Lecture Notes in Statistics 170

Springer
New York
Berlin
Heidelberg
Hong Kong
London
Milan
Paris
Tokyo

Tadeusz Caliński
Sanpei Kageyama

Block Designs: A Randomization Approach

Volume II: Design

Springer

Tadeusz Caliński
Department of Mathematical
and Statistical Methods
Agricultural University of Poznań
Wojska Polskiego 28
PL-60-637 Poznań
Poland
calinski@owl.au.poznan.pl

Sanpei Kageyama
Department of Mathematics
Graduate School of Education
Hiroshima University
1-1-1 Kagamiyama
Higashi-Hiroshima 739-8524
Japan
ksanpei@hiroshima-u.ac.jp

Library of Congress Cataloging-in-Publication Data
Caliński, T.
Block Designs: a randomization approach / Tadeusz Caliński, Sanpei Kageyama.
p. cm.—(Lecture notes in statistics; 170)
Includes bibliographical references and indexes.
Contents: v. 2. Design.
ISBN 0-387-95470-8 (v. 2: softcover: alk. paper)
1. Block designs. I. Kageyama, Sanpei, 1945- II. Title. III. Lecture notes in statistics (Springer-Verlag); v. 170.
QA279.C35 2000
519.5—dc21 00-030762

ISBN 0-387-95470-8 Printed on acid-free paper.

Printed in the United States of America.

9 8 7 6 5 4 3 2 1 SPIN 10873992

www.springer-ny.com

Springer-Verlag New York Berlin Heidelberg
A member of BertelsmannSpringer Science+Business Media GmbH

To Maria Calińska and Masako Kageyama

Preface

The book is composed of two volumes, each consisting of five chapters. In Volume I, following some statistical motivation based on a randomization model, a general theory of the analysis of experiments in block designs has been developed. In the present Volume II, the primary aim is to present methods of constructing block designs that satisfy the statistical requirements described in Volume I, particularly those considered in Chapters 3 and 4, and also to give some catalogues of plans of the designs. Thus, the constructional aspects are of predominant interest in Volume II, with a general consideration given in Chapter 6. The main design investigations are systematized by separating the material into two contents, depending on whether the designs provide unit efficiency factors for some contrasts of treatment parameters (Chapter 7) or not (Chapter 8). This distinction in classifying block designs may be essential from a practical point of view. In general, classification of block designs, whether proper or not, is based here on efficiency balance (EB) in the sense of the new terminology proposed in Section 4.4 (see, in particular, Definition 4.4.2). Most of the attention is given to connected proper designs because of their statistical advantages as described in Volume I, particularly in Chapter 3. When all contrasts are of equal importance, either the class of $(v-1;0;0)$-EB designs, i.e., orthogonal designs, or that of $(0;v-1;0)$-EB designs, i.e., efficiency balanced designs, is to be recommended in the search for an appropriate design, where v is the number of distinct treatments to be compared in the experiment. If not, that is, when the experimenter wants to estimate some of the contrasts of treatment parameters with higher and some other with lower efficiencies, the case of $(\rho_0;\rho_1,...,\rho_{m-1};0)$-EB designs, with $m=2$, 3 or more, may be of interest, where $\{\rho_\beta\}$ denote multiplicities of the relevant efficiency factors. In these designs, of several distinct efficiency factors, one of them could be equal to 1, i.e., ρ_0 could be nonzero. This may be very desirable in certain practical situations in which the experimenter wants to estimate some of the contrasts with full efficiency.

In separate chapters, resolvable designs (Chapter 9) and some special designs (Chapter 10), including nonbinary designs and some disconnected designs, are considered. Each chapter gives some general considerations on the characterization and construction of the relevant block designs. The reader can choose any chapter according to his/her experimental interest. In this sense the reader may find in each chapter design characteristics (called the parameters of the design), actual plans or methods of construction, the efficiency factors of the designs, and finally some illustrations by examples.

To avoid repetition, the term design is used to denote a connected block design, except when stated otherwise, particularly in Section 10.4, where some disconnected block designs are considered. In the terminology of Section 4.4, this means that unless otherwise stated, it will always be assumed that $\rho_m = 0$ (see again Definition 4.4.2).

As stated in the preface to Volume I, the book is aimed at an advanced audience, including students at the postgraduate level and research workers interested in designing and analyzing experiments with full understanding of the principles. Whereas in the previous volume the main interest has been in the analysis, in the present volume the emphasis is on the design. The knowledge of the rules underlying the analysis is, however, essential at the stage of choosing a proper design for the experiment to be conducted.

The help of several colleagues in the preparation of this book is appreciated. In particular thanks are due to Dr. Paweł Krajewski for his help in computing the example analyses and to Dr. Idzi Siatkowski, Dr. Takashi Seo, and Dr. Miwako Mishima for their assistance in the technical preparation of the manuscript. Last but not least, the authors are most grateful to Dr. John Kimmel, the editor at Springer, and to several reviewers for their many instructive comments and suggestions, and also for their inspiring encouragement.

Poznań, Poland
Hiroshima, Japan
May 2002

Tadeusz Caliński
Sanpei Kageyama

Comments and corrections will be welcomed. The authors' email addresses are in page iv. We plan to maintain a list of corrections for Volume I that will be found on our home pages. At the present time our home pages have the following address:

http://home.hiroshima-u.ac.jp/matedu/staff/ftp/corre.pdf

or

http://home.hiroshima-u.ac.jp/matedu/staff/kageyama-e.html

Contents

Volume I: ANALYSIS

6
Constructional Approaches and Methods

As already mentioned in the Preface to Volume I, and then detailed in its Section 1.5, in Volume II of this monograph constructional aspects of suitable designs are to be discussed, particularly with regard to some statistical concepts essential for planning of experiments. In the present chapter various methods of constructing designs will be described, along with some combinatorial and statistical properties of block designs, mainly those related to the efficiency factors of the design for estimating the corresponding basic contrasts in the intra-block analysis, as defined in Section 3.4. In general, the classification of block designs adopted here is based on the new terminology proposed in Section 4.4 (see, in particular, Definition 4.4.2). It may be helpful to recall it shortly.

A block design, defined in Section 2.1 and discussed in Section 2.2, with disconnectedness of degree $g-1$ (connected when $g=1$), is said to be ($\rho_0; \rho_1, ..., \rho_{m-1}; \rho_m$)-EB if a complete set of its $v-1$ basic contrasts can be partitioned into at most $m+1$ disjoint and nonempty subsets such that all ρ_β basic contrasts of the βth subset correspond to a common efficiency factor $\varepsilon_\beta = 1-\mu_\beta$, different for different $\beta = 0, 1, ..., m-1, m$, i.e., so that the matrix $\boldsymbol{M}_0$ (introduced in Section 4.3.2) has the spectral decomposition of the form

$$\boldsymbol{M}_0 = \boldsymbol{r}^{-\delta}\boldsymbol{N}\boldsymbol{k}^{-\delta}\boldsymbol{N}' - n^{-1}\boldsymbol{1}_v\boldsymbol{r}' = \sum_{\beta=0}^{m}\sum_{j=1}^{\rho_\beta}\mu_\beta \boldsymbol{s}_{\beta j}\boldsymbol{s}_{\beta j}'\boldsymbol{r}^\delta = \sum_{\beta=0}^{m}\mu_\beta \boldsymbol{L}_\beta, \quad (6.0.0)$$

$$\boldsymbol{L}_\beta = \sum_{j=1}^{\rho_\beta}\boldsymbol{s}_{\beta j}\boldsymbol{s}_{\beta j}'\boldsymbol{r}^\delta, \ \text{rank}(\boldsymbol{L}_\beta) = \rho_\beta, \ \beta = 0, 1, ..., m,$$

where the distinct eigenvalues $0 = \mu_0 < \mu_1 < \cdots < \mu_{m-1} < \mu_m = 1$ have the multiplicities $\rho_0 \geq 0$, $\rho_1 \geq 1$, ..., $\rho_{m-1} \geq 1$, $\rho_m = g-1 \geq 0$, respectively. This allows the C-matrix of the design (see Sections 2.2 and 3.4) to be written as

$C_1 = r^{\delta}\sum_{\beta=0}^{m-1}(1-\mu_\beta)L_\beta$. The parameters of a $(\rho_0; \rho_1, ..., \rho_{m-1}; \rho_m)$-EB design can be written as v, b, r, k, $\varepsilon_\beta = 1-\mu_\beta$, ρ_β, L_β for $\beta = 0, 1, ..., m-1, m$.

Thus, (i) ρ_0 gives the number of basic contrasts estimated in the intra-block analysis with full efficiency, i.e., not confounded with blocks, (ii) $\{\rho_1, ..., \rho_{m-1}\}$ give the numbers of basic contrasts estimated in the intra-block analysis with the corresponding distinct efficiencies $\{\varepsilon_\beta,\ \beta = 1, ..., m-1\}$ less than 1, i.e., partially confounded with blocks, and (iii) ρ_m gives the number of basic contrasts with zero efficiency, i.e., totally confounded with blocks. In a connected design $\rho_m = 0$, i.e., no basic contrast is totally confounded with blocks.

Most attention will be paid to connected proper designs because of their statistical advantages that have been revealed in Volume I, i.e., to proper $(\rho_0; \rho_1, ..., \rho_{m-1}; 0)$-EB designs, with $m = 1$, 2 or more. In these designs, of possibly several distinct efficiency factors, one of the factors may be equal to 1, i.e., ρ_0 may be nonzero. This would be very desirable in certain practical situations in which the experimenter wants to estimate, in the intra-block analysis, some of the contrasts with full efficiency. In subsequent chapters the reader will learn how to construct block designs that provide the unit efficiency factor for some contrasts (Chapter 7) or, on the contrary, to build block designs in which no contrast receives the unit efficiency (Chapter 8).

To avoid repetitions, by a design a connected block design will be meant throughout the volume, except when differently stated.

Methods of constructing block designs have appeared in the literature in abundance. Many of the most important combinatorial techniques used for constructing designs are explained by Lindner and Rodger (1997), more up to date than in an earlier book by Constantine (1987). The main methods are those of recursive construction and of direct construction. The recursive method consists in constructing designs from certain minimal designs, or in composing designs with the use of some smaller ones. The direct method was systematically treated first by Bose (1939), who introduced initial difference sets for many series of balanced incomplete block (BIB) designs using his technique of symmetrically repeated differences. To become familiar with the concept of difference sets the reader is referred to Baumert (1971) or Part IV (Chapters 12 and 13) of Colbourn and Dinitz (1996).

The direct method easily yields desirable constructions, but is applicable only for special values of design parameters. Most of the block designs have been constructed by using the notions of group, ring, Galois field, finite geometry, orthogonal Latin squares, Hadamard matrix, orthogonal array, error-correcting code, etc. For properties of a Galois field and other algebraic structures, the reader is referred to Appendix B, while for some fundamental properties of finite geometries and orthogonal Latin squares, to Appendices C and D, respectively.

6.0 Basic designs

There are several block designs which can be utilized in constructing further designs with some desirable efficiency properties. Two types of such basic designs which play a key role in these constructions are the BIB designs and the partially balanced incomplete block (PBIB) designs. These two basic classes of designs, described in the following subsections (6.0.1 and 6.0.2), can be considered as some special cases of the general $(\rho_0; \rho_1, ..., \rho_{m-1}; \rho_m)$-EB designs, the first obtained by taking $m = 2$ with $\rho_0 = 0, \rho_1 = v - 1$ and $\rho_m = 0$ (for connectedness), the other obtained for $m \geq 2$ by imposing some additional conditions, usually with $\rho_0 \geq 0$ and $\rho_m = 0$.

6.0.1 BIB designs

As already mentioned in Section 2.4.1, a BIB design is a proper binary equireplicate pairwise balanced design. This type of designs was introduced by Yates (1936a, 1936b), though their combinatorial idea was known much earlier (see Woolhouse, 1844; Kirkman, 1847). These designs are traditionally defined as follows (Raghavarao, 1971, Definition 4.3.2; see also Definition 2.4.2 in this monograph).

An arrangement of v treatments in b blocks is called a BIB design with a $v \times b$ incidence matrix $\boldsymbol{N} = [n_{ij}]$ if it meets the following conditions:

(0) Every treatment occurs at most once in a block, i.e., $n_{ij} = 0$ or 1 for all i and j.

(i) Each block contains $k(< v)$ treatments, i.e., $\sum_{i=1}^{v} n_{ij} = k$ for all j.

(ii) Every treatment occurs in exactly r blocks, i.e., $\sum_{j=1}^{b} n_{ij} = r$ for all i.

(iii) Every two treatments concur in exactly λ blocks, i.e., $\sum_{j=1}^{b} n_{ij}n_{i'j} = \lambda(> 0)$ for all i, i' $(i \neq i') = 1, 2, ..., v$.

According to the new terminology described in Section 4.4, a BIB design can be regarded as a $(0; v-1; 0)$-EB design specified by the relevant quantities v, b, r, k and by the unique efficiency factor $\varepsilon_1 = \lambda v/(rk)$, of multiplicity $\rho_1 = v - 1$, and the corresponding matrix $\boldsymbol{L}_1 = \boldsymbol{I}_v - v^{-1}\mathbf{1}_v\mathbf{1}_v'$. Hence, its C-matrix is $\boldsymbol{C}_1 = \varepsilon_1 r \boldsymbol{L}_1$ $=(\lambda v/k)(\boldsymbol{I}_v - v^{-1}\mathbf{1}_v\mathbf{1}_v')$.

The quantities v, b, r, k, λ are called the parameters of a BIB design. Though the condition (ii) is redundant, it is retained in the definition for convenience. Evident relations among the parameters are $vr = bk$ and $\lambda(v-1) = r(k-1)$. Thus, a BIB design has only three free parameters v, k, λ, as $r = \lambda(v-1)/(k-1)$ and $b = \lambda v(v-1)/[k(k-1)]$. Since r and b must be integers, necessary conditions for the existence of a BIB design are $\lambda(v-1) \equiv 0 \pmod{k-1}$ and $\lambda v(v-1) \equiv 0$ $[\text{mod } k(k-1)]$. Even if there are positive integers v, b, r, k, λ satisfying these relations, there does not always exist a BIB design with these parameters. [On the other hand, for any positive integers $v = k$ and $b = r = \lambda$ there exists a block design, the randomized block design (RBD), but this is not a BIB design.] As a very simple method of constructing BIB designs, the following can be mentioned here. The collection of all possible combinations of k treatments from the set of v treatments yields a BIB design with parameters v, $b = \binom{v}{k}$, $r = \binom{v-1}{k-1}$, k, $\lambda = \binom{v-2}{k-2}$. It is called an unreduced BIB design or a BIB design of all combination type. Since constructions of BIB designs will be discussed thoroughly in Chapter 8, only some combinatorial properties are described here.

In a BIB design with parameters v, b, r, k, λ, the concurrence (concordance) matrix (see Section 2.4.1) is $\boldsymbol{NN}' = (r - \lambda)\boldsymbol{I}_v + \lambda \boldsymbol{1}_v \boldsymbol{1}_v'$. Also it can easily be shown that

$$|\boldsymbol{NN}'| = rk(r - \lambda)^{v-1} > 0, \tag{6.0.1}$$

noting that $v > k$ implies $r > \lambda$. Hence, on account of Corollary 2.3.1(a), Fisher's inequality $b \geq v$ or, equivalently, $r \geq k$ holds. When the equalities hold, a special type of a BIB design is obtained.

Definition 6.0.1. A BIB design is said to be square or symmetric if $v = b$ (or, equivalently, $r = k$).

The term "symmetric" (or "symmetrical") is popular, although the term may be misleading as the incidence matrix need not be symmetric about the main diagonal. But the reason for calling a square BIB design symmetric may be that there is the following "symmetry" in the properties of the rows and columns of its incidence matrix, as Theorem 6.0.1 below shows: (1) any row contains k 1's; (2) any column contains k 1's; (3) any two columns both have 1's in exactly λ rows; (4) any two rows both have 1's in exactly λ columns. (How to transform the incidence matrix of a symmetric BIB design into a symmetric matrix will be seen in Remark 8.2.3.)

Theorem 6.0.1. *In a symmetric BIB design with parameters $v = b$, $r = k$, λ, any two blocks have exactly λ treatments in common.*

Proof. For such a BIB design, $\boldsymbol{NN}' = (k - \lambda)\boldsymbol{I}_v + \lambda \boldsymbol{1}_v \boldsymbol{1}_v'$ and, by (6.0.1), $\boldsymbol{N}$ is nonsingular. Now, the matrix $\boldsymbol{NN}'\boldsymbol{N} = (k - \lambda)\boldsymbol{N} + \lambda \boldsymbol{1}_v \boldsymbol{1}_v' \boldsymbol{N}$, when premultiplied on both sides by $\boldsymbol{N}^{-1}$, yields $\boldsymbol{N}'\boldsymbol{N} = (k - \lambda)\boldsymbol{I}_v + \lambda \boldsymbol{1}_v \boldsymbol{1}_v'$ [see the relations in (2.2.1)]. □

Through Theorem 6.0.1 every symmetric BIB design yields two other BIB designs (as indicated by Raghavarao, 1971, p. 65). To obtain them from a given symmetric BIB design with parameters $v = b$, $r = k$, λ, one can proceed as follows.

I. Choose a block, and remove it and its treatments from the design. Then the remaining structure is a BIB design with parameters $v_1 = v - k, b_1 = v - 1, r_1 = k, k_1 = k - \lambda$, $\lambda_1 = \lambda$, which is called a residual design.

II. Choose a block and remove it, but retain its treatments whenever else they appear in the design. Then remove all treatments not appearing in the chosen block. The remaining structure is a BIB design with parameters $v_2 = k, b_2 = v - 1, r_2 = k - 1, k_2 = \lambda$, $\lambda_2 = \lambda - 1$, which is called a derived design.

It is said that a BIB design is quasi-residual (quasi-derived) if it has the same parameters as those of a design obtained by residuation (derivation) from a symmetric BIB design. One may be interested to know whether the above method of obtaining the residual and derived designs from a symmetric BIB design has a converse. That is, is a quasi-residual (quasi-derived) design a residual (derived) design? The answer is yes if $\lambda = 1$ or 2. This means that if a quasi-residual design exists with $\lambda = 1$ or 2, then the generating symmetric design also exists. The answer is negative for $\lambda = 3$, as was shown by Bhattacharya (1945) who found a quasi-residual BIB design with parameters $v = 16, b = 24, r = 9, k = 6$, $\lambda = 3$ in which two of its blocks have four treatments in common. This means, by Theorem 6.0.1, that his BIB design cannot be embedded into the corresponding symmetric BIB design with parameters $v = b = 25, r = k = 9$, $\lambda = 3$. The reader who wishes to follow this topic further, is referred to a paper by Hall and Connor (1953) or to the book by Hall (1986).

Example 6.0.1. Consider a symmetric BIB design with parameters $v = b = 7, r = k = 3, \lambda = 1$ (see also Example 6.8.1 or No. 24 in Table 8.2 later in the book) whose incidence matrix is given by

$$\boldsymbol{N} = \begin{bmatrix} 1 & 0 & 0 & 0 & 1 & 0 & 1 \\ 1 & 1 & 0 & 0 & 0 & 1 & 0 \\ 0 & 1 & 1 & 0 & 0 & 0 & 1 \\ 1 & 0 & 1 & 1 & 0 & 0 & 0 \\ 0 & 1 & 0 & 1 & 1 & 0 & 0 \\ 0 & 0 & 1 & 0 & 1 & 1 & 0 \\ 0 & 0 & 0 & 1 & 0 & 1 & 1 \end{bmatrix}.$$

Remove the first block and its treatments from the design. Then a residual design is given by the incidence matrix

$$\boldsymbol{N}^{(1)} = \begin{bmatrix} 1 & 1 & 0 & 0 & 0 & 1 \\ 1 & 0 & 1 & 1 & 0 & 0 \\ 0 & 1 & 0 & 1 & 1 & 0 \\ 0 & 0 & 1 & 0 & 1 & 1 \end{bmatrix},$$

i.e., is a BIB design with parameters $v_1 = 4, b_1 = 6, r_1 = 3, k_1 = 2, \lambda_1 = 1$. Next, again in the original design choose the first block and remove it, but retain its treatments whenever else they appear in the design, while remove all treatments not appearing in the first block. Then a derived design is given by the incidence matrix

$$\boldsymbol{N}^{(2)} = \begin{bmatrix} 0 & 0 & 0 & 1 & 0 & 1 \\ 1 & 0 & 0 & 0 & 1 & 0 \\ 0 & 1 & 1 & 0 & 0 & 0 \end{bmatrix}$$

with parameters $v_2 = 3, b_2 = 6, r_2 = 2, k_2 = 1, \lambda_2 = 0$. In fact, it follows that

$$\boldsymbol{N}^* = \begin{bmatrix} \mathbf{1}_k & \boldsymbol{N}^{(2)} \\ \mathbf{0} & \boldsymbol{N}^{(1)} \end{bmatrix},$$

with $k = 3$, after appropriate permutation of its rows, coincides with the original incidence matrix $\boldsymbol{N}$. This structural relation among a symmetric BIB design and its residual and derived designs holds in general. Thus, the embedding problem mentioned above is to construct a symmetric BIB design $\boldsymbol{N}$ from a joint combination of a residual design $\boldsymbol{N}^{(1)}$ and a derived design $\boldsymbol{N}^{(2)}$.

Other well-known necessary conditions for the existence of a symmetric BIB design have been obtained by Schutzenberger (1949), Bruck and Ryser (1949), Shrikhande (1950), and Chowla and Ryser (1950), independently.

Theorem 6.0.2. *In a symmetric BIB design with parameters $v = b, r = k, \lambda$,*

(i) *if v is even, $k - \lambda$ is a perfect square;*
(ii) *if v is odd, the equation $x^2 = (k-\lambda)y^2 + (-1)^{(v-1)/2}\lambda z^2$ has a solution in integers, not all zero.*

Proof. When $v = b$, (6.0.1) yields $|\boldsymbol{N}|^2 = k^2(k-\lambda)^{v-1}$ which must be a square. This implies (i). For (ii) refer to the original papers or to the book by Hall (1986, p.133). □

Shrikhande's (1950) result is expressed in terms of the Hilbert norm residue symbol but is equivalent to the above. Theorem 6.0.2, nowadays called the Bruck-Chowla-Ryser theorem, is very powerful. For example, (i) shows the nonexistence of symmetric BIB designs with parameters $v = 22, k = 7, \lambda = 2$; $v = 34, k = 12, \lambda = 4$; $v = 46, k = 10, \lambda = 2$, whereas (ii) shows the nonexistence of symmetric BIB designs with parameters $v = 43, k = 7, \lambda = 1$; $v = 43, k = 15, \lambda = 3$; $v = 29, k = 8, \lambda = 2$. It is conjectured that the conditions in Theorem 6.0.2 are also sufficient.

For a BIB design several relations on its parameters can be derived. The following two of them are of particular interest when using the intra-block and the inter-block information for obtaining combined estimators of treatment parametric functions, as can be seen in Sections 3.8.4 and 3.9.2. Note that the bound (6.0.2) below is the same as in (3.8.83). If it holds, then the gain in precision of the combined estimator (3.8.81) is uniformly positive, as shown in Section 3.8.4.

Theorem 6.0.3. *In a BIB design with parameters $v(>3), b, r, k, \lambda$, the inequality*

$$\frac{k(v-1)}{v(v-3)} < \lambda \tag{6.0.2}$$

holds, except for two existing BIB designs with parameters $v = 4, b = 6, r = 3, k = 2, \lambda = 1$ and with $v = 4, b = 4, r = 3, k = 3, \lambda = 2$.

Proof. It follows that (6.0.2) is equivalent to $0 < \lambda(v-3) + (v-1)[\lambda(v-3) - k]$ $[= f(v)$, say$]$. It is clear that if $v \geq k/\lambda + 3$, then $f(v)$ is positive. Then when $v < k/\lambda + 3$, one can have $k < v < k/\lambda + 3 \leq k + 3$, because $\lambda \geq 1$. This implies that $v = k+1$ or $k+2$.

Case (I): $v = k+1$. Because it is known (Kageyama and Kuwada, 1985, Proposition 10) that there exists a BIB design with $v = k+1$ if and only if its incidence matrix is given by $b - r$ copies of $\mathbf{1}_v\mathbf{1}_v' - \boldsymbol{I}_v$, one can let $\lambda = s(v-2)$ for a positive integer s. In this case, $f(v) = sv(v-2)(v-3) - (v-1)^2 \geq$

$v(v-2)(v-3)-(v-1)^2 > 0$ for $v \geq 5$. Hence the remaining case is $v=4$, and then $f(4)=8s-9>0$ for $s \geq 2$. Thus, only the case $v=4$ and $s=1$, i.e., a BIB design with parameters $v=b=4, r=k=3$, $\lambda=2$, violates (6.0.2).

Case (II): $v = k+2$. Since it is known (Kageyama and Kuwada, 1985, Proposition 11) that there exists a BIB design with $v = k+2$ only if its parameters can be expressed as $v = k+2, b = s\binom{v}{2}$, $r = s\binom{v-1}{2}$, k, $\lambda = s\binom{v-2}{2}$ for a positive integer s, one can get $f(v) = (v-2)[sv(v-3)^2 - 2(v-1)]/2 \geq (v-2)[v(v-3)^2 - 2(v-1)]/2 > 0$ for $v \geq 5$. Hence the remaining case is $v=4$, and then $f(4) = 2(2s-3)$ for $s \geq 2$. Thus, only the case $v=4$ and $s=1$, i.e., a BIB design with parameters $v=4, b=6, r=3, k=2, \lambda=1$, violates (6.0.2). This completes the proof. □

The following theorem concerns a condition for the Yates (1940b) estimators (considered in Section 3.9.1) to secure the uniformly smaller variance (USV) property of the resulting combined estimator (3.8.81), the property considered in Section 3.9.2.

Theorem 6.0.4. *Let $d_1 = vr-b-(v-1)$, the residual degrees of freedom (d.f.) in the intra-block analysis. Then in a BIB design with parameters v, b, r, k, λ, the inequality*

$$\frac{1}{2} \leq \frac{\lambda d_1}{r(r-1)} \frac{b-3}{d_1+2} \tag{6.0.3}$$

holds, except for eight existing BIB designs with parameters: $v=b=3, r=k=2, \lambda=1$; $v=b=4, r=k=3, \lambda=2$; $v=b=5, r=k=4, \lambda=3$; $v=3, b=6, r=4, k=2, \lambda=2$; $v=3, b=9, r=6, k=2, \lambda=3$; $v=4, b=6, r=3, k=2, \lambda=1$; $v=4, b=12, r=6, k=2, \lambda=2$; $v=5, b=10, r=4, k=2, \lambda=1$.

Proof. This may be given by considering the following three cases.

(I) $b=v$, i.e., $r=k$ (a symmetric BIB design). If $k=2$, then $d_1=1$ and (6.0.3) is equivalent to $\lambda(v-3) \geq 3$, which never holds, as $\lambda(v-1)=2$ if $r=k=2$, giving $v=3$. For this, the only possibility is a BIB design with parameters $v=b=3, r=k=2$, $\lambda=1$, which does exist. When $k \geq 3$ (and hence $v \geq 4$), it can be shown that (6.0.3) is reduced to $v(v-5)(k-2)-(v+3) \geq 0$, which holds for $v \geq 6$ (requiring $k \geq 5$). Then, for $4 \leq v \leq 5$, it follows that there exist only two BIB designs, with parameters $v=b=4, r=k=3, \lambda=2$, and with $v=b=5$, $r=k=4, \lambda=3$, which violate (6.0.3).

(II) $b>v$ and $k=2$. It is clear that the parameters of such BIB designs can be expressed as v, $b=sv(v-1)/2, r=s(v-1), k=2, \lambda=s$ for a positive integer s, and such designs always exist. Hence $d_1 = (v-1)(sv-2)/2$. Then the relation (6.0.3) with these parameters reduces to $0 \leq v(v-1)s^2+(v^2-13v+6)s-2(v-9)[= f(s;v)$, say], which holds for $s \geq 4$ at any $v \geq 3$ and for $v^2-13v+6 \geq 0$ at any $s \geq 1$. Now $v^2-13v+6 < 0$ implies $v = 3,4,5,6,7,8,9,10,11,12$. Each of such v shows $f(s;v) > 0$ for $s \geq 4$. Next, when $s=1$, $f(1;v) \geq 0$ reduces to $v \geq 6$. Then for $4 \leq v \leq 5$ it follows that there exist only two BIB designs with parameters $v=4, b=6, r=3, k=2, \lambda=1$, and with $v=5, b=10$, $r=4, k=2, \lambda=1$, which violate (6.0.3). Similarly, when $s=2$ and 3, there exist only three BIB designs with parameters $v = 3, b = 6, r = 4, k = 2, \lambda = 2$; $v = 4, b = 12, r = 6, k = 2, \lambda = 2$; $v = 3, b = 9, r = 6, k = 2, \lambda = 3$, which all violate (6.0.3).

(III) $b > v$ and $k \geq 3$. It follows from $d_1 = vr - b - v + 1$ that (6.0.3) reduces to $0 \geq -(b-v)+3r-6k+3-b(k-1)[(b-6)(k-2)+r-6]/(v-1)(= g$, say). As $b/(v-1) > 1$, it follows that $g < -(b-v)-r(k-4)-3-(k-1)(k-2)(b-6)(= g'$, say), which is negative, if $k \geq 4$, because $k \geq 4$ implies $b \geq 6$. Next, when $k = 3$, $g' = -(b-v)-(b-r)-(b-9)$ which is negative if $b \geq 9$. Then, as the remaining cases, $k = 3$ and $b = 5, 6, 7, 8$ are considered. It can be seen (Table 8.2) that among them there exists only one BIB design, and for it, $g < 0$. □

Note that in Theorem 6.0.4 one should confine attention to the cases with $b > 3$ only.

6.0.2 PBIB designs

When a design is to be proper, one of the most typical and standard choices, other than a BIB design, is a PBIB design with two associate classes. Following the usual definition (see Raghavarao, 1971, Section 8.1), an incomplete block design is called a PBIB design with s associate classes (or an s-associate PBIB design) if it satisfies the following conditions:

(i) The experimental material is divided into b blocks of k experimental units each, different treatments being applied to units of the same block.

(ii) There are v ($> k$) treatments each of which occurs in r blocks.

(iii) Between any two treatments a relation of association satisfies the following requirements:

(a) Two treatments are either first, second, ... , or sth associates.

(b) Each treatment has exactly n_i ith associates, $i = 1, 2, ..., s$.

(c) Given any two treatments which are ith associates, the number of treatments which are both jth associates of the first and uth associates of the second is p^i_{ju}, and this number is independent of the pair of treatments chosen among the ith associates. It follows that $p^i_{ju} = p^i_{uj}$, $i, j, u = 1, 2, ..., s$.

(iv) Two treatments which are ith associates occur together in exactly λ_i blocks, this number being independent of the particular pair of ith associates chosen, $i = 1, 2, ..., s$.

Note that the association relations, forming the so-called "association scheme with s associate classes" for the v treatments of a PBIB design, are determined only by the conditions given in the part (iii) of the above definition, and that they do not depend on how the treatments are arranged into blocks, until the part (iv) is invoked. In this sense the above association scheme can be defined independently of a design and depends only on parameters $n_1, n_2, ..., n_s, p^i_{ju}$, $i, j, u = 1, 2, ..., s$, called parameters of the association scheme. The numbers $v, b, r, k, \lambda_1, \lambda_2, ..., \lambda_s$ are then called the parameters of the design.

Remark 6.0.1. As will be indicated in Chapter 8, among the class of all connected incomplete binary equireplicate block designs a BIB design is the most efficient design in some sense. However, unfortunately, BIB designs exist for a limited number of cases and also may require relatively large treatment replications. Thus, as designs requiring relatively small number of replications and having statistical properties similar to those of BIB designs, Bose and Nair (1939) introduced PBIB designs, also to generalize a class of lattice designs, which will be discussed in Section 9.6. Unlike in a BIB design, however, in a PBIB design, the elementary contrasts, i.e., such which concern only differences between two treatments, are not all estimated with the same variance in the intra-block analysis. For example, in a 2-associate PBIB design the variance takes one of two values, depending on whether the two treatments are first associates or second associates. This will be shown in some detail in Chapter 7 (Section 7.3).

In general, for a connected s-associate PBIB design with parameters v, b, r, k, λ_1, λ_2, ..., λ_s, it is known (see Bose and Mesner, 1959; Yamamoto and Fujii, 1963; Raghavarao, 1971, Section 8.3) that the incidence matrix $\boldsymbol{N}$ of the design satisfies the equalities

$$\begin{aligned} \boldsymbol{N}\boldsymbol{N}' &= r\boldsymbol{A}_0 + \lambda_1\boldsymbol{A}_1 + \cdots + \lambda_s\boldsymbol{A}_s \\ &= rk\boldsymbol{A}_0^{\#} + \psi_1\boldsymbol{A}_1^{\#} + \cdots + \psi_s\boldsymbol{A}_s^{\#}, \ 0 \le \psi_i < rk, \ i = 1, 2, ..., s. \end{aligned}$$

Here the $v \times v$ matrices $\boldsymbol{A}_i = (a^i_{\alpha\beta}), i = 1, 2, ..., s$, are defined as

$$a^i_{\alpha\beta} = \begin{cases} 1 & \text{if } \alpha \text{ and } \beta \text{ are } i\text{th associates,} \\ 0 & \text{otherwise.} \end{cases}$$

They are called the association matrices of the association scheme considered. From definition, these matrices are symmetrical, with constant row and column totals, those of the matrix $\boldsymbol{A}_i$ being each equal to n_i. Further, let every treatment be the 0th associate of itself and such of no other treatment. Then, evidently,

$$n_0 = 1, \ p^0_{ii} = n_i, \ p^0_{ii'} = 0, \ p^i_{0i} = 1, \ p^i_{0i'} = 0, \ \lambda_0 = r, \ i \ne i',$$

$$\boldsymbol{A}_0 = \boldsymbol{I}_v, \ \sum_{i=0}^{s} \boldsymbol{A}_i = \boldsymbol{1}_v\boldsymbol{1}_v'.$$

Also, it can be seen (see Raghavarao, 1971, p.124) that $\boldsymbol{A}_j\boldsymbol{A}_u = \sum_{i=0}^{s} p^i_{ju}\boldsymbol{A}_i$ for $j, u = 0, 1, ..., s$. [This property shows a possibility of taking association matrices as incidence matrices of some block designs, as will be seen in Theorem 8.3.7 and so on.] As to the $v \times v$ matrices $\boldsymbol{A}_i^{\#}, i = 0, 1, ..., s$, they are obtainable by appropriate linear combinations of the association matrices, and are mutually orthogonal idempotent matrices such that

$$\boldsymbol{A}_0^{\#} = \frac{1}{v}\boldsymbol{1}_v\boldsymbol{1}_v', \ \ \boldsymbol{A}_i^{\#}\boldsymbol{A}_{i'}^{\#} = \delta_{ii'}\boldsymbol{A}_i^{\#}, \ \ \sum_{i=0}^{s} \boldsymbol{A}_i^{\#} = \boldsymbol{I}_v, \ \ \text{rank}(\boldsymbol{A}_i^{\#}) = \rho_i,$$

$\delta_{ii'}$ being the Kronecker delta. Also, ψ_i, $i = 1, 2, ..., s$, are the distinct eigenvalues of $\boldsymbol{NN'}$ other than rk, their multiplicities being ρ_i, respectively.

Now, referring to Sections 4.3 and 4.4, and particularly to (4.4.1), (4.4.4) and (4.4.11), it can be seen that

$$\boldsymbol{F} = \frac{1}{r}\boldsymbol{C}_1 = \boldsymbol{I}_v - \boldsymbol{M} = \boldsymbol{I}_v - \frac{1}{rk}\boldsymbol{NN'}$$

$$= (1 - \frac{\psi_1}{rk})\boldsymbol{A}_1^{\#} + (1 - \frac{\psi_2}{rk})\boldsymbol{A}_2^{\#} + \cdots + (1 - \frac{\psi_s}{rk})\boldsymbol{A}_s^{\#}, \qquad (6.0.4)$$

which shows that a connected s-associate PBIB design is either a (ρ_0; $\rho_1, ..., \rho_{s-1}$; 0)-EB or a (0; $\rho_1, ..., \rho_s$; 0)-EB design with parameters

$$v,\ b,\ r,\ k,\ \varepsilon_\beta = 1 - \frac{\psi_\beta}{rk},\ \rho_\beta,\ \boldsymbol{L}_\beta = \boldsymbol{A}_\beta^{\#},\ \beta \in \{0, 1, ..., s\},$$

as defined in Section 4.4. Note, however, that this notation will coincide with that used in Definition 4.4.2 if the subscripts at the eigenvalues ψ_β and at the idempotent matrices $\boldsymbol{A}_\beta^{\#}$ above, $\beta = 1, 2, ..., s$, are renumbered according to the order $0 = \psi_0 < \psi_1 < \cdots < \psi_{s-1}$ or $0 < \psi_1 < \cdots < \psi_{s-1} < \psi_s$, depending on whether $\psi_0 = 0$ exists or not. If it does, then $\varepsilon_0 = 1$ with $\rho_0 > 0$, which implies that in such an s-associate PBIB design, ρ_0 basic contrasts are estimated with full efficiency. In this sense, a PBIB design with one of the distinct eigenvalues of the matrix $\boldsymbol{NN'}$ being zero, is of special importance from the practical point of view. On the other hand, for a basic contrast $\boldsymbol{c}'_{\beta j}\boldsymbol{\tau}(= r\boldsymbol{s}'_{\beta j}\boldsymbol{\tau})$ the variance of its BLUE in the intra-block analysis is $\text{Var}[(\widehat{\boldsymbol{c}'_{\beta j}\boldsymbol{\tau}})_{\text{intra}}] = \varepsilon_\beta^{-1}\sigma_1^2$ for $\beta = 0, 1, ..., s-1$ or $\beta = 1, 2, ..., s$ and $j = 1, 2, ..., \rho_\beta$ [see (4.4.7) and the comments following it].

Here mainly the case of $s = 2$ will be discussed in detail, i.e., the 2-associate PBIB designs. The following relations on parameters of these designs are well-known (see, Raghavarao, 1971, Sections 8.2–8.4; Clatworthy, 1973):

$$vr = bk, \quad v - 1 = n_1 + n_2, \quad r(k-1) = n_1\lambda_1 + n_2\lambda_2,$$

$$p_{11}^1 + p_{12}^1 = n_1 - 1, \quad p_{11}^2 + p_{12}^2 = n_1, \quad n_1 p_{12}^1 = n_2 p_{11}^2,$$

$$p_{21}^1 + p_{22}^1 = n_2, \quad p_{21}^2 + p_{22}^2 = n_2 - 1, \quad n_1 p_{22}^1 = n_2 p_{12}^2.$$

The parameters p_{ju}^i can also be written in matrix notation as

$$\boldsymbol{P}_1 = [p_{ju}^1] = \begin{bmatrix} p_{11}^1 & p_{12}^1 \\ p_{21}^1 & p_{22}^1 \end{bmatrix}, \quad \boldsymbol{P}_2 = [p_{ju}^2] = \begin{bmatrix} p_{11}^2 & p_{12}^2 \\ p_{21}^2 & p_{22}^2 \end{bmatrix}.$$

Usually $\lambda_1 \neq \lambda_2$. Otherwise the 2-associate PBIB design becomes a BIB design discussed in Section 6.0.1.

Following Connor and Clatworthy (1954), for the incidence matrix $\boldsymbol{N}$ of a connected 2-associate PBIB design with parameters v, b, r, k, λ_1, λ_2; n_1, n_2, p_{ju}^i,

$i, j, u = 1, 2$, the eigenvalues of $\boldsymbol{NN}'$ and their multiplicities are determined as follows:

$$\psi_i = r - \frac{1}{2}\{(\lambda_1 - \lambda_2)[-\gamma + (-1)^i\sqrt{\Delta}] + \lambda_1 + \lambda_2\},$$

where

$$\gamma = p_{12}^2 - p_{12}^1,\ \delta = p_{12}^1 + p_{12}^2,\ \Delta = \gamma^2 + 2\delta + 1,$$

and

$$\rho_i = \frac{n_1 + n_2}{2} + (-1)^i \frac{n_1 - n_2 + \gamma(n_1 + n_2)}{2\sqrt{\Delta}},\ i = 1, 2.$$

Note that the multiplicities ρ_i do not depend on the design parameters but only on the parameters of the association scheme. That ρ_1 and ρ_2 are to be integers, is a necessary condition for the existence of the association scheme. Also note that, to be consistent with the notation of Section 4.4, the subscript numeration corresponding to $0 = \psi_0 < \psi_1$ or $0 < \psi_1 < \psi_2$ would have to be adopted, as shown above.

Following Bose and Shimamoto (1952), it will be interesting to pay special attention to four standard association schemes with two associate classes which provide the following types of 2-associate PBIB designs: (i) Group Divisible (GD), (ii) Triangular, (iii) Latin-square type (L_i) and (iv) Cyclic (C). At first each of these association schemes will be described along with other parameters, including eigenvalues of $\boldsymbol{NN}'$, with their multiplicities, where $\boldsymbol{N}$ is the incidence matrix of the PBIB design based on the considered association scheme, in which two treatments called first associates occur together in λ_1 blocks, whereas two treatments called second associates occur together in λ_2 blocks. In this presentation the traditional notation is used, to avoid confusion with the relevant literature. [The symbol n used in (A) and (B) below should not be confused with that denoting the total number of units (plots) used in the experiment, as introduced in Section 2.2. Also the symbol m used in (A) should not be confused with that in a $(\rho_0; \rho_1, ..., \rho_{m-1}; \rho_m)$-EB design. It is hoped that it will be always possible to recognize from the context the appropriate meaning of these symbols.]

(A) A GD association scheme (Raghavarao, 1971, Definition 8.4.1). There are $v = mn$ treatments which are divided into m groups of n treatments each as follows

$$\begin{array}{ccccc} 1 & 2 & \cdots & n \\ n+1 & n+2 & \cdots & 2n \\ \vdots & \vdots & & \vdots \\ (i-1)n+1 & (i-1)n+2 & \cdots & in \\ \vdots & \vdots & & \vdots \\ (m-1)n+1 & (m-1)n+2 & \cdots & mn \end{array}$$

such that any two treatments of the same group (i.e., from the same row, as each row above represents a group) are first associates and two treatments from different groups are second associates. Then

$$n_1 = n - 1,\ n_2 = n(m-1),$$

$$\boldsymbol{P}_1 = \begin{bmatrix} n-2 & 0 \\ 0 & n(m-1) \end{bmatrix},\quad \boldsymbol{P}_2 = \begin{bmatrix} 0 & n-1 \\ n-1 & n(m-2) \end{bmatrix},$$

$$\psi_1 = rk - v\lambda_2,\ \psi_2 = r - \lambda_1,\ \rho_1 = m-1,\ \rho_2 = m(n-1),$$

$$\boldsymbol{A}_0 = \boldsymbol{I}_v,\ \boldsymbol{A}_1 = \boldsymbol{I}_m \otimes (\mathbf{1}_n\mathbf{1}_n' - \boldsymbol{I}_n),\ \boldsymbol{A}_2 = \mathbf{1}_v\mathbf{1}_v' - \boldsymbol{A}_0 - \boldsymbol{A}_1,$$

$$\boldsymbol{A}_0^{\#} = \frac{1}{v}\mathbf{1}_v\mathbf{1}_v',\ \boldsymbol{A}_1^{\#} = (\boldsymbol{I}_m - \frac{1}{m}\mathbf{1}_m\mathbf{1}_m') \otimes \frac{1}{n}\mathbf{1}_n\mathbf{1}_n',\ \boldsymbol{A}_2^{\#} = \boldsymbol{I}_m \otimes (\boldsymbol{I}_n - \frac{1}{n}\mathbf{1}_n\mathbf{1}_n'),$$

where the symbol $\otimes$ is used to denote the Kronecker product of the matrices.

GD designs are further divided into three exhaustive and mutually exclusive classes as follows. A GD design is said to be singular if $\psi_2 = 0$, semi-regular if $\psi_2 > 0$ and $\psi_1 = 0$, and regular if $\psi_2 > 0$ and $\psi_1 > 0$. In view of the efficiency of the design it may be preferable to use singular or semi-regular GD designs, because one of the eigenvalues ψ_i is then zero.

Example 6.0.2. The following incidence matrix gives a semi-regular GD design, SR6 in Clatworthy (1973), with parameters $v = 6, b = 9, r = 3, k = 2, \lambda_1 = 0$, $\lambda_2 = 1, m = 2, n = 3$, based on two groups $\{1,2,3\}, \{4,5,6\}$ of three treatments:

$$\begin{array}{c} 1 \\ 2 \\ 3 \\ 4 \\ 5 \\ 6 \end{array} \left[\begin{array}{ccc|ccc|ccc} 1 & 0 & 0 & 1 & 0 & 0 & 1 & 0 & 0 \\ 0 & 1 & 0 & 0 & 1 & 0 & 0 & 1 & 0 \\ 0 & 0 & 1 & 0 & 0 & 1 & 0 & 0 & 1 \\ 1 & 0 & 0 & 0 & 1 & 0 & 0 & 0 & 1 \\ 0 & 1 & 0 & 0 & 0 & 1 & 1 & 0 & 0 \\ 0 & 0 & 1 & 1 & 0 & 0 & 0 & 1 & 0 \end{array}\right].$$

Note that this design has three superblocks consisting of consecutive three blocks each, i.e., is resolvable (see Section 6.0.3).

Because in a GD design $\psi_1 = rk - v\lambda_2$ and $\psi_2 = r - \lambda_1$, a basic contrast represented by an eigenvector of $\boldsymbol{A}_1^{\#}$, i.e., a contrast involving not individual treatments but groups of them, obtains for its best linear unbiased estimator (BLUE) in the intra-block analysis a variance $rk\sigma_1^2/(v\lambda_2)$, whereas a basic contrast represented by an eigenvector of $\boldsymbol{A}_2^{\#}$, i.e., in particular, a contrast involving individual treatments from the same group, obtains for its intra-block BLUE a variance $rk\sigma_1^2/[r(k-1)+\lambda_1]$. This means that PBIB designs based on a GD association scheme are useful for experimental situations in which the treatments can be divided into a number of groups, such for which one wants to estimate the within group contrasts with a constant efficiency and the between group contrasts also with a constant, though different, efficiency. If a singular GD design is used, the within group contrasts are then estimated in the intra-block analysis with full efficiency, whereas in case of a semi-regular GD design, the between group contrasts are estimated there with full efficiency (see also John, 1987, Section 3.5.2).

Remark 6.0.2. A GD design may be considered as suitable for an $m \times n$ factorial experiment with two factors A and B, applied at m and n levels, respectively. Then it is instructive to write the matrix $\boldsymbol{F}$ [introduced in Section 4.4.3, here in (6.0.4)] as

$$\boldsymbol{F} = \varepsilon_1 \boldsymbol{A}^{\#}_{1\cdot} + \varepsilon_2 (\boldsymbol{A}^{\#}_{\cdot 2} + \boldsymbol{A}^{\#}_{12}),$$

where $\varepsilon_1 = 1 - (rk - v\lambda_2)/(rk) = v\lambda_2/(rk)$, $\varepsilon_2 = 1 - (r - \lambda_1)/(rk) = [r(k-1) + \lambda_1]/(rk)$, and where the following spectral decompositions hold:

$$\boldsymbol{A}^{\#}_{1\cdot} = (\boldsymbol{I}_m - \frac{1}{m}\mathbf{1}_m\mathbf{1}'_m) \otimes \frac{1}{n}\mathbf{1}_n\mathbf{1}'_n = \frac{1}{n}\sum_{j_1=1}^{m-1}(\boldsymbol{p}_{j_1} \otimes \mathbf{1}_n)(\boldsymbol{p}_{j_1} \otimes \mathbf{1}_n)',$$

where $\boldsymbol{p}'_{j_1}\boldsymbol{p}_{j'_1} = \delta_{j_1 j'_1}$ and $\mathbf{1}'_m \boldsymbol{p}_{j_1} = 0$ for $j_1, j'_1 = 1, 2, ..., m-1$,

$$\boldsymbol{A}^{\#}_{\cdot 2} = \frac{1}{m}\mathbf{1}_m\mathbf{1}'_m \otimes (\boldsymbol{I}_n - \frac{1}{n}\mathbf{1}_n\mathbf{1}'_n) = \frac{1}{m}\sum_{j_2=1}^{m-1}(\mathbf{1}_m \otimes \boldsymbol{p}_{j_2})(\mathbf{1}_m \otimes \boldsymbol{p}_{j_2})',$$

where $\boldsymbol{p}'_{j_2}\boldsymbol{p}_{j'_2} = \delta_{j_2 j'_2}$ and $\mathbf{1}'_n \boldsymbol{p}_{j_2} = 0$ for $j_2, j'_2 = 1, 2, ..., n-1$, and

$$\boldsymbol{A}^{\#}_{12} = (\boldsymbol{I}_m - \frac{1}{m}\mathbf{1}_m\mathbf{1}'_m) \otimes (\boldsymbol{I}_n - \frac{1}{n}\mathbf{1}_n\mathbf{1}'_n) = \sum_{j_1=1}^{m-1}\sum_{j_2=1}^{n-1}(\boldsymbol{p}_{j_1} \otimes \boldsymbol{p}_{j_2})(\boldsymbol{p}_{j_1} \otimes \boldsymbol{p}_{j_2})'.$$

The eigenvectors $\{\boldsymbol{p}_{j_1}\}$ of $\boldsymbol{I}_m - m^{-1}\mathbf{1}_m\mathbf{1}'_m$ and the eigenvectors $\{\boldsymbol{p}_{j_2}\}$ of $\boldsymbol{I}_n - n^{-1}\mathbf{1}_n\mathbf{1}'_n$, not uniquely determined, can be chosen in accordance with the interest of the experimenter to define relevant contrasts. For this note that, with the vector $\boldsymbol{\tau}$ of treatment parameters appropriately ordered, interesting basic contrasts of the design can be obtained from the following explanations:

(i) $\{(\boldsymbol{p}_{j_1} \otimes \mathbf{1}_n)'\boldsymbol{\tau},\ j_1 = 1, 2, ..., m-1\}$ are mutually orthogonal contrasts among the A main effects, receiving the efficiency factor ε_1;

(ii) $\{(\mathbf{1}_m \otimes \boldsymbol{p}_{j_2})'\boldsymbol{\tau},\ j_2 = 1, 2, ..., n-1\}$ are mutually orthogonal contrasts among the B main effects, receiving the efficiency factor ε_2;

(iii) $\{(\boldsymbol{p}_{j_1} \otimes \boldsymbol{p}_{j_2})'\boldsymbol{\tau},\ j_1 = 1, 2, ..., m-1,\ j_2 = 1, 2, ..., n-1\}$ are mutually orthogonal AB interaction contrasts, receiving the efficiency factor ε_2.

(To be consistent with the notation in Section 4.4, the symbols $\varepsilon_1, \varepsilon_2$ would have to be interchanged in case of $\psi_1 = rk - v\lambda_2 > r - \lambda_1 = \psi_2$. Further obvious change in the subscripts would have to be made in case of $\psi_1 = 0$ or $\psi_2 = 0$, i.e., if $\varepsilon_1 = 1$ or $\varepsilon_2 = 1$, respectively.)

(B) A triangular association scheme (Raghavarao, 1971, Definition 8.4.3). There are $v = n(n-1)/2$ treatments which are arranged in an array of n rows and n columns with the following properties:

(i) The positions in the principal diagonal (running from the upper left-hand to the lower right-hand corner) are left blank.
(ii) The $n(n-1)/2$ positions above the principal diagonal are filled by the numbers $1, 2, ..., n(n-1)/2$ corresponding to the treatments.
(iii) The $n(n-1)/2$ positions below the principal diagonal are filled such that the array is symmetric about the principal diagonal.
(iv) For any treatment i the first associates are exactly those treatments which occur in the same row or in the same column as the treatment i.

It follows then that

$$n_1 = 2(n-2),\ n_2 = (n-2)(n-3)/2,$$

$$\boldsymbol{P}_1 = \begin{bmatrix} n-2 & n-3 \\ n-3 & (n-3)(n-4)/2 \end{bmatrix},\ \boldsymbol{P}_2 = \begin{bmatrix} 4 & 2n-8 \\ 2n-8 & (n-4)(n-5)/2 \end{bmatrix},$$

$$\psi_1 = r + (n-4)\lambda_1 - (n-3)\lambda_2,\ \psi_2 = r - 2\lambda_1 + \lambda_2,$$

$$\rho_1 = n-1,\ \rho_2 = n(n-3)/2,$$

$$\boldsymbol{A}_0 = \boldsymbol{I}_v,\ \boldsymbol{A}_1,\ \boldsymbol{A}_2 = \boldsymbol{1}_v\boldsymbol{1}'_v - \boldsymbol{I}_v - \boldsymbol{A}_1,$$

$$\boldsymbol{A}_0^{\#} = \frac{1}{v}\boldsymbol{1}_v\boldsymbol{1}'_v,\ \boldsymbol{A}_1^{\#} = \frac{1}{n(n-2)}(2n\boldsymbol{I}_v + n\boldsymbol{A}_1 - 4\boldsymbol{1}_v\boldsymbol{1}'_v),$$

$$\boldsymbol{A}_2^{\#} = \frac{1}{(n-1)(n-2)}[(n-1)(n-4)\boldsymbol{I}_v - (n-1)\boldsymbol{A}_1 + 2\boldsymbol{1}_v\boldsymbol{1}'_v].$$

Note that in general $\boldsymbol{A}_1$ above is not given in a closed form in terms of n.

Example 6.0.3. Based on the triangular scheme of ten treatments

$$\begin{array}{ccccc} & 1 & 2 & 3 & 4 \\ 1 & & 5 & 6 & 7 \\ 2 & 5 & & 8 & 9 \\ 3 & 6 & 8 & & 10 \\ 4 & 7 & 9 & 10 & \end{array},$$

in which the diagonal positions are blank, the following incidence matrix gives a triangular design, T9 in Clatworthy (1973), with parameters $v = b = 10, r = k = 3, \lambda_1 = 1, \lambda_2 = 0, n = 5$:

$$\begin{array}{r} 1 \\ 2 \\ 3 \\ 4 \\ 5 \\ 6 \\ 7 \\ 8 \\ 9 \\ 10 \end{array} \begin{bmatrix} 1 & 0 & 0 & 0 & 0 & 0 & 0 & 0 & 1 & 1 \\ 1 & 0 & 1 & 0 & 1 & 0 & 0 & 0 & 0 & 0 \\ 0 & 0 & 1 & 0 & 0 & 0 & 1 & 0 & 0 & 1 \\ 0 & 0 & 0 & 0 & 1 & 0 & 1 & 0 & 1 & 0 \\ 1 & 0 & 0 & 1 & 0 & 1 & 0 & 0 & 0 & 0 \\ 0 & 0 & 0 & 0 & 0 & 1 & 0 & 1 & 0 & 1 \\ 0 & 0 & 0 & 1 & 0 & 0 & 0 & 1 & 1 & 0 \\ 0 & 1 & 1 & 0 & 0 & 1 & 0 & 0 & 0 & 0 \\ 0 & 1 & 0 & 1 & 1 & 0 & 0 & 0 & 0 & 0 \\ 0 & 1 & 0 & 0 & 0 & 0 & 1 & 1 & 0 & 0 \end{bmatrix}.$$

Here

$$A_1 = \begin{bmatrix} 0 & 1 & 1 & 1 & 1 & 1 & 1 & 0 & 0 & 0 \\ 1 & 0 & 1 & 1 & 1 & 0 & 0 & 1 & 1 & 0 \\ 1 & 1 & 0 & 1 & 0 & 1 & 0 & 1 & 0 & 1 \\ 1 & 1 & 1 & 0 & 0 & 0 & 1 & 0 & 1 & 1 \\ 1 & 1 & 0 & 0 & 0 & 1 & 1 & 1 & 1 & 0 \\ 1 & 0 & 1 & 0 & 1 & 0 & 1 & 1 & 0 & 1 \\ 1 & 0 & 0 & 1 & 1 & 1 & 0 & 0 & 1 & 1 \\ 0 & 1 & 1 & 0 & 1 & 1 & 0 & 0 & 1 & 1 \\ 0 & 1 & 0 & 1 & 1 & 0 & 1 & 1 & 0 & 1 \\ 0 & 0 & 1 & 1 & 0 & 1 & 1 & 1 & 1 & 0 \end{bmatrix},$$

with $n_1 = 6$, i.e., $\boldsymbol{A}_1\boldsymbol{1}_{10} = 6\boldsymbol{1}_{10}$.

(C) An L_i association scheme (see Raghavarao, 1971, Definition 8.4.4). There are $v = s^2$ treatments which are arranged in an $s \times s$ array and $i-2$ mutually orthogonal Latin squares of order s (see Appendix D) are superimposed. Two treatments are first associates if and only if they occur in the same row or column of the array or in positions occupied by the same letter in any of the Latin squares. Two treatments in other cases are second associates. Then

$$n_1 = i(s-1),\ n_2 = (s-i+1)(s-1),\ 2 \le i \le s,$$

$$\boldsymbol{P}_1 = \begin{bmatrix} (i-1)(i-2)+s-2 & (s-i+1)(i-1) \\ (s-i+1)(i-1) & (s-i+1)(s-i) \end{bmatrix},$$

$$\boldsymbol{P}_2 = \begin{bmatrix} i(i-1) & i(s-i) \\ i(s-i) & (s-i)(s-i-1)+s-2 \end{bmatrix},$$

$$\psi_1 = r + (s-i)\lambda_1 - (s-i+1)\lambda_2,\ \psi_2 = r - i\lambda_1 + (i-1)\lambda_2,$$
$$\rho_1 = i(s-1),\ \rho_2 = (s-i+1)(s-1).$$

In general, when $i > 2$, closed expressions of all $\boldsymbol{A}_\beta$ and $\boldsymbol{A}_\beta^{\#}$ are not provided for $\beta > 0$, because they depend on the concrete structures of $i-2$ mutually orthogonal Latin squares of order s. Hence, a special case of L_i, i.e., L_2 being the most commonly used, is stated. [In fact, out of 145 L_i designs given in Clatworthy (1973), there are 90 L_2, 33 L_3, 10 L_4, 6 L_5, 3 L_6, 2 L_7 and one L_8 designs. Originally, L_i, for $i = 4$, 5 and 6, were used by Yates [1936a, Table VIII(b)] to construct BIB designs.] That is, two treatments are first associates if and only if they occur in the same row or column of the array. Then

$$n_1 = 2(s-1),\ n_2 = (s-1)^2,$$

$$\boldsymbol{P}_1 = \begin{bmatrix} s-2 & s-1 \\ s-1 & (s-1)(s-2) \end{bmatrix},\ \boldsymbol{P}_2 = \begin{bmatrix} 2 & 2(s-2) \\ 2(s-2) & (s-2)^2 \end{bmatrix},$$

$$\psi_1 = r + (s-2)\lambda_1 - (s-1)\lambda_2, \ \psi_2 = r - 2\lambda_1 + \lambda_2,$$

$$\rho_1 = 2(s-1), \ \rho_2 = (s-1)^2,$$

$$\boldsymbol{A}_0 = \boldsymbol{I}_v, \ \boldsymbol{A}_1 = \boldsymbol{I}_s \otimes (\mathbf{1}_s\mathbf{1}_s' - \boldsymbol{I}_s) + (\mathbf{1}_s\mathbf{1}_s' - \boldsymbol{I}_s) \otimes \boldsymbol{I}_s,$$

$$\boldsymbol{A}_2 = (\mathbf{1}_s\mathbf{1}_s' - \boldsymbol{I}_s) \otimes (\mathbf{1}_s\mathbf{1}_s' - \boldsymbol{I}_s), \ \boldsymbol{A}_0^{\#} = \frac{1}{v}\mathbf{1}_v\mathbf{1}_v',$$

$$\boldsymbol{A}_1^{\#} = \frac{1}{s}\mathbf{1}_s\mathbf{1}_s' \otimes (\boldsymbol{I}_s - \frac{1}{s}\mathbf{1}_s\mathbf{1}_s') + (\boldsymbol{I}_s - \frac{1}{s}\mathbf{1}_s\mathbf{1}_s') \otimes \frac{1}{s}\mathbf{1}_s\mathbf{1}_s',$$

$$\boldsymbol{A}_2^{\#} = (\boldsymbol{I}_s - \frac{1}{s}\mathbf{1}_s\mathbf{1}_s') \otimes (\boldsymbol{I}_s - \frac{1}{s}\mathbf{1}_s\mathbf{1}_s').$$

Example 6.0.4. Based on the L_2 scheme of nine treatments, the following incidence matrix right gives an L_2 design, LS72 in Clatworthy (1973), with parameters $v = 9, b = 6, r = 4, k = 6, \lambda_1 = 3, \lambda_2 = 2, s = 3$:

$$\begin{array}{ccc} 1 & 2 & 3 \\ 4 & 5 & 6 \\ 7 & 8 & 9 \end{array}; \qquad \begin{array}{c} 1 \\ 2 \\ 3 \\ 4 \\ 5 \\ 6 \\ 7 \\ 8 \\ 9 \end{array} \begin{bmatrix} 0 & 1 & 1 & 0 & 1 & 1 \\ 0 & 1 & 1 & 1 & 0 & 1 \\ 0 & 1 & 1 & 1 & 1 & 0 \\ 1 & 0 & 1 & 0 & 1 & 1 \\ 1 & 0 & 1 & 1 & 0 & 1 \\ 1 & 0 & 1 & 1 & 1 & 0 \\ 1 & 1 & 0 & 0 & 1 & 1 \\ 1 & 1 & 0 & 1 & 0 & 1 \\ 1 & 1 & 0 & 1 & 1 & 0 \end{bmatrix}.$$

(D) A cyclic association scheme (Raghavarao, 1971, Definition 8.4.5). There are v $(= 4t + 1, t$ being an integer) treatments denoted by $0, 1, ..., v - 1$, the elements of an Abelian group M of v elements (see Appendix B). The first associates of the ith treatment are $i + d_1, i + d_2, ..., i + d_{n_1}$ (mod v), while the other treatments are its second associates. Here the elements d_i are nonzero and have to satisfy the following conditions:

(i) The elements are all different and form a subset $D = \{d_1, d_2, ..., d_{n_1}\}$ of the group M.

(ii) Among the $n_1(n_1 - 1)$ differences $d_j - d_{j'}$ $(j, j' = 1, 2, ..., n_1;\ j \neq j')$ reduced mod v, each of the elements of D occurs p_{11}^1 times, whereas each of the other nonzero elements of M occurs p_{11}^2 times.

(iii) For each d_i in D, there exists d_k in D such that $d_k = -d_i$.

Ma (1984) showed that any cyclic association scheme and the design based on it have exclusively the following parameters:

$$v = 4t + 1 (\text{a prime}), \ n_1 = n_2 = 2t,$$

$$\boldsymbol{P}_1 = \begin{bmatrix} t-1 & t \\ t & t \end{bmatrix}, \quad \boldsymbol{P}_2 = \begin{bmatrix} t & t \\ t & t-1 \end{bmatrix},$$

$$\psi_1 = r - \frac{1}{2}[(\lambda_2 - \lambda_1)\sqrt{4t+1} + \lambda_1 + \lambda_2], \quad \rho_1 = 2t,$$

$$\psi_2 = r - \frac{1}{2}[(\lambda_1 - \lambda_2)\sqrt{4t+1} + \lambda_1 + \lambda_2], \quad \rho_2 = 2t.$$

As $4t + 1$ is a prime, the two eigenvalues ψ_1 and ψ_2 are not rational. This is a point different from the previous association schemes. (See also John, 1987, Chapter 4.)

Example 6.0.5. The following incidence matrix gives a cyclic design, C2 in Clatworthy (1973), with parameters $v = 5, b = 10, r = 4, k = 2, \lambda_1 = 2,\ \lambda_2 = 0, t = 1$:

$$\left[\begin{array}{ccccc|ccccc} 1 & 0 & 0 & 1 & 0 & 1 & 0 & 0 & 1 & 0 \\ 0 & 1 & 0 & 0 & 1 & 0 & 1 & 0 & 0 & 1 \\ 1 & 0 & 1 & 0 & 0 & 1 & 0 & 1 & 0 & 0 \\ 0 & 1 & 0 & 1 & 0 & 0 & 1 & 0 & 1 & 0 \\ 0 & 0 & 1 & 0 & 1 & 0 & 0 & 1 & 0 & 1 \end{array}\right],$$

which is 2-resolvable (see Section 6.0.3).

With these descriptions of the standard association schemes with $s = 2$ associate classes, one can calculate the intra-block efficiencies and variancies of basic contrasts for the 2-associate PBIB designs based on any of the above four association schemes. This will be discussed in Chapters 7 and 8.

Though most of PBIB designs utilized in this book are of 2-associate classes, in Chapters 7 and 8 some 3-associate PBIB designs with parameters v, b, r, k, λ_1, λ_2, λ_3 will also be discussed, including their constructions, to show how designs with more numbers of distinct efficiency factors can be obtained. As standard association schemes with three associate classes, which provide 3-associate PBIB designs, the following four schemes are available: (i) Rectangular association scheme (Vartak, 1955), (ii) Group divisible 3-associate association scheme (Roy, 1953-1954), (iii) Cubic association scheme (Raghavarao and Chandrasekhararao, 1964), and (iv) Extended triangular association scheme (John, 1966). But only two of these association schemes will be picked up here, those which provide more practical and easily available 3-associate PBIB designs. These are (i) and (ii), and in what follows are presented under (E) and (F), respectively.

(E) A rectangular association scheme (Raghavarao, 1971, Section 8.12.1). There are $v = st$ treatments arranged in a rectangle of s rows and t columns. Two treatments are first associates if they appear in the same row, second associates if they appear in the same column, otherwise they are third associates. Then

$$n_1 = t - 1, \ n_2 = s - 1, \ n_3 = (s-1)(t-1),$$

$$P_1 = \begin{bmatrix} t-2 & 0 & 0 \\ 0 & 0 & s-1 \\ 0 & s-1 & (s-1)(t-2) \end{bmatrix},$$

$$P_2 = \begin{bmatrix} 0 & 0 & t-1 \\ 0 & s-2 & 0 \\ t-1 & 0 & (s-2)(t-1) \end{bmatrix},$$

$$P_3 = \begin{bmatrix} 0 & 1 & t-2 \\ 1 & 0 & s-2 \\ t-2 & s-2 & (s-2)(t-2) \end{bmatrix},$$

$$\psi_1 = r-\lambda_1+(s-1)(\lambda_2-\lambda_3),\ \psi_2 = r-\lambda_2+(t-1)(\lambda_1-\lambda_3),\ \psi_3 = r-\lambda_1-\lambda_2+\lambda_3,$$

$$\rho_1 = t-1,\ \rho_2 = s-1,\ \rho_3 = (s-1)(t-1),$$

$$\boldsymbol{A}_0 = \boldsymbol{I}_v,\ \boldsymbol{A}_1 = \boldsymbol{I}_s \otimes (\boldsymbol{1}_t\boldsymbol{1}_t' - \boldsymbol{I}_t),\ \boldsymbol{A}_2 = (\boldsymbol{1}_s\boldsymbol{1}_s' - \boldsymbol{I}_s) \otimes \boldsymbol{I}_t,$$

$$\boldsymbol{A}_3 = (\boldsymbol{1}_s\boldsymbol{1}_s' - \boldsymbol{I}_s) \otimes (\boldsymbol{1}_t\boldsymbol{1}_t' - \boldsymbol{I}_t);\ \boldsymbol{A}_0^{\#} = \frac{1}{v}\boldsymbol{1}_v\boldsymbol{1}_v',\ \boldsymbol{A}_1^{\#} = \frac{1}{s}\boldsymbol{1}_s\boldsymbol{1}_s' \otimes (\boldsymbol{I}_t - \frac{1}{t}\boldsymbol{1}_t\boldsymbol{1}_t'),$$

$$\boldsymbol{A}_2^{\#} = (\boldsymbol{I}_s - \frac{1}{s}\boldsymbol{1}_s\boldsymbol{1}_s') \otimes \frac{1}{t}\boldsymbol{1}_t\boldsymbol{1}_t',\ \boldsymbol{A}_3^{\#} = (\boldsymbol{I}_s - \frac{1}{s}\boldsymbol{1}_s\boldsymbol{1}_s') \otimes (\boldsymbol{I}_t - \frac{1}{t}\boldsymbol{1}_t\boldsymbol{1}_t').$$

Example 6.0.6. Based on the 3×2 rectangular scheme of six treatments, the following incidence matrix right gives a rectangular design, No.4 in Table 1 of Sinha, Kageyama and Singh (1993), with parameters $v = 6, b = 8, r = 4, k = 3, \lambda_1 = 0, \lambda_2 = 3, \lambda_3 = 1, s = 3, t = 2$:

$$\begin{matrix} 1 & 2 \\ 3 & 4 \\ 5 & 6 \end{matrix};\qquad \begin{matrix} 1 \\ 2 \\ 3 \\ 4 \\ 5 \\ 6 \end{matrix} \left[\begin{array}{cccc|cccc} 1 & 1 & 0 & 0 & 1 & 1 & 0 & 0 \\ 0 & 0 & 1 & 1 & 0 & 0 & 1 & 1 \\ 1 & 0 & 0 & 1 & 1 & 1 & 0 & 0 \\ 0 & 1 & 1 & 0 & 0 & 0 & 1 & 1 \\ 0 & 1 & 0 & 1 & 1 & 1 & 0 & 0 \\ 1 & 0 & 1 & 0 & 0 & 0 & 1 & 1 \end{array}\right],$$

which is 2-resolvable (see Section 6.0.3).

(F) A group divisible 3-associate association scheme (a special case of Raghavarao, 1971, Section 8.12.6). There are $v = s_1s_2s_3$ treatments each denoted by three indices (i_1, i_2, i_3), $i_1 = 1, 2, ..., s_1$; $i_2 = 1, 2, ..., s_2$; $i_3 = 1, 2, ..., s_3$. Two treatments (i_1, i_2, i_3) and (j_1, j_2, j_3) are the uth associates if only their first $3 - u$ indices are the same. Then

$$n_1 = s_3 - 1,\ n_2 = s_3(s_2 - 1),\ n_3 = s_3s_2(s_1 - 1),$$

$$P_1 = \begin{bmatrix} s_3-2 & 0 & 0 \\ 0 & s_3(s_2-1) & 0 \\ 0 & 0 & s_3s_2(s_1-1) \end{bmatrix},$$

$$P_2 = \begin{bmatrix} 0 & s_3-1 & 0 \\ s_3-1 & s_3(s_2-2) & 0 \\ 0 & 0 & s_3s_2(s_1-1) \end{bmatrix},$$

$$P_3 = \begin{bmatrix} 0 & 0 & s_3-1 \\ 0 & 0 & s_3(s_2-1) \\ s_3-1 & s_3(s_2-1) & s_3s_2(s_1-2) \end{bmatrix},$$

$$\psi_1 = r-\lambda_3+(s_3-1)(\lambda_1-\lambda_3)+s_3(s_2-1)(\lambda_2-\lambda_3),\ \psi_2 = r-\lambda_2+(s_3-1)(\lambda_1-\lambda_2),$$

$$\psi_3 = r-\lambda_1,\ \rho_1 = s_1-1,\ \rho_2 = s_1(s_2-1),\ \rho_3 = s_1s_2(s_3-1),$$

$$A_0 = I_v,\ A_1 = I_{s_1s_2} \otimes (1_{s_3}1'_{s_3} - I_{s_3}),$$

$$A_2 = I_{s_1} \otimes (1_{s_2}1'_{s_2} - I_{s_2}) \otimes 1_{s_3}1'_{s_3},\ A_3 = (1_{s_1}1'_{s_1} - I_{s_1}) \otimes 1_{s_2}1'_{s_3};$$

$$A_0^{\#} = \frac{1}{v}1_v1'_v,\ A_1^{\#} = (I_{s_1} - \frac{1}{s_1}1_{s_1}1'_{s_1}) \otimes \frac{1}{s_2s_3}1_{s_2s_3}1'_{s_2s_3},$$

$$A_2^{\#} = I_{s_1} \otimes (I_{s_2} - \frac{1}{s_2}1_{s_2}1'_{s_2}) \otimes \frac{1}{s_3}1_{s_3}1'_{s_3},\ A_3^{\#} = I_{s_1s_2} \otimes (I_{s_3} - \frac{1}{s_3}1_{s_3}1'_{s_3}).$$

Example 6.0.7. The following incidence matrix shows a 3-associate GD PBIB design with parameters $v = b = 8$, $r = k = 4$, $s_1 = s_2 = s_3 = 2$, $\lambda_1 = 2$, $\lambda_2 = 1$, $\lambda_3 = 2$, by taking the eight treatments as (1, 1, 1), (1, 1, 2), (1, 2, 1), (1, 2, 2), (2, 1, 1), (2, 1, 2), (2, 2, 1), (2, 2, 2):

$$\begin{matrix} (1,1,1) \\ (1,1,2) \\ (1,2,1) \\ (1,2,2) \\ (2,1,1) \\ (2,1,2) \\ (2,2,1) \\ (2,2,2) \end{matrix} \left[\begin{array}{cccc|cccc} 1 & 1 & 0 & 0 & 1 & 0 & 1 & 0 \\ 1 & 1 & 0 & 0 & 0 & 1 & 0 & 1 \\ 0 & 0 & 1 & 1 & 1 & 0 & 0 & 1 \\ 0 & 0 & 1 & 1 & 0 & 1 & 1 & 0 \\ 1 & 0 & 0 & 1 & 0 & 0 & 1 & 1 \\ 0 & 1 & 1 & 0 & 0 & 0 & 1 & 1 \\ 1 & 0 & 1 & 0 & 1 & 1 & 0 & 0 \\ 0 & 1 & 0 & 1 & 1 & 1 & 0 & 0 \end{array}\right],$$

which is 2-resolvable (see Section 6.0.3). Evidently, this design could well be used for a 2^3 factorial experiment, allowing the contrast between main effects of one of the factors to be estimated in the intra-block analysis with full efficiency.

From these descriptions one can easily obtain the efficiencies and variances of the intra-block BLUEs for basic contrasts of a PBIB design that is based on any of the above two association schemes.

Remark 6.0.3. There are a few cases in which the schemes degenerate into other association schemes with two associate classes. For example, in (E), if $\lambda_2 = \lambda_3$, then $\psi_1 = \psi_3$ and a GD association scheme with s groups of t treatments is obtained; if $\lambda_1 = \lambda_3$, then $\psi_2 = \psi_3$ and one gets a GD association scheme with

t groups of s treatments; if $\lambda_1 = \lambda_2$, then $\psi_1 = \psi_2$ only in the case of $s = t$ and then an L_2 association scheme results. In (F), if $\lambda_1 = \lambda_2$, then $\psi_2 = \psi_3$ and one obtains a GD association scheme with s_1 groups of $s_2 s_3$ treatments; if $\lambda_2 = \lambda_3$, then $\psi_1 = \psi_2$ and a GD association scheme with $s_1 s_2$ groups of s_3 treatments is obtained. In other cases, the schemes may not degenerate into any others. Thus, the values of the concurrences λ_i play an important role for possible reduction of the associate classes (see Kageyama, 1974b).

6.0.3 Resolvability

The concept of resolvability (see Chapter 5) introduced by Bose (1942a) was generalized to α-resolvability by Shrikhande and Raghavarao (1963, 1964). This concept can be further generalized to $(\alpha_1, \alpha_2, ..., \alpha_a)$-resolvability as follows (see Kageyama, 1976b). The α-resolvability will be mainly discussed in Chapter 9 because of its rich published results on existence and some statistical arguments in Chapter 5.

Definition 6.0.2. An equireplicate block design is said to be $(\alpha_1, \alpha_2, ..., \alpha_a)$-resolvable if its blocks can be separated into a (≥ 2) disjoint sets (called "superblocks" in Chapter 5) such that the hth set consisting of b_h blocks contains every treatment exactly α_h (≥ 1) times, i.e., the hth set forms an α_h-replication set for each treatment, $h = 1, 2, ..., a$ (another term used for a superblock is a "resolution set"). In particular, when $\alpha_1 = \alpha_2 = \cdots = \alpha_a$ ($= \alpha$, say), it is said to be α-resolvable for $\alpha \geq 1$. A 1-resolvable design is simply called a resolvable block design in the sense of Bose (1942a).

Note that this definition of α-resolvability corresponds to that of α-resolvability introduced by Shrikhande and Raghavarao (1963, 1964). In an α-resolvable BIB design with parameters v, b, $r = \alpha a, k$, λ, it follows from (6.0.1) that $v = \text{rank}(\boldsymbol{N}\boldsymbol{N}') = \text{rank}(\boldsymbol{N}) \leq b - (a-1)$, i.e., $b \geq v + a - 1$ (see also Raghavarao, 1971, p. 61). Hughes and Piper (1976) have derived the same inequality for an $(\alpha_1, \alpha_2, ..., \alpha_a)$-resolvable BIB design.

One of the earliest examples of a resolvable BIB design is the Kirkman (1850a) school girl problem formulated in 1850 and pursued further in another paper (Kirkman, 1850b). The problem was to find different row arrangements such that any two girls would be assigned to the same row exactly on one day. This can be seen as equivalent to finding a resolvable solution of a BIB design with parameters $v = 6t+3$, $b = (2t+1)(3t+1)$, $r = 3t+1$, $k = 3$, $\lambda = 1$. Kirkman himself gave some solutions and many mathematicians worked on this problem in the late 19th and early 20th century. A relevant bibliography can be found in Eckenstein (1912). However, no complete solution was known until Ray-Chaudhuri and Wilson (1971) completely solved the problem.

Example 6.0.8. The following incidence matrix $\boldsymbol{N}$ gives a resolvable BIB design with parameters $v = 9, b = 12, r = 4, k = 3, \lambda = 1$, $a = 4$ (see a design of No. 9* in Table 9.1 of Chapter 9), which is also affine resolvable with $q_1 = 0$ and $q_2 = 1$ (see Definition 6.0.3 and the comment following Theorem 6.0.5, with

$\alpha = 1$, in this section; also Bose, 1942a):

$$N = \left[\begin{array}{ccc|ccc|ccc|ccc} 0&0&1&1&0&0&1&0&0&0&1&0 \\ 1&0&0&0&0&1&1&0&0&1&0&0 \\ 0&1&0&1&0&0&0&0&1&1&0&0 \\ 0&1&0&0&1&0&1&0&0&0&0&1 \\ 0&0&1&0&1&0&0&1&0&1&0&0 \\ 0&1&0&0&0&1&0&1&0&0&1&0 \\ 1&0&0&0&1&0&0&0&1&0&1&0 \\ 1&0&0&1&0&0&0&1&0&0&0&1 \\ 0&0&1&0&0&1&0&0&1&0&0&1 \end{array}\right].$$

Example 6.0.9. Based on the L_2 scheme of nine treatments

$$\begin{array}{ccc} 1 & 2 & 3 \\ 4 & 5 & 6 \\ 7 & 8 & 9 \end{array}$$

the following incidence matrix gives a 2-resolvable L_2 design, LS1 in Clatworthy (1973), with parameters $v = 9, b = 18, r = 4, k = 2, \lambda_1 = 1, \lambda_2 = 0, s = 3, a = 2$:

$$\left[\begin{array}{ccccccccc|ccccccccc} 1&0&1&0&0&0&0&0&0&1&0&0&0&0&0&1&0&0 \\ 1&1&0&0&0&0&0&0&0&0&1&0&0&0&0&0&1&0 \\ 0&1&1&0&0&0&0&0&0&0&0&1&0&0&1&0&0&0 \\ 0&0&0&1&1&0&0&0&0&1&0&0&1&0&0&0&0&0 \\ 0&0&0&0&1&1&0&0&0&0&1&0&0&1&0&0&0&0 \\ 0&0&0&0&0&1&0&0&1&0&0&0&1&0&1&0&0&0 \\ 0&0&0&1&0&0&1&0&0&0&0&0&0&0&0&1&0&1 \\ 0&0&0&0&0&0&1&1&0&0&0&0&0&1&0&0&1&0 \\ 0&0&0&0&0&0&0&1&1&0&0&1&0&0&0&0&0&1 \end{array}\right].$$

Consider a BIB design with parameters $v = 16$, $b = 40$, $r = 15$, $k = 6$, $\lambda = 5$. This design is not resolvable (1-resolvable) as v is not divisible by k, but this design is (3,6,6)-resolvable because it can be generated by the blocks [(0, 1, 3, 8, 9, 11), (1, 2, 4, 9, 10, 12), (2, 3, 5, 10, 11, 13), (3, 4, 6, 11, 12, 14), (4, 5, 7, 12, 13, 15), (5, 6, 8, 13, 14, 0), (6, 7, 9, 14, 15, 1), (7, 8, 10, 15, 0, 2)], [(0, 1, 3, 5, 9, 12), (0, 1, 2, 3, 6, 12) mod 16], where mod n denotes cyclic development of modulo n. This example shows a new possibility of using the $(\alpha_1, \alpha_2, ..., \alpha_a)$-resolvability in BIB designs. For such practical applications, refer to John (1961) and Kageyama (1976b). Also, $(\alpha_1, \alpha_2, ..., \alpha_a)$-resolvable BIB designs are used in the construction of designs with desired properties here. For example, see Theorems 6.3.4 and 6.3.5 later as fundamental methods of construction.

The parameters v, b, r, k of a proper $(\alpha_1, \alpha_2, ..., \alpha_a)$-resolvable design satisfy the equalities

$$b = \sum_{h=1}^{a} b_h, \ r = \sum_{h=1}^{a} \alpha_h, \ vr = bk, \ v\alpha_h = kb_h, \ b\alpha_h = rb_h, \ h = 1, 2, ..., a.$$

Note that the above definition of $(\alpha_1, \alpha_2, ..., \alpha_a)$-resolvability can also be applied to a general block design which may be balanced in several possible senses (see Kageyama, 1976b). However, the definition of "affine" resolvability must be given in a different way.

Here only those $(\alpha_1, \alpha_2, ..., \alpha_a)$-resolvable block designs are considered which have a constant block size within each set (superblock). The constant block size within the hth set is denoted, as in Lemma 5.5.2, by $k_{(h)}$ for $h = 1, 2, ..., a$ (see Mukerjee and Kageyama, 1985).

Definition 6.0.3. An $(\alpha_1, \alpha_2, ..., \alpha_a)$-resolvable design with a constant block size within each set (superblock) is said to be affine $(\alpha_1, \alpha_2, ..., \alpha_a)$-resolvable if:

(i) for $h = 1, 2, ..., a$, every two distinct blocks from the hth set intersect in the same number, say q_{hh}, of treatments;

(ii) for $h \neq h' = 1, 2, ..., a$, every block from the hth set intersects every block of the h'th set in the same number, say $q_{hh'}$, of treatments.

Example 6.0.10. The following incidence matrix gives an affine (2,2,1,1)-resolvable design with parameters $v = 9, b = 12, r = 6$, $a = 4$, $k_{(1)} = k_{(2)} = 6$, $k_{(3)} = k_{(4)} = 3$, $b_1 = b_2 = b_3 = b_4 = 3$, $q_{11} = q_{22} = 3$, $q_{33} = q_{44} = 0$, $q_{12} = 4, q_{13} = q_{14} = 2$, $q_{23} = q_{24} = 2$, $q_{34} = 1$:

$$\left[\begin{array}{ccc|ccc|ccc|ccc}
0 & 1 & 1 & 0 & 1 & 1 & 1 & 0 & 0 & 1 & 0 & 0 \\
0 & 1 & 1 & 1 & 0 & 1 & 0 & 1 & 0 & 0 & 1 & 0 \\
0 & 1 & 1 & 1 & 1 & 0 & 0 & 0 & 1 & 0 & 0 & 1 \\
1 & 0 & 1 & 0 & 1 & 1 & 0 & 1 & 0 & 0 & 0 & 1 \\
1 & 0 & 1 & 1 & 0 & 1 & 0 & 0 & 1 & 1 & 0 & 0 \\
1 & 0 & 1 & 1 & 1 & 0 & 1 & 0 & 0 & 0 & 1 & 0 \\
1 & 1 & 0 & 0 & 1 & 1 & 0 & 0 & 1 & 0 & 1 & 0 \\
1 & 1 & 0 & 1 & 0 & 1 & 1 & 0 & 0 & 0 & 0 & 1 \\
1 & 1 & 0 & 1 & 1 & 0 & 0 & 1 & 0 & 1 & 0 & 0
\end{array}\right].$$

It is evident that for affine $(\alpha_1, \alpha_2, ..., \alpha_a)$-resolvable designs

$$q_{hh}(b_h - 1) = k_{(h)}(\alpha_h - 1) \ \text{ and } \ q_{hh'}b_{h'} = k_{(h)}\alpha_{h'} \ \ (h \neq h' = 1, 2, ..., a).$$

For combinatorial properties of the affine $(\alpha_1, \alpha_2, ..., \alpha_a)$-resolvability, refer to Raghavarao (1962, 1971), Shrikhande and Raghavarao (1964), Kageyama (1973b, 1973c, 1974c, 1976a, 1984), Mukerjee and Kageyama (1985), Hughes and Piper (1976), Kageyama and Tsuji (1979). Here a typical result is shown, in which the number of blocks in each set is constant, say b_0 (as in Remark 5.5.1), and all blocks are of equal size, k. This gives a characterization of attaining the bound $b \geq v + a - 1$ presented following Definition 6.0.2.

Theorem 6.0.5 (Shrikhande and Raghavarao, 1964; Raghavarao, 1971, Theorems 4.6.2 and 5.4.1). *An α-resolvable BIB design with parameters $v, b = b_0 a, r = \alpha a, k, \lambda$ is affine α-resolvable if and only if $b = v + a - 1$.*

Furthermore, note that if $\alpha_1 = \alpha_2 = \cdots = \alpha_a = \alpha$ and $k_{(h)} = k$ for all $h = 1, 2, ..., a$, then $b_1 = b_2 = \cdots = b_a = b_0$ and for an affine α-resolvable proper block design with parameters $v, b = b_0 a, r = \alpha a, k$, in the notation of Definition 6.0.3, $q_{hh} = (\alpha - 1)k/(b_0 - 1)$ $(= q_1$, say) and $q_{hh'} = \alpha k/b_0 = k^2/v$ $(= q_2$, say). It also follows from Theorem 6.0.5 that $q_1 (= q_{hh}) = k + \lambda - r$ for an affine α-resolvable BIB design with parameters v, b, r, k, λ (see Kageyama, 1973b).

In general, given a set of parameters, the construction of resolvable nonproper block designs, having some balance property, is not so simple. In particular, there is not much in the literature devoted to nonproper designs (see, e.g., Kageyama, 1988b). Ceranka, Kageyama and Mejza (1986) presented four different techniques for constructing α-resolvable C-designs (see Section 4.4.2 for this term) which may be nonproper. The four construction techniques are based on dualization, merging of treatments and dualization, complementation, and juxtaposition. These ideas will be explained in the present chapter and used throughout Volume II. For a class of proper block designs, Shrikhande (1976) gave an excellent survey of known combinatorial results on affine resolvable BIB designs.

For (affine) α-resolvable PBIB designs see also Raghavarao (1971, Sections 4.6 and 12.6), Kageyama (1977a), Kageyama and Mohan (1985) and Ghosh, Bhimani and Kageyama (1989). For affine resolvable designs constructed from orthogonal arrays see Bailey, Monod and Morgan (1995). For the statistical analysis of such designs see Ceranka (1975).

6.1 Dualization

Let $\mathcal{D}$ be a block design. Its dual is defined as a design $\mathcal{D}_*$ whose treatment labels are given by the block labels of the design $\mathcal{D}$ and whose block labels are given by the treatment labels of $\mathcal{D}$. So if the design $\mathcal{D}$ has the parameters $(v, b, r, k, \cdots)$, then its dual design $\mathcal{D}_*$ has the parameters $(v_* = b, b_* = v, r_* = k, k_* = r, \cdots)$. Consequently, if $\boldsymbol{N} = (n_{ij})$ is the $v \times b$ incidence matrix of the design $\mathcal{D}$, then the incidence matrix of $\mathcal{D}_*$ is given by the $b \times v$ matrix $\boldsymbol{N}'$, where $\boldsymbol{N}'$ is the transpose of $\boldsymbol{N}$. Thus, the dualization means the interchanging of the role of treatments and blocks in a block design.

If duals of block designs are considered, some reasonable block designs can be obtained. For example, by Theorem 6.0.1 one obtains the following result.

Corollary 6.1.1. *The dual of a symmetric BIB design is again a symmetric BIB design.*

A general result can now be given as follows.

Theorem 6.1.1. *The dual of a* $(\rho_0; \rho_1, ..., \rho_{m-1}; \rho_m)$*-EB design with parameters* $v, b, \boldsymbol{r}, \boldsymbol{k}, \varepsilon_\beta, \rho_\beta, \boldsymbol{L}_\beta$ *for* $\beta = 0, 1, ..., m-1, m$ *is again a* $(\rho_{*0}; \rho_{*1}, ..., \rho_{*m-1}; \rho_{*m})$*-EB design with parameters* $v_* = b, b_* = v, \boldsymbol{r}_* = \boldsymbol{k}, \boldsymbol{k}_* = \boldsymbol{r}, \varepsilon_{*\beta}, \rho_{*\beta}, \boldsymbol{L}_{*\beta}$*, where* $\varepsilon_{*\beta} = \varepsilon_\beta$ *for all* β*,* $\rho_{*\beta} = \rho_\beta$ *for* $\beta = 1, 2, ..., m-1, m$*, but* $\rho_{*0} = b - v + \rho_0$*, and where* $\boldsymbol{L}_{*\beta} = (1-\varepsilon_\beta)^{-1}\boldsymbol{k}^{-\delta}\boldsymbol{N}'\boldsymbol{L}_\beta\boldsymbol{r}^{-\delta}\boldsymbol{N}$ *for* $\beta = 1, 2, ..., m-1, m$*, while* $\boldsymbol{L}_{*0} = \boldsymbol{I}_b - \sum_{\beta=1}^m \boldsymbol{L}_{*\beta} - n^{-1}\boldsymbol{1}_b\boldsymbol{k}'$.

Proof. Applying Lemma 2.3.3 one can write for the original design

$$\boldsymbol{M} = \boldsymbol{r}^{-\delta}\boldsymbol{N}\boldsymbol{k}^{-\delta}\boldsymbol{N}' = \sum_{\beta=1}^{m}(1-\varepsilon_\beta)\boldsymbol{L}_\beta + \frac{1}{n}\boldsymbol{1}_v\boldsymbol{r}',$$

with $\boldsymbol{L}_\beta, \beta = 0, 1, ..., m$, defined in (4.3.7). Analogously, on account of Remark 2.3.1, one obtains for the dual design

$$\boldsymbol{M}_* = \boldsymbol{k}^{-\delta}\boldsymbol{N}'\boldsymbol{r}^{-\delta}\boldsymbol{N} = \sum_{\beta=1}^{m}(1-\varepsilon_\beta)\boldsymbol{L}_{*\beta} + \frac{1}{n}\boldsymbol{1}_b\boldsymbol{k}',$$

with $\boldsymbol{L}_{*\beta} = (1-\varepsilon_\beta)^{-1}\boldsymbol{k}^{-\delta}\boldsymbol{N}'\boldsymbol{L}_\beta\boldsymbol{r}^{-\delta}\boldsymbol{N}$ for $\beta = 1, 2, ..., m$, following from (2.3.3), (4.3.7), (4.4.1) and (4.4.3), and with $\boldsymbol{L}_{*0} = \boldsymbol{I}_b - \sum_{\beta=1}^m \boldsymbol{L}_{*\beta} - n^{-1}\boldsymbol{1}_b\boldsymbol{k}'$ following from the obvious equality $\boldsymbol{I}_b = \sum_{\beta=0}^m \boldsymbol{L}_{*\beta} + n^{-1}\boldsymbol{1}_b\boldsymbol{k}'$. The formula for ρ_{*0} follows from the equalities $v = \sum_{\beta=0}^m \rho_\beta + 1$ and $b = \sum_{\beta=0}^m \rho_{*\beta} + 1$, and the equalities $\rho_{*\beta} = \rho_\beta$ for $\beta = 1, 2, ..., m$, evident from the fact that $\boldsymbol{M}$ and $\boldsymbol{M}_*$ have exactly the same nonzero eigenvalues. □

The theorem applies whether $\rho_m = 0$ or not.

Remark 6.1.1. If all ε_β's are less than 1, i.e., $\rho_0 = 0$, then the dual design has $\rho_{*0} = b - v$. In other words, the dual of a $(0; \rho_1, ..., \rho_{m-1}; \rho_m)$-EB design is a $(\rho_{*0}; \rho_1, ..., \rho_{m-1}; \rho_m)$-EB design, with $\rho_{*0} > 0$, unless $b = v$.

From Theorem 6.1.1, the following corollary can be obtained (see also Saha, 1976, Corollary 1; Ceranka and Mejza, 1978, Corollary 2).

Corollary 6.1.2. *The dual of a* $(0; v-1; 0)$*-EB design, having the incidence matrix* $\boldsymbol{N}$*, with parameters* v*,* b $(> v)$*,* $\boldsymbol{r}$*,* $\boldsymbol{k}$*,* ε_1*,* $\rho_1 = v - 1$*,* $\boldsymbol{L}_1 = \boldsymbol{I}_v - n^{-1}\boldsymbol{1}_v\boldsymbol{r}'$ *is a* $(\rho_{*0};\ \rho_{*1};\ 0)$*-EB design with parameters* $v_* = b$*,* $b_* = v$*,* $\boldsymbol{r}_* = \boldsymbol{k}$*,* $\boldsymbol{k}_* = \boldsymbol{r}$*,* $\varepsilon_{*0} = 1$*,* $\varepsilon_{*1} = \varepsilon_1$*,* $\rho_{*0} = b - v$*,* $\rho_{*1} = v - 1$*,* $\boldsymbol{L}_{*0} = \boldsymbol{I}_b - n^{-1}\boldsymbol{1}_b\boldsymbol{k}' - \boldsymbol{L}_{*1}$*,* $\boldsymbol{L}_{*1} = (1-\varepsilon_1)^{-1}\ \boldsymbol{k}^{-\delta}\boldsymbol{N}'\boldsymbol{L}_1\boldsymbol{r}^{-\delta}\boldsymbol{N}$.

Example 6.1.1. The dual of an affine resolvable BIB design, i.e., $(0;8;0)$-EB with $\varepsilon_1 = 3/4$ and $\boldsymbol{L}_1 = \boldsymbol{I}_9 - 9^{-1}\boldsymbol{1}_9\boldsymbol{1}_9'$, with parameters $v = 9, b = 12, r = 4, k = 3, \lambda = 1, a = 4, q_1 = 0, q_2 = 1$, given in Example 6.0.8 with the incidence matrix $\boldsymbol{N}$, yields a $(3;8;0)$-EB design with parameters $v_* = 12, b_* = 9$, $r_* = 3, k_* = 4$, $\varepsilon_{*0} = 1$, $\varepsilon_{*1} = 3/4$, $\boldsymbol{L}_{*0} = (\boldsymbol{I}_4 - 4^{-1}\boldsymbol{1}_4\boldsymbol{1}_4') \otimes 3^{-1}\boldsymbol{1}_3\boldsymbol{1}_3'$, $\boldsymbol{L}_{*1} = \boldsymbol{I}_4 \otimes (\boldsymbol{I}_3 - 3^{-1}\boldsymbol{1}_3\boldsymbol{1}_3')$, whose incidence matrix is of the form

$$\boldsymbol{N}^* = \boldsymbol{N}' = \begin{array}{r} 1\\2\\3\\4\\5\\6\\7\\8\\9\\10\\11\\12 \end{array} \left[\begin{array}{ccccccccc} 0&1&0&0&0&0&1&1&0\\ 0&0&1&1&0&1&0&0&0\\ 1&0&0&0&1&0&0&0&1\\ \hline 1&0&1&0&0&0&0&1&0\\ 0&0&0&1&1&0&1&0&0\\ 0&1&0&0&0&1&0&0&1\\ \hline 1&1&0&1&0&0&0&0&0\\ 0&0&0&0&1&1&0&1&0\\ 0&0&1&0&0&0&1&0&1\\ \hline 0&1&1&0&1&0&0&0&0\\ 1&0&0&0&0&1&1&0&0\\ 0&0&0&1&0&0&0&1&1 \end{array}\right].$$

In fact, this design is a semi-regular GD design with 12 treatments based on four groups $\{1,2,3\}$, $\{4,5,6\}$, $\{7,8,9\}, \{10,11,12\}$ of three treatments (see Section 6.0.2(A) and Theorem 7.3.4).

Also, by Theorem 6.1.1 with $m = 2$, $\varepsilon_0 = 1$ and ε_1 (< 1), of respective multiplicities ρ_0 and $\rho_1 = v - 1 - \rho_0$, one can get the following corollary (see also Saha, 1976, Corollary 2).

Corollary 6.1.3. *The dual of a* $(\rho_0; \rho_1; 0)$*-EB design* $\boldsymbol{N}$ *with parameters* v, b, $\boldsymbol{r}$, $\boldsymbol{k}$, $\varepsilon_0 = 1$, ε_1, ρ_0, ρ_1, $\boldsymbol{L}_0$, $\boldsymbol{L}_1$ *is a* $(\rho_{*0}; \rho_{*1}; 0)$*-EB design with parameters* $v_* = b$, $b_* = v$, $\boldsymbol{r}_* = \boldsymbol{k}$, $\boldsymbol{k}_* = \boldsymbol{r}$, $\varepsilon_{*0} = 1$, $\varepsilon_{*1} = \varepsilon_1$, $\rho_{*0} = b - v + \rho_0$, $\rho_{*1} = v - 1 - \rho_0 = \rho_1$, $\boldsymbol{L}_{*0} = \boldsymbol{I}_b - n^{-1}\boldsymbol{1}_b\boldsymbol{k}' - \boldsymbol{L}_{*1}$, $\boldsymbol{L}_{*1} = (1 - \varepsilon_1)^{-1}\boldsymbol{k}^{-\delta}\boldsymbol{N}'\boldsymbol{L}_1\boldsymbol{r}^{-\delta}\boldsymbol{N}$.

The dualization has been a popular technique of constructing more block designs from a design. Assume that there exists a BIB design with $r = 2k + 1$ and $\lambda = 1$. Then it is easy to see that the parameters are $v = k(2k - 1)$, $b = 4k^2 - 1$, $r = 2k + 1, k = k$, $\lambda = 1$. Shrikhande (1952) has proved that the dual of this design is a 2-associate PBIB design with parameters $v_* = 4k^2 - 1$, $n_1 = 2k^2$, $p_{11}^1 = p_{11}^2 = k^2$. Hence, it obviously follows (see also Theorem 8.3.7) that an association matrix of the first associates is the incidence matrix of a symmetric BIB design with parameters $v = b = 4k^2 - 1$, $r = k = 2k^2$, $\lambda = k^2$, which is equivalent to a Hadamard matrix of order $4k^2$ [see Theorem 6.8.1(ii)]. This is given in Theorem 5.9.2 of Raghavarao (1971) in another manner.

Remark 6.1.2. The concept of dualization can be generalized by introducing a dual design $\mathcal{D}_*^{(s)}$ with respect to s-tuples for $s \geq 1$, such that if the jth block of the parent design $\mathcal{D}$ with parameters v, b, r, k includes an s-tuple of treatments, $s < k$, then the corresponding block of $\mathcal{D}_*^{(s)}$ will have the jth treatment of $\mathcal{D}_*^{(s)}$. The resulting design has parameters $v_*^{(s)} = b, b_*^{(s)} = \binom{v}{s}$, $r_*^{(s)} = \binom{k}{s}$, $k_*^{(s)}$ being the number of times s-tuples of treatments occur in the parent design, $\lambda_*^{(s)}$ being the number of s-tuples in a set of the numbers of treatments common to any two blocks in the parent design (see Kageyama and Mohan, 1984b). [Note that if the number of times s-tuples of treatments occur and the number of treatments common to any two blocks in the parent design are not constant, then the values of $k_*^{(s)}$ and $\lambda_*^{(s)}$ are also varying.] Here $\mathcal{D}_*^{(1)}$ is the usual dualization $\mathcal{D}_*$ described in this section. In particular, $\mathcal{D}_*^{(2)}$ for symmetric BIB designs and affine α-resolvable BIB designs $\mathcal{D}$ yield BIB designs and GD designs respectively (see Mohan and Kageyama, 1983). This will be used to prove Theorem 8.3.3(8)–(10) (also see a comment following Theorem 8.2.16).

6.2 Complementation

Let $\mathcal{D}$ be a binary block design. Its complement is defined as a binary block design $\tilde{\mathcal{D}}$ having the same number of treatments v and the same number of blocks b as the parent design $\mathcal{D}$ has, and whose jth block receives precisely those treatments which are not assigned to the jth block ($j = 1, 2, ..., b$) of $\mathcal{D}$. Thus, the complement of a binary block design with the $v \times b$ incidence matrix $\boldsymbol{N}$ is a design with the incidence matrix $\mathbf{1}_v\mathbf{1}_b'- \boldsymbol{N}$ and, hence, with replications $b\mathbf{1}_v - \boldsymbol{r}$ and block sizes $v\mathbf{1}_b - \boldsymbol{k}$.

Here a connected equireplicate and proper binary block design is considered as a basic design, for simplicity of argument.

Example 6.2.1. The complement of the design in Example 6.0.8 gives a BIB design with parameters $v = 9, b = 12, r = 8, k = 6, \lambda = 5$, whose incidence matrix is

$$\left[\begin{array}{ccc|ccc|ccc|ccc} 1&1&0&0&1&1&0&1&1&1&0&1\\ 0&1&1&1&1&0&0&1&1&0&1&1\\ 1&0&1&0&1&1&1&1&0&0&1&1\\ 1&0&1&1&0&1&0&1&1&1&1&0\\ 1&1&0&1&0&1&1&0&1&0&1&1\\ 1&0&1&1&1&0&1&0&1&1&0&1\\ 0&1&1&1&0&1&1&1&0&1&0&1\\ 0&1&1&0&1&1&1&0&1&1&1&0\\ 1&1&0&1&1&0&1&1&0&1&1&0 \end{array}\right].$$

The resulting design is again affine 2-resolvable (see Theorem 6.0.5 with $b_0 = 3, a = 4, \alpha = 2$, and a comment after Theorem 6.0.5 with $q_1 = 3$ and $q_2 = 4$).

Theorem 6.2.1. *If $\boldsymbol{N}$ is the $v \times b$ incidence matrix of a binary (ρ_0; ρ_1, ..., ρ_{m-1}; 0)-EB design with parameters v, b, r, k, ε_β, ρ_β, $\boldsymbol{L}_\beta$, $\beta = 0, 1, ..., m-1$, then its complement $\mathbf{1}_v\mathbf{1}_b' - \boldsymbol{N}$ is the incidence matrix of a binary ($\tilde{\rho}_0$; $\tilde{\rho}_1$, ..., $\tilde{\rho}_{m-1}$; 0)-EB design with parameters $\tilde{v} = v$, $\tilde{b} = b$, $\tilde{r} = b - r$, $\tilde{k} = v - k$, $\tilde{\varepsilon}_\beta = 1 - rk(1-\varepsilon_\beta)/[(v-k)(b-r)]$, $\tilde{\rho}_\beta = \rho_\beta$, $\tilde{\boldsymbol{L}}_\beta = \boldsymbol{L}_\beta$, provided that $v(2r-b)/(rk) < \varepsilon_\beta \leq 1$ for $\beta = 0, 1, ..., m-1$. In the extreme case of $\varepsilon_{m-1} = v(2r-b)/(rk)$, the resulting design is ($\tilde{\rho}_0$; $\tilde{\rho}_1$, ..., $\tilde{\rho}_{m-2}$; $\tilde{\rho}_{m-1}$)-EB, where $\tilde{\rho}_{m-1}$ is the multiplicity of*

$\tilde{\varepsilon}_{m-1} = 0$.

Proof. From Section 4.4 and the proof of Theorem 6.1.1 it is evident that the original design satisfies the equalities

$$\boldsymbol{M} = \frac{1}{rk}\boldsymbol{N}\boldsymbol{N}' = \sum_{\beta=1}^{m-1}(1-\varepsilon_\beta)\boldsymbol{L}_\beta + \frac{1}{v}\mathbf{1}_v\mathbf{1}_v'$$

and, on account of (4.4.11),

$$\boldsymbol{I}_v - \frac{1}{rk}\boldsymbol{N}\boldsymbol{N}' = \sum_{\beta=0}^{m-1}\varepsilon_\beta\boldsymbol{L}_\beta.$$

Now, replacing $\boldsymbol{N}$ by $\mathbf{1}_v\mathbf{1}_b' - \boldsymbol{N}$ note that, on account of (2.2.1), $(\mathbf{1}_v\mathbf{1}_b' - \boldsymbol{N})\mathbf{1}_b = (b-r)\mathbf{1}_v$ and $(\mathbf{1}_b\mathbf{1}_v' - \boldsymbol{N}')\mathbf{1}_v = (v-k)\mathbf{1}_b$. This allows to write

$$\begin{aligned}\tilde{\boldsymbol{M}} &= \frac{1}{(v-k)(b-r)}(\mathbf{1}_v\mathbf{1}_b' - \boldsymbol{N})(\mathbf{1}_b\mathbf{1}_v' - \boldsymbol{N}') \\ &= \frac{1}{(v-k)(b-r)}[(b-2r)\mathbf{1}_v\mathbf{1}_v' + \boldsymbol{N}\boldsymbol{N}'] \\ &= \sum_{\beta=1}^{m-1}(1-\tilde{\varepsilon}_\beta)\boldsymbol{L}_\beta + \frac{1}{v}\mathbf{1}_v\mathbf{1}_v'\end{aligned}$$

and

$$\boldsymbol{I}_v - \frac{1}{(v-k)(b-r)}(\mathbf{1}_v\mathbf{1}_b' - \boldsymbol{N})(\mathbf{1}_b\mathbf{1}_v' - \boldsymbol{N}') = \sum_{\beta=0}^{m-1}\tilde{\varepsilon}_\beta\boldsymbol{L}_\beta,$$

where

$$\tilde{\varepsilon}_\beta = 1 - rk(1-\varepsilon_\beta)/[(v-k)(b-r)],\ \beta = 0, 1, ..., m-1,$$

satisfying the condition $0 < \tilde{\varepsilon}_\beta \leq 1$ if and only if $v(2r-b)/(rk) < \varepsilon_\beta \leq 1$. If, however, $\varepsilon_\beta = v(2r-b)/(rk)$ for $\beta = m-1$ (the smallest ε_β), then $\tilde{\varepsilon}_{m-1} = 0$. Thus the proof is complete. □

Note that in Theorem 6.2.1, $\tilde{\varepsilon}_\beta = 1$ if and only if $\varepsilon_\beta = 1$. This theorem does not apply to the case of $\rho_m > 0$ in general, unless $v = 2k$ (i.e., $b = 2r$). In this case $\tilde{\varepsilon}_\beta = \varepsilon_\beta$ for any β.

In a class of proper, equireplicate and binary block designs, when $m = 2, 3$ or more, one can get special cases of Theorem 6.2.1 by referring to BIB designs and PBIB designs considered in Sections 6.0.1 and 6.0.2, respectively.

Corollary 6.2.1. *The complement of a 2-associate PBIB design with parameters* $v, b, r, k, \lambda_1, \lambda_2, \psi_1, \psi_2, \rho_1, \rho_2, \boldsymbol{A}_1^\#, \boldsymbol{A}_2^\#$ *is a 2-associate PBIB design, having the same association scheme as the original, with parameters* $\tilde{v} = v, \tilde{b} = b, \tilde{r} = b-r, \tilde{k} = v-k, \tilde{\lambda}_1 = b-2r+\lambda_1, \tilde{\lambda}_2 = b-2r+\lambda_2, \psi_1, \psi_2, \rho_1, \rho_2, \boldsymbol{A}_1^\#, \boldsymbol{A}_2^\#$, *i.e.,* *is a* $(\tilde{\rho}_0;\tilde{\rho}_1;0)$*-EB or* $(0;\tilde{\rho}_1,\tilde{\rho}_2;0)$*-EB design, the former if* ψ_1 *or* ψ_2 *is equal to* 0, *provided that* $0 \le \psi_i < (b-r)(v-k)$ *for* $i = 1,2$.

Proof. Let $\boldsymbol{N}$ be the incidence matrix of the original design. Then $\tilde{\boldsymbol{N}} = \mathbf{1}_v\mathbf{1}_b' - \boldsymbol{N}$ is the incidence matrix of the complementary design. Because $\boldsymbol{N}\boldsymbol{N}' = r\boldsymbol{A}_0 + \lambda_1\boldsymbol{A}_1 + \lambda_2\boldsymbol{A}_2 = rk\boldsymbol{A}_0^{\#} + \psi_1\boldsymbol{A}_1^{\#} + \psi_2\boldsymbol{A}_2^{\#}$, where $\mathbf{1}_v\mathbf{1}_v' = \boldsymbol{A}_0 + \boldsymbol{A}_1 + \boldsymbol{A}_2 = v\boldsymbol{A}_0^{\#}$, it follows that $\tilde{\boldsymbol{N}}(\tilde{\boldsymbol{N}})' = (\mathbf{1}_v\mathbf{1}_b' - \boldsymbol{N})(\mathbf{1}_b\mathbf{1}_v' - \boldsymbol{N}') = (b-2r)\mathbf{1}_v\mathbf{1}_v' + \boldsymbol{N}\boldsymbol{N}' = (b-r)\boldsymbol{A}_0 + (b-2r+\lambda_1)\boldsymbol{A}_1 + (b-2r+\lambda_2)\boldsymbol{A}_2 = (b-r)(v-k)\boldsymbol{A}_0^{\#} + \psi_1\boldsymbol{A}_1^{\#} + \psi_2\boldsymbol{A}_2^{\#}$, giving $\tilde{\varepsilon}_i = 1 - \psi_i/(\tilde{r}\tilde{k})$, positive if and only if $\psi_i < \tilde{r}\tilde{k}$ for $i = 1,2$, which completes the proof. □

When $\lambda_1 = \lambda_2 = \lambda$ in Corollary 6.2.1, one can get the following result.

Corollary 6.2.2. *The complement of a BIB design with parameters* v, b, r, k, λ, $\varepsilon_1 = \lambda v/(rk)$, $\rho_1 = v-1$, $\boldsymbol{L}_1 = \boldsymbol{I}_v - v^{-1}\mathbf{1}_v\mathbf{1}_v'$ *is a BIB design with parameters* $\tilde{v} = v$, $\tilde{b} = b$, $\tilde{r} = b-r$, $\tilde{k} = v-k$, $\tilde{\lambda} = b-2r+\lambda$, $\tilde{\varepsilon}_1 = \tilde{\lambda}\tilde{v}/(\tilde{r}\tilde{k})$, ρ_1, $\boldsymbol{L}_1$, *i.e., is a binary proper* $(0;v-1;0)$*-EB design, provided that* $v > k+1$ *or, equivalently,* $b > 2r-\lambda$ [*this following also directly from the condition* $\varepsilon_\beta > v(2r-b)/(rk)$ *in Theorem* 6.2.1].

6.3 Supplementation

The technique of supplementation is one of the standard methods of constructing "balanced" designs. It consists in adding one or more supplementary treatments to each of the blocks of the available block designs. Such experiments may be of interest in various fields of research, particularly in agriculture and medicine, where often one or more treatments are used as controls or standards. The reason is as follows: In a comparative experiment, there may be some treatment (or treatments), usually a control, whose position is logically different from the rest of the treatments. For instance, in plant breeding trials, often the available seed material is limited for the new strains as compared to standard (or check) varieties. The basic approach of constructing designs suitable for the indicated situation is to augment (supplement) any standard design (RBD, BIB, PBIB or any such design) with one or more additional treatments. In the literature of the subject, the reinforced (Das, 1958), augmented (Federer, 1956, 1961), supplemented balanced (Pearce, 1960), nearly balanced (Nigam, 1976), and orthogonally supplemented balanced (Caliński, 1971) designs have been considered, all developed for such experiments.

The incidence matrix of a supplemented block design may be written in general as

$$\boldsymbol{N}^* = \begin{bmatrix} \boldsymbol{N}_1 \\ \boldsymbol{N}_2 \end{bmatrix}, \tag{6.3.1}$$

where $\boldsymbol{N}_1$ is the incidence matrix of the basic design with $\boldsymbol{k}_1$, and $\boldsymbol{N}_2$ is that of the supplementary treatments with $\boldsymbol{k}_2$. From (6.3.1) the weighted concurrence matrix of the supplemented design is

$$\boldsymbol{N}^*(\boldsymbol{k}^*)^{-\delta}(\boldsymbol{N}^*)' = \begin{bmatrix} \boldsymbol{N}_1(\boldsymbol{k}^*)^{-\delta}\boldsymbol{N}_1' & \boldsymbol{N}_1(\boldsymbol{k}^*)^{-\delta}\boldsymbol{N}_2' \\ \boldsymbol{N}_2(\boldsymbol{k}^*)^{-\delta}\boldsymbol{N}_1' & \boldsymbol{N}_2(\boldsymbol{k}^*)^{-\delta}\boldsymbol{N}_2' \end{bmatrix}. \tag{6.3.2}$$

Here $\boldsymbol{k}^* = \boldsymbol{k}_1 + \boldsymbol{k}_2$. Depending on the pattern of the submatrices in (6.3.2), following from the structures of $\boldsymbol{N}_1$ and $\boldsymbol{N}_2$, one may have different types of balance in the resulting supplemented block designs (see Caliński and Ceranka, 1974; Puri, Nigam and Narain, 1977; Ceranka and Chudzik, 1984). This technique usually yields non-equireplicate designs. Some typical cases of (6.3.1) are discussed here. More cases will be considered in Chapter 7.

Further, let the incidence matrix $\boldsymbol{N}_i$ have $\boldsymbol{r}_i$ and n_i as the replication vector and the total number of experimental units, respectively, of the ith block design, $i = 1, 2$, and let $\boldsymbol{r}^* = [\boldsymbol{r}_1', \boldsymbol{r}_2']'$ and $n = n_1 + n_2$. In this case the following can be obtained.

Theorem 6.3.1. *Let $\boldsymbol{N}_1$ be the $v \times b$ incidence matrix of a $(0; \rho_1, ..., \rho_{m-1}; 0)$-EB design with parameters v, b, $\boldsymbol{r}_1$, $\boldsymbol{k}_1$, ε_β, ρ_β, $\boldsymbol{L}_\beta$, $\beta = 1, 2, ..., m-1$. Then*

$$\boldsymbol{N}^* = \begin{bmatrix} \boldsymbol{N}_1 \\ \boldsymbol{N}_2 \end{bmatrix} \quad \textit{with } \boldsymbol{N}_2 = \frac{1}{n}\boldsymbol{r}_2(\boldsymbol{k}^*)' \textit{ of size } t \times b$$

is the incidence matrix of a $(\rho_0^*; \rho_1^*, ..., \rho_{m^*-1}^*; 0)$-EB design, for $m^* \leq m+1$, with parameters $v^* = v + t$, $b^* = b$, $\boldsymbol{r}^* = [\boldsymbol{r}_1', \boldsymbol{r}_2']'$, $\boldsymbol{k}^* = (n/n_1)\boldsymbol{k}_1$, $\varepsilon_0^* = 1$, $\varepsilon_\beta^* = 1 - n_1(1-\varepsilon_\beta)/n$, $\rho_0^* = t$, $\rho_\beta^* = \rho_\beta$, $\boldsymbol{L}_0^* = \boldsymbol{I}_{v+t} - n^{-1}\boldsymbol{1}_{v+t}(\boldsymbol{r}^*)' - \sum_{\beta=1}^{m^*-1}\boldsymbol{L}_\beta^*$, $\boldsymbol{L}_\beta^* = \mathrm{diag}[\boldsymbol{L}_\beta : \mathbf{O}]$, $\beta = 1, 2, ..., m^*-1$.

Proof. Parameters v^*, b^*, $\boldsymbol{r}^*$ are obvious. Also $\boldsymbol{k}_2 = \boldsymbol{N}_2'\boldsymbol{1} = (n_2/n)\boldsymbol{k}^*$ and then $\boldsymbol{k}_1 = \boldsymbol{k}^* - \boldsymbol{k}_2 = \boldsymbol{k}^* - (n_2/n)\boldsymbol{k}^* = (n_1/n)\boldsymbol{k}^*$. Hence $\boldsymbol{k}^* = (n/n_1)\boldsymbol{k}_1$. It follows that the matrix $\boldsymbol{M}$ is of the form

$$\begin{bmatrix} \boldsymbol{r}_1^{-\delta} & \mathbf{O} \\ \mathbf{O} & \boldsymbol{r}_2^{-\delta} \end{bmatrix} \begin{bmatrix} \boldsymbol{N}_1(\boldsymbol{k}^*)^{-\delta}\boldsymbol{N}_1' & \boldsymbol{N}_1(\boldsymbol{k}^*)^{-\delta}\boldsymbol{N}_2' \\ \boldsymbol{N}_2(\boldsymbol{k}^*)^{-\delta}\boldsymbol{N}_1' & \boldsymbol{N}_2(\boldsymbol{k}^*)^{-\delta}\boldsymbol{N}_2' \end{bmatrix}$$

$$= \begin{bmatrix} \boldsymbol{r}_1^{-\delta}\boldsymbol{N}_1(\boldsymbol{k}^*)^{-\delta}\boldsymbol{N}_1' & n^{-1}\boldsymbol{1}_v\boldsymbol{r}_2' \\ n^{-1}\boldsymbol{1}_t\boldsymbol{r}_1' & n^{-1}\boldsymbol{1}_t\boldsymbol{r}_2' \end{bmatrix},$$

i.e.,

$$\begin{aligned} \boldsymbol{M}_0^* &= (\boldsymbol{r}^*)^{-\delta}\boldsymbol{N}^*(\boldsymbol{k}^*)^{-\delta}(\boldsymbol{N}^*)' - \frac{1}{n}\boldsymbol{1}_{v+t}(\boldsymbol{r}^*)' \\ &= \mathrm{diag}[\boldsymbol{r}_1^{-\delta}\boldsymbol{N}_1(\boldsymbol{k}^*)^{-\delta}\boldsymbol{N}_1' - \frac{1}{n}\boldsymbol{1}_v\boldsymbol{r}_1' : \mathbf{O}] \\ &= \frac{n_1}{n}\mathrm{diag}[\boldsymbol{r}_1^{-\delta}\boldsymbol{N}_1\boldsymbol{k}_1^{-\delta}\boldsymbol{N}_1' - \frac{1}{n_1}\boldsymbol{1}_v\boldsymbol{r}_1' : \mathbf{O}] \\ &= \frac{n_1}{n}\mathrm{diag}[\sum_{\beta=0}^{m-1}(1-\varepsilon_\beta)\boldsymbol{L}_\beta : \mathbf{O}] \end{aligned}$$

which, from (6.0.0), completes the proof. □

Remark 6.3.1. In Theorem 6.3.1, if one of the efficiency factors ε_β is equal to 1, i.e., $\rho_0 > 0$, then the resulting design is exactly $(\rho_0^*; \rho_1^*, ..., \rho_{m-1}^*; 0)$-EB.

In Theorem 6.3.1, if the basic design is proper, i.e., $\boldsymbol{k} = k\mathbf{1}_b$, then $\boldsymbol{k}^* = (n/b)\mathbf{1}_b$ and $\boldsymbol{N}_2 = b^{-1}\boldsymbol{r}_2\mathbf{1}_b'$. As a block design belonging to this category, a BIB, PBIB design and so on can be utilized to produce new designs. This is further discussed below.

Let $\boldsymbol{N}_1$ be the incidence matrix of some block design. Then as the first resulting design consider the following,

$$\boldsymbol{N}^* = \begin{bmatrix} \boldsymbol{N}_1 \\ \mathbf{1}_s\mathbf{1}_b' \end{bmatrix}, \tag{6.3.3}$$

i.e., with $\boldsymbol{N}_2$ in (6.3.1) representing an orthogonal design (an RBD). Suppose that $\boldsymbol{N}_1$ represents a BIB design. Then the following result is useful.

Theorem 6.3.2. *The existence of a BIB design with parameters v, b, r, k, λ implies the existence of a $(\rho_0^*; \rho_1^*; 0)$-EB design with parameters $v^* = v + s, b^* = b, \boldsymbol{r}^* = [r\mathbf{1}_v', b\mathbf{1}_s']', \boldsymbol{k}^* = k + s$, $\varepsilon_0^* = 1$, $\varepsilon_1^* = 1 - (r - \lambda)/[r(k + s)]$, $\rho_0^* = s, \rho_1^* = v - 1$, $\boldsymbol{L}_0^* = \boldsymbol{I}_{v+s} - (n^*)^{-1}\mathbf{1}_{v+s}(\boldsymbol{r}^*)' - \boldsymbol{L}_1^*$, $\boldsymbol{L}_1^* = \text{diag}[\boldsymbol{I}_v - v^{-1}\mathbf{1}_v\mathbf{1}_v' : \mathbf{O}]$, where $n^* = b(k + s)$ and $s \geq 1$.*

Proof. From (6.3.3) note that $\boldsymbol{N}^*\mathbf{1}_b = \boldsymbol{r}^*$, as given in the theorem, and that

$$\boldsymbol{N}^*(\boldsymbol{k}^*)^{-\delta}(\boldsymbol{N}^*)' = \frac{1}{k+s}\begin{bmatrix} (r-\lambda)\boldsymbol{I}_v + \lambda\mathbf{1}_v\mathbf{1}_v' & r\mathbf{1}_v\mathbf{1}_s' \\ r\mathbf{1}_s\mathbf{1}_v' & b\mathbf{1}_s\mathbf{1}_s' \end{bmatrix},$$

its eigenvalues with respect to $(\boldsymbol{r}^*)^\delta$ being $\mu_1 = \cdots = \mu_s = 0$, $\mu_{s+1} = \cdots = \mu_{v+s-1} = (r-\lambda)/[r(k+s)]$ and $\mu_{v+s} = 1$ (see Lemma 2.3.3). The corresponding eigenvectors of this matrix with respect to $(\boldsymbol{r}^*)^\delta$, when $(\boldsymbol{r}^*)^\delta$-orthonormalized and applied to the formulae (4.3.5) and (4.3.7), give

$$\boldsymbol{M}^* = \frac{r-\lambda}{r(k+s)}\boldsymbol{L}_1^* + \frac{1}{b(k+s)}\mathbf{1}_{v+s}(\boldsymbol{r}^*)',$$

where $\boldsymbol{L}_1^*$ =diag$[\boldsymbol{I}_v - v^{-1}\mathbf{1}_v\mathbf{1}_v' : \mathbf{O}]$. Now, since $\boldsymbol{L}_0^* + \boldsymbol{L}_1^* = \boldsymbol{I}_{v+s} - [b(k+s)]^{-1}\mathbf{1}_{v+s}(\boldsymbol{r}^*)'$, the matrix $\boldsymbol{L}_0^*$ is readily obtainable. The remaining results given in the theorem are obvious. □

Remark 6.3.2. In connection with Theorem 6.3.2, note (see Section 4.4) that the matrix $\boldsymbol{C}_1^*$ can be written as $\boldsymbol{C}_1^* = (\boldsymbol{r}^*)^\delta\boldsymbol{L}_0^* + \varepsilon_1^*(\boldsymbol{r}^*)^\delta\boldsymbol{L}_1^* = (\boldsymbol{r}^*)^\delta - [b(k+s)]^{-1}\boldsymbol{r}^*(\boldsymbol{r}^*)' + (\varepsilon_1^* - 1)(\boldsymbol{r}^*)^\delta\boldsymbol{L}_1^*$, and that a possible choice of its g-inverse is the matrix

$$\begin{aligned} \boldsymbol{\Omega}^* &= \left(\frac{1}{\varepsilon_1^*} - 1\right)\boldsymbol{L}_1^*(\boldsymbol{r}^*)^{-\delta} + (\boldsymbol{r}^*)^{-\delta} \\ &= \left(\frac{1}{\varepsilon_1^*} - 1\right)\begin{bmatrix} \frac{1}{r}\left(\boldsymbol{I}_v - \frac{1}{v}\mathbf{1}_v\mathbf{1}_v'\right) & \mathbf{O} \\ \mathbf{O} & \mathbf{O} \end{bmatrix} + \begin{bmatrix} \frac{1}{r}\boldsymbol{I}_v & \mathbf{O} \\ \mathbf{O} & \frac{1}{b}\boldsymbol{I}_s \end{bmatrix} \end{aligned}$$

(see Section 4.4.1). This means that there are three different variances for the BLUEs of the elementary contrasts estimated in the intra-block analysis.

A generalized version of Theorem 6.3.2 can now easily be given.

Theorem 6.3.3. *If $\boldsymbol{N}_1$ is the incidence matrix of a proper $(\rho_0; \rho_1; 0)$-EB design with parameters $v, b, \boldsymbol{r}, k$, ε_0, ε_1, ρ_0, ρ_1, $\boldsymbol{L}_0$, $\boldsymbol{L}_1$, then the incidence matrix* (6.3.3) *yields a $(\rho_0^*; \rho_1^*; 0)$-EB design with parameters $v^* = v + s, b^* = b, \boldsymbol{r}^* = [\boldsymbol{r}', b\boldsymbol{1}_s']', k^* = k + s$, $\varepsilon_0^* = 1$, $\varepsilon_1^* = (k\varepsilon_1 + s)/(k + s)$, $\rho_0^* = \rho_0 + s, \rho_1^* = \rho_1$, $\boldsymbol{L}_0^* = \boldsymbol{I}_{v+s} - [b(k + s)]^{-1}\boldsymbol{1}_{v+s}(\boldsymbol{r}^*)' - \boldsymbol{L}_1^*$, $\boldsymbol{L}_1^* = \mathrm{diag}[\boldsymbol{L}_1 : \mathbf{O}]$, where $s \geq 1$.*

In the supplemented design (6.3.3), that is of the type usually considered, every supplementary treatment has to be included once in every block. Therefore, the number of replications required for supplementary treatments becomes quite large as compared to the treatments of the basic design. This could cause some inconvenience in many experimental situations, especially when a large number of trials are to be conducted over a number of years and/or locations. This is all the more so when high costs of control treatments are involved. For such situation, Puri and Kageyama (1985) have given supplemented designs for comparative trials where the supplementary treatments need not be replicated in all blocks and at the same time every supplementary treatment occurs equal number of times with every treatment of the basic design in a block, to ensure that the contrasts involving supplementary treatments are estimated with higher efficiency. But most of these designs have at least three efficiency factors. Designs of this type will be discussed in Chapters 7 and 8 in detail.

For constructing supplemented block designs with three efficiency factors, the property of resolvability may also be useful. In the present case, for a block design having the partition of blocks like in the case of $(\alpha_1, \alpha_2, ..., \alpha_a)$-resolvability described in Section 6.0.3, one can produce the following two results to reduce the number of replications of supplementary treatments, as compared with the usual supplemented block designs discussed earlier.

Theorem 6.3.4. *If $\boldsymbol{N}_1 \equiv \boldsymbol{N}$ is the incidence matrix of an $(\alpha_1, \alpha_2, ..., \alpha_a)$-resolvable BIB design with parameters v, b, r, k, λ, then the incidence matrix*

$$\begin{bmatrix} \boldsymbol{N} \\ \mathrm{diag}[\boldsymbol{1}_s\boldsymbol{1}_{b_1}' : \boldsymbol{1}_s\boldsymbol{1}_{b_2}' : \cdots : \boldsymbol{1}_s\boldsymbol{1}_{b_a}'] \end{bmatrix} = \boldsymbol{N}^* \quad (say)$$

yields a $(\rho_0^*; \rho_1^*, \rho_2^*; 0)$-EB design with parameters $v^* = v + sa$, $b^* = b(= \sum_{h=1}^{a} b_h)$, $\boldsymbol{r}^* = [r\boldsymbol{1}_v', b_1\boldsymbol{1}_s', ..., b_a\boldsymbol{1}_s']'$, $k^* = k + s$, $\varepsilon_0^* = 1$, $\varepsilon_1^* = 1 - (r - \lambda)/[r(k + s)]$, $\varepsilon_2^* = k/(k + s)$, $\rho_0^* = a(s - 1) + 1$, $\rho_1^* = v - 1$, $\rho_2^* = a - 1$, $\boldsymbol{L}_0^* = \boldsymbol{I}_{v+sa} - [b(k + s)]^{-1}\boldsymbol{1}_{v+sa}(\boldsymbol{r}^*)' - \boldsymbol{L}_1^* - \boldsymbol{L}_2^*$, $\boldsymbol{L}_1^* = \mathrm{diag}[\boldsymbol{I}_v - v^{-1}\boldsymbol{1}_v\boldsymbol{1}_v' : \mathbf{O}]$, $\boldsymbol{L}_2^* = \mathrm{diag}[\mathbf{O} : (\boldsymbol{I}_a - b^{-1}\boldsymbol{1}_a\boldsymbol{b}') \otimes s^{-1}\boldsymbol{1}_s\boldsymbol{1}_s']$, where $\boldsymbol{b} = [b_1, b_2, ..., b_a]'$ and $s \geq 1$.

Proof. By using the $(\alpha_1, \alpha_2, ..., \alpha_a)$-resolvability for the original design, it follows that, with the incidence matrix $\boldsymbol{N}^*$,

$$\boldsymbol{M}^* = (\boldsymbol{r}^*)^{-\delta}\boldsymbol{N}^*(\boldsymbol{k}^*)^{-\delta}(\boldsymbol{N}^*)'$$

$$= \frac{1}{k+s}\begin{bmatrix} \frac{1}{r}NN' & \frac{\alpha_1}{r}\mathbf{1}_v\mathbf{1}'_s & \cdots & \frac{\alpha_a}{r}\mathbf{1}_v\mathbf{1}'_s \\ \frac{\alpha_1}{b_1}\mathbf{1}_s\mathbf{1}'_v & & & \\ \vdots & & I_a \otimes \mathbf{1}_s\mathbf{1}'_s & \\ \frac{\alpha_a}{b_a}\mathbf{1}_s\mathbf{1}'_v & & & \end{bmatrix},$$

which, by noting that $\alpha_h/b_h = r/b$, $h = 1,2,...,a$, and $(n^*)^{-1}\mathbf{1}_{v^*}(r^*)' = [b(k+s)]^{-1}\mathbf{1}_{v+sa}(r^*)'$ with $r^* = [r\mathbf{1}'_v, b_1\mathbf{1}'_s, ..., b_a\mathbf{1}'_s]'$, yields, on account of (4.3.8),

$$M_0^* = (r^*)^{-\delta}N^*(k^*)^{-\delta}(N^*)' - (n^*)^{-1}\mathbf{1}_{v^*}(r^*)'$$

$$= \frac{1}{k+s}\text{diag}[\frac{1}{r}NN' - \frac{r}{b}\mathbf{1}_v\mathbf{1}'_v : (I_a - \frac{1}{b}\mathbf{1}_a b')\otimes\mathbf{1}_s\mathbf{1}'_s], \quad (6.3.4)$$

where $b = [b_1, b_2, ..., b_a]'$. Because, from the properties of a BIB design,

$$\frac{1}{r}NN' - \frac{r}{b}\mathbf{1}_v\mathbf{1}'_v = \frac{r-\lambda}{r}\left(I_v - \frac{1}{v}\mathbf{1}_v\mathbf{1}'_v\right),$$

one can write, using the notation of the theorem, the equality (6.3.4) as $M_0^* = \mu_1 L_1^* + \mu_2 L_2^*$. This, by checking that $M_0^* L_\beta^* = \mu_\beta L_\beta^*$ for $\beta = 1, 2$, yields the required result. □

Example 6.3.1. Consider for N the resolvable BIB design, mentioned in Example 6.0.8, with parameters $v = 9, b = 12, r = 4, k = 3, \lambda = 1$, having blocks

[(2,7,8), (3,4,6), (1,5,9)], [(1,3,8), (4,5,7), (2,6,9)],
[(1,2,4), (5,6,8), (3,7,9)], [(2,3,5), (1,6,7), (4,8,9)].

Here $a = 4$ and $b_1 = b_2 = b_3 = b_4 = 3$. Then, Theorem 6.3.4 produces a $(\rho_0^*; \rho_1^*, \rho_2^*; 0)$-EB design with parameters, for $s \geq 1$, $v^* = 4s + 9$, $b^* = 12$, $r^* = [4\mathbf{1}'_9, 3\mathbf{1}'_{4s}]'$, $k^* = s + 3$, $\varepsilon_0^* = 1$, $\varepsilon_1^* = (4s+9)/[4(s+3)]$, $\varepsilon_2^* = 3/(s+3)$, $\rho_0^* = 4s - 3$, $\rho_1^* = 8$, $\rho_2^* = 3$, $L_0^* = I_{4s+9} - [12(s+3)]^{-1}\mathbf{1}_{4s+9}(r^*)' - L_1^* - L_2^*$, $L_1^* = \text{diag}[I_9 - 9^{-1}\mathbf{1}_9\mathbf{1}'_9 : \mathbf{O}]$, $L_2^* = \text{diag}[\mathbf{O} : (I_4 - 4^{-1}\mathbf{1}_4\mathbf{1}'_4) \otimes s^{-1}\mathbf{1}_s\mathbf{1}'_s]$, and with the incidence matrix for, e.g., $s = 1$ as

$$N^* = \begin{bmatrix} N \\ \text{diag}[\mathbf{1}'_3, \mathbf{1}'_3, \mathbf{1}'_3, \mathbf{1}'_3] \end{bmatrix},$$

with parameters $v^* = 13$, $b^* = 12$, $r^* = [4\mathbf{1}'_9, 3\mathbf{1}'_4]'$, $k^* = 4$, $\varepsilon_0^* = 1$, $\varepsilon_1^* = 13/16$, $\varepsilon_2^* = 3/4$, $\rho_0^* = 1$, $\rho_1^* = 8$ and $\rho_2^* = 3$.

By the same approach as in Theorem 6.3.4, one can obtain the following theorem.

Theorem 6.3.5. *If N is the incidence matrix of an $(\alpha_1, \alpha_2, ..., \alpha_a)$-resolvable 2-associate PBIB design with parameters v, $b(= \sum_{h=1}^a b_h)$, $r, k, \lambda_1, \lambda_2, \{\psi_1, \psi_2\} = \{0, \psi\}$ (for example, let here $\psi_1 = 0$), $\rho_1, \rho_2, A_1^\#, A_2^\#$, then, for $s \geq 1$,*

$$\begin{bmatrix} N \\ \text{diag}[\mathbf{1}_s\mathbf{1}'_{b_1} : \mathbf{1}_s\mathbf{1}'_{b_2} : \cdots : \mathbf{1}_s\mathbf{1}'_{b_a}] \end{bmatrix}$$

is the incidence matrix of a $(\rho_0^*; \rho_1^*, \rho_2^*; 0)$*-EB design with parameters* $v^* = v + sa$, $b^* = b$, $\boldsymbol{r}^* = [r\mathbf{1}_v', b_1\mathbf{1}_s', ..., b_a\mathbf{1}_s']'$, $\boldsymbol{k}^* = \boldsymbol{k} + s$, $\varepsilon_0^* = 1$, $\varepsilon_1^* = 1 - \psi_2/[r(k+s)]$, $\varepsilon_2^* = k/(k+s)$, $\rho_0^* = \rho_1 + a(s-1) + 1$, $\rho_1^* = \rho_2$, $\rho_2^* = a - 1$, $\boldsymbol{L}_0^* = \boldsymbol{I}_{v+sa} - [b(k+s)]^{-1}\mathbf{1}_{v+sa}(\boldsymbol{r}^*)' - \boldsymbol{L}_1^* - \boldsymbol{L}_2^*$, $\boldsymbol{L}_1^* = \text{diag}[\boldsymbol{A}_2^{\#} : \mathbf{O}]$, $\boldsymbol{L}_2^* = \text{diag}[\mathbf{O} : (\boldsymbol{I}_a - b^{-1}\mathbf{1}_a\boldsymbol{b}') \otimes s^{-1}\mathbf{1}_s\mathbf{1}_s']$, *where* $\boldsymbol{b} = [b_1, b_2, ..., b_a]'$.

Example 6.3.2. Consider for $\boldsymbol{N}$ the resolvable semi-regular GD design, given in Example 6.0.2, based on two groups $\{1,2,3\}$, $\{4,5,6\}$ of three treatments, with parameters $v = 6, b = 9, r = 3, k = 2, \lambda_1 = 0, \lambda_2 = 1, m = 2, n = 3, \psi_1 = 0$, $\psi_2 = 3$, having blocks

$$[(1,4),(2,5),(3,6)],[(1,6),(2,4),(3,5)],[(1,5),(2,6),(3,4)],$$

in which $a = 3, b_1 = b_2 = b_3 = 3$. Then Theorem 6.3.5 produces a $(\rho_0^*; \rho_1^*, \rho_2^*; 0)$-EB design with parameters, for $s \geq 1$, $v^* = 3(s+2)$, $b^* = 9$, $r^* = 3$, $k^* = s+2$, $\varepsilon_0^* = 1$, $\varepsilon_1^* = (s+1)/(s+2)$, $\varepsilon_2^* = 2/(s+2)$, $\rho_0^* = 3s-1$, $\rho_1^* = 4$, $\rho_2^* = 2$, $\boldsymbol{L}_0^* = \boldsymbol{I}_{3(s+2)} - [3(s+2)]^{-1}\mathbf{1}_{3(s+2)}\mathbf{1}_{3(s+2)}' - \boldsymbol{L}_1^* - \boldsymbol{L}_2^*$, $\boldsymbol{L}_1^* = \text{diag}[\boldsymbol{I}_2 \otimes (\boldsymbol{I}_3 - 3^{-1}\mathbf{1}_3\mathbf{1}_3') : \mathbf{O}]$, $\boldsymbol{L}_2^* = \text{diag}[\mathbf{O} : (\boldsymbol{I}_3 - 3^{-1}\mathbf{1}_3\mathbf{1}_3') \otimes s^{-1}\mathbf{1}_s\mathbf{1}_s']$. Its incidence matrix is given for, e.g., $s = 1$ as

$$\boldsymbol{N}^* = \begin{bmatrix} 1 & 0 & 0 & 1 & 0 & 0 & 1 & 0 & 0 \\ 0 & 1 & 0 & 0 & 1 & 0 & 0 & 1 & 0 \\ 0 & 0 & 1 & 0 & 0 & 1 & 0 & 0 & 1 \\ 1 & 0 & 0 & 0 & 1 & 0 & 0 & 0 & 1 \\ 0 & 1 & 0 & 0 & 0 & 1 & 1 & 0 & 0 \\ 0 & 0 & 1 & 1 & 0 & 0 & 0 & 1 & 0 \\ 1 & 1 & 1 & 0 & 0 & 0 & 0 & 0 & 0 \\ 0 & 0 & 0 & 1 & 1 & 1 & 0 & 0 & 0 \\ 0 & 0 & 0 & 0 & 0 & 0 & 1 & 1 & 1 \end{bmatrix}.$$

This design is a (2; 6; 0)-EB design with parameters $v^* = 9$, $b^* = 9$, $r^* = 3$, $k^* = 3$, $\varepsilon_0^* = 1$, $\varepsilon_1^* = 2/3$, $\rho_0^* = 2$, $\rho_1^* = 6$, $\boldsymbol{L}_0^* = \boldsymbol{I}_9 - 9^{-1}\mathbf{1}_9\mathbf{1}_9' - \boldsymbol{L}_1^*$, $\boldsymbol{L}_1^* = \text{diag}[\boldsymbol{I}_2 \otimes (\boldsymbol{I}_3 - 3^{-1}\mathbf{1}_3\mathbf{1}_3') : \boldsymbol{I}_3 - 3^{-1}\mathbf{1}_3\mathbf{1}_3']$. When $s \geq 2$, the design becomes $(\rho_0^*; \rho_1^*, \rho_2^*; 0)$-EB.

6.4 Juxtaposition

This is an obvious and useful method of constructing a block design from other designs, which also includes a copy procedure. A juxtaposed design is a union of some designs, whose incidence matrices are $\boldsymbol{N}_h$, $h = 1, 2, ..., a$, with the common number of treatments. This union is described by the incidence matrix $[\boldsymbol{N}_1 : \boldsymbol{N}_2 : \cdots : \boldsymbol{N}_a]$. An essence of the result is due to the fact that the C-matrix of a design obtained by juxtaposing some designs is given by the summation of the C-matrices of their component designs.

Theorem 6.4.1. *Let* $\boldsymbol{N}$ *be the incidence matrix of a* $(\rho_0; \rho_1, ..., \rho_{m-1}; 0)$*-EB design with parameters* v, b, $\boldsymbol{r}$, $\boldsymbol{k}$, ε_β, ρ_β, $\boldsymbol{L}_\beta$, $\beta = 0, 1, ..., m-1$. *Then any* *juxtaposition* $N^* = 1'_s \otimes N$ *is the incidence matrix of a* $(\rho_0^*; \rho_1^*, ..., \rho_{m-1}^*; 0)$*-EB design with parameters* $v^* = v$, $b^* = sb$, $r^* = sr$, $k^* = 1_s \otimes k$, $\varepsilon_\beta^* = \varepsilon_\beta, \rho_\beta^* = \rho_\beta$, $L_\beta^* = L_\beta$, $\beta = 0, 1, ..., m-1$.

Proof. Since $M^* = (r^*)^{-\delta} N^* (k^*)^{-\delta} (N^*)'$ is exactly the same as the matrix M of the original design (the same applying to the matrix M_0^*), the proof is complete. □

Theorem 6.4.2. *Let* N_h, $h = 1, 2, ..., a$, *be the incidence matrices of a s-associate PBIB designs, based on the same association scheme, with parameters* v, $b^{(h)}$, $r^{(h)}$, $k^{(h)}$, $\lambda_1^{(h)}$, $\lambda_2^{(h)}$, ... , $\lambda_s^{(h)}$, $\psi_1^{(h)}$, $\psi_2^{(h)}$, ... , $\psi_s^{(h)}$, ρ_1, ρ_2, ... , ρ_s, $A_1^\#$, $A_2^\#$, ... , $A_s^\#$. *Then* $N^* = [N_1 : N_2 : \cdots : N_a]$ *is the incidence matrix of a* $(\rho_0^*; \rho_1^*, ..., \rho_{s-1}^*; 0)$*-EB or* $(0; \rho_1^*, \rho_2^*, ..., \rho_s^*; 0)$*-EB design with parameters* $v^* = v$, $b^* = \sum_{h=1}^a b^{(h)}$, $r^* = \sum_{h=1}^a r^{(h)}$, $k^* = [k^{(1)} 1'_{b(1)}, k^{(2)} 1'_{b(2)}, ..., k^{(a)} 1'_{b(a)}]'$, $\varepsilon_t^* = 1 - (\sum_{h=1}^a \psi_t^{(h)}/k^{(h)})/r^*$, ρ_t^* $(\rho_0^* \geq 0)$ *equal to the rank of* $L_t^* = A_t^\#$, *with* $t \in \{0, 1, ..., s\}$ *corresponding to the order* $1 = \varepsilon_0^* > \varepsilon_1^* > \cdots > \varepsilon_{s-1}^* > 0$ *in the first instance or to the order* $1 > \varepsilon_1^* > \cdots > \varepsilon_{s-1}^* > \varepsilon_s^* > 0$ *in the second. Only when* $\psi_t^{(h)} = 0$ *for all* $h = 1, 2, ..., a$ *with a common fixed* t, *the case with* $\varepsilon_0^* = 1$ *and* $\rho_0^* > 0$ *occurs.*

Proof. The parameters v^*, b^*, r^*, k^* are obvious. For $N^* = [N_1 : N_2 : \cdots : N_a]$, $M^* = (r^*)^{-1} N^* (k^*)^{-\delta} (N^*)' = (r^*)^{-1} \sum_{h=1}^a {k^{(h)}}^{-1} N_h N_h'$ which, with the relation between $N_h N_h'$ and the matrices $A^\#$ in Section 6.0.2, completes the proof. □

As a special case, one can get the following result.

Corollary 6.4.1. *If* N_h, $h = 1, 2$, *are incidence matrices of two GD designs with parameters* $v = mn$ *(m groups of n treatments each),* $b^{(h)}$, $r^{(h)}$, $k^{(h)}$, $\lambda_1^{(h)}$, $\lambda_2^{(h)}$, *then the incidence matrix* $N^* = [N_1 : N_2]$ *yields a* $(\rho_0^*; \rho_1^*, \rho_2^*; 0)$*-EB design with parameters* $v^* = mn$, $b^* = b^{(1)} + b^{(2)}$, $r^* = r^{(1)} + r^{(2)}$, $k^* = [k^{(1)} 1'_{b(1)}, k^{(2)} 1'_{b(2)}]'$, $\varepsilon_1^* = 1 - [(r^{(1)} k^{(1)} - v\lambda_2^{(1)})/k^{(1)} + (r^{(2)} k^{(2)} - v\lambda_2^{(2)})/k^{(2)}]/(r^{(1)} + r^{(2)})$, $\varepsilon_2^* = 1 - [(r^{(1)} - \lambda_1^{(1)})/k^{(1)} + (r^{(2)} - \lambda_1^{(2)})/k^{(2)}]/(r^{(1)} + r^{(2)})$, $\rho_1^* = m - 1$, $\rho_2^* = m(n-1)$, $L_1^* = (I_m - m^{-1} 1_m 1'_m) \otimes n^{-1} 1_n 1'_n$, $L_2^* = I_m \otimes (I_n - n^{-1} 1_n 1'_n)$. *In particular,* (i) *when at least one of the two basic GD designs is regular, or one of them is singular whereas the other is semi-regular,* $\rho_0^* = 0$, (ii) *when they are both singular,* ε_2^* *becomes* $\varepsilon_0^* = 1$ *with* $\rho_0^* = m(n-1)$, *and* (iii) *when they are both semi-regular,* ε_1^* *becomes* $\varepsilon_0^* = 1$ *with* $\rho_0^* = m - 1$. *Thus, to be more precise, the resulting design is* $(0; \rho_1^*, \rho_2^*; 0)$*-EB in the case* (i), *possibly with the change of the subscripts, so that* $\varepsilon_1^* > \varepsilon_2^*$, *but is* $(\rho_0^*; \rho_1^*; 0)$*-EB in any of the cases* (ii) *and* (iii).

Theorem 6.4.3. *Let* N *be the* $v \times b$ *incidence matrix of a GD design with parameters* $v = mn$ *(m groups of n treatments each),* b, r, k, λ_1, λ_2. *Then the incidence matrix*

$$[N : 1_m 1'_s \otimes I_n] = N^* \ (say)$$

yields a $(\rho_0^*; \rho_1^*, \rho_2^*, \rho_3^*; 0)$*-EB design with parameters* $v^* = mn$, $b^* = b+ns$, $r^* = r+s$, $\boldsymbol{k}^* = [k\mathbf{1}'_b, m\mathbf{1}'_{ns}]'$, $\varepsilon_1^* = 1-(r-\lambda_1)/[k(r+s)]$, $\varepsilon_2^* = 1-(r-\lambda_1+sk)/[k(r+s)]$, $\varepsilon_3^* = 1-(rk-v\lambda_2)/[k(r+s)]$, $\rho_1^* = (m-1)(n-1)$, $\rho_2^* = n-1$, $\rho_3^* = m-1$, $\boldsymbol{L}_1^* = (\boldsymbol{I}_m - m^{-1}\mathbf{1}_m\mathbf{1}'_m) \otimes (\boldsymbol{I}_n - n^{-1}\mathbf{1}_n\mathbf{1}'_n)$, $\boldsymbol{L}_2^* = m^{-1}\mathbf{1}_m\mathbf{1}'_m \otimes (\boldsymbol{I}_n - n^{-1}\mathbf{1}_n\mathbf{1}'_n)$, $\boldsymbol{L}_3^* = (\boldsymbol{I}_m - m^{-1}\mathbf{1}_m\mathbf{1}'_m) \otimes n^{-1}\mathbf{1}_n\mathbf{1}'_n$ *for a positive integer* s. *In particular,* (i) *when the basic GD design is regular,* $\rho_0^* = 0$, (ii) *when it is singular,* ε_1^* *becomes* $\varepsilon_0^* = 1$ *with* $\rho_0^* = (m-1)(n-1)$, *and* (iii) *when it is semi-regular,* ε_3^* *becomes* $\varepsilon_0^* = 1$ *with* $\rho_0^* = m-1$. *Thus, in fact, the resulting design is* $(0; \rho_1^*, \rho_2^*, \rho_3^*; 0)$*-EB in the case* (i), *but is* $(\rho_0^*; \rho_1^*, \rho_2^*; 0)$*-EB in any of the cases* (ii) *and* (iii), *with appropriate renumbering of the subscripts.*

Proof. For the incidence matrix $\boldsymbol{N}^*$, it follows that

$$\begin{aligned}
\boldsymbol{M}^* &= \frac{1}{r+s}\boldsymbol{N}^*\mathrm{diag}\left[\frac{1}{k}\boldsymbol{I}_b : \frac{1}{m}\boldsymbol{I}_{ns}\right](\boldsymbol{N}^*)' \\
&= \frac{1}{k(r+s)}\boldsymbol{N}\boldsymbol{N}' + \frac{s}{r+s}\left(\frac{1}{m}\mathbf{1}_m\mathbf{1}'_m \otimes \boldsymbol{I}_n\right) \\
&= \frac{1}{k(r+s)}\left\{rk\left(\frac{1}{m}\mathbf{1}_m\mathbf{1}'_m \otimes \frac{1}{n}\mathbf{1}_n\mathbf{1}'_n\right) + (r-\lambda_1)(\boldsymbol{I}_m \otimes \left(\boldsymbol{I}_n - \frac{1}{n}\mathbf{1}_n\mathbf{1}'_n\right))\right. \\
&\quad \left. + (rk - v\lambda_2)(\left(\boldsymbol{I}_m - \frac{1}{m}\mathbf{1}_m\mathbf{1}'_m\right) \otimes \frac{1}{n}\mathbf{1}_n\mathbf{1}'_n)\right\} + \frac{s}{r+s}\left(\frac{1}{m}\mathbf{1}_m\mathbf{1}'_m \otimes \boldsymbol{I}_n\right) \\
&= \frac{r-\lambda_1}{k(r+s)}\boldsymbol{L}_1^* + \frac{r-\lambda_1+sk}{k(r+s)}\boldsymbol{L}_2^* + \frac{rk-v\lambda_2}{k(r+s)}\boldsymbol{L}_3^* + \frac{1}{v}\mathbf{1}_v\mathbf{1}'_v,
\end{aligned}$$

which, after checking the required properties of the matrices $\boldsymbol{L}_\beta^*, \beta = 1,2,3$ (as in the proof of Theorem 6.3.2), completes the proof. □

Note that since there exists a GD design with $k = m$, Theorem 6.4.3 produces also proper and equireplicate designs, as the following example shows.

Example 6.4.1. Consider a semi-regular GD design, based on four groups $\{1,2,3\}$, $\{4,5,6\}$, $\{7,8,9\}$, $\{10,11,12\}$ of three treatments, known as SR41 (Clatworthy, 1973), with parameters $v = 12$, $b = 9$, $r = 3$, $k = 4$, $\lambda_1 = 0$, $\lambda_2 = 1$, $m = 4, n = 3$, whose blocks are given by

(1, 4, 7, 10), (2, 6, 8, 10), (3, 5, 9, 10), (1, 5, 8, 11), (2, 4, 9, 11),

(3, 6, 7, 11), (1, 6, 9, 12), (3, 4, 8, 12), (2, 5, 7, 12).

Then Theorem 6.4.3 yields a $(\rho_0^*; \rho_1^*, \rho_2^*; 0)$-EB design with parameters $v^* = 12$, $b^* = 3s+9$, $r^* = s+3$, $k^* = 4$, $\varepsilon_0^* = 1$, $\varepsilon_1^* = (4s+9)/(4s+12)$, $\varepsilon_2^* = 9/(4s+12)$, $\rho_0^* = 3$, $\rho_1^* = 6$, $\rho_2^* = 2$, $\boldsymbol{L}_0^* = (\boldsymbol{I}_4 - 4^{-1}\mathbf{1}_4\mathbf{1}'_4) \otimes 3^{-1}\mathbf{1}_3\mathbf{1}'_3$, $\boldsymbol{L}_1^* = (\boldsymbol{I}_4 - 4^{-1}\mathbf{1}_4\mathbf{1}'_4) \otimes (\boldsymbol{I}_3 - 3^{-1}\mathbf{1}_3\mathbf{1}'_3)$, $\boldsymbol{L}_2^* = 4^{-1}\mathbf{1}_4\mathbf{1}'_4 \otimes (\boldsymbol{I}_3 - 3^{-1}\mathbf{1}_3\mathbf{1}'_3)$. For example, when $s = 1$, it has the parameters $v^* = b^* = 12, r^* = k^* = 4$, $\varepsilon_0^* = 1$, $\varepsilon_1^* = 13/16$,

$\varepsilon_2^* = 9/16$, $\rho_0^* = 3$, $\rho_1^* = 6$, $\rho_2^* = 2$, with the incidence matrix as

$$N^* = \begin{matrix} 1 \\ 2 \\ 3 \\ 4 \\ 5 \\ 6 \\ 7 \\ 8 \\ 9 \\ 10 \\ 11 \\ 12 \end{matrix} \begin{bmatrix} 1 & 0 & 0 & 1 & 0 & 0 & 1 & 0 & 0 & 1 & 0 & 0 \\ 0 & 1 & 0 & 0 & 1 & 0 & 0 & 0 & 1 & 0 & 1 & 0 \\ 0 & 0 & 1 & 0 & 0 & 1 & 0 & 1 & 0 & 0 & 0 & 1 \\ 1 & 0 & 0 & 0 & 1 & 0 & 0 & 1 & 0 & 1 & 0 & 0 \\ 0 & 0 & 1 & 1 & 0 & 0 & 0 & 0 & 1 & 0 & 1 & 0 \\ 0 & 1 & 0 & 0 & 0 & 1 & 1 & 0 & 0 & 0 & 0 & 1 \\ 1 & 0 & 0 & 0 & 0 & 1 & 0 & 0 & 1 & 1 & 0 & 0 \\ 0 & 1 & 0 & 1 & 0 & 0 & 0 & 1 & 0 & 0 & 1 & 0 \\ 0 & 0 & 1 & 0 & 1 & 0 & 1 & 0 & 0 & 0 & 0 & 1 \\ 1 & 1 & 1 & 0 & 0 & 0 & 0 & 0 & 0 & 1 & 0 & 0 \\ 0 & 0 & 0 & 1 & 1 & 1 & 0 & 0 & 0 & 0 & 1 & 0 \\ 0 & 0 & 0 & 0 & 0 & 0 & 1 & 1 & 1 & 0 & 0 & 1 \end{bmatrix}.$$

6.5 Merging

The effect of merging treatments was first discussed by Pearce (1971) who showed that the merging of two treatments leads to greater precision in the rest of the design, if it has any effect at all. In general, a procedure of merging treatments in a block design may produce also nonbinary block designs.

Theorem 6.5.1. *The existence of a BIB design with parameters v, b, r, k, λ implies the existence of a proper $(0; v^* - 1; 0)$-EB design with parameters $v^* = s, b^* = b, \boldsymbol{r}^* = r\boldsymbol{a}, k^* = k$, $\varepsilon^* = \lambda v/(rk)$, $\boldsymbol{L}^* = \boldsymbol{I}_s - v^{-1}\boldsymbol{1}_s\boldsymbol{a}'$, where $\boldsymbol{a} = [a_1, a_2, ..., a_s]'$ with nonnegative integers a_i such that $\sum_{i=1}^s a_i = v$ for $s < v$.*

Proof. The design is obtained by merging treatments of the starting BIB design into s mutually exclusive subsets of sizes a_1, a_2, ..., a_s. Let $\boldsymbol{N}^*$ be the incidence matrix of the design obtained by the proposed merging. Then the parameters $v^*, b^*, \boldsymbol{r}^*, k^*$ are obvious. Following Section 4.4, consider $(\boldsymbol{r}^*)^{-\delta}\boldsymbol{N}^*(\boldsymbol{k}^*)^{-\delta}(\boldsymbol{N}^*)'$ $(= \boldsymbol{M}^*$, say), where $(\boldsymbol{r}^*)^\delta = r \cdot \text{diag}[a_1, a_2, ..., a_s]$ and $(\boldsymbol{k}^*)^\delta = k\boldsymbol{I}_b$. Since the (i, i)th or (i, j)th element of $\boldsymbol{N}^*(\boldsymbol{N}^*)'$ is $a_i r + a_i(a_i - 1)\lambda$ or $\lambda a_i a_j$, respectively, it follows that

$$\boldsymbol{M}^* = \frac{1}{rk}[(r-\lambda)\boldsymbol{I}_s + \lambda\boldsymbol{1}_s\boldsymbol{a}'],\ \boldsymbol{M}_0^* = \boldsymbol{M}^* - \frac{1}{v}\boldsymbol{1}_s\boldsymbol{a}' = \frac{r-\lambda}{rk}\left(\boldsymbol{I}_s - \frac{1}{v}\boldsymbol{1}_s\boldsymbol{a}'\right)$$

which show that the resulting design is $(0; v^* - 1; 0)$-EB with $\varepsilon^* = 1 - (r - \lambda)/(rk) = \lambda v/(rk)$. Thus the proof is complete. □

Example 6.5.1. Consider a BIB design (see No. 18 in Table 8.2) with parameters $v = 6$, $b = 10$, $r = 5$, $k = 3$, $\lambda = 2$, whose incidence matrix is given by

$$N = \begin{bmatrix} 1 & 0 & 1 & 0 & 1 & 1 & 0 & 0 & 0 & 1 \\ 1 & 1 & 0 & 1 & 0 & 1 & 1 & 0 & 0 & 0 \\ 0 & 1 & 1 & 0 & 1 & 0 & 1 & 1 & 0 & 0 \\ 1 & 0 & 1 & 1 & 0 & 0 & 0 & 1 & 1 & 0 \\ 0 & 1 & 0 & 1 & 1 & 0 & 0 & 0 & 1 & 1 \\ 0 & 0 & 0 & 0 & 0 & 1 & 1 & 1 & 1 & 1 \end{bmatrix}.$$

By merging the first two consecutive treatments, one obtains the following incidence matrix,

$$N^* = \begin{bmatrix} 2 & 1 & 1 & 1 & 1 & 2 & 1 & 0 & 0 & 1 \\ 0 & 1 & 1 & 0 & 1 & 0 & 1 & 1 & 0 & 0 \\ 1 & 0 & 1 & 1 & 0 & 0 & 0 & 1 & 1 & 0 \\ 0 & 1 & 0 & 1 & 1 & 0 & 0 & 0 & 1 & 1 \\ 0 & 0 & 0 & 0 & 0 & 1 & 1 & 1 & 1 & 1 \end{bmatrix},$$

which represents a (0; 4; 0)-EB design with parameters $v^* = 5$, $b^* = 10$, $r^* = [10, 51_4']'$, $k^* = 3$, $\varepsilon_1^* = 4/5$, $\rho_1^* = 4$, $L_1^* = I_5 - 6^{-1}1_5[2,1,1,1,1]$.

Theorem 6.5.1 has been generalized to the following theorem.

Theorem 6.5.2 (Puri and Nigam, 1975b). *If there exists a $(0; v-1; 0)$-EB design with v treatments in b blocks, then the design obtained by replacing a_u ($u = 1, 2, ..., s$; $\sum_u a_u = v$) treatments by a newly defined treatment is again a $(0; s-1; 0)$-EB design with s treatments, b blocks and the same efficiency factor as that of the original design.*

In particular, when there exists an equireplicate $(0; v-1; 0)$-EB design with parameters v, b, r, $\boldsymbol{k}$, the design obtained by replacing a_u ($u = 1, 2, ..., s$; $\sum_u a_u = v$) treatments by a newly defined treatment is a $(0; s-1; 0)$-EB design with parameters s, b, $r^* = r[a_1, a_2, ..., a_s]'$, $\boldsymbol{k}$ and with the same efficiency factor as that of the original design.

Puri and Nigam (1983) extended further the above-mentioned results, especially for group divisible (GD) designs. Because binary block designs are of prime interest, binary and nonequireplicate "balanced" designs will be introduced here.

Consider a GD design with parameters $v = mn, b, r, k, \lambda_1, \lambda_2$. Let the treatments be arranged into m groups according to the following array:

$$\begin{bmatrix} 1 & 2 & \cdots & n \\ n+1 & n+2 & \cdots & 2n \\ \vdots & \vdots & & \vdots \\ (i-1)n+1 & (i-1)n+2 & \cdots & in \\ \vdots & \vdots & & \vdots \\ (m-1)n+1 & (m-1)n+2 & \cdots & mn \end{bmatrix} \begin{matrix} \cdots \text{ 1st group} \\ \cdots \text{ 2nd group} \\ \\ \cdots \ i\text{th group} \\ \\ \cdots \ m\text{th group} \end{matrix}$$

[see Section 6.0.2(A)]. The treatments of each of the m groups are divided into an arbitrary number of disjoint subgroups, s_i of the ith group. The subgroup sizes for the ith group are denoted by $a_{i1}, a_{i2}, ..., a_{is_i}$ such that $\sum_{j=1}^{s_i} a_{ij} = n$ and $1 \leq a_{ij} \leq n$. Then the treatments in each subgroup are merged to give a single treatment of the resulting design $\mathcal{D}$. It is clear that $\mathcal{D}$ has now $\sum_{i=1}^{m} s_i$ treatments with replications as $r[\boldsymbol{a}_1', \boldsymbol{a}_2', ..., \boldsymbol{a}_m']'$, where $\boldsymbol{a}_i = [a_{i1}, a_{i2}, ..., a_{is_i}]'$, $i = 1, 2, ..., m$. In this case some calculation yields the following theorem.

Theorem 6.5.3. *The existence of a GD design with parameters $v = mn, b, r, k, \lambda_1, \lambda_2$ implies the existence of a proper $(\rho_0^*; \rho_1^*, \rho_2^*; 0)$-EB design with parameters $v^* = \sum_{i=1}^{m} s_i$, $b^* = b$, $\boldsymbol{r}^* = r[\boldsymbol{a}_1', \boldsymbol{a}_2', ..., \boldsymbol{a}_m']'$, $k^* = k$, $\varepsilon_1^* = v\lambda_2/(rk)$, $\varepsilon_2^* = 1 - (r - \lambda_1)/(rk)$, $\rho_1^* = m - 1$, $\rho_2^* = v^* - m$, $\boldsymbol{L}_1^* = n^{-1}\text{diag}[\boldsymbol{1}_{s_1}\boldsymbol{a}_1' : \boldsymbol{1}_{s_2}\boldsymbol{a}_2' : \cdots : \boldsymbol{1}_{s_m}\boldsymbol{a}_m'] - (vr)^{-1}\boldsymbol{1}_{v^*}(\boldsymbol{r}^*)'$, $\boldsymbol{L}_2^* = \boldsymbol{I}_{v^*} - n^{-1}\text{diag}[\boldsymbol{1}_{s_1}\boldsymbol{a}_1' : \boldsymbol{1}_{s_2}\boldsymbol{a}_2' : \cdots : \boldsymbol{1}_{s_m}\boldsymbol{a}_m']$. When the basic GD design is regular, $\rho_0^* = 0$.*

In Theorem 6.5.3, if a singular or semi-regular GD design is taken as a starting design, then the resulting design is obviously $(\rho_0^*; \rho_1^*; 0)$-EB. These results are stated below.

Corollary 6.5.1. *The existence of a singular GD design with parameters $v = mn$, b, r, k, $\lambda_1(= r)$, λ_2 implies the existence of a proper $(\rho_0^*; \rho_1^*; 0)$-EB design with parameters $v^* = \sum_{i=1}^{m} s_i$, $b^* = b$, $\boldsymbol{r}^* = r[\boldsymbol{a}_1', \boldsymbol{a}_2', ..., \boldsymbol{a}_m']'$, $k^* = k$, $\varepsilon_0^* = 1$, $\varepsilon_1^* = v\lambda_2/(rk)$, $\rho_0^* = v^* - m$, $\rho_1^* = m - 1$, $\boldsymbol{L}_0^* = \boldsymbol{I}_{v^*} - n^{-1}\text{diag}[\boldsymbol{1}_{s_1}\boldsymbol{a}_1' : \boldsymbol{1}_{s_2}\boldsymbol{a}_2' : \cdots : \boldsymbol{1}_{s_m}\boldsymbol{a}_m']$, $\boldsymbol{L}_1^* = n^{-1}\text{diag}[\boldsymbol{1}_{s_1}\boldsymbol{a}_1' : \boldsymbol{1}_{s_2}\boldsymbol{a}_2' : \cdots : \boldsymbol{1}_{s_m}\boldsymbol{a}_m'] - (vr)^{-1}\boldsymbol{1}_{v^*}(\boldsymbol{r}^*)'$.*

Corollary 6.5.2. *The existence of a semi-regular GD design with parameters $v = mn, b, r, k, \lambda_1, \lambda_2(= rk/v)$ implies the existence of a proper $(\rho_0^*; \rho_1^*; 0)$-EB design with parameters $v^* = \sum_{i=1}^{m} s_i$, $b^* = b$, $\boldsymbol{r}^* = r[\boldsymbol{a}_1', \boldsymbol{a}_2', ..., \boldsymbol{a}_m']'$, $k^* = k$, $\varepsilon_0^* = 1$, $\varepsilon_1^* = 1 - (r - \lambda_1)/(rk)$, $\rho_0^* = m - 1$, $\rho_1^* = v^* - m$, $\boldsymbol{L}_0^* = n^{-1}\text{diag}[\boldsymbol{1}_{s_1}\boldsymbol{a}_1' : \boldsymbol{1}_{s_2}\boldsymbol{a}_2' : \cdots : \boldsymbol{1}_{s_m}\boldsymbol{a}_m'] - (vr)^{-1}\boldsymbol{1}_{v^*}(\boldsymbol{r}^*)'$, $\boldsymbol{L}_1^* = \boldsymbol{I}_{v^*} - n^{-1}\text{diag}[\boldsymbol{1}_{s_1}\boldsymbol{a}_1' : \boldsymbol{1}_{s_2}\boldsymbol{a}_2' : \cdots : \boldsymbol{1}_{s_m}\boldsymbol{a}_m']$.*

When in Theorem 6.5.3 $\lambda_1 = 0$, the resulting design is binary.

Corollary 6.5.3. *The existence of a GD design with parameters $v = mn$ (m groups of n treatments each), b, r, k, $\lambda_1 = 0$, λ_2 implies the existence of a $(\rho_0^*; \rho_1^*, \rho_2^*; 0)$-EB design with parameters $v^* = \sum_{i=1}^{m} s_i$, $b^* = b$, $\boldsymbol{r}^* = r[\boldsymbol{a}_1', \boldsymbol{a}_2', ..., \boldsymbol{a}_m']'$, $k^* = k$, $\varepsilon_1^* = 1 - (rk - v\lambda_2)/(rk)$, $\varepsilon_2^* = 1 - 1/k$, $\rho_1^* = m - 1$, $\rho_2^* = v^* - m$, $\boldsymbol{L}_1^* = n^{-1}\text{diag}[\boldsymbol{1}_{s_1}\boldsymbol{a}_1' : \boldsymbol{1}_{s_2}\boldsymbol{a}_2' : \cdots : \boldsymbol{1}_{s_m}\boldsymbol{a}_m'] - (vr)^{-1}\boldsymbol{1}_{v^*}(\boldsymbol{r}^*)'$, $\boldsymbol{L}_2^* = \boldsymbol{I}_{v^*} - n^{-1}\text{diag}[\boldsymbol{1}_{s_1}\boldsymbol{a}_1' : \boldsymbol{1}_{s_2}\boldsymbol{a}_2' : \cdots : \boldsymbol{1}_{s_m}\boldsymbol{a}_m']$.*

If the number of subgroups in each group is the same, i.e., $s_1 = s_2 = \cdots = s_m (= s$, say), and also the subgroup sizes for merging the treatments in each group are the same, $a_1, a_2, ..., a_s$, i.e., $\boldsymbol{a}_i = [a_1, a_2, ..., a_s]'$ for all i, then one obtains the following result.

Corollary 6.5.4. *The existence of a GD design with parameters $v = mn, b, r, k, \lambda_1, \lambda_2$ implies the existence of a proper $(\rho_0^*; \rho_1^*, \rho_2^*; 0)$-EB design with parameters*

$v^* = ms$, $b^* = b$, $\boldsymbol{r}^* = r\boldsymbol{1}_m \otimes \boldsymbol{a}$, $k^* = k$, $\varepsilon_1^* = v\lambda_2/(rk)$, $\varepsilon_2^* = 1 - (r - \lambda_1)/(rk)$, $\rho_0^* \geq 0$, $\rho_1^* = m - 1$, $\rho_2^* = m(s-1)$, $\boldsymbol{L}_1^* = (\boldsymbol{I}_m - m^{-1}\boldsymbol{1}_m\boldsymbol{1}_m') \otimes n^{-1}\boldsymbol{1}_s\boldsymbol{a}'$, $\boldsymbol{L}_2^* = \boldsymbol{I}_m \otimes (\boldsymbol{I}_s - n^{-1}\boldsymbol{1}_s\boldsymbol{a}')$, where $\boldsymbol{a} = [a_1, a_2, ..., a_s]'$ with $\sum_{i=1}^{s} a_i = n$. In particular, when the original GD design is singular, the resulting design is $(\rho_0^*; \rho_1^*; 0)$-EB with $\rho_0^* = m(s-1)$ and $\rho_1^* = m - 1$. On the other hand, when the original GD design is semi-regular, the resulting design is $(\rho_0^*; \rho_1^*; 0)$-EB with $\rho_0^* = m - 1$ and $\rho_1^* = m(s-1)$. Furthermore, when the original GD design is regular, the resulting design is $(0; \rho_1^*, \rho_2^*; 0)$-EB with $\rho_0^* = 0$.

Note that $\lambda_1 = 0$ cannot be assumed for a singular GD design.

6.6 Patterned method

The patterned method is a generalization of the idea of juxtaposition discussed in the previous section. The method may result in large values of design parameters, however, it is very simple and useful for producing more efficient designs.

Theorem 6.6.1. *Let $\boldsymbol{N}$ be the $v \times b$ incidence matrix of a BIB design with parameters v, b, r, k, λ. Then $\boldsymbol{1}_s\boldsymbol{1}_t' \otimes \boldsymbol{N}$ is the incidence matrix of an r-resolvable $(\rho_0^*; \rho_1^*; 0)$-EB design with parameters $v^* = sv$, $b^* = tb$, $r^* = tr$, $k^* = sk$, $\varepsilon_0^* = 1$, $\varepsilon_1^* = \lambda v/(rk)$, $\rho_0^* = v(s-1)$, $\rho_1^* = v - 1$, $\boldsymbol{L}_0^* = (\boldsymbol{I}_s - s^{-1}\boldsymbol{1}_s\boldsymbol{1}_s') \otimes \boldsymbol{I}_v$, $\boldsymbol{L}_1^* = s^{-1}\boldsymbol{1}_s\boldsymbol{1}_s' \otimes (\boldsymbol{I}_v - v^{-1}\boldsymbol{1}_v\boldsymbol{1}_v')$.*

Proof. Let $\boldsymbol{N}^* = \boldsymbol{1}_s\boldsymbol{1}_t' \otimes \boldsymbol{N}$. Then the parameters v^*, b^*, r^*, k^* are obvious. Furthermore, it follows that, from (6.0.0),

$$\begin{aligned}
\boldsymbol{M}_0^* &= \boldsymbol{M}^* - \frac{1}{v^*}\boldsymbol{1}_{v^*}\boldsymbol{1}_{v^*}' \\
&= \frac{1}{tr}\boldsymbol{N}^*(\boldsymbol{k}^*)^{-\delta}(\boldsymbol{N}^*)' - \frac{1}{sv}\boldsymbol{1}_{sv}\boldsymbol{1}_{sv}' \\
&= \frac{1}{srk}\boldsymbol{1}_s\boldsymbol{1}_s' \otimes \boldsymbol{N}\boldsymbol{N}' - \frac{1}{sv}\boldsymbol{1}_{sv}\boldsymbol{1}_{sv}' \\
&= \frac{1}{rk}\left(\frac{1}{s}\boldsymbol{1}_s\boldsymbol{1}_s'\right) \otimes \left[rk\left(\frac{1}{v}\boldsymbol{1}_v\boldsymbol{1}_v'\right) + (r-\lambda)\left(\boldsymbol{I}_v - \frac{1}{v}\boldsymbol{1}_v\boldsymbol{1}_v'\right)\right] - \frac{1}{sv}\boldsymbol{1}_{sv}\boldsymbol{1}_{sv}' \\
&= \frac{r-\lambda}{rk}\left[\frac{1}{s}\boldsymbol{1}_s\boldsymbol{1}_s' \otimes \left(\boldsymbol{I}_v - \frac{1}{v}\boldsymbol{1}_v\boldsymbol{1}_v'\right)\right]
\end{aligned}$$

which completes the proof. □

Though this design is proper, it does not belong to a category of 2-associate PBIB designs. The resulting design may become a 3-associate PBIB design.

Example 6.6.1. Consider an unreduced BIB design with parameters $v = 4$, $b = 6$, $r = 3, k = 2, \lambda = 1$, i.e., with blocks (1,2), (1,3), (1,4), (2,3), (2,4), (3,4). Then Theorem 6.6.1 produces a 3-resolvable $(\rho_0^*; \rho_1^*; 0)$-EB design with parameters $v^* = 4s$, $b^* = 6t$, $r^* = 3t$, $k^* = 2s$, $\varepsilon_0^* = 1$, $\varepsilon_1^* = 2/3$, $\rho_0^* = 4(s-1)$, $\rho_1^* = 3$, $\boldsymbol{L}_0^* = (\boldsymbol{I}_s - s^{-1}\boldsymbol{1}_s\boldsymbol{1}_s') \otimes \boldsymbol{I}_4$, $\boldsymbol{L}_1^* = s^{-1}\boldsymbol{1}_s\boldsymbol{1}_s' \otimes (\boldsymbol{I}_4 - 4^{-1}\boldsymbol{1}_4\boldsymbol{1}_4')$, and with the incidence matrix, for $s = 2$ and $t = 2$ as

$$N^* = \left[\begin{array}{cccccc|cccccc} 1&1&1&0&0&0&1&1&1&0&0&0\\ 1&0&0&1&1&0&1&0&0&1&1&0\\ 0&1&0&1&0&1&0&1&0&1&0&1\\ 0&0&1&0&1&1&0&0&1&0&1&1\\ 1&1&1&0&0&0&1&1&1&0&0&0\\ 1&0&0&1&1&0&1&0&0&1&1&0\\ 0&1&0&1&0&1&0&1&0&1&0&1\\ 0&0&1&0&1&1&0&0&1&0&1&1 \end{array}\right].$$

Theorem 6.6.2. *Let $\boldsymbol{N}$ be the $v \times b$ incidence matrix of a 2-associate PBIB design with parameters v, b, r, k, λ_1, λ_2, ψ_1, $\psi_2 = 0$, ρ_1, ρ_2, $\boldsymbol{A}_1^\#$, $\boldsymbol{A}_2^\#$. Then $\mathbf{1}_s\mathbf{1}_t' \otimes \boldsymbol{N}$ is the incidence matrix of an r-resolvable $(\rho_0^*; \rho_1^*; 0)$-EB design with parameters $v^* = sv$, $b^* = tb$, $r^* = tr$, $k^* = sk$, $\varepsilon_0^* = 1$, $\varepsilon_1^* = 1 - \psi_1/(rk)$, $\rho_0^* = v(s-1) + \rho_2$, $\rho_1^* = \rho_1$, $\boldsymbol{L}_0^* = (\boldsymbol{I}_s - s^{-1}\mathbf{1}_s\mathbf{1}_s') \otimes \boldsymbol{I}_v + s^{-1}\mathbf{1}_s\mathbf{1}_s' \otimes \boldsymbol{A}_2^\#$, $\boldsymbol{L}_1^* = s^{-1}\mathbf{1}_s\mathbf{1}_s' \otimes \boldsymbol{A}_1^\#$.*

Proof. For $\boldsymbol{N}^* = \mathbf{1}_s\mathbf{1}_t' \otimes \boldsymbol{N}$, it follows that, from (6.0.0),

$$\begin{aligned} \boldsymbol{M}^* &= \frac{1}{tr}\boldsymbol{N}^*(\boldsymbol{k}^*)^{-\delta}(\boldsymbol{N}^*)' = \frac{1}{srk}\mathbf{1}_s\mathbf{1}_s' \otimes \boldsymbol{N}\boldsymbol{N}' \\ &= \frac{1}{rk}\left[\frac{1}{s}\mathbf{1}_s\mathbf{1}_s' \otimes (rk\left(\frac{1}{v}\mathbf{1}_v\mathbf{1}_v'\right) + \psi_1\boldsymbol{A}_1^\# + \psi_2\boldsymbol{A}_2^\#)\right] \\ &= \frac{1}{sv}\mathbf{1}_{sv}\mathbf{1}_{sv}' + \frac{\psi_1}{rk}\left(\frac{1}{s}\mathbf{1}_s\mathbf{1}_s' \otimes \boldsymbol{A}_1^\#\right) + \frac{\psi_2}{rk}\left(\frac{1}{s}\mathbf{1}_s\mathbf{1}_s' \otimes \boldsymbol{A}_2^\#\right) \end{aligned}$$

and

$$\begin{aligned} \frac{1}{sv}\mathbf{1}_{sv}\mathbf{1}_{sv}' + \frac{1}{s}\mathbf{1}_s\mathbf{1}_s' \otimes \boldsymbol{A}_1^\# + \frac{1}{s}\mathbf{1}_s\mathbf{1}_s' \otimes \boldsymbol{A}_2^\# &= \frac{1}{s}\mathbf{1}_s\mathbf{1}_s' \otimes \left(\frac{1}{v}\mathbf{1}_v\mathbf{1}_v' + \boldsymbol{A}_1^\# + \boldsymbol{A}_2^\#\right) \\ &= \frac{1}{s}\mathbf{1}_s\mathbf{1}_s' \otimes \boldsymbol{I}_v \end{aligned}$$

which completes the proof. □

Compare multiplicities of $\varepsilon_0^* = 1$ in Theorems 6.6.1 and 6.6.2. The multiplicity ρ_0^* in Theorem 6.6.2 is greater than ρ_0^* in Theorem 6.6.1. This means that for the same number sv of treatments and the same number of replications, in Theorem 6.6.2 more basic contrasts are estimated with full efficiency than in Theorem 6.6.1. Hence if for given parameters such a PBIB design is available, one would rather recommend the use of Theorem 6.6.2 to construct a $(\rho_0; \rho_1; 0)$-EB design.

Theorem 6.6.3. *If $\boldsymbol{N}$ is the $v \times b$ incidence matrix of a $(\rho_0; \rho_1; 0)$-EB design with parameters $v = 2k$, b, r, k, $\varepsilon_0 = 1$, ε_1, ρ_0, ρ_1, $\boldsymbol{L}_0$, $\boldsymbol{L}_1$, then the pattern given* *by the incidence matrix*

$$N^* = \begin{bmatrix} N & 1_v 1_b' \\ 1_v 1_b' & 1_v 1_b' - N \end{bmatrix}$$

yields a $(\rho_0^*; \rho_1^*, \rho_2^*; 0)$*-EB design with parameters* $v^* = 4k$, $b^* = 4r$, $r^* = 3r$, $k^* = 3k$, $\varepsilon_0^* = 1$, $\varepsilon_1^* = (8+\varepsilon_1)/9$, $\varepsilon_2^* = 8/9$, $\rho_0^* = 2\rho_0$, $\rho_1^* = 2\rho_1$, $\rho_2^* = 1$, $L_0^* = I_2 \otimes L_0$, $L_1^* = I_2 \otimes L_1$, $L_2^* = (I_2 - 2^{-1}1_2 1_2') \otimes (2k)^{-1} 1_{2k} 1_{2k}'$, *where* $\rho_0 + \rho_1 = 2k - 1$.

Proof. The result follows from a straightforward checking of the spectral decomposition (4.3.8) of the relevant matrix M^*, with matrices L_β^* as in the theorem. □

Thus, a $(\rho_0; \rho_1; 0)$-EB design with $v = 2k$ can be used to obtain a $(\rho_0^*; \rho_1^*, \rho_2^*; 0)$-EB design with $\rho_0^* > 0$.

Remark 6.6.1. In Theorem 6.6.3, if $c_{\beta j}$, $j = 1, 2, ..., \rho_\beta$, represent the ρ_β basic contrasts of the original design N which are estimated in the intra-block analysis with efficiency ε_β, $\beta = 0, 1$, then the basic contrasts of the resulting design can be obtained from

$$\left\{\begin{bmatrix} 1 \\ 1 \end{bmatrix} \otimes c_{\beta j}\right\};\ \left\{\begin{bmatrix} 1 \\ -1 \end{bmatrix} \otimes c_{\beta j}\right\};\ \begin{bmatrix} 1 \\ -1 \end{bmatrix} \otimes 1_v,\ \beta = 0, 1;\ j = 1, 2, ..., \rho_\beta,$$

which are estimated intra-block with efficiency $\{(8+\varepsilon_\beta)/9\}, \{(8+\varepsilon_\beta)/9\}$ and $8/9$, respectively. If the above design is used as a two-factor factorial experiment with factors A and B having the numbers of levels 2 and v, respectively, then the specified basic contrasts involve the B factor main effects, the AB interactions and the A factor main effects, respectively.

Although results in this section are aimed mainly at constructing binary designs, some of them can be generalized to cover constructions of nonbinary designs, as the methods presented below show.

Theorem 6.6.4. *Let N be the $v \times b$ incidence matrix of a* $(\rho_0; \rho_1, ..., \rho_{m-1}; 0)$*-EB design with parameters* $v, b, r, k, \varepsilon_\beta, \rho_\beta, L_\beta$, $\beta = 0, 1, ..., m-1$. *Then for nonnegative integers a and c, not both* 0, *the incidence matrix*

$$N^* = \begin{bmatrix} N & a1_v 1_b' & c(1_v 1_b' - N) \\ c(1_v 1_b' - N) & N & a1_v 1_b' \\ a1_v 1_b' & c(1_v 1_b' - N) & N \end{bmatrix} \tag{6.6.1}$$

yields a block design with parameters $v^* = 3v$, $b^* = 3b$, $r^* = (a+c)b + (1-c)r$, $k^* = (a+c)v + (1-c)k$, *and such that*

(0) *when* $c = 0$ *and* $a \geq 1$, *it is a* $(\rho_0^*; \rho_1^*, ..., \rho_m^*; 0)$*-EB design with* $\varepsilon_\beta^* = 1 - (1-\varepsilon_\beta)rk/(r^*k^*)$, $\varepsilon_m^* = 1 - [rk + (r^* - 2r)(k^* - k)]/(r^*k^*)$, $\rho_\beta^* = 3\rho_\beta$, $\rho_m^* = 2$, $L_\beta^* = I_3 \otimes L_\beta$, $\beta = 0, 1, ..., m-1$, $L_m^* = (I_3 - 3^{-1}1_3 1_3') \otimes v^{-1} 1_v 1_v'$;

(i) *when* $c = 1$, *it is a* $(\rho_0^*; \rho_1^*, ..., \rho_m^*; 0)$*-EB design with* $\varepsilon_0^* = 1$, $\varepsilon_\beta^* = 1 - 3(1 - \varepsilon_\beta)rk/(r^*k^*)$, $\varepsilon_m^* = 1 - [(a^2 - a + 1)vb - 3r(v-k)]/(r^*k^*)$, $\rho_0^* = v - 1 + 2\rho_0$, $\rho_\beta^* = 2\rho_\beta$, $\rho_m^* = 2$, $\boldsymbol{L}_0^* = 3^{-1}\boldsymbol{1}_3\boldsymbol{1}_3' \otimes (\boldsymbol{I}_v - v^{-1}\boldsymbol{1}_v\boldsymbol{1}_v') + (\boldsymbol{I}_3 - 3^{-1}\boldsymbol{1}_3\boldsymbol{1}_3') \otimes \boldsymbol{L}_0$, $\boldsymbol{L}_\beta^* = (\boldsymbol{I}_3 - 3^{-1}\boldsymbol{1}_3\boldsymbol{1}_3') \otimes \boldsymbol{L}_\beta$, $\beta = 1, 2, ..., m-1$, $\boldsymbol{L}_m^* = (\boldsymbol{I}_3 - 3^{-1}\boldsymbol{1}_3\boldsymbol{1}_3') \otimes v^{-1}\boldsymbol{1}_v\boldsymbol{1}_v'$;

(ii) *when* $c \geq 2$, *it is a* $(\rho_0^*; \rho_1^*, ..., \rho_{2m-1}^*; 0)$*-EB design with* $\varepsilon_0^* = 1$, $\varepsilon_\beta^* = 1 - (1-c)^2(1 - \varepsilon_\beta)rk/(r^*k^*)$, $\varepsilon_{m-1+\beta}^* = 1 - (c^2 + c + 1)(1 - \varepsilon_\beta)rk/(r^*k^*)$, $\varepsilon_{2m-1}^* = 1 - [(c^2 + c + 1)rk + (a^2 + c^2 - ac)vb - (2c^2 + a + c - ac)vr]/(r^*k^*)$, $\rho_0^* = 3\rho_0$, $\rho_\beta^* = \rho_\beta$, $\rho_{m-1+\beta}^* = 2\rho_\beta$, $\rho_{2m-1}^* = 2$, $\boldsymbol{L}_0^* = \boldsymbol{I}_3 \otimes \boldsymbol{L}_0$, $\boldsymbol{L}_\beta^* = 3^{-1}\boldsymbol{1}_3\boldsymbol{1}_3' \otimes \boldsymbol{L}_\beta$, $\boldsymbol{L}_{m-1+\beta}^* = (\boldsymbol{I}_3 - 3^{-1}\boldsymbol{1}_3\boldsymbol{1}_3') \otimes \boldsymbol{L}_\beta$, $\beta = 1, 2, ..., m-1$, $\boldsymbol{L}_{2m-1}^* = (\boldsymbol{I}_3 - 3^{-1}\boldsymbol{1}_3\boldsymbol{1}_3') \otimes v^{-1}\boldsymbol{1}_v\boldsymbol{1}_v'$.

Proof. The parameters v^*, b^*, r^*, k^* are obvious. Noting that $\boldsymbol{N}^*(\boldsymbol{N}^*)' = \boldsymbol{I}_3 \otimes \boldsymbol{A} + (\boldsymbol{1}_3\boldsymbol{1}_3' - \boldsymbol{I}_3) \otimes \boldsymbol{B}$ with $\boldsymbol{A} = (1 + c^2)\boldsymbol{N}\boldsymbol{N}' + (a^2 + c^2)b\boldsymbol{1}_v\boldsymbol{1}_v' - 2c^2r\boldsymbol{1}_v\boldsymbol{1}_v' = (1 + c^2)rk\boldsymbol{M}_0 + [(1 + c^2)rk + (a^2 + c^2)vb - 2c^2vr]v^{-1}\boldsymbol{1}_v\boldsymbol{1}_v'$ and $\boldsymbol{B} = -c\boldsymbol{N}\boldsymbol{N}' + [(a + c - ac)r + abc]\boldsymbol{1}_v\boldsymbol{1}_v' = -rkc\boldsymbol{M}_0 + [(a + c - ac)vr + abcv - crk]v^{-1}\boldsymbol{1}_v\boldsymbol{1}_v'$, where the incidence matrix $\boldsymbol{N}$ and the matrix $\boldsymbol{M}_0$, defined in (6.0.0), are taken from the starting design, it can be shown that the latter matrix for the resulting design is of the form

$$\begin{aligned}
\boldsymbol{M}_0^* &= \frac{rk}{r^*k^*}\begin{bmatrix} 1+c^2 & -c & -c \\ -c & 1+c^2 & -c \\ -c & -c & 1+c^2 \end{bmatrix} \otimes \boldsymbol{M}_0 \\
&\quad + \frac{1}{r^*k^*}[(1 + c + c^2)rk + (a^2 + c^2 - ac)vb \\
&\quad - (a + c - ac + 2c^2)vr][(\boldsymbol{I}_3 - \frac{1}{3}\boldsymbol{1}_3\boldsymbol{1}_3') \otimes \frac{1}{v}\boldsymbol{1}_v\boldsymbol{1}_v'] \\
&= \sum_{\beta=0}^{m-1} \frac{rk}{r^*k^*}(1 - c)^2(1 - \varepsilon_\beta)\frac{1}{3}\boldsymbol{1}_3\boldsymbol{1}_3' \otimes \boldsymbol{L}_\beta \\
&\quad + \sum_{\beta=0}^{m-1} \frac{rk}{r^*k^*}(1 + c + c^2)(1 - \varepsilon_\beta)(\boldsymbol{I}_3 - \frac{1}{3}\boldsymbol{1}_3\boldsymbol{1}_3') \otimes \boldsymbol{L}_\beta \\
&\quad + \frac{1}{r^*k^*}[(1 + c + c^2)rk + (a^2 + c^2 - ac)vb \\
&\quad - (2c^2 + a + c - ac)vr](\boldsymbol{I}_3 - \frac{1}{3}\boldsymbol{1}_3\boldsymbol{1}_3') \otimes \frac{1}{v}\boldsymbol{1}_v\boldsymbol{1}_v'.
\end{aligned}$$

This completes the proof. □

Remark 6.6.2. By Theorem 6.6.4(i), particular binary designs can be obtained when taking $c = 1$ and $a = 0$ or 1. The properties of these designs are as indicated in the theorem. Further, note that the values of efficiency factors of the designs for $a = 0$ and $a = 1$ are the same.

For nonnegative integers a and c

$$N^* = \begin{bmatrix} N & a\mathbf{1}_v\mathbf{1}_b' & c\mathbf{1}_v\mathbf{1}_b' \\ c\mathbf{1}_v\mathbf{1}_b' & N & a\mathbf{1}_v\mathbf{1}_b' \\ a\mathbf{1}_v\mathbf{1}_b' & c\mathbf{1}_v\mathbf{1}_b' & N \end{bmatrix}. \tag{6.6.2}$$

Because

$$N^*(N^*)' = I_3 \otimes NN' + [(d-e)I_3 + e\mathbf{1}_3\mathbf{1}_3'] \otimes \mathbf{1}_v\mathbf{1}_v'$$

with $d = b(a^2+c^2)$ and $e = r(a+c) + abc$, the following can be obtained by a line similar to Theorem 6.6.4.

Theorem 6.6.5. *Let N be the $v \times b$ incidence matrix of a BIB design with parameters v, b, r, k, λ. Then (6.6.2) yields a $(0;\rho_1^*;\rho_2^*;0)$-EB design N^* with parameters $v^* = 3v$, $b^* = 3b$, $r^* = r + (a+c)b$, $k^* = k + (a+c)v$, $\varepsilon_1^* = 1-(r-\lambda)/(r^*k^*)$, $\varepsilon_2^* = 1-[rk+vb(a^2+c^2-ac)-(a+c)vr]/(r^*k^*)$, $\rho_1^* = 3(v-1)$, $\rho_2^* = 2$, $L_1^* = I_3 \otimes (I_v - v^{-1}\mathbf{1}_v\mathbf{1}_v')$, $L_2^* = (I_3 - 3^{-1}\mathbf{1}_3\mathbf{1}_3') \otimes v^{-1}\mathbf{1}_v\mathbf{1}_v'$.*

Theorem 6.6.6. *Let N be the $v \times b$ incidence matrix of a 2-associate PBIB design with parameters $v, b, r, k, \lambda_1, \lambda_2, \psi_1, \psi_2, \rho_1, \rho_2, A_1^{\#}, A_2^{\#}$. Then (6.6.2) is the incidence matrix of a $(\rho_0^*;\rho_1^*,\rho_2^*,\rho_3^*;0)$-EB design with parameters $v^* = 3v$, $b^* = 3b$, $r^* = r+(a+c)b$, $k^* = k+(a+c)v$, $\varepsilon_1^* = 1-\psi_1/(r^*k^*)$, $\varepsilon_2^* = 1-\psi_2/(r^*k^*)$, $\varepsilon_3^* = 1-[rk+vb(a^2+c^2-ac)-(a+c)vr]/(r^*k^*)$, $\rho_0^* \geq 0$, $\rho_1^* = 3\rho_1$, $\rho_2^* = 3\rho_2$, $\rho_3^* = 2$, $L_1^* = I_3 \otimes A_1^{\#}$, $L_2^* = I_3 \otimes A_2^{\#}$, $L_3^* = (I_3 - 3^{-1}\mathbf{1}_3\mathbf{1}_3') \otimes v^{-1}\mathbf{1}_v\mathbf{1}_v'$. In particular, when one of the ψ_1 and ψ_2 is zero, the resulting design is $(\rho_0^{**};\rho_1^{**},\rho_2^{**};0)$-EB with $\rho_0^{**} \geq 1$.*

A method of adding a new treatment to a starting design is also considered.

Theorem 6.6.7. *Let N be the $v \times b$ incidence matrix of a BIB design with parameters v, b, r, k, λ. Then for a positive integer c the incidence matrix*

$$N^* = \begin{bmatrix} N & \mathbf{1}_v\mathbf{1}_b' - N \\ c\mathbf{1}_b' & \mathbf{0}' \end{bmatrix} \tag{6.6.3}$$

yields a $(0;\rho_1^*,\rho_2^*;0)$-EB design with parameters $v^* = v+1$, $b^* = 2b$, $r^* = b[\mathbf{1}_v', c]'$, $k^* = [(k+c)\mathbf{1}_b', (v-k)\mathbf{1}_b']'$, $\varepsilon_1^* = 1-(v+c)(r-\lambda)/[b(k+c)(v-k)]$, $\varepsilon_2^* = 1 - c(b-r)/[b(k+c)]$, $\rho_1^* = v-1$, $\rho_2^* = 1$, $L_1^* = \text{diag}[I_v - v^{-1}\mathbf{1}_v\mathbf{1}_v' : 0]$ and

$$L_2^* = \frac{1}{v(v+c)} \begin{bmatrix} c\mathbf{1}_v\mathbf{1}_v' & -cv\mathbf{1}_v \\ -v\mathbf{1}_v' & v^2 \end{bmatrix}. \tag{6.6.4}$$

Proof. The parameters v^*, b^*, r^*, k^* need no explanation. Consider the matrix $M^* = (r^*)^{-\delta}N^*(k^*)^{-\delta}(N^*)'$, where $(r^*)^{\delta} = \text{diag}[bI_v : bc]$ and $(k^*)^{\delta} = \text{diag}[(k+c)I_b : (v-k)I_b]$. Then it follows that

$$M^* = \frac{1}{b}\begin{bmatrix} \frac{1}{k+c}NN' + \frac{1}{v-k}(\mathbf{1}_v\mathbf{1}_b' - N)(\mathbf{1}_b\mathbf{1}_v' - N') & \frac{cr}{k+c}\mathbf{1}_v \\ \frac{r}{k+c}\mathbf{1}_v' & \frac{cb}{k+c} \end{bmatrix}$$

$$= \frac{1}{b}\begin{bmatrix} \frac{(r-\lambda)(v+c)}{(v-k)(k+c)}\boldsymbol{I}_v + (\frac{\lambda}{k+c} + \frac{b-2r+\lambda}{v-k})\boldsymbol{1}_v\boldsymbol{1}_v' & \frac{cr}{k+c}\boldsymbol{1}_v \\ \frac{r}{k+c}\boldsymbol{1}_v' & \frac{cb}{k+c} \end{bmatrix}$$

$$= \frac{1}{b(v+c)}\boldsymbol{1}_{v+1}(\boldsymbol{r}^*)' + \frac{(v+c)(r-\lambda)}{b(k+c)(v-k)}\begin{bmatrix} \boldsymbol{I}_v - \frac{1}{v}\boldsymbol{1}_v\boldsymbol{1}_v' & 0 \\ 0 & 0 \end{bmatrix}$$

$$+\frac{c(b-r)}{bv(k+c)(v+c)}\begin{bmatrix} c\boldsymbol{1}_v\boldsymbol{1}_v' & -cv\boldsymbol{1}_v \\ -v\boldsymbol{1}_v' & v^2 \end{bmatrix}$$

which completes the proof. □

Note that in Theorem 6.6.7, $\varepsilon_1^* = \varepsilon_2^*$ holds if and only if $c = (r-\lambda)/(b-2r+\lambda)$.

Theorem 6.6.8. *Let $\boldsymbol{N}$ be the $v \times b$ incidence matrix of a $(0; v-1; 0)$-EB design with parameters $v, b, r, k, \varepsilon, \boldsymbol{L}$. Then for a positive integer c the pattern* (6.6.3) *yields a $(0; \rho_1^*, \rho_2^*; 0)$-EB design with parameters $v^* = v+1$, $b^* = 2b$, $\boldsymbol{r}^* = b[\boldsymbol{1}_v', c]'$, $\boldsymbol{k}^* = [(k+c)\boldsymbol{1}_b', (v-k)\boldsymbol{1}_b']'$, $\varepsilon_1^* = 1-k^2(v+c)(1-\varepsilon)/[v(k+c)(v-k)]$, $\varepsilon_2^* = 1-c(b-r)/[b(k+c)]$, $\rho_1^* = v-1$, $\rho_2^* = 1$, $\boldsymbol{L}_1^* = \mathrm{diag}[\boldsymbol{I}_v - v^{-1}\boldsymbol{1}_v\boldsymbol{1}_v' : 0]$, $\boldsymbol{L}_2^*$ being the same as* (6.6.4).

Proof. A procedure similar to the proof of Theorem 6.6.7 can be applied to establish this theorem. □

Remark 6.6.3. In Theorem 6.6.8, if c is chosen equal to $c = k^2v(1-\varepsilon)/[(v-k)^2 - k^2(1-\varepsilon)]$, then the resulting design becomes $(0; v^*-1; 0)$-EB.

Similarly, the following can be obtained.

Theorem 6.6.9. *The existence of a 2-associate PBIB design with parameters $v, b, r, k, \lambda_1, \lambda_2, \psi_1, \psi_2, \rho_1, \rho_2, \boldsymbol{A}_1^\#, \boldsymbol{A}_2^\#$ implies the existence of a $(\rho_0^*; \rho_1^*, \rho_2^*, \rho_3^*; 0)$-EB design with parameters $v^* = v+1$, $b^* = 2b$, $\boldsymbol{r}^* = b[\boldsymbol{1}_v', c]'$, $\boldsymbol{k}^* = [(k+c)\boldsymbol{1}_b', (v-k)\boldsymbol{1}_b']'$, $\varepsilon_\beta^* = 1-(v+c)\psi_\beta/[b(k+c)(v-k)]$, $\varepsilon_3^* = 1-c(b-r)/[b(k+c)]$, $\rho_0^* \geq 0$, $\rho_\beta^* = \rho_\beta$, $\rho_3^* = 1$, $\boldsymbol{L}_\beta^* = \mathrm{diag}[\boldsymbol{A}_\beta^\# : 0]$, $\beta = 1, 2$, $\boldsymbol{L}_3^*$ being the same as* (6.6.4). *In particular, if one of the ψ_1 and ψ_2 is zero, the resulting design is $(\rho_0^{**}; \rho_1^{**}, \rho_2^{**}; 0)$-EB with $\rho_0^{**} \geq 1$.*

Remark 6.6.4. In Theorem 6.6.9, if c is chosen equal to $c = vk\psi_\beta/[r(v-k)^2 - k\psi_\beta]$ for some β, then the resulting design is $(0; \tilde{\rho}_1, \tilde{\rho}_2; 0)$-EB.

This can also be generalized into two cases.

Theorem 6.6.10. *The existence of a $(0; \rho_1, \rho_2; 0)$-EB design with parameters v, $b, r, k, \varepsilon_0, \varepsilon_1, \rho_0, \rho_1, \boldsymbol{L}_0, \boldsymbol{L}_1$ implies the existence of a $(0; \rho_1^*, \rho_2^*, \rho_3^*; 0)$-EB design with parameters $v^* = v+1$, $b^* = 2b$, $\boldsymbol{r}^* = b[\boldsymbol{1}_v', c]'$, $\boldsymbol{k}^* = [(k+c)\boldsymbol{1}_b', (v-k)\boldsymbol{1}_b']'$, $\varepsilon_\beta^* = 1-k^2(v+c)(1-\varepsilon_\beta)/[v(k+c)(v-k)]$, $\varepsilon_3^* = 1-c(b-r)/[b(k+c)]$, $\rho_\beta^* = \rho_\beta$, $\rho_3^* = 1$, $\boldsymbol{L}_\beta^* = \mathrm{diag}[\boldsymbol{L}_\beta : 0]$, $\beta = 1, 2$, $\boldsymbol{L}_3^*$ being the same as* (6.6.4).

Proof. This follows from an approach similar to the proof of Theorem 6.6.7. □

Corollary 6.6.1. *The existence of a $(\rho_0;\rho_1;0)$-EB design with parameters v, b, r, k, $\varepsilon_0 = 1$, ε_1, ρ_0, ρ_1, $\boldsymbol{L}_0$, $\boldsymbol{L}_1$ implies the existence of a $(\rho_0^*;\rho_1^*,\rho_2^*;0)$-EB design with parameters $v^* = v+1$, $b^* = 2b$, $\boldsymbol{r}^* = b[\boldsymbol{1}_v', c]'$, $\boldsymbol{k}^* = [(k+c)\boldsymbol{1}_b', (v-k)\boldsymbol{1}_b']'$, $\varepsilon_0^* = 1$, $\varepsilon_1^* = 1-k^2(v+c)(1-\varepsilon_1)/[v(k+c)(v-k)]$, $\varepsilon_2^* = 1-c(b-r)/[b(k+c)]$, $\rho_\beta^* = \rho_\beta$, $\rho_2^* = 1$, $\boldsymbol{L}_\beta^* = \mathrm{diag}[\boldsymbol{L}_\beta : 0]$, $\beta = 0,1$, $\boldsymbol{L}_2^*$ being the same as (6.6.4).*

One can also produce nonbinary designs by simply changing the configuration.

Theorem 6.6.11. *The existence of a $(\rho_0;\rho_1,...,\rho_{m-1};0)$-EB design, having the $v\times b$ incidence matrix $\boldsymbol{N}$, with parameters v, b, $\boldsymbol{r}$, $\boldsymbol{k}$, ε_β, $\boldsymbol{L}_\beta$, $\beta = 0,1,...,m-1$, implies the existence of a $(\rho_0^*;\rho_1^*,...,\rho_{m-1}^*;0)$-EB design, having the incidence matrix $[c_1\boldsymbol{N} : c_2\boldsymbol{N}]$, with parameters $v^* = v$, $b^* = 2b$, $\boldsymbol{r}^* = (c_1+c_2)\boldsymbol{r}$, $\boldsymbol{k}^* = [c_1,c_2]'\otimes\boldsymbol{k}$, $\varepsilon_\beta^* = \varepsilon_\beta$, $\rho_\beta^* = \rho_\beta$, $\boldsymbol{L}_\beta^* = \boldsymbol{L}_\beta$, $\beta = 0,1,...,m-1$, where c_1 and c_2 are two nonnegative integers.*

Enlarging or reducing the number of blocks in a nonbinary design is now considered. Suppose there exists a $(\rho_0;\rho_1,...,\rho_{m-1};0)$-EB design whose incidence matrix can be expressed as $[c\boldsymbol{N}_1 : \boldsymbol{N}_2]$, where $\boldsymbol{N}_1$ and $\boldsymbol{N}_2$ are of sizes $v\times b_1$ and $v\times b_2$ respectively. Let there exist any two numbers c_1 and c_2 $(c_1+c_2 = c)$ such that $c_1\boldsymbol{N}_1$ and $c_2\boldsymbol{N}_1$ are matrices with nonnegative elements only, $\boldsymbol{N}_1'\boldsymbol{1}_v = \boldsymbol{k}_1$ and $\boldsymbol{N}_2'\boldsymbol{1}_v = \boldsymbol{k}_2$. Then consider $[c_1\boldsymbol{N}_1 : c_2\boldsymbol{N}_1 : \boldsymbol{N}_2]$ as a new incidence matrix. In this case the following can be obtained.

Theorem 6.6.12. *The existence of a $(\rho_0;\rho_1,...,\rho_{m-1};0)$-EB design with the $v\times b$ incidence matrix $\boldsymbol{N} = [c\boldsymbol{N}_1 : \boldsymbol{N}_2]$ and with parameters v, b, $\boldsymbol{r}$, $\boldsymbol{k} = [c\boldsymbol{k}_1', \boldsymbol{k}_2']'$, ε_β, ρ_β, $\boldsymbol{L}_\beta$, $\beta = 0,1,...,m-1$, implies the existence of a $(\rho_0^*;\rho_1^*,...,\rho_{m-1}^*;0)$-EB design with the incidence matrix $\boldsymbol{N}^* = [c_1\boldsymbol{N}_1 : c_2\boldsymbol{N}_1 : \boldsymbol{N}_2]$ and with parameters $v^* = v$, $b^* = b_1+b$, $\boldsymbol{r}^* = \boldsymbol{r}$, $\boldsymbol{k}^* = [c_1\boldsymbol{k}_1', c_2\boldsymbol{k}_1', \boldsymbol{k}_2']'$, $\varepsilon_\beta^* = \varepsilon_\beta$, $\rho_\beta^* = \rho_\beta$, $\boldsymbol{L}_\beta^* = \boldsymbol{L}_\beta$, $\beta = 0,1,...,m-1$, and vice versa.*

Proof. Since it follows that $\boldsymbol{N}$ and $\boldsymbol{N}^*$ have the same matrix $\boldsymbol{M}$, the proof is immediate. □

Theorem 6.6.12 provides a technique for reducing the number of blocks in the case of designs with repeated blocks. A large number of designs for $m-1 = 1,2,3$ with repeated blocks are available in the literature. These designs, including BIB and PBIB designs, can be used to obtain new designs with different block sizes.

Other patterned methods will be given in Chapters 7, 8 and 10. Here, a typical method of constructing nonproper and nonequireplciate $(0;\rho_1;0)$-EB designs is presented. It is based on an addition of a new treatment to some available block designs.

Theorem 6.6.13. *If $\boldsymbol{N}_h$, $h = 1,2$, are the $v\times b_h$ incidence matrices, the first of a BIB design with parameters $v, b_1, r_1, k_1, \lambda_1$, and the second of a binary $(0;v-1;0)$-EB design with parameters v, b_2, r_2, $\boldsymbol{k}$, ε_1, respectively, then for positive* *integers x and y such that*

$$\frac{x}{y} = \frac{(k_1+1)(vr_2 - b_2) - k_1 r_2 (v-1)}{(v-1)(r_1 - \lambda_1)}$$

there exists a $(0; \rho_1^*; 0)$-EB design with parameters $v^* = v+1, b^* = b_1x + b_2y$, $\boldsymbol{r}^* = [b_1x, (r_1x + r_2y)\mathbf{1}_v']'$, $\boldsymbol{k}^* = [(k_1+1)\mathbf{1}_{b_1x}', \mathbf{1}_y' \otimes \boldsymbol{k}']'$, $\varepsilon_1^* = 1 - r_2y/[(k_1+1)(r_1x+r_2y)]$, $\rho_1^* = v$, $\boldsymbol{L}_1^* = \boldsymbol{I}_{v+1} - (n^*)^{-1}\mathbf{1}_{v+1}'(\boldsymbol{r}^*)'$, where $n^* = b_1x + (r_1x + r_2y)v$, whose incidence matrix is of the form

$$\boldsymbol{N}^* = \begin{bmatrix} \mathbf{1}_{xb_1}' & \mathbf{0}' \\ \mathbf{1}_x' \otimes \boldsymbol{N}_1 & \mathbf{1}_y' \otimes \boldsymbol{N}_2 \end{bmatrix}.$$

Proof. The values of the parameters $v^*, b^*, \boldsymbol{r}^*$, $\boldsymbol{k}^*$ are obvious. Furthermore, under the given condition, it follows that

$$\boldsymbol{M}_0^* = \frac{r_2y}{(k_1+1)(r_1x+r_2y)}\left(\boldsymbol{I}_{v+1} - \frac{1}{n^*}\mathbf{1}_{v+1}'(\boldsymbol{r}^*)'\right),$$

where $n^* = b_1x + (r_1x + r_2y)v$. This completes the proof. □

Some special cases of Theorem 6.6.13 are immediately obtainable by taking, as $\boldsymbol{N}_1$ or $\boldsymbol{N}_2$, the structures $\boldsymbol{I}_v$, $\mathbf{1}_v\mathbf{1}_{b_1}'$ or $\mathbf{1}_v\mathbf{1}_v' - \boldsymbol{I}_v$. [In particular, note that $\boldsymbol{I}_v$ and $\mathbf{1}_v\mathbf{1}_b'$ can be regarded as incidence matrices of quasi-BIB designs with parameters $v = b$, $r = k = 1$, $\lambda = 0$, and with parameters v, b, $r = b$, $k = v$, $\lambda = b$, respectively.] Thus, a series of particular applications of Theorem 6.6.13 can yield nonequireplicate $(0; v^*-1; 0)$-EB designs from equireplicate $(0; v-1; 0)$-EB designs, as will be seen in Chapter 8. Finally, a special case of Theorem 6.6.13, that will be useful in Chapter 8, is as follows.

Corollary 6.6.2. *If $\boldsymbol{N}_h$, $h = 1, 2$, are the $v \times b_h$ incidence matrices of two BIB designs with parameters $v, b_h, r_h, k_h, \lambda_h$, then for positive integers x and y such that $x/y = [\lambda_2 v(k_1+1) - k_1k_2r_2]/[k_2(r_1 - \lambda_1)]$ there exists a $(0; \rho_1^*; 0)$-EB design with parameters $v^* = v+1, b^* = b_1x + b_2y$, $\boldsymbol{r}^* = [xb_1, (xr_1 + yr_2)\mathbf{1}_v']'$, $\boldsymbol{k}^* = [(k_1+1)\mathbf{1}_{xb_1}', k_2\mathbf{1}_{yb_2}']'$, $\varepsilon_1^* = 1 - r_2k_2(r_1 - \lambda_1)/\{v(k_1+1)[r_1(k_1+1)\lambda_2 - \lambda_1r_2k_2]\}$, $\rho_1^* = v$, $\boldsymbol{L}_1^* = \boldsymbol{I}_{v+1} - (n^*)^{-1}\mathbf{1}_{v+1}'(\boldsymbol{r}^*)'$, where $n^* = b_1x + (r_1x + r_2y)v$, whose incidence matrix is given by*

$$\begin{bmatrix} \mathbf{1}_{xb_1}' & \mathbf{0}' \\ \mathbf{1}_x' \otimes \boldsymbol{N}_1 & \mathbf{1}_y' \otimes \boldsymbol{N}_2 \end{bmatrix}.$$

Remark 6.6.5. In particular, when $x = y = 1$ and $\boldsymbol{N}_2 = \mathbf{1}_v\mathbf{1}_{b_1}' - \boldsymbol{N}_1$, the condition in Corollary 6.6.2 becomes $b_1 = 3r_1 - 2\lambda_1$. It has been shown (by Kageyama and Kuwada, 1985) that the parameters of a BIB design satisfying $b_1 = 3r_1 - 2\lambda_1$ can only be expressed as $v = 2k_1 + 1, b_1 = s(2k_1+1), r_1 = sk_1, k_1, \lambda_1 = s(k_1 - 1)/2$ for a positive integer s with even $s(k_1 - 1)$. Then the following applies: If $\boldsymbol{N}_1$ is the incidence matrix of a BIB design with parameters $v = 2k_1+1, b_1 = s(2k_1+1), r_1 = sk_1, k_1, \lambda_1 = s(k_1-1)/2$ for a positive integer s with even $s(k_1-1)$, then Corollary 6.6.2 shows the existence of a $(0; v^*-1; 0)$-EB design with parameters $v^* = 2k_1+2$, $b^* = 2s(2k_1+1)$, $r^* = s(2k_1+1)$, $k^* = k_1+1$, $\varepsilon_1^* = 2k_1/(2k_1+1)$, $\boldsymbol{L}_1^* = \boldsymbol{I}_{2k_1+2} - (2k_1+2)^{-1}\boldsymbol{1}_{2k_1+2}\boldsymbol{1}'_{2k_1+2}$ (see also Theorem 8.2.13).

6.7 Methods of differences

The most powerful direct method of constructing block designs is based on the method of differences. However, it may be applicable only for special values of design parameters. Various construction methods have been studied by Bose (1939, 1942b), who developed two fundamental theorems. These principles of construction can be utilized to produce block designs considered in Chapters 6 to 10. By this method one can save space for the presentation of all design blocks.

Let M be an Abelian group (a module; see Appendix B) of m elements and let there be n symbols $x_1, x_2, ..., x_n$ corresponding to each element x in M. Then altogether there are mn symbols corresponding to M, and it is said that symbols with the same subscript, say i, belong to the ith class, $i = 1, 2, ..., n$.

Suppose that a k-subset (i.e., containing k symbols), denoted by B, from the mn symbols is chosen, and further suppose that $p_i(\geq 0)$ symbols in B belong to the ith class. Then $\sum_{i=1}^n p_i = k$. Here denote by $x_i^{(1)}, x_i^{(2)}, ..., x_i^{(p_i)}$ the symbols of the ith class in B, and similarly by $y_j^{(1)}, y_j^{(2)}, ..., y_j^{(p_j)}$ those of the jth class, where all $x_i^{(1)}, x_i^{(2)}, ..., x_i^{(p_i)}, y_j^{(1)}, y_j^{(2)}, ..., y_j^{(p_j)}$ belong to M.

The $p_i(p_i-1)$ differences $x_i^{(w)} - x_i^{(w')}$ $(w \neq w' = 1, 2, ..., p_i)$ may be called pure differences of type (i, i) arising from the set B. Similarly, the p_ip_j differences $x_i^{(w)} - y_j^{(w')}$ $(w = 1, 2, ..., p_i;\ w' = 1, 2, ..., p_j, i \neq j)$ may be called mixed differences of type (i, j) arising from B. Altogether there are n types of pure differences and $n(n-1)$ types of mixed differences in gneral.

Consider a family of s sets, $B_1, B_2, ..., B_s$, of k symbols each, for which $p_{i\ell}$ denotes the number of symbols of the ith class belonging to B_ℓ, and suppose that they satisfy the following conditions:

(1) Among the $\sum_{\ell=1}^s p_{i\ell}(p_{i\ell}-1)$ pure differences of type (i, i) arising from $B_1, B_2, ..., B_s$, every nonzero element of M is repeated exactly λ times, independently of i.

(2) Among the $\sum_{\ell=1}^s p_{i\ell}p_{j\ell}$ mixed differences of type (i, j) arising from B_1, B_2, ..., B_s, every element of M is repeated exactly λ times, independently of i and j.

Then it is said (Bose, 1939) that in the s sets the differences are "symmetrically repeated, each occurring λ times". This concept will be recognized below as the same as the notion of "concurrences" introduced in Section 2.4.

Example 6.7.1. Let $M = \{0, 1, 2, 3, 4 \pmod 5\} = Z_5$ as a module of residue classes (for the definition of Z_5 see Appendix B, below Theorem B.10), i.e., $m = 5$, and let $n = 2$. This gives altogether the 10 symbols $0_1, 0_2$, $1_1, 1_2$, $2_1, 2_2$, $3_1, 3_2$, $4_1, 4_2$ (that are regarded later as 10 treatments in a design). Put $s = 3$ and consider the sets

$$B_1 = (0_1, 3_1, 0_2, 4_2),\ B_2 = (0_1, 0_2, 1_2, 3_2),\ B_3 = (1_1, 2_1, 3_1, 0_2),$$

investigating two types of differences of symbols. Note that in B_1 one has $p_1 = 2$ and $p_2 = 2$; in B_2 one has $p_1 = 1$ and $p_2 = 3$; in B_3 one has $p_1 = 3$ and $p_2 = 1$. Now one can check that the pure differences of type (1,1), i.e., $x_1^{(w)} - x_1^{(w')}$, are $0 - 3 = -3 \equiv 2 \pmod 5$, $3 - 0 = 3$ (in B_1), and $1 - 2 = -1 \equiv 4$, $2 - 1 = 1$, $1-3 = -2 \equiv 3$, $3-1 = 2$, $2-3 = -1 \equiv 4$, $3-2 = 1$ (in B_3). Here all the nonzero elements 1, 2, 3, 4 in Z_5 occur twice (i.e., corresponding to $\lambda = 2$). Similarly, it is clear that among the pure differences of type (2,2), i.e., $x_2^{(w)} - x_2^{(w')}$, each of the nonzero elements in Z_5 occurs twice, because in B_1, $0 - 4 \equiv 1$, $4 - 0 = 4$; and in B_2, $0 - 1 \equiv 4$, $1 - 0 = 1$, $0 - 3 \equiv 2$, $3 - 0 = 3$, $1 - 3 \equiv 3$, $3 - 1 = 2$. Next one can check the mixed differences of type (1,2), i.e., $x_1^{(w)} - y_2^{(w')}$, as $0 - 0 = 0, 0 - 4 = -4 \equiv 1$, $3 - 0 = 3, 3 - 4 = -1 \equiv 4$ (in B_1), $0 - 0 = 0$, $0 - 1 = -1 \equiv 4, 0 - 3 = -3 \equiv 2$ (in B_2), and $1 - 0 = 1, 2 - 0 = 2$, $3 - 0 = 3$ (in B_3).Here each of the elements 0, 1, 2, 3, 4 in Z_5 occurs twice (i.e., corresponding to $\lambda = 2$). Similarly, it can also be checked that among the mixed differences of type (2,1), i.e., $x_2^{(w)} - y_1^{(w')}$, each of the elements in Z_5 occurs twice, because in B_1, $0 - 0 = 0$, $0 - 3 \equiv 2$, $4 - 0 = 4$, $4 - 3 = 1$; in B_2, $0 - 0 = 0$, $1 - 0 = 1$, $3 - 0 = 3$; and in B_3, $0 - 1 \equiv 4$, $0 - 2 \equiv 3$, $0 - 3 \equiv 2$. Thus the differences are symmetrically repeated twice ($\lambda = 2$) in this family of three sets B_1, B_2 and B_3 above.

Example 6.7.2. Let $M = \{0, 1, 2, 3, 4, 5, 6, 7, 8\}$ i.e., $m = 9$, and let $n = 1$ (i.e., there is only one class). This gives altogether the 9 symbols $0, 1, 2, 3, 4, 5, 6$, 7, 8 (that are regarded later as nine treatments in a design). Put $s = 2$ and consider the sets

$$B_1 = (0, 1, 2, 4),\ B_2 = (0, 3, 4, 7).$$

Because there is one class, one has only to check all the differences of symbols in the sets simply. They are: from B_1, $0-1 = -1 \equiv 8 \pmod 9$, $0-2 = -2 \equiv 7$, $0-4 = -4 \equiv 5$, $1-0 = 1$, $1-2 = -1 \equiv 8$, $1-4 = -3 \equiv 6$, $2-0 = 2$, $2-1 = 1$, $2-4 = -2 \equiv 7$, $4-0 = 4$, $4-1 = 3$, $4-2 = 2$; from B_2, $0-3 = -3 \equiv 6$ (mod 9), $0-4 = -4 \equiv 5$, $0-7 = -7 \equiv 2$, $3-0 = 3$, $3-4 = -1 \equiv 8$, $3-7 = -4 \equiv 5$, $4-0 = 4$, $4-3 = 1$, $4-7 = -3 \equiv 6$, $7-0 = 7$, $7-3 = 4$, $7-4 = 3$. Here all the nonzero elements 1, 2, 3, 4, 5, 6, 7, 8 of M occur three times (i.e., corresponding to $\lambda = 3$). Thus the differences are symmetrically repeated three times ($\lambda = 3$) in this family of two sets B_1 and B_2 above.

When applying the above theory to a block design, symbols are considered as treatments (or varieties), while sets as blocks. With this in mind, the following results can be given.

Theorem 6.7.1 (First fundamental theorem; Theorem I of Bose, 1939; Theorem 5.7.1 of Raghavarao, 1971; Theorem 4.3 of Dey, 1986; Theorem 4 in Chapter 3 of Street and Street, 1987). *Let M be an Abelian group of m elements $y^{(w)}$ $(w = 1, 2, ..., m)$ and to every element let there correspond n symbols $y_u^{(w)}$, $u = 1, 2, ..., n$, so that there will be mn symbols altogether, those with the same subscript u being considered as belonging to the same class (the uth). Suppose that there exist s sets of symbols, $B_1, B_2, ..., B_s$, such that*

(i) *every set contains exactly k distinct symbols;*
(ii) *among the ks symbols occurring in the s sets, exactly r symbols belong to each of the n classes;*
(iii) *the differences arising from the s sets are symmetrically repeated, each occurring λ times.*

Then one can form other sets $B_{j\theta}$ of symbols $y_u^{(x)}$, where

$$y_u^{(x)} = y_u^{(w)} + \theta, \quad y_u^{(w)} \in B_j, \quad \theta \in M.$$

The resulting ms sets $B_{j\theta}$ $(j = 1, 2, ..., s;\ \theta = y^{(1)}, y^{(2)}, ..., y^{(m)})$ compose a BIB design with parameters $v = mn, b = ms, r, k, \lambda$.

Note that, in Theorem 6.7.1, $0 \in M$ and then one θ has to be zero. Hence in fact $B_j \equiv B_{j0}$ for $j = 1, 2, ..., s$.

The blocks $B_1, B_2, ..., B_s$ (as in Theorem 6.7.1) from which the complete solution for the design can be constructed are usually called "initial blocks" (or difference sets generating a BIB design). They are also called base blocks or starter blocks (see, for example, Street and Street, 1987).

Example 6.7.1(continued). As seen from the previous discussion in this example, it follows that the 3 sets (blocks) satisfy all the conditions of Theorem 6.7.1 with $m = 5, n = 2$ and $s = 3$. Hence one can take such sets as three initial blocks, and by Theorem 6.7.1 one can produce a BIB design with parameters $v = 10$, $b = 15$, $r = 6$, $k = 4$, $\lambda = 2$, by noting that all subscripts of 10 symbols are invariant under development mod m $(= 5)$, whose all developed blocks are given by

$$\begin{aligned}
&(0_1, 3_1, 0_2, 4_2)(= B_1), (1_1, 4_1, 1_2, 0_2), (2_1, 0_1, 2_2, 1_2),\\
&\quad (3_1, 1_1, 3_2, 2_2),\ (4_1, 2_1, 4_2, 3_2),\\
&(0_1, 0_2, 1_2, 3_2)(= B_2), (1_1, 1_2, 2_2, 4_2), (2_1, 2_2, 3_2, 0_2),\\
&\quad (3_1, 3_2, 4_2, 1_2),\ (4_1, 4_2, 0_2, 2_2),\\
&(1_1, 2_1, 3_1, 0_2)(= B_3), (2_1, 3_1, 4_1, 1_2), (3_1, 4_1, 0_1, 2_2),\\
&\quad (4_1, 0_1, 1_1, 3_2),\ (0_1, 1_1, 2_1, 4_2),
\end{aligned}$$

with 10 symbols $0_1, 0_2$, $1_1, 1_2$, $2_1, 2_2$, $3_1, 3_2$, $4_1, 4_2$ being regarded as ten treatments. The incidence matrix of the design is

$$\begin{array}{c} 0_1 \\ 0_2 \\ 1_1 \\ 1_2 \\ 2_1 \\ 2_2 \\ 3_1 \\ 3_2 \\ 4_1 \\ 4_2 \end{array} \left[\begin{array}{ccccc|cccccccccc} 1 & 0 & 1 & 0 & 0 & 1 & 0 & 0 & 0 & 0 & 0 & 0 & 1 & 1 & 1 \\ 1 & 1 & 0 & 0 & 0 & 1 & 0 & 1 & 0 & 1 & 1 & 0 & 0 & 0 & 0 \\ 0 & 1 & 0 & 1 & 0 & 0 & 1 & 0 & 0 & 0 & 1 & 0 & 0 & 1 & 1 \\ 0 & 1 & 1 & 0 & 0 & 1 & 1 & 0 & 1 & 0 & 0 & 1 & 0 & 0 & 0 \\ 0 & 0 & 1 & 0 & 1 & 0 & 0 & 1 & 0 & 0 & 1 & 1 & 0 & 0 & 1 \\ 0 & 0 & 1 & 1 & 0 & 0 & 1 & 1 & 0 & 1 & 0 & 0 & 1 & 0 & 0 \\ 1 & 0 & 0 & 1 & 0 & 0 & 0 & 0 & 1 & 0 & 1 & 1 & 1 & 0 & 0 \\ 0 & 0 & 0 & 1 & 1 & 1 & 0 & 1 & 1 & 0 & 0 & 0 & 0 & 1 & 0 \\ 0 & 1 & 0 & 0 & 1 & 0 & 0 & 0 & 0 & 1 & 0 & 1 & 1 & 1 & 0 \\ 1 & 0 & 0 & 0 & 1 & 0 & 1 & 0 & 1 & 1 & 0 & 0 & 0 & 0 & 1 \end{array}\right],$$

which shows that the design is (2,4)-resolvable. Note that the initial blocks correspond to the columns 1st, 6th and 11th of the above matrix.

Example 6.7.2(continued). As seen from the previous discussion in this example, it follows that the 2 sets (blocks) satisfy all the conditions of Theorem 6.7.1 with $m = 9, n = 1$ and $s = 2$. Hence by Theorem 6.7.1 one can produce a BIB design with parameters $v = 9$, $b = 18$, $r = 8$, $k = 4$, $\lambda = 3$, whose all developed blocks are given by

$$\begin{aligned} &(0,1,2,4)(= B_1), (1,2,3,5), (2,3,4,6), (3,4,5,7), (4,5,6,8), \\ &\quad (5,6,7,0), (6,7,8,1), (7,8,0,2), (8,0,1,3), \\ &(0,3,4,7)(= B_2), (1,4,5,8), (2,5,6,0), (3,6,7,1), (4,7,8,2), \\ &\quad (5,8,0,3), (6,0,1,4), (7,1,2,5), (8,2,3,6). \end{aligned}$$

The resulting design is described by the incidence matrix

$$\begin{array}{c} 0 \\ 1 \\ 2 \\ 3 \\ 4 \\ 5 \\ 6 \\ 7 \\ 8 \end{array} \left[\begin{array}{ccccccccc|ccccccccc} 1 & 0 & 0 & 0 & 0 & 1 & 0 & 1 & 1 & 1 & 0 & 1 & 0 & 0 & 1 & 1 & 0 & 0 \\ 1 & 1 & 0 & 0 & 0 & 0 & 1 & 0 & 1 & 0 & 1 & 0 & 1 & 0 & 0 & 1 & 1 & 0 \\ 1 & 1 & 1 & 0 & 0 & 0 & 0 & 1 & 0 & 0 & 0 & 1 & 0 & 1 & 0 & 0 & 1 & 1 \\ 0 & 1 & 1 & 1 & 0 & 0 & 0 & 0 & 1 & 1 & 0 & 0 & 1 & 0 & 1 & 0 & 0 & 1 \\ 1 & 0 & 1 & 1 & 1 & 0 & 0 & 0 & 0 & 1 & 1 & 0 & 0 & 1 & 0 & 1 & 0 & 0 \\ 0 & 1 & 0 & 1 & 1 & 1 & 0 & 0 & 0 & 0 & 1 & 1 & 0 & 0 & 1 & 0 & 1 & 0 \\ 0 & 0 & 1 & 0 & 1 & 1 & 1 & 0 & 0 & 0 & 0 & 1 & 1 & 0 & 0 & 1 & 0 & 1 \\ 0 & 0 & 0 & 1 & 0 & 1 & 1 & 1 & 0 & 1 & 0 & 0 & 1 & 1 & 0 & 0 & 1 & 0 \\ 0 & 0 & 0 & 0 & 1 & 0 & 1 & 1 & 1 & 0 & 1 & 0 & 0 & 1 & 1 & 0 & 0 & 1 \end{array}\right],$$

which shows that the design is 4-resolvable. Evidently, the initial blocks are represented by the columns 1st and 10th of the matrix.

Example 6.7.3. To illustrate Theorem 6.7.1 with $n = 1$ (i.e., $r = sk$) consider again the construction of a BIB design with parameters $v = 9$, $b = 18$, $r = 8$, $k =$

4, $\lambda = 3$. As one can see from Example 6.7.2, two initial blocks, (0, 1, 2, 4) and (0, 3, 4, 7), mod 9, generate the BIB design (see also a solution No. 37 in Table 8.2 of Chapter 8). However, another construction of the BIB design can here be shown for further understanding of the meaning of the initial blocks generating many BIB designs given in Table 8.2. This will also illustrate Theorem 8.2.11 given later. Let the 9 treatments be taken now as the elements of a Galois field GF(3^2) (see Appendix B). If x is a primitive element, i.e., $\mathrm{GF}(3^2) = \{0, x^0(\equiv x^8), x^1, ..., x^7\}$, the two initial blocks are (x^0, x^2, x^4, x^6) and (x, x^3, x^5, x^7), which can be shown to satisfy conditions (i), (ii) and (iii) of Theorem 6.7.1 with $m = 9, n = 1$ and $s = 2$. As explained in Appendix B, the elements of GF(3^2) can be represented by the polynomials $a_1x + a_2$ where $a_1, a_2 \in \{0, 1, 2\}$ (mod 3). By using the minimum function $x^2 + x + 2$ in GF(3^2) (see again Appendix B and Table B.3), the initial blocks can be written as $(1, 2x + 1, 2, x + 2)$ and $(x, 2x + 2, 2x, x + 1)$, because $x^0 = 1, x^1 = x, x^2 = 2x + 1, x^3 = 2x + 2, x^4 = 2, x^5 = 2x, x^6 = x + 2, x^7 = x + 1$. The complete solution is obtained by adding to these the polynomials $a_1x + a_2$ $(a_1, a_2 = 0, 1, 2)$ and reducing the coefficients (mod 3). This procedure can be explained more as follows. To $a_1x + a_2$ let there correspond the element a_1a_2 in $M = \{a_1a_2 : a_1, a_2 = 0, 1, 2\}$, i.e., the module of residue classes [mod (3,3)]. Then the correspondence is isomorphic. Hence one can present the treatments by the elements of M, and write the above initial blocks as (01, 21, 02, 12), (10, 22, 20, 11) and, by Theorem 6.7.1, complete the solution by adding to these the nonzero elements, 01, 02, 10, 11, 12, 20, 21, 22, of M as follows.

(01, 21, 02, 12), (02, 22, 00, 10), (00, 20, 01, 11),
(11, 01, 12, 22), (12, 02, 10, 20), (10, 00, 11, 21),
(21, 11, 22, 02), (22, 12, 20, 00), (20, 10, 21, 01);
(10, 22, 20, 11), (11, 20, 21, 12), (12, 21, 22, 10),
(20, 02, 00, 21), (21, 00, 01, 22), (22, 01, 02, 20),
(00, 12, 10, 01), (01, 10, 11, 02), (02, 11, 12, 00).

Thus, once initial blocks of a block design are found, it is easy to get all the blocks forming the design. In this sense initial blocks are needed only to present a block design through the method of differences. Thus the main problem while using the method of differences is how to find initial blocks for a given block design.

Theorem 6.7.2 (Second fundamental theorem; Theorem II of Bose, 1939; Theorem 5.7.2 of Raghavarao, 1971; Theorem 4.4 of Dey, 1986; Theorem 5 in Chapter 3 of Street and Street, 1987). *Let M be an Abelian group of m elements $y^{(w)}$ $(w = 1, 2, ..., m)$ and to every element let there correspond n symbols defined as in Theorem 6.7.1. Further, let there be adjoined a new symbol ∞. Suppose that there exist $s + t$ initial sets of symbols, $B_1, B_2, ..., B_s, B_1', B_2', ..., B_t'$, such that*

(i) *each of the sets $B_1, B_2, ..., B_s$ contains exactly k different symbols and each of the sets $B_1', B_2', ..., B_t'$ contains ∞ and $k - 1$ other distinct symbols;*

(ii) *among the ks symbols occurring in the sets* $B_1, B_2, ..., B_s$, *exactly* $mt-\lambda$ *of them belong to each of the n classes, and among the* $(k-1)t$ *finite symbols occurring in the sets* $B_1', B_2', ..., B_t'$, *exactly* λ *of them belong to each of the n classes* [*so that* $ks = n(mt-\lambda)$ *and* $(k-1)t = n\lambda$];

(iii) *the differences arising from the* $s+t$ *sets* $B_1, B_2, ..., B_s, B_1'', B_2'', ..., B_t''$ (*where* $B_j'' = B_j'$ *with* ∞ *deleted*) *are symmetrically repeated, each occurring* λ *times.*

Then other sets $B_{j\theta}$ *and* $B'_{j\theta}$ *of symbols*

$$y_u^{(x)} = y_u^{(w)} + \theta, \quad y_u^{(w)} \in B_j \text{ or } B_j', \quad \theta \in M, \quad \infty + \theta = \infty$$

can be formed. The resulting $n(s+t)$ *sets* $B_{j\theta}$ *and* $B'_{j\theta}$ *compose a BIB design with parameters* $v = mn+1, b = m(s+t), r = ms, k, \lambda$.

Note that, in Theorem 6.7.2, $0 \in M$ and then θ has to be zero. Hence in fact $B_{j0} \equiv B_j$ and $B'_{j'0} \equiv B'_{j'}$ for $j = 1, 2, ..., s$ and $j' = 1, 2, ..., t$.

Example 6.7.4. To illustrate Theorem 6.7.2 with $n = 1$, consider the construction of a BIB design with parameters $v = 6, b = 10, r = 5, k = 3, \lambda = 2$. Because 2 is a primitive element of GF(5) (see Appendix B, Definition B.3), the two initial blocks are $(0, 2^0, 2^2)$ and $(\infty, 2^0, 2^2)$, that is, (0, 1, 4) and (∞, 1, 4) (mod 5), which can be shown to satisfy the conditions (i), (ii) and (iii) of Theorem 6.7.2 with $m = 5, n = 1, s = t = 1$. The complete solution can be obtained in accordance with Theorem 6.7.2 by developing these initial blocks as follows.

$$\begin{aligned}&(0, 1, 4), (1, 2, 0), (2, 3, 1), (3, 4, 2), (4, 0, 3),\\ &(\infty, 1, 4), (\infty, 2, 0), (\infty, 3, 1), (\infty, 4, 2), (\infty, 0, 3),\end{aligned}$$

whose incidence matrix is

$$\begin{matrix} 0 \\ 1 \\ 2 \\ 3 \\ 4 \\ \infty \end{matrix} \begin{bmatrix} 1 & 1 & 0 & 0 & 1 & 0 & 1 & 0 & 0 & 1 \\ 1 & 1 & 1 & 0 & 0 & 1 & 0 & 1 & 0 & 0 \\ 0 & 1 & 1 & 1 & 0 & 0 & 1 & 0 & 1 & 0 \\ 0 & 0 & 1 & 1 & 1 & 0 & 0 & 1 & 0 & 1 \\ 1 & 0 & 0 & 1 & 1 & 1 & 0 & 0 & 1 & 0 \\ 0 & 0 & 0 & 0 & 0 & 1 & 1 & 1 & 1 & 1 \end{bmatrix} \quad (= \boldsymbol{N}, \text{ say}),$$

(see also a solution, No. 18, in Table 8.2 of Chapter 8). Note that the initial blocks are represented by the columns 1st and 6th of the matrix $\boldsymbol{N}$.

These two fundamental theorems (Theorems 6.7.1 and 6.7.2) play a specially important role in constructing BIB designs, whose properties and systematic constructions will be described in Chapter 8 (Section 8.2).

6.8 Hadamard matrix

A square matrix $\boldsymbol{H}$ of order h whose elements are $+1$ or -1 is called a Hadamard matrix of order h provided that its rows are pairwise orthogonal, in other words that $\boldsymbol{HH}' = h\boldsymbol{I}_h$.

There exist Hadamard matrices of order 1 and of order 2, but it can be shown that every other Hadamard matrix has order $4t$ for some positive integer t. Hadamard matrices of infinite many orders have been constructed, and it has been conjectured that one exists for every t (see Hall, 1986). The smallest order which is undecided is 428 ($= 4 \times 107$), see Sawada (1985). As a simple method of construction, for example, for a Hadamard matrix of order 2,

$$\boldsymbol{H} = \begin{bmatrix} +1 & +1 \\ +1 & -1 \end{bmatrix},$$

its s times Kronecker product, i.e., $\boldsymbol{H} \otimes \boldsymbol{H} \otimes \cdots \otimes \boldsymbol{H}$, is a Hadamard matrix of order 2^s for every positive integer s. For individual examples of Hadamard matrices of small orders refer to Hedayat and Wallis (1978), Hall (1986) or Seberry and Yamada (1992). More discussion and notes can be found in Colbourn and Dinitz (1996; Part IV, Chapter 24, pp. 370-377), and Hedayat, Sloane and Stufken (1999; Chapter 7).

Let a change of a row mean the replacement of all its elements -1 by $+1$ and vice versa. Then, given a Hadamard matrix, one can change every row whose first element is -1 so to obtain an equivalent matrix whose first column is composed of $+1$'s only. Similarly, every column having -1 as its first element can then be converted so that the first row is composed of $+1$'s only. A matrix in this form is said to be normalized. For example, a Hadamard matrix $\boldsymbol{H}_1$ of order 4 can be converted to $\boldsymbol{H}_2$ in a normalized form:

$$\boldsymbol{H}_1 = \begin{bmatrix} +1 & -1 & +1 & +1 \\ -1 & +1 & +1 & +1 \\ +1 & +1 & -1 & +1 \\ +1 & +1 & +1 & -1 \end{bmatrix}, \quad \boldsymbol{H}_2 = \begin{bmatrix} +1 & +1 & +1 & +1 \\ +1 & +1 & -1 & -1 \\ +1 & -1 & -1 & +1 \\ +1 & -1 & +1 & -1 \end{bmatrix}.$$

The following results are well known, for example, see Hedayat and Wallis (1978), and Kageyama and Tanaka (1981).

Theorem 6.8.1. *The existence of a Hadamard matrix of order* $4t$ $(t > 1)$ *is equivalent to the existence of two symmetric BIB designs with parameters:*
(i) $v = b = 4t - 1,\ r = k = 2t - 1,\ \lambda = t - 1;$
(ii) $v = b = 4t - 1,\ r = k = 2t,\ \lambda = t.$

Proof. Let $\boldsymbol{H}_0$ be a matrix of order $4t - 1$ obtained by deleting the first row and the first column of a normalized Hadamard matrix of order $4t$. Then the matrix $2^{-1}(\boldsymbol{1}_{4t-1}\boldsymbol{1}'_{4t-1} + \boldsymbol{H}_0)$ yields the incidence matrix of the required design (i), i.e., by converting $+1$ and -1 in $\boldsymbol{H}_0$ into 1 and 0, respectively. Similarly, the matrix

$2^{-1}(\mathbf{1}_{4t-1}\mathbf{1}'_{4t-1} - \boldsymbol{H}_0)$ yields the design (ii). □

Corollary 6.8.1. *The existence of a Hadamard matrix of order* $4t$ $(t > 1)$ *implies the existence of BIB designs with the following parameters*

$$\text{(i) } v = 2t-1, b = 4t-2, r = 2t-2, k = t-1, \lambda = t-2;$$
$$\text{(ii) } v = 2t, b = 4t-2, r = 2t-1, k = t, \lambda = t-1;$$
$$\text{(iii) } v = 2t-1, b = 4t-2, r = 2t, k = t, \lambda = t.$$

Proof. Without loss of generality, the incidence matrix of a symmetric BIB design (i) in Theorem 6.8.1 can be presented in the form

$$\begin{bmatrix} \mathbf{1}_{2t-1} & \boldsymbol{A} \\ \mathbf{0}_{2t} & \boldsymbol{B} \end{bmatrix}.$$

It is clear that $\boldsymbol{A}$ and $\boldsymbol{B}$ are the incidence matrices of the designs (i) and (ii) in Corollary 6.8.1, respectively. It also follows that $\mathbf{1}_{2t-1}\mathbf{1}'_{4t-2} - \boldsymbol{A}$ yields the design (iii). □

Remark 6.8.1. In Corollary 6.8.1, the design (i) is a derived design of a symmetric BIB design, while (ii) is a residual design of a symmetric BIB design (see Section 6.0.1).

Example 6.8.1. A relation between a Hadamard matrix and a symmetric BIB design can be illustrated as follows. Consider the following three matrices:

$$\begin{bmatrix}
+1 & +1 & +1 & +1 & +1 & +1 & +1 & +1 \\
+1 & +1 & -1 & -1 & -1 & +1 & -1 & +1 \\
+1 & +1 & +1 & -1 & -1 & -1 & +1 & -1 \\
+1 & -1 & +1 & +1 & -1 & -1 & -1 & +1 \\
+1 & +1 & -1 & +1 & +1 & -1 & -1 & -1 \\
+1 & -1 & +1 & -1 & +1 & +1 & -1 & -1 \\
+1 & -1 & -1 & +1 & -1 & +1 & +1 & -1 \\
+1 & -1 & -1 & -1 & +1 & -1 & +1 & +1
\end{bmatrix},$$

$$\begin{bmatrix}
+1 & -1 & -1 & -1 & +1 & -1 & +1 \\
+1 & +1 & -1 & -1 & -1 & +1 & -1 \\
-1 & +1 & +1 & -1 & -1 & -1 & +1 \\
+1 & -1 & +1 & +1 & -1 & -1 & -1 \\
-1 & +1 & -1 & +1 & +1 & -1 & -1 \\
-1 & -1 & +1 & -1 & +1 & +1 & -1 \\
-1 & -1 & -1 & +1 & -1 & +1 & +1
\end{bmatrix}, \quad
\begin{bmatrix}
1 & 0 & 0 & 0 & 1 & 0 & 1 \\
1 & 1 & 0 & 0 & 0 & 1 & 0 \\
0 & 1 & 1 & 0 & 0 & 0 & 1 \\
1 & 0 & 1 & 1 & 0 & 0 & 0 \\
0 & 1 & 0 & 1 & 1 & 0 & 0 \\
0 & 0 & 1 & 0 & 1 & 1 & 0 \\
0 & 0 & 0 & 1 & 0 & 1 & 1
\end{bmatrix}.$$

The first is a normalized Hadamard matrix of order 8, the second is derived by deleting the first row and first column of the first matrix, and finally the third is the incidence matrix of a BIB design with parameters $v = b = 7, r = k = 3, \lambda = 1$

obtained by converting +1 and −1 in the second matrix into 1 and 0, respectively.

Theorem 6.8.2. *The existence of a Hadamard matrix of order* $4t$ $(t > 1)$ *implies the existence of GD designs* [*see Section* 6.0.2(A)] *with the following parameters:*

(i) $v = 12t, b = 16t - 4, r = 8t - 2, k = 6t, \lambda_1 = 4t - 2, \lambda_2 = 4t - 1$. *Here* $m = 3, n = 4t$.

(ii) $v = 16t - 4, b = 12t, r = 6t, k = 8t - 2, \lambda_1 = 2t, \lambda_2 = 3t$. *Here* $m = 4t - 1, n = 4$.

(iii) $v = 16t - 4, b = 16t - 4, r = 8t - 3, k = 8t - 3, \lambda_1 = 4t - 3, \lambda_2 = 4t - 2$. *Here* $m = 4, n = 4t - 1$.

(iv) $v = 16t - 4, b = 16t - 4, r = 8t - 1, k = 8t - 1, \lambda_1 = 4t - 1, \lambda_2 = 4t$. *Here* $m = 4, n = 4t - 1$.

(v) $v = 4t, b = 8t - 4, r = 4t - 2, k = 2t, \lambda_1 = 2t - 2, \lambda_2 = 2t - 1$. *Here* $m = 2, n = 2t$.

Proof. Let $\boldsymbol{H}_1$ be a matrix of size $4t \times (4t - 1)$ obtained by deleting the first column of a normalized Hadamard matrix of order $4t$. Then the matrix $2^{-1}(\mathbf{1}_{12t}\mathbf{1}'_{16t-4} + \boldsymbol{H}_2)$ and its transpose produce incidence matrices of the designs (i) and (ii), respectively, if

$$\boldsymbol{H}_2 = \begin{bmatrix} \boldsymbol{H}_1 & \boldsymbol{H}_1 & -\boldsymbol{H}_1 & -\boldsymbol{H}_1 \\ \boldsymbol{H}_1 & -\boldsymbol{H}_1 & \boldsymbol{H}_1 & -\boldsymbol{H}_1 \\ \boldsymbol{H}_1 & -\boldsymbol{H}_1 & -\boldsymbol{H}_1 & \boldsymbol{H}_1 \end{bmatrix}.$$

Let $\boldsymbol{H}_3$ be a matrix of order $4t - 1$ obtained by deleting the first row and the first column of a normalized Hadamard matrix of order $4t$. Then the matrices $2^{-1}(\mathbf{1}_{16t-4}\mathbf{1}'_{16t-4}+\boldsymbol{H}_4)$ and $2^{-1}(\mathbf{1}_{16t-4}\mathbf{1}'_{16t-4}-\boldsymbol{H}_4)$ yield the incidence matrices of the designs (iii) and (iv), respectively, if

$$\boldsymbol{H}_4 = \begin{bmatrix} -\boldsymbol{H}_3 & \boldsymbol{H}_3 & \boldsymbol{H}_3 & \boldsymbol{H}_3 \\ \boldsymbol{H}_3 & -\boldsymbol{H}_3 & \boldsymbol{H}_3 & \boldsymbol{H}_3 \\ \boldsymbol{H}_3 & \boldsymbol{H}_3 & -\boldsymbol{H}_3 & \boldsymbol{H}_3 \\ \boldsymbol{H}_3 & \boldsymbol{H}_3 & \boldsymbol{H}_3 & \boldsymbol{H}_3 \end{bmatrix}.$$

Let $\boldsymbol{H}_5$ be the incidence matrix of a residual design (see Section 6.0.1) of a symmetric BIB design (of Theorem 6.8.1) derivable from a normalized Hadamard matrix of order $4t$. Then

$$\begin{bmatrix} \boldsymbol{H}_5 & \boldsymbol{H}_5 \\ \boldsymbol{H}_5 & \mathbf{1}_{2t}\mathbf{1}'_{4t-2} - \boldsymbol{H}_5 \end{bmatrix}$$

is the incidence matrix of the design (v). □

Theorem 6.8.3. *The existence of two Hadamard matrices of order* $4t$ *and* $4f$ *implies the existence of a semi-regular GD design with parameters* $v = 4t(4f -$

$1), b = 4f(4t-1), r = 2f(4t-1), k = 2t(4f-1), \lambda_1 = f(4t-2), \lambda_2 = f(4t-1)$. *Here* $m = 4f - 1, n = 4t$.

Proof. Let $\boldsymbol{A}$ be a normalized Hadamard matrix of order $4f$ whose first row is deleted. Let $\boldsymbol{B}$ be a normalized Hadamard matrix of order $4t$ whose first column is deleted. Then it follows that $2^{-1}(\boldsymbol{1}_{4t(4f-1)\times 4f(4t-1)} + \boldsymbol{A} \otimes \boldsymbol{B})$ is the incidence matrix of the required design. □

Finally, a BIB design with the affine resolvability can be derived as follows.

Theorem 6.8.4. *The existence of a Hadamard matrix of order* $4t$ *implies the existence of an affine resolvable BIB design with parameters* $v = 4t, b = 2(4t - 1), r = 4t - 1, k = 2t, \lambda = 2t - 1$ *for* $t \geq 1$.

Proof. It follows from Theorem 6.8.1(i) that the existence of a Hadamard matrix of order $4t$ implies the existence of a symmetric BIB design, having the incidence matrix $\boldsymbol{N}$, with parameters $v = b = 4t - 1, r = k = 2t - 1, \lambda = t - 1$. Let $\boldsymbol{N}^{*\prime} = [\boldsymbol{N}' : \boldsymbol{1}_{4t-1}]$. Then it can be shown that $[\boldsymbol{N}^* : \boldsymbol{1}_{4t}\boldsymbol{1}_{4t}' - \boldsymbol{N}^*]$ produces the required design. In fact, $4t - 1$ resolution sets (superblocks) consisting of two blocks can be formed, by taking one block from $\boldsymbol{N}^*$ and the other as its complementary block, i.e., the corresponding block from $\boldsymbol{1}_{4t}\boldsymbol{1}_{4t}' - \boldsymbol{N}^*$. The affine resolvability can be shown by Theorem 6.0.5. □

6.9 Finite geometry

With the help of a Galois field (see Appendix B) one can construct finite geometries, i.e., a finite projective t-dimensional geometry over GF(q) denoted by PG(t, q) and the corresponding affine geometry AG(t, q), which are explained in Appendices C.1 and C.2, respectively. Let

$$\phi(t, d, q) = \frac{(q^{t+1} - 1)(q^t - 1) \cdots (q^{t-d+1} - 1)}{(q^{d+1} - 1)(q^d - 1) \cdots (q - 1)},$$

with $\phi(t, -1, q) = 1$ conventionally. Then the following theorem applies (see also Raghavarao, 1971, Section 5.6).

Theorem 6.9.1. *If* q *is a prime or a prime power, then there exists a BIB design with parameters* $v = \phi(t, 0, q)$, $b = \phi(t, d, q)$, $r = \phi(t - 1, d - 1, q)$, $k = \phi(d, 0, q)$, $\lambda = \phi(t - 2, d - 2, q)$ *for* $t > d$.

Proof. To every point of PG(t, q) let there correspond a treatment. Furthermore, to every d-flat, i.e., d-dimensional subspace of PG(t, q), let there correspond a block containing all those treatments whose corresponding points occur in the d-flat. Thus, in PG(t, q), because the number of d-flats is given by $\phi(t, d, q)$, the number of d-flats including a particular point is $\phi(t - 1, d - 1, q)$, whereas the number of d-flats including two distinct points is $\phi(t - 2, d - 2, q)$ (see Theorem C.1.1, and Corollaries C.1.2.1 and C.1.2.2 in Appendix C), the proof is complete. □

Note that a BIB design resulting from Theorem 6.9.1 is denoted by PG(t, q) : d. Special cases of Theorem 6.9.1 are provided for some t and d.

Corollary 6.9.1. *When q is a prime or a prime power, there exist BIB designs with the following parameters*

(i) $v = b = q^2 + q + 1, r = k = q + 1, \lambda = 1$;
(ii) $v = b = (q + 1)(q^2 + 1), r = k = q^2 + q + 1, \lambda = q + 1$;
(iii) $v = (q+1)(q^2+1), b = (q^2+1)(q^2+q+1), r = q^2+q+1, k = q+1, \lambda = 1$.

Proof. The designs correspond, (i) to $t = 2, d = 1$, i.e., PG$(2, q)$: 1, (ii) to $t = 3, d = 2$, i.e., PG$(3, q)$: 2, and (iii) to $t = 3, d = 1$, i.e., PG$(3, q)$: 1, respectively, of the BIB design of Theorem 6.9.1. □

Example 6.9.1. By Corollary 6.9.1 (i), with $q = 2$, in a PG(2,2) there are 7 points and 7 lines consisting of 3 points each such that each point is included in 3 lines and any two distinct lines have only one point in common. Let 7 points be $1, 2, 3, 4, \infty_1, \infty_2, \infty_3$. Then 7 such lines of 3 points each are given by $\{1, 2, \infty_1\}$, $\{3, 4, \infty_1\}$, $\{1, 3, \infty_2\}$, $\{2, 4, \infty_2\}$, $\{1, 4, \infty_3\}$, $\{2, 3, \infty_3\}$, $\{\infty_1, \infty_2, \infty_3\}$, which yield a BIB design with parameters $v = b = 7, r = k = 3, \lambda = 1$, by regarding points and lines in the PG(2,2) as treatments and blocks, respectively (see also Appendix C.3). Note (see Appendix C.2) that $\infty_i, i = 1, 2, 3$, are called points at infinity and $\{\infty_1, \infty_2, \infty_3\}$ is called a line at infinity.

Similarly, to every point of AG(t, q) let there correspond a treatment. Furthermore, to every d-flat let there correspond a block containing all the treatments whose corresponding points form the d-flats. In this case the following applies. Note that the parallelism in d-flats constitutes resolution sets for the resolvability, i.e., one replicate of every treatment.

Theorem 6.9.2. *If q is a prime or a prime power, there exists a resolvable BIB design with parameters $v = q^t$, $b = q^{t-d}\phi(t-1, d-1, q)$, $r = \phi(t-1, d-1, q)$, $k = q^d$, $\lambda = \phi(t-2, d-2, q)$ for $t > d$.*

Note that a BIB design resulting from Theorem 6.9.2 is denoted by AG(t, q) : d. By the use of an relation between PG(t, q) and AG(t, q) (see Appendix C), from every BIB design given by Theorem 6.9.1 it is possible to obtain a corresponding BIB design given by Theorem 6.9.2 by cutting out those treatments which correspond to points at infinity (like ∞_i in Example 6.9.1), together with all those blocks which consists of such treatments only.

Special cases of Theorem 6.9.2 are provided for some t and d.

Corollary 6.9.2. *When q is a prime or a prime power, there exist resolvable BIB designs with the following parameters*

(i) $v = q^2, b = q(q + 1), r = q + 1, k = q, \lambda = 1$;
(ii) $v = q^3, b = q(q^2 + q + 1), r = q^2 + q + 1, k = q^2, \lambda = q + 1$;

(iii) $v = q^3, b = q^2(q^2+q+1), r = q^2+q+1, k = q, \lambda = 1$.

Proof. The designs correspond, (i) to $t = 2, d = 1$, i.e., AG$(2,q)$: 1, (ii) to $t = 3, d = 2$, i.e., AG$(3,q)$: 2, and (iii) to $t = 3, d = 1$, i.e., AG$(3,q)$: 1, respectively, of the BIB design of Theorem 6.9.2. □

Example 6.9.2. By Corollary 6.9.2 (i) with $q = 2$, in an AG(2,2) there are 4 points and 6 lines consisting of 2 points each such that each point is included in 3 lines and any two distinct lines have only one point in common. This affine plane is provided by deleting three points $\infty_1, \infty_2, \infty_3$ in a PG(2,2) of Example 6.9.1. That is, 6 lines of 2 points each are given by

$$\{1,2\},\{3,4\},\{1,3\},\{2,4\},\{1,4\},\{2,3\},$$

which yield an affine resolvable BIB design with parameters $v = 4, b = 6, r = 3, k = 2, \lambda = 1$, in which three resolution sets of 2 blocks each are formed by every two blocks from the beginning, while any two blocks from different sets have exactly one common treatment (see Section 6.0.3 and Appendix C.3).

Examples 6.9.1 and 6.9.2 illustrate the fact (see Appendix C.2) that an affine plane can be obtained from a projective plane by excluding all the points at infinity and a line at infinity.

Remark 6.9.1. The two series given in Corollaries 6.9.1(i) and 6.9.2(i) are especially called orthogonal series of Yates.

As a generalization of Example 6.9.2, the following result can be obtained.

Corollary 6.9.3. *There exists an affine resolvable BIB design with parameters* $v = q^2$, $b = q(q+1)$, $r = q+1$, $k = q$, $\lambda = 1$ *for* q *being a prime or a prime power.*

Proof. The resolvable BIB design with parameters $v = q^2$, $b = q(q+1)$, $r = q+1$, $k = q$, $\lambda = 1$ given in Corollary 6.9.2(i) satisfies $b = v+r-1$. Hence the required result follows by Theorem 6.0.5 in Section 6.0.3. □

Such BIB designs of Corollary 6.9.3 are called balanced lattice design (see, e.g., John, 1987, Section 3.4.2; John and Williams, 1995, Section 4.2).

6.10 Orthogonal Latin squares

Some detailed arguments on Latin squares can be found in Appendix D. A Latin square of order h is a square arrangement of h symbols in h rows and h columns such that every symbol occurs once in each row and once in each column. If in two Latin squares of the same order, when superimposed on one another, every ordered pair of symbols occurs exactly once, then the two Latin squares are said to be orthogonal. If in a set of Latin squares every pair of them is orthogonal, then the set is called a set of mutually orthogonal Latin squares (MOLS). In particular, a set of $h-1$ MOLS of order h is called a complete set of the MOLS. For example, the following three sets consisting of two, three and four Latin squares

$$\begin{bmatrix} 0 & 1 & 2 \\ 2 & 0 & 1 \\ 1 & 2 & 0 \end{bmatrix}, \begin{bmatrix} 0 & 2 & 1 \\ 2 & 1 & 0 \\ 1 & 0 & 2 \end{bmatrix};$$

$$\begin{bmatrix} 0 & 1 & 2 & 3 \\ 1 & 0 & 3 & 2 \\ 2 & 3 & 0 & 1 \\ 3 & 2 & 1 & 0 \end{bmatrix}, \begin{bmatrix} 0 & 1 & 2 & 3 \\ 2 & 3 & 0 & 1 \\ 3 & 2 & 1 & 0 \\ 1 & 0 & 3 & 2 \end{bmatrix}, \begin{bmatrix} 0 & 1 & 2 & 3 \\ 3 & 2 & 1 & 0 \\ 1 & 0 & 3 & 2 \\ 2 & 3 & 0 & 1 \end{bmatrix};$$

$$\begin{bmatrix} 0 & 1 & 2 & 3 & 4 \\ 1 & 2 & 3 & 4 & 0 \\ 2 & 3 & 4 & 0 & 1 \\ 3 & 4 & 0 & 1 & 2 \\ 4 & 0 & 1 & 2 & 3 \end{bmatrix}, \begin{bmatrix} 0 & 1 & 2 & 3 & 4 \\ 2 & 3 & 4 & 0 & 1 \\ 4 & 0 & 1 & 2 & 3 \\ 1 & 2 & 3 & 4 & 0 \\ 3 & 4 & 0 & 1 & 2 \end{bmatrix}, \begin{bmatrix} 0 & 1 & 2 & 3 & 4 \\ 3 & 4 & 0 & 1 & 2 \\ 1 & 2 & 3 & 4 & 0 \\ 4 & 0 & 1 & 2 & 3 \\ 2 & 3 & 4 & 0 & 1 \end{bmatrix},$$

$$\begin{bmatrix} 0 & 1 & 2 & 3 & 4 \\ 4 & 0 & 1 & 2 & 3 \\ 3 & 4 & 0 & 1 & 2 \\ 2 & 3 & 4 & 0 & 1 \\ 1 & 2 & 3 & 4 & 0 \end{bmatrix};$$

form complete sets of MOLS of order 3, of order 4 and of order 5, respectively.

Now it is specially again shown that the two series given in Corollaries 6.9.1(i) and 6.9.2(i) are also constructed by use of a complete set of MOLS of order q [see Hall (1986) together with Appendices C and D; Raghavarao, 1971, Section 5.9].

Theorem 6.10.1. *The existence of a complete set of MOLS of order q implies the existence of BIB designs with parameters*

(i) $v = q^2, b = q(q+1), r = q+1, k = q, \ \lambda = 1$;
(ii) $v = b = q^2 + q + 1, r = k = q + 1, \ \lambda = 1.$

Proof. Let q^2 symbols be arranged in a $q \times q$ square L as

$$L = \begin{bmatrix} 1 & 2 & \cdots & q \\ q+1 & q+2 & \cdots & 2q \\ \vdots & \vdots & & \vdots \\ (q-1)q+1 & (q-1)q+2 & \cdots & q^2 \end{bmatrix}.$$

These q^2 symbols in L are regarded as treatments of the designs. Further, let $L_1, L_2, ..., L_{q-1}$ form a complete set of MOLS of order q. Then it follows from the property of MOLS that the following procedure can yield the required design (i) with $q(q+1)$ blocks:

(1) The blocks B_j, $j = 1, 2, ..., q$, are sets formed by q symbols in each row of L.

(2) The blocks B_j, $j = q+1, q+2, ..., 2q$, are sets formed by q symbols in each column of L.

(3) The remaining $q(q-1)$ blocks $B_{(\alpha+1)q+\beta+1}$, $\alpha = 1, 2, ..., q-1$, $\beta = 0, 1, ..., q-1$, are sets formed by q symbols of L (in the same positions) corresponding to the symbol β of L_α.

In the above solution of the design (i), one could have more new $q+1$ symbols, $q^2+1, q^2+2, ..., q^2+q+1$, such that to the above all blocks given in (1) add a symbol q^2+1, to (2) add a symbol q^2+2, and to (3) add a symbol $q^2+(\alpha+2)$ for each of $\alpha = 1, 2, ..., q-1$. Finally, a new block consisting of the new symbols $q^2+1, q^2+2, ..., q^2+q+1$ is added. Hence it follows that the new q^2+q+1 blocks of size $q+1$ each give the required design (ii). □

Note that the solution of a BIB design (i) given in the proof of Theorem 6.10.1 is resolvable.

Remark 6.10.1. When q is a prime or a prime power, there exists a complete set of MOLS of order q (see Appendix D), and hence the two series of BIB designs given in Theorem 6.10.1 can always be constructed. On the other hand, when q is a composite number, it is unknown whether or not there exists a complete set of MOLS of order q.

Example 6.10.1. It is to illustrate that the above-mentioned complete set,

$$L_1 = \begin{bmatrix} 0 & 1 & 2 & 3 \\ 1 & 0 & 3 & 2 \\ 2 & 3 & 0 & 1 \\ 3 & 2 & 1 & 0 \end{bmatrix}, \quad L_2 = \begin{bmatrix} 0 & 1 & 2 & 3 \\ 2 & 3 & 0 & 1 \\ 3 & 2 & 1 & 0 \\ 1 & 0 & 3 & 2 \end{bmatrix}, \quad L_3 = \begin{bmatrix} 0 & 1 & 2 & 3 \\ 3 & 2 & 1 & 0 \\ 1 & 0 & 3 & 2 \\ 2 & 3 & 0 & 1 \end{bmatrix},$$

of MOLS of order 4 produces BIB designs with parameters (i) $v = 16, b = 20, r = 5, k = 4, \lambda = 1$; and with (ii) $v = 21, b = 21, r = 5, k = 5, \lambda = 1$. First consider a 4×4 array as

$$L = \begin{bmatrix} 1 & 2 & 3 & 4 \\ 5 & 6 & 7 & 8 \\ 9 & 10 & 11 & 12 \\ 13 & 14 & 15 & 16 \end{bmatrix}.$$

Then, following the procedure given in the proof of Theorem 6.10.1, 20 blocks are formed for the design (i) as follows:

By the procedure (1), four blocks $B_1 = (1, 2, 3, 4), B_2 = (5, 6, 7, 8), B_3 = (9, 10, 11, 12)$, $B_4 = (13, 14, 15, 16)$ can be taken through L,

By the procedure (2), four blocks $B_5 = (1, 5, 9, 13), B_6 = (2, 6, 10, 14), B_7 = (3, 7, 11, 15), B_8 = (4, 8, 12, 16)$ can be taken through L,

By taking the procedure (3) for the above L_1, one can form four blocks $B_9 =$ (1, 6, 11, 16) (from symbols of L corresponding to the symbol 0 in L_1), $B_{10} =$ (2, 5, 12, 15) (from symbols of L corresponding to the symbol 1 in L_1), $B_{11} =$ (3, 8, 9, 14) (from symbols of L corresponding to the symbol 2 in L_1), $B_{12} =$ (4, 7, 10, 13) (from symbols of L corresponding to the symbol 3 in L_1),

By the procedure (3) for the above L_2, four blocks $B_{13} =$ (1, 7, 12, 14) (from symbols of L corresponding to the symbol 0 in L_2), $B_{14} =$ (2, 8, 11, 13) (from symbols of L corresponding to the symbol 1 in L_2), $B_{15} =$ (3, 5, 10, 16) (from symbols of L corresponding to the symbol 2 in L_2), $B_{16} =$ (4, 6, 9, 15) (from symbols of L corresponding to the symbol 3 in L_2) can be formed,

Finally, by the procedure (3) for the above L_3, four blocks $B_{17} =$ (1, 8, 10, 15) (from symbols of L corresponding to the symbol 0 in L_3), $B_{18} =$ (2, 7, 9, 16) (from symbols of L corresponding to the symbol 1 in L_3), $B_{19} =$ (3, 6, 12, 13) (from symbols of L corresponding to the symbol 2 in L_3), $B_{20} =$ (4, 5, 11, 14) (from symbols of L corresponding to the symbol 3 in L_3) can be formed.

Thus, the incidence matrix of the design (i) with these blocks is given by

$$\begin{array}{c} 1\\2\\3\\4\\5\\6\\7\\8\\9\\10\\11\\12\\13\\14\\15\\16 \end{array}
\left[\begin{array}{cccc|cccc|cccc|cccc|cccc}
1&0&0&0&1&0&0&0&1&0&0&0&1&0&0&0&1&0&0&0\\
1&0&0&0&0&1&0&0&0&1&0&0&0&1&0&0&0&1&0&0\\
1&0&0&0&0&0&1&0&0&0&1&0&0&0&1&0&0&0&1&0\\
1&0&0&0&0&0&0&1&0&0&0&1&0&0&0&1&0&0&0&1\\
0&1&0&0&1&0&0&0&0&1&0&0&0&0&1&0&0&0&0&1\\
0&1&0&0&0&1&0&0&1&0&0&0&0&0&0&1&0&0&1&0\\
0&1&0&0&0&0&1&0&0&0&0&1&1&0&0&0&0&1&0&0\\
0&1&0&0&0&0&0&1&0&0&1&0&0&1&0&0&1&0&0&0\\
0&0&1&0&1&0&0&0&0&0&1&0&0&0&0&1&0&1&0&0\\
0&0&1&0&0&1&0&0&0&0&0&1&0&0&1&0&1&0&0&0\\
0&0&1&0&0&0&1&0&1&0&0&0&0&1&0&0&0&0&0&1\\
0&0&1&0&0&0&0&1&0&1&0&0&1&0&0&0&0&0&1&0\\
0&0&0&1&1&0&0&0&0&0&0&1&0&1&0&0&0&0&1&0\\
0&0&0&1&0&1&0&0&0&0&1&0&1&0&0&0&0&0&0&1\\
0&0&0&1&0&0&1&0&0&1&0&0&0&0&0&1&1&0&0&0\\
0&0&0&1&0&0&0&1&1&0&0&0&0&0&1&0&0&1&0&0
\end{array}\right],$$

which is resolvable. Furthermore, the incidence matrix of the design (ii) can be given, after adding five new treatments, 17,18,19,20,21, and one new block consisting of them, to the design (i), by

1	1	0	0	0	1	0	0	0	1	0	0	0	1	0	0	0	1	0	0	0	0
2	1	0	0	0	0	1	0	0	0	1	0	0	0	1	0	0	0	1	0	0	0
3	1	0	0	0	0	0	1	0	0	0	1	0	0	0	1	0	0	0	1	0	0
4	1	0	0	0	0	0	0	1	0	0	0	1	0	0	0	1	0	0	0	1	0
5	0	1	0	0	1	0	0	0	0	1	0	0	0	0	1	0	0	0	0	1	0
6	0	1	0	0	0	1	0	0	1	0	0	0	0	0	0	1	0	0	1	0	0
7	0	1	0	0	0	0	1	0	0	0	0	1	1	0	0	0	0	1	0	0	0
8	0	1	0	0	0	0	0	1	0	0	1	0	0	1	0	0	1	0	0	0	0
9	0	0	1	0	1	0	0	0	0	0	1	0	0	0	0	1	0	1	0	0	0
10	0	0	1	0	0	1	0	0	0	0	0	1	0	0	1	0	1	0	0	0	0
11	0	0	1	0	0	0	1	0	1	0	0	0	0	1	0	0	0	0	0	1	0
12	0	0	1	0	0	0	0	1	0	1	0	0	1	0	0	0	0	0	1	0	0
13	0	0	0	1	1	0	0	0	0	0	0	1	0	1	0	0	0	0	1	0	0
14	0	0	0	1	0	1	0	0	0	0	1	0	1	0	0	0	0	0	0	1	0
15	0	0	0	1	0	0	1	0	0	1	0	0	0	0	0	1	1	0	0	0	0
16	0	0	0	1	0	0	0	1	1	0	0	0	0	0	1	0	0	1	0	0	0
17	1	1	1	1	0	0	0	0	0	0	0	0	0	0	0	0	0	0	0	0	1
18	0	0	0	0	1	1	1	1	0	0	0	0	0	0	0	0	0	0	0	0	1
19	0	0	0	0	0	0	0	0	1	1	1	1	0	0	0	0	0	0	0	0	1
20	0	0	0	0	0	0	0	0	0	0	0	0	1	1	1	1	0	0	0	0	1
21	0	0	0	0	0	0	0	0	0	0	0	0	0	0	0	0	1	1	1	1	1

.

A similar explanation can be seen in Matrix C.1 in Appendix C, Section C.3.

As another application of MOLS related to Theorem 6.10.1 the following theorem can be shown. For the existence of MOLS, refer to Appendix D, and Colbourn and Dinitz (1996; pp. 111-172).

Theorem 6.10.2 (Shrikhande and Singh, 1962). *The existence of* $s-2$ *MOLS of order* $2s$ *implies the existence of a symmetric BIB design with parameters* $v = b = 4s^2$, $r = k = s(2s-1)$, $\lambda = s(s-1)$.

Proof. It follows from Bose and Shimamoto (1952) that the existence of $s-2$ MOLS of order $2s$ implies the existence of a 2-associate association scheme with $v = 4s^2$, $n_1 = s(2s-1)$, $p^1_{11} = p^2_{11} = s(s-1)$. Hence, it can obviously be seen (as a similar approach see Theorem 8.3.7) that an association matrix of the first associates of this association scheme is the incidence matrix of the required design. □

7
Designs with Full Efficiency for Some Contrasts

The purpose of this chapter is to present methods of constructing connected block designs that provide the unit efficiency factor for some contrasts of treatment parameters. In the terminology introduced in Chapter 4 (Section 4.4) and recalled at the beginning of Chapter 6, this means that various cases of $(\rho_0; \rho_1, ..., \rho_{m-1}; 0)$-EB designs, with $m = 1, 2, 3$ and more, for which $\rho_0 \geq 1$, will be of interest. The chapter begins with a general consideration on such designs (Section 7.1), by recalling relevant results discussed in Volume I and providing several corresponding optimality results. Next, a large number of methods of construction are given for designs separated into four classes, those covering designs that are (i) proper and equireplicate, (ii) proper and nonequireplicate, (iii) nonproper and equireplicate and (iv) nonproper and nonequireplicate, first for $m = 1$ (Section 7.2), then for $m = 2$ (Section 7.3), then for $m = 3$ (Section 7.4), and finally for $m > 3$ (Section 7.5).

7.1 General consideration

In a $(\rho_0; \rho_1, ..., \rho_{m-1}; \rho_m)$-EB design, as defined in Chapter 4 (Definition 4.4.2), there are possibly m distinct nonzero efficiency factors of the design for estimating the corresponding basic contrasts in the intra-block analysis, i.e., under the submodel $\boldsymbol{y}_1 = \boldsymbol{\phi}_1 \boldsymbol{y}$, with $\mathrm{E}(\boldsymbol{y}_1) = \boldsymbol{\phi}_1 \boldsymbol{\Delta}' \boldsymbol{\tau}$ and $\mathrm{Cov}(\boldsymbol{y}_1) = \boldsymbol{\phi}_1 \sigma_1^2$, where $\boldsymbol{\phi}_1 = \boldsymbol{I}_n - \boldsymbol{D}' \boldsymbol{k}^{-\delta} \boldsymbol{D}'$ (see Section 3.2.1). Therefore, if $\rho_0 \geq 1$ and $m \geq 2$, not all basic contrasts can be estimated there with the same efficiency, however, some of them can be estimated with full efficiency. Designs of such possibility will be discussed, along with their properties and constructions. If not otherwise stated, connected designs will be considered, this being denoted by $\rho_m = 0$.

As explained in Section 4.4 (Section 4.4.1 in particular), a $(\rho_0; \rho_1, ..., \rho_{m-1}; 0)$-

EB design is a connected block design specified by the parameters v, b, $\boldsymbol{r}$, $\boldsymbol{k}$, $\varepsilon_\beta = 1 - \mu_\beta$, ρ_β, $\boldsymbol{L}_\beta$ for $\beta = 0, 1, ..., m-1, m$, where ρ_0 gives the number of basic contrasts estimated in the intra-block analysis (i.e., under the intra-block submodel of Section 3.2.1) with full efficiency, $\varepsilon_0 = 1$, whereas $\rho_1, ..., \rho_{m-1}$ give the numbers of basic contrasts estimated in that analysis with efficiencies $(1 >)\ \varepsilon_1 > \cdots > \varepsilon_{m-1}$, respectively, and where (as defined in Section 3.4)

$$\boldsymbol{L}_\beta = \boldsymbol{S}_\beta \boldsymbol{S}'_\beta \boldsymbol{r}^\delta, \quad \text{with} \quad \boldsymbol{S}_\beta = [\boldsymbol{s}_{\beta 1} : \boldsymbol{s}_{\beta 2} : \cdots : \boldsymbol{s}_{\beta \rho_\beta}], \tag{7.1.1}$$

the latter representing the βth subset of basic contrasts, $\{\boldsymbol{c}'_{\beta j}\boldsymbol{\tau} = \boldsymbol{s}'_{\beta j}\boldsymbol{r}^\delta\boldsymbol{\tau}, j = 1, 2, ..., \rho_\beta\}$, those corresponding to a common efficiency factor ε_β. Thus (see also Section 4.4.3), a connected block design, denoted as $(\rho_0; \rho_1, ..., \rho_{m-1}; 0)$-EB , is well characterized from the statistical point of view by the triples $\{\mu_\beta, \rho_\beta, \boldsymbol{L}_\beta\}$ for $\beta = 0, 1, ..., m-1$, in which the idempotent matrices $\{\boldsymbol{L}_\beta\}$, their ranks $\{\rho_\beta\}$ and the corresponding eigenvalues $\{\mu_\beta = 1 - \varepsilon_\beta\}$ are obtainable from the spectral decomposition of the matrix $\boldsymbol{M}_0 = \boldsymbol{r}^{-\delta}\boldsymbol{N}\boldsymbol{k}^{-\delta}\boldsymbol{N}' - n^{-1}\boldsymbol{1}_v\boldsymbol{r}'$ (introduced in Section 4.3.2), i.e., from

$$\boldsymbol{M}_0 = \sum_{\beta=0}^{m-1} \mu_\beta \boldsymbol{S}_\beta \boldsymbol{S}'_\beta \boldsymbol{r}^\delta = \sum_{\beta=0}^{m-1} \mu_\beta \boldsymbol{L}_\beta, \quad \text{with} \quad \mu_\beta = 1 - \varepsilon_\beta, \tag{7.1.2}$$

where $\{\boldsymbol{L}_\beta\}$, defined as above, satisfy the conditions

$$\boldsymbol{M}_0 \boldsymbol{L}_\beta = \mu_\beta \boldsymbol{L}_\beta, \quad \text{with} \quad \mu_0 = 0, \quad 0 < \mu_\beta < 1 \quad \text{otherwise},$$

and have the properties

$$\boldsymbol{L}_\beta^2 = \boldsymbol{L}_\beta, \quad \boldsymbol{L}_\beta \boldsymbol{L}_{\beta'} = \boldsymbol{O} \ \text{ if } \ \beta \neq \beta', \ \text{ for } \ \beta, \beta' = 0, 1, ..., m-1,$$

$$\sum_{\beta=0}^{m-1} \boldsymbol{L}_\beta = \boldsymbol{I}_v - n^{-1}\boldsymbol{1}_v\boldsymbol{r}',$$

and

$$\text{rank}(\boldsymbol{L}_\beta) = \rho_\beta, \quad \text{giving} \quad \sum_{\beta=0}^{m-1} \rho_\beta = v - 1. \tag{7.1.3}$$

Correspondingly to the above decomposition of $\boldsymbol{M}_0$, also the matrix $\boldsymbol{C}_1 = \boldsymbol{r}^\delta - \boldsymbol{N}\boldsymbol{k}^{-\delta}\boldsymbol{N}'$ (introduced in Section 3.2.1) and the matrix $\boldsymbol{F} = \boldsymbol{r}^{-\delta/2}\boldsymbol{C}_1\boldsymbol{r}^{-\delta/2}$ (used in Section 4.4.3) can be written in their decomposed forms as

$$\boldsymbol{C}_1 = \sum_{\beta=0}^{m-1} \varepsilon_\beta \boldsymbol{r}^\delta \boldsymbol{L}_\beta \quad \text{and} \quad \boldsymbol{F} = \sum_{\beta=0}^{m-1} \varepsilon_\beta \boldsymbol{r}^{\delta/2} \boldsymbol{L}_\beta \boldsymbol{r}^{-\delta/2}, \tag{7.1.4}$$

respectively, where the matrices $\{\boldsymbol{L}_\beta\}$ may conveniently be written either as

$$\boldsymbol{L}_\beta = \sum_{j=1}^{\rho_\beta} \boldsymbol{s}_{\beta j}\boldsymbol{s}'_{\beta j}\boldsymbol{r}^\delta, \quad \text{or as} \quad \boldsymbol{L}_\beta = \boldsymbol{r}^{-\delta/2}\sum_{j=1}^{\rho_\beta} \boldsymbol{p}_{\beta j}\boldsymbol{p}'_{\beta j}\boldsymbol{r}^{\delta/2},$$

with the vectors $\{s_{\beta j}\}$ obtainable from the equality

$$C_1 s_{\beta j} = \varepsilon_\beta r^\delta s_{\beta j}, \quad \varepsilon_\beta = 1 - \mu_\beta,$$

and satisfying the conditions

$$s'_{\beta j} r^\delta 1_v = 0 \text{ and } s'_{\beta j} r^\delta s_{\beta' j'} = \begin{cases} 1 & \text{if } \beta = \beta' \text{ and } j = j', \\ 0 & \text{otherwise,} \end{cases}$$

or the vectors $\{p_{\beta j}\}$ obtainable from

$$F p_{\beta j} = \varepsilon_\beta p_{\beta j}, \quad \varepsilon_\beta = 1 - \mu_\beta,$$

and satisfying the conditions

$$p'_{\beta j} r^{\delta/2} 1_v = 0 \text{ and } p'_{\beta j} p_{\beta' j'} = \begin{cases} 1 & \text{if } \beta = \beta' \text{ and } j = j', \\ 0 & \text{otherwise,} \end{cases}$$

and, hence, obtainable directly as $p_{\beta j} = r^{\delta/2} s_{\beta j}$, for $\beta = 0, 1, ..., m-1$ and $j = 1, 2, ..., \rho_\beta$.

Now, the following properties of a $(\rho_0; \rho_1, ..., \rho_{m-1}; 0)$-EB design, characterized by the triples $\{\mu_\beta = 1 - \varepsilon_\beta, \rho_\beta, L_\beta\}$ for $\beta = 0, 1, ..., m-1$, can easily be recalled (from Sections 3.4 and 4.4).

(1) If, in accordance with (7.1.1), $S_\beta = [s_{\beta 1} : s_{\beta 2} : \cdots : s_{\beta \rho_\beta}]$ is such that $L_\beta S_\beta = S_\beta$, then each of the ρ_β basic contrasts $\{c'_{\beta j}\tau = s'_{\beta j} r^\delta \tau, j = 1, 2, ..., \rho_\beta\}$ of the design, its βth subset, is estimated in the intra-block analysis with the same efficiency $\varepsilon_\beta = 1 - \mu_\beta$, obtainable from (7.1.2) or (7.1.4), for any $\beta = 0, 1, ..., m-1$.

(2) The average efficiency factor of the design, defined in Remark 3.4.1(c), is equal to the harmonic mean of the efficiency factors $\{\varepsilon_\beta\}$, i.e.,

$$\varepsilon = \left(\frac{1}{v-1} \sum_{\beta=0}^{m-1} \rho_\beta \varepsilon_\beta^{-1} \right)^{-1} = \frac{v-1}{\sum_{\beta=0}^{m-1} \rho_\beta \varepsilon_\beta^{-1}} \quad \left(\leq \frac{v-1}{\sum_\beta \rho_\beta} = 1 \right).$$

It becomes 1 if and only if $\rho_1 = \rho_2 = \cdots = \rho_{m-1} = 0$, i.e., if $\rho_0 = v - 1$.

(3) The covariance (dispersion) matrix of the intra-block BLUEs of the ρ_β basic contrasts from the βth subset, i.e., of $S'_\beta r^\delta \tau$, is given by $(\sigma_1^2/\varepsilon_\beta) I_{\rho_\beta}$, and that of contrasts $A'_\beta S'_\beta r^\delta \tau$, where A_β is some matrix of ρ_β rows, is given by $(\sigma_1^2/\varepsilon_\beta) A'_\beta A_\beta$, for any $\beta = 0, 1, ..., m-1$, as it follows from Corollary 3.4.2. (For the definition of σ_1^2 see Section 3.2.1.)

(4) The matrix $r^\delta L_\beta$ appearing in (7.1.4) spans the subspace, of dimension ρ_β, of all contrasts which are estimated in the intra-block analysis with the efficiency ε_β, i.e., contrasts of the type $s'_\beta L'_\beta r^\delta \tau$ for some s_β (such that $L_\beta s_\beta \neq 0$), $\beta = 0, 1, ..., m-1$. The variance of the intra-block BLUE of any such contrast is given in (4.4.7), i.e., is of the form $\varepsilon_\beta^{-1} s'_\beta r^\delta L_\beta s_\beta \sigma_1^2$, reducing to $(r/\varepsilon_\beta) s'_\beta L_\beta s_\beta \sigma_1^2$

if the design is equireplicate, with $r_1 = r_2 = \cdots = r_v = r$. In particular, if $\boldsymbol{L}_\beta \boldsymbol{s}_\beta = \boldsymbol{s}_\beta$, the variance is simplified to $\varepsilon_\beta^{-1}\boldsymbol{s}'_\beta \boldsymbol{r}^\delta \boldsymbol{s}_\beta \sigma_1^2$ or $(r/\varepsilon_\beta)\boldsymbol{s}'_\beta \boldsymbol{s}_\beta \sigma_1^2$, respectively.

(5) More generally, for any set of contrasts $\boldsymbol{A}'\boldsymbol{S}'\boldsymbol{r}^\delta\boldsymbol{\tau}$, where $\boldsymbol{S} = [\boldsymbol{S}_0 : \boldsymbol{S}_1 : \cdots : \boldsymbol{S}_{m-1}]$, the covariance matrix of their intra-block BLUEs is of the form

$$\text{Cov}[(\widehat{\boldsymbol{A}'\boldsymbol{S}'\boldsymbol{r}^\delta\boldsymbol{\tau}})_{\text{intra}}] = \boldsymbol{A}'\boldsymbol{\varepsilon}^{-\delta}\boldsymbol{A}\sigma_1^2, \tag{7.1.5}$$

where $\boldsymbol{\varepsilon}^{-\delta}$ is the inverse of $\boldsymbol{\varepsilon}^\delta = \text{diag}[\boldsymbol{I}_{\rho_0} : \varepsilon_1\boldsymbol{I}_{\rho_1} : \cdots : \varepsilon_{m-1}\boldsymbol{I}_{\rho_{m-1}}]$, as it follows from Theorem 3.4.2. Alternatively, for any set of contrasts $\boldsymbol{U}'\boldsymbol{\tau}$, the covariance matrix of their intra-block BLUEs can be written as

$$\text{Cov}[(\widehat{\boldsymbol{U}'\boldsymbol{\tau}})_{\text{intra}}] = \boldsymbol{U}'\boldsymbol{C}_1^-\boldsymbol{U}\sigma_1^2,$$

where $\boldsymbol{C}_1^-$ is a generalized inverse (g-inverse) of the matrix $\boldsymbol{C}_1$, as it follows directly from Section 3.2.1, by noting that in case of a connected block design any set of contrasts $\boldsymbol{U}'\boldsymbol{\tau}$ satisfies the condition $\boldsymbol{U} = \boldsymbol{C}_1\boldsymbol{S}$ for some matrix $\boldsymbol{S}$. As a g-inverse of $\boldsymbol{C}_1$, one can take the matrix $\sum_{\beta=0}^{m-1}\varepsilon_\beta^{-1}\boldsymbol{L}_\beta\boldsymbol{r}^{-\delta}$, or sometimes a more convenient matrix $\boldsymbol{\Omega}$, given in Section 4.4.1 as

$$\boldsymbol{\Omega} = (\boldsymbol{C}_1 + n^{-1}\boldsymbol{r}\boldsymbol{r}')^{-1} = \left(\sum_{\beta=1}^{m-1}\frac{1-\varepsilon_\beta}{\varepsilon_\beta}\boldsymbol{L}_\beta + \boldsymbol{I}_v\right)\boldsymbol{r}^{-\delta}. \tag{7.1.6}$$

In particular, the variance of the intra-block BLUE of any contrast $\boldsymbol{c}'\boldsymbol{\tau}$, is obtainable from (7.1.6) as

$$\text{Var}[(\widehat{\boldsymbol{c}'\boldsymbol{\tau}})_{\text{intra}}] = \left(\sum_{\beta=1}^{m-1}\frac{1-\varepsilon_\beta}{\varepsilon_\beta}\boldsymbol{c}'\boldsymbol{L}_\beta\boldsymbol{r}^{-\delta}\boldsymbol{c} + \boldsymbol{c}'\boldsymbol{r}^{-\delta}\boldsymbol{c}\right)\sigma_1^2.$$

In connection with the property (5), it may be worth noting that if $\boldsymbol{A}$ is a $(v-1)\times(v-1)$ orthogonal matrix, then, from (7.1.5),

$$\text{tr}\{\sigma_1^{-2}\text{Cov}[(\widehat{\boldsymbol{A}'\boldsymbol{S}'\boldsymbol{r}^\delta\boldsymbol{\tau}})_{\text{intra}}]\} = \text{tr}(\boldsymbol{A}'\boldsymbol{\varepsilon}^{-\delta}\boldsymbol{A}) = \text{tr}(\boldsymbol{\varepsilon}^{-\delta}) = \sum_{\beta=0}^{m-1}\rho_\beta\varepsilon_\beta^{-1}, \tag{7.1.7}$$

which on account of the property (2) can equivalently be written as

$$\text{tr}\{\text{Cov}[(\widehat{\boldsymbol{A}'\boldsymbol{S}'\boldsymbol{r}^\delta\boldsymbol{\tau}})_{\text{intra}}]\} = (v-1)\sigma_1^2/\varepsilon,$$

where ε is the average efficiency factor of the design. This, by comparing with the property (3), shows that the above term used for ε is fully justified. In fact, the trace appearing in (7.1.7) is used in one of the commonly applied optimality criteria, viz., in defining the A-optimality of a design. More precisely, a design is said to be A-optimal for estimating in the intra-block analysis the basic contrasts $\boldsymbol{S}'\boldsymbol{r}^\delta\boldsymbol{\tau}$ if it minimizes, in the considered class of designs, $\text{tr}(\boldsymbol{\varepsilon}^{-\delta})$, i.e., maximizes the harmonic mean of the efficiency factors, ε, in that class. Evidently, if a design is A-optimal for $\boldsymbol{S}'\boldsymbol{r}^{\delta}\boldsymbol{\tau}$, it is also such for $\boldsymbol{A}'\boldsymbol{S}'\boldsymbol{r}^{\delta}\boldsymbol{\tau}$, provided that $\boldsymbol{A}$ is orthogonal. According to two other of the most common optimality criteria, one has either to minimize the determinant of $\boldsymbol{\varepsilon}^{-\delta}$, i.e., to maximize the geometric mean of the efficiency factors, to obtain a D-optimal design, or to minimize the maximum eigenvalue of $\boldsymbol{\varepsilon}^{-\delta}$, i.e., to maximize the smallest efficiency factor, ε_{m-1}, to obtain an E-optimal design (see, e.g., Kiefer, 1959, Section 4A; John, 1987, Section 2.4; Cheng and Bailey, 1991). However, note that in the application of these optimality criteria it is implicitly assumed that σ_1^2 is fixed for the class of designs considered, whether or not it is known (see a discussion in the classic paper by Kiefer, 1959, p. 294).

Applying the optimality criteria mentioned above, i.e., the minimization of $\text{tr}(\boldsymbol{\varepsilon}^{-\delta})$ for A-optimality, of $|\boldsymbol{\varepsilon}^{-\delta}|$ for D-optimality and of the maximum eigenvalue of $\boldsymbol{\varepsilon}^{-\delta}$ for E-optimality, the following results can be obtained for $(\rho_0; \rho_1, ..., \rho_{m-1}; 0)$-EB designs when considering estimation of their basic contrasts under the intra-block submodel $\boldsymbol{y}_1 = \boldsymbol{\phi}_1\boldsymbol{y}$, i.e., as in Section 3.4.

Lemma 7.1.1. *Let the estimation be considered under the intra-block submodel. Then a $(v-1;0;0)$-EB design is A-, D- and E-optimal for a complete set of its basic contrasts in the class of all $(\rho_0; \rho_1, ..., \rho_{m-1}; 0)$-EB designs with the same treatment replications,* $\boldsymbol{r}$.

Proof. This result follows from the inequalities $\{\varepsilon_i \leq 1, i = 1, 2, ..., v-1\}$, which hold for any $(\rho_0; \rho_1, ..., \rho_{m-1}; 0)$-EB design, and become equalities for a $(v-1;0;0)$-EB design. (For a more general result see Corsten, Ceranka and Mejza, 1984.) □

Lemma 7.1.2. *Let the estimation be considered under the intra-block submodel. Then a $(\rho_0^*; \rho_1^*; 0)$-EB design is A-, D- and E-optimal for a complete set of its basic contrasts in the class of all $(\rho_0; \rho_1, ..., \rho_{m-1}; 0)$-EB designs with the same treatment replications, $\boldsymbol{r}$, and with constant both* $\text{rank}(\boldsymbol{M}_0)$ *and* $\text{tr}(\boldsymbol{M}_0)$.

Proof. First note that, from (7.1.2) and (7.1.3), for a $(\rho_0; \rho_1, ..., \rho_{m-1}; 0)$-EB design, $\text{rank}(\boldsymbol{M}_0) = v - 1 - \rho_0$ and

$$\text{tr}(\boldsymbol{M}_0) = \sum_{\beta=1}^{m-1} \rho_\beta \mu_\beta = v - 1 - \rho_0 - \sum_{\beta=1}^{m-1} \rho_\beta \varepsilon_\beta.$$

Hence, constant $\text{rank}(\boldsymbol{M}_0)$ and $\text{tr}(\boldsymbol{M}_0)$ imply for a $(\rho_0^*; \rho_1^*; 0)$-EB design that in the considered class of designs $\rho_0^* = \rho_0$, $\rho_1^* = v - 1 - \rho_0$ and $\rho_1^*\varepsilon_1^* = \sum_{\beta=1}^{m-1} \rho_\beta \varepsilon_\beta$. Furthermore, in that class

$$\text{tr}(\boldsymbol{\varepsilon}^{-\delta}) = \rho_0 + \sum_{\beta=1}^{m-1} \rho_\beta \varepsilon_\beta^{-1}, \quad |\boldsymbol{\varepsilon}^{-\delta}| = (\prod_{\beta=1}^{m-1} \varepsilon_\beta^{\rho_\beta})^{-1}$$

and the maximum eigenvalue of $\boldsymbol{\varepsilon}^{-\delta}$ is ε_{m-1}^{-1}, these being reduced for a $(\rho_0^*; \rho_1^*; 0)$-

EB design to

$$\text{tr}(\boldsymbol{\varepsilon}^{-\delta}) = \rho_0 + \rho_1^*(\varepsilon_1^*)^{-1}, \quad |\boldsymbol{\varepsilon}^{-\delta}| = (\varepsilon_1^*)^{-\rho_1^*}$$

and the maximum eigenvalue of $\boldsymbol{\varepsilon}^{-\delta}$ to $(\varepsilon_1^*)^{-1}$, respectively. Thus, the optimality of a $(\rho_0^*; \rho_1^*; 0)$-EB design follows from the inequalities:

$$\rho_1^*/\varepsilon_1^* \leq \sum_{\beta=1}^{m-1} \rho_\beta/\varepsilon_\beta \quad \text{for its } A\text{-optimality},$$

$$-\rho_1^*\log(\varepsilon_1^*) \leq -\sum_{\beta=1}^{m-1} \rho_\beta \log(\varepsilon_\beta) \quad \text{for its } D\text{-optimality},$$

$$(\varepsilon_1^*)^{-1} \leq (\varepsilon_{m-1})^{-1} \quad \text{for its } E\text{-optimality},$$

which can easily be verified from the known properties of the arithmetic, geometric and harmonic means. (This lemma can also be proved applying Theorem 2.1 of Cheng and Bailey, 1991.) □

Lemma 7.1.3. *Let the estimation be considered under the intra-block submodel. Then a* $(0; v-1; 0)$*-EB design is A-, D- and E-optimal for a complete set of its basic contrasts in the class of all* $(0; \rho_1, ..., \rho_{m-1}; 0)$*-EB designs with the same treatment replications,* $\boldsymbol{r}$*, and with a constant* $\text{tr}(\boldsymbol{M}_0)$.

Proof. This result follows immediately from Lemma 7.1.2. (For some D- and E-optimality results see also Mukerjee and Saha, 1990.) □

In connection with Lemmas 7.1.2 and 7.1.3, note that always $\text{rank}(\boldsymbol{M}_0) = \text{rank}(\boldsymbol{N}) - 1$. Also, it may be interesting to note that in a binary proper connected block design $\text{tr}(\boldsymbol{M}_0) = (v-k)/k$, whereas in a binary equireplicate connected block design $\text{tr}(\boldsymbol{M}_0) = (b-r)/r$, where k is the common size of all b blocks for a proper design, and r is the common replication of all v treatments in an equireplicate design. Thus, when $\boldsymbol{r}$ is fixed, a constant $\text{tr}(\boldsymbol{M}_0)$ means a constant k in the first case and it means a constant b in the second case.

Now, a question of interest is whether these optimality results obtained for estimation under the intra-block submodel $\boldsymbol{y}_1 = \boldsymbol{\phi}_1\boldsymbol{y}$, with a simple covariance matrix of the form $\text{Cov}(\boldsymbol{y}_1) = \boldsymbol{\phi}_1\sigma_1^2$, on which the intra-block analysis in Section 3.2.1 is based, can be extended to the situation of estimation under the overall randomization model (3.1.1), i.e., the model $\boldsymbol{y} = \boldsymbol{\Delta}'\boldsymbol{\tau} + \boldsymbol{D}'\boldsymbol{\beta} + \boldsymbol{\eta} + \boldsymbol{e}$ derived in Section 3.1.1. It appears that a straightforward extension is possible for proper connected block designs, for which the covariance matrix of $\boldsymbol{y}$ is of the form

$$\text{Cov}(\boldsymbol{y}) = \boldsymbol{\phi}_1\sigma_1^2 + \boldsymbol{\phi}_2\sigma_2^2 + \boldsymbol{\phi}_3\sigma_3^2, \tag{7.1.8}$$

with the matrices $\boldsymbol{\phi}_1, \boldsymbol{\phi}_2, \boldsymbol{\phi}_3$ and the variances $\sigma_1^2, \sigma_2^2, \sigma_3^2$ defined in Section 3.5, i.e., designs that are generally balanced, in the sense of Definition 3.6.1 (see also Bogacka and Mejza, 1994).

In fact, a simple extension of Lemmas 7.1.1–7.1.3 can be obtained when assuming that the ratio $\delta = \sigma_2^2/\sigma_1^2$ is for the considered class of designs fixed, known and satisfies the inequality $\delta \geq 1$. To see this, note that if δ is known then, from Corollary 3.8.2, for any proper connected block design the covariance matrix of the BLUEs of a set of contrasts $\boldsymbol{A}'\boldsymbol{S}'\boldsymbol{r}^{\delta}\boldsymbol{\tau}$, where $\boldsymbol{S} = [\boldsymbol{S}_0 : \boldsymbol{S}_1 : \cdots : \boldsymbol{S}_{m-1}]$, is under the model (3.1.1), with (7.1.8), of the form

$$\mathrm{Cov}(\widehat{\boldsymbol{A}'\boldsymbol{S}'\boldsymbol{r}^{\delta}\boldsymbol{\tau}}) = \boldsymbol{A}'\mathrm{diag}\left[\boldsymbol{I}_{\rho_0} : \frac{w_{11}}{\varepsilon_{11}}\boldsymbol{I}_{\rho_1} : \cdots : \frac{w_{1,m-1}}{\varepsilon_{1,m-1}}\boldsymbol{I}_{\rho_{m-1}}\right]\boldsymbol{A}\sigma_1^2, \qquad (7.1.9)$$

where the weights $\{w_{1\beta}\}$ are defined [see (3.8.16)] as

$$w_{1\beta} = \frac{\varepsilon_{1\beta}\sigma_2^2}{\varepsilon_{1\beta}\sigma_2^2 + \varepsilon_{2\beta}\sigma_1^2} = \frac{\varepsilon_{1\beta}\delta}{\varepsilon_{1\beta}\delta + \varepsilon_{2\beta}}, \quad \beta = 1, 2, ..., m-1, \qquad (7.1.10)$$

with $\varepsilon_{1\beta} = \varepsilon_\beta$ (in the notation of the intra-block analysis) and $\varepsilon_{2\beta} = 1 - \varepsilon_\beta$. Hence, the following general result can be proved.

Theorem 7.1.1. *Let the estimation be considered under the overall randomization model with the covariance matrix of the form* (7.1.8), *for which it is assumed that the ratio* $\delta = \sigma_2^2/\sigma_1^2$ *is known and satisfies the condition* $\delta \geq 1$. *Then a* $(\rho_0^*; \rho_1^*; 0)$*-EB design is A-, D- and E-optimal for a complete set of its basic contrasts in the class of all proper* $(\rho_0; \rho_1, ..., \rho_{m-1}; 0)$*-EB designs with the same treatment replications,* $\boldsymbol{r}$, *and with constant both* $\mathrm{rank}(\boldsymbol{M}_0)$ *and* $\mathrm{tr}(\boldsymbol{M}_0)$.

Proof. As noticed in the proof of Lemma 7.1.2, constant $\mathrm{rank}(\boldsymbol{M}_0)$ and $\mathrm{tr}(\boldsymbol{M}_0)$ imply for a $(\rho_0^*; \rho_1^*; 0)$-EB design that in the considered class of designs the equalities $\rho_0^* = \rho_0$, $\rho_1^* = v - 1 - \rho_0$ and $\rho_1^*\varepsilon_1^* = \sum_{\beta=1}^{m-1}\rho_\beta\varepsilon_\beta$ hold. The last two, in turn, imply that $\rho_1^*\varepsilon_{11}^*/w_{11}^* = \sum_{\beta=1}^{m-1}\rho_\beta\varepsilon_{1\beta}/w_{1\beta}$, as can easily be proved by noting that $\varepsilon_{1\beta}/w_{1\beta} = (\varepsilon_{1\beta}\delta + \varepsilon_{2\beta})/\delta = \varepsilon_\beta + (1 - \varepsilon_\beta)/\delta = (1 - \delta^{-1})\varepsilon_\beta + \delta^{-1}$. Furthermore, from (7.1.9), in the considered class

$$\mathrm{tr}[\sigma_1^{-2}\mathrm{Cov}(\widehat{\boldsymbol{S}'\boldsymbol{r}^{\delta}\boldsymbol{\tau}})] = \rho_0 + \sum_{\beta=1}^{m-1}\rho_\beta w_{1\beta}/\varepsilon_{1\beta},$$

$$|\sigma_1^{-2}\mathrm{Cov}(\widehat{\boldsymbol{S}'\boldsymbol{r}^{\delta}\boldsymbol{\tau}})| = [\prod_{\beta=1}^{m-1}(\varepsilon_{1\beta}/w_{1\beta})^{\rho_\beta}]^{-1}$$

and the maximum eigenvalue of $\sigma_1^{-2}\mathrm{Cov}(\widehat{\boldsymbol{S}'\boldsymbol{r}^{\delta}\boldsymbol{\tau}})$ is $w_{1,m-1}/\varepsilon_{1,m-1}$ (≥ 1 if $\delta \geq 1$), these being reduced for a $(\rho_0^*; \rho_1^*; 0)$-EB design to

$$\mathrm{tr}[\sigma_1^{-2}\mathrm{Cov}(\widehat{\boldsymbol{S}'\boldsymbol{r}^{\delta}\boldsymbol{\tau}})] = \rho_0 + \rho_1^* w_{11}^*/\varepsilon_{11}^*, \quad |\sigma_1^{-2}\mathrm{Cov}(\widehat{\boldsymbol{S}'\boldsymbol{r}^{\delta}\boldsymbol{\tau}})| = (\varepsilon_{11}^*/w_{11}^*)^{-\rho_1^*}$$

and the maximum eigenvalue of $\sigma_1^{-2}\mathrm{Cov}(\widehat{\boldsymbol{S}'\boldsymbol{r}^{\delta}\boldsymbol{\tau}})$ to $w_{11}^*/\varepsilon_{11}^*$ (≥ 1 if $\delta \geq 1$), respectively. Thus, the optimality of a $(\rho_0^*; \rho_1^*; 0)$-EB design follows from the inequalities:

$$\rho_1^* w_{11}^*/\varepsilon_{11}^* \leq \sum_{\beta=1}^{m-1} \rho_\beta w_{1\beta}/\varepsilon_{1\beta} \quad \text{for its } A\text{-optimality},$$

$$-\rho_1^* \log(\varepsilon_{11}^*/w_{11}^*) \leq -\sum_{\beta=1}^{m-1} \rho_\beta \log(\varepsilon_{1\beta}/w_{1\beta}) \quad \text{for its } D\text{-optimality},$$

$$w_{11}^*/\varepsilon_{11}^* \leq w_{1,m-1}/\varepsilon_{1,m-1} \quad \text{for its } E\text{-optimality},$$

which can be obtained using the same argument as in the proof of Lemma 7.1.2.

Alternatively, this theorem can be proved applying Theorem 2.1 of Cheng and Bailey (1991). To see this, replace there n by $v-1$ and x_i by ε_{1i}/w_{1i} for $i = 1, 2, ..., v-1$, and note that for any design in the considered class the condition of constant

$$\sum_{i=1}^{v-1} \frac{\varepsilon_{1i}}{w_{1i}} = \rho_0 + \sum_{\beta=1}^{m-1} \rho_\beta \frac{\varepsilon_{1\beta}}{w_{1\beta}} = \rho_0 + (1-\delta^{-1}) \sum_{\beta=1}^{m-1} \rho_\beta \varepsilon_\beta + \delta^{-1}(v-1-\rho_0)$$

is satisfied, because $\text{rank}(\boldsymbol{M}_0) = v-1-\rho_0$ and $\text{tr}(\boldsymbol{M}_0) = v-1-\rho_0-\sum_{\beta=1}^{m-1} \rho_\beta \varepsilon_\beta$ are constant in that class. Furthermore, note that a $(\rho_0^*; \rho_1^*; 0)$-EB design, with $\rho_0^* = \rho_0$ and $\rho_1^* = v-1-\rho_0$, satisfies the following conditions:

(i) $\varepsilon_{1i}^*/w_{1i}^* > 0$ for $i = 1, 2, ..., v-1$ (if $\delta > 0$),

(ii) there are two distinct values among $\{\varepsilon_{1i}^*/w_{1i}^*\}$,

(iii) $\sum_{i=1}^{v-1} (\varepsilon_{1i}^*/w_{1i}^*)^2 = \rho_0 + \rho_1^*(\varepsilon_{11}^*/w_{11}^*)^2 \leq \rho_0 + \sum_{i=\rho_0+1}^{v-1} (\varepsilon_{1i}/w_{1i})^2$,

(iv) $\sum_{i=1}^{v-1} \varepsilon_{1i}^*/w_{1i}^* = \rho_0 + \rho_1^* \varepsilon_{11}^*/w_{11}^* = \rho_0 + \sum_{i=\rho_0+1}^{v-1} \varepsilon_{1i}/w_{1i}$.

Hence, following the same reasoning as in Cheng and Bailey (1991), it can be concluded that Theorem 7.1.1 is proved. □

Corollary 7.1.1. *If in Theorem* 7.1.1 *the conditions of constant* $\text{rank}(\boldsymbol{M}_0)$ *and constant* $\text{tr}(\boldsymbol{M}_0)$ *are removed, then the theorem extends the result of Lemma* 7.1.1 *to estimation under the overall randomized model with known* $\delta = \sigma_2^2/\sigma_1^2$, *provided that attention is confined to proper designs only, and that* $\delta \geq 1$.

Proof. This follows from the fact that $\varepsilon_{1\beta}/w_{1\beta} = \varepsilon_\beta + \delta^{-1}(1-\varepsilon_\beta) \leq 1$ if and only if $1 \leq \delta$. □

Corollary 7.1.2. *If in Theorem* 7.1.1 *the condition of a constant* $\text{rank}(\boldsymbol{M}_0)$ *is replaced by a constant* $\text{rank}(\boldsymbol{M}_0) = v-1$, *then the theorem extends the result of Lemma* 7.1.3 *to estimation under the overall randomization model with known* $\delta = \sigma_2^2/\sigma_1^2$, *provided that attention is confined to proper designs only, whether or not* $\delta \geq 1$.

Proof. This result follows immediately from Theorem 7.1.1, when taking $\rho_0 = 0$. (See also Bogacka and Mejza, 1994, Corollary 2.) □

Remark 7.1.1. In connection with Theorem 7.1.1 it is interesting to note that the assumption of $\delta = \sigma_2^2/\sigma_1^2 \geq 1$ is not necessary to prove the theorem. It would be sufficient to assume that $\delta > 0$. However, the condition $\delta \geq 1$ is essential for the ρ_0 basic contrasts, chosen so to be estimated in the intra-block analysis with full efficiency, to retain their advantageous position when estimated in the overall analysis combining information from the intra-block and the inter-block stratum (as described in Section 3.8.1), in the sense of being estimated with a precision not smaller than the precision of estimating the remaining basic contrasts. To see these precisions, note that, under the overall randomization model with (7.1.8),

$$\text{Var}(\widehat{\mathbf{c}'_{\beta j}\boldsymbol{\tau}}) = \text{Var}(\widehat{\mathbf{s}'_{\beta j}\mathbf{r}^{\delta}\boldsymbol{\tau}}) = \sigma_1^2 w_{1\beta}/\varepsilon_{1\beta},$$

with $w_{1\beta}/\varepsilon_{1\beta} \geq 1$ if and only if $\delta = \sigma_2^2/\sigma_1^2 \geq 1$ for $\beta = 1, 2, ..., m-1$, whereas for $\beta = 0$,

$$\text{Var}(\widehat{\mathbf{c}'_{0j}\boldsymbol{\tau}}) = \text{Var}(\widehat{\mathbf{s}'_{0j}\mathbf{r}^{\delta}\boldsymbol{\tau}}) = \sigma_1^2,$$

as it follows from (7.1.9) and (7.1.10).

The above results concerning the optimality properties under the overall randomization model are applicable to proper block designs only, i.e., those which are generally balanced, and under the condition that the ratio $\delta = \sigma_2^2/\sigma_1^2$ for the stratum variances appearing in the covariance matrix (7.1.8) is known, as it is assumed in Section 3.8.1. As to the assumption that $\delta \geq 1$, it reflects the usual aim of designing and conducting an experiment so that the intra-block heterogeneity of experimental units is reduced as far as possible. [See the definitions of σ_1^2 and σ_2^2 in (3.5.7) and the discussion in Section 4.5.] The more successfully, in this sense, the grouping of units into blocks is performed, the higher the value of δ will be. Thus, in searching for an optimal design, one has to be prepared to use it in the experiment in such a way that the ratio δ becomes not smaller than 1, and possibly higher.

Another problem, one has to be aware of, is that in practice the true value of δ is usually not known and is to be estimated. Replacing the unknown δ by its estimate $\hat{\delta}$ changes the properties of the estimators used in estimating the basic contrasts, by expanding their variances, as can be seen in Section 3.8.3. This may cause doubts whether there is any gain in precision resulting from the recovery of inter-block information over the estimation in the intra-block analysis, when instead of the true value of δ its estimate is used. This has been discussed in Section 3.8.4, where an approximate measure of this gain is given, in (3.8.82), and where it has been shown, e.g., that for a BIB design with $v > 3$ this measure is uniformly (for all values of w_{11}) positive if and only if the inequality

$$(v-1)^2/[v(v-3)] < r(k-1)/k \tag{7.1.11}$$

holds. More generally, as can be shown, this measure of precision is uniformly positive for a proper $(\rho_0; \rho_1; 0)$-EB design if and only if

$$\max\{\rho_1^2, (b-1)(\rho_1 - 2) + 2\rho_1\} < (n-v)(\rho_1 - 2), \tag{7.1.12}$$

provided that $\rho_1 > 2$. In the case of $\rho_1 = v - 1 > 2$, the inequality (7.1.12) is reduced to

$$(b-1)(v-3) + 2(v-1) < (n-v)(v-3),$$

and this is equivalent to (7.1.11) for a proper equireplicate design.

A practical conclusion from the discussion above is that in the search for an optimal connected proper block design one can rely on the results given in Lemmas 7.1.1 – 7.1.3, and then recover in the analysis the inter-block information, i.e., use the combined estimators presented in Section 3.8.3, whenever a gain in precision can be expected according to the conditions given in Section 3.8.4. For a proper $(\rho_0; \rho_1; 0)$-EB design, if chosen, this means to employ in the analysis the combined estimators when the design satisfies the inequality (7.1.12). However, one has to remember that the desirable optimality of a chosen design will be preserved, when going from the intra-block analysis to the analysis combining the intra-block and inter-block information, if the grouping of experimental units into blocks is sufficiently successful for increasing the ratio $\delta = \sigma_2^2/\sigma_1^2$ over 1.

Unfortunately, the conclusions given above apply to proper block designs only. Whether and how much they can be extended to nonproper designs remains, at present, an open question. Nevertheless, when confined to intra-block analysis, one can use the results of Lemmas 7.1.1–7.1.3 for proper as well as for nonproper block designs.

As already mentioned at the beginning, the subject of consideration in this chapter is the class of $(\rho_0; \rho_1, ..., \rho_{m-1}; 0)$-EB designs in which $\rho_0 \geq 1$, i.e., designs in which one of the distinct eigenvalues of the matrix $\boldsymbol{C}_1$ with respect to $\boldsymbol{r}^\delta$ is 1. This eigenvalue is denoted in (7.1.4) by ε_0 and its multiplicity by ρ_0. The exposition of these designs will be systematized by separating the material into four sections, depending on whether the number of distinct eigenvalues, i.e., distinct efficiency factors, m, is 1, 2, 3, or more. Although of main interest here are the $(\rho_0; \rho_1; 0)$-EB designs, $(\rho_0; \rho_1, ..., \rho_{m-1}; 0)$-EB designs with $m > 2$ may be, in certain experimental situations, of some interest from the practical point of view. Throughout the chapter, whether it is mentioned or not, the term efficiency factor will be used in the sense of the efficiency for estimating in the intra-block analysis (i.e., as in Section 3.4).

7.2 Orthogonal designs

In the extreme case, when $m = 1$ and $\rho_0 = v-1$, i.e., the design is $(v-1; 0; 0)$-EB, every contrast of treatment parameters is estimated in the intra-block analysis with the same efficiency ε $(= \varepsilon_1 = \cdots = \varepsilon_{v-1}) = 1$.

On the other hand, it is known (from Corollaries 2.3.3 and 2.3.4) that for a connected block design the unique ε, with multiplicity $v - 1$, is equal to 1 if and only if $\boldsymbol{N} = n^{-1}\boldsymbol{r}\boldsymbol{k}'$ (describing an orthogonal design), which for a binary design is $\boldsymbol{N} = \boldsymbol{1}_v\boldsymbol{1}_b'$ (describing the randomized block design, RBD). Then one obtains the following result.

Theorem 7.2.1. *A binary* $(v-1;0;0)$*-EB design is the randomized block design (RBD).*

On account of Theorem 7.2.1, as a connected binary block design with full efficiency, one can have only the RBD, which is the most widely used among all experimental designs. In agricultural field experimentation, e.g., the RBD can be used when the experimental field is subdivided into blocks of experimental units (plots) in such a way that the units within each block are more homogeneous than within the area as a whole (see Pearce, 1983, p. 41), and the number of units within each block is equal to the number of treatments being investigated. As described in Sections 1.3 and 1.4, within a block all the treatments are randomly allocated to the experimental units, this being repeated over all chosen blocks, with a separate randomization for each block, and with an independent randomization or random selection of the blocks themselves. This is one of the methods aimed at controlling variability of experimental units. The design is extremely popular in industrial and agricultural research, commonly used in laboratory, field, greenhouse, and other experiments, but also in educational, medical, marketing and sociological experimentation. For more discussion on the statistical implications of the randomizations involved see Section 1.4, also Caliński (1996).

As it follows from Lemma 7.1.1, a connected orthogonal block design, given by $\boldsymbol{N} = n^{-1}\boldsymbol{r}\boldsymbol{k}'$, is A-, D- and E-optimal in the class of all connected block designs with the same treatment replications . Moreover, as shown by Corsten et al. (1984), a connected block design is uniformly optimal, in the sense of minimizing the variance of the intra-block BLUE for any contrast of treatment parameters, if and only if the design is orthogonal. Consequently, the RBD satisfies all the above optimality criteria in the class of connected binary equireplicate block designs with the same r, a common replication for all v treatments. Furthermore, on account of Corollary 7.1.1, these optimality results are valid not only when considering the estimation of contrasts under the intra-block submodel (3.2.2) but also when considering it under the overall randomization model (3.1.1), provided that $\delta = \sigma_2^2/\sigma_1^2$ is constant in the considered class of designs, and is known to be not smaller than 1. Also note that in case of an orthogonal connected block design, the best linear unbiased estimation of contrasts of treatment parameters under the submodel (3.2.2) coincides with that under the overall model (3.1.1), as it follows from Corollary 3.2.5(a) and Remark 3.2.7.

Finally, one has to remember that the above optimality results apply under the assumption that the variance ratio δ is known to be constant in the class of designs considered. Such assumption may be not realistic when the intra-block heterogeneity of experimental units cannot be controlled to the same extent independently of the block sizes. In practice, it is usually easier to reduce this heterogeneity when the blocks include smaller number of units (i.e., plots in an agricultural field experiment). Therefore, often it becomes desirable to look for an optimal design within a class of block designs with a fixed block size. This means to restrict attention to proper $(\rho_0;\rho_1,...,\rho_{m-1};0)$-EB designs with a chosen constant block size k, such which allows to keep the ratio δ consistently well above 1. If further reducing the class to binary designs, a constant $\mathrm{tr}(\boldsymbol{M}_0)$ in Theorem 7.1.1 will mean a constant value of $(v-k)/k$ (see the discussion following Lemma 7.1.3). One would then arrive at the RBD only for $v = k$. Because in many cases the number of treatments to be investigated may be larger than a reasonably chosen block size k, other than connected orthogonal block designs are of interest. They will be considered in the following sections.

7.3 Designs with two efficiency factors

As pointed out by Pearce (1983, Section 1.6), in many practical situations all the contrasts may not be of equal importance for the experimenter, and it may be desirable to estimate and test in the intra-block analysis only the most important contrasts with full efficiency, allowing all the other to be estimated in that analysis with equal but lower efficiency. To meet such circumstances, $(\rho_0;\rho_1;0)$-EB designs will be discussed in this section, giving some characterizations and constructions of these designs, with $\rho_0 \geq 1$. Similar considerations for the case of $m = 3$ will be presented in Section 7.4.

Thus, the interest here is in designs in which the number of distinct efficiency factors is $m = 2$ and one of them is equal to one. In such designs not all of the basic contrasts can be estimated with the same efficiency, but some of them, ρ_0 in number, can be estimated in the intra-block analysis with full efficiency. This means that any of these designs can be called orthogonal for the set of contrasts receiving the unit efficiency factor [see also Remark 3.4.1(a)]. Note, however, that such a design does not belong to the class of connected orthogonal block designs considered in Section 7.2, unless $\rho_0 = v - 1$.

Let $\varepsilon_0 = 1$ with multiplicity ρ_0. Then the other eigenvalue ε_1 has the multiplicity $\rho_1 = v - 1 - \rho_0$. In this case the average efficiency factor of the design can be given by

$$\varepsilon = \frac{(v-1)\varepsilon_1}{(v-1)\varepsilon_1 + (1-\varepsilon_1)\rho_1} = 1 - \frac{(1-\varepsilon_1)\rho_1}{(v-1)\varepsilon_1 + (1-\varepsilon_1)\rho_1}, \tag{7.3.1}$$

which depends only on the eigenvalue ε_1 (< 1) and its multiplicity.

From Lemma 7.1.2, it is known that a $(\rho_0^*;\rho_1^*;0)$-EB design is A-, D- and E-optimal for estimating in the intra-block analysis its basic contrasts, among all $(\rho_0;\rho_1,...,\rho_{m-1};0)$-EB designs with the same treatment replications, $\boldsymbol{r} = [r_1, r_2, ..., r_v]'$, and with constant values of both $\mathrm{rank}(\boldsymbol{M}_0)$ and $\mathrm{tr}(\boldsymbol{M}_0)$. More explicitly, this means that the design is optimal, with regard to the three optimality criteria, in the class of all connected block designs with constant vector $\boldsymbol{r}$ and the values of ρ_0 and $\mathrm{tr}(\boldsymbol{r}^{-\delta}\boldsymbol{C}_1) = \rho_0 + \sum_{\beta=1}^{m-1}\varepsilon_\beta\rho_\beta$. Note that the last condition can be replaced by that of constant value of $\mathrm{tr}(\boldsymbol{r}^{-\delta}\boldsymbol{N}\boldsymbol{k}^{-\delta}\boldsymbol{N}')$.

Now some methods of construction will be described. At first, by Theorem 6.4.1, once a $(\rho_0;\rho_1;0)$-EB design is available, a few of its copies yield a $(\rho_0^*;\rho_1^*;0)$-EB design having the same properties concerning efficiency factors. More methods can be given. A general procedure of constructing $(\rho_0;\rho_1;0)$-EB designs with parameters v, b, $\boldsymbol{r}$, $\boldsymbol{k}$ and the relevant incidence matrix $\boldsymbol{N}$ consists in finding such $\boldsymbol{N}$ which provides $\varepsilon_0 = 1$, ε_1, $\boldsymbol{L}_0$ and $\boldsymbol{L}_1$ that satisfy (7.1.4) with $m = 2$ and $\rho_0 \geq 1$, i.e., gives the decomposition $\boldsymbol{C}_1 = \boldsymbol{r}^\delta \boldsymbol{L}_0 + \varepsilon_1 \boldsymbol{r}^\delta \boldsymbol{L}_1$ or, equivalently,

$$\boldsymbol{M}_0 = (1 - \varepsilon_1)\boldsymbol{L}_1,$$

where $\boldsymbol{M}_0$ is as defined in Section 4.3.2 and recalled in Section 7.1. From this standpoint, the concepts and methods introduced by Caliński (1971), and further developed by Saha (1976), Puri and Nigam (1977a, 1977b), Ceranka (1983), Ceranka and Kozłowska (1983, 1984), and others, may be utilized to construct $(\rho_0; \rho_1; 0)$-EB designs. Recall Definition 4.3.1 and the discussion on simple partially efficiency-balanced [PEB(s)] designs, or C-designs, in Section 4.4.2.

For example, concerning the technique of dualization (see Section 6.1), Corollary 6.1.3 shows that the dual of a $(\rho_0; \rho_1; 0)$-EB design is again a $(\rho_0^*; \rho_1^*; 0)$-EB design with $\varepsilon_0^* = 1$ and $\varepsilon_1^* = \varepsilon_1$, i.e., with the same property regarding the efficiency factors, provided that $b > v - \rho_0$. By (7.3.1) and Corollary 6.1.3, the average efficiency factor of the dual design is given by

$$\varepsilon^* = \frac{(b-1)\varepsilon_1}{(b-1)\varepsilon_1 + (1-\varepsilon_1)\rho_1}. \tag{7.3.2}$$

From (7.3.1), (7.3.2) and Corollary 6.1.3, the following observations can be obtained (see also Ceranka, 1983, Corollaries 3.6, 3.9 and 3.10).

Remark 7.3.1. The average efficiency factor of a $(\rho_0; \rho_1; 0)$-EB design is the same as that of its dual if and only if $v = b$.

Remark 7.3.2. The average efficiency factor of the dual of a $(\rho_0; \rho_1; 0)$-EB design is greater than that of the original design if and only if $b > v$.

Remark 7.3.3. The average efficiency factor of the dual of a $(\rho_0; \rho_1; 0)$-EB design is smaller than that of the original design if and only if $b < v$. It becomes equal to ε_1 if and only if $b = v - \rho_0$ (i.e., $\rho_1 = b - 1$), which is the extreme case of the dual, being then $(0; v^* - 1; 0)$-EB.

In this section, a large number of procedures of constructing designs with two efficiency factors, in which one of them is equal to one, will be presented. Using these procedures, one can choose or construct a plan suitable to a purpose of an experiment without any difficulty.

7.3.1 Proper designs that are equireplicate

Practical considerations, such as, e.g., related to crop variety trials, dictate that in most of the designs used for agricultural field experiments all treatments are to be equally replicated, though they may be exceptions from this. A wide choice of binary equireplicate block designs with various possible block sizes is open to the experimenter. Usually, blocks of equal sizes, i.e., proper designs, are used.

From the considerations in Sections 4.4 and 7.1, the role of the matrices $\{\boldsymbol{L}_\beta\}$ and the corresponding efficiency factors $\{\varepsilon_\beta\}$ in the characterization of an equireplicate block design should become apparent. If the experimental problem can be expressed in terms of contrasts of treatment parameters, and the contrasts can be classified into subsets (classes) of different importance, then in searching for an appropriate design one should look for such which provides matrices $\{\boldsymbol{L}_\beta\}$ spanning subspaces of contrasts that cover those of real experimental interest, and that are associated with efficiency factors well reflecting the classification of contrasts according to their importance. In particular, it is desirable that the subspace spanned by the matrix $\boldsymbol{L}_0$ includes all contrasts of the highest importance.

If the contrasts of experimental interest are classified into two subsets only, one of them composed of contrasts requiring estimation in the intra-block analysis with full efficiency, then the search for a suitable (connected) design should be confined to $(\rho_0; \rho_1; 0)$-EB designs.

Consider first a proper equireplicate $(\rho_0; \rho_1; 0)$-EB design with parameters v, b, r, k. Recall, from (7.1.2), (7.1.3) and (7.1.4) that for any such design

$$\boldsymbol{F} = r^{-1}\boldsymbol{C}_1 = \boldsymbol{L}_0 + \varepsilon_1\boldsymbol{L}_1,$$

where

$$\boldsymbol{L}_\beta = \sum_{j=1}^{\rho_\beta} \boldsymbol{p}_{\beta j}\boldsymbol{p}'_{\beta j}, \quad \boldsymbol{p}'_{\beta j}\boldsymbol{p}_{\beta j'} = \delta_{jj'}, \quad \boldsymbol{L}_\beta\boldsymbol{L}_{\beta'} = \delta_{\beta\beta'}\boldsymbol{L}_\beta,$$

and

$$\text{rank}(\boldsymbol{L}_\beta) = \rho_\beta, \ \beta = 0, 1, \ \rho_0 + \rho_1 = v - 1.$$

Also note that the average efficiency factor of the design is $\varepsilon = (v-1)/(\rho_0 + \rho_1/\varepsilon_1)$, whereas any ρ_β basic contrasts of the βth subset, $\{\boldsymbol{c}'_{\beta j}\boldsymbol{\tau} = r^{1/2}\boldsymbol{p}'_{\beta j}\boldsymbol{\tau}$, $j = 1, 2, ..., \rho_\beta\}$, are estimated in the intra-block analysis with the same efficiency $\varepsilon_\beta, \beta = 0, 1$. Moreover, from Sections 4.4.3 and 7.1(4), note that $\boldsymbol{L}_0$ spans the subspace of all contrasts estimated in the intra-block analysis with full efficiency, $\varepsilon_0 = 1$, i.e., contrasts of the type $\boldsymbol{p}'_0\boldsymbol{L}'_0\boldsymbol{\tau}$ for some $\boldsymbol{p}_0$, whereas $\boldsymbol{L}_1$ spans the subspace of all contrasts which are estimated in that analysis with a lower efficiency, $\varepsilon_1 < 1$, i.e., contrasts of the type $\boldsymbol{p}'_1\boldsymbol{L}'_1\boldsymbol{\tau}$ for some $\boldsymbol{p}_1$. The dimensions of these subspaces are ρ_0 and ρ_1, respectively, giving $\rho_0 + \rho_1 = v - 1$ since only connected designs are considered here. The intra-block BLUEs and their variances for these contrasts are given by $(\widehat{\boldsymbol{p}'_\beta\boldsymbol{L}'_\beta\boldsymbol{\tau}})_{\text{intra}} = (\varepsilon_\beta r)^{-1}\boldsymbol{p}'_\beta\boldsymbol{L}_\beta\boldsymbol{Q}_1$ and $\text{Var}[(\widehat{\boldsymbol{p}'_\beta\boldsymbol{L}'_\beta\boldsymbol{\tau}})_{\text{intra}}] = (\varepsilon_\beta r)^{-1}\boldsymbol{p}'_\beta\boldsymbol{L}_\beta\boldsymbol{p}_\beta\sigma_1^2$, respectively, where $\beta = 0, 1$. [See formulae (4.4.12) and (4.4.13).]

As to the optimality of a proper equireplicate $(\rho_0^*; \rho_1^*; 0)$-EB design, it can be recalled (from Lemma 7.1.2) that it is A-, D- and E-optimal for the intra-block estimation of its basic contrasts, in the class of all proper equireplicate $(\rho_0; \rho_1, ..., \rho_{m-1}; 0)$-EB designs with the same treatment replications, a common r here, and with constant both $\text{rank}(\boldsymbol{M}_0)$ and $\text{tr}(\boldsymbol{M}_0)$. Here, the last two conditions mean that in the considered class both ρ_0 and $(rk)^{-1}\mathrm{tr}(\boldsymbol{NN}')$ are constant, the latter reducing to v/k if the class is confined to binary designs.

Now the problem is how to construct such $(\rho_0; \rho_1; 0)$-EB designs.

7.3.1.1 PBIB designs

When a proper equireplicate binary design is to be used, one of the most standard choices is a partially balanced incomplete block (PBIB) design with two associate classes, described in Section 6.0.2. As shown there, a connected 2-associate PBIB design with parameters $v, b, r, k, \lambda_1, \lambda_2, \psi_1, \psi_2, \rho_1, \rho_2, \boldsymbol{A}_i, \boldsymbol{A}_i^{\#}, i = 0, 1, 2$, is $(\rho_0; \rho_1; 0)$-EB, provided that $\psi_1 = 0$ or $\psi_2 = 0$. Letting ψ_i, ρ_i and $\boldsymbol{A}_i^{\#}, i = 1, 2$, to be renumbered, with $\beta = 0, 1$, so that $\psi_0 = 0$ and $\psi_1 > 0$, with multiplicities ρ_0 and ρ_1, respectively, it will be convenient to write $\varepsilon_0 = 1 - (rk)^{-1}\psi_0 = 1$, $\varepsilon_1 = 1 - (rk)^{-1}\psi_1$ and, correspondingly, $\boldsymbol{L}_0 = \boldsymbol{A}_0^{\#}$, $\boldsymbol{L}_1 = \boldsymbol{A}_1^{\#}$, in accordance with the decomposition

$$
\begin{aligned}
\frac{1}{rk}\boldsymbol{NN}' &= \frac{\psi_0}{rk}\boldsymbol{A}_0^{\#} + \frac{\psi_1}{rk}\boldsymbol{A}_1^{\#} + \frac{1}{v}\mathbf{1}_v\mathbf{1}_v' \\
&= (1-\varepsilon_0)\boldsymbol{L}_0 + (1-\varepsilon_1)\boldsymbol{L}_1 + \frac{1}{v}\mathbf{1}_v\mathbf{1}_v',
\end{aligned}
$$

comparable with (2.3.1) and (4.3.8) [here with (7.1.2)]. In this sense, a PBIB design with one of the eigenvalues of $\boldsymbol{NN}'$ being zero, i.e., with the rank of the incidence matrix $\boldsymbol{N}$ smaller than v, is of special interest, as allowing some of its basic contrasts to be estimated in the intra-block analysis with full efficiency. To be consistent with the notation already adopted in this chapter, from now onward, when presenting parameters of a PBIB design the symbols $\{\varepsilon_\beta, \rho_\beta, \boldsymbol{L}_\beta\}$, instead of $\{\psi_i, \rho_i, \boldsymbol{A}_i^{\#}\}$, will be used.

Some methods of constructing such PBIB designs are worth to be presented. First, however, note that Clatworthy (1973) has tabulated comprehensively more than 800 actual plans of 2-associate PBIB designs with treatment replications and block sizes within the scope of $2 \leq r, k \leq 10$. These tables are a revised and enlarged version of the earlier tables given by Bose, Clatworthy and Shrikhande (1954). The reader will find in Clatworthy's book (1973) also a number of methods of construction. [Sinha (1991a) has prepared a list of new GD designs found after 1973.] Here, focusing attention on PBIB designs with one of the ψ_1 and ψ_2 being zero, some construction methods of such designs are described.

(1) Singular GD designs. This design is characterized by $r - \lambda_1 = 0$ [see Section 6.0.2(A)]. In this case the following theorem is obvious.

Theorem 7.3.1 (Bose and Connor, 1952; Raghavarao, 1971, Theorem 8.5.1). *The existence of a BIB design with parameters v, b, r, k, λ is equivalent to the existence of a singular GD design with parameters $v^* = vn$ (v groups of n treatments), $b^* = b$, $r^* = r$, $k^* = nk$, $\lambda_1^* = r$, $\lambda_2^* = \lambda$, i.e., of a $(\rho_0^*; \rho_1^*; 0)$-EB design with $\varepsilon_0^* = 1$, $\varepsilon_1^* = \lambda v/(rk)$, $\rho_0^* = v(n-1)$, $\rho_1^* = v - 1$, $\boldsymbol{L}_0^* = \boldsymbol{I}_v \otimes (\boldsymbol{I}_n - n^{-1}\mathbf{1}_n\mathbf{1}_n')$, $\boldsymbol{L}_1^* = (\boldsymbol{I}_v - v^{-1}\mathbf{1}_v\mathbf{1}_v') \otimes n^{-1}\mathbf{1}_n\mathbf{1}_n'$.* [The symbol n here should not be confused with n denoting the total number of experimental units used in the experiment, as defined in Section 2.2. Also see the remark made just before Section 6.0.2(A). Further, note that m has in the notation of this book special meaning (see Definition 4.4.2).]

Proof. This result follows from a procedure consisting in replacing each treatment of the BIB design by a group of n treatments. (See also Raghavarao, 1971, Theorem 8.5.1 and its proof.) □

The essence of Theorem 7.3.1 is that any singular GD design can be derived from a relevant BIB design, provided it exists. Hence, as a companion result to this theorem, the following result is useful.

Theorem 7.3.2 (Kageyama and Tsuji, 1977, Theorem 2.1). *A GD design with parameters $v = mn$ (m groups of n treatments each), b, r, k, λ_1, λ_2 is singular if and only if k/n is an integer and every block contains exactly k/n groups of treatments of the association scheme.*

In a singular GD design with parameters $v^*, b^*, r^*, k^*, \lambda_1^*, \lambda_2^*$ considered in Theorem 7.3.1, any $v(n-1)$ basic contrasts among n treatments belonging to the same of the v groups (i.e., being the first associates) are estimated with full efficiency and the variance σ_1^2 each, whereas any $v-1$ basic contrasts among the v different groups of treatments (i.e., being second associates) are estimated with the efficiency $\varepsilon_1^* = \lambda v/(rk)$ and the variance $rk\sigma_1^2/(\lambda v)$ each, when referring, as usual, to the intra-block analysis.

(2) Semi-regular GD designs. Here $rk = v\lambda_2$ and $r - \lambda_1 > 0$. The following characterization result concerning the block structure of these designs is useful.

Theorem 7.3.3 (Bose and Connor, 1952). *A GD design with parameters $v = mn$, b, r, k, λ_1, λ_2 is semi-regular if and only if k/m is an integer and every block contains exactly k/m treatments from each group of the association scheme. If the condition is satisfied, then the design is $(\rho_0; \rho_1; 0)$-EB with $\varepsilon_0 = 1$, $\varepsilon_1 = [r(k-1)+\lambda_1]/(rk)$, $\rho_0 = m-1$, $\rho_1 = m(n-1)/2$, $\boldsymbol{L}_0 = (\boldsymbol{I}_m - m^{-1}\mathbf{1}_m\mathbf{1}_m')\otimes n^{-1}\mathbf{1}_n\mathbf{1}_n'$ and $\boldsymbol{L}_1 = \boldsymbol{I}_m \otimes (\boldsymbol{I}_n - n^{-1}\mathbf{1}_n\mathbf{1}_n')$.*

For a proof see Raghavarao (1971, Theorem 8.5.6). (Although only the necessity part is proved there, it is easy to see from this proof that the condition of the theorem is also sufficient for making a GD design to become semi-regular.) See also Kageyama and Tsuji (1977, Theorem 2.2).

Another useful result is the following, originally noticed by Bose, Shrikhande and Bhattacharya (1953, p. 182).

Theorem 7.3.4 (Raghavarao, 1971, Theorem 10.3.3). *The dual of an affine α-resolvable BIB design is a semi-regular GD design.*

The theorem is obvious from the definitions of affine α-resolvable BIB designs (Definitions 6.0.2 and 6.0.3) and GD designs [Section 6.0.2(A)].

A particular application of Theorem 7.3.4 can be seen when considering an affine resolvable BIB design with parameters $v^* = q^2$, $b^* = q(q+1)$, $r^* = q+1$, $k^* = q$, $\lambda^* = 1$, which exists if q is a prime or a prime power (see Corollary 6.9.3). By taking the dual of the design one obtains a semi-regular GD design characterized as follows.

Corollary 7.3.1 (Raghavarao, 1971, Theorem 8.5.8 with $t = 0$). *If q is a prime or a prime power, there exists a semi-regular GD design with parameters $v = (q+1)q$, $b = q^2$, $r = q$, $k = q+1$, $\lambda_1 = 0$, $\lambda_2 = 1$; $m = q+1, n = q$, i.e., a $(\rho_0; \rho_1; 0)$-EB design with $\varepsilon_0 = 1$, $\varepsilon_1 = q/(q+1)$, $\rho_0 = q$, $\rho_1 = q^2 - 1$, $\boldsymbol{L}_0 = [\boldsymbol{I}_{q+1} - (q+1)^{-1}\mathbf{1}_{q+1}\mathbf{1}'_{q+1}] \otimes q^{-1}\mathbf{1}_q\mathbf{1}'_q$, $\boldsymbol{L}_1 = \boldsymbol{I}_{q+1} \otimes (\boldsymbol{I}_q - q^{-1}\mathbf{1}_q\mathbf{1}'_q)$.*

In this semi-regular GD design, any q basic contrasts among the $q+1$ different groups of treatments (i.e., being second associates) are estimated in the intra-block analysis with full efficiency and the variance σ_1^2 each, whereas any q^2-1 basic contrasts among treatments belonging to the same group (i.e., being first associates) are estimated in that analysis with the efficiency $q/(q+1)$ and the variance $(q+1)\sigma_1^2/q$ each.

Theorem 7.3.5. *Let $\boldsymbol{N}$ be the $v \times b$ incidence matrix of a BIB design with parameters $v = 2k$, $b = 2r$, r, k, λ. Then the incidence matrix*

$$\boldsymbol{N}^* = \begin{bmatrix} \boldsymbol{N} & \boldsymbol{N} \\ \boldsymbol{N} & \mathbf{1}_v\mathbf{1}'_b - \boldsymbol{N} \end{bmatrix} \tag{7.3.3}$$

yields an r-resolvable semi-regular GD design with parameters $v^* = 4k$, $b^* = 4r$, $r^* = 2r$, $k^* = 2k$, $\lambda_1^* = 2\lambda$, $\lambda_2^* = r$, $m^* = 2, n^* = 2k$, i.e., a $(\rho_0^*; \rho_1^*; 0)$-EB design with $\varepsilon_0^* = 1$, $\varepsilon_1^* = 1 - (r-\lambda)/(2rk)$, $\rho_0^* = 1$, $\rho_1^* = 2(v-1)$, $\boldsymbol{L}_0^* = (\boldsymbol{I}_2 - 2^{-1}\mathbf{1}_2\mathbf{1}'_2) \otimes v^{-1}\mathbf{1}_v\mathbf{1}'_v$, $\boldsymbol{L}_1^* = \boldsymbol{I}_2 \otimes (\boldsymbol{I}_v - v^{-1}\mathbf{1}_v\mathbf{1}'_v)$.

Proof. In the resulting design with the incidence matrix (7.3.3), let the first v treatments form one group and the other v treatments form the other group. Then there are $m^* = 2$ groups of $n^* = v = 2k$ treatments each. Because for a BIB design $\boldsymbol{N}\boldsymbol{N}' = (r-\lambda)\boldsymbol{I}_v + \lambda\mathbf{1}_v\mathbf{1}'_v$ (see Section 6.0.1), it follows from (7.3.3) that

$$\boldsymbol{N}^*(\boldsymbol{N}^*)' = \begin{bmatrix} 2(r-\lambda)\boldsymbol{I}_v + 2\lambda\mathbf{1}_v\mathbf{1}'_v & r\mathbf{1}_v\mathbf{1}'_v \\ r\mathbf{1}_v\mathbf{1}'_v & 2(r-\lambda)\boldsymbol{I}_v + 2\lambda\mathbf{1}_v\mathbf{1}'_v \end{bmatrix},$$

which shows that the resulting design is a semi-regular GD design with the indicated parameters. The matrices $\boldsymbol{L}_0^*$ and $\boldsymbol{L}_1^*$ are readily obtainable from the general description of this class of designs, given in Section 6.0.2(A), by taking $\boldsymbol{L}_0^* = \boldsymbol{A}_1^{\#}$ and $\boldsymbol{L}_1^* = \boldsymbol{A}_2^{\#}$ there, with m replaced by $m^* = 2$ and n by $n^* = v$. It can easily be checked that (see Section 7.1)

$$\boldsymbol{M}_0^* = \frac{1}{r^*k^*}\boldsymbol{N}^*(\boldsymbol{N}^*)' - \frac{1}{v^*}\mathbf{1}_{v^*}\mathbf{1}'_{v^*} = \mu_1^*\boldsymbol{L}_1^*,$$

with $\mu_1^* = (r-\lambda)/(2rk)$, $\boldsymbol{L}_1^* = \boldsymbol{I}_2 \otimes (\boldsymbol{I}_v - v^{-1}\mathbf{1}_v\mathbf{1}'_v)$, and that $\boldsymbol{M}_0^*\boldsymbol{L}_0^* = \mathbf{O}$ for $\boldsymbol{L}_0^* = (\boldsymbol{I}_2 - 2^{-1}\mathbf{1}_2\mathbf{1}'_2) \otimes v^{-1}\mathbf{1}_v\mathbf{1}'_v$, which completes the proof. □

For the subsequent discussion, it is also interesting to note that the design obtained in Theorem 7.3.5 gives

$$\begin{aligned} \boldsymbol{F}^* &= \frac{1}{r^*}\boldsymbol{C}_1^* = \boldsymbol{L}_0^* + \varepsilon_1^*\boldsymbol{L}_1^* \\ &= \left(\boldsymbol{I}_2 - \frac{1}{2}\mathbf{1}_2\mathbf{1}_2'\right)\otimes\frac{1}{v}\mathbf{1}_v\mathbf{1}_v' + \left(1 - \frac{r-\lambda}{2rk}\right)\boldsymbol{I}_2\otimes\left(\boldsymbol{I}_v - \frac{1}{v}\mathbf{1}_v\mathbf{1}_v'\right). \end{aligned}$$

Example 7.3.1. Employing an unreduced BIB design with parameters $v = 4$, $b = 6$, $r = 3, k = 2, \lambda = 1$, Theorem 7.3.5 produces a 3-resolvable $(\rho_0^*; \rho_1^*; 0)$-EB design with parameters $v^* = 8$, $b^* = 12$, $r^* = 6$, $k^* = 4$, $\varepsilon_0^* = 1$, $\varepsilon_1^* = 5/6$, $\rho_0^* = 1$, $\rho_1^* = 6$, $\boldsymbol{L}_0^* = (\boldsymbol{I}_2 - 2^{-1}\mathbf{1}_2\mathbf{1}_2')\otimes 4^{-1}\mathbf{1}_4\mathbf{1}_4'$, $\boldsymbol{L}_1^* = \boldsymbol{I}_2\otimes(\boldsymbol{I}_4 - 4^{-1}\mathbf{1}_4\mathbf{1}_4')$, and with the incidence matrix

$$\boldsymbol{N}^* = \left[\begin{array}{cccccc|cccccc} 1&1&1&0&0&0&1&1&1&0&0&0\\ 1&0&0&1&1&0&1&0&0&1&1&0\\ 0&1&0&1&0&1&0&1&0&1&0&1\\ 0&0&1&0&1&1&0&0&1&0&1&1\\ 1&1&1&0&0&0&0&0&0&1&1&1\\ 1&0&0&1&1&0&0&1&1&0&0&1\\ 0&1&0&1&0&1&1&0&1&0&1&0\\ 0&0&1&0&1&1&1&1&0&1&0&0 \end{array}\right].$$

Here any contrast of the type $\boldsymbol{p}_0'(\boldsymbol{L}_0^*)'\boldsymbol{\tau}$, for some $\boldsymbol{p}_0$, is estimated in the intra-block analysis with full efficiency and the variance $6^{-1}\boldsymbol{p}_0'\boldsymbol{L}_0^*\boldsymbol{p}_0\sigma_1^2$, whereas any contrast of the type $\boldsymbol{p}_1'(\boldsymbol{L}_1^*)'\boldsymbol{\tau}$, for some $\boldsymbol{p}_1$, is estimated in that analysis with the efficiency 5/6 and the variance $5^{-1}\boldsymbol{p}_1'\boldsymbol{L}_1^*\boldsymbol{p}_1\sigma_1^2$. In terms of basic contrasts, this means that a basic contrast represented by an eigenvector of the matrix $\boldsymbol{L}_0^*$ corresponding to its nonzero eigenvalue is estimated in the intra-block analysis with full efficiency and the variance σ_1^2, whereas any 6 basic contrasts represented by eigenvectors of the matrix $\boldsymbol{L}_1^*$ corresponding to its nonzero eigenvalue are estimated in that analysis with the efficiency 5/6 and the variance $6\sigma_1^2/5$ each. As described in Theorem 7.3.5, this is a semi-regular GD design with 8 treatments divided into two groups of 4 members each. Therefore, one could also say that the basic contrast estimated with full efficiency in the intra-block analysis compares the two groups, whereas the remaining 6 basic contrasts can be chosen as the within group comparisons. The average efficiency factor of the design is 35/41 = 0.8537. The $\boldsymbol{\Omega}$ matrix (7.1.6) obtained for this design is of the form

$$\boldsymbol{\Omega}^* = \frac{1}{6}\left[\frac{1-5/6}{5/6}\boldsymbol{I}_2\otimes\left(\boldsymbol{I}_4 - \frac{1}{4}\mathbf{1}_4\mathbf{1}_4'\right) + \boldsymbol{I}_2\otimes\boldsymbol{I}_4\right] = \frac{1}{120}\boldsymbol{I}_2\otimes(24\boldsymbol{I}_4 - \mathbf{1}_4\mathbf{1}_4').$$

It shows, for example, that elementary contrasts are estimated with two different variances, $2\sigma_1^2/5$ and $23\sigma_1^2/60$, in the intra-block analysis. In particular, the first variance corresponds to the intra-block BLUEs of elementary contrasts between treatments from the same group. In fact, any such contrast is of the type $\boldsymbol{c}_1'\boldsymbol{\tau}$, where $\boldsymbol{c}_1 = \boldsymbol{L}_1^*\boldsymbol{c}_1$, so that the variance of its intra-block BLUE is $(\varepsilon_1 r)^{-1}\boldsymbol{c}_1'\boldsymbol{c}_1\sigma_1^2$.

However, note that the two variances concerning elementary contrasts are almost equal.

Theorem 6.6.4 with $a = 0$ and $c = 1$ yields the following result when the basic design is a BIB design with $b = 3(r - \lambda)$.

Theorem 7.3.6. *Let $\boldsymbol{N}$ be the $v \times b$ incidence matrix of a BIB design with parameters $v, b = 3(r - \lambda)$, r, k, λ. Then the incidence matrix*

$$\boldsymbol{N}^* = \begin{bmatrix} \boldsymbol{N} & \boldsymbol{O} & \boldsymbol{1}_v\boldsymbol{1}_b' - \boldsymbol{N} \\ \boldsymbol{1}_v\boldsymbol{1}_b' - \boldsymbol{N} & \boldsymbol{N} & \boldsymbol{O} \\ \boldsymbol{O} & \boldsymbol{1}_v\boldsymbol{1}_b' - \boldsymbol{N} & \boldsymbol{N} \end{bmatrix}$$

yields a semi-regular GD design with parameters $v^* = 3v$, $b^* = 3b$, $r^* = b$, $k^* = v$, $\lambda_1^* = 0$, $\lambda_2^* = r - \lambda$, $m^* = v$, $n^* = 3$, i.e., a $(\rho_0^*; \rho_1^*; 0)$-EB design with $\varepsilon_0^* = 1$, $\varepsilon_1^* = 1 - 1/v$, $\rho_0^* = v - 1$, $\rho_1^* = 2v$, $\boldsymbol{L}_0^* = 3^{-1}\boldsymbol{1}_3\boldsymbol{1}_3' \otimes (\boldsymbol{I}_v - v^{-1}\boldsymbol{1}_v\boldsymbol{1}_v')$, $\boldsymbol{L}_1^* = (\boldsymbol{I}_3 - 3^{-1}\boldsymbol{1}_3\boldsymbol{1}_3') \otimes \boldsymbol{I}_v$.

Proof. For the $3v$ treaments in the resulting design with the above incidence matrix, let every ith treatment from the original BIB design form a group of three treatments each, as $\{i, v+i, 2v+i\}$, $i = 1, 2, ..., v$. [Note that the numbering (arrangement) of the present treatments is different from that of treatments in the usual GD association scheme mentioned in Section 6.0.2(A).] Then there are $m^* = v$ groups of $n^* = 3$ treatments each. Now when $b = 3(r - \lambda)$ it can easily be shown that the resulting design is a semi-regular GD design with the indicated parameters. Hence a procedure similar to that used in the proof of Theorem 7.3.5 completes the proof. □

Note that the design obtained in Theorem 7.3.6 is a suitable design for $3 \times v$ factorial experiment, when one wants to estimate contrasts among the main effects of the second factor with full efficiency.

Example 7.3.2. Taking a doubled unreduced BIB design with parameters $v = 3$, $b = 6$, $r = 4$, $k = 2$, $\lambda = 2$, Theorem 7.3.6 produces a $(2; 6; 0)$-EB design with parameters $v^* = 9$, $b^* = 18$, $r^* = 6$, $k^* = 3$, $\varepsilon_0^* = 1$, $\varepsilon_1^* = 2/3$, $\rho_0^* = 2$, $\rho_1^* = 6$, $\boldsymbol{L}_0^* = 3^{-1}\boldsymbol{1}_3\boldsymbol{1}_3' \otimes (\boldsymbol{I}_3 - 3^{-1}\boldsymbol{1}_3\boldsymbol{1}_3')$, $\boldsymbol{L}_1^* = (\boldsymbol{I}_3 - 3^{-1}\boldsymbol{1}_3\boldsymbol{1}_3') \otimes \boldsymbol{I}_3$, its incidence matrix being

$$\begin{bmatrix} 1&1&1&1&0&0&0&0&0&0&0&0&0&0&0&0&1&1 \\ 1&1&0&0&1&1&0&0&0&0&0&0&0&0&1&1&0&0 \\ 0&0&1&1&1&1&0&0&0&0&0&0&1&1&0&0&0&0 \\ 0&0&0&0&1&1&1&1&1&1&0&0&0&0&0&0&0&0 \\ 0&0&1&1&0&0&1&1&0&0&1&1&0&0&0&0&0&0 \\ 1&1&0&0&0&0&0&0&1&1&1&1&0&0&0&0&0&0 \\ 0&0&0&0&0&0&0&0&0&0&1&1&1&1&1&1&0&0 \\ 0&0&0&0&0&0&0&0&1&1&0&0&1&1&0&0&1&1 \\ 0&0&0&0&0&0&1&1&0&0&0&0&0&0&1&1&1&1 \end{bmatrix}.$$

Here the three groups of three treatments of the GD association scheme are given by $\{1,4,7\}$, $\{2,5,8\}$, $\{3,6,9\}$, as it follows from Theorem 7.3.6. [Note that to comply with the notation in Section 6.0.2(A), the treatments are to be renumbered to give the usual three groups $\{1,2,3\}$, $\{4,5,6\}$, $\{7,8,9\}$, i.e., to place the original treatments in the order 1, 4, 7, 2, 5, 8, 3, 6, 9 through some permutations of rows of the original incidence matrix. In this sense, further, note that the present forms of $\boldsymbol{L}_\beta^*$ are different from those obtained from $\boldsymbol{A}_\beta^\#$ given in Section 6.0.2(A).] The average efficiency factor of the design is $8/11 = 0.7273$. Furthermore, the matrix $\boldsymbol{\Omega}$ obtained for the resulting design is

$$\begin{aligned}\boldsymbol{\Omega}^* &= \frac{1}{6}\left[\frac{1-2/3}{2/3}\left(\boldsymbol{I}_3 - \frac{1}{3}\boldsymbol{1}_3\boldsymbol{1}_3'\right)\otimes\boldsymbol{I}_3 + \boldsymbol{I}_3\otimes\boldsymbol{I}_3\right] \\ &= \frac{1}{36}(9\boldsymbol{I}_3 - \boldsymbol{1}_3\boldsymbol{1}_3')\otimes\boldsymbol{I}_3,\end{aligned}$$

in which, for example, elementary contrasts are estimated in the intra-block analysis with two different variances, $4\sigma_1^2/9$ and $\sigma_1^2/2$.

Another method of obtaining semi-regular GD designs is the procedure of merging treatments in GD designs, as proposed by Puri and Nigam (1983) and explained in Section 6.5. Since of prime interest are binary block designs, as a basic design consider a GD design with parameters $v = mn, b, r, k$, $\lambda_1 = 0$ and positive λ_2. Suppose that treatments within each of the m groups are divided into s disjoint subgroups, i.e., n/s is an integer. Then merging, in each of the m groups, the $n/s, n/s, ..., n/s$ consecutive not concurring treatments in each of the s subgroups yields by Corollary 6.5.2 the following result.

Corollary 7.3.2. *The existence of a semi-regular GD design with parameters* $v = mn$ *(m groups of n treatments each),* b, r, k, $\lambda_1 = 0$, λ_2 *implies the existence of a semi-regular GD design with parameters* $v^* = ms$, $b^* = b$, $r^* = nr/s$, $k^* = k$, $\lambda_1^* = 0$, $\lambda_2^* = (n/s)^2\lambda_2$, *i.e., a* $(\rho_0^*;\rho_1^*;0)$*-EB design with* $\varepsilon_0^* = 1$, $\varepsilon_1^* = (k-1)/k$, $\rho_0^* = m-1$, $\rho_1^* = m(s-1)$, $\boldsymbol{L}_0^* = (\boldsymbol{I}_m - m^{-1}\boldsymbol{1}_m\boldsymbol{1}_m')\otimes s^{-1}\boldsymbol{1}_s\boldsymbol{1}_s'$, $\boldsymbol{L}_1^* = \boldsymbol{I}_m\otimes(\boldsymbol{I}_s - s^{-1}\boldsymbol{1}_s\boldsymbol{1}_s')$, *provided that n/s is an integer.*

Example 7.3.3. If a semi-regular GD design given in Corollary 7.3.1 is taken, a merging approach of Corollary 7.3.2 produces a semi-regular GD design with parameters $v^* = (q+1)s$, $b^* = q^2$, $r^* = q^2/s$, $k^* = q+1$, $\lambda_1^* = 0$, $\lambda_2^* = (q/s)^2$, i.e., a $(\rho_0^*;\rho_1^*;0)$-EB design with parameters $\varepsilon_0^* = 1$, $\varepsilon_1^* = q/(q+1)$, $\rho_0^* = q$, $\rho_1^* = (q+1)(s-1)$, $\boldsymbol{L}_0^* = [\boldsymbol{I}_{q+1} - (q+1)^{-1}\boldsymbol{1}_{q+1}\boldsymbol{1}_{q+1}']\otimes s^{-1}\boldsymbol{1}_s\boldsymbol{1}_s'$, $\boldsymbol{L}_1^* = \boldsymbol{I}_{q+1}\otimes(\boldsymbol{I}_s - s^{-1}\boldsymbol{1}_s\boldsymbol{1}_s')$, if s is an integer such that q/s is an integer, provided q is a prime or a prime power. Now let $q = 4$ and $s = 2$. Then one obtains a semi-regular GD design with $v^* = 10$ treatments consisting of 5 groups of 2 members each, i.e., a (4; 5; 0)-EB design with parameters $\varepsilon_0^* = 1$, $\varepsilon_1^* = 4/5$, $\rho_0^* = 4$, $\rho_1^* = 5$, $\boldsymbol{L}_0^* = (\boldsymbol{I}_5 - 5^{-1}\boldsymbol{1}_5\boldsymbol{1}_5')\otimes 2^{-1}\boldsymbol{1}_2\boldsymbol{1}_2'$, $\boldsymbol{L}_1^* = \boldsymbol{I}_5\otimes(\boldsymbol{I}_2 - 2^{-1}\boldsymbol{1}_2\boldsymbol{1}_2')$, and with the incidence matrix $\boldsymbol{N}^*$ as given later. To see this, one can take as the starting semi-regular GD design that with parameters $v = 20$, $b = 16$, $r = 4$, $k = 5$, $\lambda_1 = 0$, $\lambda_2 = 1$, $m = 5, n = 4$, based on five groups $\{1,2,3,4\}$, $\{5,6,7,8\}$,

$\{9, 10, 11, 12\}$, $\{13, 14, 15, 16\}$, $\{17, 18, 19, 20\}$ of four treatments, i.e., the design SR58 from Clatworthy (1973), and with the incidence matrix

$$
\boldsymbol{N} = \left[\begin{array}{cccccccccccccccc}
0 & 0 & 0 & 0 & 1 & 0 & 0 & 0 & 0 & 1 & 0 & 0 & 0 & 0 & 1 & 1 \\
0 & 1 & 1 & 0 & 0 & 0 & 0 & 0 & 0 & 0 & 1 & 0 & 0 & 1 & 0 & 0 \\
0 & 0 & 0 & 0 & 0 & 1 & 0 & 0 & 1 & 0 & 0 & 1 & 1 & 0 & 0 & 0 \\
1 & 0 & 0 & 1 & 0 & 0 & 1 & 1 & 0 & 0 & 0 & 0 & 0 & 0 & 0 & 0 \\
\hline
0 & 0 & 0 & 1 & 0 & 0 & 0 & 0 & 1 & 0 & 0 & 0 & 0 & 1 & 0 & 1 \\
1 & 1 & 0 & 0 & 0 & 0 & 0 & 0 & 0 & 0 & 0 & 0 & 1 & 0 & 1 & 0 \\
0 & 0 & 0 & 0 & 0 & 0 & 0 & 1 & 0 & 1 & 1 & 1 & 0 & 0 & 0 & 0 \\
0 & 0 & 1 & 0 & 1 & 1 & 1 & 0 & 0 & 0 & 0 & 0 & 0 & 0 & 0 & 0 \\
\hline
0 & 0 & 1 & 0 & 0 & 0 & 0 & 1 & 0 & 0 & 0 & 0 & 1 & 0 & 0 & 1 \\
1 & 0 & 0 & 0 & 1 & 0 & 0 & 0 & 0 & 0 & 0 & 1 & 0 & 1 & 0 & 0 \\
0 & 0 & 0 & 0 & 0 & 0 & 1 & 0 & 1 & 0 & 1 & 0 & 0 & 0 & 1 & 0 \\
0 & 1 & 0 & 1 & 0 & 1 & 0 & 0 & 0 & 1 & 0 & 0 & 0 & 0 & 0 & 0 \\
\hline
0 & 1 & 0 & 0 & 0 & 0 & 1 & 0 & 0 & 0 & 0 & 1 & 0 & 0 & 0 & 1 \\
0 & 0 & 0 & 1 & 1 & 0 & 0 & 0 & 0 & 0 & 1 & 0 & 1 & 0 & 0 & 0 \\
0 & 0 & 0 & 0 & 0 & 1 & 0 & 1 & 0 & 0 & 0 & 0 & 0 & 1 & 1 & 0 \\
1 & 0 & 1 & 0 & 0 & 0 & 0 & 0 & 1 & 1 & 0 & 0 & 0 & 0 & 0 & 0 \\
\hline
1 & 0 & 0 & 0 & 0 & 1 & 0 & 0 & 0 & 0 & 1 & 0 & 0 & 0 & 0 & 1 \\
0 & 0 & 1 & 1 & 0 & 0 & 0 & 0 & 0 & 0 & 0 & 1 & 0 & 0 & 1 & 0 \\
0 & 0 & 0 & 0 & 0 & 0 & 1 & 0 & 0 & 1 & 0 & 0 & 1 & 1 & 0 & 0 \\
0 & 1 & 0 & 0 & 1 & 0 & 0 & 1 & 1 & 0 & 0 & 0 & 0 & 0 & 0 & 0
\end{array}\right].
$$

In each of the 5 groups, merging the first two and the last two consecutive treatments, yields the following incidence matrix

$$
\boldsymbol{N}^* = \begin{bmatrix}
0 & 1 & 1 & 0 & 1 & 0 & 0 & 0 & 0 & 1 & 1 & 0 & 0 & 1 & 1 & 1 \\
1 & 0 & 0 & 1 & 0 & 1 & 1 & 1 & 1 & 0 & 0 & 1 & 1 & 0 & 0 & 0 \\
1 & 1 & 0 & 1 & 0 & 0 & 0 & 0 & 1 & 0 & 0 & 0 & 1 & 1 & 1 & 1 \\
0 & 0 & 1 & 0 & 1 & 1 & 1 & 1 & 0 & 1 & 1 & 1 & 0 & 0 & 0 & 0 \\
1 & 0 & 1 & 0 & 1 & 0 & 0 & 1 & 0 & 0 & 0 & 1 & 1 & 1 & 0 & 1 \\
0 & 1 & 0 & 1 & 0 & 1 & 1 & 0 & 1 & 1 & 1 & 0 & 0 & 0 & 1 & 0 \\
0 & 1 & 0 & 1 & 1 & 0 & 1 & 0 & 0 & 0 & 1 & 1 & 1 & 0 & 0 & 1 \\
1 & 0 & 1 & 0 & 0 & 1 & 0 & 1 & 1 & 1 & 0 & 0 & 0 & 1 & 1 & 0 \\
1 & 0 & 1 & 1 & 0 & 1 & 0 & 0 & 0 & 0 & 1 & 1 & 0 & 0 & 1 & 1 \\
0 & 1 & 0 & 0 & 1 & 0 & 1 & 1 & 1 & 1 & 0 & 0 & 1 & 1 & 0 & 0
\end{bmatrix},
$$

which shows the existence of the required $(4; 5; 0)$-EB design. In fact, this is a semi-regular GD design with parameters $v^* = 10$, $b^* = 16$, $r^* = 8$, $k^* = 5$, $\lambda_1^* = 0$, $\lambda_2^* = 4$, $m^* = m = 5, n^* = s = 2$, i.e., the design SR54 from Clatworthy (1973). Here, the average efficiency factor of the design is $36/41 = 0.8780$. The matrix $\boldsymbol{\Omega}$ is obtained as

$$
\boldsymbol{\Omega}^* = \frac{1}{64}\boldsymbol{I}_5 \otimes (10\boldsymbol{I}_2 - \mathbf{1}_2\mathbf{1}_2').
$$

Hence, for example, elementary contrasts obtain two different variances $10\sigma_1^2/32$ and $9\sigma_1^2/32$, when estimated in the intra-block analysis. This design may well be used for 5×2 factorial experiment, with factors A and B applied at 5 and 2 levels respectively, when the researcher is interested in obtaining the intra-block BLUEs of contrasts involving the A factor main effects with full efficiency.

Finally, another method of obtaining a semi-regular GD design from another semi-regular GD design can be explained. This method has been noted by Kageyama (1985b), also to clarify the structural classifications of the designs. This method is based on the property given in Theorem 7.3.3. First select s of the m groups, and then delete the sn treatments in the s selected groups from the b original blocks. Then it follows from Theorem 7.3.3 that the remaining structure yields a semi-regular GD design with parameters $v^* = v - sn$, $b^* = b$, $r^* = r$, $k^* = k - sk/m$, $\lambda_1^* = \lambda_1$, $\lambda_2^* = \lambda_2$ for an integer s such that $1 \le s \le m-2$. Note that this procedure preserves a property of the α-resolvability. Any of the 111 available semi-regular GD designs, presented in Clatworthy (1973), can be obtained from one of 20 semi-regular GD designs, with the appropriate s, which are regarded as representatives of each class, as shown by Kageyama (1985b). Thus, in a semi-regular GD design even if some groups of treatments are totally missing in an experiment, the property of being a semi-regular GD design can be preserved for the remaining structure.

(3) Triangular designs with $\psi_1 = r + (n-4)\lambda_1 - (n-3)\lambda_2 = 0$, i.e., the rank of its incidence matrix is smaller than v (see Lemma 2.3.3 and Section 6.0.2). Any such design can be characterized as follows. Let $\boldsymbol{A}_1$ be an association matrix on the first associates of the association scheme considered (see Section 6.0.2).

Theorem 7.3.7 (Raghavarao, 1960b; Kageyama and Tsuji, 1977). *In a triangular design with parameters* $v = n(n-1)/2$, b, r, k, λ_1, λ_2, *the equality* $r + (n-4)\lambda_1 - (n-3)\lambda_2 = 0$ *holds if and only if* $2k/n$ *is an integer and every block of the design contains exactly* $2k/n$ *treatments from each of the* n *rows of the association scheme. If the condition is satisfied, then the design is* $(\rho_0; \rho_1; 0)$*-EB with* $\varepsilon_0 = 1$, $\varepsilon_1 = [r(k-1) + 2\lambda_1 - \lambda_2]/(rk)$, $\rho_0 = n-1$, $\rho_1 = n(n-3)/2$, $\boldsymbol{L}_0 = [n(n-2)]^{-1}(2n\boldsymbol{I}_v + n\boldsymbol{A}_1 - 4\boldsymbol{1}_v\boldsymbol{1}_v')$ *and* $\boldsymbol{L}_1 = [(n-1)(n-2)]^{-1}\{(n-1)(n-4)\boldsymbol{I}_v - (n-1)\boldsymbol{A}_1 + 2\boldsymbol{1}_v\boldsymbol{1}_v'\}$.

A method of constructing a triangular design with $\psi_1 = 0$ is obtainable from the following theorem.

Theorem 7.3.8 (Shrikhande, 1960). *The dual of a BIB design with parameters* $v = (n-1)(n-2)/2$, $b = n(n-1)/2$, $r = n$, $k = n-2$, $\lambda = 2$, *for any* $n \neq 8$, *yields a triangular design with parameters* $v_* = n(n-1)/2$, $b_* = (n-1)(n-2)/2$, $r_* = n-2$, $k_* = n$, $\lambda_{*1} = 1$, $\lambda_{*2} = 2$, *i.e., a* $(\rho_{*0}; \rho_{*1}; 0)$*-EB design with* $\varepsilon_{*0} = 1$, $\varepsilon_{*1} = (n-1)/n$, $\rho_{*0} = n-1$, $\rho_{*1} = n(n-3)/2$, $\boldsymbol{L}_{*0} = [n(n-2)]^{-1}(2n\boldsymbol{I}_{v^*} + n\boldsymbol{A}_1 - 4\boldsymbol{1}_{v^*}\boldsymbol{1}_{v^*}')$ *and* $\boldsymbol{L}_{*1} = [(n-1)(n-2)]^{-1}\{(n-1)(n-4)\boldsymbol{I}_{v^*} - (n-1)\boldsymbol{A}_1 + 2\boldsymbol{1}_{v^*}\boldsymbol{1}_{v^*}'\}$.

Proof. The incidence matrix $\boldsymbol{N}$ of a BIB design with parameters as assumed satisfies the equality $\boldsymbol{NN}' = (n-2)\boldsymbol{I}_v + 2\mathbf{1}_v\mathbf{1}_v'$ (see Section 6.0.1). Now taking the $b \times b$ matrix

$$\boldsymbol{A}_1 = (n-4)\boldsymbol{I}_b + 2\mathbf{1}_b\mathbf{1}_b' - \boldsymbol{N}'\boldsymbol{N},$$

where $b = n(n-1)/2$, it can be seen that it is symmetric, with zero diagonal elements and with the remaining elements being some integers. Since $\mathbf{1}_b'\boldsymbol{A}_1\mathbf{1}_b = \text{tr}(\boldsymbol{A}_1^2)$, these integers are 0 and 1 only. Also it has the property $\boldsymbol{A}_1\mathbf{1}_b = 2(n-2)\mathbf{1}_b$. Thus, $\boldsymbol{A}_1$ can be regarded as an association matrix (see Section 6.0.2). Moreover, as it holds that $\boldsymbol{N}'\boldsymbol{N} = (n-2)\boldsymbol{I}_b + \boldsymbol{A}_1 + 2(\mathbf{1}_b\mathbf{1}_b' - \boldsymbol{I}_b - \boldsymbol{A}_1)$, the matrix $\boldsymbol{A}_1$ may be seen as the first association matrix of an association scheme with two associate classes, appropriate for a design having $\boldsymbol{N}'$ as the incidence matrix. In fact, as in addition the three equalities

$$\boldsymbol{A}_1^2 = 2(n-2)\boldsymbol{I}_b + (n-2)\boldsymbol{A}_1 + 4(\mathbf{1}_b\mathbf{1}_b' - \boldsymbol{I}_b - \boldsymbol{A}_1),$$

$$\boldsymbol{A}_1(\mathbf{1}_b\mathbf{1}_b' - \boldsymbol{I}_b - \boldsymbol{A}_1) = (n-3)\boldsymbol{A}_1 + 2(n-4)(\mathbf{1}_b\mathbf{1}_b' - \boldsymbol{I}_b - \boldsymbol{A}_1)$$

and

$$(\mathbf{1}_b\mathbf{1}_b' - \boldsymbol{I}_b - \boldsymbol{A}_1)^2 = \frac{(n-2)(n-3)}{2}\boldsymbol{I}_b + \frac{(n-3)(n-4)}{2}\boldsymbol{A}_1 + \frac{(n-4)(n-5)}{2}(\mathbf{1}_b\mathbf{1}_b' - \boldsymbol{I}_b - \boldsymbol{A}_1)$$

hold, the considered association scheme has evidently the parameters

$$n_1 = 2(n-2), n_2 = (n-2)(n-3)/2, p_{11}^1 = n-2, p_{11}^2 = 4,$$

$$p_{12}^1 = n-3, p_{12}^2 = 2(n-4), p_{22}^1 = (n-3)(n-4)/2, p_{22}^2 = (n-4)(n-5)/2.$$

These are those of a triangular association scheme [see Section 6.0.2 (B)]. This implies that the dual of the BIB design is a triangular design with the parameters as stated, provided however that $n \neq 8$ (as for $n = 8$ the determination by the parameters of this association scheme is not unique; see Raghavarao, 1971, Sections 8.7.2 and 8.7.3). □

Example 7.3.4. A BIB design with parameters $v = 6, b = 10, r = 5, k = 3, \lambda = 2$ with the incidence matrix $\boldsymbol{N}$, given in Example 6.7.3 satisfies Theorem 7.3.8 when $n = 5$. Then, by dualizing it, one obtains a triangular design with parameters $v_* = 10$, $b_* = 6$, $r_* = 3$, $k_* = 5$, $\lambda_{*1} = 1$, $\lambda_{*2} = 2$ having blocks, (1, 2, 5, 7, 10), (2, 3, 6, 8, 1), (3, 4, 7, 9, 2), (4, 5, 8, 10, 3), (5, 6, 9, 1, 4), (6, 7, 8, 9, 10), whose incidence matrix is given by $\boldsymbol{N}'$. Note that in this case the triangular association scheme is, for example, of the form

$$\begin{array}{ccccc} & 1 & 4 & 8 & 7 \\ 1 & & 10 & 9 & 3 \\ 4 & 10 & & 2 & 6 \\ 8 & 9 & 2 & & 5 \\ 7 & 3 & 6 & 5 & \end{array} .$$

[Further, note that the order of these treatments is different from that given in Section 6.0.2(B).] From Theorem 7.3.8, this is a $(\rho_{*0}; \rho_{*1}; 0)$-EB design with $\varepsilon_{*0} = 1$, $\varepsilon_{*1} = 4/5$, $\rho_{*0} = 4$, $\rho_{*1} = 5$, $\boldsymbol{L}_{*0} = \boldsymbol{I}_{10} + (2/5)\mathbf{1}_{10}\mathbf{1}'_{10} - 3^{-1}\boldsymbol{N}'\boldsymbol{N}$ and $\boldsymbol{L}_{*1} = -2^{-1}\mathbf{1}_{10}\mathbf{1}'_{10} + 3^{-1}\boldsymbol{N}'\boldsymbol{N}$, where

$$\boldsymbol{N}'\boldsymbol{N} = \begin{bmatrix} 3 & 2 & 1 & 1 & 2 & 2 & 1 & 1 & 1 & 1 \\ 2 & 3 & 2 & 1 & 1 & 1 & 2 & 1 & 1 & 1 \\ 1 & 2 & 3 & 2 & 1 & 1 & 1 & 2 & 1 & 1 \\ 1 & 1 & 2 & 3 & 2 & 1 & 1 & 1 & 2 & 1 \\ 2 & 1 & 1 & 2 & 3 & 1 & 1 & 1 & 1 & 2 \\ 2 & 1 & 1 & 1 & 1 & 3 & 1 & 2 & 2 & 1 \\ 1 & 2 & 1 & 1 & 1 & 1 & 3 & 1 & 2 & 2 \\ 1 & 1 & 2 & 1 & 1 & 2 & 1 & 3 & 1 & 2 \\ 1 & 1 & 1 & 2 & 1 & 2 & 2 & 1 & 3 & 1 \\ 1 & 1 & 1 & 1 & 2 & 1 & 2 & 2 & 1 & 3 \end{bmatrix}.$$

(4) Triangular designs with $\psi_2 = r - 2\lambda_1 + \lambda_2 = 0$, i.e., the rank of its incidence matrix is smaller than v. There is a simple method of constructing such designs. Just by writing down the rows of the triangular association scheme [see again Section 6.0.2(B)] as blocks of a design, the following theorem can be obtained. Note that the design given in the theorem can also be obtained by dualizing an unreduced BIB design with parameters $v^* = n, b^* = n(n-1)/2, r^* = n-1$, $k^* = 2, \lambda^* = 1$.

Theorem 7.3.9 (Raghavarao, 1971, Theorem 8.8.1). *There always exists a triangular design with parameters* $v = n(n-1)/2$, $b = n$, $r = 2$, $k = n-1$, $\lambda_1 = 1$, $\lambda_2 = 0$, *i.e., a* $(\rho_0; \rho_1; 0)$*-EB design with* $\varepsilon_0 = 1$, $\varepsilon_1 = n/[2(n-1)]$, $\rho_0 = n(n-3)/2$, $\rho_1 = n-1$, $\boldsymbol{L}_0 = \boldsymbol{I}_v + 2[(n-1)(n-2)]^{-1}\mathbf{1}_v\mathbf{1}'_v - (n-2)^{-1}\boldsymbol{N}'\boldsymbol{N}$ *and* $\boldsymbol{L}_1 = (n-2)^{-1}[\boldsymbol{N}'\boldsymbol{N} - (4/n)\mathbf{1}_v\mathbf{1}'_v]$, *where* $\boldsymbol{N}$ *is the incidence matrix of the original BIB design to be dualized.*

Proof. It follows from Corollary 6.1.2, or by the same idea as in the proof of Theorem 7.3.8 (see Raghavarao, 1971, Theorems 8.7.1, 8.7.2 and 8.8.1 together). □

Example 7.3.5. Consider an unreduced BIB design with parameters $v = 4, b = 6, r = 3, k = 2, \lambda = 1$ with the incidence matrix

$$\boldsymbol{N} = \begin{bmatrix} 1 & 1 & 1 & 0 & 0 & 0 \\ 1 & 0 & 0 & 1 & 1 & 0 \\ 0 & 1 & 0 & 1 & 0 & 1 \\ 0 & 0 & 1 & 0 & 1 & 1 \end{bmatrix}.$$

Then, by dualizing it, one obtains a triangular design with parameters $v_* = 6$, $b_* = 4$, $r_* = 2$, $k_* = 3$, $\lambda_{*1} = 1$, $\lambda_{*2} = 0$ having blocks, (1, 2, 3), (1, 4, 5), (2, 4,

6), (3, 5, 6), whose incidence matrix is given by $\boldsymbol{N}'$. Note that in this case the triangular association scheme is of the form

$$\begin{array}{cccc} & 1 & 2 & 3 \\ 1 & & 4 & 5 \\ 2 & 4 & & 6 \\ 3 & 5 & 6 & \end{array}.$$

From Theorem 7.3.9, this is a $(\rho_{*0}; \rho_{*1}; 0)$-EB design with $\varepsilon_{*0} = 1$, $\varepsilon_{*1} = 2/3$, $\rho_{*0} = 2$, $\rho_{*1} = 3$, $\boldsymbol{L}_{*0} = \boldsymbol{I}_6 + 3^{-1}\mathbf{1}_6\mathbf{1}_6' - 2^{-1}\boldsymbol{N}'\boldsymbol{N}$ and $\boldsymbol{L}_{*1} = 2^{-1}(\boldsymbol{N}'\boldsymbol{N} - \mathbf{1}_6\mathbf{1}_6')$, where

$$\boldsymbol{N}'\boldsymbol{N} = \begin{bmatrix} 2 & 1 & 1 & 1 & 1 & 0 \\ 1 & 2 & 1 & 1 & 0 & 1 \\ 1 & 1 & 2 & 0 & 1 & 1 \\ 1 & 1 & 0 & 2 & 1 & 1 \\ 1 & 0 & 1 & 1 & 2 & 1 \\ 0 & 1 & 1 & 1 & 1 & 2 \end{bmatrix}.$$

Here, when grouping the treatments into the subsets, $\{1, 6\}$, $\{2, 5\}$ and $\{3, 4\}$, the contrasts between these subsets will be estimated with full efficiency, and those within the subsets with the efficiency 2/3, in the intra-block analysis.

(5) L_2 designs with $\psi_1 = r + (s-2)\lambda_1 - (s-1)\lambda_2 = 0$, i.e., the rank of its incidence matrix is smaller than v (see Lemma 2.3.3 and Section 6.0.2). There is a characterization theorem for these designs as follows.

Theorem 7.3.10 (Raghavarao, 1960b; Kageyama and Tsuji, 1977). *In an L_2 design with parameters $v = s^2$, b, r, k, λ_1, λ_2, the equality $r + (s-2)\lambda_1 - (s-1)\lambda_2 = 0$ holds if and only if k/s is an integer and every block of the design contains exactly k/s treatments from each of the s rows (or columns) of the association scheme. If the condition is satisfied, then the design is $(\rho_0; \rho_1; 0)$-EB with $\varepsilon_0 = 1$, $\varepsilon_1 = [r(k-1) + 2\lambda_1 - \lambda_2]/(rk)$, $\rho_0 = 2(s-1)$, $\rho_1 = (s-1)^2$, $\boldsymbol{L}_0 = s^{-1}\mathbf{1}_s\mathbf{1}_s' \otimes (\boldsymbol{I}_s - s^{-1}\mathbf{1}_s\mathbf{1}_s') + (\boldsymbol{I}_s - s^{-1}\mathbf{1}_s\mathbf{1}_s') \otimes s^{-1}\mathbf{1}_s\mathbf{1}_s'$ and $\boldsymbol{L}_1 = (\boldsymbol{I}_s - s^{-1}\mathbf{1}_s\mathbf{1}_s') \otimes (\boldsymbol{I}_s - s^{-1}\mathbf{1}_s\mathbf{1}_s')$.*

By use of a complete set of mutually orthogonal Latin squares of order s (see Appendix D), the following theorem can be shown.

Theorem 7.3.11 (Raghavarao, 1971, Theorem 8.10.1). *When s is a prime or a prime power, there exists an L_2 design with parameters $v = s^2$, $b = s(s-1)$, $r = s-1$, $k = s$, $\lambda_1 = 0$, $\lambda_2 = 1$, i.e., a $(\rho_0; \rho_1; 0)$-EB design with $\varepsilon_0 = 1$, $\varepsilon_1 = (s-2)/(s-1)$, $\rho_0 = 2(s-1)$, $\rho_1 = (s-1)^2$, $\boldsymbol{L}_0 = s^{-1}\mathbf{1}_s\mathbf{1}_s' \otimes (\boldsymbol{I}_s - s^{-1}\mathbf{1}_s\mathbf{1}_s') + (\boldsymbol{I}_s - s^{-1}\mathbf{1}_s\mathbf{1}_s') \otimes s^{-1}\mathbf{1}_s\mathbf{1}_s'$ and $\boldsymbol{L}_1 = (\boldsymbol{I}_s - s^{-1}\mathbf{1}_s\mathbf{1}_s') \otimes (\boldsymbol{I}_s - s^{-1}\mathbf{1}_s\mathbf{1}_s')$.*

In a design resulting from Theorem 7.3.11, when s is relatively large, $(s-2)/(s-1) \approx 1$, i.e., one obtains a design with almost one efficiency factor, providing nearly full efficiency for all basic contrasts.

(6) L_2 designs with $\psi_2 = r - 2\lambda_1 + \lambda_2 = 0$, i.e., the rank of its incidence

matrix is smaller than v. Clatworthy (1967) shows the following result by taking blocks formed by combining all possible pairs of rows and all possible pairs of columns in an $s \times s$ array.

Theorem 7.3.12 (Clatworthy, 1967; Raghavarao, 1971, Theorem 8.10.5). *There always exists an* L_2 *design with parameters* $v = s^2$, $b = s(s-1)$, $r = 2(s-1)$, $k = 2s$, $\lambda_1 = s$, $\lambda_2 = 2$, *i.e., a* $(\rho_0; \rho_1; 0)$*-EB design with* $\varepsilon_0 = 1$, $\varepsilon_1 = (3s-2)/[4(s-1)]$, $\rho_0 = (s-1)^2$, $\rho_1 = 2(s-1)$, $\boldsymbol{L}_0 = (\boldsymbol{I}_s - s^{-1}\mathbf{1}_s\mathbf{1}_s') \otimes (\boldsymbol{I}_s - s^{-1}\mathbf{1}_s\mathbf{1}_s')$ *and* $\boldsymbol{L}_1 = s^{-1}\mathbf{1}_s\mathbf{1}_s' \otimes (\boldsymbol{I}_s - s^{-1}\mathbf{1}_s\mathbf{1}_s') + (\boldsymbol{I}_s - s^{-1}\mathbf{1}_s\mathbf{1}_s') \otimes s^{-1}\mathbf{1}_s\mathbf{1}_s'$.

In all the cases (1) to (6) of 2-associate PBIB designs considered above, it is to be remembered that not only the basic contrasts based on the relevant matrix $\boldsymbol{L}_\beta$ are estimated in the intra-block analysis with the efficiency ε_β, but also other contrasts of the form $\boldsymbol{p}_\beta'\boldsymbol{L}_\beta'\boldsymbol{\tau}$ are estimated in that analysis with the same efficiency ε_β and with the variance $(\varepsilon_\beta r)^{-1}\boldsymbol{p}_\beta'\boldsymbol{L}_\beta\boldsymbol{p}_\beta\sigma_1^2$, $\beta = 0, 1$. This means that the suitability of any of the discussed designs for use in an experiment depends very much on the structures of the matrices $\boldsymbol{L}_\beta$ and their correspondence to the experimental purposes. For example, any of the L_2 designs considered in Theorem 7.3.11 or in Theorem 7.3.12 may be suitable for an s^2 factorial experiment, i.e., an experiment involving two factors each at s levels. This is due to the structures of the matrices $\boldsymbol{L}_0$ and $\boldsymbol{L}_1$ of these designs, as indicated in the theorems. Evidently, in case of the L_2 design in Theorem 7.3.11 contrasts of the type $(\mathbf{1}_s' \otimes \boldsymbol{c}')\boldsymbol{L}_0'\boldsymbol{\tau} = (\mathbf{1}_s' \otimes \boldsymbol{c}')\boldsymbol{\tau}$, with any vector $\boldsymbol{c}$ such that $\boldsymbol{c}'\mathbf{1}_s = 0$, as well as those of the type $(\boldsymbol{c}' \otimes \mathbf{1}_s')\boldsymbol{L}_0'\boldsymbol{\tau} = (\boldsymbol{c}' \otimes \mathbf{1}_s')\boldsymbol{\tau}$, with such $\boldsymbol{c}$, are interesting from the point of view of comparing the main effects of the s levels of each of the two factors. The efficiency factor for any of these contrasts estimated in the intra-block analysis is $\varepsilon_0 = 1$ and the relevant variance is of the form $r^{-1}s\boldsymbol{c}'\boldsymbol{c}\sigma_1^2$ [see the formula (4.4.13)]. Other contrasts, of the type $(\boldsymbol{c}_1' \otimes \boldsymbol{c}_2')\boldsymbol{L}_1'\boldsymbol{\tau} = (\boldsymbol{c}_1' \otimes \boldsymbol{c}_2')\boldsymbol{\tau}$, with any vectors $\boldsymbol{c}_1$ and $\boldsymbol{c}_2$ such that $\boldsymbol{c}_1'\mathbf{1}_s = \boldsymbol{c}_2'\mathbf{1}_s = 0$, are interesting from the point of view of investigating the two-factor interactions. The efficiency factor for any of these contrasts is $\varepsilon_1 = (s-2)/(s-1)$ and the relevant variance is of the form $(\varepsilon_1 r)^{-1}\boldsymbol{c}_1'\boldsymbol{c}_1\boldsymbol{c}_2'\boldsymbol{c}_2\sigma_1^2$. Thus, if one wants to have the main effect contrasts estimated with full efficiency, the L_2 designs considered in Theorem 7.3.11 can be recommended. On the contrary, if the interaction contrasts are to be estimated with full efficiency, the L_2 designs from Theorem 7.3.12 become preferable, as can easily be seen.

7.3.1.2 Other than PBIB designs

As mentioned in Section 6.0.2, an association scheme does not always exist for any parameters. In fact, there may be a possibility of getting a $(\rho_0; \rho_1; 0)$-EB design, which is equireplicate and proper, but not necessarily one of the 2-associate PBIB designs. A large number of designs in this case can be constructed through the methods described in Chapter 6, particularly the patterned method (Section 6.6), as can be seen from Theorems 6.6.1 and 6.6.2. Further results will be given here. One way of obtaining classes of $(\rho_0; \rho_1; 0)$-EB designs is offered by the following theorem.

Theorem 7.3.13. *Let $\boldsymbol{N}$ be the $v \times b$ incidence matrix of a 2-associate PBIB design with parameters $v = 2k$, b, r, k, λ_1, λ_2, $\varepsilon_0 = 1$, ε_1, ρ_0, ρ_1, $\boldsymbol{L}_0$, $\boldsymbol{L}_1$. Then the incidence pattern (7.3.3) yields an r-resolvable $(\rho_0^*; \rho_1^*; 0)$-EB design with parameters $v^* = 4k$, $b^* = 4r$, $r^* = 2r$, $k^* = 2k$, $\varepsilon_0^* = 1$, $\varepsilon_1^* = (1+\varepsilon_1)/2$, $\rho_0^* = 2\rho_0 + 1$, $\rho_1^* = 2\rho_1$, $\boldsymbol{L}_0^* = \boldsymbol{I}_2 \otimes \boldsymbol{L}_0 + (\boldsymbol{I}_2 - 2^{-1}\mathbf{1}_2\mathbf{1}_2') \otimes v^{-1}\mathbf{1}_v\mathbf{1}_v'$, $\boldsymbol{L}_1^* = \boldsymbol{I}_2 \otimes \boldsymbol{L}_1$.*

Proof. The proof is entirely similar to that of Theorem 7.3.5. It is left as an exercise to the reader. □

Example 7.3.6. Consider a singular GD design, S6 (in the table of singular GD designs prepared by Clatworthy, 1973), based on four groups $\{1,2\}$, $\{3,4\}$, $\{5,6\}$, $\{7,8\}$ of two treatments each, with parameters $v = 8$, $b = 6$, $r = 3$, $k = 4$, $\lambda_1 = 3$, $\lambda_2 = 1$; $m = 4$, $n = 2$; $\varepsilon_0 = 1$, $\varepsilon_1 = 2/3$, $\rho_0 = 4$, $\rho_1 = 3$, $\boldsymbol{L}_0 = \boldsymbol{I}_4 \otimes (\boldsymbol{I}_2 - 2^{-1}\mathbf{1}_2\mathbf{1}_2')$, $\boldsymbol{L}_1 = (\boldsymbol{I}_4 - 4^{-1}\mathbf{1}_4\mathbf{1}_4') \otimes 2^{-1}\mathbf{1}_2\mathbf{1}_2'$, whose blocks are (1, 2, 3, 4), (5, 6, 7, 8), (1, 2, 5, 6), (3, 4, 7, 8), (1, 2, 7, 8), (3, 4, 5, 6), giving an 8×6 incidence matrix $\boldsymbol{N}$. Theorem 7.3.13 yields a 3-resolvable $(\rho_0^*; \rho_1^*; 0)$-EB design with parameters $v^* = 16$, $b^* = 12$, $r^* = 6$, $k^* = 8$, $\varepsilon_0^* = 1$, $\varepsilon_1^* = 5/6$, $\rho_0^* = 9$, $\rho_1^* = 6$, $\boldsymbol{L}_0^* = \boldsymbol{I}_8 \otimes (\boldsymbol{I}_2 - 2^{-1}\mathbf{1}_2\mathbf{1}_2') + (\boldsymbol{I}_2 - 2^{-1}\mathbf{1}_2\mathbf{1}_2') \otimes 8^{-1}\mathbf{1}_8\mathbf{1}_8'$, $\boldsymbol{L}_1^* = \boldsymbol{I}_2 \otimes (\boldsymbol{I}_4 - 4^{-1}\mathbf{1}_4\mathbf{1}_4') \otimes 2^{-1}\mathbf{1}_2\mathbf{1}_2'$, and with the incidence matrix

$$\boldsymbol{N}^* = \begin{bmatrix} \boldsymbol{N} & \boldsymbol{N} \\ \boldsymbol{N} & \mathbf{1}_8\mathbf{1}_6' - \boldsymbol{N} \end{bmatrix} = \left[\begin{array}{cccccc|cccccc}
1&0&1&0&1&0&1&0&1&0&1&0\\
1&0&1&0&1&0&1&0&1&0&1&0\\
1&0&0&1&0&1&1&0&0&1&0&1\\
1&0&0&1&0&1&1&0&0&1&0&1\\
0&1&1&0&0&1&0&1&1&0&0&1\\
0&1&1&0&0&1&0&1&1&0&0&1\\
0&1&0&1&1&0&0&1&0&1&1&0\\
0&1&0&1&1&0&0&1&0&1&1&0\\
1&0&1&0&1&0&0&1&0&1&0&1\\
1&0&1&0&1&0&0&1&0&1&0&1\\
1&0&0&1&0&1&0&1&1&0&1&0\\
1&0&0&1&0&1&0&1&1&0&1&0\\
0&1&1&0&0&1&1&0&0&1&1&0\\
0&1&1&0&0&1&1&0&0&1&1&0\\
0&1&0&1&1&0&1&0&1&0&0&1\\
0&1&0&1&1&0&1&0&1&0&0&1
\end{array}\right].$$

Here, the average efficiency factor of the design is $25/27 = 0.9259$. The $\boldsymbol{\Omega}$ matrix for the resulting design is

$$\boldsymbol{\Omega}^* = \frac{1}{240}\boldsymbol{I}_2 \otimes [\boldsymbol{I}_4 \otimes (40\boldsymbol{I}_2 + 31\mathbf{1}_2\mathbf{1}_2') + (\boldsymbol{I}_4 - \mathbf{1}_4\mathbf{1}_4') \otimes \mathbf{1}_2\mathbf{1}_2'].$$

Hence, for example, elementary contrasts obtain three different variances, $\sigma_1^2/3$, $43\sigma_1^2/120$ and $11\sigma_1^2/30$, when estimated in the intra-block analysis. The smallest,

$\sigma_1^2/3$, concerns elementary contrasts belonging to the subspace spanned by the matrix $\boldsymbol{L}_0^*$, e.g., those between treatments 1st and 2nd, 3rd and 4th, etc.

Theorem 6.6.2 and Corollary 7.3.1 produce the following result.

Corollary 7.3.3. *If q is a prime or a prime power, there exists a q-resolvable $(\rho_0;\rho_1;0)$-EB design with parameters $v = sq(q+1)$, $b = tq^2$, $r = tq$, $k = s(q+1)$, $\varepsilon_0 = 1$, $\varepsilon_1 = 1 - 1/(q+1)$, $\rho_0 = q(q+1)(s-1)+q$, $\rho_1 = q^2-1$, $\boldsymbol{L}_0 = (\boldsymbol{I}_s - s^{-1}\mathbf{1}_s\mathbf{1}_s') \otimes \boldsymbol{I}_{mn} + s^{-1}\mathbf{1}_s\mathbf{1}_s' \otimes (\boldsymbol{I}_m - m^{-1}\mathbf{1}_m\mathbf{1}_m') \otimes n^{-1}\mathbf{1}_n\mathbf{1}_n'$, $\boldsymbol{L}_1 = s^{-1}\mathbf{1}_s\mathbf{1}_s' \otimes \boldsymbol{I}_m \otimes (\boldsymbol{I}_n - n^{-1}\mathbf{1}_n\mathbf{1}_n')$ for positive integers s and t, where $m = q+1$ and $n = q$.*

Example 7.3.7. In an unreduced BIB design with parameters $v = 4, b = 6$, $r = 3, k = 2, \lambda = 1$, a rearrangement of the blocks into [(1,2), (3,4)], [(1,3), (2,4)], [(1,4), (2,3)] gives an affine resolvable solution with an evident incidence matrix $\boldsymbol{N}$. By taking the dual of the design, one can obtain a semi-regular GD design, as shown in Corollary 7.3.1, with $\boldsymbol{N}^* = \boldsymbol{N}'$ and parameters $v^* = 6, b^* = 4, r^* = 2$, $k^* = 3$, $\lambda_1^* = 0, \lambda_2^* = 1$, $m^* = 3, n^* = 2$, $\varepsilon_0^* = 1$, $\varepsilon_1^* = 2/3$, $\rho_0^* = 2, \rho_1^* = 3$, following from $q = 2$. Then Corollary 7.3.3 yields a 2-resolvable $(\tilde{\rho}_0;\tilde{\rho}_1;0)$-EB design with parameters $\tilde{v} = 6s$, $\tilde{b} = 4t$, $\tilde{r} = 2t$, $\tilde{k} = 3s$, $\tilde{\varepsilon}_0 = 1$, $\tilde{\varepsilon}_1 = 2/3$, $\tilde{\rho}_0 = 6s-4$, $\tilde{\rho}_1 = 3$, $\tilde{\boldsymbol{L}}_0 = (\boldsymbol{I}_s - s^{-1}\mathbf{1}_s\mathbf{1}_s') \otimes \boldsymbol{I}_6 + s^{-1}\mathbf{1}_s\mathbf{1}_s' \otimes (\boldsymbol{I}_3 - 3^{-1}\mathbf{1}_3\mathbf{1}_3') \otimes 2^{-1}\mathbf{1}_2\mathbf{1}_2'$, $\tilde{\boldsymbol{L}}_1 = s^{-1}\mathbf{1}_s\mathbf{1}_s' \otimes \boldsymbol{I}_3 \otimes (\boldsymbol{I}_2 - 2^{-1}\mathbf{1}_2\mathbf{1}_2')$, and with the incidence matrix given, for $s = 2$ and $t = 2$, as

$$\tilde{\boldsymbol{N}} = \mathbf{1}_2\mathbf{1}_2' \otimes \boldsymbol{N}^* = \left[\begin{array}{cccc|cccc} 1&1&0&0&1&1&0&0\\ 0&0&1&1&0&0&1&1\\ 1&0&1&0&1&0&1&0\\ 0&1&0&1&0&1&0&1\\ 1&0&0&1&1&0&0&1\\ 0&1&1&0&0&1&1&0\\ 1&1&0&0&1&1&0&0\\ 0&0&1&1&0&0&1&1\\ 1&0&1&0&1&0&1&0\\ 0&1&0&1&0&1&0&1\\ 1&0&0&1&1&0&0&1\\ 0&1&1&0&0&1&1&0 \end{array}\right], \quad \begin{array}{l} \tilde{v} = 12,\ \tilde{b} = 8, \\ \tilde{r} = 4,\ \tilde{k} = 6, \\ \tilde{\varepsilon}_0 = 1,\ \tilde{\varepsilon}_1 = 2/3, \\ \tilde{\rho}_0 = 8,\ \tilde{\rho}_1 = 3. \end{array}$$

Here, the average efficiency factor of the design is $22/25 = 0.88$. The matrix $\boldsymbol{\Omega}$ is obtained as

$$\tilde{\boldsymbol{\Omega}} = \frac{1}{32}\left[\boldsymbol{I}_2 \otimes \boldsymbol{I}_3 \otimes (10\boldsymbol{I}_2 - \mathbf{1}_2\mathbf{1}_2') + (\mathbf{1}_2\mathbf{1}_2' - \boldsymbol{I}_2) \otimes \boldsymbol{I}_3 \otimes (2\boldsymbol{I}_2 - \mathbf{1}_2\mathbf{1}_2')\right].$$

Hence, for example, an elementary contrast is estimated in the intra-block analysis with one of three different variances $5\sigma_1^2/8$, $9\sigma_1^2/16$ and $\sigma_1^2/2$. The latter applies to elementary contrasts represented by eigenvectors of $\tilde{\boldsymbol{L}}_0$ corresponding to its unit (= 1) eigenvalue, e.g., to contrasts between treatments 1st and 7th, 2nd and 8th, etc.

The design constructed in Example 7.3.7 can well be used for a three-factor experiment with factors A, B and C applied at s (= 2), 3 and 2 levels, respectively ($\tilde{v} = s \times 3 \times 2$). The structures of the matrices $\tilde{\boldsymbol{L}}_0$ and $\tilde{\boldsymbol{L}}_1$ show that under this design any contrast involving the A factor main effects, any contrast involving the B factor main effects and any contrast involving the AB, AC and ABC interactions will be estimated in the intra-block analysis with full efficiency, whereas a contrast involving the C factor main effects and any contrast involving the BC interactions will be estimated in that analysis with a lower efficiency, $\tilde{\varepsilon}_1 = 2/3$.

With regard to the complementation method (Section 6.2), a general result is given in Theorem 6.2.1. For the present section, its version concerning $(\rho_0; \rho_1; 0)$-EB design is as follows.

Corollary 7.3.4. *If $\boldsymbol{N}$ is the $v \times b$ incidence matrix of a binary $(\rho_0; \rho_1; 0)$-EB design with parameters $v, b, r, k, \varepsilon_\beta$, ρ_β, $\boldsymbol{L}_\beta, \beta = 0, 1$, then its complement $\mathbf{1}_v\mathbf{1}_b' - \boldsymbol{N}$ is the incidence matrix of a binary $(\tilde{\rho}_0; \tilde{\rho}_1; 0)$-EB design with parameters $\tilde{v} = v, \tilde{b} = b, \tilde{r} = b - r, \tilde{k} = v - k$, $\tilde{\varepsilon}_\beta = 1 - rk(1 - \varepsilon_\beta)/[(v - k)(b - r)]$, $\tilde{\rho}_\beta = \rho_\beta, \tilde{\boldsymbol{L}}_\beta = \boldsymbol{L}_\beta$, provided that $v(2r - b)/(rk) < \varepsilon_\beta \leq 1$ for $\beta = 0, 1$. In the extreme case of $v(2r - b)/(rk) = \varepsilon_1$, the resulting design is $(\rho_0; 0; \rho_1)$-EB.*

Note that, in Corollary 7.3.4, $\tilde{\varepsilon}_\beta = 1$ if and only if $\varepsilon_\beta = 1$.

As to the dualization method, recall (see Section 6.1) that this procedure means the interchanging of the role of blocks and treatments in a block design. Applications of Corollaries 6.1.2 and 6.1.3 will now be presented, by dualizing designs with 1 or 2 distinct efficiency factors.

Because a BIB design with parameters v, b, r, k, λ has

$$\varepsilon_1 = \frac{\lambda v}{rk}\ (< 1), \quad \rho_1 = v - 1, \quad \boldsymbol{L}_1 = \boldsymbol{I}_v - \frac{1}{v}\mathbf{1}_v\mathbf{1}_v',$$

i.e., is $(0; v-1; 0)$-EB (see Section 6.0.1), the following corollary can be obtained.

Corollary 7.3.5. *The dual of a BIB design with parameters v, $b(> v)$, r, k, λ is a $(\rho_0^*; \rho_1^*; 0)$-EB design with parameters $v^* = b$, $b^* = v$, $r^* = k$, $k^* = r$, $\varepsilon_0^* = 1$, $\varepsilon_1^* = \lambda v/(rk)$, $\rho_0^* = b - v$, $\rho_1^* = v - 1$, $\boldsymbol{L}_0^* = \boldsymbol{I}_b - b^{-1}\mathbf{1}_b\mathbf{1}_b' - \boldsymbol{L}_1^*$, $\boldsymbol{L}_1^* = (r - \lambda)^{-1}\boldsymbol{N}'(\boldsymbol{I}_v - v^{-1}\mathbf{1}_v\mathbf{1}_v')\boldsymbol{N}$, where $\boldsymbol{N}$ is the incidence matrix of the dualized BIB design.*

Because a singular or semi-regular GD design with parameters $v = mn, b, r, k, \lambda_1, \lambda_2$ has the properties [described in Section 6.0.2(A)] that make it $(\rho_0; \rho_1; 0)$-EB with $\varepsilon_0 = 1$, $\varepsilon_1 = \lambda_2 v/(rk)$, $\rho_0 = m(n-1)$, $\rho_1 = m - 1$, $\boldsymbol{L}_0 = \boldsymbol{I}_m \otimes (\boldsymbol{I}_n - n^{-1}\mathbf{1}_n\mathbf{1}_n')$, $\boldsymbol{L}_1 = (\boldsymbol{I}_m - m^{-1}\mathbf{1}_m\mathbf{1}_m') \otimes n^{-1}\mathbf{1}_n\mathbf{1}_n'$, if it is singular, or with $\varepsilon_0 = 1$, $\varepsilon_1 = [r(k-1) + \lambda_1]/(rk)$, $\rho_0 = m - 1$, $\rho_1 = m(n-1)$, $\boldsymbol{L}_0 = (\boldsymbol{I}_m - m^{-1}\mathbf{1}_m\mathbf{1}_m') \otimes n^{-1}\mathbf{1}_n\mathbf{1}_n'$, $\boldsymbol{L}_1 = \boldsymbol{I}_m \otimes (\boldsymbol{I}_n - n^{-1}\mathbf{1}_n\mathbf{1}_n')$, if it is semi-regular, one gets the following procedure (see also Puri, Nigam and Kageyama, 1987, Corollary 3.4).

Corollary 7.3.6. *The dual of a singular GD design is a (k/n)-resolvable $(\rho_0^*;\rho_1^*;0)$-EB design with parameters $v^* = b$, $b^* = v = mn$, $r^* = k$, $k^* = r$, $\varepsilon_0^* = 1$, $\varepsilon_1^* = v\lambda_2/(rk)$, $\rho_0^* = b - m$, $\rho_1^* = m - 1$, $\boldsymbol{L}_0^* = \boldsymbol{I}_b - b^{-1}\mathbf{1}_b\mathbf{1}_b' - \boldsymbol{L}_1^*$, $\boldsymbol{L}_1^* = (rk - v\lambda_2)^{-1}\boldsymbol{N}'[(\boldsymbol{I}_m - m^{-1}\mathbf{1}_m\mathbf{1}_m') \otimes n^{-1}\mathbf{1}_n\mathbf{1}_n']\boldsymbol{N}$, where $\boldsymbol{N}$ is the incidence matrix of the original singular GD design.*

Corollary 7.3.7. *The dual of a semi-regular GD design is a (k/m)-resolvable $(\rho_0^*;\rho_1^*;0)$-EB design with parameters $v^* = b$, $b^* = mn$, $r^* = k$, $k^* = r$, $\varepsilon_0^* = 1$, $\varepsilon_1^* = 1-(r-\lambda_1)/(rk)$, $\rho_0^* = b-m(n-1)-1$, $\rho_1^* = m(n-1)$, $\boldsymbol{L}_0^* = \boldsymbol{I}_b - b^{-1}\mathbf{1}_b\mathbf{1}_b' - \boldsymbol{L}_1^*$, $\boldsymbol{L}_1^* = (r - \lambda_1)^{-1}\boldsymbol{N}'[\boldsymbol{I}_m \otimes (\boldsymbol{I}_n - n^{-1}\mathbf{1}_n\mathbf{1}_n')]\boldsymbol{N}$, where $\boldsymbol{N}$ is the incidence matrix of the original semi-regular GD design.*

As to the α-resolvable properties of designs given by Corollaries 7.3.6 and 7.3.7, they will be discussed in Lemmas 9.3.1 and 9.3.2, which will give the final Corollaries 9.3.2 and 9.3.3.

Similarly, by dualization of other 2-associate PBIB designs with one of the eigenvalues of $\boldsymbol{NN}'$ being zero, discussed in Section 6.0.2 and in the present section, various $(\rho_0;\rho_1;0)$-EB designs with $\rho_0 \geq 1$ can be obtained.

In conclusion, by the procedure of dualization the following two useful results have been observed (see Corollaries 6.1.2 and 6.1.3), which will be used specially in Section 7.3.4:

(1) The dual of a $(0; v-1; 0)$-EB design yields a $(\rho_0^*;\rho_1^*;0)$-EB design with $\rho_0^* = b - v$ and $\rho_1^* = \rho_1 = v - 1$.

(2) The dual of a $(\rho_0;\rho_1;0)$-EB design with $\rho_0 \geq 1$ is again a $(\rho_0^*;\rho_1^*;0)$-EB design with $\rho_0^* = b - v + \rho_0$ and $\rho_1^* = \rho_1$.

7.3.2 Proper designs that are nonequireplicate

As explained in Section 7.1, in a $(\rho_0;\rho_1;0)$-EB design with parameters $v, b, \boldsymbol{r}, \boldsymbol{k}, \varepsilon_\beta$, ρ_β, $\boldsymbol{L}_\beta$ for $\beta = 0,1$,

$$\boldsymbol{C}_1 = \boldsymbol{r}^\delta \boldsymbol{L}_0 + \varepsilon_1 \boldsymbol{r}^\delta \boldsymbol{L}_1,$$

$$\boldsymbol{C}_1 \boldsymbol{s}_{\beta j} = \varepsilon_\beta \boldsymbol{r}^\delta \boldsymbol{s}_{\beta j},\ j = 1,2,...,\rho_\beta,$$

$$\boldsymbol{L}_\beta = \sum_{j=1}^{\rho_\beta} \boldsymbol{s}_{\beta j}\boldsymbol{s}_{\beta j}'\boldsymbol{r}^\delta,\quad \boldsymbol{L}_0 + \boldsymbol{L}_1 = \boldsymbol{I}_v - \frac{1}{n}\mathbf{1}_v\boldsymbol{r}'.$$

As in Section 4.4.3 [see also Section 7.1(4)], any contrast given by $\boldsymbol{s}_\beta'\boldsymbol{L}_\beta'\boldsymbol{r}^\delta\boldsymbol{\tau}$ for some $\boldsymbol{s}_\beta$ ($\boldsymbol{L}_\beta\boldsymbol{s}_\beta \neq \mathbf{0}$) obtains the intra-block BLUE of the form

$$(\widehat{\boldsymbol{s}_\beta'\boldsymbol{L}_\beta'\boldsymbol{r}^\delta\boldsymbol{\tau}})_{\text{intra}} = \varepsilon_\beta^{-1}\boldsymbol{s}_\beta'\boldsymbol{L}_\beta'\boldsymbol{Q}_1,$$

with the variance

$$\text{Var}[(\widehat{\boldsymbol{s}_\beta'\boldsymbol{L}_\beta'\boldsymbol{r}^\delta\boldsymbol{\tau}})_{\text{intra}}] = \varepsilon_\beta^{-1}\boldsymbol{s}_\beta'\boldsymbol{r}^\delta\boldsymbol{L}_\beta\boldsymbol{s}_\beta\sigma_1^2,$$

$\beta = 0, 1$. This variance is reduced to $\varepsilon_\beta^{-1}\sigma_1^2$ if s_β represents one of the basic contrasts corresponding to the common efficiency factor ε_β.

Unequal replication numbers often appear in supplemented block designs, i.e., designs developed by adding supplementary treatments to those already appearing in available block designs. The technique of supplementation is described in Section 6.3, main constructional results being given in Theorems 6.3.1 – 6.3.5 (see also Ceranka, 1983, Section 6). In particular, consider the following pattern,

$$\boldsymbol{N}^* = \begin{bmatrix} \boldsymbol{N} \\ \mathbf{1}_s\mathbf{1}_b' \end{bmatrix}, \tag{7.3.4}$$

which is the same as (6.3.3).

Through Theorem 6.3.1 with a BIB design $\boldsymbol{N}$, a $(\rho_0; \rho_1; 0)$-EB design can be obtained (for example, see Theorem 6.3.2). Of course, instead of a BIB design one can utilize a special type of 2-associate PBIB designs. Theorem 6.3.3 implies the following result.

Corollary 7.3.8. *The existence of a 2-associate PBIB design with parameters* v, b, r, k, λ_1, λ_2, $\varepsilon_0 = 1$, $\varepsilon_1 < 1$, ρ_0, ρ_1, $\boldsymbol{L}_0$, $\boldsymbol{L}_1$ *implies the existence of a* $(\rho_0^*; \rho_1^*; 0)$*-EB design* (7.3.4) *with parameters* $v^* = v+s$, $b^* = b$, $\boldsymbol{r}^* = [r\mathbf{1}_v', b\mathbf{1}_s']'$, $k^* = k+s$, $\varepsilon_0^* = 1$, $\varepsilon_1^* = (k\varepsilon_1 + s)/(k+s)$, $\rho_0^* = \rho_0 + s$, $\rho_1^* = \rho_1$, $\boldsymbol{L}_0^* = \boldsymbol{I}_{v+s} - [b(k+s)]^{-1}\mathbf{1}_{v+s}(\boldsymbol{r}^*)' - \boldsymbol{L}_1^*$, $\boldsymbol{L}_1^* = \operatorname{diag}[\boldsymbol{L}_1 : \mathbf{O}]$ *for* $s \geq 1$.

For any 2-associate PBIB design in which the rank of its $\boldsymbol{NN}'$ is smaller than v, the number of treatments, Corollary 7.3.8 produces a $(\rho_0; \rho_1; 0)$-EB design in which $\rho_0 \geq 1$. For example, if a semi-regular GD design is taken as a basic design in (7.3.4), then the following result can be obtained by referring to Sections 6.0.2(A) and 7.3.1.1(2).

Example 7.3.8. The existence of a semi-regular GD design with parameters $v = mn$, b, r, k, λ_1, λ_2 implies the existence of a $(\rho_0^*; \rho_1^*; 0)$-EB design (7.3.4) with parameters $v^* = v + s$, $b^* = b$, $\boldsymbol{r}^* = [r\mathbf{1}_v', b\mathbf{1}_s']'$, $k^* = k+s$, $\varepsilon_0^* = 1$, $\varepsilon_1^* = 1 - (r - \lambda_1)/[r(k+s)]$, $\rho_0^* = m + s - 1$, $\rho_1^* = m(n-1)$, $\boldsymbol{L}_0^* = \operatorname{diag}[\boldsymbol{I}_m \otimes n^{-1}\mathbf{1}_n\mathbf{1}_n' : \boldsymbol{I}_s] - [b(k+s)]^{-1}\mathbf{1}_{v+s}(\boldsymbol{r}^*)'$, $\boldsymbol{L}_1^* = \operatorname{diag}[\boldsymbol{I}_m \otimes (\boldsymbol{I}_n - n^{-1}\mathbf{1}_n\mathbf{1}_n') : \mathbf{O}]$ for $s \geq 1$. The resulting design gives, by (7.1.6),

$$\boldsymbol{\Omega}^* = \left(\frac{1}{\varepsilon_1^*} - 1\right)\begin{bmatrix} \frac{1}{r}\boldsymbol{I}_m \otimes (\boldsymbol{I}_n - \frac{1}{n}\mathbf{1}_n\mathbf{1}_n') & \mathbf{O} \\ \mathbf{O} & \mathbf{O} \end{bmatrix} + \begin{bmatrix} \frac{1}{r}\boldsymbol{I}_v & \mathbf{O} \\ \mathbf{O} & \frac{1}{b}\boldsymbol{I}_s \end{bmatrix},$$

which shows that in this design elementary contrasts are estimated in the intrablock analysis with four different variances.

Example 7.3.9. The existence of a singular GD design with parameters $v = mn$, b, r, k, λ_1, λ_2 implies the existence of a $(\rho_0^*; \rho_1^*; 0)$-EB design (7.3.4) with parameters $v^* = v+s$, $b^* = b$, $\boldsymbol{r}^* = [r\mathbf{1}_v', b\mathbf{1}_s']'$, $k^* = k+s$, $\varepsilon_0^* = 1$, $\varepsilon_1^* = 1-(rk - v\lambda_2)/[r(k+s)]$, $\rho_0^* = v+s-m$, $\rho_1^* = m-1$, $\boldsymbol{L}_0^* = \boldsymbol{I}_{v+s} - [b(k+s)]^{-1}\mathbf{1}_{v+s}(\boldsymbol{r}^*)' - \boldsymbol{L}_1^*$, $\boldsymbol{L}_1^* = \operatorname{diag}[(\boldsymbol{I}_m - m^{-1}\mathbf{1}_m\mathbf{1}_m') \otimes n^{-1}\mathbf{1}_n\mathbf{1}_n' : \mathbf{O}]$ for $s \geq 1$. The matrix $\boldsymbol{\Omega}$ for the resulting design is of the form

$$\boldsymbol{\Omega}^* = \left(\frac{1}{\varepsilon_1^*} - 1\right)\begin{bmatrix} \frac{1}{r}(\boldsymbol{I}_m - \frac{1}{m}\mathbf{1}_m\mathbf{1}_m') \otimes \frac{1}{n}\mathbf{1}_n\mathbf{1}_n' & \mathbf{O} \\ \mathbf{O} & \mathbf{O} \end{bmatrix} + \begin{bmatrix} \frac{1}{r}\boldsymbol{I}_v & \mathbf{O} \\ \mathbf{O} & \frac{1}{b}\boldsymbol{I}_s \end{bmatrix}.$$

Hence, for example, elementary contrasts obtain four different variances in the intra-block analysis.

These results can be derived similarly as Theorem 6.3.1, by noting that a 2-associate PBIB design is $(\rho_0; \rho_1; 0)$-EB if the rank of its incidence matrix $\boldsymbol{N}$ is smaller than v [see Lemma 2.3.3 and Section 6.0.2(A)].

As a special case of (7.3.4) consider the pattern

$$\boldsymbol{N}^* = \begin{bmatrix} \boldsymbol{I}_b \\ \mathbf{1}_s\mathbf{1}_b' \end{bmatrix}, \tag{7.3.5}$$

which yields the following observation (see also Singh and Dey, 1979; Ceranka and Chudzik, 1984).

Corollary 7.3.9. *There always exists a $(\rho_0^*; \rho_1^*; 0)$-EB design with parameters $v^* = b + s$, $b^* = b$, $\boldsymbol{r}^* = [\mathbf{1}_b', b\mathbf{1}_s']'$, $k^* = s + 1$, $\varepsilon_0^* = 1$, $\varepsilon_1^* = s/(s+1)$, $\rho_0^* = s$, $\rho_1^* = b - 1$, $\boldsymbol{L}_0^* = \boldsymbol{I}_{b+s} - [b(s+1)]^{-1}\mathbf{1}_{b+s}(\boldsymbol{r}^*)' - \boldsymbol{L}_1^*$, $\boldsymbol{L}_1^* = \operatorname{diag}[\boldsymbol{I}_b - b^{-1}\mathbf{1}_b\mathbf{1}_b' : \mathbf{O}]$. Its incidence matrix is given by* (7.3.5).

In this corollary, the matrix $\boldsymbol{\Omega}$ can be obtained as

$$\boldsymbol{\Omega}^* = \left(\frac{1}{\varepsilon_1^*} - 1\right)\begin{bmatrix} \boldsymbol{I}_b - \frac{1}{b}\mathbf{1}_b\mathbf{1}_b' & \mathbf{O} \\ \mathbf{O} & \mathbf{O} \end{bmatrix} + \begin{bmatrix} \boldsymbol{I}_b & \mathbf{O} \\ \mathbf{O} & \frac{1}{b}\boldsymbol{I}_s \end{bmatrix}.$$

Hence, elementary contrasts can be estimated in the intra-block analysis with three different variances. Note that Corollary 7.3.9 can be regarded as a special (extreme) case of Theorem 6.3.2 when $v = b, r = k = 1$ and $\lambda = 0$.

Another pattern of supplementation is that considered by Ceranka (1983, p. 44), a special case of which is analysed in the following theorem.

Theorem 7.3.14. *If $\boldsymbol{N}$ is the $v \times b$ incidence matrix of a BIB design with parameters v, b, r, k, λ, then the incidence matrix*

$$\boldsymbol{N}^* = \begin{bmatrix} \boldsymbol{N}' \\ \boldsymbol{I}_v \otimes \mathbf{1}_\alpha \end{bmatrix}$$

yields a $(\rho_0^*; \rho_1^*; 0)$-EB design with parameters $v^* = b + v\alpha$, $b^* = v$, $\boldsymbol{r}^* = [k\mathbf{1}_b', \mathbf{1}_{v\alpha}']'$, $k^* = r + \alpha$, $\varepsilon_0^* = 1$, $\varepsilon_1^* = \lambda v/[k(r+\alpha)]$, $\rho_0^* = b + v(\alpha - 1)$, $\rho_1^* = v - 1$,

$$\boldsymbol{L}_0^* = \boldsymbol{I}_{b+v\alpha} - \frac{1}{v(r+\alpha)}\mathbf{1}_{b+v\alpha}(\boldsymbol{r}^*)' - \boldsymbol{L}_1^*$$

and

$$\boldsymbol{L}_1^* = \frac{k}{r - \lambda + \alpha k}\begin{bmatrix} \frac{1}{k}\boldsymbol{N}'(\boldsymbol{I}_v - \frac{1}{v}\mathbf{1}_v\mathbf{1}_v')\boldsymbol{N} & \frac{1}{k}\boldsymbol{N}'\{(\boldsymbol{I}_v - \frac{1}{v}\mathbf{1}_v\mathbf{1}_v') \otimes \mathbf{1}_\alpha'\} \\ \{(\boldsymbol{I}_v - \frac{1}{v}\mathbf{1}_v\mathbf{1}_v') \otimes \mathbf{1}_\alpha\}\boldsymbol{N} & (\boldsymbol{I}_v - \frac{1}{v}\mathbf{1}_v\mathbf{1}_v') \otimes \mathbf{1}_\alpha\mathbf{1}_\alpha' \end{bmatrix}.$$

Proof. As pointed out in Section 5 of Puri and Kageyama (1985), one can first consider some juxtaposition of the incidence matrix, i.e., $(\boldsymbol{N}^*)' = [\boldsymbol{N} : \boldsymbol{I}_v \otimes \boldsymbol{1}'_\alpha]$, and then prove that the resulting design is a $(\rho_0; \rho_1; 0)$-EB design in which $\rho_0 \geq 0$. After that, taking the dual of this design, i.e., $[(\boldsymbol{N}^*)']' = \boldsymbol{N}^*$, one can complete the proof by utilizing Corollary 6.1.2 or Corollary 6.1.3.

For $\boldsymbol{N}^{**} = (\boldsymbol{N}^*)'$, where $\boldsymbol{N}$ is of a BIB design, the matrix $\boldsymbol{M}_0$ is given by

$$\begin{aligned}
\boldsymbol{M}_0^{**} &= \frac{1}{r+\alpha}[\boldsymbol{N} : \boldsymbol{I}_v \otimes \boldsymbol{1}'_\alpha] \begin{bmatrix} \frac{1}{k}\boldsymbol{I}_b & \mathbf{O} \\ \mathbf{O} & \boldsymbol{I}_s \end{bmatrix} \begin{bmatrix} \boldsymbol{N}' \\ \boldsymbol{I}_v \otimes \boldsymbol{1}_\alpha \end{bmatrix} - \frac{1}{v}\boldsymbol{1}_v\boldsymbol{1}'_v \\
&= \frac{1}{r+\alpha}\left(\frac{1}{k}\boldsymbol{N}\boldsymbol{N}' + \alpha \boldsymbol{I}_v\right) - \frac{1}{v}\boldsymbol{1}_v\boldsymbol{1}'_v \\
&= \frac{r-\lambda+\alpha k}{k(r+\alpha)}\left(\boldsymbol{I}_v - \frac{1}{v}\boldsymbol{1}_v\boldsymbol{1}'_v\right),
\end{aligned}$$

since $\boldsymbol{N}\boldsymbol{N}' = (r-\lambda)\boldsymbol{I}_v + \lambda\boldsymbol{1}_v\boldsymbol{1}'_v$ (see Section 6.0.1). This, by referring to Section 4.4, shows that $\boldsymbol{N}^{**} = [\boldsymbol{N} : \boldsymbol{I}_v \otimes \boldsymbol{1}'_\alpha]$ is the incidence matrix of a $(0; \rho_1^{**}; 0)$-EB design in which

$$\varepsilon_1^{**} = 1 - \frac{r-\lambda+\alpha k}{k(r+\alpha)} = \frac{\lambda v}{k(r+\alpha)} \quad (< 1),$$

$$\boldsymbol{L}_1^{**} = \boldsymbol{I}_v - \frac{1}{v}\boldsymbol{1}_v\boldsymbol{1}'_v, \; \rho_1^{**} = v - 1.$$

Then Corollary 6.1.2 clearly yields the required result. In fact, from this corollary and the result above,

$$\begin{aligned}
\boldsymbol{L}_1^* &= \frac{k(r+\alpha)}{r-\lambda+\alpha k} \begin{bmatrix} \frac{1}{k}\boldsymbol{I}_b & \mathbf{O} \\ \mathbf{O} & \boldsymbol{I}_{v\alpha} \end{bmatrix} \begin{bmatrix} \boldsymbol{N}' \\ \boldsymbol{I}_v \otimes \boldsymbol{1}_\alpha \end{bmatrix} \frac{1}{r+\alpha}(\boldsymbol{I}_v - \frac{1}{v}\boldsymbol{1}_v\boldsymbol{1}'_v)[\boldsymbol{N} : \boldsymbol{I}_v \otimes \boldsymbol{1}'_\alpha] \\
&= \frac{k}{r-\lambda+\alpha k} \begin{bmatrix} \frac{1}{k}\boldsymbol{N}'(\boldsymbol{I}_v - \frac{1}{v}\boldsymbol{1}_v\boldsymbol{1}'_v)\boldsymbol{N} & \frac{1}{k}\boldsymbol{N}'\{(\boldsymbol{I}_v - \frac{1}{v}\boldsymbol{1}_v\boldsymbol{1}'_v) \otimes \boldsymbol{1}'_\alpha\} \\ \{(\boldsymbol{I}_v - \frac{1}{v}\boldsymbol{1}_v\boldsymbol{1}'_v) \otimes \boldsymbol{1}_\alpha\}\boldsymbol{N} & (\boldsymbol{I}_v - \frac{1}{v}\boldsymbol{1}_v\boldsymbol{1}'_v) \otimes \boldsymbol{1}_\alpha\boldsymbol{1}'_\alpha \end{bmatrix}
\end{aligned}$$

and

$$(\boldsymbol{L}_1^*)^2 = \boldsymbol{L}_1^*, \quad \operatorname{tr}(\boldsymbol{L}_1^*) = v - 1.$$

Thus, the proof is complete. (As a similar approach see Theorem 6.4.3.) □

Remark 7.3.4. Theorem 7.3.14 can be proved directly by noting that

$$\boldsymbol{M}_0^* = \frac{1}{r+\alpha} \begin{bmatrix} \frac{1}{k}\boldsymbol{N}'\boldsymbol{L}_1\boldsymbol{N} & \frac{1}{k}\boldsymbol{N}'(\boldsymbol{L}_1 \otimes \boldsymbol{1}'_\alpha) \\ (\boldsymbol{L}_1 \otimes \boldsymbol{1}_\alpha)\boldsymbol{N} & \boldsymbol{L}_1 \otimes \boldsymbol{1}_\alpha\boldsymbol{1}'_\alpha \end{bmatrix}, \quad \boldsymbol{L}_1 = \boldsymbol{I}_v - \frac{1}{v}\boldsymbol{1}_v\boldsymbol{1}'_v,$$

and that the matrix

$$\frac{k}{r-\lambda+k\alpha} \begin{bmatrix} \frac{1}{k}\boldsymbol{N}'\boldsymbol{L}_1\boldsymbol{N} & \frac{1}{k}\boldsymbol{N}'(\boldsymbol{L}_1 \otimes \boldsymbol{1}'_\alpha) \\ (\boldsymbol{L}_1 \otimes \boldsymbol{1}_\alpha)\boldsymbol{N} & \boldsymbol{L}_1 \otimes \boldsymbol{1}_\alpha\boldsymbol{1}'_\alpha \end{bmatrix}$$

is idempotent, and hence can be taken as $\boldsymbol{L}_1^*$.

The matrix $\boldsymbol{\Omega}$ for the design resulting from Theorem 7.3.14 is obtained as

$$\left(\frac{1}{\varepsilon_1^*}-1\right)\frac{k}{r-\lambda+\alpha k}\begin{bmatrix}\frac{1}{k^2}\boldsymbol{N}'(\boldsymbol{I}_v-\frac{1}{v}\mathbf{1}_v\mathbf{1}_v')\boldsymbol{N} & \frac{1}{k}\boldsymbol{N}'\{(\boldsymbol{I}_v-\frac{1}{v}\mathbf{1}_v\mathbf{1}_v')\otimes\mathbf{1}_\alpha'\} \\ \frac{1}{k}\{(\boldsymbol{I}_v-\frac{1}{v}\mathbf{1}_v\mathbf{1}_v')\otimes\mathbf{1}_\alpha\}\boldsymbol{N} & (\boldsymbol{I}_v-\frac{1}{v}\mathbf{1}_v\mathbf{1}_v')\otimes\mathbf{1}_\alpha\mathbf{1}_\alpha'\end{bmatrix}$$

$$+\begin{bmatrix}\frac{1}{k}\boldsymbol{I}_b & \mathbf{O} \\ \mathbf{O} & \boldsymbol{I}_{v\alpha}\end{bmatrix}.$$

This means that in estimating elementary contrasts the number of different variances for their estimators depends on the structure of $\boldsymbol{N}$.

One can also adopt Theorem 7.3.14 to a design of a more simple pattern. In particular, when the basic design is an RBD, the following result applies.

Corollary 7.3.10. *A supplemented design with the incidence matrix*

$$\boldsymbol{N}^*=\begin{bmatrix}\mathbf{1}_v\mathbf{1}_b' \\ \boldsymbol{I}_b\otimes\mathbf{1}_\alpha\end{bmatrix}$$

is a $(\rho_0^*;\rho_1^*;0)$*-EB design with parameters* $v^*=v+s, b^*=b, \boldsymbol{r}^*=[b\mathbf{1}_v',\mathbf{1}_s']', k^*=v+\alpha$, $\varepsilon_0^*=1$, $\varepsilon_1^*=v/(v+\alpha)$, $\rho_0^*=v+s-b$, $\rho_1^*=b-1$, $\boldsymbol{L}_0^*=\boldsymbol{I}_{v+s}-[b(v+\alpha)]^{-1}\mathbf{1}_{v+s}(\boldsymbol{r}^*)'-\boldsymbol{L}_1^*$, $\boldsymbol{L}_1^*=\operatorname{diag}[\mathbf{O}:(\boldsymbol{I}_b-b^{-1}\mathbf{1}_b\mathbf{1}_b')\otimes\alpha^{-1}\mathbf{1}_\alpha\mathbf{1}_\alpha']$, *where* $s=b\alpha$.

Proof. In Theorem 7.3.14, let $\boldsymbol{N}=\mathbf{1}_b\mathbf{1}_v'$, i.e., take for $\boldsymbol{N}$, instead of a BIB design, the dual of an RBD. Then interchange b and v to obtain $k=b$ and $r=\lambda=v$, which leads to the required result. {As in Remark 7.3.4, a direct proof is given by noting that

$$\boldsymbol{M}_0^*=\frac{\alpha}{v+\alpha}\operatorname{diag}[\mathbf{O}:(\boldsymbol{I}_b-\frac{1}{b}\mathbf{1}_b\mathbf{1}_b')\otimes\frac{1}{\alpha}\mathbf{1}_\alpha\mathbf{1}_\alpha']$$

in the present case.} □

In Corollary 7.3.10, the matrix $\boldsymbol{\Omega}$ is, on account of (7.1.6), given simply by

$$\left(\frac{1}{\varepsilon_1^*}-1\right)\boldsymbol{L}_1^*(\boldsymbol{r}^*)^{-\delta}+(\boldsymbol{r}^*)^{-\delta}=\frac{s}{v}\begin{bmatrix}\mathbf{O} & \mathbf{O} \\ \mathbf{O} & \left(\boldsymbol{I}_b-\frac{1}{b}\mathbf{1}_b\mathbf{1}_b'\right)\otimes\frac{1}{\alpha}\mathbf{1}_\alpha\mathbf{1}_\alpha'\end{bmatrix}+\begin{bmatrix}\frac{1}{b}\boldsymbol{I}_v & \mathbf{O} \\ \mathbf{O} & \boldsymbol{I}_s\end{bmatrix},$$

which shows that elementary contrasts are estimated in the intra-block analysis with four different variances.

The design of Corollary 7.3.10 can be explained as follows. There are $s=b\alpha$ new treatments (e.g., breeding strains) available for a single replication, which are to be compared with v standard treatments (e.g., established varieties) available for b replications. Augment the basic RBD, comprising the standard treatments, by supplementing each block with α new treatments such that each new treatment appears in one block only. Here the only basic contrasts that are confounded with block differences are those between the new treatments belonging to different blocks. They are estimated in the intra-block analysis with efficiency $v/(v+\alpha)$, while any other basic contrasts are estimated there with full efficiency.

Next, another supplemented design can be obtained by a straightforward use of Theorem 7.3.14, with $\boldsymbol{N}$ confined to a symmetric BIB design (see Theorem 6.0.1 and Corollary 6.1.1) and $\alpha = 1$. In fact, referring to Theorem 7.3.14 and noting that, with $\boldsymbol{N}'$ replaced by $\boldsymbol{N}$, $\boldsymbol{N}'(\boldsymbol{I}_v - v^{-1}\mathbf{1}_v\mathbf{1}_v')\boldsymbol{N} = (k-\lambda)(\boldsymbol{I}_v - v^{-1}\mathbf{1}_v\mathbf{1}_v')$ and $\boldsymbol{N}'(\boldsymbol{I}_v - v^{-1}\mathbf{1}_v\mathbf{1}_v') = \boldsymbol{N} - (k/v)\mathbf{1}_v\mathbf{1}_v'$, the following corollary can be obtained.

Corollary 7.3.11. *If $\boldsymbol{N}$ is the $v \times b$ incidence matrix of a symmetric BIB design with parameters $v = b$, $r = k$, λ, then the incidence matrix*

$$\boldsymbol{N}^* = \begin{bmatrix} \boldsymbol{N} \\ \boldsymbol{I}_v \end{bmatrix}$$

yields a $(\rho_0^*;\rho_1^*;0)$-EB design with parameters $v^* = 2v$, $b^* = v$, $\boldsymbol{r}^* = [r\mathbf{1}_v', \mathbf{1}_v']'$, $k^* = k+1$, $\varepsilon_0^* = 1$, $\varepsilon_1^* = \lambda v/[k(k+1)]$, $\rho_0^* = v$, $\rho_1^* = v-1$,

$$\boldsymbol{L}_0^* = \boldsymbol{I}_{2v} - \frac{1}{v(k+1)}\mathbf{1}_{2v}(\boldsymbol{r}^*)' - \boldsymbol{L}_1^*$$

and

$$\boldsymbol{L}_1^* = \frac{1}{2k-\lambda}\begin{bmatrix} (k-\lambda)(\boldsymbol{I}_v - \frac{1}{v}\mathbf{1}_v\mathbf{1}_v') & \boldsymbol{N} - \frac{k}{v}\mathbf{1}_v\mathbf{1}_v' \\ k(\boldsymbol{N}' - \frac{k}{v}\mathbf{1}_v\mathbf{1}_v') & k(\boldsymbol{I}_v - \frac{1}{v}\mathbf{1}_v\mathbf{1}_v') \end{bmatrix},$$

giving

$$\boldsymbol{\Omega}^* = \frac{k(k+1)-\lambda k}{\lambda v(2k-\lambda)}\begin{bmatrix} \frac{k-\lambda}{k}(\boldsymbol{I}_v - \frac{1}{v}\mathbf{1}_v\mathbf{1}_v') & \boldsymbol{N} - \frac{k}{v}\mathbf{1}_v\mathbf{1}_v' \\ \boldsymbol{N}' - \frac{k}{v}\mathbf{1}_v\mathbf{1}_v' & k(\boldsymbol{I}_v - \frac{1}{v}\mathbf{1}_v\mathbf{1}_v') \end{bmatrix} + \begin{bmatrix} \frac{1}{k}\boldsymbol{I}_v & \mathbf{O} \\ \mathbf{O} & \boldsymbol{I}_v \end{bmatrix}.$$

Because a large number of symmetric BIB designs are available, as will be seen in Chapter 8 (Section 8.2), one can produce many $(\rho_0;\rho_1;0)$-EB designs in which v basic contrasts (among $2v$ treatments) are estimated in the intra-block analysis with full efficiency. Note that in fact under the design considered in Corollary 7.3.11 each contrast $\boldsymbol{s}_0'(\boldsymbol{r}^*)^{\delta}\boldsymbol{\tau}$ such that $\boldsymbol{s}_0' = [\boldsymbol{s}', -\boldsymbol{s}'\boldsymbol{N}]$, for any $v \times 1$ vector $\boldsymbol{s}$, will be estimated in the intra-block analysis with full efficiency.

Finally, one can also consider the merging method as suitable for constructing $(\rho_0;\rho_1;0)$-EB designs that are proper and nonequireplicate. Theorems 6.5.1 and 6.5.3 are useful for this purpose. Recall that Corollary 6.5.2 leads to Corollary 7.3.2 when $\boldsymbol{a}_i = [a_1, a_2, ..., a_s]'$ for all i, with $a_1 = a_2 = \cdots = a_s = n/s$.

In Corollary 6.5.2, if a semi-regular GD design is taken as a basic design, then the resulting design has one of the efficiency factors equal to one. The assumption $\lambda_1 = 0$ is only needed to produce a binary design. Without such assumption, Theorem 6.5.3 is valid but the resulting designs become nonbinary.

Example 7.3.10. Consider a semi-regular GD design, SR58 in Clatworthy (1973), with parameters $v = 20, b = 16$, $r = 4$, $k = 5$, $\lambda_1 = 0$, $\lambda_2 = 1$, $m = 5$, $n = 4$, having blocks (4, 7, 11, 13, 19), (1, 8, 11, 15, 17), (2, 5, 12, 15, 19), (1, 6, 9, 16, 19), (1, 5, 10, 13, 20), (4, 5, 9, 14, 17), (2, 8, 9, 13, 18), (3, 6, 12, 13, 17), (2, 7, 10, 16, 17), (2, 6, 11, 14, 20), (4, 6, 10, 15, 18), (3, 8, 10, 14, 19), (1, 7, 12, 14, 18), (3, 5, 11, 16, 18), (3, 7, 9, 15, 20), (4, 8, 12, 16, 20), where the five treatment groups of the association scheme are $\{1,2,3,4\}$, $\{5,6,7,8\}$, $\{9,10,11,12\}$, $\{13,14,15,16\}$, $\{17,18,19,20\}$. Two kinds of merging, following Corollary 6.5.2, will now be considered.

(1) The first design, $\mathcal{D}_1$, is obtained by merging treatment numbers $\{11,12\}$, $\{15,16\}$ and $\{19,20\}$, i.e., by forming subgroups of sizes $\boldsymbol{a}_1 = [1,1,1,1]' = \boldsymbol{a}_2$, $\boldsymbol{a}_3 = [1,1,2]' = \boldsymbol{a}_4 = \boldsymbol{a}_5$, respectively. Then the design $\mathcal{D}_1$, through Corollary 6.5.2, is a (4; 12; 0)-EB design with parameters $v^* = 17$, $b^* = 16$, $\boldsymbol{r}^* = 4[\boldsymbol{a}_1', \boldsymbol{a}_2', \boldsymbol{a}_3', \boldsymbol{a}_4', \boldsymbol{a}_5']'$ (of elements 4 or 8), $k^* = 5$, $\varepsilon_0^* = 1$, $\varepsilon_1^* = 4/5$, $\rho_0^* = 4$, $\rho_1^* = 12$, $\boldsymbol{L}_0^* = 4^{-1}\mathrm{diag}[\mathbf{1}_4\mathbf{1}_4' : \mathbf{1}_4\mathbf{1}_4' : \mathbf{1}_3[1,1,2] : \mathbf{1}_3[1,1,2] : \mathbf{1}_3[1,1,2]] - (80)^{-1}\mathbf{1}_{17}(\boldsymbol{r}^*)'$, $\boldsymbol{L}_1^* = \boldsymbol{I}_{17} - 4^{-1}\mathrm{diag}[\mathbf{1}_4\mathbf{1}_4' : \mathbf{1}_4\mathbf{1}_4' : \mathbf{1}_3[1,1,2] : \mathbf{1}_3[1,1,2] : \mathbf{1}_3[1,1,2]]$. The average efficiency factor of $\mathcal{D}_1$ is $16/19 = 0.8421$. Note that in the original semi-regular GD design it is $76/91 = 0.8352$. Furthermore, the matrix $\boldsymbol{\Omega}$ is given here by

$$\boldsymbol{\Omega}^* = \frac{1}{64}\mathrm{diag}\Big[\boldsymbol{I}_2 \otimes (20\boldsymbol{I}_4 - \mathbf{1}_4\mathbf{1}_4') : \boldsymbol{I}_3 \otimes (\mathrm{diag}[20,20,10] - \mathbf{1}_3\mathbf{1}_3')\Big],$$

which shows, for example, that there are five different variances for elementary contrasts estimated in the intra-block analysis. The incidence matrix of $\mathcal{D}_1$ is given by

$$\boldsymbol{N}^* = \begin{bmatrix}
0 & 1 & 0 & 1 & 1 & 0 & 0 & 0 & 0 & 0 & 0 & 0 & 1 & 0 & 0 & 0 \\
0 & 0 & 1 & 0 & 0 & 0 & 1 & 0 & 1 & 1 & 0 & 0 & 0 & 0 & 0 & 0 \\
0 & 0 & 0 & 0 & 0 & 0 & 0 & 1 & 0 & 0 & 0 & 1 & 0 & 1 & 1 & 0 \\
1 & 0 & 0 & 0 & 0 & 1 & 0 & 0 & 0 & 0 & 1 & 0 & 0 & 0 & 0 & 1 \\
0 & 0 & 1 & 0 & 1 & 1 & 0 & 0 & 0 & 0 & 0 & 0 & 0 & 1 & 0 & 0 \\
0 & 0 & 0 & 1 & 0 & 0 & 0 & 1 & 0 & 1 & 1 & 0 & 0 & 0 & 0 & 0 \\
1 & 0 & 0 & 0 & 0 & 0 & 0 & 0 & 1 & 0 & 0 & 0 & 1 & 0 & 1 & 0 \\
0 & 1 & 0 & 0 & 0 & 0 & 1 & 0 & 0 & 0 & 0 & 1 & 0 & 0 & 0 & 1 \\
0 & 0 & 0 & 1 & 0 & 1 & 1 & 0 & 0 & 0 & 0 & 0 & 0 & 0 & 1 & 0 \\
0 & 0 & 0 & 0 & 1 & 0 & 0 & 0 & 1 & 0 & 1 & 1 & 0 & 0 & 0 & 0 \\
1 & 1 & 1 & 0 & 0 & 0 & 0 & 1 & 0 & 1 & 0 & 0 & 1 & 1 & 0 & 1 \\
1 & 0 & 0 & 0 & 1 & 0 & 1 & 1 & 0 & 0 & 0 & 0 & 0 & 0 & 0 & 0 \\
0 & 0 & 0 & 0 & 0 & 1 & 0 & 0 & 0 & 1 & 0 & 1 & 1 & 0 & 0 & 0 \\
0 & 1 & 1 & 1 & 0 & 0 & 0 & 0 & 1 & 0 & 1 & 0 & 0 & 1 & 1 & 1 \\
0 & 1 & 0 & 0 & 0 & 1 & 0 & 1 & 1 & 0 & 0 & 0 & 0 & 0 & 0 & 0 \\
0 & 0 & 0 & 0 & 0 & 0 & 1 & 0 & 0 & 0 & 1 & 0 & 1 & 1 & 0 & 0 \\
1 & 0 & 1 & 1 & 1 & 0 & 0 & 0 & 0 & 1 & 0 & 1 & 0 & 0 & 1 & 1
\end{bmatrix}.$$

(2) The second design, $\mathcal{D}_2$, whose incidence matrix is given by

$$\boldsymbol{N}^* = \begin{bmatrix} 0&1&0&1&1&0&0&0&0&0&0&0&1&0&0&0 \\ 0&0&1&0&0&0&1&0&1&1&0&0&0&0&0&0 \\ 1&0&0&0&0&1&0&1&0&0&1&1&0&1&1&1 \\ 0&0&1&0&1&1&0&0&0&0&0&0&0&1&0&0 \\ 0&0&0&1&0&0&0&1&0&1&1&0&0&0&0&0 \\ 1&1&0&0&0&0&1&0&1&0&0&1&1&0&1&1 \\ 0&0&0&1&0&1&1&0&0&0&0&0&0&0&1&0 \\ 0&0&0&0&1&0&0&0&1&0&1&1&0&0&0&0 \\ 1&1&1&0&0&0&0&1&0&1&0&0&1&1&0&1 \\ 1&0&0&0&1&0&1&1&0&0&0&0&0&0&0&0 \\ 0&0&0&0&0&1&0&0&0&1&0&1&1&0&0&0 \\ 0&1&1&1&0&0&0&0&1&0&1&0&0&1&1&1 \\ 0&1&0&0&0&1&0&1&1&0&0&0&0&0&0&0 \\ 0&0&0&0&0&0&1&0&0&0&1&0&1&1&0&0 \\ 1&0&1&1&1&0&0&0&0&1&0&1&0&0&1&1 \end{bmatrix},$$

is obtained by merging treatment numbers $\{3,4\}$, $\{7,8\}$, $\{11,12\}$, $\{15,16\}$ and $\{19,20\}$, respectively, i.e., by using $\boldsymbol{a}_1 = \boldsymbol{a}_2 = \boldsymbol{a}_3 = \boldsymbol{a}_4 = \boldsymbol{a}_5 = [1,1,2]'$. Then the design $\mathcal{D}_2$, through Corollary 6.5.2 with $s = 3$, is a $(4;10;0)$-EB design with parameters $v^* = 15$, $b^* = 16$, $\boldsymbol{r}^* = 4\boldsymbol{1}_5 \otimes [1,1,2]'$, $k^* = 5$, $\varepsilon_0^* = 1$, $\varepsilon_1^* = 4/5$, $\rho_0^* = 4$, $\rho_1^* = 10$, $\boldsymbol{L}_0^* = (\boldsymbol{I}_5 - 5^{-1}\boldsymbol{1}_5\boldsymbol{1}_5') \otimes 4^{-1}\boldsymbol{1}_3[1,1,2]$, $\boldsymbol{L}_1^* = \boldsymbol{I}_5 \otimes (\boldsymbol{I}_3 - 4^{-1}\boldsymbol{1}_3[1,1,2])$. The average efficiency factor of $\mathcal{D}_2$ is $28/33 = 0.8485$. Furthermore, the matrix $\boldsymbol{\Omega}$ is given by

$$\boldsymbol{\Omega}^* = \frac{1}{64}\boldsymbol{I}_5 \otimes (\mathrm{diag}[20,20,10] - \boldsymbol{1}_3\boldsymbol{1}_3'),$$

which provides five different variances, $5\sigma_1^2/8$, $15\sigma_1^2/32$, $19\sigma_1^2/32$, $7\sigma_1^2/16$, $9\sigma_1^2/32$, for elementary contrasts estimated in the intra-block analysis. In this example, the average efficiency factors of the design increase with the number of treatments merged, though the efficiency factors ε_1 and ε_2, and the total number of experimental units remain unchanged with the merging of treatments.

Incidentally, on account of Theorem 6.1.1, the dual design of any of the nonproper and equireplicate $(\rho_0;\rho_1;0)$-EB designs, which will be discussed in Section 7.3.3, may produce proper and nonequireplicate designs with two efficiency factors belonging to the present section.

7.3.3 Nonproper designs that are equireplicate

PBIB designs are equireplicate and proper. In some practical situations, however, nonproper block designs may be required (see also Section 8.2.2), though they are less recommendable from the statistical point of view. Here, binary block designs with block sizes, $k_1, k_2, \ldots, k_b$ (≥ 2), that are not all equal, will be considered.

Continuing to be interested in designs that allow to estimate some basic contrasts with full efficiency, attention will now be focused on methods of constructing nonproper equireplicate $(\rho_0;\rho_1;0)$-EB designs with $\rho_0 \geq 1$. Not so many methods of constructing such designs are available in the literature. Some of them require large replication numbers. The problem of constructing a $(\rho_0;\rho_1;0)$-EB design with $r_1 = r_2 = \cdots = r_v = r$ and $k_1, k_2, ..., k_b$ not all equal is related to finding some incidence matrix $\boldsymbol{N}$ such for which the matrix

$$\frac{1}{r}\boldsymbol{C}_1 = \boldsymbol{I}_v - \frac{1}{r}\boldsymbol{N}\boldsymbol{k}^{-\delta}\boldsymbol{N}'$$

can be written as $r^{-1}\boldsymbol{C}_1 = \boldsymbol{L}_0 + \varepsilon_1\boldsymbol{L}_1$, following from the spectral decomposition of the matrix $r^{-1}\boldsymbol{N}\boldsymbol{k}^{-\delta}\boldsymbol{N}' - v^{-1}\mathbf{1}_v\mathbf{1}_v'$ $(= \boldsymbol{M}_0)$, as shown in Section 4.4 and recalled in (7.1.2). This observation may be used for checking whether any suggested construction satisfies conditions of the required design.

The statistical implication of nonproper designs are discussed in Section 3. Under the inter-block submodel (3.2.4), Theorem 3.2.2 gives a necessary and sufficient condition for the existence of the inter-block BLUE of a given contrast. An equivalent condition for the existence of the inter-block BLUE for any conformable contrast is shown in Corollary 3.2.2. If the design is proper, then this condition is satisfied automatically. Thus, among nonproper designs one would recommend designs which satisfy Corollary 3.2.2.

As shown in Section 6.4, a union (juxtaposition) of some designs with a common number of treatments may yield a new design.

From the results in Corollary 6.4.1, the following two corollaries can be obtained.

Corollary 7.3.12. *The existence of two semi-regular GD designs with parameters* $v = mn$, $b^{(h)}$, $r^{(h)}$, $k^{(h)}$, $\lambda_1^{(h)}$, $\lambda_2^{(h)}$, $h = 1, 2$, *implies the existence of a* $(\rho_0^*;\rho_1^*;0)$*-EB design with parameters* $v^* = mn$, $b^* = b^{(1)} + b^{(2)}$, $r^* = r^{(1)} + r^{(2)}$, $\boldsymbol{k}^* = [k^{(1)}\mathbf{1}'_{b^{(1)}}, k^{(2)}\mathbf{1}'_{b^{(2)}}]'$, $\varepsilon_0^* = 1$, $\varepsilon_1^* = 1 - [(r^{(1)} - \lambda_1^{(1)})/k^{(1)} + (r^{(2)} - \lambda_1^{(2)})/k^{(2)}]/(r^{(1)} + r^{(2)})$, $\rho_0^* = m - 1$, $\rho_1^* = m(n-1)$, $\boldsymbol{L}_0^* = (\boldsymbol{I}_m - m^{-1}\mathbf{1}_m\mathbf{1}_m') \otimes n^{-1}\mathbf{1}_n\mathbf{1}_n'$, $\boldsymbol{L}_1^* = \boldsymbol{I}_m \otimes (\boldsymbol{I}_n - n^{-1}\mathbf{1}_n\mathbf{1}_n')$.

Note that designs obtainable by Corollary 7.3.12 satisfy Corollary 3.2.2 on account of Corollary 3.2.3, as the subdesigns used in the union are proper and equireplicate each.

Example 7.3.11. Consider two semi-regular GD designs, based on two groups $\{1,2,3\}, \{4,5,6\}$ of three treatments, SR6 and SR35 in Clatworthy (1973), with the following parameters

$v = 6, b = 9, r = 3, k = 2, \lambda_1 = 0, \lambda_2 = 1; m = 2, n = 3,$

$v = 6, b = 9, r = 6, k = 4, \lambda_1 = 3, \lambda_2 = 4; m = 2, n = 3,$

whose blocks are given by

[(1,4), (2,5), (3,6)], [(1,6), (2,4), (3,5)], [(1,5), (2,6), (3,4)]

and

[(1,2,4,5), (1,3,4,6), (2,3,5,6)], [(1,2,5,6), (1,3,4,5), (2,3,4,6)],
[(1,2,4,6), (1,3,5,6), (2,3,4,5)],

respectively. The first design is resolvable, whereas the second is 2-resolvable. Juxtaposition of these blocks yields a $(1; 4; 0)$-EB design with parameters $v^* = 6$, $b^* = 18$, $r^* = 9$, $k_j^* = 2$ or 4, $\varepsilon_0^* = 1$, $\varepsilon_1^* = 3/4$, $\rho_0^* = 1$, $\rho_1^* = 4$, $\boldsymbol{L}_0^* = (\boldsymbol{I}_2 - 2^{-1}\mathbf{1}_2\mathbf{1}_2') \otimes 3^{-1}\mathbf{1}_3\mathbf{1}_3'$, $\boldsymbol{L}_1^* = \boldsymbol{I}_2 \otimes (\boldsymbol{I}_3 - 3^{-1}\mathbf{1}_3\mathbf{1}_3')$, and with the incidence matrix

$$\boldsymbol{N}^* = \begin{bmatrix} 1&0&0&1&0&0&1&0&0&1&1&0&1&1&0&1&1&0 \\ 0&1&0&0&1&0&0&1&0&1&0&0&1&0&1&1&0&1 \\ 0&0&1&0&0&1&0&0&1&0&1&1&0&1&1&0&1&1 \\ 1&0&0&0&1&0&0&0&1&1&1&0&0&1&1&1&0&1 \\ 0&1&0&0&0&1&1&0&0&1&0&1&1&1&0&0&1&1 \\ 0&0&1&1&0&0&0&1&0&0&1&1&1&0&1&1&1&0 \end{bmatrix},$$

which satisfies Corollary 3.2.2. Here, the average efficiency factor of the design is $15/19 = 0.7895$. The variances of the intra-block BLUEs of basic contrasts are σ_1^2 or $4\sigma_1^2/3$, depending on whether they correspond to $\boldsymbol{L}_0^*$ or $\boldsymbol{L}_1^*$, respectively [see Section 7.1(3)]. Furthermore, for the evaluation of the variance of the intra-block BLUE of any contrast, the matrix $\boldsymbol{\Omega}$ is

$$\boldsymbol{\Omega}^* = \frac{1}{81}\boldsymbol{I}_2 \otimes (12\boldsymbol{I}_3 - \mathbf{1}_3\mathbf{1}_3').$$

Hence, an elementary contrast $\boldsymbol{c}_1'\boldsymbol{\tau}$ involving treatments from the same group (i.e., $\{1,2,3\}$ or $\{4,5,6\}$) obtains a variance $8\sigma_1^2/27 = 0.296296\sigma_1^2$, whereas an elementary contrast $\boldsymbol{c}_2'\boldsymbol{\tau}$ involving treatments from different groups has a variance $22\sigma_1^2/81 = 0.271605\sigma_1^2$, when estimated in the intra-block analysis. Here, for example, one can take $\boldsymbol{c}_1 = [1,-1,0,0,0,0]'$ and $\boldsymbol{c}_2 = [1,0,0,-1,0,0]'$, respectively.

The designs obtained in Example 7.3.11 may be interesting, in particular, for a factorial experiment with two factors, A and B, applied at 2 and 3 levels, respectively, if of main interest is the comparison between A factor main effects.

Corollary 7.3.13. *The existence of two singular GD designs with parameters* $v = mn$, $b^{(h)}$, $r^{(h)}$, $k^{(h)}$, $\lambda_1^{(h)}$, $\lambda_2^{(h)}$, $h = 1,2$, *implies the existence of a* $(\rho_0^*; \rho_1^*; 0)$*-EB design with parameters* $v^* = mn$, $b^* = b^{(1)} + b^{(2)}$, $r^* = r^{(1)} + r^{(2)}$, $\boldsymbol{k}^* = [k^{(1)}\mathbf{1}_{b^{(1)}}', k^{(2)}\mathbf{1}_{b^{(2)}}']'$, $\varepsilon_0^* = 1$, $\varepsilon_1^* = 1 - [(r^{(1)}k^{(1)} - v\lambda_2^{(1)})/k^{(1)} + (r^{(2)}k^{(2)} - v\lambda_2^{(2)})/k^{(2)}]/(r^{(1)} + r^{(2)})$, $\rho_0^* = m(n-1)$, $\rho_1^* = m-1$, $\boldsymbol{L}_0^* = \boldsymbol{I}_m \otimes (\boldsymbol{I}_n - n^{-1}\mathbf{1}_n\mathbf{1}_n')$, $\boldsymbol{L}_1^* = (\boldsymbol{I}_m - m^{-1}\mathbf{1}_m\mathbf{1}_m') \otimes n^{-1}\mathbf{1}_n\mathbf{1}_n'$.

Note again that Corollary 7.3.13 provides designs that, through Corollary 3.2.3, satisfy Corollary 3.2.2 to obtain also the inter-block BLUE of any contrast partially confounded with blocks.

Example 7.3.12. Take two singular GD designs, based on four groups $\{1,2\}$,

$\{3, 4\}, \{5, 6\}, \{7, 8\}$ of two treatments, S6 and S18 in Clatworthy (1973), with respective parameters

$v = 8, b = 6, r = 3, k = 4, \lambda_1 = 3, \lambda_2 = 1; m = 4, n = 2,$
$v = 8, b = 4, r = 3, k = 6, \lambda_1 = 3, \lambda_2 = 2; m = 4, n = 2,$

whose blocks are given by

$$[(1,2,3,4), (5,6,7,8)], [(1,2,5,6), (3,4,7,8)], [(1,2,7,8), (3,4,5,6)]$$

and

$$(1,2,3,4,5,6), (3,4,5,6,7,8), (1,2,5,6,7,8), (1,2,3,4,7,8),$$

respectively. The first design is affine resolvable. Corollary 7.3.13 yields a 3-resolvable [more precisely, a (1,1,1,3)-resolvable] $(4; 3; 0)$-EB design with parameters $v^* = 8$, $b^* = 10$, $r^* = 6$, $k_j^* = 4$ or 6, $\varepsilon_0^* = 1$, $\varepsilon_1^* = 7/9$, $\rho_0^* = 4$, $\rho_1^* = 3$, $\boldsymbol{L}_0^* = \boldsymbol{I}_4 \otimes (\boldsymbol{I}_2 - 2^{-1}\mathbf{1}_2\mathbf{1}_2')$, $\boldsymbol{L}_1^* = (\boldsymbol{I}_4 - 4^{-1}\mathbf{1}_4\mathbf{1}_4') \otimes 2^{-1}\mathbf{1}_2\mathbf{1}_2'$, and with the incidence matrix $\boldsymbol{N}^*$ below, which satisfies Corollary 3.2.3 and, hence, Corollary 3.2.2. Here, the average efficiency factor of the design is $49/55 = 0.8909$. The variances of the intra-block BLUEs of basic contrasts are σ_1^2 or $9\sigma_1^2/7$, depending on whether they correspond to $\boldsymbol{L}_0^*$ or $\boldsymbol{L}_1^*$. The matrix $\boldsymbol{\Omega}$ for the resulting design is

$$\boldsymbol{\Omega}^* = \frac{1}{168}\left[\boldsymbol{I}_4 \otimes (28\boldsymbol{I}_2 + 31\mathbf{1}_2\mathbf{1}_2') + (\boldsymbol{I}_4 - \mathbf{1}_4\mathbf{1}_4') \otimes \mathbf{1}_2\mathbf{1}_2'\right].$$

Thus an elementary contrast involving treatments from the same group (given as $\{1, 2\}, \{3, 4\}, \{5.6\}$ or $\{7, 8\}$) obtains the variance $\sigma_1^2/3 = 0.333333\sigma_1^2$, whereas that involving treatments from different groups obtains the variance $8\sigma_1^2/21 = 0.380952\sigma_1^2$, when estimated in the intra-block analysis. Now,

$$\boldsymbol{N}^* = \left[\begin{array}{cccccc|cccc} 1&0&1&0&1&0&1&0&1&1\\ 1&0&1&0&1&0&1&0&1&1\\ 1&0&0&1&0&1&1&1&0&1\\ 1&0&0&1&0&1&1&1&0&1\\ 0&1&1&0&0&1&1&1&1&0\\ 0&1&1&0&0&1&1&1&1&0\\ 0&1&0&1&1&0&0&1&1&1\\ 0&1&0&1&1&0&0&1&1&1 \end{array}\right].$$

As a special case of Corollary 6.4.1, a useful pattern is considered in the following corollary, where as one of the singular GD designs that having the incidence matrix $\boldsymbol{I}_m \otimes \mathbf{1}_n\mathbf{1}_s'$ is used.

Corollary 7.3.14. *If $\boldsymbol{N}$ is the $v \times b$ incidence matrix of a singular GD design with parameters $v = mn$ (m groups of n treatments each), b, r, k, λ_1, λ_2, $\varepsilon_0 = 1$, $\varepsilon_1 = \lambda_2 v/(rk)$, $\rho_0 = m(n-1)$, $\rho_1 = m - 1$, $\boldsymbol{L}_0 = \boldsymbol{I}_m \otimes (\boldsymbol{I}_n - n^{-1}\mathbf{1}_n\mathbf{1}_n')$, $\boldsymbol{L}_1 = (\boldsymbol{I}_m - m^{-1}\mathbf{1}_m\mathbf{1}_m') \otimes n^{-1}\mathbf{1}_n\mathbf{1}_n'$, then $[\boldsymbol{N} : \boldsymbol{I}_m \otimes \mathbf{1}_n\mathbf{1}_s']$ is the incidence matrix of a $(\rho_0^*; \rho_1^*; 0)$-EB design with parameters $v^* = mn$, $b^* = b + ms$, $r^* = r + s$,*

$\boldsymbol{k}^* = [k\mathbf{1}'_b, n\mathbf{1}'_{ms}]'$, $\varepsilon_0^* = 1$, $\varepsilon_1^* = 1 - [k(r+s) - v\lambda_2]/[k(r+s)]$, $\rho_0^* = m(n-1)$, $\rho_1^* = m-1$, $\boldsymbol{L}_0^* = \boldsymbol{I}_m \otimes (\boldsymbol{I}_n - n^{-1}\mathbf{1}_n\mathbf{1}'_n)$, $\boldsymbol{L}_1^* = (\boldsymbol{I}_m - m^{-1}\mathbf{1}_m\mathbf{1}'_m) \otimes n^{-1}\mathbf{1}_n\mathbf{1}'_n$.

Note that designs obtainable by Corollary 7.3.14 and the following Corollaries 7.3.15 to 7.3.18 satisfy Corollary 3.2.2 on account of Corollary 3.2.3.

Since the existence of a singular GD design is equivalent to the existence of a BIB design (see Theorem 7.3.1), Corollary 7.3.14 shows that the existence of any BIB design implies the existence of a $(\rho_0; \rho_1; 0)$-EB design with $\rho_0 \geq 1$ and the properties considered here. Furthermore, since singular GD designs are also available in Clatworthy (1973) in abundance, such plans can be used to derive those designs.

In order to apply the idea of Theorem 6.4.2 to other 2-associate PBIB designs, one can utilize Corollary 6.2.1. Then the following result is obtained.

Corollary 7.3.15. *If $\boldsymbol{N}$ is the $v \times b$ incidence matrix of a 2-associate PBIB design with parameters v, b, r, k, λ_1, λ_2, ε_0, $\varepsilon_1 < 1$, ρ_0, ρ_1, $\boldsymbol{L}_0$, $\boldsymbol{L}_1$, then $[\boldsymbol{N} : \mathbf{1}_v\mathbf{1}'_b - \boldsymbol{N}]$ is the incidence matrix of a $(\rho_0^*; \rho_1^*; 0)$-EB design with parameters $v^* = v$, $b^* = 2b$, $r^* = b$, $\boldsymbol{k}^* = [k\mathbf{1}'_b, (v-k)\mathbf{1}'_b]'$, $\varepsilon_0^* = 1$, $\varepsilon_1^* = 1 - k(1-\varepsilon_1)/(v-k)$, $\rho_0^* = \rho_0$, $\rho_1^* = \rho_1$, $\boldsymbol{L}_0^* = \boldsymbol{L}_0$, $\boldsymbol{L}_1^* = \boldsymbol{L}_1$.*

This corollary has a wide coverage of application. Using PBIB designs discussed in this section, one can produce various kinds of $(\rho_0; \rho_1; 0)$-EB designs. For example, by Theorem 7.3.9 and Corollary 7.3.15, one can obtain the following corollary which utilizes a triangular association scheme [see Section 6.0.2(B)].

Corollary 7.3.16. *For $n \geq 4$, there exists a $(\rho_0; \rho_2; 0)$-EB design with parameters $v = n(n-1)/2$, $b = 2n$, $r = n$, $\boldsymbol{k} = [(n-1)\mathbf{1}'_n, [(n-1)(n-2)/2]\mathbf{1}'_n]'$, $\varepsilon_0 = 1$, $\varepsilon_1 = (n-2)/(n-1)$, $\rho_0 = n(n-3)/2$, $\rho_1 = n-1$, $\boldsymbol{L}_0 = [(n-1)(n-2)]^{-1}[(n-1)(n-4)\boldsymbol{I}_v - (n-1)\boldsymbol{A}_1 + 2\mathbf{1}_v\mathbf{1}'_v]$, $\boldsymbol{L}_1 = [n(n-2)]^{-1}(2n\boldsymbol{I}_v + n\boldsymbol{A}_1 - 4\mathbf{1}_v\mathbf{1}'_v)$, where $\boldsymbol{A}_1$ is the first association matrix of a triangular association scheme.*

The average efficiency factor of the design given in Corollary 7.3.16 is $(n+1)(n-2)^2/[n(n-2)(n-3)+2(n-1)^2]$.

More juxtaposition methods of constructing $(\rho_0; \rho_1; 0)$-EB designs can be given. In particular, since $\mathbf{1}_v\mathbf{1}'_v$ has two distinct eigenvalues v and 0 with respective multiplicities 1 and $v-1$, one can present the following result.

Corollary 7.3.17. *If $\boldsymbol{N}$ is the $v \times b$ incidence matrix of a 2-associate PBIB design with parameters v, b, r, k, λ_1, λ_2, $\varepsilon_0 = 1$, $\varepsilon_1 < 1$, ρ_0, ρ_1, $\boldsymbol{L}_0$, $\boldsymbol{L}_1$, then $[\boldsymbol{N} : \mathbf{1}_v\mathbf{1}'_p]$ for any integer p is the incidence matrix of a $(\rho_0^*; \rho_1^*; 0)$-EB design with parameters $v^* = v$, $b^* = b+p$, $r^* = r+p$, $\boldsymbol{k}^* = [k\mathbf{1}'_b, v\mathbf{1}'_p]'$, $\varepsilon_0^* = 1$, $\varepsilon_1^* = 1 - r(1-\varepsilon_1)/(r+p)$, $\rho_0^* = \rho_0$, $\rho_1^* = \rho_1$, $\boldsymbol{L}_0^* = \boldsymbol{L}_0$, $\boldsymbol{L}_1^* = \boldsymbol{L}_1$.*

This corollary, together with Theorem 7.3.12, yields the following observation [see Section 6.0.2(C)].

Corollary 7.3.18. *For any integers $s \geq 2$ and $p \geq 1$, there exists a $(\rho_0; \rho_1; 0)$-EB design with parameters $v = s^2$, $b = s(s-1)+p$, $r = 2(s-1)+p$, $\boldsymbol{k} =$*

$[2s\mathbf{1}'_{s(s-1)}, s^2\mathbf{1}'_p]'$, $\varepsilon_0 = 1$, $\varepsilon_1 = 1-(s-2)/[2(2s-2+p)]$, $\rho_0 = (s-1)^2$, $\rho_1 = 2(s-1)$, $\boldsymbol{L}_0 = (\boldsymbol{I}_s - s^{-1}\mathbf{1}_s\mathbf{1}'_s)\otimes(\boldsymbol{I}_s - s^{-1}\mathbf{1}_s\mathbf{1}'_s)$, $\boldsymbol{L}_1 = s^{-1}\mathbf{1}_s\mathbf{1}'_s\otimes(\boldsymbol{I}_s - s^{-1}\mathbf{1}_s\mathbf{1}'_s)$ $+ (\boldsymbol{I}_s - s^{-1}\mathbf{1}_s\mathbf{1}'_s)\otimes s^{-1}\mathbf{1}_s\mathbf{1}'_s$.

Note that ε_1 increases to 1 with p increasing. If the design from Corollary 7.3.18 is used, then the average efficiency factor of the design is $(s+1)(2s+p-2)/[2s^2+(p-1)s+p]$.

7.3.4 Nonproper designs that are nonequireplicate

For nonproper and nonequireplicate $(\rho_0;\rho_1;0)$-EB designs no systematic method of construction is available in the literature. Usually such designs are less recommendable than those considered in Section 7.3.2, and are not often used in practice. But, there may be some situations where one has to use such designs because of certain restrictions. In this section, Corollary 6.1.2 and then Theorem 6.4.1 or Corollary 6.1.3 will be utilized to construct various kinds of designs of this type. The main role in the construction of designs in this section will be played by Theorem 6.6.13, which provides designs that are to be dualized.

Corollary 6.1.2 (concerning dual designs) can be used when $b > v$ in the original $(0;v-1;0)$-EB design. This will be discussed in Section 8.2.2. Recall that in a $(0;v-1;0)$-EB design $\boldsymbol{L}_1 = \boldsymbol{I}_v - n^{-1}\mathbf{1}_v\boldsymbol{r}'$ (see Section 4.4.2).

Corollary 6.1.2 and Theorem 6.6.13 yield the following result.

Theorem 7.3.15. *If $\boldsymbol{N}_h$, $h=1,2$, are the $v\times b_h$ incidence matrices, the first of a BIB design, with parameters v, b_1, r_1, k_1, λ_1, and the second of a $(0;v-1;0)$-EB design, with parameters v, b_2, r_2, $\boldsymbol{k}$, ε $(\equiv \varepsilon_1)$, respectively, then for positive integers x and y such that*

$$\frac{x}{y} = \frac{(k_1+1)(vr_2-b_2)-k_1r_2(v-1)}{(v-1)(r_1-\lambda_1)}$$

there exists a $(\rho_0^*;\rho_1^*;0)$-EB design $\boldsymbol{N}^*$ with parameters $v^* = b_1x+b_2y$, $b^* = v+1$, $\boldsymbol{r}^* = [(k_1+1)\mathbf{1}'_{b_1x}, \mathbf{1}'_y\otimes\boldsymbol{k}']'$, $\boldsymbol{k}^* = [b_1x, (r_1x+r_2y)\mathbf{1}'_v]'$, $\varepsilon_0^* = 1$, $\varepsilon_1^* = 1-r_2y/[(k_1+1)(r_1x+r_2y)]$, $\rho_0^* = b_1x+b_2y-v-1$, $\rho_1^* = v$, $\boldsymbol{L}_0^* = \boldsymbol{I}_{v^*} - [(\boldsymbol{k}^*)'\mathbf{1}_{v+1}]^{-1}\mathbf{1}_{v^*}(\boldsymbol{r}^*)' - \boldsymbol{L}_1^*$, $\boldsymbol{L}_1^* = (1-\varepsilon_1^*)^{-1}(\boldsymbol{r}^*)^{-\delta}\boldsymbol{N}^*\{\boldsymbol{I}_{v+1} - [(\boldsymbol{k}^*)'\mathbf{1}_{v+1}]^{-1}\mathbf{1}_{v+1}(\boldsymbol{k}^*)'\}(\boldsymbol{k}^*)^{-\delta}(\boldsymbol{N}^*)'$, whose incidence matrix is given by

$$\boldsymbol{N}^* = \begin{bmatrix} \mathbf{1}'_{xb_1} & \mathbf{0}' \\ \mathbf{1}'_x\otimes\boldsymbol{N}_1 & \mathbf{1}'_y\otimes\boldsymbol{N}_2 \end{bmatrix}'. \tag{7.3.6}$$

Example 7.3.13. For an unreduced BIB design with parameters $v=5, b=10$, $r=4, k=2$, $\lambda=1$, and a $(0;4;0)$-EB design with parameters $v=5, b=15$,

$r = 9$, $\boldsymbol{k} = [4\mathbf{1}_6', 2\mathbf{1}_6', 3\mathbf{1}_3']'$, $\varepsilon = 5/6$, whose incidence matrix is given by

$$\boldsymbol{N}_2 = \begin{bmatrix} 0&0&1&1&1&1&1&1&1&1&1&0&0&0&0 \\ 1&1&0&0&1&1&1&1&1&0&0&1&1&0&0 \\ 1&1&1&1&0&0&1&1&1&0&0&0&0&1&1 \\ 1&1&1&1&1&1&0&0&0&1&0&1&0&1&0 \\ 1&1&1&1&1&1&0&0&0&0&1&0&1&0&1 \end{bmatrix},$$

Theorem 7.3.15 produces a $(\rho_0^*;\rho_1^*;0)$-EB design with parameters $v^* = 10x+15y$, $b^* = 6$, $\boldsymbol{r}^* = [3\mathbf{1}_{10x}', \mathbf{1}_y' \otimes \boldsymbol{k}']'$, $\boldsymbol{k}^* = [10x, (4x+9y)\mathbf{1}_5']'$, $\varepsilon_0^* = 1$, $\varepsilon_1^* = 4/5$, $\rho_0^* = 10x+15y-6$, $\rho_1^* = 5$, $\boldsymbol{L}_0^* = \boldsymbol{I}_{10x+15y} - [5(6x+9y)]^{-1}\mathbf{1}_{10x+15y}(\boldsymbol{r}^*)' - \boldsymbol{L}_1^*$, $\boldsymbol{L}_1^* = 5(\boldsymbol{r}^*)^{-\delta}\boldsymbol{N}^*\{\boldsymbol{I}_6 - [5(6x+9y)]^{-1}\mathbf{1}_6(\boldsymbol{k}^*)'\}(\boldsymbol{k}^*)^{-\delta}(\boldsymbol{N}^*)'$ for positive integers x and y such that $x/y = 3/2$.

Theorem 7.3.15 with $\boldsymbol{N}_1 = \boldsymbol{I}_v$ yields the following result.

Corollary 7.3.19. *If there exists an equireplicate $(0; v-1; 0)$-EB design with parameters v, b, r, $\boldsymbol{k}$, ε, and with the incidence matrix $\boldsymbol{N}$, then for positive integers x and y such that $x/y = [(v+1)r - 2b]/(v-1)$ there exists a $(\rho_0^*;\rho_1^*;0)$-EB design $\boldsymbol{N}^*$ with parameters $v^* = vx + by$, $b^* = v+1$, $\boldsymbol{r}^* = [2\mathbf{1}_{xv}', \mathbf{1}_y' \otimes \boldsymbol{k}']'$, $\boldsymbol{k}^* = [vx, (x+yr)\mathbf{1}_v']'$, $\varepsilon_0^* = 1$, $\varepsilon_1^* = 1 - r(v-1)/[4(vr-b)]$, $\rho_0^* = vx+by-v-1$, $\rho_1^* = v$, $\boldsymbol{L}_0^* = \boldsymbol{I}_{v^*} - [v(2x+yr)]^{-1}\mathbf{1}_{v^*}(\boldsymbol{r}^*)' - \boldsymbol{L}_1^*$, $\boldsymbol{L}_1^* = (1-\varepsilon_1^*)^{-1}(\boldsymbol{r}^*)^{-\delta}\boldsymbol{N}^*\{\boldsymbol{I}_{v+1} - [v(2x+yr)]^{-1}\mathbf{1}_{v+1}(\boldsymbol{k}^*)'\}(\boldsymbol{k}^*)^{-\delta}(\boldsymbol{N}^*)'$, whose incidence matrix is given by*

$$\boldsymbol{N}^* = \begin{bmatrix} \mathbf{1}_{xv}' & \mathbf{0}' \\ \mathbf{1}_x' \otimes \boldsymbol{I}_v & \mathbf{1}_y' \otimes \boldsymbol{N} \end{bmatrix}'.$$

Example 7.3.14. For a $(0; v-1; 0)$-EB design with parameters $v = 6, b = 18$, $r = 8, \boldsymbol{k} = [4\mathbf{1}_6', 2\mathbf{1}_{12}']'$, $\varepsilon = 3/4$, whose incidence matrix is given by

$$\boldsymbol{N} = \begin{bmatrix} 0&0&1&1&1&1&1&1&0&0&0&0&1&1&0&0&0&0 \\ 1&1&0&0&1&1&1&0&1&0&0&0&0&0&1&1&0&0 \\ 1&1&1&1&0&0&0&1&1&0&0&0&0&0&0&0&1&1 \\ 1&1&1&1&0&0&0&0&0&1&1&0&1&0&1&0&0&0 \\ 1&1&0&0&1&1&0&0&0&1&0&1&0&1&0&0&1&0 \\ 0&0&1&1&1&1&0&0&0&0&1&1&0&0&0&1&0&1 \end{bmatrix},$$

Corollary 7.3.19 produces a $(\rho_0^*;\rho_1^*;0)$-EB design with parameters $v^* = 6x + 18y$, $b^* = 7$, $\boldsymbol{r}^* = [2\mathbf{1}_{xv}', \mathbf{1}_y' \otimes \boldsymbol{k}']'$, $\boldsymbol{k}^* = [6x, (x+8y)\mathbf{1}_6']'$, $\varepsilon_0^* = 1$, $\varepsilon_1^* = 2/3$, $\rho_0^* = 6x+18y-7$, $\rho_1^* = 6$, $\boldsymbol{L}_0^* = \boldsymbol{I}_{v^*} - [12(x+4y)]^{-1}\mathbf{1}_{v^*}(\boldsymbol{r}^*)' - \boldsymbol{L}_1^*$, $\boldsymbol{L}_1^* = 3(\boldsymbol{r}^*)^{-\delta}\boldsymbol{N}^*\{\boldsymbol{I}_7 - [12(x+4y)]^{-1}\mathbf{1}_7(\boldsymbol{k}^*)'\}(\boldsymbol{k}^*)^{-\delta}(\boldsymbol{N}^*)'$, for positive integers x and y such that $x = 4y$. For example, for $x = 4$ and $y = 1$, one can get a $(35; 6; 0)$-EB design with parameters $v^* = 42$, $b^* = 7$, $\boldsymbol{r}^* = [2\mathbf{1}_{24}', 4\mathbf{1}_6', 2\mathbf{1}_{12}']'$, $\boldsymbol{k}^* = [24, 12\mathbf{1}_6']'$, $\varepsilon_0^* = 1$, $\varepsilon_1^* = 2/3$, $\rho_0^* = 35$, $\rho_1^* = 6$, $\boldsymbol{L}_0^* = \boldsymbol{I}_{42} - (96)^{-1}\mathbf{1}_{42}(\boldsymbol{r}^*)' - \boldsymbol{L}_1^*$, $\boldsymbol{L}_1^* =$

$3(\boldsymbol{r}^*)^{-\delta}\boldsymbol{N}^*\{\boldsymbol{I}_7 - (96)^{-1}\mathbf{1}_7(\boldsymbol{k}^*)'\}(\boldsymbol{k}^*)^{-\delta}(\boldsymbol{N}^*)'$, whose incidence matrix is of the form

$$\boldsymbol{N}^* = \begin{bmatrix} \mathbf{1}_{24} & \mathbf{1}_4 \otimes \boldsymbol{I}_6 \\ \mathbf{0} & \boldsymbol{N}' \end{bmatrix}.$$

The matrix $\boldsymbol{\Omega}$ for the resulting design is

$$\boldsymbol{\Omega}^* = \left(\frac{1}{2}\boldsymbol{L}_1^* + \boldsymbol{I}_{42}\right)\mathrm{diag}\left[\frac{1}{2}\boldsymbol{I}_{24} : \frac{1}{4}\boldsymbol{I}_6 : \frac{1}{2}\boldsymbol{I}_{12}\right].$$

Theorem 7.3.15 with $\boldsymbol{N}_1 = \mathbf{1}_v, x = m$ and $y = 1$ yields the following result.

Corollary 7.3.20. *If there exists an equireplicate* $(0; v-1; 0)$*-EB design with parameters* v, b, r, $\boldsymbol{k}$, ε, *and with the incidence matrix* $\boldsymbol{N}$ *such that* $b = 2vr/(v+1)$, *then for a positive integer* m *there exists a* $(\rho_0^*; \rho_1^*; 0)$*-EB design* $\boldsymbol{N}^*$ *with parameters* $v^* = m + b$, $b^* = v + 1$, $\boldsymbol{r}^* = [(v+1)\mathbf{1}'_m, \boldsymbol{k}']'$, $\boldsymbol{k}^* = [m, (m+r)\mathbf{1}'_v]'$, $\varepsilon_0^* = 1$, $\varepsilon_1^* = 1 - r/[(v+1)(m+r)]$, $\rho_0^* = m + b - v - 1$, $\rho_1^* = v$, $\boldsymbol{L}_0^* = \boldsymbol{I}_{v^*} - [m(v+1)+vr]^{-1}\mathbf{1}_{v^*}(\boldsymbol{r}^*)' - \boldsymbol{L}_1^*$, $\boldsymbol{L}_1^* = (1-\varepsilon_1^*)^{-1}(\boldsymbol{r}^*)^{-\delta}\boldsymbol{N}^*\{\boldsymbol{I}_{v+1} - [m(v+1)+vr]^{-1}\mathbf{1}_{v+1}(\boldsymbol{k}^*)'\}(\boldsymbol{k}^*)^{-\delta}(\boldsymbol{N}^*)'$, *whose incidence matrix is given by*

$$\boldsymbol{N}^* = \begin{bmatrix} \mathbf{1}'_m & \mathbf{0}' \\ \mathbf{1}_v\mathbf{1}'_m & \boldsymbol{N} \end{bmatrix}'. \tag{7.3.7}$$

Example 7.3.15. For a $(0;4;0)$-EB design with parameters $v = 5, b = 15$, $r = 9, \boldsymbol{k} = [4\mathbf{1}'_6, 3\mathbf{1}'_3, 2\mathbf{1}'_6]'$, $\varepsilon = 5/6$ (see Example 7.3.13), Corollary 7.3.20 produces a $(\rho_0^*; \rho_1^*; 0)$-EB design with parameters $v^* = m + 15$, $b^* = 6$, $\boldsymbol{r}^* = [6\mathbf{1}'_m, 4\mathbf{1}'_6, 3\mathbf{1}'_3, 2\mathbf{1}'_6]'$, $\boldsymbol{k}^* = [m, (m+9)\mathbf{1}'_5]'$, $\varepsilon_0^* = 1$, $\varepsilon_1^* = (2m+15)/(2m+18)$, $\rho_0^* = m + 9$, $\rho_1^* = 5$, $\boldsymbol{L}_0^* = \boldsymbol{I}_{v^*} - (6m+45)^{-1}\mathbf{1}_{v^*}(\boldsymbol{r}^*)' - \boldsymbol{L}_1^*$, $\boldsymbol{L}_1^* = [(2m+18)/3](\boldsymbol{r}^*)^{-\delta}\boldsymbol{N}^*\{\boldsymbol{I}_6 - (6m+45)^{-1}\mathbf{1}_6(\boldsymbol{k}^*)'\}(\boldsymbol{k}^*)^{-\delta}(\boldsymbol{N}^*)'$, for any positive integer m. For example, for $m = 2$, one can get a $(11;5;0)$-EB design with parameters $v^* = 17$, $b^* = 6$, $\boldsymbol{r}^* = [6\mathbf{1}'_2, 4\mathbf{1}'_6, 3\mathbf{1}'_3, 2\mathbf{1}'_6]'$, $\boldsymbol{k}^* = [2, 11\mathbf{1}'_5]'$, $\varepsilon_0^* = 1$, $\varepsilon_1^* = 19/22$, $\rho_0^* = 11$, $\rho_1^* = 5$, $\boldsymbol{L}_0^* = \boldsymbol{I}_{17} - (57)^{-1}\mathbf{1}_{17}(\boldsymbol{r}^*)' - \boldsymbol{L}_1^*$, $\boldsymbol{L}_1^* = (22/3)(\boldsymbol{r}^*)^{-\delta}\boldsymbol{N}^*\{\boldsymbol{I}_6 - (57)^{-1}\mathbf{1}_6(\boldsymbol{k}^*)'\}(\boldsymbol{k}^*)^{-\delta}(\boldsymbol{N}^*)'$, whose incidence matrix is of the form

$$\boldsymbol{N}^* = \begin{bmatrix} \mathbf{1}_2 & \mathbf{1}_2\mathbf{1}'_5 \\ \mathbf{0} & \boldsymbol{N}' \end{bmatrix}.$$

The matrix $\boldsymbol{\Omega}$ for the resulting design is

$$\boldsymbol{\Omega}^* = \left(\frac{3}{19}\boldsymbol{L}_1^* + \boldsymbol{I}_{17}\right)\mathrm{diag}\left[\frac{1}{6}\boldsymbol{I}_2 : \frac{1}{4}\boldsymbol{I}_6 : \frac{1}{3}\boldsymbol{I}_3 : \frac{1}{2}\boldsymbol{I}_6\right].$$

Theorem 7.3.15 with $\boldsymbol{N}_1 = \mathbf{1}_v\mathbf{1}'_v - \boldsymbol{I}_v$ yields the following result.

Corollary 7.3.21. *If there exists an equireplicate* $(0; v-1; 0)$*-EB design with parameters* v, b, r, $\boldsymbol{k}$, ε, *and with the incidence matrix* $\boldsymbol{N}$, *then for positive* *integers x and y such that $x/y = [(2v-1)r - bv]/(v-1)$ there exists a $(\rho_0^*; \rho_1^*; 0)$-EB design $\boldsymbol{N}^*$ with parameters $v^* = xv + yb$, $b^* = v+1$, $\boldsymbol{r}^* = [v\mathbf{1}'_{xv}, \mathbf{1}'_y \otimes \boldsymbol{k}']'$, $\boldsymbol{k}^* = [xv, [x(v-1)+yr]\mathbf{1}'_v]'$, $\varepsilon_0^* = 1$, $\varepsilon_1^* = 1 - r/[v^2(2r-b)]$, $\rho_0^* = xv + yb - v - 1$, $\rho_1^* = v$, $\boldsymbol{L}_0^* = \boldsymbol{I}_{v^*} - [v(xv+yr)]^{-1}\mathbf{1}_{v^*}(\boldsymbol{r}^*)' - \boldsymbol{L}_1^*$, $\boldsymbol{L}_1^* = (1-\varepsilon_1^*)^{-1}(\boldsymbol{r}^*)^{-\delta}\boldsymbol{N}^*\{\boldsymbol{I}_{v+1} - [v(xv+yr)]^{-1}\mathbf{1}_{v+1}(\boldsymbol{k}^*)'\}(\boldsymbol{k}^*)^{-\delta}(\boldsymbol{N}^*)'$, whose incidence matrix is given by*

$$\boldsymbol{N}^* = \begin{bmatrix} \mathbf{1}'_{xv} & \mathbf{0}' \\ \mathbf{1}'_x \otimes (\mathbf{1}_v\mathbf{1}'_v - \boldsymbol{I}_v) & \mathbf{1}'_y \otimes \boldsymbol{N} \end{bmatrix}'. \tag{7.3.8}$$

Example 7.3.16. For a $(0;4;0)$-EB design with parameters $v = 5, b = 15, r = 9$, $\boldsymbol{k} = [4\mathbf{1}'_6, 3\mathbf{1}'_3, 2\mathbf{1}'_6]'$, $\varepsilon = 5/6$ (see Example 7.3.13 for the incidence matrix $\boldsymbol{N}$ of this design), Corollary 7.3.21 produces a $(\rho_0^*; \rho_1^*; 0)$-EB design with parameters $v^* = 5x + 15y$, $b^* = 6$, $\boldsymbol{r}^* = [5\mathbf{1}'_{5x}, \mathbf{1}'_y \otimes \boldsymbol{k}']'$, $\boldsymbol{k}^* = [5x, (4x+9y)\mathbf{1}'_5]'$, $\varepsilon_0^* = 1$, $\varepsilon_1^* = 22/25$, $\rho_0^* = 5x + 15y - 6$, $\rho_1^* = 5$, $\boldsymbol{L}_0^* = \boldsymbol{I}_{v^*} - [5(5x+9y)]^{-1}\mathbf{1}_{v^*}(\boldsymbol{r}^*)' - \boldsymbol{L}_1^*$, $\boldsymbol{L}_1^* = (25/3)(\boldsymbol{r}^*)^{-\delta}\boldsymbol{N}^*\{\boldsymbol{I}_6 - [5(5x+9y)]^{-1}\mathbf{1}_6(\boldsymbol{k}^*)'\}(\boldsymbol{k}^*)^{-\delta}(\boldsymbol{N}^*)'$ for positive integers x and y such that $x/y = 3/2$. For example, for $x = 3$ and $y = 2$, one can get a $(39;5;0)$-EB design with parameters $v^* = 45$, $b^* = 6$, $\boldsymbol{r}^* = [5\mathbf{1}'_{15}, \mathbf{1}'_2 \otimes \boldsymbol{k}']'$, $\boldsymbol{k}^* = [15, 30\mathbf{1}'_5]'$, $\varepsilon_0^* = 1$, $\varepsilon_1^* = 22/25$, $\rho_0^* = 39$, $\rho_1^* = 5$, $\boldsymbol{L}_0^* = \boldsymbol{I}_{45} - (165)^{-1}\mathbf{1}_{45}(\boldsymbol{r}^*)' - \boldsymbol{L}_1^*$, $\boldsymbol{L}_1^* = (25/3)(\boldsymbol{r}^*)^{-\delta}\boldsymbol{N}^*\{\boldsymbol{I}_6 - (165)^{-1}\mathbf{1}_6(\boldsymbol{k}^*)'\}(\boldsymbol{k}^*)^{-\delta}(\boldsymbol{N}^*)'$, whose incidence matrix is of the form

$$\boldsymbol{N}^* = \begin{bmatrix} \mathbf{1}_{15} & \mathbf{1}_3 \otimes (\mathbf{1}_5\mathbf{1}'_5 - \boldsymbol{I}_5) \\ \mathbf{0} & \mathbf{1}_2 \otimes \boldsymbol{N}' \end{bmatrix}.$$

The matrix $\boldsymbol{\Omega}$ is obtained as

$$\boldsymbol{\Omega}^* = \left(\frac{3}{22}\boldsymbol{L}_1^* + \boldsymbol{I}_{45}\right)\mathrm{diag}\left[\frac{1}{5}\boldsymbol{I}_{15} : \frac{1}{4}\boldsymbol{I}_6 : \frac{1}{3}\boldsymbol{I}_3 : \frac{1}{2}\boldsymbol{I}_6 : \frac{1}{4}\boldsymbol{I}_6 : \frac{1}{3}\boldsymbol{I}_3 : \frac{1}{2}\boldsymbol{I}_6\right].$$

Corollaries 6.1.2 and 6.6.2, or directly Theorem 7.3.15, yield the following result.

Corollary 7.3.22. *If $\boldsymbol{N}_h, h = 1,2$, are the $v \times b_h$ incidence matrices of two BIB designs with parameters v, b_h, r_h, k_h, λ_h, then for positive integers x and y such that $x/y = [\lambda_2 v(k_1+1) - k_1k_2r_2]/[k_2(r_1 - \lambda_1)]$ there exists a $(\rho_0^*; \rho_1^*; 0)$-EB design $\boldsymbol{N}^*$ with parameters $v^* = xb_1 + yb_2$, $b^* = v+1$, $\boldsymbol{r}^* = [(k_1+1)\mathbf{1}'_{xb_1}, k_2\mathbf{1}'_{yb_2}]'$, $\boldsymbol{k}^* = [xb_1, (xr_1 + yr_2)\mathbf{1}'_v]'$, $\varepsilon_0^* = 1$, $\varepsilon_1^* = 1 - r_2k_2(r_1 - \lambda_1)/\{v(k_1+1)[r_1(k_1+1)\lambda_2 - \lambda_1r_2k_2]\}$, $\rho_0^* = xb_1 + yb_2 - v - 1$, $\rho_1^* = v$, $\boldsymbol{L}_0^* = \boldsymbol{I}_{v^*} - [xb_1(k_1+1) + yvr_2]^{-1}\mathbf{1}_{v^*}(\boldsymbol{r}^*)' - \boldsymbol{L}_1^*$, $\boldsymbol{L}_1^* = (1-\varepsilon_1^*)^{-1}(\boldsymbol{r}^*)^{-\delta}\boldsymbol{N}^*\{\boldsymbol{I}_{v+1} - [xb_1(k_1+1) + yvr_2]^{-1}\mathbf{1}_{v+1}(\boldsymbol{k}^*)'\}(\boldsymbol{k}^*)^{-\delta}(\boldsymbol{N}^*)'$, whose incidence matrix is given by (7.3.6).*

Example 7.3.17. For a BIB design with parameters $v = 6, b = 10, r = 5, k = 3$,

$\lambda = 2$ (see Example 6.7.3), whose incidence matrix is given by

$$N = \begin{bmatrix} 1 & 1 & 0 & 0 & 1 & 0 & 1 & 0 & 0 & 1 \\ 1 & 1 & 1 & 0 & 0 & 1 & 0 & 1 & 0 & 0 \\ 0 & 1 & 1 & 1 & 0 & 0 & 1 & 0 & 1 & 0 \\ 0 & 0 & 1 & 1 & 1 & 0 & 0 & 1 & 0 & 1 \\ 1 & 0 & 0 & 1 & 1 & 1 & 0 & 0 & 1 & 0 \\ 0 & 0 & 0 & 0 & 0 & 1 & 1 & 1 & 1 & 1 \end{bmatrix},$$

Corollary 7.3.22 with $\boldsymbol{N}_1 = \boldsymbol{N}_2 = \boldsymbol{N}$ presents a $(\rho_0^*; \rho_1^*; 0)$-EB design, whose incidence matrix $\boldsymbol{N}^*$, with parameters $v^* = 10(x+y)$, $b^* = 7$, $\boldsymbol{r}^* = [4\boldsymbol{1}'_{10x}, 3\boldsymbol{1}'_{10y}]'$, $\boldsymbol{k}^* = [10x, 5(x+y)\boldsymbol{1}'_6]'$, $\varepsilon_0^* = 1$, $\varepsilon_1^* = 13/16$, $\rho_0^* = 10(x+y) - 7$, $\rho_1^* = 6$, $\boldsymbol{L}_0^* = \boldsymbol{I}_{v^*} - [10(4x+3y)]^{-1}\boldsymbol{1}_{v^*}(\boldsymbol{r}^*)' - \boldsymbol{L}_1^*$, $\boldsymbol{L}_1^* = (16/3)(\boldsymbol{r}^*)^{-\delta}\boldsymbol{N}^*\{\boldsymbol{I}_7 - [10(4x+3y)]^{-1}\boldsymbol{1}_7(\boldsymbol{k}^*)'\}(\boldsymbol{k}^*)^{-\delta}(\boldsymbol{N}^*)'$, for positive integers x and y such that $x/y = 1/3$. For example, for $x = 1$ and $y = 3$, one can get a $(33; 6; 0)$-EB design with parameters $v^* = 40$, $b^* = 7$, $\boldsymbol{r}^* = [4\boldsymbol{1}'_{10}, 3\boldsymbol{1}'_{30}]'$, $\boldsymbol{k}^* = [10, 20\boldsymbol{1}'_6]'$, $\varepsilon_0^* = 1$, $\varepsilon_1^* = 13/16$, $\rho_0^* = 33$, $\rho_1^* = 6$, $\boldsymbol{L}_0^* = \boldsymbol{I}_{40} - (130)^{-1}\boldsymbol{1}_{40}(\boldsymbol{r}^*)' - \boldsymbol{L}_1^*$, $\boldsymbol{L}_1^* = (16/3)(\boldsymbol{r}^*)^{-\delta}\boldsymbol{N}^*\{\boldsymbol{I}_7 - (130)^{-1}\boldsymbol{1}_7(\boldsymbol{k}^*)'\}(\boldsymbol{k}^*)^{-\delta}(\boldsymbol{N}^*)'$, whose incidence matrix is of the form

$$\boldsymbol{N}^* = \begin{bmatrix} \boldsymbol{1}_{10} & \boldsymbol{N}' \\ \boldsymbol{0} & \boldsymbol{1}_3 \otimes \boldsymbol{N}' \end{bmatrix}.$$

The matrix $\boldsymbol{\Omega}$ for the resulting design is

$$\boldsymbol{\Omega}^* = \left(\frac{3}{13}\boldsymbol{L}_1^* + \boldsymbol{I}_{40}\right)\operatorname{diag}\left[\frac{1}{4}\boldsymbol{I}_{10} : \frac{1}{3}\boldsymbol{I}_{30}\right].$$

Corollary 7.3.22 with $\boldsymbol{N}_1 = \boldsymbol{I}_v$ yields the following result.

Corollary 7.3.23. *If $\boldsymbol{N}$ is the $v \times b$ incidence matrix of a BIB design with parameters v, b, r, k, λ, then for positive integers x and y such that $x/y = 2\lambda v/k - r$ there exists a $(\rho_0^*; \rho_1^*; 0)$-EB design $\boldsymbol{N}^*$ with parameters $v^* = xv + yb$, $b^* = v + 1$, $\boldsymbol{r}^* = [2\boldsymbol{1}'_{xv}, k\boldsymbol{1}'_{yb}]'$, $\boldsymbol{k}^* = [xv, (yr + x)\boldsymbol{1}'_v]'$, $\varepsilon_0^* = 1$, $\varepsilon_1^* = 1 - rk/(4\lambda v)$, $\rho_0^* = xv + yb - v - 1$, $\rho_1^* = v$, $\boldsymbol{L}_0^* = \boldsymbol{I}_{v^*} - [v(2x + yr)]^{-1}\boldsymbol{1}_{v^*}(\boldsymbol{r}^*)' - \boldsymbol{L}_1^*$, $\boldsymbol{L}_1^* = (1 - \varepsilon_1^*)^{-1}(\boldsymbol{r}^*)^{-\delta}\boldsymbol{N}^*\{\boldsymbol{I}_{v+1} - [v(2x + yr)]^{-1}\boldsymbol{1}_{v+1}(\boldsymbol{k}^*)'\}(\boldsymbol{k}^*)^{-\delta}(\boldsymbol{N}^*)'$, whose incidence matrix is given by*

$$\boldsymbol{N}^* = \begin{bmatrix} \boldsymbol{1}'_{xv} & \boldsymbol{0}' \\ \boldsymbol{1}'_x \otimes \boldsymbol{I}_v & \boldsymbol{1}'_y \otimes \boldsymbol{N} \end{bmatrix}'.$$

Example 7.3.18. When $2\lambda + 1$ is a prime or a prime power, there exists a BIB design with parameters $v = 2(\lambda + 1)$, $b = 2(2\lambda + 1)$, $r = 2\lambda + 1$, $k = \lambda + 1$, λ (for the existence see Theorem 8.2.8). Then Corollary 7.3.23 produces a $(\rho_0^*; \rho_1^*; 0)$-EB design with parameters $v^* = 2(\lambda + 1)x + 2(2\lambda + 1)y$, $b^* = 2\lambda + 3$, $\boldsymbol{r}^* = [2\boldsymbol{1}'_{xv}, (\lambda + 1)\boldsymbol{1}'_{yb}]'$, $\boldsymbol{k}^* = [2x(\lambda + 1), \{x + (2\lambda + 1)y\}\boldsymbol{1}'_{2(\lambda+1)}]'$, $\varepsilon_0^* = 1$, $\varepsilon_1^* = (6\lambda - 1)/(8\lambda)$, $\rho_0^* = 2(\lambda+1)x + 2(2\lambda+1)y - 2\lambda - 3$, $\rho_1^* = 2(\lambda+1)$, $\boldsymbol{L}_0^* = \boldsymbol{I}_{v^*} -$

$\{2(\lambda+1)[2x+(2\lambda+1)y]\}^{-1}\mathbf{1}_{v^*}(\boldsymbol{r}^*)' - \boldsymbol{L}_1^*$, $\boldsymbol{L}_1^* = [8\lambda/(2\lambda+1)](\boldsymbol{r}^*)^{-\delta}\boldsymbol{N}^*\{\boldsymbol{I}_{2\lambda+3} - \{2(\lambda+1)[2x+(2\lambda+1)y]\}^{-1}\mathbf{1}_{2\lambda+3}(\boldsymbol{k}^*)'\}(\boldsymbol{k}^*)^{-\delta}(\boldsymbol{N}^*)'$, for positive integers x and y such that $x/y = 2\lambda - 1$.

By Corollary 7.3.20 the following result can be obtained.

Corollary 7.3.24. *If $\boldsymbol{N}$ is the $v \times b$ incidence matrix of a BIB design with parameters v, b, r, k, λ such that $r = 2\lambda$, then for a positive integer m there exists a $(\rho_0^*; \rho_1^*; 0)$-EB design $\boldsymbol{N}^*$ with parameters $v^* = b + m$, $b^* = v + 1$, $\boldsymbol{r}^* = [(v+1)\mathbf{1}_m', k\mathbf{1}_b']'$, $\boldsymbol{k}^* = [m, (r+m)\mathbf{1}_v']'$, $\varepsilon_0^* = 1$, $\varepsilon_1^* = 1 - r/[(v+1)(r+m)]$, $\rho_0^* = b + m - v - 1$, $\boldsymbol{L}_0^* = \boldsymbol{I}_{v^*} - [v(r+m)+m]^{-1}\mathbf{1}_{v^*}(\boldsymbol{r}^*)' - \boldsymbol{L}_1^*$, $\boldsymbol{L}_1^* = (1 - \varepsilon_1^*)^{-1}(\boldsymbol{r}^*)^{-\delta}\boldsymbol{N}^*\{\boldsymbol{I}_{v+1} - [v(r+m)+m]^{-1}\mathbf{1}_{v+1}(\boldsymbol{k}^*)'\}(\boldsymbol{k}^*)^{-\delta}(\boldsymbol{N}^*)'$, $\rho_1^* = v$, whose incidence matrix is given by* (7.3.7).

Corollary 7.3.22 with $\boldsymbol{N}_2 = \mathbf{1}_v\mathbf{1}_p'$ yields the following result.

Corollary 7.3.25. *If $\boldsymbol{N}$ is the $v \times b$ incidence matrix of a BIB design with parameters v, b, r, k, λ, then for positive integers x and y such that $x/y = p/(r - \lambda)$ there exists a $(\rho_0^*; \rho_1^*; 0)$-EB design $\boldsymbol{N}^*$ with parameters $v^* = xb + yp$, $b^* = v+1$, $\boldsymbol{r}^* = [(k+1)\mathbf{1}_{xb}', v\mathbf{1}_{yp}']'$, $\boldsymbol{k}^* = [xb, (xr+yp)\mathbf{1}_v']'$, $\varepsilon_0^* = 1$, $\varepsilon_1^* = 1-(r-\lambda)/[(k+1)(2r-\lambda)]$, $\rho_0^* = xb+yp-v-1$, $\rho_1^* = v$, $\boldsymbol{L}_0^* = \boldsymbol{I}_{v^*} - [xb(k+1)+ypv]^{-1}\mathbf{1}_{v^*}(\boldsymbol{r}^*)' - \boldsymbol{L}_1^*$, $\boldsymbol{L}_1^* = (1 - \varepsilon_1^*)^{-1}(\boldsymbol{r}^*)^{-\delta}\boldsymbol{N}^*\{\boldsymbol{I}_{v+1} - [xb(k+1)+ypv]^{-1}\mathbf{1}_{v+1}(\boldsymbol{k}^*)'\}(\boldsymbol{k}^*)^{-\delta}(\boldsymbol{N}^*)'$, whose incidence matrix is given by*

$$\boldsymbol{N}^* = \begin{bmatrix} \mathbf{1}_{xb}' & \mathbf{0}' \\ \mathbf{1}_x' \otimes \boldsymbol{N} & \mathbf{1}_y' \otimes \mathbf{1}_v\mathbf{1}_p' \end{bmatrix}'.$$

Example 7.3.19. For a BIB design with parameters $v = b = 7$, $r = k = 3$, $\lambda = 1$ (see Example 6.0.1), whose incidence matrix is given by

$$\boldsymbol{N} = \begin{bmatrix} 1 & 0 & 0 & 0 & 1 & 0 & 1 \\ 1 & 1 & 0 & 0 & 0 & 1 & 0 \\ 0 & 1 & 1 & 0 & 0 & 0 & 1 \\ 1 & 0 & 1 & 1 & 0 & 0 & 0 \\ 0 & 1 & 0 & 1 & 1 & 0 & 0 \\ 0 & 0 & 1 & 0 & 1 & 1 & 0 \\ 0 & 0 & 0 & 1 & 0 & 1 & 1 \end{bmatrix},$$

Corollary 7.3.25 produces a $(\rho_0^*; \rho_1^*; 0)$-EB design with parameters $v^* = 7x + py$, $b^* = 8$, $\boldsymbol{r}^* = [4\mathbf{1}_{7x}', 7\mathbf{1}_{yp}']'$, $\boldsymbol{k}^* = [7x, (3x+py)\mathbf{1}_7']'$, $\varepsilon_0^* = 1$, $\varepsilon_1^* = 9/10$, $\rho_0^* = 7x + py - 8$, $\rho_1^* = 7$, $\boldsymbol{L}_0^* = \boldsymbol{I}_{v^*} - [7(4x+py)]^{-1}\mathbf{1}_{v^*}(\boldsymbol{r}^*)' - \boldsymbol{L}_1^*$, $\boldsymbol{L}_1^* = 10(\boldsymbol{r}^*)^{-\delta}\boldsymbol{N}^*\{\boldsymbol{I}_8 - [7(4x+py)]^{-1}\mathbf{1}_8(\boldsymbol{k}^*)'\}(\boldsymbol{k}^*)^{-\delta}(\boldsymbol{N}^*)'$ for positive integers x, y and p satisfying $x/y = p/2$. For example, for $x = p = 2$ and $y = 2$, one can get a $(10; 7; 0)$-EB design with parameters $v^* = 18$, $b^* = 8$, $\boldsymbol{r}^* = [4\mathbf{1}_{14}', 7\mathbf{1}_4']'$, $\boldsymbol{k}^* = [14, 10\mathbf{1}_7']'$, $\varepsilon_0^* = 1$, $\varepsilon_1^* = 9/10$, $\rho_0^* = 10$, $\rho_1^* = 7$, $\boldsymbol{L}_0^* = \boldsymbol{I}_{18} - [(\boldsymbol{k}^*)'\mathbf{1}_8]^{-1}\mathbf{1}_{18}(\boldsymbol{r}^*)' - \boldsymbol{L}_1^*$, $\boldsymbol{L}_1^* =$

$10(\boldsymbol{r}^*)^{-\delta}\boldsymbol{N}^*\{\boldsymbol{I}_8-[(\boldsymbol{k}^*)'\boldsymbol{1}_8]^{-1}\boldsymbol{1}_8(\boldsymbol{k}^*)'\}(\boldsymbol{k}^*)^{-\delta}(\boldsymbol{N}^*)'$, whose incidence matrix is of the form

$$\boldsymbol{N}^* = \begin{bmatrix} \boldsymbol{1}_{14} & \boldsymbol{1}_2 \otimes \boldsymbol{N}' \\ \boldsymbol{0} & \boldsymbol{1}_2 \otimes \boldsymbol{1}_2\boldsymbol{1}_2' \end{bmatrix}.$$

The matrix $\boldsymbol{\Omega}$ for the resulting design is

$$\boldsymbol{\Omega}^* = \left(\frac{1}{9}\boldsymbol{L}_1^* + \boldsymbol{I}_{18}\right)\operatorname{diag}\left[\frac{1}{4}\boldsymbol{I}_{14} : \frac{1}{7}\boldsymbol{I}_4\right].$$

Using directly Corollary 7.3.21, or Corollary 7.3.22 with $\boldsymbol{N}_1 = \boldsymbol{1}_v\boldsymbol{1}_v' - \boldsymbol{I}_v$, the following result can be obtained.

Corollary 7.3.26. *If $\boldsymbol{N}$ is the $v \times b$ incidence matrix of a BIB design with parameters v, b, r, k, λ, then for positive integers x and y such that $x/y = \lambda v^2/k - r(v-1)$ there exists a $(\rho_0^*; \rho_1^*; 0)$-EB design $\boldsymbol{N}^*$ with parameters $v^* = xv + yb$, $b^* = v+1$, $\boldsymbol{r}^* = [v\boldsymbol{1}_{xv}', k\boldsymbol{1}_{yb}']'$, $\boldsymbol{k}^* = [xv, [x(v-1)+yr]\boldsymbol{1}_v']'$, $\varepsilon_0^* = 1$, $\varepsilon_1^* = 1 - k/[v^2(2k-v)]$, $\rho_0^* = xv + yb - v - 1$, $\rho_1^* = v$, $\boldsymbol{L}_0^* = \boldsymbol{I}_{v^*} - [v(xv+yr)]^{-1}\boldsymbol{1}_{v^*}(\boldsymbol{r}^*)' - \boldsymbol{L}_1^*$, $\boldsymbol{L}_1^* = (1-\varepsilon_1^*)^{-1}(\boldsymbol{r}^*)^{-\delta}\boldsymbol{N}^*\{\boldsymbol{I}_{v+1} - [v(xv+yr)]^{-1}\boldsymbol{1}_{v+1}(\boldsymbol{k}^*)'\}(\boldsymbol{k}^*)^{-\delta}(\boldsymbol{N}^*)'$, whose incidence matrix is given by* (7.3.8).

Example 7.3.20. There exists a BIB design with parameters $v = b = 4\lambda - 1$, $r = k = 2\lambda$ and λ, if $4\lambda - 1$ is a prime or a prime power (for the existence see Theorem 8.2.10 with $m = 1$). Then Corollaries 7.3.24 and 7.3.26 produce, respectively, $(\rho_0^*; \rho_1^*; 0)$-EB designs with parameters

(i) $v^* = 4\lambda - 1 + m$, $b^* = 4\lambda$, $\boldsymbol{r}^* = [4\lambda\boldsymbol{1}_m', 2\lambda\boldsymbol{1}_{4\lambda-1}']'$, $\boldsymbol{k}^* = [m, (2\lambda+m)\boldsymbol{1}_{4\lambda-1}']'$, $\varepsilon_0^* = 1$, $\varepsilon_1^* = 1 - 1/(4\lambda+2m)$, $\rho_0^* = m - 1$, $\rho_1^* = 4\lambda - 1$, $\boldsymbol{L}_0^* = \boldsymbol{I}_{v^*} - [2\lambda(4\lambda - 1 + 2m)]^{-1}\boldsymbol{1}_{v^*}(\boldsymbol{r}^*)' - \boldsymbol{L}_1^*$, $\boldsymbol{L}_1^* = (4\lambda + 2m)(\boldsymbol{r}^*)^{-\delta}\boldsymbol{N}^*\{\boldsymbol{I}_{4\lambda} - [2\lambda(4\lambda - 1 + 2m)]^{-1}\boldsymbol{1}_{4\lambda}(\boldsymbol{k}^*)'\}(\boldsymbol{k}^*)^{-\delta}(\boldsymbol{N}^*)'$ for any positive integer m (≥ 2);

(ii) $v^* = (4\lambda - 1)(x + y)$, $b^* = 4\lambda$, $\boldsymbol{r}^* = [(4\lambda - 1)\boldsymbol{1}_{x(4\lambda-1)}', 2\lambda\boldsymbol{1}_{y(4\lambda-1)}']'$, $\boldsymbol{k}^* = [x(4\lambda - 1), \{2x(2\lambda - 1) + 2y\lambda\}\boldsymbol{1}_{4\lambda-1}']'$, $\varepsilon_0^* = 1$, $\varepsilon_1^* = 1 - 4\lambda/[(4\lambda - 1)(8\lambda - 1)]$, $\rho_0^* = (4\lambda - 1)(x + y) - 4\lambda$, $\rho_1^* = 4\lambda - 1$, $\boldsymbol{L}_0^* = \boldsymbol{I}_{v^*} - \{(4\lambda - 1)[x(4\lambda - 1) + 2y\lambda]\}^{-1}\boldsymbol{1}_{v^*}(\boldsymbol{r}^*)' - \boldsymbol{L}_1^*$, $\boldsymbol{L}_1^* = [(4\lambda - 1)(8\lambda - 1)/(4\lambda)](\boldsymbol{r}^*)^{-\delta}\boldsymbol{N}^*\{\boldsymbol{I}_{4\lambda} - \{(4\lambda - 1)[x(4\lambda - 1) + 2y\lambda]\}^{-1}\boldsymbol{1}_{4\lambda}(\boldsymbol{k}^*)'\}(\boldsymbol{k}^*)^{-\delta}(\boldsymbol{N}^*)'$, for positive integers x and y such that $x/y = 1/2$.

Corollary 7.3.22 with $\boldsymbol{N}_2 = \boldsymbol{1}_v\boldsymbol{1}_v' - \boldsymbol{I}_v$ yields the following result.

Corollary 7.3.27. *If $\boldsymbol{N}$ is the $v \times b$ incidence matrix of a BIB design with parameters v, b, r, k, λ, then for positive integers x and y such that $x/y = (v^2 - 2v - k)/[(v-1)(r-\lambda)]$ there exists a $(\rho_0^*; \rho_1^*; 0)$-EB design $\boldsymbol{N}^*$ with parameters $v^* = xb + yv$, $b^* = v+1$, $\boldsymbol{r}^* = [(k+1)\boldsymbol{1}_{xb}', (v-1)\boldsymbol{1}_{yv}']'$, $\boldsymbol{k}^* = [xb, [xr + y(v-1)]\boldsymbol{1}_v']'$, $\varepsilon_0^* = 1$, $\varepsilon_1^* = 1 - (v-1)^2(r-\lambda)/[vr(k+1)(2v-k-3)]$, $\rho_0^* = xb + (y-1)v - 1$, $\rho_1^* = v$, $\boldsymbol{L}_0^* = \boldsymbol{I}_{v^*} - [xb(k+1) + yv(v-1)]^{-1}\boldsymbol{1}_{v^*}(\boldsymbol{r}^*)' - \boldsymbol{L}_1^*$, $\boldsymbol{L}_1^* =$*

$(1-\varepsilon_1^*)^{-1}(\boldsymbol{r}^*)^{-\delta}\boldsymbol{N}^*\{\boldsymbol{I}_{v+1}-[xb(k+1)+yv(v-1)]^{-1}\boldsymbol{1}_{v+1}(\boldsymbol{k}^*)'\}(\boldsymbol{k}^*)^{-\delta}(\boldsymbol{N}^*)'$, *whose incidence matrix is given by*

$$\boldsymbol{N}^*=\begin{bmatrix}\boldsymbol{1}'_{xb} & \boldsymbol{0}'\\ \boldsymbol{1}'_x\otimes\boldsymbol{N} & \boldsymbol{1}'_y\otimes(\boldsymbol{1}_v\boldsymbol{1}'_v-\boldsymbol{I}_v)\end{bmatrix}'.$$

Example 7.3.21. For a BIB design with parameters $v=9, b=12, r=4, k=3$, $\lambda=1$, whose incidence matrix is given in Example 6.0.8, Corollary 7.3.27 produces a $(\rho_0^*;\rho_1^*;0)$-EB design with parameters $v^*=3(4x+3y)$, $b^*=10$, $\boldsymbol{r}^*=[4\boldsymbol{1}'_{12x},8\boldsymbol{1}'_{9y}]'$, $\boldsymbol{k}^*=[12x,4(x+2y)\boldsymbol{1}'_9]'$, $\varepsilon_0^*=1$, $\varepsilon_1^*=8/9$, $\rho_0^*=3(4x+3y)-10$, $\rho_1^*=9$, $\boldsymbol{L}_0^*=\boldsymbol{I}_{v^*}-(48x+72y)^{-1}\boldsymbol{1}_{v^*}(\boldsymbol{r}^*)'-\boldsymbol{L}_1^*$, $\boldsymbol{L}_1^*=9(\boldsymbol{r}^*)^{-\delta}\boldsymbol{N}^*[\boldsymbol{I}_{10}-(48x+72y)^{-1}\boldsymbol{1}_{10}(\boldsymbol{k}^*)'](\boldsymbol{k}^*)^{-\delta}(\boldsymbol{N}^*)'$ for positive integers x and y satisfying $2x=5y$. For example, for $x=5$ and $y=2$, one can get a $(68;9;0)$-EB design with parameters $v^*=78$, $b^*=10$, $\boldsymbol{r}^*=[4\boldsymbol{1}'_{60},8\boldsymbol{1}'_{18}]'$, $\boldsymbol{k}^*=[60,36\boldsymbol{1}'_9]'$, $\varepsilon_0^*=1$, $\varepsilon_1^*=8/9$, $\rho_0^*=68$, $\rho_1^*=9$, $\boldsymbol{L}_0^*=\boldsymbol{I}_{78}-(384)^{-1}\boldsymbol{1}_{78}(\boldsymbol{r}^*)'-\boldsymbol{L}_1^*$, $\boldsymbol{L}_1^*=9(\boldsymbol{r}^*)^{-\delta}\boldsymbol{N}^*[\boldsymbol{I}_{10}-(384)^{-1}\boldsymbol{1}_{10}(\boldsymbol{k}^*)'](\boldsymbol{k}^*)^{-\delta}(\boldsymbol{N}^*)'$, whose incidence matrix is of the form

$$\boldsymbol{N}^*=\begin{bmatrix}\boldsymbol{1}_{60} & \boldsymbol{1}_5\otimes\boldsymbol{N}'\\ \boldsymbol{0} & \boldsymbol{1}_2\otimes(\boldsymbol{1}_9\boldsymbol{1}'_9-\boldsymbol{I}_9)\end{bmatrix}.$$

The matrix $\boldsymbol{\Omega}$ for the resulting design is

$$\boldsymbol{\Omega}^*=\left(\frac{1}{8}\boldsymbol{L}_1^*+\boldsymbol{I}_{78}\right)\text{diag}\left[\frac{1}{4}\boldsymbol{I}_{60}:\frac{1}{8}\boldsymbol{I}_{18}\right].$$

Using Corollary 7.3.23 with $\boldsymbol{N}=\boldsymbol{1}_v\boldsymbol{1}'_s$ and $y=1$, the following result can be obtained.

Corollary 7.3.28. *There exists a* $(\rho_0^*;\rho_1^*;0)$*-EB design* $\boldsymbol{N}^*$ *with parameters* $v^*=s(v+1)$, $b^*=v+1$, $\boldsymbol{r}^*=[2\boldsymbol{1}'_{sv},v\boldsymbol{1}'_s]'$, $\boldsymbol{k}^*=[sv,2s\boldsymbol{1}'_v]'$, $\varepsilon_0^*=1$, $\varepsilon_1^*=3/4$, $\rho_0^*=(s-1)(v+1)$, $\rho_1^*=v$, $\boldsymbol{L}_0^*=\boldsymbol{I}_{s(v+1)}-(3sv)^{-1}\boldsymbol{1}_{s(v+1)}(\boldsymbol{r}^*)'-\boldsymbol{L}_1^*$, $\boldsymbol{L}_1^*=4(\boldsymbol{r}^*)^{-\delta}\boldsymbol{N}^*[\boldsymbol{I}_{v+1}-(3sv)^{-1}\boldsymbol{1}_{v+1}(\boldsymbol{k}^*)'](\boldsymbol{k}^*)^{-\delta}(\boldsymbol{N}^*)'$, *whose incidence matrix is given by*

$$\boldsymbol{N}^*=\begin{bmatrix}\boldsymbol{1}'_{sv} & \boldsymbol{0}'\\ \boldsymbol{1}'_s\otimes\boldsymbol{I}_v & \boldsymbol{1}_v\boldsymbol{1}'_s\end{bmatrix}',$$

where s is any positive integer ≥ 2.

Remark 7.3.5. In the designs constructed here, it should be noted that the number of treatments is greater than the number of blocks. This is an attractive property for designing experiments with large numbers of treatments, as, e.g., in plant breeding research.

Corollary 6.1.3 or Theorem 6.4.1 can also be used to construct a $(\rho_0^*;\ \rho_1^*;\ 0)$-EB design from another $(\rho_0;\ \rho_1;\ 0)$-EB design, that may be nonproper and nonequireplicate, through its dualization or its juxtaposition. In this sense $(\rho_0^*;$

ρ_1^*; 0)-EB designs constructed here can further be used to construct more such designs. But this routine method will not be discussed here.

7.3.5 Illustration

In the present Section 7.3, a large number of procedures of constructing designs with two efficiency factors, one of them being equal to one, have been presented. Hence, one can easily choose or construct a design suitable for an experiment in which certain contrasts of treatment parameters are estimated in the intra-block analysis with full efficiency. Here, a real example of such experiment will be presented, together with the analysis of the experimental data, worked in detail. The analysis will cover not only the intra-block analysis but also the inter-block analysis and, finally, a combination of both of them. Estimation and testing hypotheses will be considered.

Example 7.3.22. Ceranka (1975, 1983) has analyzed data from a plant-breeding field experiment with 25 breeding strains and 2 standard varieties of sunflower compared in a block design based on the incidence matrix $\boldsymbol{N}^*$ of the type (7.3.4), i.e.,

$$\boldsymbol{N}^* = \begin{bmatrix} \boldsymbol{N} \\ \mathbf{1}_s\mathbf{1}_b' \end{bmatrix},$$

with

$$\boldsymbol{N} = [\boldsymbol{N}_1 : \boldsymbol{N}_2 : \boldsymbol{N}_3 : \boldsymbol{N}_4 : \boldsymbol{N}_5 : \boldsymbol{N}_6],$$

where

$$\boldsymbol{N}_1' = \begin{bmatrix} 1&1&1&1&1&0&0&0&0&0&0&0&0&0&0&0&0&0&0&0&0&0&0&0&0 \\ 0&0&0&0&0&1&1&1&1&1&0&0&0&0&0&0&0&0&0&0&0&0&0&0&0 \\ 0&0&0&0&0&0&0&0&0&0&1&1&1&1&1&0&0&0&0&0&0&0&0&0&0 \\ 0&0&0&0&0&0&0&0&0&0&0&0&0&0&0&1&1&1&1&1&0&0&0&0&0 \\ 0&0&0&0&0&0&0&0&0&0&0&0&0&0&0&0&0&0&0&0&1&1&1&1&1 \end{bmatrix},$$

$$\boldsymbol{N}_2' = \begin{bmatrix} 1&0&0&0&0&1&0&0&0&0&1&0&0&0&0&1&0&0&0&0&1&0&0&0&0 \\ 0&1&0&0&0&0&1&0&0&0&0&1&0&0&0&0&1&0&0&0&0&1&0&0&0 \\ 0&0&1&0&0&0&0&1&0&0&0&0&1&0&0&0&0&1&0&0&0&0&1&0&0 \\ 0&0&0&1&0&0&0&0&1&0&0&0&0&1&0&0&0&0&1&0&0&0&0&1&0 \\ 0&0&0&0&1&0&0&0&0&1&0&0&0&0&1&0&0&0&0&1&0&0&0&0&1 \end{bmatrix},$$

$$\boldsymbol{N}_3' = \begin{bmatrix} 1&0&0&0&0&0&1&0&0&0&0&0&1&0&0&0&0&0&1&0&0&0&0&0&1 \\ 0&1&0&0&0&0&0&1&0&0&0&0&0&1&0&0&0&0&0&1&1&0&0&0&0 \\ 0&0&1&0&0&0&0&0&1&0&0&0&0&0&1&1&0&0&0&0&0&1&0&0&0 \\ 0&0&0&1&0&0&0&0&0&1&1&0&0&0&0&0&1&0&0&0&0&0&1&0&0 \\ 0&0&0&0&1&1&0&0&0&0&0&1&0&0&0&0&0&1&0&0&0&0&0&1&0 \end{bmatrix},$$

$$\boldsymbol{N}_4' = \begin{bmatrix} 1&0&0&0&0&0&0&1&0&0&0&0&0&0&1&0&1&0&0&0&0&0&0&1&0 \\ 0&1&0&0&0&0&0&0&1&0&1&0&0&0&0&0&0&1&0&0&0&0&0&0&1 \\ 0&0&1&0&0&0&0&0&0&1&0&1&0&0&0&0&0&0&1&0&1&0&0&0&0 \\ 0&0&0&1&0&1&0&0&0&0&0&0&1&0&0&0&0&0&0&1&0&1&0&0&0 \\ 0&0&0&0&1&0&1&0&0&0&0&0&0&1&0&1&0&0&0&0&0&0&1&0&0 \end{bmatrix},$$

$$N_5' = \begin{bmatrix} 1&0&0&0&0&0&0&0&1&0&0&1&0&0&0&0&0&0&0&1&0&0&1&0&0 \\ 0&1&0&0&0&0&0&0&0&1&0&0&1&0&0&1&0&0&0&0&0&0&0&1&0 \\ 0&0&1&0&0&1&0&0&0&0&0&0&0&1&0&0&1&0&0&0&0&0&0&0&1 \\ 0&0&0&1&0&0&1&0&0&0&0&0&0&0&1&0&0&1&0&0&1&0&0&0&0 \\ 0&0&0&0&1&0&0&1&0&0&1&0&0&0&0&0&0&0&1&0&0&1&0&0&0 \end{bmatrix},$$

$$N_6' = \begin{bmatrix} 1&0&0&0&0&0&0&0&0&1&0&0&0&1&0&0&0&1&0&0&0&1&0&0&0 \\ 0&1&0&0&0&1&0&0&0&0&0&0&0&0&1&0&0&0&1&0&0&0&1&0&0 \\ 0&0&1&0&0&0&1&0&0&0&1&0&0&0&0&0&0&0&0&1&0&0&0&1&0 \\ 0&0&0&1&0&0&0&1&0&0&0&1&0&0&0&1&0&0&0&0&0&0&0&0&1 \\ 0&0&0&0&1&0&0&0&1&0&0&0&1&0&0&0&1&0&0&0&1&0&0&0&0 \end{bmatrix},$$

and with

$$\mathbf{1}_s\mathbf{1}_b' = [\mathbf{1}_2\mathbf{1}_5' : \mathbf{1}_2\mathbf{1}_5' : \mathbf{1}_2\mathbf{1}_5' : \mathbf{1}_2\mathbf{1}_5' : \mathbf{1}_2\mathbf{1}_5' : \mathbf{1}_2\mathbf{1}_5'].$$

It can be seen that the 25×30 incidence matrix $\boldsymbol{N}$ of the basic design represents a BIB design, with $v = 25$ treatments (here breeding strains) replicated $r = 6$ times, every two of the treatments concurring in exactly $\lambda = 1$ block and each of the $b = 30$ blocks being of size $k = 5$. Also note that, according to Theorem 6.10.1(i), this BIB design can be derived from a complete set of MOLS of order $q = 5$ (as defined in Appendix D). More precisely, let $q^2 = 25$ symbols be arranged in a 5×5 square L as

$$L = \begin{bmatrix} 1 & 2 & 3 & 4 & 5 \\ 6 & 7 & 8 & 9 & 10 \\ 11 & 12 & 13 & 14 & 15 \\ 16 & 17 & 18 & 19 & 20 \\ 21 & 22 & 23 & 24 & 25 \end{bmatrix},$$

and regard the symbols in L as treatment labels of the design. Further, let L_1, L_2, L_3, L_4 represent a complete set of MOLS of order $q = 5$. Then one can easily check that the blocks described by $\boldsymbol{N}_1$ correspond to the rows of L, i.e., that all treatments indicated by the jth column of $\boldsymbol{N}_1$ have their labels appearing in the jth row of L ($j = 1, 2, ..., 5$). Similary, the blocks described by $\boldsymbol{N}_2$ correspond to the columns of L. To explain the composition of blocks described by $\boldsymbol{N}_3$, one has to superimpose the Latin square L_1 on the array L, and then note that all treatments belonging to the same block given by $\boldsymbol{N}_3$ have their labels corresponding to the same letter in the superimposed L_1. In the same way, the compositions of blocks given by $\boldsymbol{N}_4$, $\boldsymbol{N}_5$ and $\boldsymbol{N}_6$ can be detected by superimposing the remaining mutually orthogonal Latin squares L_2, L_3 and L_4, respectively. Of course, the exact correspondence will depend on how the complete set of MOLS is ordered. In this example, it is as originally used by Yates (1936a).

In addition, it may be noted that the 2×30 incidence matrix $\mathbf{1}_s\mathbf{1}_b'$ of the supplementary treatments represents an RBD, with $s = 2$ treatments (here standard varieties) replicated $b = 30$ times.

Thus, on account of Theorem 6.3.2, the resulting design given by the 27×30 incidence matrix $\boldsymbol{N}^*$ (see above) is a $(\rho_0^*; \rho_1^*; 0)$-EB design with parameters $v^* =$

$v + s = 27, b^* = b = 30, \boldsymbol{r}^* = [r\mathbf{1}'_v, b\mathbf{1}'_s]' = [6\mathbf{1}'_{25}, 30\mathbf{1}'_2]', k^* = k + s = 7,\ \varepsilon_0^* = 1,$ $\varepsilon_1^* = 1 - (r - \lambda)/[r(k + s)] = 1 - 5/42 = 37/42\ (= 0.880952),\ \rho_0^* = s = 2, \rho_1^* =$ $v - 1 = 24,\ \boldsymbol{L}_1^* = \mathrm{diag}[\boldsymbol{I}_v - v^{-1}\mathbf{1}_v\mathbf{1}'_v : \mathbf{O}],\ \boldsymbol{L}_0^* = \boldsymbol{I}_{v+s} - (n^*)^{-1}\mathbf{1}_{v+s}(\boldsymbol{r}^*)' - \boldsymbol{L}_1^*,$ where $n^* = b(k + s) = 30(7) = 210$, i.e.,

$$\boldsymbol{L}_1^* = \begin{bmatrix} \boldsymbol{I}_{25} - (25)^{-1}\mathbf{1}_{25}\mathbf{1}'_{25} & \mathbf{O} \\ \mathbf{O} & \mathbf{O} \end{bmatrix}$$

and

$$\boldsymbol{L}_0^* = \frac{1}{175}\begin{bmatrix} 21\mathbf{1}_{25}\mathbf{1}'_{25} & -25\mathbf{1}_{25}\mathbf{1}'_2 \\ -5\mathbf{1}_2\mathbf{1}'_{25} & 175\boldsymbol{I}_2 - 25\mathbf{1}_2\mathbf{1}'_2 \end{bmatrix}.$$

To complete the description of the design and its use in the analyzed experiment, the assignment of treatments to experimental units (plots) through randomization is to be explained. Here, it will be assumed that the randomization of blocks and plots within the blocks have been implemented according to the procedure described in Section 3.1.1. (Thus, the fact that the BIB design represented by $\boldsymbol{N}$ is resolvable, as already noted in Section 6.10, will be ignored here; this will be taken into account when reconsidering this example in Chapter 9.) This assumption means that the order in which the columns of the matrix $\boldsymbol{N}^*$ are assigned to the real blocks of plots, formed in the experimental field, has been chosen at random, and that for each block the order in which the treatments indicated by 1's in the assigned column of $\boldsymbol{N}^*$ are then assigned to the plots of the block has also been chosen at random.

The plant trait observed on the experimental units, and taken here for the analysis, is the average diameter of capitulum (head) in centimeters. The individual plot observations, ordered according to the order of blocks in the incidence matrix $\boldsymbol{N}^*$ of the design, are given in Table 7.1.

To perform the intra-block and the inter-block analysis, first note that because the design is a $(\rho_0^*; \rho_1^*; 0)$-EB design, its matrix $\boldsymbol{C}_1$ can simply be obtained here, applying the formula (4.4.5), as

$$\boldsymbol{C}_1^* = \begin{bmatrix} \boldsymbol{C}_{1,11}^* & \boldsymbol{C}_{1,12}^* \\ \boldsymbol{C}_{1,21}^* & \boldsymbol{C}_{1,22}^* \end{bmatrix},$$

with

$$\boldsymbol{C}_{1,11}^* = \frac{1}{7}\left(37\boldsymbol{I}_{25} - \mathbf{1}_{25}\mathbf{1}'_{25}\right), \quad \boldsymbol{C}_{1,12}^* = -\frac{6}{7}\mathbf{1}_{25}\mathbf{1}'_2,$$

$$\boldsymbol{C}_{1,21}^* = -\frac{6}{7}\mathbf{1}_2\mathbf{1}'_{25} \quad \text{and} \quad \boldsymbol{C}_{1,22}^* = \frac{1}{7}\left(210\boldsymbol{I}_2 - 30\mathbf{1}_2\mathbf{1}'_2\right).$$

A suitable choice of a g-inverse of $\boldsymbol{C}_1^*$ is the 27×27 matrix (see Section 4.4.1 and Remark 6.3.2)

$$\boldsymbol{\Omega}^* = \begin{bmatrix} \boldsymbol{\Omega}_{1,11}^* & \boldsymbol{\Omega}_{1,12}^* \\ \boldsymbol{\Omega}_{1,21}^* & \boldsymbol{\Omega}_{1,22}^* \end{bmatrix},$$

Table 7.1
Experimental data, observations of the average diameter of capitulum (head) of sunflower in centimeters, analyzed in Example 7.3.22

Block	Strain	Observation	Block	Strain	Observation	Block	Strain	Observation
1	27	17.4	11	13	16.6	21	1	15.3
1	1	17.9	11	19	16.5	21	12	16.5
1	4	17.3	11	27	16.3	21	23	14.3
1	5	17.7	11	26	14.1	21	27	16.2
1	2	16.5	11	1	15.0	21	26	15.2
1	26	15.7	11	25	16.7	21	20	17.1
1	3	16.6	11	7	15.3	21	9	17.1
2	10	18.1	12	20	19.6	22	10	15.0
2	8	19.0	12	27	16.9	22	16	15.1
2	26	18.5	12	2	17.6	22	26	14.6
2	27	18.5	12	21	17.5	22	2	18.0
2	7	17.1	12	26	18.2	22	13	16.8
2	9	18.1	12	8	19.0	22	24	15.6
2	6	18.3	12	14	17.9	22	27	16.8
3	14	16.1	13	3	15.4	23	14	15.7
3	11	17.1	13	15	17.1	23	25	13.8
3	26	15.9	13	22	17.9	23	26	14.5
3	13	16.9	13	27	16.0	23	3	16.2
3	27	18.9	13	9	15.9	23	27	15.7
3	15	17.1	13	16	17.6	23	6	16.5
3	12	17.6	13	26	17.3	23	17	16.6
4	18	19.0	14	23	16.7	24	4	15.8
4	20	19.6	14	4	19.7	24	18	16.1
4	16	17.8	14	11	19.0	24	27	16.4
4	27	16.1	14	27	17.2	24	7	15.9
4	26	16.5	14	10	17.7	24	15	15.9
4	17	17.1	14	26	15.3	24	26	14.0
4	19	18.5	14	17	16.5	24	21	16.2
5	22	17.4	15	24	16.0	25	8	15.4
5	24	17.3	15	12	17.0	25	11	15.5
5	27	18.3	15	5	19.1	25	22	14.8
5	25	15.5	15	27	16.1	25	27	16.4
5	23	16.1	15	18	19.4	25	26	16.0
5	21	17.4	15	26	19.8	25	5	17.2
5	26	18.0	15	6	17.5	25	19	14.5
6	11	21.0	16	15	11.4	26	1	12.1
6	21	16.0	16	26	11.2	26	18	13.4
6	27	18.6	16	27	13.0	26	26	10.7
6	1	19.0	16	24	10.4	26	22	11.0
6	16	19.0	16	8	10.7	26	27	12.3
6	26	16.0	16	1	13.0	26	10	13.2
6	6	18.2	16	17	11.8	26	14	14.1
7	12	16.6	17	9	14.2	27	6	12.1
7	22	17.6	17	18	13.9	27	15	13.3
7	2	17.5	17	2	13.1	27	23	13.0
7	26	16.1	17	11	12.4	27	27	14.1
7	27	17.0	17	27	13.0	27	26	11.3
7	17	16.5	17	25	10.5	27	2	12.2
7	7	18.0	17	26	9.9	27	19	12.3
8	3	17.9	18	10	12.8	28	20	16.8
8	13	15.8	18	12	12.5	28	24	13.8
8	23	16.7	18	27	12.6	28	3	12.3
8	27	18.6	18	26	12.2	28	27	13.5
8	26	16.2	18	3	12.5	28	11	14.3
8	18	17.5	18	19	12.7	28	26	12.9
8	8	18.0	18	21	12.2	28	7	14.9
9	14	17.0	19	26	11.8	29	12	12.5
9	26	16.3	19	22	11.9	29	16	12.5
9	9	17.7	19	13	12.0	29	26	11.3
9	4	16.1	19	27	11.4	29	25	11.5
9	27	19.3	19	4	12.6	29	8	12.5
9	24	15.8	19	6	12.8	29	27	14.4
9	19	17.6	19	20	14.5	29	4	11.8
10	10	15.8	20	14	12.6	30	9	13.3
10	20	19.1	20	16	12.7	30	13	13.8
10	26	18.2	20	23	10.8	30	27	14.1
10	25	17.4	20	7	11.2	30	21	11.6
10	15	18.0	20	26	10.5	30	5	12.8
10	27	16.8	20	5	12.2	30	17	13.2
10	5	18.8	20	27	11.5	30	26	12.5

with

$$\boldsymbol{\Omega}_{11}^* = \frac{1}{1110}(210\boldsymbol{I}_{25} - \mathbf{1}_{25}\mathbf{1}_{25}'), \quad \boldsymbol{\Omega}_{12}^* = \mathbf{O},$$

$$\boldsymbol{\Omega}_{21}^* = \mathbf{O} \quad \text{and} \quad \boldsymbol{\Omega}_{22}^* = \frac{37}{1110}\boldsymbol{I}_2 = \frac{1}{30}\boldsymbol{I}_2.$$

This matrix is of full rank, $v + s = 27$, i.e., is nonsingular. Furthermore, note that the vector $\boldsymbol{Q}_1$ is here of the form $\boldsymbol{Q}_1^* = \boldsymbol{\Delta}^*\boldsymbol{y} - (k^*)^{-1}\boldsymbol{N}^*\boldsymbol{D}^*\boldsymbol{y}$, where $\boldsymbol{\Delta}^*\boldsymbol{y}$ is the vector of treatment totals and $\boldsymbol{D}^*\boldsymbol{y}$ is that of block totals, giving

$$\begin{aligned}\boldsymbol{Q}_1^* = \; [&1.243, \;\; 1.714, \; -2.243, \;\; 1.186, \;\; 4.843,\\ &0.457, \; -0.129, \;\; 1.343, \;\; 2.786, \; -1.686,\\ &4.329, \; -0.086, \;\; 0.329, \;\; 1.486, \;\; 1.229,\\ &1.914, \; -0.843, \;\; 5.829, \;\; 0.600, \;\; 10.614,\\ -&3.900, \; -0.843, \; -4.443, \; -4.900, \; -5.614, \; -23.957, \;\; 8.743]'.\end{aligned}$$

From these results, applying the formulae given in Section 3.2.1, the intra-block analysis of variance is obtainable as follows. [For $(\boldsymbol{C}_1^*)^-$ one can take either the matrix $\boldsymbol{\Omega}^*$ or any other g-inverse of $\boldsymbol{C}_1^*$.]

Source	Degrees of freedom	Sum of squares	Mean square	F
Treatments	$v^* - 1 = 26$	80.523	3.097	3.369
Residuals	$n^* - b^* - v^* + 1 = 154$	141.586	$s_1^2 = 0.919390$	
Total	$n^* - b^* = 180$	222.109		

For this value of F, the corresponding P value is about 10^{-6}.

As to the inter-block analysis, one needs the matrix $\boldsymbol{C}_2$ and the vector $\boldsymbol{Q}_2$. The first can be obtained here [see (3.5.15) and (3.5.16)] as

$$\boldsymbol{C}_2^* = (k^*)^{-1}\boldsymbol{N}^*(\boldsymbol{N}^*)' - (n^*)^{-1}\boldsymbol{r}^*(\boldsymbol{r}^*)' = (\boldsymbol{r}^*)^\delta - \boldsymbol{C}_1^* - (n^*)^{-1}\boldsymbol{r}^*(\boldsymbol{r}^*)'$$

$$= \begin{bmatrix} \boldsymbol{C}_{2,11}^* & \mathbf{O} \\ \mathbf{O} & \mathbf{O} \end{bmatrix},$$

with

$$\boldsymbol{C}_{2,11}^* = \frac{5}{7}\left(\boldsymbol{I}_{25} - \frac{1}{25}\mathbf{1}_{25}\mathbf{1}_{25}'\right).$$

It appears that a possible simple choice of a g-inverse of $\boldsymbol{C}_2^*$ is the 27×27 matrix

$$\frac{7}{5}\begin{bmatrix} \boldsymbol{I}_{25} & \mathbf{O} \\ \mathbf{O} & \mathbf{O} \end{bmatrix}.$$

Also note that the rank of $\boldsymbol{C}_2^*$ is 24, and that here the vector $\boldsymbol{Q}_2$ is of the form $\boldsymbol{Q}_2^* = (k^*)^{-1}\boldsymbol{N}^*\boldsymbol{D}^*\boldsymbol{y} - (n^*)^{-1}\boldsymbol{r}^*(\mathbf{1}_{n^*})'\boldsymbol{y}$, which gives

$$
\begin{aligned}
\boldsymbol{Q}_2^* = [&-1.8743, \quad 0.2543, \quad 0.2114, \ -0.8171, \quad 0.0257,\\
&2.0114, \ -0.4029, \quad 0.3257, \quad 0.5829, \quad 1.3543,\\
&2.0400, \ -0.1457, \ -1.3600, \ -1.0171, \ -1.3600,\\
&-0.1457, \ -0.3886, \quad 0.5400, \ -1.4314, \quad 3.1543,\\
&1.8686, \ -1.4886, \ -0.8886, \quad 0.8686, \ -1.9171, \ 0.0000, \ 0.0000]'.
\end{aligned}
$$

These results, used in the formulae given in Section 3.2.2, yield the inter-block analysis of variance, which can be presented as

Source	Degrees of freedom	Sum of squares	Mean square	F
Treatments	$v^* - \rho_0^* - 1 = 24$	60.318	2.513	0.012
Residuals	$b^* - v^* + \rho_0^* = 5$	1007.506	$s_2^2 = 201.501$	
Total	$b^* - 1 = 29$	1067.824		

With this value of the statistic F, the tested null hypothesis cannot be rejected.

The results of the F test followed from the intra-block and the inter-block analysis show that only the hypothesis $H_{01} : \mathrm{E}(\boldsymbol{y}_1) = \boldsymbol{0}$ (see Section 3.2.1) is to be rejected for the present example, while there is no evidence for rejecting the hypothesis $H_{02} : \mathrm{E}(\boldsymbol{y}_2) = \boldsymbol{0}$ (see Section 3.2.2). Certainly, the rejection of H_{01} is sufficient for rejecting the overall null hypothesis H_0 defined in Section 3.8.5, i.e., the hypothesis $H_0 = H_{01} \cap H_{02}$. In fact, it may be noted that, for the present design, H_{01} is equivalent to $\tau_1 = \tau_2 = \cdots = \tau_{27}$, whereas H_{02} is equivalent to $\tau_1 = \tau_2 = \cdots = \tau_{25}$, which means that $H_0 = H_{01}$. Thus, it can be concluded that, in this example, the treatment (strain) parameters are almost surely not constant within the whole experiment.

Now, after rejecting H_0, it may be interesting to ask which contrasts in the treatment parameters $\boldsymbol{\tau}$ are responsible for the rejection. To answer this question, it will be useful to choose a set of basic contrasts, estimate them as described in Sections 3.5 and 3.6 and then proceed as shown in Section 3.8.5.

Suppose that from the subject of the research point of view, the researcher has chosen basic contrasts represented by the following vectors:

$$
\begin{aligned}
\boldsymbol{s}_{01} &= [\,\boldsymbol{0}\boldsymbol{1}_{25}', \ -1, \ 1]'/\sqrt{60},\\
\boldsymbol{s}_{02} &= [-2\boldsymbol{1}_{25}', \ 5\boldsymbol{1}_2']'/\sqrt{2100},\\
\boldsymbol{s}_{11} &= [-1, \ 1, \ \boldsymbol{0}\boldsymbol{1}_{25}']'/\sqrt{12},\\
\boldsymbol{s}_{12} &= [\,\boldsymbol{0}\boldsymbol{1}_2', \ -1, \ 1, \ \boldsymbol{0}\boldsymbol{1}_{23}']'/\sqrt{12},\\
\boldsymbol{s}_{13} &= [-\boldsymbol{1}_2', \ \boldsymbol{1}_2', \ \boldsymbol{0}\boldsymbol{1}_{23}']'/\sqrt{24},\\
\boldsymbol{s}_{14} &= [-\boldsymbol{1}_4', \ 4, \ \boldsymbol{0}\boldsymbol{1}_{22}']'/\sqrt{120},\\
\boldsymbol{s}_{15} &= [\,\boldsymbol{0}\boldsymbol{1}_5', \ -1, \ 1, \ \boldsymbol{0}\boldsymbol{1}_{20}']'/\sqrt{12},\\
\boldsymbol{s}_{16} &= [\,\boldsymbol{0}\boldsymbol{1}_7', \ -1, \ 1, \ \boldsymbol{0}\boldsymbol{1}_{18}']'/\sqrt{12},\\
\boldsymbol{s}_{17} &= [\,\boldsymbol{0}\boldsymbol{1}_5', \ -\boldsymbol{1}_2', \ \boldsymbol{1}_2', \ \boldsymbol{0}\boldsymbol{1}_{18}']'/\sqrt{24},
\end{aligned}
$$

$$
\begin{aligned}
s_{1\,8} &= [\,0 1'_5,\ -1'_4,\ 4,\ 0 1'_{17}]'/\sqrt{120},\\
s_{1\,9} &= [\,0 1'_{10},\ -1,\ 1,\ 0 1'_{15}]'/\sqrt{12},\\
s_{1\,10} &= [\,0 1'_{12},\ -1,\ 1,\ 0 1'_{13}]'/\sqrt{12},\\
s_{1\,11} &= [\,0 1'_{10},\ -1'_2,\ 1'_2,\ 0 1'_{13}]'/\sqrt{24},\\
s_{1\,12} &= [\,0 1'_{10},\ -1'_4,\ 4,\ 0 1'_{12}]'/\sqrt{120},\\
s_{1\,13} &= [\,0 1'_{15},\ -1,\ 1,\ 0 1'_{10}]'/\sqrt{12},\\
s_{1\,14} &= [\,0 1'_{17},\ -1,\ 1,\ 0 1'_{8}]'/\sqrt{12},\\
s_{1\,15} &= [\,0 1'_{15},\ -1'_2,\ 1'_2,\ 0 1'_{8}]'/\sqrt{24},\\
s_{1\,16} &= [\,0 1'_{15},\ -1'_4,\ 4,\ 0 1'_{7}]'/\sqrt{120},\\
s_{1\,17} &= [\,0 1'_{20},\ -1,\ 1,\ 0 1'_{5}]'/\sqrt{12},\\
s_{1\,18} &= [\,0 1'_{22},\ -1,\ 1,\ 0 1'_{3}]'/\sqrt{12},\\
s_{1\,19} &= [\,0 1'_{20},\ -1'_2,\ 1'_2,\ 0 1'_{3}]'/\sqrt{24},\\
s_{1\,20} &= [\,0 1'_{20},\ -1'_4,\ 4,\ 0 1'_{2}]'/\sqrt{120},\\
s_{1\,21} &= [-1'_5,\ 1'_5,\ 0 1'_{17}]'/\sqrt{60},\\
s_{1\,22} &= [\,0 1'_{10},\ -1'_5,\ 1'_5,\ 0 1'_{7}]'/\sqrt{60},\\
s_{1\,23} &= [-1'_{10},\ 1'_{10},\ 0 1'_{7}]'/\sqrt{120},\\
s_{1\,24} &= [-1'_{20},\ 4 1'_5,\ 0 1'_{2}]'/\sqrt{600},
\end{aligned}
$$

It can easily be checked that, as required (see Section 7.1),

$$
\begin{aligned}
C_1^* s_{0j} &= (r^*)^{\delta} s_{0j} && \text{for } j = 1, 2\ (= \rho_0^*),\\
C_1^* s_{1j} &= (37/42)(r^*)^{\delta} s_{1j} && \text{for } j = 1, 2, ..., 24\ (= \rho_1^*),
\end{aligned}
$$

and that the vectors $s_{0\,1}$, $s_{0\,2}$, $s_{1\,1}$ $s_{1\,2}$, ..., $s_{1\,24}$ and $s_{v^*} = (n^*)^{-1/2} 1_{v^*}$ are all $(r^*)^{\delta}$-orthonormal, which confirms that the parametric functions $\{c'_{\beta j}\tau = s'_{\beta j}(r^*)^{\delta}\tau,\ j = 1, 2, ..., \rho_\beta^*,\ \beta = 0, 1,\}$ form a complete set of basic contrasts (see Lemma 2.3.3 and Definition 3.4.1).

Referring to Theorem 3.6.1, it is now possible to obtain for these contrasts the within stratum BLUEs and their variances. Note that, for this example, the intra-block BLUEs and their variances are obtainable by the formulae

$$
(\widehat{c'_{0j}\tau})_1 = (\varepsilon_{10}^*)^{-1} s'_{0j} Q_1^* \quad \text{and} \quad \mathrm{Var}[(\widehat{c'_{0j}\tau})_1] = (\varepsilon_{10}^*)^{-1}\sigma_1^2 \quad \text{for } j = 1, 2,
$$

$$
(\widehat{c'_{1j}\tau})_1 = (\varepsilon_{11}^*)^{-1} s'_{1j} Q_1^* \quad \text{and} \quad \mathrm{Var}[(\widehat{c'_{1j}\tau})_1] = (\varepsilon_{11}^*)^{-1}\sigma_1^2 \quad \text{for } j = 1, 2, ..., 24,
$$

whereas the inter-block BLUEs and their variances can be obtained by the formulae

$$
(\widehat{c'_{1j}\tau})_2 = (\varepsilon_{21}^*)^{-1} s'_{1j} Q_2^* \quad \text{and} \quad \mathrm{Var}[(\widehat{c'_{1j}\tau})_2] = (\varepsilon_{21}^*)^{-1}\sigma_2^2 \quad \text{for } j = 1, 2, ..., 24,
$$

with $\varepsilon_{1\beta}^* = \varepsilon_\beta^*$ and $\varepsilon_{2\beta}^* = 1 - \varepsilon_\beta^*$ for $\beta = 0, 1$. Unbiased estimates of the unknown values of σ_1^2 and σ_2^2 are provided by the intra-block residual mean square, s_1^2, and the inter-block residual mean square, s_2^2, respectively. Certainly, the coefficients

$\varepsilon_{1\beta}^*$, $\beta = 0, 1$, are the efficiency factors of the analyzed design for the corresponding basic contrasts when estimated in the intra-block analysis, here $\varepsilon_{10}^* = \varepsilon_0^* = 1$ and $\varepsilon_{11}^* = \varepsilon_1^* = 37/42$, whereas ε_{21}^* is the efficiency factor of the design for the corresponding basic contrasts when estimated in the inter-block analysis, here $\varepsilon_{21}^* = 1 - \varepsilon_1^* = 5/42$. In particular, it should be noted that the unique BLUEs, the intra-block BLUEs for the first two basic contrasts, for which $\varepsilon_{10}^* = 1$, are simultaneously the BLUEs of those contrasts under the overall randomization model (3.1.1), as it follows from Theorem 3.5.2.

To test the hypotheses on the nullity of the individual basic contrasts considered here, the F test statistics given in (3.8.91) in relation to the intra-block analysis and those given in (3.8.92) in relation to the inter-block analysis are to be used. Evidently, for $\beta = 0$ those F statistics are unique, obtainable in the intra-block analysis only. Note that in any case, the F statistic is equal to the square of the relevant estimate of the contrast divided by its estimated variance. Results of the intra-block and the inter-block estimation and testing obtained for the considered basic contrasts are presented in Table 7.2. Examining the results there, first note that the equalities (3.8.89) and (3.8.90) are really satisfied by the estimates of the considered basic contrasts, i.e., that

$$(\boldsymbol{Q}_1^*)'(\boldsymbol{C}_1^*)^-\boldsymbol{Q}_1^* = \sum_{j=1}^{2}(\widehat{\boldsymbol{c}_{0j}'\boldsymbol{\tau}})_1^2 + \varepsilon_{11}^*\sum_{j=1}^{24}(\widehat{\boldsymbol{c}_{1j}'\boldsymbol{\tau}})_1^2 = 80.5229$$

and

$$(\boldsymbol{Q}_2^*)'(\boldsymbol{C}_2^*)^-\boldsymbol{Q}_2^* = \varepsilon_{21}^*\sum_{j=1}^{24}(\widehat{\boldsymbol{c}_{1j}'\boldsymbol{\tau}})_2^2 = 60.3183.$$

It may also be checked that the average value of the intra-block individual F statistics is equal to the value of F_1 defined in (3.8.86), and the average value of the inter-block individual F statistics is equal to the value of F_2 in (3.8.87), i.e., that

$$F_1 = (26)^{-1}\Big(\sum_{j=1}^{2} F_{1,0j} + \sum_{j=1}^{24} F_{1,1j}\Big) = 3.369,$$

where

$$F_{1,\beta j} = \varepsilon_{1\beta}^*(\widehat{\boldsymbol{c}_{\beta j}'\boldsymbol{\tau}})_1^2/s_1^2 \quad \text{for } j = 1, 2, ..., \rho_\beta,\ \beta = 0, 1,$$

and

$$F_2 = (24)^{-1}\sum_{j=1}^{24} F_{2,1j} = 0.012,$$

where

$$F_{2,1j} = \varepsilon_{21}^*(\widehat{\boldsymbol{c}_{1j}'\boldsymbol{\tau}})_2^2/s_2^2 \quad \text{for } j = 1, 2, ..., \rho_1\ (= 24).$$

An essential conclusion from Table 7.2 is that only the contrasts denoted by $j = 1, 2$ for $\beta = 0$ and $j = 16, 24$ for $\beta = 1$ can be considered responsible for rejecting the hypothesis H_{01}, and so the hypothesis H_0. Note that this means that the two standard varieties, denoted as strains 26th and 27th, are significantly different

Table 7.2
Intra-block and inter-block estimates of the basic contrasts, together with their estimated variances and the relevant F statistics, obtained in Example 7.3.22

Basic contrast	Intra-block estimate	Estimated variance	Intra-block F	Inter-block estimate	Estimated variance	Inter-block F
$c'_{01}\tau$	4.22	0.919	19.384			
$c'_{02}\tau$	−2.32	0.919	5.875			
$c'_{11}\tau$	0.15	1.044	0.023	5.16	1692.61	0.016
$c'_{12}\tau$	1.12	1.044	1.209	−2.49	1692.61	0.004
$c'_{13}\tau$	−0.93	1.044	0.829	1.74	1692.61	0.002
$c'_{14}\tau$	1.81	1.044	3.141	1.79	1692.61	0.002
$c'_{15}\tau$	−0.19	1.044	0.035	−5.85	1692.61	0.020
$c'_{16}\tau$	0.47	1.044	0.214	0.62	1692.61	0.000
$c'_{17}\tau$	0.88	1.044	0.743	−1.20	1692.61	0.001
$c'_{18}\tau$	−1.16	1.044	1.291	2.22	1692.61	0.003
$c'_{19}\tau$	−1.45	1.044	2.005	−5.30	1692.61	0.017
$c'_{1\,10}\tau$	0.38	1.044	0.138	0.83	1692.61	0.000
$c'_{1\,11}\tau$	−0.56	1.044	0.303	−7.32	1692.61	0.032
$c'_{1\,12}\tau$	−0.12	1.044	0.013	−3.80	1692.61	0.009
$c'_{1\,13}\tau$	−0.90	1.044	0.782	−0.59	1692.61	0.000
$c'_{1\,14}\tau$	−1.71	1.044	2.813	−4.78	1692.61	0.014
$c'_{1\,15}\tau$	1.24	1.044	1.476	−0.61	1692.61	0.000
$c'_{1\,16}\tau$	3.62	1.044	12.573	10.77	1692.61	0.069
$c'_{1\,17}\tau$	1.00	1.044	0.962	−8.14	1692.61	0.039
$c'_{1\,18}\tau$	−0.15	1.044	0.022	4.26	1692.61	0.011
$c'_{1\,19}\tau$	−1.07	1.044	1.089	−0.69	1692.61	0.000
$c'_{1\,20}\tau$	−0.87	1.044	0.721	−6.16	1692.61	0.022
$c'_{1\,21}\tau$	−0.58	1.044	0.325	6.58	1692.61	0.026
$c'_{1\,22}\tau$	1.59	1.044	2.413	3.87	1692.61	0.009
$c'_{1\,23}\tau$	1.65	1.044	2.596	−1.37	1692.61	0.001
$c'_{1\,24}\tau$	−5.27	1.044	26.609	−2.67	1692.61	0.004
Critical values		$F_{0.05;1,154}$ $F_{0.01;1,154}$	$= 3.90$ $= 6.80$		$F_{0.05;1,5}$ $F_{0.01;1,5}$	$= 6.61$ $= 16.26$

with regard to the analyzed trait, that they are jointly different from the average of the 25 breeding strains, that among the latter there are significant differences between the strains 16th, 17th, 18th, 19th on one side and the strain 20th on the other, and that on the average there is a significant difference between the first twenty and the remaining five breeding strains. Also note that the results of the inter-block individual F tests confirm that the hypothesis H_{02} cannot be rejected in the inter-block analysis of variance. However, one has to note that the inter-block F test of this hypothesis is rather poor, as based only on 5 d.f. for residuals.

Because for twenty four of the considered basic contrasts two kinds of estimates and tests have been obtained, using the intra-block and the inter-block information separately, it might now be interesting to perform for them the combined estimation and testing, based on the information from both of these strata. As to the combined estimators, the formula (3.8.96) is applicable. To test the hypotheses $H_{0,1j} : \mathbf{c}'_{1j}\boldsymbol{\tau} = 0$ for $j = 1, 2, ..., 24$, the formula (3.8.98) can be applied. These formulae, however, involve estimates of the weights $w_{1\beta} = w_1 = \varepsilon^*_{11}\sigma^2_2/(\varepsilon^*_{11}\sigma^2_2 + \varepsilon^*_{21}\sigma^2_1)$ and $w_{2\beta} = w_2 = 1 - w_1$, defined in (3.8.16), here for $\beta = 1$ only. To obtain these estimates, it is necessary first to estimate the stratum variances σ^2_1 and σ^2_2, defined in (3.5.7). Computationally, this estimation can be accomplished by employing the Nelder (1968) iterative procedure described in Section 3.8.2 (equivalent to the corresponding REML and iterated MINQUE procedure). When entering this procedure with preliminary estimates of the stratum variances obtained by the Yates–Rao estimators (described in Section 3.9.1), i.e., with $\hat{\sigma}^2_{1(YR)} = 0.919388$ and $\hat{\sigma}^2_{2(YR)} = 40.1833$, then the procedure provides the following results for the quantities appearing in the formulae (3.8.35) and (3.8.36) of the adopted Nelder estimators of the stratum variances:

Iteration cycle	1	2	3
d'_1	154.07398	154.08114	154.08111
d'_2	28.92602	28.91886	28.91889
$\hat{w}_1$	0.99692	0.99662	0.99662
$\hat{w}_2$	0.00308	0.00338	0.00338
$\hat{\sigma}^2_1 \equiv \hat{\sigma}^2_{1(N)}$	0.91897	0.91893	0.91893
$\hat{\sigma}^2_2 \equiv \hat{\sigma}^2_{2(N)}$	36.60999	36.61799	36.61797

Coinciding results have also been obtained by employing the REML estimation procedure provided by GENSTAT 5 (1996) software, here running through 8 iteration cycles to reach convergence with the same precision.

Now, with $\hat{\zeta}_j = \hat{\zeta} = 0.0002776$ (equal for $j = 1, 2, ..., 24$) obtained according to the formula following (3.8.78), it can be seen that the condition (3.8.84) is satisfied, as $\hat{w}_2/\hat{w}_1 = 0.0033915$. Hence, the estimated approximate variance (3.8.77) of the empirical combined estimator (3.8.81) can be expected to be smaller than the estimated variance of the intra-block BLUE for any of the basic contrasts $\mathbf{c}'_{1j}\boldsymbol{\tau}, j = 1, 2, ..., 24$. In fact, from (3.8.77), it follows that

$$\widehat{\mathrm{Var}}(\widetilde{\mathbf{c}'_{1j}\boldsymbol{\tau}}) \cong (\varepsilon^*_1)^{-1}\hat{w}_1(1 + \hat{\zeta})\hat{\sigma}^2_1 = 1.13161\hat{\sigma}^2_1 \; (= 1.03987),$$

which is smaller than

$$\widehat{\mathrm{Var}}[(\widehat{\mathbf{c}'_{1j}\boldsymbol{\tau}})_1] = (\varepsilon^*_1)^{-1}\hat{\sigma}^2_1 = 1.13514\hat{\sigma}^2_1 \; (= 1.04363, \;\; \text{if } \hat{\sigma}^2_1 = s^2_1),$$

obtained from (3.4.2) for the intra-block BLUE (see Table 7.2). Furthermore, applying the formulae (3.8.96) [(3.8.81)], i.e., the empirical estimator

$$\widetilde{c'_{1j}\tau} = \hat{w}_1(\widehat{c'_{1j}\tau})_1 + \hat{w}_2(\widehat{c'_{1j}\tau})_2,$$

(3.8.98), i.e., the corresponding F statistic

$$F_{1j} = [\hat{w}_1(1+\hat{\zeta})]^{-1}\varepsilon_1^*(\widetilde{c'_{1j}\tau})^2/s_1^2 \quad \text{for} \quad j = 1, 2, ..., 24,$$

with $s_1^2 = 0.919390$ from the intra-block anlysis, and (3.8.101), i.e.,

$$d_{(j)} = \frac{2}{2 + \dfrac{3\zeta(w_2 - 3w_1\zeta)^2}{(1+\zeta)^2 w_1 w_2}},$$

the following results have been obtained:

$\widetilde{c'_{1\,1}\tau} =$	0.171,	$F_{1\,1} =$	0.028,	$P =$	0.8668,
$\widetilde{c'_{1\,2}\tau} =$	1.111,	$F_{1\,2} =$	1.187,	$P =$	0.2776,
$\widetilde{c'_{1\,3}\tau} =$	−0.921,	$F_{1\,3} =$	0.816,	$P =$	0.3679,
$\widetilde{c'_{1\,4}\tau} =$	1.810,	$F_{1\,4} =$	3.150,	$P =$	0.0779,
$\widetilde{c'_{1\,5}\tau} =$	−0.211,	$F_{1\,5} =$	0.043,	$P =$	0.8363,
$\widetilde{c'_{1\,6}\tau} =$	0.473,	$F_{1\,6} =$	0.215,	$P =$	0.6433,
$\widetilde{c'_{1\,7}\tau} =$	0.873,	$F_{1\,7} =$	0.733,	$P =$	0.3931,
$\widetilde{c'_{1\,8}\tau} =$	−1.149,	$F_{1\,8} =$	1.269,	$P =$	0.2617,
$\widetilde{c'_{1\,9}\tau} =$	−1.460,	$F_{1\,9} =$	2.047,	$P =$	0.1545,
$\widetilde{c'_{1\,10}\tau} =$	0.381,	$F_{1\,10} =$	0.139,	$P =$	0.7095,
$\widetilde{c'_{1\,11}\tau} =$	−0.586,	$F_{1\,11} =$	0.330,	$P =$	0.5667,
$\widetilde{c'_{1\,12}\tau} =$	−0.131,	$F_{1\,12} =$	0.016,	$P =$	0.8981,
$\widetilde{c'_{1\,13}\tau} =$	−0.902,	$F_{1\,13} =$	0.783,	$P =$	0.3777,
$\widetilde{c'_{1\,14}\tau} =$	−1.724,	$F_{1\,14} =$	2.856,	$P =$	0.0931,
$\widetilde{c'_{1\,15}\tau} =$	1.235,	$F_{1\,15} =$	1.466,	$P =$	0.2278,
$\widetilde{c'_{1\,16}\tau} =$	3.647,	$F_{1\,16} =$	12.781,	$P =$	0.0005,
$\widetilde{c'_{1\,17}\tau} =$	0.971,	$F_{1\,17} =$	0.906,	$P =$	0.3427,
$\widetilde{c'_{1\,18}\tau} =$	−0.135,	$F_{1\,18} =$	0.017,	$P =$	0.8950,
$\widetilde{c'_{1\,19}\tau} =$	−1.065,	$F_{1\,19} =$	1.089,	$P =$	0.2983,
$\widetilde{c'_{1\,20}\tau} =$	−0.885,	$F_{1\,20} =$	0.753,	$P =$	0.3868,
$\widetilde{c'_{1\,21}\tau} =$	−0.558,	$F_{1\,21} =$	0.299,	$P =$	0.5853,
$\widetilde{c'_{1\,22}\tau} =$	1.595,	$F_{1\,22} =$	2.444,	$P =$	0.1200,
$\widetilde{c'_{1\,23}\tau} =$	1.636,	$F_{1\,23} =$	2.572,	$P =$	0.1108,
$\widetilde{c'_{1\,24}\tau} =$	−5.261,	$F_{1\,24} =$	26.603,	$P =$	0.0000.

As shown in Section 3.8.5 in connection with (3.8.98), under the hypothesis $H_{0,1j} : \boldsymbol{c}'_{1j}\boldsymbol{\tau} = 0$, the distribution of the applied statistic F_{1j} can be approximated by the central F distribution with the degrees of freedom (d.f.) $d_{(j)}$ given above and $d_1 = n^* - b^* - v^* + 1$. For this example, $\hat{d}_{(j)} = 0.999999$, obtained with w_1, w_2 and ζ replaced by their estimates, whereas $d_1 = 154$, as already used in the intra-block analysis. With these d.f., the relevant P values have been obtained as given above. They are below 0.05 only for the basic contrasts denoted by $j = 16, 24$ (for $\beta = 1$), confirming that the relevant hypotheses $H_{0,1j}$, for $j = 16, 24$, can surely be rejected. The rejection of $H_{0,0j}$ for $j = 1, 2$ need not be cofirmed, because there is no inter-block information for these two contrasts.

Also, it may be noted that the statistic

$$F_{(1)} = (24)^{-1} \sum_{j=1}^{24} F_{1j} = 2.6060,$$

can be used to test the intersection hypothesis

$$H_{0(1)} = \bigcap_{j=1}^{24} H_{0,1j} : \boldsymbol{U}'_{(1)}\boldsymbol{\tau} = \boldsymbol{0}, \quad \text{where} \quad \boldsymbol{U}_{(1)} = [\boldsymbol{c}_{1\,1} : \boldsymbol{c}_{1\,2} : \cdots : \boldsymbol{c}_{1\,24}].$$

It follows, from the results given in Sections 3.7.4 and 3.8.5, that the distribution of the statistic $F_{(1)}$ can be approximated by the central F distribution with the d.f. d and d_1, where d given in (3.7.74) can here be written equivalently as

$$d = \frac{2\rho_1}{2 + \dfrac{\zeta(w_2 - 3w_1\zeta)^2}{(1+\zeta)^2 w_1 w_2}(2 + \rho_1)}.$$

In this example, $\rho_1 = \rho_1^* = 24, \hat{d} = 23.9998$ and $d_1 = 154$. With these d.f., the relevant P value for $F_{(1)} = 2.6060$ is 0.000223, showing that the hypothesis $H_{0(1)}$ can definitely be rejected.

Finally, to complete the analysis, it may be interesting to obtain, from the combined estimates of basic contrasts, the empirical estimator of $\boldsymbol{\tau}$, together with its covariance matrix. For this, the formula (3.8.52) can be used, which here gets the form

$$\tilde{\boldsymbol{\tau}} = (\boldsymbol{r}^*)^{-\delta}[\sum_{j=1}^{2} \boldsymbol{c}_{0j}(\widehat{\boldsymbol{c}'_{0j}\boldsymbol{\tau}}) + \sum_{j=1}^{24} \boldsymbol{c}_{1j}(\widetilde{\boldsymbol{c}'_{1j}\boldsymbol{\tau}})] + (n^*)^{-1}\boldsymbol{1}_{v^*}\boldsymbol{1}'_{n^*}\boldsymbol{y},$$

where $(\boldsymbol{r}^*)^{-\delta} = \text{diag}[(1/6)\boldsymbol{I}_{25} : (1/30)\boldsymbol{I}_2]$, $n^* = 210$, $v^* = 27$, and where $\widehat{\boldsymbol{c}'_{0j}\boldsymbol{\tau}}$, $j = 1, 2$, are the intra-block estimators of the first two basic contrasts (see Table 7.2) and $\widetilde{\boldsymbol{c}'_{1j}\boldsymbol{\tau}}$, $j = 1, 2, ..., 24$, are the empirical combined estimators of the remaining basic contrasts chosen in the present example. With the estimates of these contrasts already obtained above, the vector $\tilde{\boldsymbol{\tau}}$ becomes

$$\tilde{\boldsymbol{\tau}} = [15.701,\ 15.800,\ 15.053,\ 15.695,\ 16.388,\\ 15.571,\ 15.449,\ 15.730,\ 16.003,\ 15.164,\\ 16.301,\ 15.458,\ 15.531,\ 15.751,\ 15.700,\\ 15.836,\ 15.314,\ 16.577,\ 15.582,\ 17.491,\\ 14.749,\ 15.309,\ 14.633,\ 14.555,\ 14.408,\ 14.690,\ 15.780]'.$$

These estimates of treatment parameters coincide exactly with the results obtained directly from the GENSTAT 5 (1996) software (employed here as in Example 3.7.3). The covariance matrix of $\tilde{\boldsymbol{\tau}}$, as considered in Theorem 3.8.2, is not readily obtainable, but can be approximated using the formula

$$\mathrm{Cov}(\tilde{\boldsymbol{\tau}}) \cong \sigma_1^2 \boldsymbol{H}_0 + \sigma_1^2 \frac{w_1}{\varepsilon_1^*}(1+\zeta)\boldsymbol{H}_1 + \sigma_3^2 (n^*)^{-1} \mathbf{1}_{v^*}\mathbf{1}'_{v^*},$$

following from (3.8.21) and the derivation in Section 3.8.4 that leads to (3.8.76), the matrices $\boldsymbol{H}_0$ and $\boldsymbol{H}_1$ being here of the forms

$$\boldsymbol{H}_0 = (\boldsymbol{r}^*)^{-\delta} \sum_{j=1}^{2} \boldsymbol{c}_{0j}\boldsymbol{c}'_{0j} (\boldsymbol{r}^*)^{-\delta} = \frac{1}{210}\begin{bmatrix} \frac{2}{5}\mathbf{1}_{25}\mathbf{1}'_{25} & -\mathbf{1}_{25}\mathbf{1}'_{2} \\ -\mathbf{1}_{2}\mathbf{1}'_{25} & 7\boldsymbol{I}_2 - \mathbf{1}_2\mathbf{1}'_2 \end{bmatrix} = \boldsymbol{L}_0^*(\boldsymbol{r}^*)^{-\delta}$$

and

$$\boldsymbol{H}_1 = (\boldsymbol{r}^*)^{-\delta} \sum_{j=1}^{24} \boldsymbol{c}_{1j}\boldsymbol{c}'_{1j} (\boldsymbol{r}^*)^{-\delta} = \frac{1}{6}\begin{bmatrix} \boldsymbol{I}_{25} - \frac{1}{25}\mathbf{1}_{25}\mathbf{1}'_{25} & \mathbf{O} \\ \mathbf{O} & \mathbf{O} \end{bmatrix} = \boldsymbol{L}_1^*(\boldsymbol{r}^*)^{-\delta}.$$

The above approximation of $\mathrm{Cov}(\tilde{\boldsymbol{\tau}})$ shows that for any contrast $\boldsymbol{c}'\boldsymbol{\tau}$, the variance of its empirical estimator $\widetilde{\boldsymbol{c}'\boldsymbol{\tau}} = \boldsymbol{c}'\tilde{\boldsymbol{\tau}}$ can be approximated as

$$\mathrm{Var}(\widetilde{\boldsymbol{c}'\boldsymbol{\tau}}) = \boldsymbol{c}'\mathrm{Cov}(\tilde{\boldsymbol{\tau}})\boldsymbol{c} \cong [\boldsymbol{c}'\boldsymbol{H}_0\boldsymbol{c} + \frac{w_1}{\varepsilon_1^*}(1+\zeta)\boldsymbol{c}'\boldsymbol{H}_1\boldsymbol{c}]\sigma_1^2,$$

the last term in the above formula for $\mathrm{Cov}(\tilde{\boldsymbol{\tau}})$ being not involved, because of $\boldsymbol{c}'\mathbf{1}_{v^*} = 0$ for a contrast. To estimate this variance, the unknown values of w_1 and ζ are to be replaced by their estimates, here $\hat{w}_1 = 0.99662$ and $\hat{\zeta} = 0.0002776$. If, for example, the researcher is interested to compare the first strain with the first standard variety, i.e., to estimate the contrast $\tau_1 - \tau_{26}$, he (she) obtains $\widetilde{\tau_1 - \tau_{26}} = 15.701 - 14.690 = 1.011$, with its estimated variance

$$\widehat{\mathrm{Var}}(\widetilde{\tau_1 - \tau_{26}}) \cong [\frac{1}{25} + \frac{\hat{w}_1}{\varepsilon_1^*}(1+\hat{\zeta})\frac{4}{25}]\hat{\sigma}_1^2 = 0.203137.$$

If in this estimation the correction factor $1+\hat{\zeta} = 1.0002776$ (derived in Section 3.8.4) is ignored, the result is 0.203091, which coincides with that obtained from the GENSTAT 5 (1996) software. In fact, in this example, $\hat{\zeta} = 0.0002776$ is too small to make any essential difference. Nevertheless, as shown in Sections 3.8.3 and 3.8.4, one has to be aware that the replacement of the unknown stratum variances σ_1^2 and σ_2^2 by their estimates increases the variance of any contrast that is not estimated in the intra-block analysis with full efficiency. It may also be noted that if the estimation of the above contrast is restricted to the intra-block analysis, its estimated variance is

$$\mathrm{Var}[(\widehat{\tau_1 - \tau_{26}})_1] = (\frac{1}{25} + \frac{1}{\varepsilon_1^*}\frac{4}{25})s_1^2 = 0.203757.$$

It may be noted that the small difference between the two variances obtained here can be explained by the fact that both the average efficiency factor of the design, $\varepsilon^* = 0.88909$, and the variance ratio, $\hat{\delta} = \hat{\sigma}_2^2/\hat{\sigma}_1^2 = 39.848$, have attained high values in this example. This indicates that the experiment was designed and conducted very successfully (see the discussion in Section 3.8.1).

7.4 Designs with three efficiency factors

In some experimental situations, the existence of three distinct efficiency factors for estimating contrasts in the intra-block analysis may be allowed. Even in this case, designs in which some basic contrasts of interest are estimated with full efficiency will usually be preferred. Thus, $(\rho_0; \rho_1, \rho_2; 0)$-EB designs with $\rho_0 \geq 1$ will be treated in this section. Four different classes of such designs are discussed.

7.4.1 Proper designs that are equireplicate

General properties of an equireplicate $(\rho_0; \rho_1, ..., \rho_{m-1}; 0)$-EB design have been considered in Section 4.4.3 and also in Section 7.1. Here, a proper equireplicate $(\rho_0; \rho_1, \rho_2; 0)$-EB design with parameters v, b, r, k will be of interest. Recall [see (7.1.3) and (7.1.4)] that for any such design

$$\boldsymbol{F} = r^{-1}\boldsymbol{C}_1 = \boldsymbol{L}_0 + \varepsilon_1\boldsymbol{L}_1 + \varepsilon_2\boldsymbol{L}_2,$$

where

$$\boldsymbol{L}_\beta = \frac{1}{r}\sum_{j=1}^{\rho_\beta} \boldsymbol{s}_{\beta j}\boldsymbol{s}'_{\beta j} = \sum_{j=1}^{\rho_\beta} \boldsymbol{p}_{\beta j}\boldsymbol{p}'_{\beta j}, \ \boldsymbol{p}'_{\beta j}\boldsymbol{p}_{\beta j'} = \delta_{jj'}, \ \boldsymbol{L}_\beta\boldsymbol{L}_{\beta'} = \delta_{\beta\beta'}\boldsymbol{L}_\beta,$$

$$\mathrm{rank}(\boldsymbol{L}_\beta) = \rho_\beta, \ \beta = 0, 1, 2, \ \rho_0 + \rho_1 + \rho_2 = v - 1,$$

$$\boldsymbol{\Omega} = \frac{1}{r}\left(\boldsymbol{L}_0 + \frac{1}{\varepsilon_1}\boldsymbol{L}_1 + \frac{1}{\varepsilon_2}\boldsymbol{L}_2 + \frac{1}{v}\boldsymbol{1}_v\boldsymbol{1}'_v\right),$$

and the average efficiency factor of the design is $\varepsilon = (v-1)/(\rho_0 + \rho_1/\varepsilon_1 + \rho_2/\varepsilon_2)$, whereas any ρ_β basic contrasts of the βth subset $\{\boldsymbol{c}'_{\beta j}\boldsymbol{\tau} = r^{1/2}\boldsymbol{p}'_{\beta j}\boldsymbol{\tau}$, $j = 1, 2, ..., \rho_\beta\}$ are estimated in the intra-block analysis with the same efficiency $\varepsilon_\beta, \beta = 0, 1, 2$. Moreover, note that $\boldsymbol{L}_0$ spans the subspace of all contrasts estimated in the intra-block analysis with full efficiency, $\varepsilon_0 = 1$, whereas $\boldsymbol{L}_1$ or $\boldsymbol{L}_2$ spans the subspace of all contrasts which are estimated in that analysis with a lower efficiency, $\varepsilon_1 < 1$ or $\varepsilon_2 < 1$, respectively. The dimensions of these subspaces are ρ_0, ρ_1 and ρ_2, respectively. These numbers sum to $v-1$ since only connected designs are considered.

Now methods of constructing such designs will be considered. One of the most standard choices for this purpose is a 3-associate PBIB design, the definition of which can be found in Section 6.0.2.

7.4.1.1 PBIB designs

Recall that a connected 3-associate PBIB design with parameters v, b, r, k, λ_1, λ_2, λ_3 is either $(0; \rho_1, \rho_2, \rho_3; 0)$-EB or $(\rho_0; \rho_1, \rho_2; 0)$-EB with $\rho_0 \geq 1$. The latter occurs if one of the eigenvalues of $\boldsymbol{NN}'$ other than rk is zero, i.e., when the incidence matrix $\boldsymbol{N}$ of the PBIB design has a rank smaller than v (see Section 6.0.2).

For standard association schemes with three associate classes, which necessarily provide 3-associate PBIB designs, see Section 6.0.2. Now some methods of constructing such PBIB designs will be presented. Several methods are available in the literature. For the present purpose, PBIB designs with $\rho_0 \geq 1$, i.e., those for which $\varepsilon_0 = 1$ exists, will be considered here. Before that, it will be useful to give some characterization results on such PBIB designs. These results are helpful if one wants to know appropriate structure of PBIB designs satisfying such properties. First, let a rectangular association scheme on st treatments [see Section 6.0.2(E)] be considered.

Theorem 7.4.1 (Kageyama and Tsuji, 1977). *In a rectangular PBIB design with parameters $v = st, b, r, k$, λ_1, λ_2, λ_3, ψ_1, ψ_2, ψ_3 [defined in Section 6.0.2(E)], the following holds:*

(i) $\psi_1 = 0$ *if and only if k/t is an integer and every block of the design contains k/t treatments from each of the t columns of the association scheme.*

(ii) $\psi_2 = 0$ *if and only if k/s is an integer and every block of the design contains k/s treatments from each of the s rows of the association scheme.*

(iii) $\psi_3 = 0$ *if and only if* $\boldsymbol{N}'[s(\boldsymbol{I}_s \otimes \mathbf{1}_t\mathbf{1}_t') + t(\mathbf{1}_s\mathbf{1}_s' \otimes \boldsymbol{I}_t) - ts\boldsymbol{I}_v] = k\mathbf{1}_b\mathbf{1}_v'$ *for the incidence matrix $\boldsymbol{N}$ of the PBIB design.*

(iv) *When s and t are relatively prime, there does not exist a rectangular PBIB design with $v = st$ treatments satisfying $\psi_1 = \psi_2 = 0$.*

A rectangular PBIB design satisfying (i), (ii) or (iii) in Theorem 7.4.1 produces a $(\rho_0;\ \rho_1,\ \rho_2; 0)$-EB design. Note that the design of Example 6.0.6 satisfies Theorem 7.4.1(ii). A method of constructing designs that satisfy Theorem 7.4.1(i) can be introduced as follows. In fact, this follows from Theorem 6.6.4 with $a = 0$ and $c = 1$.

Theorem 7.4.2 (Bhagwandas, Kageyama and Banerjee, 1985). *Let $\boldsymbol{N}$ be the $v \times b$ incidence matrix of a BIB design with parameters v, b, r, k, λ. Then the* *incidence matrix*

$$\begin{bmatrix} N & O & 1_v1_b' - N \\ 1_v1_b' - N & N & O \\ O & 1_v1_b' - N & N \end{bmatrix} \tag{7.4.1}$$

yields a rectangular PBIB design with parameters $v^* = 3v$, $b^* = 3b$, $r^* = b$, $k^* = v$, $\lambda_1^* = b-2(r-\lambda)$, $\lambda_2^* = 0$, $\lambda_3^* = r-\lambda$, $n_1^* = v-1$, $n_2^* = 2$, $n_3^* = 2(v-1)$, *i.e.*, $(\rho_0^*; \rho_1^*, \rho_2^*; 0)$*-EB with* $\varepsilon_0^* = 1$, $\varepsilon_1^* = 3(r-\lambda)(v-1)/(bv)$, $\varepsilon_2^* = [bv-3(r-\lambda)]/(bv)$, $\rho_0^* = v-1$, $\rho_1^* = 2$, $\rho_2^* = 2(v-1)$, $L_0^* = 3^{-1}1_31_3' \otimes (I_v - v^{-1}1_v1_v')$, $L_1^* = (I_3 - 3^{-1}1_31_3') \otimes v^{-1}1_v1_v'$, $L_2^* = (I_3 - 3^{-1}1_31_3') \otimes (I_v - v^{-1}1_v1_v')$.

Proof. The proof is straightforward by noting that an association scheme is given in a $3 \times v$ rectangular array [see Section 6.0.2(E)] as in

$$\begin{matrix} 1 & 2 & \cdots & v \\ v+1 & v+2 & \cdots & 2v \\ 2v+1 & 2v+2 & \cdots & 3v \end{matrix} .$$

□

Even in this case one can get a $(\rho_0^*; \rho_1^*; 0)$-EB design, as in Theorem 7.3.6, when a basic BIB design satisfies the condition $b = 3(r-\lambda)$.

Because a large number of BIB designs are available (see, for example, Table 8.2), it is easy to illustrate Theorem 7.4.2. The reader is advised to try it as an exercise. Another 3-associate PBIB design is illustrated by the following example.

Example 7.4.1. For 3-associate GD PBIB designs [see Section 6.0.2(F)], Raghavarao (1960a) gave 8 numerical constructions in the range of $r, k \leq 10$. Among them, two designs have one of the ε_β equal to one, as shown below.

(1) Example 6.0.7 shows such a design with parameters $v = b = 8$, $r = k = 4$, $s_1 = s_2 = s_3 = 2$, $\lambda_1 = 2$, $\lambda_2 = 1$, $\lambda_3 = 2$, whose incidence matrix is given by

$$\begin{matrix} (1,1,1) \\ (1,1,2) \\ (1,2,1) \\ (1,2,2) \\ (2,1,1) \\ (2,1,2) \\ (2,2,1) \\ (2,2,2) \end{matrix} \left[\begin{array}{cccc|cccc} 1 & 1 & 0 & 0 & 1 & 0 & 1 & 0 \\ 1 & 1 & 0 & 0 & 0 & 1 & 0 & 1 \\ 0 & 0 & 1 & 1 & 1 & 0 & 0 & 1 \\ 0 & 0 & 1 & 1 & 0 & 1 & 1 & 0 \\ 1 & 0 & 0 & 1 & 0 & 0 & 1 & 1 \\ 0 & 1 & 1 & 0 & 0 & 0 & 1 & 1 \\ 1 & 0 & 1 & 0 & 1 & 1 & 0 & 0 \\ 0 & 1 & 0 & 1 & 1 & 1 & 0 & 0 \end{array}\right], \quad \begin{matrix} \varepsilon_0 = 1, & \rho_0 = 1, \\ \varepsilon_1 = 7/8, & \rho_1 = 4, \\ \varepsilon_2 = 3/4, & \rho_2 = 2. \end{matrix}$$

Here, (1, 1, 1), (1, 1, 2), (1, 2, 1), (1, 2, 2), (2, 1, 1), (2, 1, 2), (2, 2, 1), (2, 2, 2) denote the 8 treatments, possibly in a factorial structure. This design gives also a 2-resolvable $(\rho_0^*; \rho_1^*, \rho_2^*; 0)$-EB design with parameters $\varepsilon_0^* = 1$, $\varepsilon_1^* = 7/8$, $\varepsilon_2^* = 3/4$, $\rho_0^* = 1$, $\rho_1^* = 4$, $\rho_2^* = 2$, $L_0^* = (I_2 - 2^{-1}1_21_2') \otimes 4^{-1}1_41_4'$, $L_1^* = I_4 \otimes (I_2 - 2^{-1}1_21_2')$, $L_2^* = I_2 \otimes (I_2 - 2^{-1}1_21_2') \otimes 2^{-1}1_21_2'$, giving

$$\Omega^* = \frac{1}{336} I_2 \otimes [I_2 \otimes (96I_2 + 1_21_2') - 7(1_21_2' - I_2) \otimes 1_21_2'].$$

Here, the average efficiency factor of the design is $147/173 = 0.8497$. The variances of the intra-block BLUEs of basic contrasts are σ_1^2, $8\sigma_1^2/7$ or $4\sigma_1^2/3$, depending on whether they correspond to $\boldsymbol{L}_0^*$, $\boldsymbol{L}_1^*$ or $\boldsymbol{L}_2^*$. Furthermore, for example, elementary contrasts are estimated in the intra-block analysis with three different variances, $4\sigma_1^2/7$, $13\sigma_1^2/21$ and $97\sigma_1^2/168$.

(2) A design with parameters $v = b = 8$, $r = k = 6$, $s_1 = s_2 = s_3 = 2$, $\lambda_1 = 4$, $\lambda_2 = 5$, $\lambda_3 = 4$. Taking the same treatments as those of (1), one can get a design below. This yields also a $(\rho_0^*; \rho_1^*, \rho_2^*; 0)$-EB design with parameters $\varepsilon_0^* = 1$, $\varepsilon_1^* = 17/18$, $\varepsilon_2^* = 8/9$, $\rho_0^* = 2$, $\rho_1^* = 4$, $\rho_2^* = 1$, $\boldsymbol{L}_1^*$ is the same as in (1), but $\boldsymbol{L}_0^*$ and $\boldsymbol{L}_2^*$ are interchanged. Here, the average efficiency factor of the design is $952/1001 = 0.9510$. The variances of the intra-block BLUEs of basic contrasts are σ_1^2, $18\sigma_1^2/17$ or $9\sigma_1^2/8$, depending on whether they correspond to $\boldsymbol{L}_0^*$, $\boldsymbol{L}_1^*$ or $\boldsymbol{L}_2^*$. Furthermore, elementary contrasts are estimated in the intra-block analysis with three different variances, $6\sigma_1^2/17$, $35\sigma_1^2/102$, $577\sigma_1^2/1632$, because in this case the matrx $\boldsymbol{\Omega}$ is obtained as

$$\boldsymbol{\Omega}^* = \frac{1}{6528}\Big[\boldsymbol{I}_4 \otimes (1152\boldsymbol{I}_2 - 151\boldsymbol{1}_2\boldsymbol{1}_2') + 17\boldsymbol{I}_2 \otimes (\boldsymbol{1}_2\boldsymbol{1}_2' - \boldsymbol{I}_2) \otimes \boldsymbol{1}_2\boldsymbol{1}_2' + 17(\boldsymbol{I}_2 - \boldsymbol{1}_2\boldsymbol{1}_2') \otimes \boldsymbol{1}_4\boldsymbol{1}_4'\Big].$$

The incidence matrix of the design is given by

$$\begin{bmatrix} 1&1&1&1&1&1&0&0\\ 1&1&1&1&0&0&1&1\\ 1&1&1&1&1&0&0&1\\ 1&1&1&1&0&1&1&0\\ 1&1&0&0&1&1&1&1\\ 0&0&1&1&1&1&1&1\\ 1&0&1&0&1&1&1&1\\ 0&1&0&1&1&1&1&1 \end{bmatrix}, \qquad \begin{array}{ll} \varepsilon_0 = 1, & \rho_0 = 2,\\ \varepsilon_1 = 17/18, & \rho_1 = 4,\\ \varepsilon_2 = 8/9, & \rho_2 = 1. \end{array}$$

Remark 7.4.1. Referring to Theorem 6.6.4 and Remark 6.6.2 with $a = c = 1$, one obtains a pattern similar to that in Theorem 7.4.2. Now, let $\boldsymbol{N}$ be the $v \times b$ incidence matrix of a BIB design with parameters v, b, r, k, λ. Then

$$\begin{bmatrix} \boldsymbol{N} & \boldsymbol{1}_v\boldsymbol{1}_b' & \boldsymbol{1}_v\boldsymbol{1}_b' - \boldsymbol{N}\\ \boldsymbol{1}_v\boldsymbol{1}_b' - \boldsymbol{N} & \boldsymbol{N} & \boldsymbol{1}_v\boldsymbol{1}_b'\\ \boldsymbol{1}_v\boldsymbol{1}_b' & \boldsymbol{1}_v\boldsymbol{1}_b' - \boldsymbol{N} & \boldsymbol{N} \end{bmatrix} \tag{7.4.2}$$

is the incidence matrix of a rectangular PBIB design with parameters $v^* = 3v$, $b^* = 3b$, $r^* = 2b$, $k^* = 2v$, $\lambda_1^* = 2(b-r+\lambda)$, $\lambda_2^* = b$, $\lambda_3^* = b+r-\lambda$, other quantities are the same as those in Theorem 7.4.2. This pattern will be discussed again in Theorem 7.4.8. Some analogy of this pattern may yield other rectangular PBIB designs (see Kageyama and Mohan, 1984a, Theorem 2).

Many other methods of constructing rectangular designs can be found in the literature. See, for example, Suen (1989), Gupta and Mukerjee (1989), Sinha

(1991b), Sinha, Kageyama and Singh (1993), Kageyama and Miao (1995a), and Sinha, Singh, Kageyama and Singh (2002). Extensive lists of rectangular designs in the range of parameters $2 \le r, k \le 10$ are given by Sinha, Kageyama and Singh (1993) and Sinha, Singh, Kageyama and Singh (2002).

7.4.1.2 Other than PBIB designs

PBIB designs are equireplicate and proper. Furthermore, they impose some association relations among the treatments of the design (as described in Section 6.0.2). Similarly as in Section 7.3.1.2, without demanding any of such association schemes, binary equireplicate and proper $(\rho_0; \rho_1, \rho_2; 0)$-EB designs, that are not PBIB designs, will now be considered.

The construction of such a design consists actually in finding some incidence matrix $\boldsymbol{N}$ such for which the matrix $r^{-1}\boldsymbol{C}_1 = \boldsymbol{I}_v - (rk)^{-1}\boldsymbol{N}\boldsymbol{N}'$ can be written as $r^{-1}\boldsymbol{C}_1 = \boldsymbol{L}_0 + \varepsilon_1\boldsymbol{L}_1 + \varepsilon_2\boldsymbol{L}_2$, following from the spectral decomposition of the matrix $(rk)^{-1}\boldsymbol{N}\boldsymbol{N}' (= \boldsymbol{M})$, as recalled (from Sections 4.3.2 and 4.4.3) at the beginning of Section 7.4.1.

As a complementation method, Theorem 6.2.1 can yield the following result.

Corollary 7.4.1. *If $\boldsymbol{N}$ is the $v \times b$ incidence matrix of a binary $(\rho_0; \rho_1, \rho_2; 0)$-EB design with parameters v, b, r, k, ε_β, ρ_β, $\boldsymbol{L}_\beta$, $\beta = 0, 1, 2$, then its complement $\boldsymbol{1}_v\boldsymbol{1}_b' - \boldsymbol{N}$ is the incidence matrix of a $(\tilde{\rho}_0; \tilde{\rho}_1, \tilde{\rho}_2; 0)$-EB design with parameters $\tilde{v} = v$, $\tilde{b} = b$, $\tilde{r} = b - r$, $\tilde{k} = v - k$, $\tilde{\varepsilon}_\beta = 1 - rk(1 - \varepsilon_\beta)/[(v-k)(b-r)]$, $\tilde{\rho}_\beta = \rho_\beta$, $\tilde{\boldsymbol{L}}_\beta = \boldsymbol{L}_\beta$, provided that $v(2r - b)/(rk) < \varepsilon_\beta \le 1$ for $\beta = 0, 1, 2$. In the extreme case of $v(2r - b)/(rk) = \varepsilon_2$, the resulting design is $(\rho_0; \rho_1; \rho_2)$-EB.*

Note that $\tilde{\varepsilon}_\beta = 1$ if and only if $\varepsilon_\beta = 1$. Once there exists a $(\rho_0; \rho_1, \rho_2; 0)$-EB design with $\rho_0 \ge 1$, its complement produces a design that is again $(\rho_0; \rho_1, \rho_2; 0)$-EB under the condition above. This technique will be utilized later frequently.

As a juxtaposition method, from the cases (ii) and (iii) in Theorem 6.4.3, the following corollaries can be obtained.

Corollary 7.4.2. *The existence of a singular GD design with parameters $v = mn$, b, r, $k = m$, λ_1, λ_2 implies the existence of a $(\rho_0^*; \rho_1^*, \rho_2^*; 0)$-EB design with parameters $v^* = mn$, $b^* = b + ns$, $r^* = r + s$, $k^* = k$, $\varepsilon_0^* = 1$, $\varepsilon_1^* = r/(r+s)$, $\varepsilon_2^* = 1 - (rk - v\lambda_2)/[k(r+s)]$, $\rho_0^* = (m-1)(n-1)$, $\rho_1^* = n - 1$, $\rho_2^* = m - 1$, $\boldsymbol{L}_0^* = (\boldsymbol{I}_m - m^{-1}\boldsymbol{1}_m\boldsymbol{1}_m') \otimes (\boldsymbol{I}_n - n^{-1}\boldsymbol{1}_n\boldsymbol{1}_n')$, $\boldsymbol{L}_1^* = m^{-1}\boldsymbol{1}_m\boldsymbol{1}_m' \otimes (\boldsymbol{I}_n - n^{-1}\boldsymbol{1}_n\boldsymbol{1}_n')$, $\boldsymbol{L}_2^* = (\boldsymbol{I}_m - m^{-1}\boldsymbol{1}_m\boldsymbol{1}_m') \otimes n^{-1}\boldsymbol{1}_n\boldsymbol{1}_n'$ for a positive integer s.*

Corollary 7.4.3. *The existence of a semi-regular GD design with parameters $v = mn$, b, r, $k = m$, λ_1, λ_2 implies the existence of a $(\rho_0^*; \rho_1^*, \rho_2^*; 0)$-EB design with parameters $v^* = mn$, $b^* = b + ns$, $r^* = r + s$, $k^* = k$, $\varepsilon_0^* = 1$, $\varepsilon_1^* = 1 - (r - \lambda_1)/[k(r+s)]$, $\varepsilon_2^* = 1 - (r - \lambda_1 + sk)/[k(r+s)]$, $\rho_0^* = m - 1$, $\rho_1^* = (m-1)(n-1)$, $\rho_2^* = n - 1$, $\boldsymbol{L}_0^* = (\boldsymbol{I}_m - m^{-1}\boldsymbol{1}_m\boldsymbol{1}_m') \otimes n^{-1}\boldsymbol{1}_n\boldsymbol{1}_n'$, $\boldsymbol{L}_1^* = (\boldsymbol{I}_m - m^{-1}\boldsymbol{1}_m\boldsymbol{1}_m') \otimes (\boldsymbol{I}_n - n^{-1}\boldsymbol{1}_n\boldsymbol{1}_n')$, $\boldsymbol{L}_2^* = m^{-1}\boldsymbol{1}_m\boldsymbol{1}_m' \otimes (\boldsymbol{I}_n - n^{-1}\boldsymbol{1}_n\boldsymbol{1}_n')$ for a positive integer s.*

Example 7.4.2 (a continuation of Example 6.4.1). There exists a $(3; 6, 2; 0)$-

EB design with parameters $v^* = b^* = 12$, $r^* = k^* = 4$, $\varepsilon_0^* = 1$, $\varepsilon_1^* = 13/16$, $\varepsilon_2^* = 9/16$, $\rho_0^* = 3$, $\rho_1^* = 6$, $\rho_2^* = 2$, $\boldsymbol{L}_0^* = (\boldsymbol{I}_4 - 4^{-1}\mathbf{1}_4\mathbf{1}_4') \otimes 3^{-1}\mathbf{1}_3\mathbf{1}_3'$, $\boldsymbol{L}_1^* = (\boldsymbol{I}_4 - 4^{-1}\mathbf{1}_4\mathbf{1}_4') \otimes (\boldsymbol{I}_3 - 3^{-1}\mathbf{1}_3\mathbf{1}_3')$, $\boldsymbol{L}_2^* = 4^{-1}\mathbf{1}_4\mathbf{1}_4' \otimes (\boldsymbol{I}_3 - 3^{-1}\mathbf{1}_3\mathbf{1}_3')$. For it,

$$\boldsymbol{\Omega}^* = \frac{1}{1404}\left[351\boldsymbol{I}_4 \otimes \boldsymbol{I}_3 + (27\boldsymbol{I}_4 + 16\mathbf{1}_4\mathbf{1}_4') \otimes (3\boldsymbol{I}_3 - \mathbf{1}_3\mathbf{1}_3')\right].$$

Here any 3 basic contrasts represented by eigenvectors of $\boldsymbol{L}_0^*$ corresponding to its nonzero eigenvalue are estimated in the intra-block analysis with full efficiency, whereas all 6 (or 2) basic contrasts represented by eigenvectors of $\boldsymbol{L}_1^*$ (or $\boldsymbol{L}_2^*$) corresponding to its nonzero eigenvalue are estimated with the efficiency 13/16 (or 9/16). The average efficiency factor ε^* of the design is $1287/1631 = 0.7891$. The variances of the intra-block BLUEs of the basic contrasts are σ_1^2, $16\sigma_1^2/13$, or $16\sigma_1^2/9$, depending on whether they correspond to $\boldsymbol{L}_0^*, \boldsymbol{L}_1^*$ or $\boldsymbol{L}_2^*$. Moreover, there are three different variances $197\sigma_1^2/351$, $421\sigma_1^2/702$ and $135\sigma_1^2/234$ for the BLUEs of the elementary contrasts estimated in the intra-block analysis.

Theorem 7.4.3. *If $\boldsymbol{N}$ is the incidence matrix of a binary $(\rho_0; \rho_1, \rho_2; 0)$-EB design with parameters v, b, r, k, ε_β, ρ_β, $\boldsymbol{L}_\beta$, $\beta = 0, 1, 2$, then the incidence matrix $[\boldsymbol{N} : \mathbf{1}_v\mathbf{1}_b' - \boldsymbol{N}]$ with $v = 2k$ yields a $(\rho_0^*; \rho_1^*, \rho_2^*; 0)$-EB design with parameters $v^* = v$, $b^* = 2b$, $r^* = b$, $k^* = k$, $\varepsilon_\beta^* = \varepsilon_\beta$, $\rho_\beta^* = \rho_\beta$, $\boldsymbol{L}_\beta^* = \boldsymbol{L}_\beta$, $\beta = 0, 1, 2$.*

Proof. Since

$$\frac{1}{b}[\boldsymbol{N} : \mathbf{1}_v\mathbf{1}_b' - \boldsymbol{N}]\mathrm{diag}\left[\frac{1}{k}\boldsymbol{I}_b : \frac{1}{v-k}\boldsymbol{I}_b\right][\boldsymbol{N} : \mathbf{1}_v\mathbf{1}_b' - \boldsymbol{N}]'$$

$$= \frac{1}{bk}\boldsymbol{N}\boldsymbol{N}' + \frac{1}{bk}(\mathbf{1}_v\mathbf{1}_b' - \boldsymbol{N})(\mathbf{1}_v\mathbf{1}_b' - \boldsymbol{N})',$$

the assumption and Corollary 7.4.1 complete the proof, since $v = 2k$, i.e., $b = 2r$. □

Note that a property of $\varepsilon_\beta^* = 1$ corresponds to $\varepsilon_\beta = 1$ in the original design.

A supplementation approach is also applicable here, as the following shows, though usually such technique yields nonequireplicate designs.

Theorem 7.4.4. *If $\boldsymbol{N}$ is the incidence matrix of an α-resolvable 2-associate PBIB design with parameters v, $b = b_0 a$, $r = \alpha a$, k, λ_1, λ_2, $\varepsilon_0 = 1$, $\varepsilon_1 (< 1)$, ρ_0, ρ_1, $\boldsymbol{L}_0$, $\boldsymbol{L}_1$, and if, in addition, the equality $r = b_0$ holds, then for $s \geq 1$*

$$\boldsymbol{N}^* = \begin{bmatrix} \boldsymbol{N} \\ \boldsymbol{I}_a \otimes \mathbf{1}_s\mathbf{1}_{b_0}' \end{bmatrix}$$

is the incidence matrix of a $(\rho_0^*; \rho_1^*, \rho_2^*; 0)$-EB design with parameters $v^* = v + sa, b^* = b, r^* = r, k^* = k + s$, $\varepsilon_0^* = 1$, $\varepsilon_1^* = (s + k\varepsilon_1)/(k + s), \varepsilon_2^* = k/(k + s)$, $\rho_0^* = \rho_0 + a(s-1) + 1$, $\rho_1^* = \rho_1, \rho_2^* = a - 1$, $\boldsymbol{L}_0^* = \boldsymbol{I}_{v^*} - (v^*)^{-1}\mathbf{1}_{v^*}\mathbf{1}_{v^*}' - \boldsymbol{L}_1^* - \boldsymbol{L}_2^*$, $\boldsymbol{L}_1^* = \mathrm{diag}[\boldsymbol{L}_1 : \mathbf{O}]$, $\boldsymbol{L}_2^* = \mathrm{diag}[\mathbf{O} : (\boldsymbol{I}_a - a^{-1}\mathbf{1}_a\mathbf{1}_a') \otimes s^{-1}\mathbf{1}_s\mathbf{1}_s']$.

Proof. Using the α-resolvability for the original design with the incidence matrix

$\boldsymbol{N}$, it follows that

$$\begin{aligned}\boldsymbol{M}_0^* &= \frac{1}{r(k+s)}\boldsymbol{N}^*(\boldsymbol{N}^*)' - \frac{1}{v^*}\mathbf{1}_{v^*}\mathbf{1}_{v^*}' \\ &= \frac{1}{r(k+s)}\begin{bmatrix} \boldsymbol{N}\boldsymbol{N}' & \alpha\mathbf{1}_v\mathbf{1}_{sa}' \\ \alpha\mathbf{1}_{sa}\mathbf{1}_v' & b_0\boldsymbol{I}_a\otimes\mathbf{1}_s\mathbf{1}_s' \end{bmatrix} - \frac{1}{v^*}\mathbf{1}_{v^*}\mathbf{1}_{v^*}' \\ &= \frac{1}{k+s}\begin{bmatrix} k(1-\varepsilon_1)\boldsymbol{L}_1 + \frac{k}{v}\mathbf{1}_v\mathbf{1}_v' & \frac{r}{b}\mathbf{1}_v\mathbf{1}_{sa}' \\ \frac{r}{b}\mathbf{1}_{sa}\mathbf{1}_v' & \boldsymbol{I}_a\otimes\mathbf{1}_s\mathbf{1}_s' \end{bmatrix} - \frac{1}{v^*}\mathbf{1}_{v^*}\mathbf{1}_{v^*}'\end{aligned}$$

which, by applying Lemma 2.3.3 and the notation of Section 4.4, completes the proof, as can be seen from Sections 6.0.2 and 6.0.3. □

Remark 7.4.2. In Theorem 7.4.4, the matrix $\boldsymbol{\Omega}$ is obtainable as

$$\boldsymbol{\Omega}^* = \text{diag}\left[\frac{k(1-\varepsilon_1)}{r(s+k\varepsilon_1)}\boldsymbol{L}_1 : \frac{s}{rk}\left(\boldsymbol{I}_a - \frac{1}{a}\mathbf{1}_a\mathbf{1}_a'\right)\otimes\frac{1}{s}\mathbf{1}_s\mathbf{1}_s'\right] + \frac{1}{r}\boldsymbol{I}_{v+sa}.$$

The restriction $r = b_0$ is here needed to get an equireplicate design. Otherwise the restriction is not necessary.

Example 7.4.3. Consider a 3-resolvable semi-regular GD design, based on two groups $\{1,2,3\},\{4,5,6\}$ of three treatments, SR8 in Clatworthy (1973), with parameters $v = 6, b = 27, r = 9, k = 2, \lambda_1 = 0, \lambda_2 = 3, m = 2, n = 3, \alpha = 3, b_0 = 9, a = 3$, having the blocks

[(1,4), (2,5), (3,6), (1,4), (2,5), (3,6), (1,4), (2,5), (3,6)],
[(1,5), (3,4), (2,6), (1,5), (3,4), (2,6), (1,5), (3,4), (2,6)],
[(1,6), (2,4), (3,5), (1,6), (2,4), (3,5), (1,6), (2,4), (3,5)],

that form the incidence matrix, say, $\boldsymbol{N}$. In this case $\varepsilon_0 = 1$, $\varepsilon_1 = 1/2$, $\rho_0 = 1, \rho_1 = 4$, $\boldsymbol{L}_0 = (\boldsymbol{I}_2 - 2^{-1}\mathbf{1}_2\mathbf{1}_2')\otimes 3^{-1}\mathbf{1}_3\mathbf{1}_3'$, $\boldsymbol{L}_1 = \boldsymbol{I}_2\otimes(\boldsymbol{I}_3 - 3^{-1}\mathbf{1}_3\mathbf{1}_3')$. Then, for $s \geq 1$, Theorem 7.4.4 produces a $(\rho_0^*;\rho_1^*,\rho_2^*;0)$-EB design with parameters $v^* = 3s+6, b^* = 27, r^* = 9, k^* = s+2, \varepsilon_0^* = 1, \varepsilon_1^* = (s+1)/(s+2), \varepsilon_2^* = 2/(s+2)$, $\rho_0^* = 3s-1$, $\rho_1^* = 4, \rho_2^* = 2$, $\boldsymbol{L}_0^* = \boldsymbol{I}_{3s+6} - [3(s+2)]^{-1}\mathbf{1}_{3s+6}\mathbf{1}_{3s+6}' - \text{diag}[\boldsymbol{I}_2\otimes(\boldsymbol{I}_3 - 3^{-1}\mathbf{1}_3\mathbf{1}_3') : (\boldsymbol{I}_3 - 3^{-1}\mathbf{1}_3\mathbf{1}_3')\otimes s^{-1}\mathbf{1}_s\mathbf{1}_s']$, $\boldsymbol{L}_1^* = \text{diag}[\boldsymbol{I}_2\otimes(\boldsymbol{I}_3 - 3^{-1}\mathbf{1}_3\mathbf{1}_3') : \mathbf{O}]$, $\boldsymbol{L}_2^* = \text{diag}[\mathbf{O} : (\boldsymbol{I}_3 - 3^{-1}\mathbf{1}_3\mathbf{1}_3')\otimes s^{-1}\mathbf{1}_s\mathbf{1}_s']$. For example, when $s = 7$, it has the incidence matrix

$$\begin{bmatrix} \boldsymbol{N} \\ \boldsymbol{I}_3\otimes\mathbf{1}_7\mathbf{1}_9' \end{bmatrix},$$

with parameters $v^* = b^* = 27, r^* = k^* = 9$, $\varepsilon_0^* = 1$, $\varepsilon_1^* = 8/9$, $\varepsilon_2^* = 2/9$, $\rho_0^* = 20$, $\rho_1^* = 4$, $\rho_2^* = 2$, $\boldsymbol{L}_0^* = \boldsymbol{I}_{27} - (27)^{-1}\mathbf{1}_{27}\mathbf{1}_{27}' - \text{diag}[\boldsymbol{I}_2\otimes(\boldsymbol{I}_3 - 3^{-1}\mathbf{1}_3\mathbf{1}_3') : (\boldsymbol{I}_3 - 3^{-1}\mathbf{1}_3\mathbf{1}_3')\otimes 7^{-1}\mathbf{1}_7\mathbf{1}_7']$, $\boldsymbol{L}_1^* = \text{diag}[\boldsymbol{I}_2\otimes(\boldsymbol{I}_3 - 3^{-1}\mathbf{1}_3\mathbf{1}_3') : \mathbf{O}]$, $\boldsymbol{L}_2^* = \text{diag}[\mathbf{O} : (\boldsymbol{I}_3 - 3^{-1}\mathbf{1}_3\mathbf{1}_3')\otimes 7^{-1}\mathbf{1}_7\mathbf{1}_7']$, giving

$$\boldsymbol{\Omega}^* = \frac{1}{108}\text{diag}\Big[\boldsymbol{I}_2\otimes[(13+\frac{1}{2})\boldsymbol{I}_3 - \frac{1}{2}\mathbf{1}_3\mathbf{1}_3'] : \boldsymbol{I}_3\otimes(12\boldsymbol{I}_7 + 4\mathbf{1}_7\mathbf{1}_7')$$

$$-2(\mathbf{1}_3\mathbf{1}_3' - \boldsymbol{I}_3) \otimes \mathbf{1}_7\mathbf{1}_7'\Big].$$

Here, the average efficiency factor of the design is $52/67 = 0.7761$. The variances of the intra-block BLUEs of basic contrasts are σ_1, $9\sigma_1^2/8$ or $9\sigma_1^2/2$, depending on whether they correspond to $\boldsymbol{L}_0^*$, $\boldsymbol{L}_1^*$ or $\boldsymbol{L}_2^*$, respectively. Furthermore, for example, elementary contrasts are estimated in the intra-block analysis with five different variances, $2\sigma_1^2/9$, $13\sigma_1^2/54$, $\sigma_1^2/4$, $29\sigma_1^2/108$ and $\sigma_1^2/3$.

Several patterned arrangements of incidence matrices of available designs may yield $(\rho_0; \rho_1, \rho_2; 0)$-EB designs that are suitable for the purpose of estimating some basic contrasts with full efficiency. Most of the results here are from Puri, Mehta and Kageyama (1987). Theorem 6.6.3 implies the following theorem.

Theorem 7.4.5. *For the $v \times b$ incidence matrix $\boldsymbol{N}$ of a 2-associate PBIB design with parameters $v = 2k$, b, r, k, λ_1, λ_2, $\varepsilon_0 = 1$, $\varepsilon_1 < 1$, ρ_0, ρ_1, $\boldsymbol{L}_0$, $\boldsymbol{L}_1$, the incidence matrix*

$$\boldsymbol{N}^* = \begin{bmatrix} \boldsymbol{N} & \mathbf{1}_v\mathbf{1}_b' \\ \mathbf{1}_v\mathbf{1}_b' & \mathbf{1}_v\mathbf{1}_b' - \boldsymbol{N} \end{bmatrix}$$

yields a $(\rho_0^*; \rho_1^*, \rho_2^*; 0)$-EB design with parameters $v^* = 4k$, $b^* = 4r$, $r^* = 3r$, $k^* = 3k$, $\varepsilon_0^* = 1$, $\varepsilon_1^* = (\varepsilon_1 + 8)/9$, $\varepsilon_2^* = 8/9$, $\rho_0^* = 2\rho_0$, $\rho_1^* = 2\rho_1$, $\rho_2^* = 1$, $\boldsymbol{L}_0^* = \boldsymbol{I}_2 \otimes \boldsymbol{L}_0$, $\boldsymbol{L}_1^* = \boldsymbol{I}_2 \otimes \boldsymbol{L}_1$, $\boldsymbol{L}_2^* = (\boldsymbol{I}_2 - 2^{-1}\mathbf{1}_2\mathbf{1}_2') \otimes (2k)^{-1}\mathbf{1}_{2k}\mathbf{1}_{2k}'$.

Proof. As the basic PBIB design is $(\rho_0; \rho_1; 0)$-EB, Theorem 6.6.3 applies. □

Thus, from a 2-associate PBIB design with $v = 2k$ in which the rank of its incidence matrix $\boldsymbol{N}$ is smaller than v (see Section 6.0.2), it is possible to obtain a $(\rho_0; \rho_1, \rho_2; 0)$-EB design in which some basic contrasts can be estimated in the intra-block analysis with full efficiency.

Example 7.4.4. Take a semi-regular GD design, based on three groups $\{1, 2\}$, $\{3, 4\}$, $\{5, 6\}$ of two treatments, SR18 in Clatworthy (1973), with parameters $v = 6, b = 4$, $r = 2, k = 3, \lambda_1 = 0$, $\lambda_2 = 1$, $m = 3, n = 2$, $\varepsilon_0 = 1$, $\varepsilon_1 = 2/3$, $\rho_0 = 2$, $\rho_1 = 3$, $\boldsymbol{L}_0 = (\boldsymbol{I}_3 - 3^{-1}\mathbf{1}_3\mathbf{1}_3') \otimes 2^{-1}\mathbf{1}_2\mathbf{1}_2'$, $\boldsymbol{L}_1 = \boldsymbol{I}_3 \otimes (\boldsymbol{I}_2 - 2^{-1}\mathbf{1}_2\mathbf{1}_2')$, having the blocks (1, 3, 5), (1, 4, 6), (2, 3, 6), (2, 4, 5). Then Theorem 7.4.5 produces a $(\rho_0^*; \rho_1^*, \rho_2^*; 0)$-EB design with parameters $v^* = 12$, $b^* = 8$, $r^* = 6$, $k^* = 9$, $\varepsilon_0^* = 1$, $\varepsilon_1^* = 26/27$, $\varepsilon_2^* = 8/9$, $\rho_0^* = 4$, $\rho_1^* = 6$, $\rho_2^* = 1$, $\boldsymbol{L}_0^* = \boldsymbol{I}_2 \otimes (\boldsymbol{I}_3 - 3^{-1}\mathbf{1}_3\mathbf{1}_3') \otimes 2^{-1}\mathbf{1}_2\mathbf{1}_2'$, $\boldsymbol{L}_1^* = \boldsymbol{I}_2 \otimes \boldsymbol{I}_3 \otimes (\boldsymbol{I}_2 - 2^{-1}\mathbf{1}_2\mathbf{1}_2')$, $\boldsymbol{L}_2^* = (\boldsymbol{I}_2 - 2^{-1}\mathbf{1}_2\mathbf{1}_2') \otimes 6^{-1}\mathbf{1}_6\mathbf{1}_6'$. Here, the average efficiency factor of the design is $1144/1181 = 0.9687$. The variances of the intra-block BLUEs of basic contrasts are σ_1^2, $27\sigma_1^2/26$ or $9\sigma_1^2/8$, depending on whether they correspond to $\boldsymbol{L}_0^*$, $\boldsymbol{L}_1^*$ or $\boldsymbol{L}_2^*$. Furthermore, the matrix $\boldsymbol{\Omega}$ for the resulting design has been obtained as

$$\boldsymbol{\Omega}^* = \frac{1}{7488}\Big[\boldsymbol{I}_6 \otimes (1296\boldsymbol{I}_2 - 111\mathbf{1}_2\mathbf{1}_2') + \boldsymbol{I}_2 \otimes (\mathbf{1}_3\mathbf{1}_3' - \boldsymbol{I}_3) \otimes 131\mathbf{1}_2\mathbf{1}_2'$$

$$+13(\boldsymbol{I}_2 - \mathbf{1}_2\mathbf{1}_2') \otimes \mathbf{1}_6\mathbf{1}_6'\Big]$$

which, for example, implies that elementary contrasts are estimated in the intra-block analysis with three different variances, $9\sigma_1^2/26, 53\sigma_1^2/156$ and $649\sigma_1^2/1872$. In fact, the incidence matrix of the design is given by

$$N^* = \begin{bmatrix} 1 & 1 & 0 & 0 & 1 & 1 & 1 & 1 \\ 0 & 0 & 1 & 1 & 1 & 1 & 1 & 1 \\ 1 & 0 & 1 & 0 & 1 & 1 & 1 & 1 \\ 0 & 1 & 0 & 1 & 1 & 1 & 1 & 1 \\ 1 & 0 & 0 & 1 & 1 & 1 & 1 & 1 \\ 0 & 1 & 1 & 0 & 1 & 1 & 1 & 1 \\ 1 & 1 & 1 & 1 & 0 & 0 & 1 & 1 \\ 1 & 1 & 1 & 1 & 1 & 1 & 0 & 0 \\ 1 & 1 & 1 & 1 & 0 & 1 & 0 & 1 \\ 1 & 1 & 1 & 1 & 1 & 0 & 1 & 0 \\ 1 & 1 & 1 & 1 & 0 & 1 & 1 & 0 \\ 1 & 1 & 1 & 1 & 1 & 0 & 0 & 1 \end{bmatrix}.$$

Another version of Theorem 6.6.2 can lead to the following result.

Theorem 7.4.6. *Let N be the $v \times b$ incidence matrix of a 2-associate PBIB design with parameters v, b, r, k, λ_1, λ_2, ε_β, ρ_β, L_β, $\beta = 1, 2$. Then $\mathbf{1}_s\mathbf{1}_t' \otimes N$ is the incidence matrix of a $(\rho_0^*; \rho_1^*, \rho_2^*; 0)$-EB design with parameters $v^* = sv$, $b^* = tb$, $r^* = tr$, $k^* = sk$, $\varepsilon_0^* = 1$, $\varepsilon_\beta^* = \varepsilon_\beta$, $\rho_0^* = (s-1)v$, $\rho_\beta^* = \rho_\beta$, $L_0^* = (I_s - s^{-1}\mathbf{1}_s\mathbf{1}_s') \otimes I_v$, $L_\beta^* = s^{-1}\mathbf{1}_s\mathbf{1}_s' \otimes L_\beta$ for positive integers $s \geq 2$ and t.*

Proof. Following the proof of Theorem 6.6.2 and its slight modification, the proof can be accomplished. □

Because 2-associate PBIB designs are available in abundance, Theorem 7.4.6 is very convenient to construct $(\rho_0; \rho_1, \rho_2; 0)$-EB designs with $\rho_0 \geq 1$. Here $(s-1)v$ basic contrasts of the zeroth subset (i.e., represented by eigenvectors of L_0^*) are estimated in the intra-block analysis with full efficiency.

Corollary 7.4.4. *The existence of a GD design with parameters $v = mn$, b, r, k, λ_1, λ_2 implies the existence of a $(\rho_0^*; \rho_1^*, \rho_2^*; 0)$-EB design with parameters $v^* = mns$, $b^* = tb$, $r^* = tr$, $k^* = sk$, $\varepsilon_0^* = 1$, $\varepsilon_1^* = 1 - (rk - v\lambda_2)/(rk)$, $\varepsilon_2^* = 1 - (r - \lambda_1)/(rk)$, $\rho_0^* = mn(s-1)$, $\rho_1^* = m - 1$, $\rho_2^* = m(n-1)$, $L_0^* = (I_s - s^{-1}\mathbf{1}_s\mathbf{1}_s') \otimes I_{mn}$, $L_1^* = s^{-1}\mathbf{1}_s\mathbf{1}_s' \otimes (I_m - m^{-1}\mathbf{1}_m\mathbf{1}_m') \otimes n^{-1}\mathbf{1}_n\mathbf{1}_n'$, $L_2^* = s^{-1}\mathbf{1}_s\mathbf{1}_s' \otimes I_m \otimes (I_n - n^{-1}\mathbf{1}_n\mathbf{1}_n')$ for positive integers $s \geq 2$ and t.*

Remark 7.4.3. (1) A design resulting from Theorem 7.4.6 can effectively be used for a two-factor factorial experiment with factors A and B applied at s and v levels, respectively. Here, all contrasts among the A factor main effects and AB interactions will be estimated with full efficiency, whereas contrasts among the B factor main effects will be estimated with the efficiency $\varepsilon_\beta^*, \beta = 1, 2$, in the intra-block analysis. (2) A design considered in Corollary 7.4.4 can also be used for a three-factor factorial experiment with factors A, B and C applied at s, m and n levels, respectively. In it, all contrasts among the A factor main effects, AB interactions, AC interactions and ABC interactions will be estimated with full efficiency, those among the B factor main effects with the efficiency ε_1^* and those among the C factor main effects and BC interactions with the efficiency ε_2^*. (3) In a design given by Corollary 7.4.4, $mn(s-1)$ basic contrasts will be estimated in the intra-block analysis with full efficiency. Furthermore, if a singular GD design is taken as a basic design, $mn(s-1)+m(n-1)$ basic contrasts will be estimated in the intra-block analysis with full efficiency, whereas if a semi-regular GD design is taken as a basic design, $mn(s-1)+m-1$ basic contrasts will be estimated there with full efficiency.

Some other patterns can be seen below.

Theorem 7.4.7. *If $\boldsymbol{N}$ is the $v \times b$ incidence matrix of a 2-associate PBIB design with parameters $v = 2k$, $b = 2r$, r, k, λ_1, λ_2, ε_β, ρ_β, $\boldsymbol{L}_\beta$, $\beta = 1, 2$, then the pattern (7.3.3) yields a $(\rho_0^*; \rho_1^*, \rho_2^*; 0)$-EB design with parameters $v^* = 4k$, $b^* = 4r$, $r^* = 2r$, $k^* = 2k$, $\varepsilon_0^* = 1$, $\varepsilon_\beta^* = (1+\varepsilon_\beta)/2$, $\rho_0^* = 1$, $\rho_\beta^* = 2\rho_\beta$, $\boldsymbol{L}_0^* = (\boldsymbol{I}_2 - 2^{-1}\boldsymbol{1}_2\boldsymbol{1}_2') \otimes (2k)^{-1}\boldsymbol{1}_{2k}\boldsymbol{1}_{2k}'$, $\boldsymbol{L}_\beta^* = \boldsymbol{I}_2 \otimes \boldsymbol{L}_\beta$.*

Proof. Take a procedure similar to that used in the proof of Theorem 7.3.5. □

Theorem 6.6.4 with $a = c = 1$ yields the following result when the basic design is a 2-associate PBIB design whose incidence matrix has a rank smaller than v (see Section 6.0.2).

Theorem 7.4.8. *If $\boldsymbol{N}$ is the $v \times b$ incidence matrix of a 2-associate PBIB design with parameters v, b, r, k, λ_1, λ_2, $\varepsilon_0 = 1$, $\varepsilon_1 < 1$, ρ_0, ρ_1, $\boldsymbol{L}_0$, $\boldsymbol{L}_1$, then the incidence matrix (7.4.2) yields a $(\rho_0^*; \rho_1^*, \rho_2^*; 0)$-EB design with parameters $v^* = 3v$, $b^* = 3b$, $r^* = 2b$, $k^* = 2v$, $\varepsilon_0^* = 1$, $\varepsilon_1^* = 1 - 3rk(1-\varepsilon_1)/(4bv)$, $\varepsilon_2^* = 3[bv + r(v-k)]/(4bv)$, $\rho_0^* = 2\rho_0 + v - 1$, $\rho_1^* = 2\rho_1$, $\rho_2^* = 2$, $\boldsymbol{L}_0^* = (\boldsymbol{I}_3 - 3^{-1}\boldsymbol{1}_3\boldsymbol{1}_3') \otimes \boldsymbol{L}_0 + 3^{-1}\boldsymbol{1}_3\boldsymbol{1}_3' \otimes (\boldsymbol{I}_v - v^{-1}\boldsymbol{1}_v\boldsymbol{1}_v')$, $\boldsymbol{L}_1^* = (\boldsymbol{I}_3 - 3^{-1}\boldsymbol{1}_3\boldsymbol{1}_3') \otimes \boldsymbol{L}_1$, $\boldsymbol{L}_2^* = (\boldsymbol{I}_3 - 3^{-1}\boldsymbol{1}_3\boldsymbol{1}_3') \otimes v^{-1}\boldsymbol{1}_v\boldsymbol{1}_v'$.*

To reduce replication numbers and block sizes of a design resulting from Theorem 7.4.8, the following structure may be considered. In fact, this can be derived from Theorem 6.6.4 with $a = 0$ and $c = 1$.

Theorem 7.4.9. *Let $\boldsymbol{N}$ be the $v \times b$ incidence matrix of a 2-associate PBIB design with parameters v, b, r, k, λ_1, λ_2, $\varepsilon_0 = 1$, $\varepsilon_1 < 1$, ρ_0, ρ_1, $\boldsymbol{L}_0$, $\boldsymbol{L}_1$. Then the incidence matrix (7.4.1) yields a $(\rho_0^*; \rho_1^*, \rho_2^*; 0)$-EB design with parameters $v^* = 3v$, $b^* = 3b$, $r^* = b$, $k^* = v$, $\varepsilon_0^* = 1$, $\varepsilon_1^* = 1 - 3rk(1-\varepsilon_1)/(bv)$, $\varepsilon_2^* = 3r(v-k)/(bv)$, $\rho_0^* = 2\rho_0 + v - 1$, $\rho_1^* = 2\rho_1$, $\rho_2^* = 2$, $\boldsymbol{L}_0^* = (\boldsymbol{I}_3 - 3^{-1}\boldsymbol{1}_3\boldsymbol{1}_3') \otimes \boldsymbol{L}_0 + 3^{-1}\boldsymbol{1}_3\boldsymbol{1}_3' \otimes (\boldsymbol{I}_v - v^{-1}\boldsymbol{1}_v\boldsymbol{1}_v')$, $\boldsymbol{L}_1^* = (\boldsymbol{I}_3 - 3^{-1}\boldsymbol{1}_3\boldsymbol{1}_3') \otimes \boldsymbol{L}_1$, $\boldsymbol{L}_2^* = (\boldsymbol{I}_3 - 3^{-1}\boldsymbol{1}_3\boldsymbol{1}_3') \otimes v^{-1}\boldsymbol{1}_v\boldsymbol{1}_v'$.*

Example 7.4.5. Take a semi-regular GD design, based on two groups $\{1, 2\}$, $\{3, 4\}$ of two treatments, SR1 in Clatworthy (1973), with parameters $v = 4$, $b = 4$, $r = 2$, $k = 2$, $\lambda_1 = 0, \lambda_2 = 1$, $m = n = 2$, $\varepsilon_0 = 1$, $\varepsilon_1 = 1/2$, $\rho_0 = 1$, $\rho_1 = 2$,

$\boldsymbol{L}_0 = (\boldsymbol{I}_2 - 2^{-1}\mathbf{1}_2\mathbf{1}_2') \otimes 2^{-1}\mathbf{1}_2\mathbf{1}_2'$, $\boldsymbol{L}_1 = \boldsymbol{I}_2 \otimes (\boldsymbol{I}_2 - 2^{-1}\mathbf{1}_2\mathbf{1}_2')$, whose blocks are given by (1,3), (2,4), (1,4), (2,3). Then Theorem 7.4.9 produces a $(\rho_0^*; \rho_1^*, \rho_2^*; 0)$-EB design with parameters $v^* = 12$, $b^* = 12$, $r^* = 4$, $k^* = 4$, $\varepsilon_0^* = 1$, $\varepsilon_1^* = 3/4$, $\varepsilon_2^* = 5/8$, $\rho_0^* = 5$, $\rho_1^* = 2$, $\rho_2^* = 4$, $\boldsymbol{L}_0^* = (\boldsymbol{I}_3 - 3^{-1}\mathbf{1}_3\mathbf{1}_3') \otimes (\boldsymbol{I}_2 - 2^{-1}\mathbf{1}_2\mathbf{1}_2') \otimes 2^{-1}\mathbf{1}_2\mathbf{1}_2' + 3^{-1}\mathbf{1}_3\mathbf{1}_3' \otimes (\boldsymbol{I}_4 - 4^{-1}\mathbf{1}_4\mathbf{1}_4')$, $\boldsymbol{L}_1^* = (\boldsymbol{I}_3 - 3^{-1}\mathbf{1}_3\mathbf{1}_3') \otimes 4^{-1}\mathbf{1}_4\mathbf{1}_4'$, $\boldsymbol{L}_2^* = (\boldsymbol{I}_3 - 3^{-1}\mathbf{1}_3\mathbf{1}_3') \otimes \boldsymbol{I}_2 \otimes (\boldsymbol{I}_2 - 2^{-1}\mathbf{1}_2\mathbf{1}_2')$, giving

$$\boldsymbol{\Omega}^* = \frac{1}{4}\left[\frac{3}{5}(\boldsymbol{I}_3 - \frac{1}{3}\mathbf{1}_3\mathbf{1}_3') \otimes \boldsymbol{I}_2 \otimes (\boldsymbol{I}_2 - \frac{1}{2}\mathbf{1}_2\mathbf{1}_2') + \frac{1}{3}(\boldsymbol{I}_3 - \frac{1}{3}\mathbf{1}_3\mathbf{1}_3') \otimes \frac{1}{4}\mathbf{1}_4\mathbf{1}_4' + \boldsymbol{I}_{12}\right].$$

Here, the average efficiency factor of the design is 165/211 = 0.7820. The variances of the intra-block BLUEs of basic contrasts are σ_1^2, $4\sigma_1^2/3$ or $8\sigma_1^2/5$, depending on whether they correspond to $\boldsymbol{L}_0^*$, $\boldsymbol{L}_1^*$ or $\boldsymbol{L}_2^*$. In fact, the incidence matrix of the design is given by

$$\begin{bmatrix} 1&0&1&0&0&0&0&0&0&1&0&1 \\ 0&1&0&1&0&0&0&0&1&0&1&0 \\ 1&0&0&1&0&0&0&0&0&1&1&0 \\ 0&1&1&0&0&0&0&0&1&0&0&1 \\ 0&1&0&1&1&0&1&0&0&0&0&0 \\ 1&0&1&0&0&1&0&1&0&0&0&0 \\ 0&1&1&0&1&0&0&1&0&0&0&0 \\ 1&0&0&1&0&1&1&0&0&0&0&0 \\ 0&0&0&0&0&1&0&1&1&0&1&0 \\ 0&0&0&0&1&0&1&0&0&1&0&1 \\ 0&0&0&0&0&1&1&0&1&0&0&1 \\ 0&0&0&0&1&0&0&1&0&1&1&0 \end{bmatrix}.$$

Theorem 6.6.4 with $a = 1$ and $c = 0$ [see also the pattern (6.6.2) with $a = 1$ and $c = 0$, i.e., Theorem 6.6.6] gives the following result when the basic design is a 2-associate PBIB design whose incidence matrix has a rank smaller than v (see Section 6.0.2).

Theorem 7.4.10. *Let $\boldsymbol{N}$ be the $v \times b$ incidence matrix of a 2-associate PBIB design with parameters v, b, r, k, λ_1, λ_2, $\varepsilon_0 = 1$, $\varepsilon_1 < 1$, ρ_0, ρ_1, $\boldsymbol{L}_0$, $\boldsymbol{L}_1$. Then*

$$\boldsymbol{N}^* = \begin{bmatrix} \boldsymbol{N} & \mathbf{1}_v\mathbf{1}_b' & \boldsymbol{O} \\ \boldsymbol{O} & \boldsymbol{N} & \mathbf{1}_v\mathbf{1}_b' \\ \mathbf{1}_v\mathbf{1}_b' & \boldsymbol{O} & \boldsymbol{N} \end{bmatrix}$$

is the incidence matrix of a $(\rho_0^*; \rho_1^*, \rho_2^*; 0)$-EB design with parameters $v^* = 3v$, $b^* = 3b$, $r^* = b + r$, $k^* = v + k$, $\varepsilon_0^* = 1$, $\varepsilon_1^* = 1 - rk(1 - \varepsilon_1)/[(b + r)(v + k)]$,

$\varepsilon_2^* = 1 - [vb - r(v-k)]/[(b+r)(v+k)]$, $\rho_0^* = 3\rho_0$, $\rho_1^* = 3\rho_1$, $\rho_2^* = 2$, $\boldsymbol{L}_0^* = \boldsymbol{I}_3 \otimes \boldsymbol{L}_0$, $\boldsymbol{L}_1^* = \boldsymbol{I}_3 \otimes \boldsymbol{L}_1$, $\boldsymbol{L}_2^* = (\boldsymbol{I}_3 - 3^{-1}\mathbf{1}_3\mathbf{1}_3') \otimes v^{-1}\mathbf{1}_v\mathbf{1}_v'$.

Thus, when a 2-associate PBIB design with one of the eigenvalues of its incidence matrix being zero, i.e., one of ε_β is 1, is taken, one can obtain designs in which some basic contrasts are estimated in the intra-block analysis with full efficiency. For example, consider a semi-regular GD design resulting from Corollary 7.3.1. Then Theorem 7.4.10 yields the following result.

Corollary 7.4.5. *If q is a prime or a prime power, there exists a $(\rho_0;\rho_1,\rho_2;0)$-EB design with parameters $v = 3q(q+1)$, $b = 3q^2$, $r = q(q+1)$, $k = (q+1)^2$, $\varepsilon_0 = 1$, $\varepsilon_1 = 1 - 1/(q+1)^3$, $\varepsilon_2 = 3q/(q+1)^2$, $\rho_0 = 3q$, $\rho_1 = 3(q^2-1)$, $\rho_2 = 2$, $\boldsymbol{L}_0 = \boldsymbol{I}_3 \otimes [\boldsymbol{I}_{q+1} - (q+1)^{-1}\mathbf{1}_{q+1}\mathbf{1}_{q+1}'] \otimes q^{-1}\mathbf{1}_q\mathbf{1}_q'$, $\boldsymbol{L}_1 = \boldsymbol{I}_3 \otimes \boldsymbol{I}_{q+1} \otimes (\boldsymbol{I}_q - q^{-1}\mathbf{1}_q\mathbf{1}_q')$, $\boldsymbol{L}_2 = (\boldsymbol{I}_3 - 3^{-1}\mathbf{1}_3\mathbf{1}_3') \otimes [q(q+1)]^{-1}\mathbf{1}_{q(q+1)}\mathbf{1}_{q(q+1)}'$.*

Finally, consider the pattern (6.6.2) for some nonnegative integers a and c. [Note that when $a = 1$ and $c = 0$, this pattern coincides with the pattern used in Theorem 7.4.10.] By taking $\boldsymbol{N}$ above as the incidence matrix of a 2-associate PBIB design, the following theorem can be obtained. This is a direct application of Theorem 6.6.6 for $\psi_1 = 0$, i.e., $\varepsilon_0 = 1$.

Theorem 7.4.11. *The existence of a 2-associate PBIB design with parameters v, b, r, k, λ_1, λ_2, $\varepsilon_0 = 1$, $\varepsilon_1 < 1$, ρ_0, ρ_1, $\boldsymbol{L}_0$, $\boldsymbol{L}_1$ implies the existence of a $(\rho_0^*;\rho_1^*,\rho_2^*;0)$-EB design with parameters $v^* = 3v$, $b^* = 3b$, $r^* = r + (a+c)b$, $k^* = k + (a+c)v$, $\varepsilon_0^* = 1$, $\varepsilon_1^* = 1 - rk(1-\varepsilon_1)/(r^*k^*)$, $\varepsilon_2^* = 1 - [rk + vb(a^2 + c^2 - ac) - (a+c)vr]/(r^*k^*)$, $\rho_0^* = 3\rho_0$, $\rho_1^* = 3\rho_1$, $\rho_2^* = 2$, $\boldsymbol{L}_0^* = \boldsymbol{I}_3 \otimes \boldsymbol{L}_0$, $\boldsymbol{L}_1^* = \boldsymbol{I}_3 \otimes \boldsymbol{L}_1$, $\boldsymbol{L}_2^* = (\boldsymbol{I}_3 - 3^{-1}\mathbf{1}_3\mathbf{1}_3') \otimes v^{-1}\mathbf{1}_v\mathbf{1}_v'$.*

For $a, c \in \{0, 1\}$, one obtains binary designs.

7.4.2 Proper designs that are nonequireplicate

Recall that $(\rho_0;\rho_1,\rho_2;0)$-EB designs have three distinct efficiency factors that are eigenvalues of $\boldsymbol{C}_1$ with respect to $\boldsymbol{r}^\delta$ [i.e., with $m = 3$ as in (7.1.4)]. Proper nonequireplicate $(\rho_0;\rho_1,\rho_2;0)$-EB designs will be considered here by following the same approach as those in Sections 7.3.2 and 7.3.4.

Applications of Theorem 6.1.1 give immediately the following results.

Corollary 7.4.6. *The dual of a $(0;\rho_1,\rho_2;0)$-EB design $\boldsymbol{N}$ with parameters v, b, r, $\boldsymbol{k}$, $\varepsilon_\beta(<1)$, ρ_β, $\boldsymbol{L}_\beta$, $\beta = 1, 2$, is a $(\rho_0^*;\rho_1^*,\rho_2^*;0)$-EB design with parameters $v^* = b$, $b^* = v$, $\boldsymbol{r}^* = \boldsymbol{k}$, $\boldsymbol{k}^* = r$, $\varepsilon_0^* = 1$, $\varepsilon_\beta^* = \varepsilon_\beta$, $\rho_0^* = b - v$, $\rho_\beta^* = \rho_\beta$, $\boldsymbol{L}_0^* = \boldsymbol{I}_b - n^{-1}\mathbf{1}_b\boldsymbol{k}' - \boldsymbol{L}_1^* - \boldsymbol{L}_2^*$, $\boldsymbol{L}_\beta^* = [r(1-\varepsilon_\beta)]^{-1}\boldsymbol{k}^{-\delta}\boldsymbol{N}'\boldsymbol{L}_\beta\boldsymbol{N}$, where $n = \boldsymbol{k}'\mathbf{1}_b$.*

Corollary 7.4.7. *The dual of a $(\rho_0;\rho_1,\rho_2;0)$-EB design $\boldsymbol{N}$ with parameters v, b, r, $\boldsymbol{k}$, $\varepsilon_0 = 1$, $\varepsilon_\beta < 1$, ρ_0, ρ_β, $\boldsymbol{L}_0$, $\boldsymbol{L}_\beta$, $\beta = 1, 2$, is a $(\rho_0^*;\rho_1^*,\rho_2^*;0)$-EB design with parameters $v^* = b$, $b^* = v$, $\boldsymbol{r}^* = \boldsymbol{k}$, $\boldsymbol{k}^* = r$, $\varepsilon_0^* = 1$, $\varepsilon_\beta^* = \varepsilon_\beta$, $\rho_0^* = b - v + \rho_0$, $\rho_\beta^* = \rho_\beta$, $\boldsymbol{L}_0^* = \boldsymbol{I}_b - n^{-1}\mathbf{1}_b\boldsymbol{k}' - \boldsymbol{L}_1^* - \boldsymbol{L}_2^*$, $\boldsymbol{L}_\beta^* = [r(1-\varepsilon_\beta)]^{-1}\boldsymbol{k}^{-\delta}\boldsymbol{N}'\boldsymbol{L}_\beta\boldsymbol{N}$, where*

$n = \boldsymbol{k}'\boldsymbol{1}_b$.

Thus, the dualization approach will be useful to construct $(\rho_0; \rho_1, \rho_2; 0)$-EB designs suitable for the present purpose. For example, the dualization of a design resulting from Corollary 6.4.1 yields the following result.

Corollary 7.4.8. *The existence of two GD designs with parameters* $v = mn$, $b^{(h)}$, $r^{(h)}$, $k^{(h)}$, $\lambda_1^{(h)}$, $\lambda_2^{(h)}$, $h = 1, 2$, *implies the existence of a* $(\rho_0^*; \rho_1^*, \rho_2^*; 0)$*-EB design with parameters* $v^* = b^{(1)} + b^{(2)}$, $b^* = mn$, $\boldsymbol{r}^* = [k^{(1)}\boldsymbol{1}'_{b^{(1)}}, k^{(2)}\boldsymbol{1}'_{b^{(2)}}]'$, $k^* = r^{(1)} + r^{(2)}$, $\varepsilon_0^* = 1$, $\varepsilon_1^* = 1 - [(r^{(1)}k^{(1)} - v\lambda_2^{(1)})/k^{(1)} + (r^{(2)}k^{(2)} - v\lambda_2^{(2)})/k^{(2)}]/(r^{(1)} + r^{(2)})$, $\varepsilon_2^* = 1 - [(r^{(1)} - \lambda_1^{(1)})/k^{(1)} + (r^{(2)} - \lambda_1^{(2)})/k^{(2)}]/(r^{(1)} + r^{(2)})$, $\rho_0^* = b^{(1)} + b^{(2)} - mn$, $\rho_1^* = m - 1$, $\rho_2^* = m(n-1)$, $\boldsymbol{L}_0^* = \boldsymbol{I}_{b^{(1)}+b^{(2)}} - [mn(r^{(1)} + r^{(2)})]^{-1}\boldsymbol{1}_{b^{(1)}+b^{(2)}}\boldsymbol{k}' - \boldsymbol{L}_1^* - \boldsymbol{L}_2^*$, $\boldsymbol{L}_1^* = [(r^{(1)}k^{(1)} - v\lambda_2^{(1)})/k^{(1)} + (r^{(2)}k^{(2)} - v\lambda_2^{(2)})/k^{(2)}]^{-1}\boldsymbol{k}^{-\delta}\boldsymbol{N}'[(\boldsymbol{I}_m - m^{-1}\boldsymbol{1}_m\boldsymbol{1}'_m) \otimes n^{-1}\boldsymbol{1}_n\boldsymbol{1}'_n]\boldsymbol{N}$, $\boldsymbol{L}_2^* = [(r^{(1)} - \lambda_1^{(1)})/k^{(1)} + (r^{(2)} - \lambda_1^{(2)})/k^{(2)}]^{-1}\boldsymbol{k}^{-\delta}\boldsymbol{N}'[\boldsymbol{I}_m \otimes (\boldsymbol{I}_n - n^{-1}\boldsymbol{1}_n\boldsymbol{1}'_n)]\boldsymbol{N}$, *where* $\boldsymbol{k}^\delta = (\boldsymbol{r}^*)^\delta$, $\boldsymbol{k} = \boldsymbol{k}^\delta\boldsymbol{1}_{b^{(1)}+b^{(2)}}$ *and* $\boldsymbol{N} = [\boldsymbol{N}_1 : \boldsymbol{N}_2]$, *with* $\boldsymbol{N}_1$ *and* $\boldsymbol{N}_2$ *representing the incidence matrices of the original two GD designs.*

Remark 7.4.4. As an application of Corollary 7.4.8, one cannot take two semi-regular (or singular) GD designs for the present purpose, because in such a case $\varepsilon_1^* = 1$ (or $\varepsilon_2^* = 1$) and then $\boldsymbol{L}_1^*$ (or $\boldsymbol{L}_2^*$) is not defined as in Corollaries 7.4.6, 7.4.7 and 7.4.8. Other choices of designs possibly used as two basic GD designs in Corollary 7.4.8 are the following four combinations: (i) a singular GD and a semi-regular GD design; (ii) a singular GD and a regular GD design; (iii) a semi-regular GD and a regular GD design; (iv) a regular GD and a regular GD design.

Example 7.4.6. Take two GD designs, based on four groups {1, 2}, {3, 4}, {5, 6}, {7, 8} of two treatments, from Clatworthy (1973):

(1) (Singular type, S18) $v = 8, b = 4, r = 3$, $k = 6$, $\lambda_1 = 3$, $\lambda_2 = 2$, $m = 4, n = 2$ with blocks, (1,2,3,4,5,6), (3,4,5,6,7,8), (1,2,5,6,7,8), (1,2,3,4,7,8);

(2) (Semi-regular type, SR36) $v = b = 8$, $r = k = 4$, $\lambda_1 = 0$, $\lambda_2 = 2$, $m = 4, n = 2$ with blocks, [(1,3,5,7), (2,4,6,8)], [(1,3,6,8), (2,4,5,7)], [(1,4,5,8), (2,3,6,7)], [(1,4,6,7), (2,3,5,8)], in an affine resolvable form.

Then Corollary 7.4.8 produces a $(\rho_0^*; \rho_1^*, \rho_2^*; 0)$-EB design with parameters $v^* = 12$, $b^* = 8$, $\boldsymbol{r}^* = [6\boldsymbol{1}'_4, 4\boldsymbol{1}'_8]'$, $k^* = 7$, $\varepsilon_0^* = 1$, $\varepsilon_1^* = 20/21$, $\varepsilon_2^* = 6/7$, $\rho_0^* = 4$, $\rho_1^* = 3$, $\rho_2^* = 4$, $\boldsymbol{L}_0^* = \boldsymbol{I}_{12} - (56)^{-1}\boldsymbol{1}_{12}(\boldsymbol{r}^*)' - \boldsymbol{L}_1^* - \boldsymbol{L}_2^*$, $\boldsymbol{L}_1^* = 3(\boldsymbol{r}^*)^{-\delta}\boldsymbol{N}'[(\boldsymbol{I}_4 - 4^{-1}\boldsymbol{1}_4\boldsymbol{1}'_4) \otimes 2^{-1}\boldsymbol{1}_2\boldsymbol{1}'_2]\boldsymbol{N} = 3\text{diag}[3^{-1}(\boldsymbol{I}_2 - 2^{-1}\boldsymbol{1}_2\boldsymbol{1}'_2) \otimes (\boldsymbol{I}_2 - 2^{-1}\boldsymbol{1}_2\boldsymbol{1}'_2) : \boldsymbol{I}_2 \otimes (\boldsymbol{I}_2 - 2^{-1}\boldsymbol{1}_2\boldsymbol{1}'_2) : \mathbf{O}]$, $\boldsymbol{L}_2^* = (\boldsymbol{r}^*)^{-\delta}\boldsymbol{N}'[\boldsymbol{I}_4 \otimes (\boldsymbol{I}_2 - 2^{-1}\boldsymbol{1}_2\boldsymbol{1}'_2)]\boldsymbol{N} = \text{diag}[6^{-1}[\boldsymbol{I}_4 + (\boldsymbol{1}_2\boldsymbol{1}'_2 - \boldsymbol{I}_2) \otimes (\boldsymbol{I}_2 - \boldsymbol{1}_2\boldsymbol{1}'_2)] : \mathbf{O} : \boldsymbol{I}_2 \otimes (\boldsymbol{I}_2 - 2^{-1}\boldsymbol{1}_2\boldsymbol{1}'_2)]$, and with the incidence matrix

N^* as

$$N^* = N' = \begin{bmatrix} 1 & 1 & 1 & 1 & 1 & 1 & 0 & 0 \\ 0 & 0 & 1 & 1 & 1 & 1 & 1 & 1 \\ 1 & 1 & 0 & 0 & 1 & 1 & 1 & 1 \\ 1 & 1 & 1 & 1 & 0 & 0 & 1 & 1 \\ 1 & 0 & 1 & 0 & 1 & 0 & 1 & 0 \\ 0 & 1 & 0 & 1 & 0 & 1 & 0 & 1 \\ 1 & 0 & 1 & 0 & 0 & 1 & 0 & 1 \\ 0 & 1 & 0 & 1 & 1 & 0 & 1 & 0 \\ 1 & 0 & 0 & 1 & 1 & 0 & 0 & 1 \\ 0 & 1 & 1 & 0 & 0 & 1 & 1 & 0 \\ 1 & 0 & 0 & 1 & 0 & 1 & 1 & 0 \\ 0 & 1 & 1 & 0 & 1 & 0 & 0 & 1 \end{bmatrix}.$$

Note that the efficiency factors are $\varepsilon_0^* = 1$ (with $\rho_0^* = 4$), $\varepsilon_1^* = 20/21 = 0.9524$ (with $\rho_1^* = 3$), $\varepsilon_2^* = 6/7 = 0.8571$ (with $\rho_2^* = 4$), and the average efficiency factor of the design is given by $\varepsilon^* = 660/709 = 0.9309$. These facts may reveal that the plan in the example is highly efficient. Furthermore, the matrix $\boldsymbol{\Omega}$ is obtained as

$$\frac{1}{4}\text{diag}\left[\boldsymbol{I}_2 \otimes \frac{1}{1080}\begin{bmatrix} 749 & -9 \\ -9 & 749 \end{bmatrix} - (\boldsymbol{1}_2\boldsymbol{1}_2' - \boldsymbol{I}_2) \otimes \frac{1}{1080}\begin{bmatrix} 9 & 11 \\ 11 & 9 \end{bmatrix} : \right.$$

$$\left. \boldsymbol{I}_4 \otimes \frac{1}{40}\begin{bmatrix} 43 & -3 \\ -3 & 43 \end{bmatrix} \right],$$

which implies that elementary contrasts are estimated in the intra-block analysis with seven different variances.

Because GD designs are available in abundance (see Clatworthy, 1973), a large number of $(\rho_0; \rho_1, \rho_2; 0)$-EB designs with $\rho_0 \geq 1$ can be produced by Corollary 7.4.8.

Corollary 7.4.7 shows that the dual of a $(\rho_0; \rho_1, \rho_2; 0)$-EB design with $\rho_0 \geq 1$ is again a $(\rho_0^*; \rho_1^*, \rho_2^*; 0)$-EB design with $\rho_0^* = b - v + \rho_0 \geq 1$. Hence the multiplicity of $\varepsilon_0^* = 1$ increases when $b > v$ in the original design. Thus, a procedure of taking the dual of such designs may be recommendable to get designs with larger values of ρ_0, if possible. Recall that a similar property of a $(\rho_0; \rho_1; 0)$-EB design is also described in Remark 7.3.2 or at the end of Section 7.3.1.

A simple juxtaposition method of constructing $(\rho_0; \rho_1, \rho_2; 0)$-EB designs is presented by Theorem 6.4.1 as follows.

Corollary 7.4.9. *If $\boldsymbol{N}$ is the $v \times b$ incidence matrix of a $(\rho_0; \rho_1, \rho_2; 0)$-EB design with parameters v, b, $\boldsymbol{r}$, $\boldsymbol{k}$, $\varepsilon_0 = 1$, ε_1, ε_2, ρ_0, ρ_1, ρ_2, $\boldsymbol{L}_0$, $\boldsymbol{L}_1$, $\boldsymbol{L}_2$, then its juxtaposition $\boldsymbol{1}_s' \otimes \boldsymbol{N}$ is the incidence matrix of a $(\rho_0^*; \rho_1^*, \rho_2^*; 0)$-EB design with parameters $v^* = v$, $b^* = sb$, $\boldsymbol{r}^* = s\boldsymbol{r}$, $\boldsymbol{k}^* = \boldsymbol{k}$, $\varepsilon_\beta^* = \varepsilon_\beta$, $\rho_\beta^* = \rho_\beta$, $\boldsymbol{L}_\beta^* = \boldsymbol{L}_\beta$,*

$\beta = 0, 1, 2$, *for a positive integer* s.

Note that the efficiency factors in both the original design and the resulting design are the same.

For the incidence matrix $\boldsymbol{N}$ of a BIB design the following pattern

$$\boldsymbol{N}^* = \begin{bmatrix} \boldsymbol{N} \\ \boldsymbol{I}_b \end{bmatrix} \tag{7.4.3}$$

is considered. This design is likely to be very useful in plant breeding trials where a sufficient material for new strains to have more than one (or two) observations may not be available, but there may not be any limitation for the standard varieties, and where the number of the standard varieties is more than one. The following is due to Singh and Dey (1979), but another simple proof can be found in Puri and Kageyama (1985). By recalling the argument on partial efficiency balance in Section 4.4, the following theorem can be given.

Theorem 7.4.12 (Singh and Dey, 1979). *The existence of a BIB design with parameters* v, b, r, k, λ *implies the existence of a* $(\rho_0^*; \rho_1^*, \rho_2^*; 0)$*-EB design with the incidence matrix* (7.4.3) *and with parameters* $v^* = v + b$, $b^* = b$, $\boldsymbol{r}^* = [r\boldsymbol{1}_v', \boldsymbol{1}_b']'$, $k^* = k+1$, $\varepsilon_0^* = 1$, $\varepsilon_1^* = 1-(2r-\lambda)/[r(k+1)]$, $\varepsilon_2^* = k/(k+1)$, $\rho_0^* = v$, $\rho_1^* = v-1$, $\rho_2^* = b - v$,

$$\boldsymbol{L}_0^* = \boldsymbol{I}_{v+b} - \frac{1}{b(k+1)}\boldsymbol{1}_{v+b}(\boldsymbol{r}^*)' - \boldsymbol{L}_1^* - \boldsymbol{L}_2^*,$$

$$\boldsymbol{L}_1^* = \frac{1}{2r-\lambda}\begin{bmatrix} (r-\lambda)(\boldsymbol{I}_v - \frac{1}{v}\boldsymbol{1}_v\boldsymbol{1}_v') & \boldsymbol{N} - \frac{k}{v}\boldsymbol{1}_v\boldsymbol{1}_b' \\ r(\boldsymbol{N}' - \frac{k}{v}\boldsymbol{1}_b\boldsymbol{1}_v') & \frac{r}{r-\lambda}(\boldsymbol{N}'\boldsymbol{N} - \frac{k^2}{v}\boldsymbol{1}_b\boldsymbol{1}_b') \end{bmatrix}$$

and

$$\boldsymbol{L}_2^* = \frac{1}{r-\lambda}\mathrm{diag}\left[\mathbf{O} : (r-\lambda)\boldsymbol{I}_b - \boldsymbol{N}'\boldsymbol{N} + \frac{\lambda k}{r}\boldsymbol{1}_b\boldsymbol{1}_b'\right].$$

Because an abundance of BIB designs are available in the literature (for example, see Chapter 8), Theorem 7.4.12 produces a large number of $(\rho_0^*; \rho_1^*, \rho_2^*; 0)$-EB designs. Note that when the basic design is a symmetric BIB design, Theorem 7.4.12 yields a $(\rho_0^*; \rho_1^*; 0)$-EB design, i.e., $\boldsymbol{L}_2^* = \mathbf{O}$ with $\rho_2^* = 0$ (see also Theorem 7.3.15 for another construction).

When $4t + 1$ is a prime or a prime power, there exists a BIB design with parameters $v = 4t+1$, $b = 8t+2$, $r = 4t$, $k = 2t$, $\lambda = 2t-1$ (for the existence see Theorem 8.2.11 along with Theorem 6.7.1). Then Theorem 7.4.12 can produce the following result.

Corollary 7.4.10. *When* $4t+1$ *is a prime or a prime power, there exists a* $(\rho_0; \rho_1, \rho_2; 0)$*-EB design with parameters* $v = 12t+3$, $b = 8t+2$, $\boldsymbol{r} = [4t\boldsymbol{1}_{4t+1}', \boldsymbol{1}_{8t+2}']'$, $k = 2t+1$, $\varepsilon_0 = 1$, $\varepsilon_1 = (2t-1)(4t+1)/[4t(2t+1)]$, $\varepsilon_2 = (2t+1)/(2t+2)$, $\rho_0 = 4t+1$, $\rho_1 = 4t$, $\rho_2 = 4t+1$,

$$\boldsymbol{L}_0 = \boldsymbol{I}_{12t+3} - \frac{1}{2(2t+1)(4t+1)}\boldsymbol{1}_{12t+3}\boldsymbol{r}' - \boldsymbol{L}_1 - \boldsymbol{L}_2,$$

$$L_1 = \frac{1}{6t+1}\begin{bmatrix} (2t+1)(I_{4t+1} - \frac{1}{4t+1}1_{4t+1}1'_{4t+1}) & N - \frac{2t}{4t+1}1_{4t+1}1'_{8t+2} \\ 4t(N' - \frac{2t}{4t+1}1_{8t+2}1'_{4t+1}) & \frac{4t}{2t+1}(N'N - \frac{4t^2}{4t+1}1_{8t+2}1'_{8t+2}) \end{bmatrix}$$

and

$$L_2 = \frac{1}{2t+1}\text{diag}\left[O : (2t+1)I_{8t+2} - N'N + \frac{2t-1}{2}1_{8t+2}1'_{8t+2}\right].$$

Theorems 6.3.4 and 6.3.5 can slightly be generalized, by adding new s (≥ 1) treatments to all the b_1^* blocks of the first a_1 sets, the next s new treatments to all the b_2^* blocks of the second a_2 sets, and so on, where $a = \sum_{h=1}^{q} a_h$ and

$$b_h^* = b_{a_1+\cdots+a_{h-1}+1} + \cdots + b_{a_1+\cdots+a_h}, \quad h = 1, 2, ..., q,$$

with $a_0 = 0$. It is obvious that the method of adding sq new treatments to the original $(\alpha_1, \alpha_2, ..., \alpha_a)$-resolvable block design with the incidence matrix N to form

$$N^* = \begin{bmatrix} N \\ \text{diag}[1_s 1'_{b_1^*} : 1_s 1'_{b_2^*} : \cdots : 1_s 1'_{b_q^*}] \end{bmatrix} \tag{7.4.4}$$

yields an $(\alpha_1^*, \alpha_2^*, ..., \alpha_q^*)$-resolvability for N, where

$$\alpha_h^* = \alpha_{a_1+\cdots+a_{h-1}+1} + \cdots + \alpha_{a_1+\cdots+a_h}, \quad h = 1, 2, ..., q.$$

Noting that $\alpha_h^*/r = b_h^*/b$, $h = 1, 2, ..., q$, the lines similar to Theorems 6.3.4 and 6.3.5 lead to Theorems 7.4.13 and 7.4.14 below.

Theorem 7.4.13. *If N is the incidence matrix of an $(\alpha_1, \alpha_2, ..., \alpha_a)$-resolvable BIB design with parameters v, b, r, k, λ, then (7.4.4) is the incidence matrix of a $(\rho_0^*; \rho_1^*, \rho_2^*; 0)$-EB design with parameters $v^* = v + sq$, $b^* = b$, $r^* = [r1'_v, b_1^* 1'_s, ..., b_q^* 1'_s]'$, $k^* = k + s$, $\varepsilon_0^* = 1$, $\varepsilon_1^* = 1 - (r - \lambda)/[r(k+s)]$, $\varepsilon_2^* = k/(k+s)$, $\rho_0^* = q(s-1)+1$, $\rho_1^* = v - 1$, $\rho_2^* = q - 1$, $L_0^* = I_{v+sq} - [b(k+s)]^{-1}1_{v+sq}(r^*)' - L_1^* - L_2^*$, $L_1^* = \text{diag}[I_v - v^{-1}1_v1'_v : O]$, $L_2^* = \text{diag}[O : [I_q - b^{-1}1_q(b^*)'] \otimes s^{-1}1_s1'_s]$, where $b^* = [b_1^*, b_2^*, ..., b_q^*]'$ and $s \geq 1$.*

Example 7.4.7. Consider for N the (3,6,6)-resolvable BIB design, mentioned in Section 6.0.3, with parameters $v = 16, b = 8 + 16 + 16 = 40, r = 15, k = 6, \lambda = 5$, having blocks [(0,1,3,8,9,11), (1,2,4,9,10,12), (2,3,5,10,11,13), (3,4,6,11,12,14), (4,5,7,12,13,15), (5,6,8,13,14,0), (6,7,9,14,15,1), (7,8,10,15,0,2)], [(0,1,3,5,9,12), (0,1,2,3,6,12) mod 16], the incidence matrix

$$N^* = \begin{bmatrix} N \\ \text{diag}[1'_8, 1'_{16}, 1'_{16}] \end{bmatrix},$$

produces a $(1; 15, 2; 0)$-EB design with parameters $v^* = 19$, $b^* = 40$, $r^* = [15 1'_{16}, 8, 16 1'_2]'$, $k^* = 7$, $\varepsilon_0^* = 1$, $\varepsilon_1^* = 19/21$, $\varepsilon_2^* = 6/7$, $\rho_0^* = 1$, $\rho_1^* = 15$, $\rho_2^* = 2$, $L_0^* = I_{19} - (280)^{-1}1_{19}(r^*)' - L_1^* - L_2^*$, $L_1^* = \text{diag}[I_{16} - (16)^{-1}1_{16}1'_{16} : O]$, $L_2^* = \text{diag}[O : I_3 - 5^{-1}1_3[1, 2, 2]]$. Also note that the average efficiency factor of the design is $\varepsilon^* = 1026/1135 = 0.9040$ [see Remark 3.4.1(c)]. In this case, a g-inverse of $\boldsymbol{C}_1^*$ is given by

$$\boldsymbol{\Omega}^* = \frac{1}{5}\mathrm{diag}\left[\frac{7}{19}\boldsymbol{I}_{16} - \frac{1}{456}\mathbf{1}_{16}\mathbf{1}_{16}' : \frac{35}{48}\boldsymbol{I}_3 - \frac{1}{48}\mathbf{1}_3\mathbf{1}_3'\right].$$

This shows that there may be five different variances for the BLUEs of the elementary contrasts estimated in the intra-block analysis.

Example 7.4.8. In the (3,6,6)-resolvable BIB design (with the incidence matrix $\boldsymbol{N}$) described in Example 7.4.7, if $a_1 = 1$ and $a_2 = 2$ (with $q = 2$) are taken, then Theorem 7.4.13 yields a $(\rho_0^*; \rho_1^*, \rho_2^*; 0)$-EB design with parameters $v^* = 2s + 16$, $b^* = 40$, $\boldsymbol{r}^* = [15\mathbf{1}_{16}', 8\mathbf{1}_s', 32\mathbf{1}_s']'$, $k^* = s + 6$, $\varepsilon_0^* = 1$, $\varepsilon_1^* = (3s + 16)/(3s + 18)$, $\varepsilon_2^* = 6/(s + 6)$, $\rho_0^* = 2s - 1$, $\rho_1^* = 15$, $\rho_2^* = 1$, $\boldsymbol{L}_0^* = \boldsymbol{I}_{2s+16} - [40(s + 6)]^{-1}\mathbf{1}_{2s+16}(\boldsymbol{r}^*)' - \boldsymbol{L}_1^* - \boldsymbol{L}_2^*$, $\boldsymbol{L}_1^* = \mathrm{diag}[\boldsymbol{I}_{16} - (16)^{-1}\mathbf{1}_{16}\mathbf{1}_{16}' : \mathbf{O}]$, $\boldsymbol{L}_2^* = \mathrm{diag}[\mathbf{O} : (\boldsymbol{I}_2 - 5^{-1}\mathbf{1}_2[1, 4]) \otimes s^{-1}\mathbf{1}_s\mathbf{1}_s']$, and with the incidence matrix, for $s = 2$, as

$$\boldsymbol{N}^* = \begin{bmatrix} \boldsymbol{N} \\ \mathrm{diag}[\mathbf{1}_2\mathbf{1}_8' : \mathbf{1}_2\mathbf{1}_{32}'] \end{bmatrix},$$

giving

$$\boldsymbol{\Omega}^* = \mathrm{diag}\left[\frac{1}{2640}(16\boldsymbol{I}_{16} - \mathbf{1}_{16}\mathbf{1}_{16}') + \frac{1}{15}\boldsymbol{I}_{16} : \begin{bmatrix} \frac{17}{120} & \frac{1}{60} & -\frac{1}{240} & -\frac{1}{240} \\ \frac{1}{60} & \frac{17}{120} & -\frac{1}{240} & -\frac{1}{240} \\ -\frac{1}{240} & -\frac{1}{240} & \frac{31}{960} & \frac{1}{960} \\ -\frac{1}{240} & -\frac{1}{240} & \frac{1}{960} & \frac{31}{960} \end{bmatrix}\right].$$

Here the design has parameters as $v^* = 20$, $b^* = 40$, $\boldsymbol{r}^* = [15\mathbf{1}_{16}', 8\mathbf{1}_2', 32\mathbf{1}_2']'$, $k^* = 8$, $\varepsilon_0^* = 1$, $\varepsilon_1^* = 11/12$, $\varepsilon_2^* = 3/4$, $\rho_0^* = 3$, $\rho_1^* = 15$, $\rho_2^* = 1$. There may be six different variances among elementary contrasts in the intra-block analysis.

Example 7.4.9. On the other hand, in the same (3,6,6)-resolvable BIB design (with the incidence matrix $\boldsymbol{N}$) as described in Example 7.4.7, if $a_1 = 2$ and $a_2 = 1$ (with $q = 2$) are taken, then Theorem 7.4.13 yields a $(\rho_0^*; \rho_1^*, \rho_2^*; 0)$-EB design with parameters $v^* = 2s + 16$, $b^* = 40$, $\boldsymbol{r}^* = [15\mathbf{1}_{16}', 24\mathbf{1}_s', 16\mathbf{1}_s']'$, $k^* = s + 6$, $\varepsilon_0^* = 1$, $\varepsilon_1^* = (3s + 16)/(3s + 18)$, $\varepsilon_2^* = 6/(s + 6)$, $\rho_0^* = 2s - 1$, $\rho_1^* = 15$, $\rho_2^* = 1$, $\boldsymbol{L}_0^* = \boldsymbol{I}_{2s+16} - [40(s + 6)]^{-1}\mathbf{1}_{2s+16}(\boldsymbol{r}^*)' - \boldsymbol{L}_1^* - \boldsymbol{L}_2^*$, $\boldsymbol{L}_1^* = \mathrm{diag}[\boldsymbol{I}_{16} - (16)^{-1}\mathbf{1}_{16}\mathbf{1}_{16}' : \mathbf{O}]$, $\boldsymbol{L}_2^* = \mathrm{diag}[\mathbf{O} : (\boldsymbol{I}_2 - 5^{-1}\mathbf{1}_2[3, 2]) \otimes s^{-1}\mathbf{1}_s\mathbf{1}_s']$, and with the incidence matrix, for $s = 2$, as

$$\boldsymbol{N}^* = \begin{bmatrix} \boldsymbol{N} \\ \mathrm{diag}[\mathbf{1}_2\mathbf{1}_{24}' : \mathbf{1}_2\mathbf{1}_{16}'] \end{bmatrix},$$

giving

$$\boldsymbol{\Omega}^* = \mathrm{diag}\left[\frac{1}{2640}(16\boldsymbol{I}_{16} - \mathbf{1}_{16}\mathbf{1}_{16}') + \frac{1}{15}\boldsymbol{I}_{16} : \begin{bmatrix} \frac{2}{45} & \frac{1}{360} & -\frac{1}{240} & -\frac{1}{240} \\ \frac{1}{360} & \frac{2}{45} & -\frac{1}{240} & -\frac{1}{240} \\ -\frac{1}{240} & -\frac{1}{240} & \frac{11}{160} & \frac{1}{160} \\ -\frac{1}{240} & -\frac{1}{240} & \frac{1}{160} & \frac{11}{160} \end{bmatrix}\right].$$

In this case, there may be six different variances among elementary contrasts in the intra-block analysis.

Note that the two designs used in Examples 7.4.8 and 7.4.9 have different replication numbers, but they have the same efficiency factor.

Example 7.4.10 (a continuation of Example 6.3.2). If one takes $s = 2$ in the design considered in Example 6.3.2 with $a = 3$ and $b_1 = b_2 = b_3 = 3$, then one can get a $(5; 4, 2; 0)$-EB design with parameters $v^* = 12$, $b^* = 9$, $r^* = 3$, $k^* = 4$, $\varepsilon_0^* = 1$, $\varepsilon_1^* = 3/4$, $\varepsilon_2^* = 1/2$, $\rho_0^* = 5$, $\rho_1^* = 4$, $\rho_2^* = 2$, $\boldsymbol{L}_0^* = \boldsymbol{I}_{12} - (12)^{-1}\mathbf{1}_{12}\mathbf{1}_{12}' - \boldsymbol{L}_1^* - \boldsymbol{L}_2^*$, $\boldsymbol{L}_1^* = \text{diag}[\boldsymbol{I}_2 \otimes (\boldsymbol{I}_3 - 3^{-1}\mathbf{1}_3\mathbf{1}_3') : \mathbf{O}]$, $\boldsymbol{L}_2^* = \text{diag}[\mathbf{O} : (\boldsymbol{I}_3 - 3^{-1}\mathbf{1}_3\mathbf{1}_3') \otimes 2^{-1}\mathbf{1}_2\mathbf{1}_2']$, whose incidence matrix is of the form

$$\boldsymbol{N}^* = \begin{bmatrix} 1 & 0 & 0 & 1 & 0 & 0 & 1 & 0 & 0 \\ 0 & 1 & 0 & 0 & 1 & 0 & 0 & 1 & 0 \\ 0 & 0 & 1 & 0 & 0 & 1 & 0 & 0 & 1 \\ 1 & 0 & 0 & 0 & 1 & 0 & 0 & 0 & 1 \\ 0 & 1 & 0 & 0 & 0 & 1 & 1 & 0 & 0 \\ 0 & 0 & 1 & 1 & 0 & 0 & 0 & 1 & 0 \\ 1 & 1 & 1 & 0 & 0 & 0 & 0 & 0 & 0 \\ 1 & 1 & 1 & 0 & 0 & 0 & 0 & 0 & 0 \\ 0 & 0 & 0 & 1 & 1 & 1 & 0 & 0 & 0 \\ 0 & 0 & 0 & 1 & 1 & 1 & 0 & 0 & 0 \\ 0 & 0 & 0 & 0 & 0 & 0 & 1 & 1 & 1 \\ 0 & 0 & 0 & 0 & 0 & 0 & 1 & 1 & 1 \end{bmatrix}.$$

Also note that the average efficiency factor of the design is $\varepsilon^* = 33/43 = 0.7674$. In this case, a g-inverse of $\boldsymbol{C}_1^*$ is given by

$$\boldsymbol{\Omega}^* = \frac{1}{3}\left(\boldsymbol{I}_{12} + \text{diag}[\frac{1}{3}\boldsymbol{I}_2 \otimes (\boldsymbol{I}_3 - \frac{1}{3}\mathbf{1}_3\mathbf{1}_3') : (\boldsymbol{I}_3 - \frac{1}{3}\mathbf{1}_3\mathbf{1}_3') \otimes \frac{1}{2}\mathbf{1}_2\mathbf{1}_2']\right).$$

There may be five different variances in the intra-block analysis for the BLUEs of the elementary contrasts.

Theorem 7.4.14. *If $\boldsymbol{N}$ is the incidence matrix of an $(\alpha_1, \alpha_2, ..., \alpha_a)$-resolvable 2-associate PBIB design with parameters $v, b, r, k, \lambda_1, \lambda_2, \varepsilon_0 = 1, \varepsilon_1 < 1, \rho_0, \rho_1, \boldsymbol{L}_0, \boldsymbol{L}_1$, then the incidence matrix* (7.4.4) *yields a $(\rho_0^*; \rho_1^*, \rho_2^*; 0)$-EB design with parameters $v^* = v + sq$, $b^* = b$, $\boldsymbol{r}^* = [r\mathbf{1}_v', b_1^*\mathbf{1}_s', ..., b_q^*\mathbf{1}_s']'$, $k^* = k + s$, $\varepsilon_0^* = 1$, $\varepsilon_1^* = 1 - k(1 - \varepsilon_1)/(k + s)$, $\varepsilon_2^* = k/(k + s)$, $\rho_0^* = \rho_0 + q(s - 1) + 1$, $\rho_1^* = \rho_1$, $\rho_2^* = q - 1$, $\boldsymbol{L}_0^* = \boldsymbol{I}_{v+sq} - [b(k + s)]^{-1}\mathbf{1}_{v+sq}(\boldsymbol{r}^*)' - \boldsymbol{L}_1^* - \boldsymbol{L}_2^*$, $\boldsymbol{L}_1^* = \text{diag}[\boldsymbol{L}_1 : \mathbf{O}]$, $\boldsymbol{L}_2^* = \text{diag}[\mathbf{O} : [\boldsymbol{I}_q - b^{-1}\mathbf{1}_q(\boldsymbol{b}^*)'] \otimes s^{-1}\mathbf{1}_s\mathbf{1}_s']$, where $\boldsymbol{b}^* = [b_1^*, b_2^*, ..., b_q^*]'$ and $s \geq 1$.*

Example 7.4.11. Consider a (1,2,2)-resolvable semi-regular GD design, based on three groups $\{1, 2, 3, 4, 5\}, \{6, 7, 8, 9, 10\}, \{11, 12, 13, 14, 15\}$ of five treatments, SR28 in Clatworthy (1973), with parameters $v = 15, b = 25, r = 5, k = 3, \lambda_1 = 0, \lambda_2 = 1, m = 3, n = 5$, having blocks

[(1,6,11),(2,9,13),(5,10,12),(3,7,15),(4,8,14)], [(2,7,14),(5,8,11),(3,6,13),
(4,9,12),(1,10,15),(3,8,12),(4,6,15),(1,9,14), (2,10,11),(5,7,13)],
[(4,10,13),(1,7,12),(2,8,15),(5,6,14),(3,9,11), (5,9,15),(3,10,14),
(4,7,11),(1,8,13),(2,6,12)],

in which $a = 3, b_1 = 5, b_2 = b_3 = 10, \alpha_1 = 1, \alpha_2 = \alpha_3 = 2$ that form the incidence matrix, $\boldsymbol{N}$, say. Then Theorem 6.3.5 and Theorem 7.4.14 with $a_1 = 2, a_2 = 1$ and $q = 2$ yield, respectively, the following two designs.

(1) A $(\rho_1^*; \rho_2^*, \rho_3^*; 0)$-EB design with parameters $v^* = 3s + 15$, $b^* = 25$, $\boldsymbol{r}^* = [5\boldsymbol{1}_{15}', 5\boldsymbol{1}_s', 10\boldsymbol{1}_{2s}']'$, $k^* = s + 3$, $\varepsilon_0^* = 1$, $\varepsilon_1^* = (s+2)/(s+3)$, $\varepsilon_2^* = 3/(s+3)$, $\rho_0^* = 3s$, $\rho_1^* = 12$, $\rho_2^* = 2$, $\boldsymbol{L}_0^* = \boldsymbol{I}_{3s+15} - [25(s+3)]^{-1}\boldsymbol{1}_{3s+15}(\boldsymbol{r}^*)' - \boldsymbol{L}_1^* - \boldsymbol{L}_2^*$, $\boldsymbol{L}_1^* = \text{diag}[\boldsymbol{I}_3 \otimes (\boldsymbol{I}_5 - 5^{-1}\boldsymbol{1}_5\boldsymbol{1}_5') : \mathbf{O}]$, $\boldsymbol{L}_2^* = \text{diag}[\mathbf{O} : (\boldsymbol{I}_3 - 5^{-1}\boldsymbol{1}_3[1,2,2]) \otimes s^{-1}\boldsymbol{1}_s\boldsymbol{1}_s']$, and with the incidence matrix, for example, when $s = 2$, as

$$\boldsymbol{N}^* = \begin{bmatrix} \boldsymbol{N} \\ \text{diag}[\boldsymbol{1}_2\boldsymbol{1}_5' : \boldsymbol{1}_2\boldsymbol{1}_{10}' : \boldsymbol{1}_2\boldsymbol{1}_{10}'] \end{bmatrix},$$

giving

$$\boldsymbol{\Omega}^* = \text{diag}\left[\frac{1}{25}\boldsymbol{I}_3 \otimes (\frac{25}{4}\boldsymbol{I}_5 - \frac{1}{4}\boldsymbol{1}_5\boldsymbol{1}_5') : \frac{1}{5}[\boldsymbol{I}_3 \otimes (\boldsymbol{I}_2 + \frac{4}{15}\boldsymbol{1}_2\boldsymbol{1}_2') + \frac{1}{15}(\boldsymbol{I}_3 - \boldsymbol{1}_3\boldsymbol{1}_3') \otimes \boldsymbol{1}_2\boldsymbol{1}_2']\right],$$

with parameters $v^* = 21$, $b^* = 25$, $\boldsymbol{r}^* = [5\boldsymbol{1}_{17}', 10\boldsymbol{1}_4']'$, $k^* = 5$, $\varepsilon_0^* = 1$, $\varepsilon_1^* = 4/5$, $\varepsilon_2^* = 3/5$, $\rho_0^* = 6$, $\rho_1^* = 12$, $\rho_2^* = 2$, $\varepsilon^* = 60/73 = 0.8219$. Note that there may be seven different variances among elementary contrasts in the intra-block analysis.

(2) A $(\rho_0^*; \rho_1^*, \rho_2^*; 0)$-EB design with parameters $v^* = 2s + 15$, $b^* = 25$, $\boldsymbol{r}^* = [5\boldsymbol{1}_{15}', 15\boldsymbol{1}_s', 10\boldsymbol{1}_s']'$, $k^* = s + 3$, $\varepsilon_0^* = 1$, $\varepsilon_1^* = (s+2)/(s+3)$, $\varepsilon_2^* = 3/(s+3)$, $\rho_0^* = 2s + 1$, $\rho_1^* = 12$, $\rho_2^* = 1$, $\boldsymbol{L}_0^* = \boldsymbol{I}_{2s+15} - [25(s+3)]^{-1}\boldsymbol{1}_{2s+15}(\boldsymbol{r}^*)' - \boldsymbol{L}_1^* - \boldsymbol{L}_2^*$, $\boldsymbol{L}_1^* = \text{diag}[\boldsymbol{I}_3 \otimes (\boldsymbol{I}_5 - 5^{-1}\boldsymbol{1}_5\boldsymbol{1}_5') : \mathbf{O}]$, $\boldsymbol{L}_2^* = \text{diag}[\mathbf{O} : (\boldsymbol{I}_2 - 5^{-1}\boldsymbol{1}_2[3,2]) \otimes s^{-1}\boldsymbol{1}_s\boldsymbol{1}_s']$, and with the incidence matrix, for example, when $s = 2$, as

$$\boldsymbol{N}^* = \begin{bmatrix} \boldsymbol{N} \\ \text{diag}[\boldsymbol{1}_2\boldsymbol{1}_{15}' : \boldsymbol{1}_2\boldsymbol{1}_{10}'] \end{bmatrix},$$

giving

$$\boldsymbol{\Omega}^* = \text{diag}\left[\frac{1}{25}\boldsymbol{I}_3 \otimes (\frac{25}{4}\boldsymbol{I}_5 - \frac{1}{4}\boldsymbol{1}_5\boldsymbol{1}_5') : \begin{bmatrix} \frac{17}{225} & \frac{2}{225} & -\frac{1}{75} & -\frac{1}{75} \\ \frac{2}{225} & \frac{17}{225} & -\frac{1}{75} & -\frac{1}{75} \\ -\frac{1}{75} & -\frac{1}{75} & \frac{6}{50} & \frac{1}{50} \\ -\frac{1}{75} & -\frac{1}{75} & \frac{1}{50} & \frac{6}{50} \end{bmatrix}\right],$$

with parameters $v^* = 19$, $b^* = 25$, $\boldsymbol{r}^* = [5\boldsymbol{1}_{15}', 15\boldsymbol{1}_2', 10\boldsymbol{1}_2']'$, $k^* = 5$, $\varepsilon_0^* = 1$, $\varepsilon_1^* = 4/5$, $\varepsilon_2^* = 3/5$, $\rho_0^* = 5$, $\rho_1^* = 12$, $\rho_2^* = 1$, $\varepsilon^* = 54/65 = 0.8308$. Note that there may be six different variances among elementary contrasts in the intra-block analysis. Further, note that both the designs (1) and (2) above have the same b^* and k^*, and the same efficiency factors with different multiplicities, but the average efficiency factors of the designs are almost the same.

Consider an α-resolvable block design with parameters v, b, r, k, in which $b = b_0 a, r = \alpha a, vr = bk, v\alpha = kb_0$, by letting $b_1 = b_2 = \cdots = b_a$ ($= b_0$, say) and $\alpha_1 = \alpha_2 = \cdots = \alpha_a$ ($= \alpha$, say) in the $(\alpha_1, \alpha_2, ..., \alpha_a)$-resolvability (see Section 6.0.3). Theorem 6.3.4 leads to the following corollaries.

Corollary 7.4.11. *If $\boldsymbol{N}$ is the incidence matrix of an α-resolvable BIB design with parameters $v, b = b_0 a, r = \alpha a, k, \lambda$, then the incidence matrix*

$$\boldsymbol{N}^* = \begin{bmatrix} \boldsymbol{N} \\ \boldsymbol{I}_a \otimes \mathbf{1}_s \mathbf{1}'_{b_0} \end{bmatrix} \tag{7.4.5}$$

yields a $(\rho_0^*; \rho_1^*, \rho_2^*; 0)$-EB design with parameters $v^* = v + sa$, $b^* = b$, $\boldsymbol{r}^* = [r\mathbf{1}'_v, b_0\mathbf{1}'_{sa}]'$, $k^* = k+s$, $\varepsilon_0^* = 1$, $\varepsilon_1^* = 1 - (r-\lambda)/[r(k+s)]$, $\varepsilon_2^* = k/(k+s)$, $\rho_0^* = a(s-1)+1$, $\rho_1^* = v-1$, $\rho_2^* = a-1$, $\boldsymbol{L}_0^* = \boldsymbol{I}_{v+sa} - [b(k+s)]^{-1}\mathbf{1}_{v+sa}(\boldsymbol{r}^*)' - \boldsymbol{L}_1^* - \boldsymbol{L}_2^*$, $\boldsymbol{L}_1^* = \mathrm{diag}[\boldsymbol{I}_v - v^{-1}\mathbf{1}_v\mathbf{1}'_v : \mathbf{O}]$, $\boldsymbol{L}_2^* = \mathrm{diag}[\mathbf{O} : (\boldsymbol{I}_a - a^{-1}\mathbf{1}_a\mathbf{1}'_a) \otimes s^{-1}\mathbf{1}_s\mathbf{1}'_s]$ for $s \geq 1$.

Corollary 7.4.12. *The existence of an α-resolvable BIB design with parameters $v, b = b_0 a, r = \alpha a, k, \lambda$, implies, through the incidence matrix*

$$\boldsymbol{N}^* = \begin{bmatrix} \boldsymbol{N} \\ \mathrm{diag}[\mathbf{1}_s\mathbf{1}'_{a_1 b_0} : \mathbf{1}_s\mathbf{1}'_{a_2 b_0} : \cdots : \mathbf{1}_s\mathbf{1}'_{a_q b_0}] \end{bmatrix}, \tag{7.4.6}$$

the existence of a $(\rho_0^*; \rho_1^*, \rho_2^*; 0)$-EB design with parameters $v^* = v + sq$, $b^* = b$, $\boldsymbol{r}^* = [r\mathbf{1}'_v, a_1 b_0 \mathbf{1}'_s, ..., a_q b_0 \mathbf{1}'_s]'$, $k^* = k + s$, $\varepsilon_0^* = 1$, $\varepsilon_1^* = 1 - (r - \lambda)/[r(k+s)]$, $\varepsilon_2^* = k/(k+s)$, $\rho_0^* = q(s-1)+1$, $\rho_1^* = v - 1$, $\rho_2^* = q - 1$, $\boldsymbol{L}_0^* = \boldsymbol{I}_{v+sq} - [b(k+s)]^{-1}\mathbf{1}_{v+sq}(\boldsymbol{r}^*)' - \boldsymbol{L}_1^* - \boldsymbol{L}_2^*$, $\boldsymbol{L}_1^* = \mathrm{diag}[\boldsymbol{I}_v - v^{-1}\mathbf{1}_v\mathbf{1}'_v : \mathbf{O}]$, $\boldsymbol{L}_2^* = \mathrm{diag}[\mathbf{O} : (\boldsymbol{I}_q - a^{-1}\mathbf{1}_q\boldsymbol{a}') \otimes s^{-1}\mathbf{1}_s\mathbf{1}'_s]$, where $\boldsymbol{a} = [a_1, a_2, ..., a_q]'$ and $s \geq 1$.

Note that there is a relation $a = \sum_{h=1}^{q} a_h$ between (7.4.5) and (7.4.6). Then $q \leq a$.

Thus, the properties of resolvability of designs have been used to construct $(\rho_0; \rho_1, \rho_2; 0)$-EB designs with $\rho_0 \geq 1$. A large number of α-resolvable BIB designs with $\alpha \geq 1$ are available in the literature. For example, refer to Bose (1947), Kageyama (1972b, 1973b, 1976b), Mohan (1980), Kageyama and Mohan (1983), and Shrikhande and Raghavarao (1963). In particular, Kageyama (1972b) and Kageyama and Mohan (1983) have tabulated practical parameter combinations for such BIB designs (along with their solutions) under some restrictions on the range of parameters.

It is known (see Shrikhande and Raghavarao, 1963) that if $4u - 1$ is a prime or a prime power, there exists a $(2u-1)$-resolvable BIB design with parameters $v = (4u-1)^2$, $b = 4u(4u-1)$, $r = 4u(2u-1)$, $k = (2u-1)(4u-1)$, $\lambda = u(4u-3)$, $b_0 = 4u - 1$, $a = 4u$. In this case Corollaries 7.4.11 and 7.4.12 produce the following $(\rho_0^*; \rho_1^*, \rho_2^*; 0)$-EB designs, respectively:

(1) $v^* = (4u-1)^2 + 4us, b^* = 4u(4u-1)$, $\boldsymbol{r}^* = [4u(2u-1)\boldsymbol{1}'_{(4u-1)^2}, (4u-1)\boldsymbol{1}'_{4us}]'$, $k^* = (2u-1)(4u-1)+s$, $\varepsilon_0^* = 1$, $\varepsilon_1^* = 1-(4u-1)/\{4(2u-1)[(2u-1)(4u-1)+s]\}$, $\varepsilon_2^* = (2u-1)(4u-1)/[(2u-1)(4u-1)+s]$, $\rho_0^* = 4u(s-1)+1$, $\rho_1^* = 8u(2u-1)$, $\rho_2^* = 4u-1$, $\boldsymbol{L}_0^* = \boldsymbol{I}_{(4u-1)^2+4us} - \{4u(4u-1)[(2u-1)(4u-1)+s]\}^{-1}\boldsymbol{1}_{(4u-1)^2+4us}(\boldsymbol{r}^*)' - \boldsymbol{L}_1^* - \boldsymbol{L}_2^*$, $\boldsymbol{L}_1^* = \operatorname{diag}[\boldsymbol{I}_{(4u-1)^2} - (4u-1)^{-2}\boldsymbol{1}_{(4u-1)^2}\boldsymbol{1}'_{(4u-1)^2} : \mathbf{O}]$, $\boldsymbol{L}_2^* = \operatorname{diag}[\mathbf{O} : [\boldsymbol{I}_{4u} - (4u)^{-1}\boldsymbol{1}_{4u}\boldsymbol{1}'_{4u}] \otimes s^{-1}\boldsymbol{1}_s\boldsymbol{1}'_s]$ for $s \geq 1$.

(2) for $q \geq 1$, $s \geq 1$ and $\sum_{h=1}^{q} a_h = 4u$, $v^* = (4u-1)^2 + sq, b^* = 4u(4u-1)$, $\boldsymbol{r}^* = [4u(2u-1)\boldsymbol{1}'_{(4u-1)^2}, (4u-1)a_1\boldsymbol{1}'_s, ..., (4u-1)a_q\boldsymbol{1}'_s]'$, $k^* = (2u-1)(4u-1)+s$, ε_β^* as above, $\rho_0^* = q(s-1)+1$, $\rho_1^* = 8u(2u-1)$, $\rho_2^* = q-1$, $\boldsymbol{L}_0^* = \boldsymbol{I}_{(4u-1)^2+sq} - \{4u(4u-1)[(2u-1)(4u-1)+s]\}^{-1}\boldsymbol{1}_{(4u-1)^2+sq}(\boldsymbol{r}^*)' - \boldsymbol{L}_1^* - \boldsymbol{L}_2^*$, $\boldsymbol{L}_1^* = \operatorname{diag}[\boldsymbol{I}_{(4u-1)^2} - (4u-1)^{-2}\boldsymbol{1}_{(4u-1)^2}\boldsymbol{1}'_{(4u-1)^2} : \mathbf{O}]$, $\boldsymbol{L}_2^* = \operatorname{diag}[\mathbf{O} : (\boldsymbol{I}_q - (4u)^{-1}\boldsymbol{1}_q[a_1, a_2, ..., a_q]) \otimes s^{-1}\boldsymbol{1}_s\boldsymbol{1}'_s]$.

As other α-resolvable BIB designs, the following designs can be listed. Such designs can produce $(\rho_0; \rho_1, \rho_2; 0)$-EB designs by Corollaries 7.4.11 and 7.4.12.

(i) for $4u-1$ being a prime or a prime power, there exists a resolvable BIB design with parameters $v = 4u$, $b = 2(4u-1)$, $r = 4u-1$, $k = 2u$, $\lambda = 2u-1$ (Sprott, 1956),

(ii) for $2k-1$ being a prime or a prime power, there exists a resolvable BIB design with parameters $v = 2k$, $b = 4(2k-1)$, $r = 2(2k-1)$, k, $\lambda = 2(k-1)$ (Sprott, 1956),

(iii) for q being a prime or a prime power, there exists a resolvable BIB design with parameters $v = q^t$, $b = q^{t-d}\phi(t-1, d-1, q)$, $r = \phi(t-1, d-1, q)$, $k = q^d$, $\lambda = \phi(t-2, d-2, q)$, where $\phi(t, d, q) = (q^{t+1}-1)(q^t-1)\cdots(q^{t-d+1}-1)/(q^{d+1}-1)(q^d-1)\cdots(q-1)$ for $t > d \geq 1$, discussed in Theorem 6.9.2,

(iv) if s and $p = s^2+s+1$ are both primes or prime powers, there exists an $(s+1)$-resolvable BIB design with parameters $v = p^2$, $b = p(p+1)$, $r = (p+1)(s+1)$, $k = p(s+1)$, $\lambda = s+p+1$, $b_0 = p$, $a = p+1$ (Shrikhande and Raghavarao, 1963).

Theorems 6.3.5 and 7.4.14 lead to the following two observations.

Corollary 7.4.13. *If $\boldsymbol{N}$ is the $v \times b$ incidence matrix of an α-resolvable 2-associate PBIB design with parameters v, $b = b_0a$, $r = \alpha a$, k, λ_1, λ_2, $\varepsilon_0 = 1$, $\varepsilon_1 < 1$, ρ_0, ρ_1, $\boldsymbol{L}_0$, $\boldsymbol{L}_1$, then the incidence matrix*

$$\boldsymbol{N}^* = \begin{bmatrix} \boldsymbol{N} \\ \boldsymbol{I}_a \otimes \boldsymbol{1}_s\boldsymbol{1}'_{b_0} \end{bmatrix}$$

yields a $(\rho_0^*; \rho_1^*, \rho_2^*; 0)$-EB design with parameters $v^* = v + sa$, $b^* = b$, $\boldsymbol{r}^* = [r\boldsymbol{1}'_v, b_0\boldsymbol{1}'_{sa}]'$, $k^* = k+s$, $\varepsilon_0^* = 1$, $\varepsilon_1^* = 1-k(1-\varepsilon_1)/(k+s)$, $\varepsilon_2^* = k/(k+s)$, $\rho_0^* = \rho_0 + a(s-1)+1$, $\rho_1^* = \rho_1$, $\rho_2^* = a-1$, $\boldsymbol{L}_0^* = \boldsymbol{I}_{v+sa} - [b(k+s)]^{-1}\boldsymbol{1}_{v+sa}(\boldsymbol{r}^*)' - \boldsymbol{L}_1^* - \boldsymbol{L}_2^*$, $\boldsymbol{L}_1^* = \operatorname{diag}[\boldsymbol{L}_1 : \mathbf{O}]$, $\boldsymbol{L}_2^* = \operatorname{diag}[\mathbf{O} : (\boldsymbol{I}_a - a^{-1}\boldsymbol{1}_a\boldsymbol{1}'_a) \otimes s^{-1}\boldsymbol{1}_s\boldsymbol{1}'_s]$ for $s \geq 1$.

Corollary 7.4.14. *If $\boldsymbol{N}$ is the $v \times b$ incidence matrix of an α-resolvable 2-associate PBIB design with parameters v, $b = b_0a$, $r = \alpha a$, k, λ_1, λ_2, $\varepsilon_0 = 1$,*

$\varepsilon_1 < 1$, ρ_0, ρ_1, $\boldsymbol{L}_0$, $\boldsymbol{L}_1$, *then the form* (7.4.6) *yields a* $(\rho_0^*; \rho_1^*, \rho_2^*; 0)$*-EB design with parameters* $v^* = v + sq$, $b^* = b$, $\boldsymbol{r}^* = [r\boldsymbol{1}'_v, a_1 b_0 \boldsymbol{1}'_s, ..., a_q b_0 \boldsymbol{1}'_s]'$, $k^* = k + s$, $\varepsilon_0^* = 1$, $\varepsilon_1^* = 1 - k(1-\varepsilon_1)/(k+s)$, $\varepsilon_2^* = k/(k+s)$, $\rho_0^* = \rho_0 + q(s-1)+1$, $\rho_1^* = \rho_1$, $\rho_2^* = q-1$, $\boldsymbol{L}_0^* = \boldsymbol{I}_{v+sq} - [b(k+s)]^{-1}\boldsymbol{1}_{v+sq}(\boldsymbol{r}^*)' - \boldsymbol{L}_1^* - \boldsymbol{L}_2^*$, $\boldsymbol{L}_1^* = \text{diag}[\boldsymbol{L}_1 : \mathbf{O}]$, $\boldsymbol{L}_2^* = \text{diag}[\mathbf{O} : (\boldsymbol{I}_q - a^{-1}\boldsymbol{1}_q[a_1, a_2, ..., a_q]) \otimes s^{-1}\boldsymbol{1}_s\boldsymbol{1}'_s]$, *where* $a = \sum_{h=1}^q a_h$ *and* $s \geq 1$.

Example 7.4.12. A semi-regular GD design, based on three groups $\{1,2\}, \{3,4\}, \{5,6\}$ of two treatments, SR19 in Clatworthy (1973), with parameters $v = 6$, $b = 8, r = 4$, $k = 3$, $\lambda_1 = 0$, $\lambda_2 = 2$, $m = 3, n = 2$, also discussed in Example 6.3.2, is resolvable with blocks

[(1,3,5), (2,4,6)], [(1,3,6), (2,4,5)], [(1,4,5), (2,3,6)], [(1,4,6), (2,3,5)],

$\alpha = 1, b_0 = 2$, $a = 4$,

or 2-resolvable with blocks

[(1,3,5), (2,4,6), (1,3,6), (2,4,5)], [(1,4,5), (2,3,6), (1,4,6), (2,3,5)],

$\alpha = 2, b_0 = 4, a = 2$.

Then Corollary 7.4.14 yields two $(\rho_0^*; \rho_1^*, \rho_2^*; 0)$-EB designs with the parameters as follows. For $s \geq 1$

(1) $v^* = 4s + 6$, $b^* = 8$, $\boldsymbol{r}^* = [4\boldsymbol{1}'_6, 2\boldsymbol{1}'_{4s}]'$, $k^* = s+3$, $\varepsilon_0^* = 1$, $\varepsilon_1^* = (s+2)/(s+3)$, $\varepsilon_2^* = 3/(s+3)$, $\rho_0^* = 4s - 1$, $\rho_1^* = 3$, $\rho_2^* = 3$, $\boldsymbol{L}_0^* = \boldsymbol{I}_{4s+6} - [8(s+3)]^{-1}\boldsymbol{1}_{4s+6}(\boldsymbol{r}^*)' - \boldsymbol{L}_1^* - \boldsymbol{L}_2^*$, $\boldsymbol{L}_1^* = \text{diag}[\boldsymbol{I}_3 \otimes (\boldsymbol{I}_2 - 2^{-1}\boldsymbol{1}_2\boldsymbol{1}'_2) : \mathbf{O}]$, $\boldsymbol{L}_2^* = \text{diag}[\mathbf{O} : (\boldsymbol{I}_4 - 4^{-1}\boldsymbol{1}_4\boldsymbol{1}'_4) \otimes s^{-1}\boldsymbol{1}_s\boldsymbol{1}'_s]$;

(2) $v^* = 2s+6$, $b^* = 8$, $r^* = 4$, $k^* = s+3$, $\varepsilon_0^* = 1$, $\varepsilon_1^* = (s+2)/(s+3)$, $\varepsilon_2^* = 3/(s+3)$, $\rho_0^* = 2s+1$, $\rho_1^* = 3$, $\rho_2^* = 1$, $\boldsymbol{L}_0^* = \boldsymbol{I}_{2s+6} - [2(s+3)]^{-1}\boldsymbol{1}_{2s+6}\boldsymbol{1}'_{2s+6} - \boldsymbol{L}_1^* - \boldsymbol{L}_2^*$, $\boldsymbol{L}_1^* = \text{diag}[\boldsymbol{I}_3 \otimes (\boldsymbol{I}_2 - 2^{-1}\boldsymbol{1}_2\boldsymbol{1}'_2) : \mathbf{O}]$, $\boldsymbol{L}_2^* = \text{diag}[\mathbf{O} : (\boldsymbol{I}_2 - 2^{-1}\boldsymbol{1}_2\boldsymbol{1}'_2) \otimes s^{-1}\boldsymbol{1}_s\boldsymbol{1}'_s]$.

The incidence matrices and $\boldsymbol{\Omega}$ matrices of the designs (1) and (2) above are, for $s = 2$, as follows. For (1), the incidence matrix is

$$\boldsymbol{N}^* = \begin{bmatrix} 1 & 0 & 1 & 0 & 1 & 0 & 1 & 0 \\ 0 & 1 & 0 & 1 & 0 & 1 & 0 & 1 \\ 1 & 0 & 1 & 0 & 0 & 1 & 0 & 1 \\ 0 & 1 & 0 & 1 & 1 & 0 & 1 & 0 \\ 1 & 0 & 0 & 1 & 1 & 0 & 0 & 1 \\ 0 & 1 & 1 & 0 & 0 & 1 & 1 & 0 \\ 1 & 1 & 0 & 0 & 0 & 0 & 0 & 0 \\ 1 & 1 & 0 & 0 & 0 & 0 & 0 & 0 \\ 0 & 0 & 1 & 1 & 0 & 0 & 0 & 0 \\ 0 & 0 & 1 & 1 & 0 & 0 & 0 & 0 \\ 0 & 0 & 0 & 0 & 1 & 1 & 0 & 0 \\ 0 & 0 & 0 & 0 & 1 & 1 & 0 & 0 \\ 0 & 0 & 0 & 0 & 0 & 0 & 1 & 1 \\ 0 & 0 & 0 & 0 & 0 & 0 & 1 & 1 \end{bmatrix}, \quad \begin{array}{l} v^* = 14,\ b^* = 8, \\ \boldsymbol{r}^* = [4\boldsymbol{1}'_6, 2\boldsymbol{1}'_8]',\ k^* = 5, \\ \varepsilon_0^* = 1,\ \rho_0^* = 7, \\ \varepsilon_1^* = 4/5,\ \rho_1^* = 3, \\ \varepsilon_2^* = 3/5,\ \rho_2^* = 3, \\ \varepsilon^* = 52/63 = 0.8254, \end{array}$$

giving

$$\Omega^* = \text{diag}\left[\frac{1}{32}\boldsymbol{I}_3 \otimes (10\boldsymbol{I}_2 - \mathbf{1}_2\mathbf{1}_2') : \frac{1}{8}[\boldsymbol{I}_4 \otimes (4\boldsymbol{I}_2 + \mathbf{1}_2\mathbf{1}_2') - (\mathbf{1}_4\mathbf{1}_4' - \boldsymbol{I}_4) \otimes \frac{1}{3}\mathbf{1}_2\mathbf{1}_2']\right];$$

For (2), it is

$$\boldsymbol{N}^* = \begin{bmatrix} 1 & 0 & 1 & 0 & 1 & 0 & 1 & 0 \\ 0 & 1 & 0 & 1 & 0 & 1 & 0 & 1 \\ 1 & 0 & 1 & 0 & 0 & 1 & 0 & 1 \\ 0 & 1 & 0 & 1 & 1 & 0 & 1 & 0 \\ 1 & 0 & 0 & 1 & 1 & 0 & 0 & 1 \\ 0 & 1 & 1 & 0 & 0 & 1 & 1 & 0 \\ 1 & 1 & 1 & 1 & 0 & 0 & 0 & 0 \\ 1 & 1 & 1 & 1 & 0 & 0 & 0 & 0 \\ 0 & 0 & 0 & 0 & 1 & 1 & 1 & 1 \\ 0 & 0 & 0 & 0 & 1 & 1 & 1 & 1 \end{bmatrix}, \quad \begin{array}{l} v^* = 10,\ b^* = 8, \\ r^* = 4,\ k^* = 5, \\ \varepsilon_0^* = 1,\ \rho_0^* = 5, \\ \varepsilon_1^* = 4/5,\ \rho_1^* = 3, \\ \varepsilon_2^* = 3/5,\ \rho_2^* = 1, \\ \varepsilon^* = 108/125 = 0.864, \end{array}$$

giving

$$\Omega^* = \text{diag}\left[\frac{1}{32}\boldsymbol{I}_3 \otimes (10\boldsymbol{I}_2 - \mathbf{1}_2\mathbf{1}_2') : \frac{1}{24}[\boldsymbol{I}_2 \otimes (6\boldsymbol{I}_2 + \mathbf{1}_2\mathbf{1}_2') - (\mathbf{1}_2\mathbf{1}_2' - \boldsymbol{I}_2) \otimes \mathbf{1}_2\mathbf{1}_2']\right].$$

In both cases, there may be five different variances among elementary contrasts in the intra-block analysis.

As a block design having full replication for the supplementary treatments, the pattern (6.3.3) or (7.3.4) can be considered, i.e., $[\boldsymbol{N}' : \mathbf{1}_b\mathbf{1}_s']'$, with the basic incidence matrix $\boldsymbol{N}$. One of the results is already described in Corollary 7.3.8, which can also yield $(\rho_0; \rho_1, \rho_2; 0)$-EB designs with $\rho_0 \geq 1$, from 2-associate PBIB designs whose incidence matrices have a rank v (see Section 6.0.2). This can be stated as follows.

Corollary 7.4.15. *The existence of a 2-associate PBIB design with parameters v, b, r, k, λ_1, λ_2, $\varepsilon_\beta(< 1)$, ρ_β, $\boldsymbol{L}_\beta$, $\beta = 1, 2$, implies the existence of a $(\rho_0^*; \rho_1^*, \rho_2^*; 0)$-EB design (7.3.4) with parameters $v^* = v + s$, $b^* = b$, $\boldsymbol{r}^* = [r\mathbf{1}_v', b\mathbf{1}_s']'$, $k^* = k + s$, $\varepsilon_0^* = 1$, $\varepsilon_\beta^* = 1 - k(1 - \varepsilon_\beta)/(k + s)$, $\rho_0^* = s$, $\rho_\beta^* = \rho_\beta$, $\boldsymbol{L}_0^* = \boldsymbol{I}_{v+s} - [b(k+s)]^{-1}\mathbf{1}_{v+s}(\boldsymbol{r}^*)' - \boldsymbol{L}_1^* - \boldsymbol{L}_2^*$, $\boldsymbol{L}_\beta^* = \text{diag}[\boldsymbol{L}_\beta : \mathbf{O}]$ for $s \geq 1$.*

For example, as an application of Corollary 7.4.15 among GD designs one can take any regular GD design, which will be given in Corollary 8.3.1, as follows.

Corollary 7.4.16. *When q is a prime or a prime power, there exists a $(\rho_0;$ $\rho_1, \rho_2; 0)$-EB design with parameters $v = q^2 + s - 1$, $b = q^2 - 1$, $\boldsymbol{r} = [q\mathbf{1}_{q^2-1}',$ $(q^2 - 1)\mathbf{1}_s']'$, $k = q + s$, $\varepsilon_0 = 1$, $\varepsilon_1 = 1 - 1/[q(q + s)]$, $\varepsilon_2 = 1 - 1/(q + s)$, $\rho_0 = s$, $\rho_1 = q$, $\rho_2 = (q + 1)(q - 2)$, $\boldsymbol{L}_0 = \boldsymbol{I}_{q^2+s-1} - [(q^2 - 1)(q + s)]^{-1}\mathbf{1}_{q^2+s-1}\boldsymbol{r}'$*

$-\boldsymbol{L}_1-\boldsymbol{L}_2$, $\boldsymbol{L}_1=\mathrm{diag}[[\boldsymbol{I}_{q+1}-(q+1)^{-1}\mathbf{1}_{q+1}\mathbf{1}'_{q+1}]\otimes(q-1)^{-1}\mathbf{1}_{q-1}\mathbf{1}'_{q-1}:\mathbf{O}]$, $\boldsymbol{L}_2=\mathrm{diag}[\boldsymbol{I}_{q+1}\otimes[\boldsymbol{I}_{q-1}-(q-1)^{-1}\mathbf{1}_{q-1}\mathbf{1}'_{q-1}]:\mathbf{O}]$ *for* $s\geq 1$.

Example 7.4.13. Consider a regular GD design, based on four groups $\{1,2\}$, $\{3,4\}$, $\{5,6\}$, $\{7,8\}$ of two treatments, with parameters $v=b=8, r=k=3$, $\lambda_1=0$, $\lambda_2=1$, $m=4$, $n=2$, whose blocks are given by (3,6,8), (4,5,7), (1,5,8), (2,6,7), (1,3,7), (2,4,8), (2,3,5), (1,4,6) (see Corollary 8.3.1 with $q=3$ later; also R54 in Clatworthy, 1973). Then Corollary 7.4.16 with $q=3$ produces a $(\rho_0^*;\rho_1^*,\rho_2^*;0)$-EB design with parameters $v^*=s+8$, $b^*=8$, $\boldsymbol{r}^*=[3\mathbf{1}'_8,8\mathbf{1}'_s]'$, $k^*=s+3$, $\varepsilon_0^*=1$, $\varepsilon_1^*=(3s+8)/(3s+9)$, $\varepsilon_2^*=(s+2)/(s+3)$, $\rho_0^*=s$, $\rho_1^*=3$, $\rho_2^*=4$, $\boldsymbol{L}_0^*=\boldsymbol{I}_{s+8}-[8(s+3)]^{-1}\mathbf{1}_{s+8}(\boldsymbol{r}^*)'-\boldsymbol{L}_1^*-\boldsymbol{L}_2^*$, $\boldsymbol{L}_1^*=\mathrm{diag}[(\boldsymbol{I}_4-4^{-1}\mathbf{1}_4\mathbf{1}'_4)\otimes 2^{-1}\mathbf{1}_2\mathbf{1}'_2:\mathbf{O}]$, $\boldsymbol{L}_2^*=\mathrm{diag}[\boldsymbol{I}_4\otimes(\boldsymbol{I}_2-2^{-1}\mathbf{1}_2\mathbf{1}'_2):\mathbf{O}]$, and with the incidence matrix, for $s=3$, as

$$\boldsymbol{N}^*=\begin{bmatrix}0&0&1&0&1&0&0&1\\0&0&0&1&0&1&1&0\\1&0&0&0&1&0&1&0\\0&1&0&0&0&1&0&1\\0&1&1&0&0&0&1&0\\1&0&0&1&0&0&0&1\\0&1&0&1&1&0&0&0\\1&0&1&0&0&1&0&0\\1&1&1&1&1&1&1&1\\1&1&1&1&1&1&1&1\\1&1&1&1&1&1&1&1\end{bmatrix},\quad \begin{array}{l}v^*=11,\ b^*=8,\\ \boldsymbol{r}^*=[3\mathbf{1}'_8,8\mathbf{1}'_3]',\ k^*=6,\\ \varepsilon_0^*=1,\ \rho_0^*=3,\\ \varepsilon_1^*=17/18,\ \rho_1^*=3,\\ \varepsilon_2^*=5/6,\ \rho_2^*=4,\\ \varepsilon^*=850/933=0.9110.\end{array}$$

Furthermore, the matrix $\boldsymbol{\Omega}$ is obtained as

$$\boldsymbol{\Omega}^*=\mathrm{diag}\left[\boldsymbol{I}_4\otimes\left(\frac{816}{2040}\boldsymbol{I}_2-\frac{53}{2040}\mathbf{1}_2\mathbf{1}'_2\right)-\frac{1}{408}(\mathbf{1}_4\mathbf{1}'_4-\boldsymbol{I}_4)\otimes\mathbf{1}_2\mathbf{1}'_2:\frac{1}{8}\boldsymbol{I}_3\right],$$

which implies that there may be four different variances among elementary contrasts in the intra-block analysis.

Some series of regular GD designs, as will be seen in Theorems 8.3.1 and 8.3.3, can produce more series of $(\rho_0;\rho_1,\rho_2;0)$-EB designs by Corollary 7.4.15.

Now 3-associate PBIB designs discussed in Section 6.0.2 will be used to produce $(\rho_0;\rho_1,\rho_2;0)$-EB designs which are of the form (7.3.4).

Theorem 7.4.15. *The existence of a* 3*-associate PBIB design with parameters* v, b, r, k, λ_1, λ_2, λ_3, $\varepsilon_0=1$, $\varepsilon_\beta<1$, ρ_0, ρ_β, $\boldsymbol{L}_0$, $\boldsymbol{L}_\beta$, $\beta=1,2$, *implies the existence of a* $(\rho_0^*;\rho_1^*,\rho_2^*;0)$*-EB design* (7.3.4) *with parameters* $v^*=v+s$, $b^*=b$, $\boldsymbol{r}^*=[r\mathbf{1}'_v,b\mathbf{1}'_s]'$, $k^*=k+s$, $\varepsilon_0^*=1$, $\varepsilon_\beta^*=1-k(1-\varepsilon_\beta)/(k+s)$, $\rho_0^*=\rho_0+s$, $\rho_\beta^*=\rho_\beta$, $\boldsymbol{L}_0^*=\boldsymbol{I}_{v+s}-[b(k+s)]^{-1}\mathbf{1}_{v+s}(\boldsymbol{r}^*)'-\boldsymbol{L}_1^*-\boldsymbol{L}_2^*$, $\boldsymbol{L}_\beta^*=\mathrm{diag}[\boldsymbol{L}_\beta:\mathbf{O}]$ *for* $s\geq 1$.

By using the rectangular PBIB design discussed in Theorem 7.4.2 with $\varepsilon_0=$

1, the following corollary can be obtained from Theorem 7.4.15.

Corollary 7.4.17. *The existence of a BIB design with parameters v, b, r, k, λ implies the existence of a $(\rho_0^*; \rho_1^*, \rho_2^*; 0)$-EB design with parameters $v^* = 3v + s$, $b^* = 3b$, $\boldsymbol{r}^* = [b\mathbf{1}'_{3v}, 3b\mathbf{1}'_s]'$, $k^* = v + s$, $\varepsilon_0^* = 1$, $\varepsilon_1^* = 1 - \{b + (v-1)[b - 3(r - \lambda)]\}/[b(v+s)]$, $\varepsilon_2^* = 1 - 3(r-\lambda)/[b(v+s)]$, $\rho_0^* = v + s - 1$, $\rho_1^* = 2$, $\rho_2^* = 2(v-1)$, $\boldsymbol{L}_0^* = \boldsymbol{I}_{3v+s} - [3b(v+s)]^{-1}\mathbf{1}_{3v+s}(\boldsymbol{r}^*)' - \boldsymbol{L}_1^* - \boldsymbol{L}_2^*$, $\boldsymbol{L}_1^* = \mathrm{diag}[(\mathbf{I}_3 - 3^{-1}\mathbf{1}_3\mathbf{1}'_3) \otimes v^{-1}\mathbf{1}_v\mathbf{1}'_v : \mathbf{O}]$, $\boldsymbol{L}_2^* = \mathrm{diag}[(\boldsymbol{I}_3 - 3^{-1}\mathbf{1}_3\mathbf{1}'_3) \otimes (\boldsymbol{I}_v - v^{-1}\mathbf{1}_v\mathbf{1}'_v) : \mathbf{O}]$.*

Note that, in Corollary 7.4.17, $\varepsilon_1^* = \varepsilon_2^*$ if and only if $b = 3(r - \lambda)$. A characterization of a BIB design satisfying $b = 3(r-\lambda)$ can be seen in Kageyama and Kuwada (1985).

For other incidence patterns discussed in the present chapter, a 3-associate PBIB design, whose incidence matrix has a rank smaller than v (see Section 6.0.2), can also be used to produce $(\rho_0;\ \rho_1, \rho_2;\ 0)$-EB designs, where $\boldsymbol{N}$ is the incidence matrix of the PBIB design. But, these are left to the reader as possible exercises. For example, see Theorem 7.4.2.

7.4.3 Nonproper designs that are equireplicate

As mentioned in Section 7.3.3, in some practical situations nonproper block designs may be required. Binary block designs with block sizes, $k_1, k_2, ..., k_b$ (≥ 2), that are not all equal, will now be considered. Not so many methods of constructing such designs are available. Note that all the designs obtainable in this section satisfy Corollary 3.2.2, on account of Corollary 3.2.3, to secure the inter-block BLUE of a contrast partially confounded with blocks.

A juxtaposition method of combining some designs with a common number of treatments is standard. Theorem 6.4.3 can yield the following result.

Corollary 7.4.18. *The existence of a singular GD design with parameters $v = mn$, b, r, k, λ_1, λ_2 implies the existence of a $(\rho_0^*; \rho_1^*, \rho_2^*; 0)$-EB design with parameters $v^* = mn$, $b^* = b + ns$, $r^* = r + s$, $\boldsymbol{k}^* = [k\mathbf{1}'_b, m\mathbf{1}'_{ns}]'$, $\varepsilon_0^* = 1$, $\varepsilon_1^* = r/(r+s)$, $\varepsilon_2^* = 1 - (rk - v\lambda_2)/[k(r+s)]$, $\rho_0^* = (m-1)(n-1)$, $\rho_1^* = n-1$, $\rho_2^* = m-1$, $\boldsymbol{L}_0^* = (\boldsymbol{I}_m - m^{-1}\mathbf{1}_m\mathbf{1}'_m) \otimes (\boldsymbol{I}_n - n^{-1}\mathbf{1}_n\mathbf{1}'_n)$, $\boldsymbol{L}_1^* = m^{-1}\mathbf{1}_m\mathbf{1}'_m \otimes (\boldsymbol{I}_n - n^{-1}\mathbf{1}_n\mathbf{1}'_n)$, $\boldsymbol{L}_2^* = (\boldsymbol{I}_m - m^{-1}\mathbf{1}_m\mathbf{1}'_m) \otimes n^{-1}\mathbf{1}_n\mathbf{1}'_n$ for a positive integer s.*

Example 7.4.14. Consider a singular GD design, based on four groups $\{1, 2, 3\}$, $\{4, 5, 6\}$, $\{7, 8, 9\}$, $\{10, 11, 12\}$ of three treatments, known as S27 (Clatworthy, 1973), with parameters $v = 12$, $b = 6$, $r = 3$, $k = 6$, $\lambda_1 = 3$, $\lambda_2 = 1$, $m = 4$, $n = 3$, whose blocks are given by

[(1,2,3,4,5,6), (7,8,9,10,11,12)], [(1,2,3,7,8,9), (4,5,6,10,11,12)],

[(1,2,3,10,11,12), (4,5,6,7,8,9)]

in an affine resolvable form. Corollary 7.4.18 produces a $(\rho_0^*; \rho_1^*, \rho_2^*; 0)$-EB design with parameters $v^* = 12$, $b^* = 3s + 6$, $r^* = s + 3$, $\boldsymbol{k}^* = [6\mathbf{1}'_6, 4\mathbf{1}'_{3s}]'$, $\varepsilon_0^* = 1$, $\varepsilon_1^* = (s+2)/(s+3)$, $\varepsilon_2^* = 3/(s+3)$, $\rho_0^* = 6$, $\rho_1^* = 3$, $\rho_2^* = 2$, $\boldsymbol{L}_0^* = (\boldsymbol{I}_4 - 4^{-1}\mathbf{1}_4\mathbf{1}'_4) \otimes$

$(\boldsymbol{I}_3 - 3^{-1}\mathbf{1}_3\mathbf{1}_3')$, $\boldsymbol{L}_1^* = (\boldsymbol{I}_4 - 4^{-1}\mathbf{1}_4\mathbf{1}_4') \otimes 3^{-1}\mathbf{1}_3\mathbf{1}_3'$, $\boldsymbol{L}_2^* = 4^{-1}\mathbf{1}_4\mathbf{1}_4' \otimes (\boldsymbol{I}_3 - 3^{-1}\mathbf{1}_3\mathbf{1}_3')$, and with the incidence matrix, for $s = 2$, as

$$\left[\begin{array}{cc|cc|cc|ccc|ccc}
1 & 0 & 1 & 0 & 1 & 0 & 1 & 0 & 0 & 1 & 0 & 0 \\
1 & 0 & 1 & 0 & 1 & 0 & 0 & 1 & 0 & 0 & 1 & 0 \\
1 & 0 & 1 & 0 & 1 & 0 & 0 & 0 & 1 & 0 & 0 & 1 \\
1 & 0 & 0 & 1 & 0 & 1 & 1 & 0 & 0 & 1 & 0 & 0 \\
1 & 0 & 0 & 1 & 0 & 1 & 0 & 1 & 0 & 0 & 1 & 0 \\
1 & 0 & 0 & 1 & 0 & 1 & 0 & 0 & 1 & 0 & 0 & 1 \\
0 & 1 & 1 & 0 & 0 & 1 & 1 & 0 & 0 & 1 & 0 & 0 \\
0 & 1 & 1 & 0 & 0 & 1 & 0 & 1 & 0 & 0 & 1 & 0 \\
0 & 1 & 1 & 0 & 0 & 1 & 0 & 0 & 1 & 0 & 0 & 1 \\
0 & 1 & 0 & 1 & 1 & 0 & 1 & 0 & 0 & 1 & 0 & 0 \\
0 & 1 & 0 & 1 & 1 & 0 & 0 & 1 & 0 & 0 & 1 & 0 \\
0 & 1 & 0 & 1 & 1 & 0 & 0 & 0 & 1 & 0 & 0 & 1
\end{array}\right], \qquad
\begin{array}{l}
v^* = b^* = 12,\ r^* = 5, \\
\boldsymbol{k}^* = [6\mathbf{1}_6', 4\mathbf{1}_6']', \\
\varepsilon_0^* = 1,\ \rho_0^* = 6, \\
\varepsilon_1^* = 4/5,\ \rho_1^* = 3, \\
\varepsilon_2^* = 3/5,\ \rho_2^* = 2.
\end{array}$$

Note that this solution is resolvable. Here, the average efficiency factor of the design is $132/157 = 0.8408$. The variances of the intra-block BLUEs of the basic contrasts are σ_1^2, $5\sigma_1^2/4$ or $5\sigma_1^2/3$, depending on whether they correspond to $\boldsymbol{L}_0^*$, $\boldsymbol{L}_1^*$ or $\boldsymbol{L}_2^*$. Furthermore, the matrix $\boldsymbol{\Omega}$ is obtained as

$$\boldsymbol{\Omega}^* = \frac{1}{720}\left[\boldsymbol{I}_4 \otimes (168\boldsymbol{I}_3 + \mathbf{1}_3\mathbf{1}_3') + (\mathbf{1}_4\mathbf{1}_4' - \boldsymbol{I}_4) \otimes (24\boldsymbol{I}_3 - 11\mathbf{1}_3\mathbf{1}_3')\right].$$

Hence, for example, elementary contrasts are estimated in the intra-block analysis with three different variances, $7\sigma_1^2/15$, $\sigma_1^2/2$ and $13\sigma_1^2/30$.

Corollary 7.4.19. *The existence of a semi-regular GD design with parameters* $v = mn$, b, r, k, λ_1, λ_2 *implies the existence of a* $(\rho_0^*; \rho_1^*, \rho_2^*; 0)$*-EB design with parameters* $v^* = mn$, $b^* = b + ns$, $r^* = r + s$, $\boldsymbol{k}^* = [k\mathbf{1}_b', m\mathbf{1}_{ns}']'$, $\varepsilon_0^* = 1$, $\varepsilon_1^* = 1 - (r - \lambda_1)/[k(r+s)]$, $\varepsilon_2^* = 1 - (r - \lambda_1 + sk)/[k(r+s)]$, $\rho_0^* = m - 1$, $\rho_1^* = (m-1)(n-1)$, $\rho_2^* = n - 1$, $\boldsymbol{L}_0^* = (\boldsymbol{I}_m - m^{-1}\mathbf{1}_m\mathbf{1}_m') \otimes n^{-1}\mathbf{1}_n\mathbf{1}_n'$, $\boldsymbol{L}_1^* = (\boldsymbol{I}_m - m^{-1}\mathbf{1}_m\mathbf{1}_m') \otimes (\boldsymbol{I}_n - n^{-1}\mathbf{1}_n\mathbf{1}_n')$, $\boldsymbol{L}_2^* = m^{-1}\mathbf{1}_m\mathbf{1}_m' \otimes (\boldsymbol{I}_n - n^{-1}\mathbf{1}_n\mathbf{1}_n')$ *for a positive integer* s.

By an approach similar to Theorem 6.4.3, the following can be obtained.

Corollary 7.4.20. *If* $\boldsymbol{N}$ *is the* $v \times b$ *incidence matrix of a singular GD design with parameters* $v = mn$, b, r, k, λ_1, λ_2, *then for positive integers* s_1 *and* s_2

$$[\boldsymbol{N} : \boldsymbol{I}_m \otimes \mathbf{1}_n\mathbf{1}_{s_1}' : \mathbf{1}_m\mathbf{1}_{s_2}' \otimes \boldsymbol{I}_n]$$

is the incidence matrix of a $(\rho_0^*; \rho_1^*, \rho_2^*; 0)$*-EB design with parameters* $v^* = mn$, $b^* = b + ms_1 + ns_2$, $r^* = r + s_1 + s_2$, $\boldsymbol{k}^* = [k\mathbf{1}_b', n\mathbf{1}_{ms_1}', m\mathbf{1}_{ns_2}']'$, $\varepsilon_0^* = 1$, $\varepsilon_1^* = 1 - s_2/(r + s_1 + s_2)$, $\varepsilon_2^* = 1 - (rk - v\lambda_2 + s_1 k)/[k(r + s_1 + s_2)]$, $\rho_0^* = (m-1)(n-1)$, $\rho_1^* = n - 1$, $\rho_2^* = m - 1$, $\boldsymbol{L}_0^* = (\boldsymbol{I}_m - m^{-1}\mathbf{1}_m\mathbf{1}_m') \otimes (\boldsymbol{I}_n - n^{-1}\mathbf{1}_n\mathbf{1}_n')$, $\boldsymbol{L}_1^* =$

$m^{-1}\mathbf{1}_m\mathbf{1}'_m \otimes (\boldsymbol{I}_n - n^{-1}\mathbf{1}_n\mathbf{1}'_n)$, $\boldsymbol{L}_2^* = (\boldsymbol{I}_m - m^{-1}\mathbf{1}_m\mathbf{1}'_m) \otimes n^{-1}\mathbf{1}_n\mathbf{1}'_n$.

Example 7.4.15. Consider a singular GD design, based on five groups {1, 2}, {3, 4}, {5, 6}, {7, 8}, {9, 10} of two treatments, known as S9 (Clatworthy, 1973), with parameters $v = b = 10$, $r = k = 4$, $\lambda_1 = 4$, $\lambda_2 = 1$, $m = 5$, $n = 2$, whose blocks are given by

[(1,2,3,4), (3,4,5,6), (5,6,7,8), (7,8,9,10), (1,2,9,10)],
[(1,2,5,6), (3,4,7,8), (5,6,9,10), (1,2,7,8), (3,4,9,10)]

in a 2-resolvable form. Then Corollary 7.4.20 produces a 2-resolvable $(\rho_0^*; \rho_1^*, \rho_2^*; 0)$-EB design with parameters $v^* = 10$, $b^* = 10 + 5s_1 + 2s_2$, $r^* = 4 + s_1 + s_2$, $\boldsymbol{k}^* = [4\mathbf{1}'_{10}, 2\mathbf{1}'_{5s_1}, 5\mathbf{1}'_{2s_2}]'$, $\varepsilon_0^* = 1$, $\varepsilon_1^* = (s_1+4)/(s_1+s_2+4)$, $\varepsilon_2^* = (2s_2+5)/[2(s_1+s_2+4)]$, $\rho_0^* = 4$, $\rho_1^* = 1$, $\rho_2^* = 4$, $\boldsymbol{L}_0^* = (\boldsymbol{I}_5 - 5^{-1}\mathbf{1}_5\mathbf{1}'_5) \otimes (\boldsymbol{I}_2 - 2^{-1}\mathbf{1}_2\mathbf{1}'_2)$, $\boldsymbol{L}_1^* = 5^{-1}\mathbf{1}_5\mathbf{1}'_5 \otimes (\boldsymbol{I}_2 - 2^{-1}\mathbf{1}_2\mathbf{1}'_2)$, $\boldsymbol{L}_2^* = (\boldsymbol{I}_5 - 5^{-1}\mathbf{1}_5\mathbf{1}'_5) \otimes 2^{-1}\mathbf{1}_2\mathbf{1}'_2$, and with the incidence matrix, for $s_1 = s_2 = 1$, as

$$\left[\begin{array}{ccccc|ccccc|ccccccc}
1&0&0&0&1&1&0&0&1&0&1&0&0&0&0&1&0\\
1&0&0&0&1&1&0&0&1&0&1&0&0&0&0&0&1\\
1&1&0&0&0&0&1&0&0&1&0&1&0&0&0&1&0\\
1&1&0&0&0&0&1&0&0&1&0&1&0&0&0&0&1\\
0&1&1&0&0&1&0&1&0&0&0&0&1&0&0&1&0\\
0&1&1&0&0&1&0&1&0&0&0&0&1&0&0&0&1\\
0&0&1&1&0&0&1&0&1&0&0&0&0&1&0&1&0\\
0&0&1&1&0&0&1&0&1&0&0&0&0&1&0&0&1\\
0&0&0&1&1&0&0&1&0&1&0&0&0&0&1&1&0\\
0&0&0&1&1&0&0&1&0&1&0&0&0&0&1&0&1
\end{array}\right], \quad
\begin{array}{l}
v^* = 10,\ b^* = 17, r^* = 6,\\
\boldsymbol{k}^* = [4\mathbf{1}'_{10}, 2\mathbf{1}'_5, 5\mathbf{1}'_2]',\\
\varepsilon_0^* = 1,\ \rho_0^* = 4,\\
\varepsilon_1^* = 5/6,\ \rho_1^* = 1,\\
\varepsilon_2^* = 7/12,\ \rho_2^* = 4.
\end{array}$$

Here, the average efficiency factor of the design is $315/422 = 0.7464$. The variances of the intra-block BLUEs of the basic contrasts are σ_1^2, $6\sigma_1^2/5$ or $12\sigma_1^2/7$, depending on whether they correspond to $\boldsymbol{L}_0^*$, $\boldsymbol{L}_1^*$ or $\boldsymbol{L}_2^*$. Furthermore, the matrix $\boldsymbol{\Omega}$ is obtained as

$$\boldsymbol{\Omega}^* = \frac{1}{1050}\left[\boldsymbol{I}_5 \otimes (182\boldsymbol{I}_2 + 91\mathbf{1}_2\mathbf{1}'_2) + (\mathbf{1}_5\mathbf{1}'_5 - \boldsymbol{I}_5) \otimes (7\boldsymbol{I}_2 - 161\mathbf{1}_2\mathbf{1}'_2)\right].$$

Hence, elementary contrasts are estimated in the intra-block analysis with three different variances, $26\sigma_1^2/75$, $8\sigma_1^2/21$ and $69\sigma_1^2/175$.

Using 3-associate PBIB designs discussed in Section 6.0.2, an argument similar to that in Theorem 6.4.2 can lead to the following results.

Theorem 7.4.16. *Let* $\boldsymbol{N}_1$ *and* $\boldsymbol{N}_2$ *be the* $v \times b^{(h)}$ *incidence matrices of two 3-associate PBIB designs with parameters* v, $b^{(h)}$, $r^{(h)}$, $k^{(h)}$, $\lambda_1^{(h)}$, $\lambda_2^{(h)}$, $\lambda_3^{(h)}$, $\varepsilon_\beta^{(h)}$, $h = 1, 2$, ρ_β, $\boldsymbol{L}_\beta$, $\beta = 1, 2, 3$. *Then the incidence matrix* $[\boldsymbol{N}_1 : \boldsymbol{N}_2]$ *yields a* $(\rho_0^*; \rho_1^*, \rho_2^*, \rho_3^*; 0)$*-EB design with parameters* $v^* = v$, $b^* = b^{(1)} + b^{(2)}$, $r^* = r^{(1)} + r^{(2)}$, $\boldsymbol{k}^* = [k^{(1)}\mathbf{1}'_{b^{(1)}}, k^{(2)}\mathbf{1}'_{b^{(2)}}]'$, $\varepsilon_\beta^* = (r^{(1)}\varepsilon_\beta^{(1)} + r^{(2)}\varepsilon_\beta^{(2)})/(r^{(1)} + r^{(2)})$,

$\rho_0^* \geq 0$, $\rho_\beta^* = \rho_\beta$, $\boldsymbol{L}_\beta^* = \boldsymbol{L}_\beta$. *In particular, if* $\varepsilon_\beta^{(h)} < 1$ *for all* $\beta = 1, 2, 3$ *and some* $h(= 1, 2)$, *then* $\rho_0^* = 0$.

Corollary 7.4.21. *The existence of two rectangular PBIB designs with parameters* $v = st$, $b^{(h)}$, $r^{(h)}$, $k^{(h)}$, $\lambda_1^{(h)}$, $\lambda_2^{(h)}$, $\lambda_3^{(h)}$, $\varepsilon_0^{(h)} = 1$, $\varepsilon_1^{(h)}$, $\varepsilon_2^{(h)}$, $h = 1, 2$, *implies the existence of a* $(\rho_0^*; \rho_1^*, \rho_2^*; 0)$*-EB design with parameters* $v^* = st$, $b^* = b^{(1)} + b^{(2)}$, $r^* = r^{(1)} + r^{(2)}$, $\boldsymbol{k}^* = [k^{(1)}\boldsymbol{1}'_{b^{(1)}}, k^{(2)}\boldsymbol{1}'_{b^{(2)}}]'$, $\varepsilon_0^* = 1$, $\varepsilon_1^* = 1 - \{[r^{(1)} - \lambda_2^{(1)} + (t-1)(\lambda_1^{(1)} - \lambda_3^{(1)})]/k^{(1)} + [r^{(2)} - \lambda_2^{(2)} + (t-1)(\lambda_1^{(2)} - \lambda_3^{(2)})]/k^{(2)}\}/(r^{(1)} + r^{(2)})$, $\varepsilon_2^* = 1 - [(r^{(1)} - \lambda_1^{(1)} - \lambda_2^{(1)} + \lambda_3^{(1)})/k^{(1)} + (r^{(2)} - \lambda_1^{(2)} - \lambda_2^{(2)} + \lambda_3^{(2)})/k^{(2)}]/(r^{(1)} + r^{(2)})$, $\rho_0^* = t - 1$, $\rho_1^* = s - 1$, $\rho_2^* = (s-1)(t-1)$, $\boldsymbol{L}_0^* = s^{-1}\boldsymbol{1}_s\boldsymbol{1}'_s \otimes (\boldsymbol{I}_t - t^{-1}\boldsymbol{1}_t\boldsymbol{1}'_t)$, $\boldsymbol{L}_1^* = (\boldsymbol{I}_s - s^{-1}\boldsymbol{1}_s\boldsymbol{1}'_s) \otimes t^{-1}\boldsymbol{1}_t\boldsymbol{1}'_t$, $\boldsymbol{L}_2^* = (\boldsymbol{I}_s - s^{-1}\boldsymbol{1}_s\boldsymbol{1}'_s) \otimes (\boldsymbol{I}_t - t^{-1}\boldsymbol{1}_t\boldsymbol{1}'_t)$.

For the existence of a rectangular PBIB design with $\varepsilon_0 = 1$ to use Corollary 7.4.21, see Theorem 7.4.2.

Corollary 7.4.22. *The existence of two* 3*-associate GD designs with parameters* $v = s_1s_2s_3$, $b^{(h)}$, $r^{(h)}$, $k^{(h)}$, $\lambda_1^{(h)}$, $\lambda_2^{(h)}$, $\lambda_3^{(h)}$, $h = 1, 2$, *implies the existence of a* $(\rho_0^*; \rho_1^*, \rho_2^*, \rho_3^*; 0)$*-EB design with parameters* $v^* = s_1s_2s_3$, $b^* = b^{(1)} + b^{(2)}$, $r^* = r^{(1)} + r^{(2)}$, $\boldsymbol{k}^* = [k^{(1)}\boldsymbol{1}'_{b^{(1)}}, k^{(2)}\boldsymbol{1}'_{b^{(2)}}]'$, $\varepsilon_1^* = 1 - \{[r^{(1)} - \lambda_3^{(1)} + (s_3 - 1)(\lambda_1^{(1)} - \lambda_3^{(1)}) + s_3(s_2 - 1)(\lambda_2^{(1)} - \lambda_3^{(1)})]/k^{(1)} + [r^{(2)} - \lambda_3^{(2)} + (s_3 - 1)(\lambda_1^{(2)} - \lambda_3^{(2)}) + s_3(s_2 - 1)(\lambda_2^{(2)} - \lambda_3^{(2)})]/k^{(2)}\}/(r^{(1)} + r^{(2)})$, $\varepsilon_2^* = 1 - \{[r^{(1)} - \lambda_2^{(1)} + (s_3 - 1)(\lambda_1^{(1)} - \lambda_2^{(1)})]/k^{(1)} + [r^{(2)} - \lambda_2^{(2)} + (s_3 - 1)(\lambda_1^{(2)} - \lambda_2^{(2)})]/k^{(2)}\}/(r^{(1)} + r^{(2)})$, $\varepsilon_3^* = 1 - [(r^{(1)} - \lambda_1^{(1)})/k^{(1)} + (r^{(2)} - \lambda_1^{(2)})/k^{(2)}]/(r^{(1)} + r^{(2)})$, $\rho_0^* \geq 0$, $\rho_1^* = s_1 - 1$, $\rho_2^* = s_1(s_2 - 1)$, $\rho_3^* = s_1s_2(s_3 - 1)$, $\boldsymbol{L}_1^* = [\boldsymbol{I}_{s_1} - (s_1)^{-1}\boldsymbol{1}_{s_1}\boldsymbol{1}'_{s_1}] \otimes (s_2s_3)^{-1}\boldsymbol{1}_{s_2}\boldsymbol{1}'_{s_3}$, $\boldsymbol{L}_2^* = \boldsymbol{I}_{s_1} \otimes [\boldsymbol{I}_{s_2} - (s_2)^{-1}\boldsymbol{1}_{s_2}\boldsymbol{1}'_{s_2}] \otimes (s_3)^{-1}\boldsymbol{1}_{s_3}\boldsymbol{1}'_{s_3}$, $\boldsymbol{L}_3^* = \boldsymbol{I}_{s_1s_2} \otimes [\boldsymbol{I}_{s_3} - (s_3)^{-1}\boldsymbol{1}_{s_3}\boldsymbol{1}'_{s_3}]$.

Juxtaposing incidence matrices of designs is one of the most standard techniques. As an example, a juxtaposition of a design and its complement is considered in the following theorem (see Theorem 7.4.3).

Theorem 7.4.17. *If* $\boldsymbol{N}$ *is the* $v \times b$ *incidence matrix of a binary* $(\rho_0; \rho_1, \rho_2; 0)$*-EB design with parameters* v, b, r, k, ε_β, ρ_β, $\boldsymbol{L}_\beta$, $\beta = 0, 1, 2$, *then the incidence matrix* $[\boldsymbol{N} : \boldsymbol{1}_v\boldsymbol{1}'_b - \boldsymbol{N}]$ *yields a* $(\rho_0^*; \rho_1^*, \rho_2^*; 0)$*-EB design with parameters* $v^* = v$, $b^* = 2b$, $r^* = b$, $\boldsymbol{k}^* = [k\boldsymbol{1}'_b, (v-k)\boldsymbol{1}'_b]'$, $\varepsilon_\beta^* = 1 - k(1 - \varepsilon_\beta)/(v - k)$, $\rho_\beta^* = \rho_\beta$, $\boldsymbol{L}_\beta^* = \boldsymbol{L}_\beta$, $\beta = 0, 1, 2$.

Note that the property of $\varepsilon_\beta^* = 1$ corresponds to $\varepsilon_\beta = 1$.

Other useful methods of construction are not readily available. The reader may be advised to try to find new methods.

7.4.4 Nonproper designs that are nonequireplicate

Nonproper and nonequireplicate $(\rho_0; \rho_1, \rho_2; 0)$-EB designs are less recommendable than those considered in Section 7.4.2 and not often used in practice. But, there may be some situations where one has to use such designs because of seveal restrictions in experimentation.

A method of dualization is appropriate for the purpose as in Theorem 6.1.1, which immediately yields the following results.

Corollary 7.4.23. *The dual of a* $(0; \rho_1, \rho_2; 0)$*-EB design* $\boldsymbol{N}$ *with parameters* v, b, $\boldsymbol{r}$, $\boldsymbol{k}$, $\varepsilon_\beta(<1)$, ρ_β, $\boldsymbol{L}_\beta$, $\beta = 1, 2$, *is a* $(\rho_0^*; \rho_1^*, \rho_2^*; 0)$*-EB design with parameters* $v^* = b$, $b^* = v$, $\boldsymbol{r}^* = \boldsymbol{k}$, $\boldsymbol{k}^* = \boldsymbol{r}$, $\varepsilon_0^* = 1$, $\varepsilon_\beta^* = \varepsilon_\beta$, $\rho_0^* = b - \rho_1 - \rho_2 - 1$, $\rho_\beta^* = \rho_\beta$, $\boldsymbol{L}_0^* = \boldsymbol{I}_b - n^{-1}\boldsymbol{1}_b\boldsymbol{k}' - \boldsymbol{L}_1^* - \boldsymbol{L}_2^*$, $\boldsymbol{L}_\beta^* = (1 - \varepsilon_\beta)^{-1}\boldsymbol{k}^{-\delta}\boldsymbol{N}'\boldsymbol{L}_\beta\boldsymbol{r}^{-\delta}\boldsymbol{N}$, *in which* $n = \boldsymbol{k}'\boldsymbol{1}_b$.

Corollary 7.4.24. *The dual of a* $(\rho_0; \rho_1, \rho_2; 0)$*-EB design* $\boldsymbol{N}$ *with parameters* v, b, $\boldsymbol{r}$, $\boldsymbol{k}$, $\varepsilon_0 = 1$, $\varepsilon_\beta < 1$, ρ_0, ρ_β, $\boldsymbol{L}_0$, $\boldsymbol{L}_\beta$, $\beta = 1, 2$, *is a* $(\rho_0^*; \rho_1^*, \rho_2^*; 0)$*-EB design with parameters* $v^* = b$, $b^* = v$, $\boldsymbol{r}^* = \boldsymbol{k}$, $\boldsymbol{k}^* = \boldsymbol{r}$, $\varepsilon_0^* = 1$, $\varepsilon_\beta^* = \varepsilon_\beta$, $\rho_0^* = b - v + \rho_0$, $\rho_\beta^* = \rho_\beta$, $\boldsymbol{L}_0^* = \boldsymbol{I}_b - n^{-1}\boldsymbol{1}_b\boldsymbol{k}' - \boldsymbol{L}_1^* - \boldsymbol{L}_2^*$, $\boldsymbol{L}_\beta^* = (1 - \varepsilon_\beta)^{-1}\boldsymbol{k}^{-\delta}\boldsymbol{N}'\boldsymbol{L}_\beta\boldsymbol{r}^{-\delta}\boldsymbol{N}$, *in which* $n = \boldsymbol{k}'\boldsymbol{1}_b$.

A simple juxtaposition method is also useful. By Theorem 6.4.1, the following corollary is obtained.

Corollary 7.4.25. *If* $\boldsymbol{N}$ *is the* $v \times b$ *incidence matrix of a* $(\rho_0; \rho_1, \rho_2; 0)$*-EB design with parameters* v, b, $\boldsymbol{r}$, $\boldsymbol{k}$, ε_β, ρ_β, $\boldsymbol{L}_\beta$, $\beta = 0, 1, 2$, *then its juxtaposition* $\boldsymbol{1}_s' \otimes \boldsymbol{N}$ *is the incidence matrix of a* $(\rho_0; \rho_1, \rho_2; 0)$*-EB design with parameters* $v^* = v$, $b^* = sb$, $\boldsymbol{r}^* = s\boldsymbol{r}$, $\boldsymbol{k}^* = \boldsymbol{1}_s \otimes \boldsymbol{k}$, $\varepsilon_\beta^* = \varepsilon_\beta$, $\rho_\beta^* = \rho_\beta$, $\boldsymbol{L}_\beta^* = \boldsymbol{L}_\beta$, *where* $\varepsilon_0 = 1$.

Once there are nonproper and nonequireplicate $(\rho_0; \rho_1, \rho_2; 0)$-EB designs, through Corollaries 7.4.23 – 7.4.25, a large number of designs suitable for the present purpose can be obtained.

In order to use Corollary 7.4.23, the existence of nonproper and nonequireplicate $(0; \rho_1, \rho_2; 0)$-EB designs will be found in Section 8.3.4 (Theorems 8.3.8 and 8.3.9), whereas for Corollaries 7.4.24 and 7.4.25, the existence of a nonproper and nonequireplicate $(\rho_0; \rho_1, \rho_2; 0)$-EB design can be discussed as below.

Consider at first a design with parameters $v^* = 6, b^* = 6, \boldsymbol{r}^* = [6\boldsymbol{1}_4', 3\boldsymbol{1}_2']'$, $\boldsymbol{k}^* = [6, 6, 4, 4, 5, 5]'$, and with the incidence matrix

$$\begin{bmatrix} 1 & 1 & 1 & 1 & 1 & 1 \\ 1 & 1 & 1 & 1 & 1 & 1 \\ 1 & 1 & 1 & 1 & 1 & 1 \\ 1 & 1 & 1 & 1 & 1 & 1 \\ 1 & 1 & 0 & 0 & 1 & 0 \\ 1 & 1 & 0 & 0 & 0 & 1 \end{bmatrix}$$

(see Pearce, 1983, p.102; also Example 3.2.2 in Volume I). After some calculation, it can be shown that this is a $(3; 1, 1; 0)$-EB design with parameters $\varepsilon_0^* = 1$, $\varepsilon_1^* = 14/15$, $\varepsilon_2^* = 8/9$, $\rho_0^* = 3$, $\rho_1^* = 1$, $\rho_2^* = 1$, $\boldsymbol{L}_0^* = \text{diag}[\boldsymbol{I}_4 - 4^{-1}\boldsymbol{1}_4\boldsymbol{1}_4' : \mathbf{O}]$,

$\boldsymbol{L}_1^* = \text{diag}[\mathbf{O} : \boldsymbol{I}_2 - 2^{-1}\mathbf{1}_2\mathbf{1}_2'],$

$$\boldsymbol{L}_2^* = \frac{1}{20}\begin{bmatrix} \mathbf{1}_4\mathbf{1}_4' & -2\mathbf{1}_4\mathbf{1}_2' \\ -4\mathbf{1}_2\mathbf{1}_4' & 8\mathbf{1}_2\mathbf{1}_2' \end{bmatrix}.$$

Here $\boldsymbol{L}_0^* + \boldsymbol{L}_1^* + \boldsymbol{L}_2^* = \boldsymbol{I}_6 - (30)^{-1}\mathbf{1}_6(\boldsymbol{r}^*)'$. Note that this design does not satisfy the condition of Corollary 3.2.2. The last example can be generalized as follows, by evaluating the matrix $\boldsymbol{M}_0$.

Theorem 7.4.18. *For positive integers s, x, y, z such that $s > x, y > z$, the matrix*

$$\begin{bmatrix} & & \mathbf{1}_s\mathbf{1}_t' & \\ \mathbf{1}_2\mathbf{1}_x' & 0\mathbf{1}_2\mathbf{1}_y' & \mathbf{1}_z' \otimes [1,0]' & \mathbf{1}_z' \otimes [0,1]' \end{bmatrix}, \quad t = x + y + 2z,$$

is the incidence matrix of a $(\rho_0; \rho_1, \rho_2; 0)$-EB design with parameters $v = s + 2$, $b = t$, $\boldsymbol{r} = [t\mathbf{1}_s', (x+z)\mathbf{1}_2']'$, $\boldsymbol{k} = [(s+2)\mathbf{1}_x', s\mathbf{1}_y', (s+1)\mathbf{1}_{2z}']'$, $\varepsilon_0 = 1$, $\varepsilon_1 = 1 - z/[(s+1)(x+z)]$, $\varepsilon_2 = 1 - [2(s+1)(y+z)(x+z) - sz(y+z) - z(s+2)(x+z)]/[t(s+1)(s+2)(x+z)]$, $\rho_0 = s - 1$, $\rho_1 = 1$, $\rho_2 = 1$, $\boldsymbol{L}_0 = \text{diag}[\boldsymbol{I}_s - s^{-1}\mathbf{1}_s\mathbf{1}_s' : \mathbf{O}]$, $\boldsymbol{L}_1 = \text{diag}[\mathbf{O} : \boldsymbol{I}_2 - 2^{-1}\mathbf{1}_2\mathbf{1}_2']$, $\boldsymbol{L}_2 = \boldsymbol{I}_{s+2} - [st + 2(x+z)]^{-1}\mathbf{1}_{s+2}[t\mathbf{1}_s', (x+z)\mathbf{1}_2']' - \boldsymbol{L}_0 - \boldsymbol{L}_1$.

In fact, the previous example is a special case of Theorem 7.4.18 when $s = 4$, $x = 2$, $y = 2$ and $z = 1$.

Theorem 7.4.19. *For positive integers s, t such that $s > t$, the incidence matrix*

$$\begin{bmatrix} \mathbf{1}_s\mathbf{1}_t' & 0\mathbf{1}_s\mathbf{1}_t' \\ 0\mathbf{1}_{s-t}\mathbf{1}_t' & \mathbf{1}_{s-t}\mathbf{1}_t' \\ \boldsymbol{I}_t & \mathbf{1}_t\mathbf{1}_t' \end{bmatrix}$$

yields a $(\rho_0; \rho_1, \rho_2; 0)$-EB design with parameters $v = 2s$, $b = 2t$, $\boldsymbol{r} = [t\mathbf{1}_{2s-t}', (t+1)\mathbf{1}_t']'$, $\boldsymbol{k} = [(s+1)\mathbf{1}_t', s\mathbf{1}_t']'$, $\varepsilon_0 = 1$, $\varepsilon_1 = 1 - 1/[(s+1)(t+1)]$, $\varepsilon_2 = t(2s+1)/[s(s+1)(t+1)]$, $\rho_0 = 2s - t - 1$, $\rho_1 = t - 1$, $\rho_2 = 1$, $\boldsymbol{L}_0 = \boldsymbol{I}_{2s} - [t(2s+1)]^{-1}\mathbf{1}_{2s}\boldsymbol{r}' - \boldsymbol{L}_1 - \boldsymbol{L}_2$, $\boldsymbol{L}_1 = \text{diag}[\mathbf{O} : \boldsymbol{I}_t - t^{-1}\mathbf{1}_t\mathbf{1}_t']$,

$$\boldsymbol{L}_2 = \frac{t+1}{y}\begin{bmatrix} \ell_{11}\mathbf{1}_s\mathbf{1}_s' & \ell_{12}\mathbf{1}_s\mathbf{1}_{s-t}' & \ell_{13}\mathbf{1}_s\mathbf{1}_t' \\ \ell_{21}\mathbf{1}_{s-t}\mathbf{1}_s' & \ell_{22}\mathbf{1}_{s-t}\mathbf{1}_{s-t}' & \ell_{23}\mathbf{1}_{s-t}\mathbf{1}_t' \\ \ell_{31}\mathbf{1}_t\mathbf{1}_s' & \ell_{32}\mathbf{1}_t\mathbf{1}_{s-t}' & \ell_{33}\mathbf{1}_t\mathbf{1}_t' \end{bmatrix},$$

where

$$\ell_{11} = s^2t(t+1),\ \ell_{12} = -st(s+1)(t+1),\ \ell_{13} = -s(t+1)[t(s+1) - s],$$

$$\ell_{21} = -st(s+1)(t+1),\ \ell_{22} = t(s+1)^2(t+1),\ \ell_{23} = (s+1)(t+1)[t(s+1) - s],$$

$$\ell_{31} = -st[t(s+1) - s],\ \ell_{32} = t(s+1)[t(s+1) - s],\ \ell_{33} = [t(s+1) - s]^2,$$
$$y = \{t(t+1)\{s^2[2s(t+1) - t] + (s-t)(3s+1)\}\}^{1/2}.$$

Proof. This follows from a direct application of Lemma 2.3.3 and Remark 2.3.1 to find eigenvalues ε_β, multiplicities ρ_β and matrices $\boldsymbol{L}_\beta$ (see Example 2.3.1). □

A special case of Theorem 7.4.19 when $s = 3$ and $t = 2$ can be seen in Pearce (1983, p. 117), and as Example 3.2.3 in Volume I. Furthermore, on account of Remark 3.2.4(a), note that Theorem 7.4.19 produces designs which satisfy the condition of Corollary 3.2.2.

7.5 Designs with more efficiency factors

In the previous sections the constructions of $(\rho_0; \rho_1, ..., \rho_{m-1}; 0)$-EB designs have been discussed mainly for $m - 1 \leq 3$, that is, designs with up to three distinct efficiency factors. The present section is devoted to the existence of designs with more efficiency factors though it is not sure whether an experimenter would like to use such plans in practice. However, such existence can also be found in Chapter 6 and the previous sections of this chapter.

7.5.1 Proper designs that are equireplicate

As described in Section 6.0.2, a PBIB design itself is such, as shown in the following theorem.

Theorem 7.5.1. *A connected s-associate PBIB design with parameters* v, b, r, k, λ_1, λ_2, ... , λ_s, $\varepsilon_0 = 1$, ε_β, ρ_0, ρ_β, $\boldsymbol{L}_0$, $\boldsymbol{L}_\beta$, $\beta = 1, 2, ..., s-1$, *is a* $(\rho_0^*; \rho_1^*, ..., \rho_{s-1}^*; 0)$*-EB design with parameters* $v^* = v$, $b^* = b$, $r^* = r$, $k^* = k$, $\varepsilon_0^* = 1$, $\varepsilon_\beta^* = \varepsilon_\beta$, $\rho_0^* = \rho_0$, $\rho_\beta^* = \rho_\beta$, $\boldsymbol{L}_0^* = \boldsymbol{L}_0$, $\boldsymbol{L}_\beta^* = \boldsymbol{L}_\beta$.

Note that some series of s-associate PBIB designs for $s \geq 2$ can be found in Sprott (1955) without mentioning association schemes but with initial blocks.

By Theorem 6.1.1 (dualization) with a PBIB design, the following corollary can be obtained.

Corollary 7.5.1. *The dual of an s-associate PBIB design with parameters* v, $b(> v)$, r, k, λ_1, λ_2, ... , λ_s, ε_β, ρ_β, $\boldsymbol{L}_\beta$, $\beta = 1, 2, ..., s$, *is a* $(\rho_0^*; \rho_1^*, ..., \rho_s^*; \rho_{s+1}^*)$*-EB design with parameters* $v^* = b$, $b^* = v$, $r^* = k$, $k^* = r$, $\varepsilon_0^* = 1$, $\varepsilon_\beta^* = \varepsilon_\beta$, $\rho_0^* = b - v$, $\rho_\beta^* = \rho_\beta$, $\boldsymbol{L}_0^* = \boldsymbol{I}_b - \sum_{\beta=1}^{s} \boldsymbol{L}_\beta^* - b^{-1}\boldsymbol{1}_b\boldsymbol{1}_b'$, $\boldsymbol{L}_\beta^* = [rk(1-\varepsilon_\beta)]^{-1}\boldsymbol{N}'\boldsymbol{L}_\beta\boldsymbol{N}$, *where* $\boldsymbol{N}$ *is the incidence matrix of the original PBIB design.*

In Corollary 7.5.1, if the rank of its incidence matrix $\boldsymbol{N}$ is smaller than v [see Lemma 2.3.3 and Section 6.0.2], the value of ρ_0^* increases and accordingly the number of efficiency factors less than 1 decreases.

By Theorem 6.2.1 (complementation) and Theorem 7.5.1, one can obtain more $(\rho_0^*; \rho_1^*, ... , \rho_{s-1}^*; 0)$-EB designs.

Theorem 7.4.3 can also be generalized as follows.

Corollary 7.5.2. *If* $\boldsymbol{N}$ *is the incidence matrix of a binary* $(\rho_0; \rho_1, ..., \rho_{m-1}; 0)$*-EB design with parameters* v, b, r, k, ε_β, ρ_β, $\boldsymbol{L}_\beta$, $\beta = 0, 1, ..., m-1$, *then the* *incidence matrix* $[\boldsymbol{N} : \mathbf{1}_v\mathbf{1}'_b - \boldsymbol{N}]$ *with* $v = 2k$ *yields a binary* $(\rho^*_0; \rho^*_1, ..., \rho^*_{m-1}; 0)$-*EB design with parameters* $v^* = v$, $b^* = 2b$, $r^* = b$, $k^* = k$, $\varepsilon^*_\beta = \varepsilon_\beta$, $\rho^*_\beta = \rho_\beta$, $\boldsymbol{L}^*_\beta = \boldsymbol{L}_\beta$.

Theorem 7.4.4 with $\varepsilon_1 < 1$ and $\varepsilon_2 < 1$ in a 2-associate PBIB design can also be stated as follows.

Theorem 7.5.2. *The existence of an* α-*resolvable* 2-*associate PBIB design with parameters* v, $b = b_0 a$, $r = \alpha a$, k, λ_1, λ_2, $\varepsilon_\beta(< 1)$, ρ_β, $\boldsymbol{L}_\beta$, $\beta = 1, 2$, *for* $r = b_0$, *implies the existence of a* $(\rho^*_0; \rho^*_1, \rho^*_2, \rho^*_3; 0)$-*EB design with parameters* $v^* = v + sa$, $b^* = b$, $r^* = r$, $k^* = k+s$, $\varepsilon^*_0 = 1$, $\varepsilon^*_\beta = (k\varepsilon_\beta + s)/(k+s)$, $\varepsilon^*_3 = k/(k+s)$, $\rho^*_0 = a(s-1)+1$, $\rho^*_\beta = \rho_\beta$, $\rho^*_3 = a-1$, $\boldsymbol{L}^*_0 = \boldsymbol{I}_{v^*} - r[b(k+s)]^{-1}\mathbf{1}_{v^*}\mathbf{1}'_{v^*} - \boldsymbol{L}^*_1 - \boldsymbol{L}^*_2 - \boldsymbol{L}^*_3$, $\boldsymbol{L}^*_\beta = \mathrm{diag}[\boldsymbol{L}_\beta : \mathbf{O}]$, $\boldsymbol{L}^*_3 = \mathrm{diag}[\mathbf{O} : (\boldsymbol{I}_a - a^{-1}\mathbf{1}_a\mathbf{1}'_a) \otimes s^{-1}\mathbf{1}_s\mathbf{1}'_s]$ *for* $s \geq 1$.

Theorem 7.4.8 with $\varepsilon_1 < 1$ and $\varepsilon_2 < 1$ in a 2-associate PBIB design can also be given in the following form.

Theorem 7.5.3. *The existence of a* 2-*associate PBIB design with parameters* v, b, r, k, λ_1, λ_2, $\varepsilon_\beta(< 1)$, ρ_β, $\boldsymbol{L}_\beta$, $\beta = 1, 2$, *implies the existence of a* $(\rho^*_0; \rho^*_1, \rho^*_2, \rho^*_3; 0)$-*EB design with parameters* $v^* = 3v$, $b^* = 3b$, $r^* = 2b$, $k^* = 2v$, $\varepsilon^*_0 = 1$, $\varepsilon^*_\beta = 1 - 3rk(1 - \varepsilon_\beta)/(4bv)$, $\varepsilon^*_3 = 3[bv + r(v-k)]/(4bv)$, $\rho^*_0 = v - 1$, $\rho^*_\beta = 2\rho_\beta$, $\rho^*_3 = 2$, $\boldsymbol{L}^*_0 = 3^{-1}\mathbf{1}_3\mathbf{1}'_3 \otimes (\boldsymbol{I}_v - v^{-1}\mathbf{1}_v\mathbf{1}'_v)$, $\boldsymbol{L}^*_\beta = (\boldsymbol{I}_3 - 3^{-1}\mathbf{1}_3\mathbf{1}'_3) \otimes \boldsymbol{L}_\beta$, $\boldsymbol{L}^*_3 = (\boldsymbol{I}_3 - 3^{-1}\mathbf{1}_3\mathbf{1}'_3) \otimes v^{-1}\mathbf{1}_v\mathbf{1}'_v$.

Theorem 7.4.9 with $\varepsilon_1 < 1$ and $\varepsilon_2 < 1$ in a 2-associate PBIB design can also be stated as below.

Theorem 7.5.4. *The existence of a* 2-*associate PBIB design with parameters* v, b, r, k, λ_1, λ_2, $\varepsilon_\beta(< 1)$, ρ_β, $\boldsymbol{L}_\beta$, $\beta = 1, 2$, *implies the existence of a* $(\rho^*_0; \rho^*_1, \rho^*_2, \rho^*_3; 0)$-*EB design with parameters* $v^* = 3v$, $b^* = 3b$, $r^* = b$, $k^* = v$, $\varepsilon^*_0 = 1$, $\varepsilon^*_\beta = 1 - 3rk(1 - \varepsilon_\beta)/(bv)$, $\varepsilon^*_3 = 3r(v-k)/(bv)$, $\rho^*_0 = v - 1$, $\rho^*_\beta = 2\rho_\beta$, $\rho^*_3 = 2$, $\boldsymbol{L}^*_0 = 3^{-1}\mathbf{1}_3\mathbf{1}'_3 \otimes (\boldsymbol{I}_v - v^{-1}\mathbf{1}_v\mathbf{1}'_v)$, $\boldsymbol{L}^*_\beta = (\boldsymbol{I}_3 - 3^{-1}\mathbf{1}_3\mathbf{1}'_3) \otimes \boldsymbol{L}_\beta$, $\boldsymbol{L}^*_3 = (\boldsymbol{I}_3 - 3^{-1}\mathbf{1}_3\mathbf{1}'_3) \otimes v^{-1}\mathbf{1}_v\mathbf{1}'_v$.

Note that the replication numbers and block sizes between Theorems 7.5.3 and 7.5.4 are different. As one can guess from Theorems 7.5.3 and 7.5.4, Theorem 6.6.4 with $a = c = 1$ gives equireplicate and proper $(\rho^*_0; \rho^*_1, ..., \rho^*_{m^*-1}; 0)$-EB designs.

7.5.2 Proper designs that are nonequireplicate

As a generalization of Theorem 6.3.2 (supplementation) obtained by taking a PBIB design instead of a BIB design as a starting design in (6.3.3) in which $s = t$, i.e., $[\boldsymbol{N}'_1 : \mathbf{1}_b\mathbf{1}'_t]'$, the following theorem can be obtained through the structure of the matrix $\boldsymbol{M}_0$.

Theorem 7.5.5. *The existence of an* s-*associate PBIB design with parameters* v, b, r, k, λ_1, λ_2, ... , λ_s, ε_β, ρ_β, $\boldsymbol{L}_\beta$, $\beta = 1, 2, ..., s$, *implies the existence of a*

$(\rho_0^*; \rho_1^*, ..., \rho_s^*; 0)$*-EB design with parameters* $v^* = v + t$, $b^* = b$, $\boldsymbol{r}^* = [r\mathbf{1}_v', b\mathbf{1}_t']'$, $k^* = k + t$, $\varepsilon_0^* = 1$, $\varepsilon_\beta^* = 1 - k(1 - \varepsilon_\beta)/(k + t)$, $\rho_0^* = t$, $\rho_\beta^* = \rho_\beta$, $\boldsymbol{L}_0^* = \boldsymbol{I}_{v+t} - [b(k+t)]^{-1}\mathbf{1}_{v+t}(\boldsymbol{r}^*)' - \sum_{\beta=1}^{s} \boldsymbol{L}_\beta^*$, $\boldsymbol{L}_\beta^* = \text{diag}[\boldsymbol{L}_\beta : \mathbf{O}]$ *for* $t \geq 1$.

Theorem 6.3.5 with $\varepsilon_1 < 1$ and $\varepsilon_2 < 1$ in a 2-associate PBIB design can also be described as follows.

Theorem 7.5.6. *The existence of an* $(\alpha_1, \alpha_2, ..., \alpha_a)$*-resolvable 2-associate PBIB design with parameters* v, $b = \sum_{h=1}^{a} b_h$, $r = \sum_{h=1}^{a} \alpha_h$, k, λ_1, λ_2, ε_β (< 1), ρ_β, $\boldsymbol{L}_\beta$, $\beta = 1, 2$, *implies the existence of a* $(\rho_0^*; \rho_1^*, \rho_2^*, \rho_3^*; 0)$*-EB design with parameters* $v^* = v + ta$, $b^* = b$, $\boldsymbol{r}^* = [r\mathbf{1}_v', b_1\mathbf{1}_t', ..., b_a\mathbf{1}_t']'$, $k^* = k + t$, $\varepsilon_0^* = 1$, $\varepsilon_\beta^* = 1 - k(1 - \varepsilon_\beta)/(k+t)$, $\varepsilon_3^* = k/(k+t)$, $\rho_0^* = a(t-1)+1$, $\rho_\beta^* = \rho_\beta$, $\rho_3^* = a - 1$, $\boldsymbol{L}_0^* = \boldsymbol{I}_{v+ta} - [b(k+t)]^{-1}\mathbf{1}_{v+ta}(\boldsymbol{r}^*)' - \boldsymbol{L}_1^* - \boldsymbol{L}_2^* - \boldsymbol{L}_3^*$, $\boldsymbol{L}_\beta^* = \text{diag}[\boldsymbol{L}_\beta : \mathbf{O}]$, $\boldsymbol{L}_3^* = \text{diag}[\mathbf{O} : (\boldsymbol{I}_a - b^{-1}\mathbf{1}_a\boldsymbol{b}') \otimes t^{-1}\mathbf{1}_t\mathbf{1}_t']$, *where* $\boldsymbol{b} = [b_1, b_2, ..., b_a]'$ *and* $t \geq 1$.

In Theorem 7.5.6, when an s-associate PBIB design as a basic design is taken, one can obtain a $(\rho_0^*; \rho_1^*, ..., \rho_s^*, \rho_{s+1}^*; 0)$-EB design (when all $\varepsilon_\beta < 1$) or $(\rho_0^*; \rho_1^*, ..., \rho_{s'}^*; 0)$-EB design (when some ε_β are one) for $s' \leq s$.

Theorem 7.5.2 can also be stated as below by relaxing the condition $r = b_0$ (see also Theorem 7.4.4 and Remark 7.4.2).

Theorem 7.5.7. *The existence of an* α*-resolvable 2-associate PBIB design with parameters* v, $b = b_0a$, $r = \alpha a$, k, λ_1, λ_2, $\varepsilon_\beta(< 1)$, ρ_β, $\boldsymbol{L}_\beta$, $\beta = 1, 2$, *implies the existence of a* $(\rho_0^*; \rho_1^*, \rho_2^*, \rho_3^*; 0)$*-EB design with parameters* $v^* = v + sa$, $b^* = b$, $\boldsymbol{r}^* = [r\mathbf{1}_v', b_0\mathbf{1}_{sa}]'$, $k^* = k + s$, $\varepsilon_0^* = 1$, $\varepsilon_\beta^* = 1 - k(1 - \varepsilon_\beta)/(k+s)$, $\varepsilon_3^* = k/(k+s)$, $\rho_0^* = a(s-1)+1$, $\rho_\beta^* = \rho_\beta$, $\rho_3^* = a - 1$, $\boldsymbol{L}_0^* = \boldsymbol{I}_{v^*} - r[b(k+s)]^{-1}\mathbf{1}_{v^*}\mathbf{1}_{v^*}' - \boldsymbol{L}_1^* - \boldsymbol{L}_2^* - \boldsymbol{L}_3^*$, $\boldsymbol{L}_\beta^* = \text{diag}[\boldsymbol{L}_\beta : \mathbf{O}]$, $\boldsymbol{L}_3^* = \text{diag}[\mathbf{O} : (\boldsymbol{I}_a - a^{-1}\mathbf{1}_a\mathbf{1}_a') \otimes s^{-1}\mathbf{1}_s\mathbf{1}_s']$ *for* $s \geq 1$.

In Theorem 7.4.14, when all ε_β are less than one, the following theorem can be obtained.

Theorem 7.5.8. *The existence of an* $(\alpha_1, \alpha_2, ..., \alpha_a)$*-resolvable 2-associate PBIB design with parameters* v, $b = \sum_{h=1}^{q} b_h^*$, $r = \sum_{h=1}^{a} \alpha_h$, k, λ_1, λ_2, ε_β (< 1), ρ_β, $\boldsymbol{L}_\beta$, $\beta = 1, 2$, *implies the existence of a* $(\rho_0^*; \rho_1^*, \rho_2^*, \rho_3^*; 0)$*-EB design with parameters* $v^* = v + sq$, $b^* = b$, $\boldsymbol{r}^* = [r\mathbf{1}_v', b_1^*\mathbf{1}_s', ..., b_q^*\mathbf{1}_s']'$, $k^* = k + s$, $\varepsilon_0^* = 1$, $\varepsilon_\beta^* = 1 - k(1 - \varepsilon_\beta)/(k+s)$, $\varepsilon_3^* = k/(k+s)$, $\rho_0^* = q(s-1)+1$, $\rho_\beta^* = \rho_\beta$, $\rho_3^* = q - 1$, $\boldsymbol{L}_0^* = \boldsymbol{I}_{v+sq} - [b(k+s)]^{-1}\mathbf{1}_{v+sq}(\boldsymbol{r}^*)' - \boldsymbol{L}_1^* - \boldsymbol{L}_2^* - \boldsymbol{L}_3^*$, $\boldsymbol{L}_\beta^* = \text{diag}[\boldsymbol{L}_\beta : \mathbf{O}]$, $\boldsymbol{L}_3^* = \text{diag}[\mathbf{O} : [\boldsymbol{I}_q - b^{-1}\mathbf{1}_q(\boldsymbol{b}^*)'] \otimes s^{-1}\mathbf{1}_s\mathbf{1}_s']$, *where* $\boldsymbol{b}^* = [b_1^*, b_2^*, ..., b_q^*]'$ *and* $s \geq 1$.

On the other hand, Corollary 7.4.13 with $\varepsilon_1 < 1$ and $\varepsilon_2 < 1$ in a 2-associate PBIB design can produce the following result.

Corollary 7.5.3. *The existence of an* α*-resolvable 2-associate PBIB design with parameters* v, $b = b_0a$, $r = \alpha a$, k, λ_1, λ_2, $\varepsilon_\beta(< 1)$, ρ_β, $\boldsymbol{L}_\beta$, $\beta = 1, 2$, *implies the existence of a* $(\rho_0^*; \rho_1^*, \rho_2^*, \rho_3^*; 0)$*-EB design with parameters* $v^* = v + sa$, $b^* = b$, $\boldsymbol{r}^* = [r\mathbf{1}_v', b_0\mathbf{1}_{sa}']'$, $k^* = k + s$, $\varepsilon_0^* = 1$, $\varepsilon_\beta^* = 1 - k(1 - \varepsilon_\beta)/(k+s)$, $\varepsilon_3^* = k/(k+s)$, $\rho_0^* = a(s-1)+1$, $\rho_\beta^* = \rho_\beta$, $\rho_3^* = a - 1$, $\boldsymbol{L}_0^* = \boldsymbol{I}_{v+sa} - [b(k+s)]^{-1}\mathbf{1}_{v+sa}(\boldsymbol{r}^*)' -$

$\boldsymbol{L}_1^* - \boldsymbol{L}_2^* - \boldsymbol{L}_3^*$, $\boldsymbol{L}_\beta^* = \text{diag}[\boldsymbol{L}_\beta : \mathbf{O}]$, $\boldsymbol{L}_3^* = \text{diag}[\mathbf{O} : (\boldsymbol{I}_a - a^{-1}\mathbf{1}_a\mathbf{1}_a') \otimes s^{-1}\mathbf{1}_s\mathbf{1}_s']$ *for* $s \geq 1$.

Corollary 7.4.14 with $\varepsilon_1 < 1$ and $\varepsilon_2 < 1$ in a 2-associate PBIB design can also be extended as follows.

Theorem 7.5.9. *The existence of an* α*-resolvable* 2*-associate PBIB design with parameters* v, $b = b_0 a$, $r = \alpha a$, k, λ_1, λ_2, $\varepsilon_\beta(< 1)$, ρ_β, $\boldsymbol{L}_\beta$, $\beta = 1, 2$, *implies the existence of a* $(\rho_0^*; \rho_1^*, \rho_2^*, \rho_3^*; 0)$*-EB design with parameters* $v^* = v + sq$, $b^* = b$, $\boldsymbol{r}^* = [r\mathbf{1}_v', a_1 b_0 \mathbf{1}_s', ..., a_q b_0 \mathbf{1}_s']'$, $k^* = k + s$, $\varepsilon_0^* = 1$, $\varepsilon_\beta^* = 1 - k(1 - \varepsilon_\beta)/(k + s)$, $\varepsilon_3^* = k/(k + s)$, $\rho_0^* = q(s - 1) + 1$, $\rho_\beta^* = \rho_\beta$, $\rho_3^* = q - 1$, $\boldsymbol{L}_0^* = \boldsymbol{I}_{v+sq} - [b(k + s)]^{-1}\mathbf{1}_{v+sq}(\boldsymbol{r}^*)' - \boldsymbol{L}_1^* - \boldsymbol{L}_2^* - \boldsymbol{L}_3^*$, $\boldsymbol{L}_\beta^* = \text{diag}[\boldsymbol{L}_\beta : \mathbf{O}]$, $\boldsymbol{L}_3^* = \text{diag}[\mathbf{O} : (\boldsymbol{I}_q - a^{-1}\mathbf{1}_q[a_1, a_2, ..., a_q]) \otimes s^{-1}\mathbf{1}_s\mathbf{1}_s']$, *where* $a = \sum_{h=1}^q a_h$ *and* $s \geq 1$.

By Theorem 7.4.15 with $\varepsilon_1 < 1$ and $\varepsilon_2 < 1$ in a 2-associate PBIB design the following theorem can be obtained.

Theorem 7.5.10. *The existence of a* 3*-associate PBIB design with parameters* v, b, r, k, λ_1, λ_2, λ_3, $\varepsilon_\beta(< 1)$, ρ_β, $\boldsymbol{L}_\beta$, $\beta = 1, 2, 3$, *implies the existence of a* $(\rho_0^*; \rho_1^*, \rho_2^*, \rho_3^*; 0)$*-EB design with parameters* $v^* = v + s$, $b^* = b$, $\boldsymbol{r}^* = [r\mathbf{1}_v', b\mathbf{1}_s']'$, $k^* = k + s$, $\varepsilon_0^* = 1$, $\varepsilon_\beta^* = 1 - k(1 - \varepsilon_\beta)/(k + s)$, $\rho_0^* = s$, $\rho_\beta^* = \rho_\beta$, $\boldsymbol{L}_0^* = \boldsymbol{I}_{v+s} - [b(k + s)]^{-1}\mathbf{1}_{v+s}(\boldsymbol{r}^*)' - \boldsymbol{L}_1^* - \boldsymbol{L}_2^* - \boldsymbol{L}_3^*$, $\boldsymbol{L}_\beta^* = \text{diag}[\boldsymbol{L}_\beta : \mathbf{O}]$ *for* $s \geq 1$.

7.5.3 Nonproper designs that are equireplicate

By Theorem 6.1.1 the dualization of designs resulting from Section 7.5.2 easily yields nonproper and equireplicate designs with more efficiency factors which are suitable for the present purpose. Note that all the designs obtainable in this section satisfy Corollary 3.2.2, on account of Corollary 3.2.3, to get the BLUE of a contrast.

Next, by Theorem 6.4.2 (juxtaposition) the following corollary can be obtained.

Corollary 7.5.4. *Let* $\boldsymbol{N}_h, h = 1, 2, ..., a$, *be the* $v \times b^{(h)}$ *incidence matrices of a* s*-associate PBIB designs, based on the same association scheme, with parameters* v, $b^{(h)}$, $r^{(h)}$, $k^{(h)}$, $\lambda_1^{(h)}$, $\lambda_2^{(h)}$, ... , $\lambda_s^{(h)}$, $\varepsilon_1^{(h)}$, $\varepsilon_2^{(h)}$, ... , $\varepsilon_s^{(h)}$, ρ_1, ρ_2, ... , ρ_s, $\boldsymbol{L}_1$, $\boldsymbol{L}_2$, ... , $\boldsymbol{L}_s$. *Then the incidence matrix* $[\boldsymbol{N}_1 : \boldsymbol{N}_2 : \cdots : \boldsymbol{N}_a]$ *yields a* $(\rho_0^*; \rho_1^*, ..., \rho_{s-1}^*; 0)$*-EB design with parameters* $v^* = v$, $b^* = \sum_{h=1}^a b^{(h)}$, $r^* = \sum_{h=1}^a r^{(h)}$, $\boldsymbol{k}^* = [k^{(1)}\mathbf{1}_{b^{(1)}}', k^{(2)}\mathbf{1}_{b^{(2)}}', ..., k^{(a)}\mathbf{1}_{b^{(a)}}']'$, $\varepsilon_0^* = 1$, $\varepsilon_\beta^* = 1 - [\sum_{h=1}^a r^{(h)}(1 - \varepsilon_{\beta+1}^{(h)})]/r^*$, $\beta = 1, 2, ..., s - 1$, $\rho_0^* = \rho_1$, $\rho_\beta^* = \rho_{\beta+1}$, $\boldsymbol{L}_0^* = \boldsymbol{L}_1$, $\boldsymbol{L}_\beta^* = \boldsymbol{L}_{\beta+1}$, *provided that* $\varepsilon_1^{(h)} = 1$ *for all* $h = 1, 2, ..., a$.

Theorem 7.4.17 can also be generalized as follows.

Theorem 7.5.11. *If* $\boldsymbol{N}$ *is the incidence matrix of a binary* $(\rho_0; \rho_1, ..., \rho_{m-1}; 0)$*-EB design with parameters* v, b, r, k, ε_β, ρ_β, $\boldsymbol{L}_\beta$, $\beta = 0, 1, ..., m - 1$, *then the incidence matrix* $[\boldsymbol{N} : \mathbf{1}_v\mathbf{1}_b' - \boldsymbol{N}]$ *yields a binary* $(\rho_0; \rho_1, ..., \rho_{m-1}; 0)$*-EB design* *with parameters* $v^* = v$, $b^* = 2b$, $r^* = b$, $\boldsymbol{k}^* = [k\mathbf{1}'_b, (v-k)\mathbf{1}'_b]'$, $\varepsilon^*_\beta = 1 - k(1 - \varepsilon_\beta)/(v-k)$, $\rho^*_\beta = \rho_\beta$, $\boldsymbol{L}^*_\beta = \boldsymbol{L}_\beta$, $\beta = 0, 1, ..., m-1$.

7.5.4 Nonproper designs that are nonequireplicate

At first, once there exists a nonproper and nonequireplicate $(\rho_0; \rho_1, ..., \rho_{m-1}; \rho_m)$-EB design with $\rho_0 \geq 0$, by Theorem 6.1.1 (dualization) one can get other $(\rho^*_0; \rho^*_1, ..., \rho^*_{m-1}; \rho^*_m)$-EB designs with $\rho^*_0 = b - v + \rho_0$. Next, by Theorem 6.4.1 (juxtaposition) one can get more such designs.

It is not so difficult to find an individual plan in this category by use of the practical procedure of Lemma 2.3.3 and Remark 2.3.1. The following is a series of such designs.

Theorem 7.5.12. *For positive integers s, t such that $s > t$, the matrix*

$$\begin{bmatrix} \mathbf{1}_s\mathbf{1}'_t & \mathbf{O} \\ \boldsymbol{I}_t & \mathbf{1}_t\mathbf{1}'_t \\ \mathbf{1}_t\mathbf{1}'_t & \boldsymbol{I}_t \\ \mathbf{0}' & \mathbf{1}'_t \end{bmatrix}$$

is the incidence matrix of a $(\rho_0; \rho_1, \rho_2, \rho_3; 0)$-EB design with parameters $v = s + 2t + 1$, $b = 2t$, $\boldsymbol{r} = [t\mathbf{1}'_s, (t+1)\mathbf{1}'_{2t}, t]'$, $\boldsymbol{k} = [(s+t+1)\mathbf{1}'_t, (t+2)\mathbf{1}'_t]'$, $\varepsilon_0 = 1$, $\varepsilon_1 = 1 - 1/[(t+1)(s+t+1)]$, $\varepsilon_2 = 1 - 1/[(t+1)(t+2)]$, $\varepsilon_3 = 2t(2t+s+3)/[(t+1)(t+2)(s+t+1)]$, $\rho_0 = s+1$, $\rho_1 = t-1$, $\rho_2 = t-1$, $\rho_3 = 1$, $\boldsymbol{L}_0 = \boldsymbol{I}_v - [t(s+2t+3)]^{-1}\mathbf{1}_v\boldsymbol{r}' - \boldsymbol{L}_1 - \boldsymbol{L}_2 - \boldsymbol{L}_3$, $\boldsymbol{L}_1 = \text{diag}[\mathbf{0}_s\mathbf{0}'_s : \boldsymbol{I}_t - t^{-1}\mathbf{1}_t\mathbf{1}'_t : \mathbf{0}_{t+1}\mathbf{0}'_{t+1}]$, $\boldsymbol{L}_2 = \text{diag}[\mathbf{0}_{s+t}\mathbf{0}'_{s+t} : \boldsymbol{I}_t - t^{-1}\mathbf{1}_t\mathbf{1}'_t : 0]$,

$$\boldsymbol{L}_3 = \begin{bmatrix} tx^2\mathbf{1}_s\mathbf{1}'_s & xy(t+1)\mathbf{1}_s\mathbf{1}'_t & xz(t+1)\mathbf{1}_s\mathbf{1}'_t & txu\mathbf{1}_s \\ txy\mathbf{1}_t\mathbf{1}'_s & y^2(t+1)\mathbf{1}_t\mathbf{1}'_t & yz(t+1)\mathbf{1}_t\mathbf{1}'_t & tyu\mathbf{1}_t \\ txz\mathbf{1}_t\mathbf{1}'_s & yz(t+1)\mathbf{1}_t\mathbf{1}'_t & z^2(t+1)\mathbf{1}_t\mathbf{1}'_t & tzu\mathbf{1}_t \\ txu\mathbf{1}'_s & yu(t+1)\mathbf{1}'_t & zu(t+1)\mathbf{1}'_t & tu^2 \end{bmatrix},$$

where $x = \mp(t+2)/[(t+s+1)\sqrt{w}]$, $y = \pm(st + t^2 - 2)/[(t+1)(s+t+1)\sqrt{w}]$, $z = \mp(t^2 + t - s - 1)/[(t+1)(s+t+1)\sqrt{w}]$, $u = \pm 1/\sqrt{w}$, $w = t[2t^4 + 3(s+1)t^3 + (s^2 + 5s - 2)t^2 + (s^2 + 6s + 1)t + 2(s^2 + 4s + 3)]/[(t+1)(s+t+1)^2]$.

Proof. This follows from the direct application of Lemma 2.3.3 and Remark 2.3.1 to find eigenvalues ε_β, multiplicities ρ_β and matrices $\boldsymbol{L}_\beta$. □

For example, a special case of Theorem 7.5.12 when $s = 3$ and $t = 2$ can be given as below, i.e., as a (4;1,1,1;0)-EB design with the incidence matrix and parameters of the form

$$\begin{bmatrix} 1 & 1 & 0 & 0 \\ 1 & 1 & 0 & 0 \\ 1 & 1 & 0 & 0 \\ 1 & 0 & 1 & 1 \\ 0 & 1 & 1 & 1 \\ 1 & 1 & 1 & 0 \\ 1 & 1 & 0 & 1 \\ 0 & 0 & 1 & 1 \end{bmatrix} \quad \begin{aligned} & v = 8,\ b = 4,\ \boldsymbol{r} = [2\mathbf{1}_3', 3\mathbf{1}_4', 2]',\ \boldsymbol{k} = [6,6,4,4]', \\ & \varepsilon_0 = 1,\ \varepsilon_1 = 17/18,\ \varepsilon_2 = 11/12,\ \varepsilon_3 = 5/9, \\ & \rho_0 = 4,\ \rho_1 = 1,\ \rho_2 = 1,\ \rho_3 = 1, \\ & \boldsymbol{L}_0 = \boldsymbol{I}_8 - (20)^{-1}\mathbf{1}_8\boldsymbol{r}' - \boldsymbol{L}_1 - \boldsymbol{L}_2 - \boldsymbol{L}_3, \\ & \boldsymbol{L}_1 = \mathrm{diag}[\mathbf{0}_3\mathbf{1}_3' : \boldsymbol{I}_2 - 2^{-1}\mathbf{1}_2\mathbf{1}_2' : \mathbf{0}_3\mathbf{1}_3'], \\ & \boldsymbol{L}_2 = \mathrm{diag}[\mathbf{0}_5\mathbf{1}_5' : \boldsymbol{I}_2 - 2^{-1}\mathbf{1}_2\mathbf{1}_2' : 0], \end{aligned}$$

$$\boldsymbol{L}_3 = \frac{1}{160}\begin{bmatrix} 24\mathbf{1}_3\mathbf{1}_3' & -24\mathbf{1}_3\mathbf{1}_2' & 6\mathbf{1}_3\mathbf{1}_2' & -36\mathbf{1}_3 \\ -16\mathbf{1}_2\mathbf{1}_3' & 16\mathbf{1}_2\mathbf{1}_2' & -4\mathbf{1}_2\mathbf{1}_2' & 24\mathbf{1}_2 \\ 4\mathbf{1}_2\mathbf{1}_3' & -4\mathbf{1}_2\mathbf{1}_2' & \mathbf{1}_2\mathbf{1}_2' & -6\mathbf{1}_2 \\ -36\mathbf{1}_3' & 36\mathbf{1}_2' & -9\mathbf{1}_2' & 54 \end{bmatrix}.$$

8

Designs with No Full Efficiency

This chapter is devoted to methods of constructing block designs that do not provide unit efficiency factors for any contrasts of treatment parameters. Taking into account the practical point of view, the cases of $(0; \rho_1, \rho_2, ..., \rho_{m-1}; 0)$-EB designs, with $m = 2, 3$ and more, will be considered. At first, a general consideration on such designs is presented in Section 8.1, by recalling relevant results discussed in Volume I and by providing some corresponding optimality results. Next, a large number of methods of construction are given for designs separated into four classes, covering designs that are (i) proper and equireplicate, (ii) proper and nonequireplicate, (iii) nonproper and equireplicate and (iv) nonproper and nonequireplicate, first for $m = 2$ (Section 8.2), then for $m = 3$ (Section 8.3), for $m = 4$ (Section 8.4), and finally for $m > 4$ (Section 8.5).

8.1 General consideration

A general consideration for a $(\rho_0; \rho_1, ..., \rho_{m-1}; 0)$-EB design with parameters v, b, $\boldsymbol{r}$, $\boldsymbol{k}$, ε_β, ρ_β, $\boldsymbol{L}_\beta$, $\beta = 0, 1, ..., m-1$, in which some basic contrasts are estimated in the intra-block analysis with full efficiency, i.e., in which $\rho_0 \geq 1$, is given in Section 7.1. In the present chapter, the discussion is confined to constructions of designs in which none of the contrasts can be estimated there with full efficiency, i.e., of designs with $\rho_0 = 0$. The latter means that $\varepsilon_\beta < 1$ for all β.

As can be recalled from Sections 3.4 and 4.4, the following properties hold for the $(0; \rho_1, \rho_2, ..., \rho_{m-1}; 0)$-EB designs.

(1) Each of the ρ_β basic contrasts of the design belonging to the βth subset, $\{\boldsymbol{c}'_{\beta j}\boldsymbol{\tau} = \boldsymbol{s}'_{\beta j}\boldsymbol{r}^{\delta}\boldsymbol{\tau}, j = 1, 2, ..., \rho_\beta\}$, is estimated in the intra-block analysis with the same efficiency $\varepsilon_\beta(< 1), \beta = 1, 2, ..., m-1$.

(2) The average efficiency factor of the design is given by the harmonic mean

$$\varepsilon = (v-1)/(\sum_{\beta=1}^{m-1} \rho_\beta/\varepsilon_\beta).$$

(3) The variance of the intra-block BLUE of any basic contrast from the βth subset is given by $\sigma_1^2/\varepsilon_\beta$, for $\beta = 1, 2, ..., m-1$, where $\sigma_1^2 = \sigma_U^2 + \sigma_e^2$, as in (3.4.2).

(4) Any contrast belonging to $\mathcal{C}(\boldsymbol{r}^\delta \boldsymbol{L}_\beta)$, the column space of the matrix $\boldsymbol{r}^\delta \boldsymbol{L}_\beta$, of dimension ρ_β, i.e., a contrast of the type $\boldsymbol{s}'_\beta \boldsymbol{L}'_\beta \boldsymbol{r}^\delta \boldsymbol{\tau}$, for any conformable vector $\boldsymbol{s}_\beta$, is estimated in the intra-block analysis with the efficiency ε_β, and the variance of its intra-block BLUE is given by $\varepsilon_\beta^{-1} \boldsymbol{s}'_\beta \boldsymbol{r}^\delta \boldsymbol{L}_\beta \boldsymbol{s}_\beta \sigma_1^2$, for $\beta = 1, 2, ..., m-1$ (this being reduced to $\varepsilon_\beta^{-1} \boldsymbol{s}'_\beta \boldsymbol{r}^\delta \boldsymbol{s}_\beta \sigma_1^2$ if $\boldsymbol{s}_\beta = \boldsymbol{L}_\beta \boldsymbol{s}_\beta$).

(5) The properties given in Section 7.1(5) apply here as well, except that the matrix $\boldsymbol{S}$ there is to be replaced here by $\boldsymbol{S} = [\boldsymbol{S}_1 : \boldsymbol{S}_2 : \cdots : \boldsymbol{S}_{m-1}]$, $\boldsymbol{\varepsilon}^\delta$ by $\boldsymbol{\varepsilon}^\delta = \text{diag}[\varepsilon_1 \boldsymbol{I}_{\rho_1} : \varepsilon_2 \boldsymbol{I}_{\rho_2} : \cdots : \varepsilon_{m-1} \boldsymbol{I}_{\rho_{m-1}}]$ and $\sum_{\beta=0}^{m-1} \varepsilon_\beta^{-1} \boldsymbol{L}_\beta \boldsymbol{r}^{-\delta}$ by $\sum_{\beta=1}^{m-1} \varepsilon_\beta^{-1} \boldsymbol{L}_\beta \boldsymbol{r}^{-\delta}$.

Subsequently, the formula (7.1.7) applies here also, but $\sum_{\beta=0}^{m-1} \rho_\beta \varepsilon_\beta^{-1}$ is to be replaced by $\sum_{\beta=1}^{m-1} \rho_\beta \varepsilon_\beta^{-1}$.

As to the optimality criteria, note that Lemma 7.1.3 is of particular interest here. Consequently, Theorem 7.1.1 can be reformulated here as follows.

Theorem 8.1.1. *Let the estimation be considered under the overall randomization model with the covariance matrix of the form* (7.1.8), *for which it is assumed that the ratio* $\delta = \sigma_2^2/\sigma_1^2$ *is known and positive. Then a* $(0; v-1; 0)$*-EB design is A-, D- and E-optimal for a complete set of its basic contrasts in the class of all proper* $(0; \rho_1, \rho_2, ..., \rho_{m-1}; 0)$*-EB designs with the same treatment replications,* $\boldsymbol{r}$, *and with constant both* $\text{rank}(\boldsymbol{M}_0) = v-1$ *and* $\text{tr}(\boldsymbol{M}_0)$.

Proof. See Corollary 7.1.2 and its proof, also Remark 7.1.1. □

Referring now to the problem caused by the replacement of an unknown true value of δ by its estimate $\hat{\delta}$, discussed in Section 3.8.4 and recalled in Section 7.1, it should be noted that for a proper $(0; v-1; 0)$-EB design, with $v > 3$, a gain in precision resulting from the recovery of inter-block information over the estimation in the intra-block analysis can be expected to be achieved uniformly if and only if the inequality

$$2(v-1) < (v-3)(n-v-b+1) \tag{8.1.1}$$

holds, as it follows from (7.1.12) for $\rho_1 = v-1 > 2$.

Thus, it will be advisable, if choosing a proper $(0; v-1; 0)$-EB design, to employ in the analysis the combined estimators of Section 3.8.3 whenever the design satisfies the inequality (8.1.1). The remaining conclusions from the discussion in Section 7.1 are valid also here, as far as they apply to the $\rho_0 = 0$ cases.

Similarly as in Chapter 7, the investigation of $(0; \rho_1, \rho_2, ..., \rho_{m-1}; 0)$-EB designs will be systematized here by separating the material into four sections, depending on the number, $m-1 = 1$, 2, 3, or more, of distinct efficiency factors. Such separation may be interesting from a practical point of view. As in Chapter 7, the term efficiency factor will be used throughout the present chapter in the sense of the efficiency for estimating in the intra-block analysis.

8.2 Designs with one efficiency factor

When all basic contrasts (defined in Section 3.4) are of equal importance for the researcher, a $(0; v-1; 0)$-EB design is to be recommended. In this section, at first general characteristics of such designs are described. Then their constructions will be considered.

As explained in Section 4.4, in a $(0; v-1; 0)$-EB design the number of nonzero distinct eigenvalues of $\boldsymbol{C}_1$ with respect to $\boldsymbol{r}^\delta$ is one, i.e., $\boldsymbol{C}_1 = \varepsilon_1(\boldsymbol{r}^\delta - n^{-1}\boldsymbol{r}\boldsymbol{r}')$, and hence every contrast of treatment parameters is estimated in the intra-block analysis with the same efficiency, given by ε_1. In this sense, such design has been called a (totally) balanced design by Jones (1959) and Caliński (1971), or an efficiency-balanced design by Puri and Nigam (1975a, 1975b) and Williams (1975) (see Theorem 2.4.3). It is also evident that in case of a $(0; v-1; 0)$-EB design, every basic contrast is estimated in the intra-block analysis with the same variance σ_1^2/ε_1 (see Sections 3.4 and 8.1). Moreover, since in this case $\boldsymbol{L}_1 = \boldsymbol{I}_v - n^{-1}\boldsymbol{1}_v\boldsymbol{r}'$ (see Section 4.4.2), it follows from (4.4.7) that for any vector $\boldsymbol{s}$ such that $\boldsymbol{r}'\boldsymbol{s} = 0$, the intra-block BLUE of $\boldsymbol{c}'\boldsymbol{\tau} = \boldsymbol{s}'\boldsymbol{L}_1\boldsymbol{r}^\delta\boldsymbol{\tau} = \boldsymbol{s}'\boldsymbol{r}^\delta\boldsymbol{\tau}$ has the variance $\varepsilon_1^{-1}\boldsymbol{s}'\boldsymbol{r}^\delta\boldsymbol{s}\sigma_1^2$. In particular, the variance of the intra-block BLUE of the elementary contrast between the ith and i'th treatment parameters is of the form $(1/r_i + 1/r_{i'})(\sigma_1^2/\varepsilon_1)$, but after $\boldsymbol{r}^{-\delta}$-normalization of the contrast, its variance becomes σ_1^2/ε_1, the same for any pair of treatments. Thus, the main advantage of $(0; v-1; 0)$-EB designs is that to every contrast corresponds the same efficiency, and also the same precision when the contrasts are $\boldsymbol{r}^{-\delta}$-normalized (i.e., such that $\boldsymbol{c}'\boldsymbol{r}^{-\delta}\boldsymbol{c} = \boldsymbol{s}'\boldsymbol{r}^\delta\boldsymbol{s} = 1$ if $\boldsymbol{c}'\boldsymbol{\tau} = \boldsymbol{s}'\boldsymbol{r}^\delta\boldsymbol{\tau}$ is the contrast).

In a $(0; v-1; 0)$-EB design, the equalities

$$\boldsymbol{N}\boldsymbol{k}^{-\delta}\boldsymbol{N}' = (1-\varepsilon_1)\boldsymbol{r}^\delta + \frac{\varepsilon_1}{n}\boldsymbol{r}\boldsymbol{r}', \quad n = \boldsymbol{r}'\boldsymbol{1}_v, \tag{8.2.1}$$

and

$$\boldsymbol{C}_1 = \boldsymbol{r}^\delta - \boldsymbol{N}\boldsymbol{k}^{-\delta}\boldsymbol{N}' = \varepsilon_1\left(\boldsymbol{r}^\delta - \frac{1}{n}\boldsymbol{r}\boldsymbol{r}'\right), \; 0 < \varepsilon_1 < 1, \; \text{rank}(\boldsymbol{C}_1) = v-1, \tag{8.2.2}$$

hold on account of (2.3.1), (2.3.2) and the equality $\sum_{i=1}^{v-1}\boldsymbol{s}_i\boldsymbol{s}_i' = \boldsymbol{r}^{-\delta} - n^{-1}\boldsymbol{1}_v\boldsymbol{1}_v'$. This, through $\boldsymbol{N}\boldsymbol{k}^{-\delta}\boldsymbol{N}'\boldsymbol{1}_v = \boldsymbol{r}$, yields the following result (corresponding to Definition 2.4.5).

Theorem 8.2.1. *A block design with parameters v, b, $\boldsymbol{r} = [r_1, r_2, ..., r_v]'$, $\boldsymbol{k}$, having the incidence matrix $\boldsymbol{N}$, is a $(0; v-1; 0)$-EB design if and only if the*

(i,i')th element of $\boldsymbol{N}\boldsymbol{k}^{-\delta}\boldsymbol{N}'$ is proportional to $r_i r_{i'}$ for all $i,i'(i \neq i') = 1,2,...,v$.

When a design is binary, $\mathrm{tr}(\boldsymbol{N}\boldsymbol{k}^{-\delta}\boldsymbol{N}') = b$ in general. Then taking the trace of both sides of (8.2.2) shows the following observation.

Theorem 8.2.2. *In a binary $(0; v-1; 0)$-EB design with parameters v, b, $\boldsymbol{r} = [r_1, r_2, ..., r_v]'$, $\boldsymbol{k} = [k_1, k_2, ..., k_b]'$, the unique efficiency factor ε_1 is given by*

$$\varepsilon_1 = n(n-b)\bigg/\left(n^2 - \sum_{i=1}^{v} r_i^2\right),$$

where $n = \sum_i r_i = \sum_j k_j$.

Remark 8.2.1. (a) Mere specification of parameters v, b, r_i, k_j determines the value of ε_1 provided the design is binary. In general, however, (8.2.2) implies that

$$\varepsilon_1 = [v - \mathrm{tr}(\boldsymbol{r}^{-\delta}\boldsymbol{N}\boldsymbol{k}^{-\delta}\boldsymbol{N}')]/(v-1).$$

(b) Because in a $(0; v-1; 0)$-EB design $\varepsilon < 1$ (by definition), it follows from Theorem 8.2.2 that if the design is binary, then $r_i < b$ for at least one i. Thus, a binary $(0; v-1; 0)$-EB design is an incomplete block design.

On the other hand, because (8.2.1) is also true for a $(v-1; 0; 0)$-EB design, with ε_1 replaced by $\varepsilon_0 = 1$, it follows that if a connected orthogonal block design is binary, then $r_i = b$ for any i, i.e., the design is a randomized block design (RBD) (see Theorem 7.2.1). Otherwise, the following theorem can be obtained.

Theorem 8.2.3. *A binary proper block design is $(0; v-1; 0)$-EB if and only if it is a balanced incomplete block (BIB) design.*

Proof. When a binary proper block design with parameters $v, b, \boldsymbol{r} = [r_1, r_2, ..., r_v]'$, $k(< v)$, is $(0; v-1; 0)$-EB, the relation (8.2.1) holds. Now, comparing elements on both sides of (8.2.1) yields $r_i/k = (1-\varepsilon_1)r_i + \varepsilon_1 r_i^2/n$ and $\sum_{j=1}^{b} n_{ij}n_{i'j}/k = \varepsilon_1 r_i r_{i'}/n$ for all $i, i'(i \neq i') = 1, 2, ..., v$, which show that the design is equireplicate and of equal concurrences (see Section 2.4.1). This shows, by Definition 2.4.2, that the design is a BIB design. The converse is obvious from Theorem 8.2.1. □

Theorem 8.2.4. *In a $(0; v-1; 0)$-EB design with parameters $v, b, \boldsymbol{r} = [r_1, r_2, ..., r_v]'$, $\boldsymbol{k}$, the inequality $b \geq v$ (Fisher's inequality) holds.*

Proof. The determinant of (8.2.1) can be found to be equal to $r_1 r_2 \cdots r_v(1-\varepsilon_1)^{v-1}$, which is nonzero, as $0 < \varepsilon_1 < 1$. That is, $\boldsymbol{N}\boldsymbol{k}^{-\delta}\boldsymbol{N}'$ is nonsingular which implies that $v = \mathrm{rank}(\boldsymbol{N}\boldsymbol{k}^{-\delta}\boldsymbol{N}') = \mathrm{rank}(\boldsymbol{N}) \leq b$. See Definition 2.3.1. □

Theorem 8.2.4 shows that the number of blocks in a $(0; v-1; 0)$-EB design is always greater than or equal to the number of treatments, which is a condition originally established by Fisher (1940) for a BIB design. However, even nonequireplicate cases (i.e., not like in a BIB design) can be treated in the class of $(0; v-1; 0)$-EB designs. This point is an advantage if one is interested to use a $(0; v-1; 0)$-EB design.

Remark 8.2.2. In a binary equireplicate $(0; v-1; 0)$-EB design with parameters $v, b, r, \boldsymbol{k}$, one obtains $|\boldsymbol{N}\boldsymbol{k}^{-\delta}\boldsymbol{N}'| = r[(b-r)/(v-1)]^{v-1}$. Thus, as in this case $b > r$ [from Remark 8.2.1 (b)], it follows again that $b \geq v$.

As a characterization of the saturated case of the inequality in Remark 8.2.2, the following theorem can be presented.

Theorem 8.2.5. *A binary equireplicate* $(0; v-1; 0)$*-EB design with parameters* $v = b$, r, $\boldsymbol{k}$ *is a BIB design.*

Proof. It follows from (8.2.1) that

$$(\boldsymbol{N}\boldsymbol{k}^{-\delta}\boldsymbol{N}')^{-1} = \frac{1}{r(1-\varepsilon_1)}\boldsymbol{I}_v - \frac{\varepsilon_1}{vr(1-\varepsilon_1)}\boldsymbol{1}_v\boldsymbol{1}_v',$$

i.e., when $b = v$ and $\boldsymbol{k} = [k_1, k_2, ..., k_b]'$,

$$\boldsymbol{k}^{\delta} = \frac{1}{r(1-\varepsilon_1)}\boldsymbol{N}'\boldsymbol{N} - \frac{\varepsilon_1}{vr(1-\varepsilon_1)}\boldsymbol{k}\boldsymbol{k}'. \tag{8.2.3}$$

Comparing the diagonal elements on both sides of (8.2.3) yields, as $n_{ij} = 0$ or 1, $k_j = [r(1-\varepsilon_1)]^{-1}k_j - \varepsilon_1[vr(1-\varepsilon_1)]^{-1}k_j^2$. This means that the design is proper, which, with Theorem 8.2.3, completes the proof. □

An interesting characterization of the case when the number of treatments equals the number of blocks is also mentioned below. Compare this with Corollary 6.1.2.

Theorem 8.2.6. *A necessary and sufficient condition for the dual of a* $(0; v-1; 0)$*-EB design with parameters* $v, b, r, \boldsymbol{k}$ *to be also a* $(0; v-1; 0)$*-EB design, is that* $v = b$. *The efficiency factor is invariant.*

Proof. This result can be obtained directly from Lemma 2.3.3 on account of Remark 2.3.1. Also refer to the original papers of Ceranka and Mejza (1978), and of Kageyama (1980). □

Finally, it may be useful to specify the estimation properties, recalled as (1) to (5) in Section 8.1, for the $(0; v-1; 0)$-EB designs. Because, as it follows from (8.2.2), for any design belonging to this class, $\boldsymbol{C}_1 = \varepsilon_1\boldsymbol{r}^{\delta}\boldsymbol{L}_1$ with $\boldsymbol{L}_1 = \boldsymbol{I}_v - n^{-1}\boldsymbol{1}_v\boldsymbol{r}'$, any set of $v-1$ $\boldsymbol{r}^{\delta}$-orthonormal vectors $\{\boldsymbol{s}_j\}$ represents a complete set of basic contrasts of the design (see Definition 3.4.1). Hence, the following properties hold for any $(0; v-1; 0)$-EB design.

(1) Each of the basic contrasts of the design, $\{\boldsymbol{c}_j'\boldsymbol{\tau} = \boldsymbol{s}_j'\boldsymbol{r}^{\delta}\boldsymbol{\tau}, j = 1, 2, ..., v-1\}$, is estimated in the intra-block analysis with the same efficiency ε_1 (< 1).

(2) The average efficiency factor of the design is given by $\varepsilon = \varepsilon_1$.

(3) The variance of the intra-block BLUE of any basic contrast of the design is given by σ_1^2/ε_1, where $\sigma_1^2 = \sigma_U^2 + \sigma_e^2$ (see Theorem 3.4.1).

(4) Any vector $\boldsymbol{s}$ representing a contrast $\boldsymbol{c}'\boldsymbol{\tau} = \boldsymbol{s}'\boldsymbol{r}^{\delta}\boldsymbol{\tau}$ satisfies the condition $\boldsymbol{L}_1\boldsymbol{s} = \boldsymbol{s}$, which implies that $\boldsymbol{c}'\boldsymbol{\tau}$ is estimated in the intra-block analysis with the efficiency ε_1, and the variance of its intra-block BLUE is given by $\varepsilon_1^{-1}\boldsymbol{s}'\boldsymbol{r}^{\delta}\boldsymbol{s}\sigma_1^2$. (Note that $\boldsymbol{c}'\boldsymbol{\tau} = \boldsymbol{s}'\boldsymbol{r}^{\delta}\boldsymbol{\tau}$ is a basic contrast if $\boldsymbol{c}'\boldsymbol{1}_v = \boldsymbol{s}'\boldsymbol{r} = 0$ and $\boldsymbol{c}'\boldsymbol{r}^{-\delta}\boldsymbol{c} = \boldsymbol{s}'\boldsymbol{r}^{\delta}\boldsymbol{s} = 1$.)

(5) More generally, for any set of contrasts $\boldsymbol{A}'\boldsymbol{S}'\boldsymbol{r}^{\delta}\boldsymbol{\tau}$, where $\boldsymbol{S} = [\boldsymbol{s}_1 : \boldsymbol{s}_2 : \cdots : \boldsymbol{s}_{v-1}]$ is such that $\boldsymbol{S}'\boldsymbol{r}^{\delta}\boldsymbol{S} = \boldsymbol{I}_{v-1}$, the covariance matrix of their intra-block BLUEs is of the form $\text{Cov}[(\widehat{\boldsymbol{A}'\boldsymbol{S}'\boldsymbol{r}^{\delta}\boldsymbol{\tau}})_{\text{intra}}] = \varepsilon_1^{-1}\boldsymbol{A}'\boldsymbol{A}\sigma_1^2$ (see Theorem 3.4.2). Alternatively, for any set of contrasts $\boldsymbol{U}'\boldsymbol{\tau}$, the covariance matrix of their intra-block BLUEs can be written as $\text{Cov}[(\widehat{\boldsymbol{U}'\boldsymbol{\tau}})_{\text{intra}}] = \boldsymbol{U}'\boldsymbol{C}_1^{-}\boldsymbol{U}\sigma_1^2 = \varepsilon_1^{-1}\boldsymbol{U}'\boldsymbol{r}^{-\delta}\boldsymbol{U}\sigma_1^2$.

Thus, in particular, an elementary contrast $\boldsymbol{c}'\boldsymbol{\tau}$ with $\boldsymbol{c} = [0, ..., 0, 1, 0, ..., 0, -1, 0, ..., 0]'$, where 1 occurs in the ith position and -1 occurs in the i'th position, gets

$$\text{Var}[(\widehat{\boldsymbol{c}'\boldsymbol{\tau}})_{\text{intra}}] = \text{Var}[(\widehat{\tau_i - \tau_{i'}})_{\text{intra}}] = \frac{1}{\varepsilon_1}\left(\frac{1}{r_i} + \frac{1}{r_{i'}}\right)\sigma_1^2,$$

which shows that although all elementary contrasts are estimated in the intra-block analysis with the same efficiency, ε_1, the variances of their intra-block BLUEs may be different, unless the design is equireplicate (see also the discussion in Section 4.1). However, if the elementary contrasts are normalized so to become basic contrasts, i.e., $\boldsymbol{c}'\boldsymbol{\tau}$ above is replaced by $\boldsymbol{c}'\boldsymbol{\tau}/\sqrt{r_i{}^{-1} + r_{i'}{}^{-1}}$, then

$$\text{Var}\left[\left(\widehat{\frac{\tau_i - \tau_{i'}}{\sqrt{r_i{}^{-1} + r_{i'}{}^{-1}}}}\right)_{\text{intra}}\right] = \frac{\sigma_1^2}{\varepsilon_1},$$

which is a constant. This point shows an advantage itself of the designs described in the following Sections 8.2.1 and 8.2.2. Further note, from the above form of $\text{Var}[(\widehat{\boldsymbol{c}'\boldsymbol{\tau}})_{\text{intra}}]$, that for a $(0; v-1; 0)$-EB design with a high efficiency factor ε_1, the variance of the intra-block BLUE of an elementary contrast is likely to become small.

8.2.1 Proper designs

Among binary proper $(0; v-1; 0)$-EB designs, there does not exist a nonequireplicate design, on account of Theorem 8.2.3. Therefore, only equireplicate designs will be considered in this section.

Practical considerations, such as, e.g., related to crop variety trials, dictate that in most of the designs used for agricultural field experiments all treatments are equally replicated, though they may be exceptions from this. A wide choice of binary block designs with various possible block sizes is open to the experimenter. Usually, blocks of equal sizes, i.e., proper designs, are used. As examples of their use for seed orchards, see Freeman (1967, 1969).

As a binary orthogonal block design, on account of Theorem 7.2.1, one can have only the RBD, which is the most widely used among all experimental designs. See again Section 7.2.

In practical experimentation, however, if the number of treatments is large and the experimental material cannot be subdivided into large blocks of sufficiently homogenous units, so to accommodate all treatments in each block, incomplete blocks are to be used. The problem of design, is to select sets of treatments to be applied to the units of different blocks. The selection is a subject of combinatorial considerations, because apart from the question of precision, it is desirable to have a design such that the statistical analysis of experimental data is as simple as possible, and such that its results can easily be interpreted. With the availability of computers, it does not seem that a simplicity in the statistical analysis is really very attractive from a practical point of view. It is, however, still an important property from the interpretational point of view. Thus, incomplete proper block designs of certain desirable patterns will be considered here.

When a binary $(0; v-1; 0)$-EB design with parameters $v, b, r, k (< v)$ is considered, it follows from (8.2.1) and Theorem 8.2.2 that $\boldsymbol{N}\boldsymbol{N}' = (r-\lambda)\boldsymbol{I}_v + \lambda \boldsymbol{1}_v\boldsymbol{1}_v'$, where $\lambda = \varepsilon_1 rk/v = r(k-1)/(v-1)$. Thus, the $v \times b$ incidence matrix $\boldsymbol{N} = [n_{ij}]$ of the design satisfies the traditional definition of a BIB design with parameters v, b, r, k, λ, as given in Definition 2.4.2. Recall that there are two basic relations $vr = bk$ and $\lambda(v-1) = r(k-1)$ satisfied by any BIB design.

The use of BIB designs in statistical planning of experiments was first initiated by Yates (1936a, 1936b). Fisher and Yates (1963) in their tables listed all BIB designs with $r \leq 10$ whose solutions were known to them, and explained their use. Though interest in these designs was greatly enhanced by their use in the statistical approach to experimentation, starting from 1936, examples of them appeared much earlier in mathematical literature, as reported by Woolhouse (1844) and Kirkman (1847).

For a BIB design with parameters v, b, r, k, λ, it follows from (8.2.2) and the pattern of $\boldsymbol{N}\boldsymbol{N}'$ that $\boldsymbol{C}_1 = (\lambda v/k)(\boldsymbol{I}_v - v^{-1}\boldsymbol{1}_v\boldsymbol{1}_v')$. When such design is used, any elementary contrast (i.e., a pairwise comparison of treatment parameters) can be estimated in the intra-block analysis with the same variance $2k\sigma_1^2/(\lambda v)$ ($> 2\sigma_1^2/r$), independently of the pair, where σ_1^2 denotes the combined error variance in that analysis [as defined in (3.4.2)]. This is due to the fact, evident from Theorem 8.2.2, that the common efficiency factor is here given by

$$\varepsilon_1 = \frac{\lambda v}{rk} = \frac{v(k-1)}{k(v-1)} = \frac{1-1/k}{1-1/v} < 1.$$

Evidently, ε_1 increases with k. [In the limit, when $k = v$, i.e., for an RBD, ε_1 becomes $\varepsilon_0 = 1$.] Furthermore, it can be shown that the most efficient binary incomplete block design for given v, b, r, k is a BIB design (if it exists). It is so, because a BIB design attains an upper bound on the average efficiency factor of a block design (see Remark 3.4.1; Section 8.1(2); Raghavarao, 1971, Section 4.5; Shah and Sinha, 1989, Section 3.3.1). In general, it is known (Kiefer, 1975) that binary designs whose C-matrix has diagonal elements all equal and also off-diagonal elements all equal (i.e., is completely symmetric in this sense) have several optimum properties (also see Fedorov, 1972; Shah and Sinha, 1989,

Chapters 2 and 3; Pukelsheim, 1993; Liski, Mandal, Shah and Sinha, 2002), and that among the class of binary proper designs for fixed v, k and n (being the total number of experimental units), a $(0; v-1; 0)$-EB design is the most efficient design in the sense of the average efficiency factor (see Caliński, Ceranka and Mejza, 1980, Lemma 3 with $\boldsymbol{X}$ replaced by $\boldsymbol{r}^{\delta}$).

Thus, under a BIB design the estimation of contrasts in the intra-block analysis becomes very simple and uniform. So if a BIB design is available for the experimenting conditions, and the researcher wants to estimate all elementary contrasts with the same precision, this design should be used.

Note that the condition (8.1.1), for a gain in precision resulting from the recovery of inter-block information over the estimation in the intra-block analysis to be uniformly positive (in the sense discussed in Section 3.8.4), is in the case of a BIB design equivalent to (6.0.2). Thus, Theorem 6.0.3 becomes interesting here.

With respect to the choice of a BIB design, even if one only asks for which values of v, b, r, k, λ such design can be constructed, this question is still far from being solved in general, although this problem of construction has a long history, going back to Woolhouse (1844). Some general results have been obtained for fixed small values of k. On the other hand, e.g., in crop variety trials, when seed is in short supply or some other economy is enforced, the trials sometimes have to be conducted in only a few replications. It is important in these circumstances to use the most efficient possible designs, like the BIB designs. However, some experience indicates that the user's needs are not always likely to comply with some preconceived opinions of what is a usual range of solutions, say $r \leq 10$ (as in standard catalogues, here in Table 8.2). In fact, recent developments in communication theory and electronic research have extended interest in incomplete block designs well beyond the conventional range of tabulation. Thus, general solutions for families of designs have some definite practical attraction.

Methods of constructing BIB designs have appeared in the literature in abundance. As mentioned in Chapter 6, the main methods are recursive construction and direct construction.

From now on, several families of BIB designs, i.e., binary proper and equireplicate $(0; v-1; 0)$-EB designs with $\varepsilon_1 = \lambda v/(rk)$, $\rho_1 = v-1$ and $\boldsymbol{L}_1 = \boldsymbol{I}_v - v^{-1}\mathbf{1}_v\mathbf{1}_v'$, will be presented along with their initial blocks (difference sets). They may be found in Bose (1939, 1942b, 1947), Hall (1986) or Sprott (1954, 1956), and can also be checked straightforwardly by use of methods of differences, as mentioned in Section 6.7 (i.e., applying Theorems 6.7.1 and 6.7.2). For a Galois field, a module, and other algebraic terms see Appendix B.

Before going further, it will be useful to recall the method of differences, as presented in Section 6.7. Consider a BIB design with 7 treatments, 7 blocks of size 3 each, each treatment having 3 replications, and such that any pair of treatments occurs together once in a block. Its blocks are given by

(0,1,3), (1,2,4), (2,3,5), (3,4,6), (4,5,0), (5,6,1), (6,0,2).

Here the cyclic procedure of construction is to get a block by adding 1 to each element in the previous block and reducing modulo 7 when necessary. Hence, the full design can be generated from one of the blocks called initial blocks. Usually as an initial block the lowest case of values (elements) is taken. Thus, in the above design as an initial block (0,1,3) is taken for the cyclic development. This example shows that the cyclic development of an initial block is among the easiest ways of constructing BIB designs.

Theorem 8.2.7 (Raghavarao, 1971, Theorem 5.8.1). *A BIB design with parameters $v = 6t + 3$, $b = (3t + 1)(2t + 1), r = 3t + 1, k = 3, \lambda = 1$ can be constructed from the $3t + 1$ initial blocks*

$$(1_1, 2t_1, 0_2), (2_1, (2t-1)_1, 0_2), ..., (t_1, (t+1)_1, 0_2),$$

$$(1_2, 2t_2, 0_3), (2_2, (2t-1)_2, 0_3), ..., (t_2, (t+1)_2, 0_3),$$

$$(1_3, 2t_3, 0_1), (2_3, (2t-1)_3, 0_1), ..., (t_3, (t+1)_3, 0_1),$$

$$(0_1, 0_2, 0_3),$$

i.e., $(0; 6t+2; 0)$*-EB with* $\varepsilon_1 = (2t+1)/(3t+1)$, $\boldsymbol{L}_1 = \boldsymbol{I}_{6t+3} - (6t+3)^{-1}\boldsymbol{1}_{6t+3}\boldsymbol{1}'_{6t+3}$.

Proof. Let $M = \{0, 1, ..., 2t \pmod{2t + 1}\}$ as an Abelian group, i.e., a module of residue classes (mod $2t + 1$). Here to every element u of M there correspond 3 treatments u_1, u_2, u_3, so that there are $6t + 3$ treatments. Consider the pairs $(1, 2t)$, $(2, 2t - 1)$, ... , $(t, t + 1)$. The two differences arising from the ith pair $(i, 2t+1-i)$ are $2t+1-2i$ and $2i-(2t+1) \equiv 2i \pmod{2t+1}$. Since $i = 1, 2, ..., t$, one can get as differences all the nonzero elements of M. This means that in the $3t + 1$ initial blocks described in the theorem the pure differences (see Section 6.7) of the type (1,1), (2,2) and (3,3) are repeated just once. Similarly, it can be shown that every element of M occurs just once among the mixed differences of all types. Hence by Theorem 6.7.1 the set of initial blocks above gives the required BIB design. In fact, the complete solution can be written down by adding to the initial blocks the elements of M, and keeping the subscripts invariant. □

Example 8.2.1. For $t = 1$ and $M = \{0, 1, 2 \pmod 3\}$, Theorem 8.2.7 provides a BIB design with parameters $v = 9, b = 12, r = 4, k = 3, \lambda = 1$, i.e.,$(0; 8; 0)$-EB with $\varepsilon_1 = 3/4, \rho_1 = 8, \boldsymbol{L}_1 = \boldsymbol{I}_9 - 9^{-1}\boldsymbol{1}_9\boldsymbol{1}'_9$. The four initial blocks are $(1_1, 2_1, 0_2)$, $(1_2, 2_2, 0_3)$, $(1_3, 2_3, 0_1)$, $(0_1, 0_2, 0_3)$. It is easy to check the symmetrically repeated property of differences holding for $\lambda = 1$, as discussed in Example 6.7.1. Thus, the conditions of Theorem 6.7.1 are satisfied, with $m = 3$, $n = 3$, $s = 4$, $k = 3$, and hence one can get the required BIB design. The complete solution of the design is given by

$$(1_1, 2_1, 0_2), (2_1, 0_1, 1_2), (0_1, 1_1, 2_2),$$

$$(1_2, 2_2, 0_3), (2_2, 0_2, 1_3), (0_2, 1_2, 2_3),$$

$$(1_3, 2_3, 0_1), (2_3, 0_3, 1_1), (0_3, 1_3, 2_1),$$

$$(0_1, 0_2, 0_3), (1_1, 1_2, 1_3), (2_1, 2_2, 2_3).$$

Theorem 8.2.8. *If $2\lambda+1$ is a prime or a prime power and $\lambda > 1$, then a BIB design with parameters $v = 2(\lambda+1)$, $b = 2(2\lambda+1), r = 2\lambda+1$, $k = \lambda+1$, λ can be constructed from the two initial blocks*

$$(0, x^0, x^2, ..., x^{2\lambda-2}), \quad (\infty, x^0, x^2, ..., x^{2\lambda-2}),$$

where x is a primitive element of Galois field GF$(2\lambda+1)$, i.e., $(0; 2\lambda+1; 0)$-EB with $\varepsilon_1 = 2\lambda/(2\lambda+1)$, $\boldsymbol{L}_1 = \boldsymbol{I}_{2(\lambda+1)} - (2\lambda+2)^{-1}\mathbf{1}_{2(\lambda+1)}\mathbf{1}'_{2(\lambda+1)}$.

Proof. It follows (see Bose, 1947) that the differences in the two initial blocks without ∞, $(0, x^0, x^2, ..., x^{2\lambda-2})$, $(x^0, x^2, ..., x^{2\lambda-2})$, are symmetrically repeated, each occurring λ times. Then apply Theorem 6.7.2 to get the required design. □

Example 8.2.2. When $\lambda = 2$, i.e., $2\lambda+1 = 5$ is a prime, $x = 2$ is a primitive element (i.e., $2^0 = 1$, $2^1 = 2$, $2^2 = 4$, $2^3 = 8 \equiv 3$, $2^4 = 16 \equiv 1$) of GF(5). By Theorem 8.2.8, two initial blocks $(0, 1, 4)$, $(\infty, 1, 4)$ produce a BIB design with parameters $v = 6$, $b = 10$, $r = 5$, $k = 3$, $\lambda = 2$ of No. 18 in Table 8.2, i.e., $(0; 5; 0)$-EB with $\varepsilon_1 = 4/5, \rho_1 = 5, \boldsymbol{L}_1 = \boldsymbol{I}_6 - 6^{-1}\mathbf{1}_6\mathbf{1}'_6$. Because (0, 1, 4) produces 6 differences as $0-1 \equiv 4$, $0-4 \equiv 1$, $1-0 = 1$, $1-4 \equiv 2$, $4-0 = 4$, $4-1 = 3$, whereas $(\infty, 1, 4)$ with ∞ deleted produces 2 differences as $1-4 \equiv 2$, $4-1 = 3$. Thus in the differences nonzero elements 1, 2, 3, 4 occur twice, i.e., the differences are symmetrically repeated, each occurring $\lambda = 2$ times. Thus, the conditions of Theorem 6.7.2 are satisfied, with $m = 5, n = 1$, $s = 1$, $t = 1$, $k = 3$, and hence one can get the required BIB design.

Theorem 8.2.9 (Sprott, 1954, Theorem 3.1; Dey, 1986, Theorem 4.17). *If $2m(2\lambda+1)+1$ is a prime or a prime power, then a BIB design with parameters $v = 2m(2\lambda+1)+1, b = mv, r = m(2\lambda+1)$, $k = 2\lambda+1$, λ can be constructed from the m initial blocks*

$$(x^i, x^{i+2m}, ..., x^{i+4\lambda m}), \quad i = 0, 1, ..., m-1,$$

where x is a primitive element of GF$(2m(2\lambda+1)+1)$, i.e., $(0; 2m(2\lambda+1); 0)$-EB with $\varepsilon_1 = \lambda[2m(2\lambda+1)+1]/[m(2\lambda+1)^2]$, $\boldsymbol{L}_1 = \boldsymbol{I}_{2m(2\lambda+1)+1} - [2m(2\lambda+1)+1]^{-1}\mathbf{1}_{2m(2\lambda+1)+1}\mathbf{1}'_{2m(2\lambda+1)+1}$.

When $m = 1$, Theorem 8.2.9 implies the following result. Also refer to Theorem 6.8.1(i).

Corollary 8.2.1 (Raghavarao, 1971, Theorem 5.7.4). *When $4\lambda+3$ is a prime or a prime power, a symmetric BIB design with parameters $v = b = 4\lambda+3$, $r = k = 2\lambda+1$, λ can be constructed from an initial block $(x^0, x^2, ..., x^{4\lambda})$, where x is a primitive element of GF$(4\lambda+3)$, i.e., $(0; 4\lambda+2; 0)$-EB with $\varepsilon_1 = \lambda(4\lambda+3)/(2\lambda+1)^2$, $\boldsymbol{L}_1 = \boldsymbol{I}_{4\lambda+3} - (4\lambda+3)^{-1}\mathbf{1}_{4\lambda+3}\mathbf{1}'_{4\lambda+3}$.*

As an example of this corollary when $\lambda = 1$, see the following remark.

Remark 8.2.3. In Corollary 8.2.1 with $\lambda = 1$, one gets a symmetric BIB design with parameters $v = b = 7, r = k = 3$, $\lambda = 1$ having an initial block (1, 2, 4) mod 7 (see No. 24 in Table 8.2). Then the complete solution is usually given by (1, 2, 4), (2, 3, 5), (3, 4, 6), (4, 5, 0), (5, 6, 1), (6, 0, 2), (0, 1, 3). On the other hand, by adding 6, 5, 4, 3, 2, 1 (mod 7) to the initial block (1, 2, 4), one gets (1, 2, 4), (0, 1, 3), (6, 0, 2), (5, 6, 1), (4, 5, 0), (3, 4, 6), (2, 3, 5), which gives a design whose incidence matrix is symmetric, as

$$\begin{bmatrix} 0 & 1 & 1 & 0 & 1 & 0 & 0 \\ 1 & 1 & 0 & 1 & 0 & 0 & 0 \\ 1 & 0 & 1 & 0 & 0 & 0 & 1 \\ 0 & 1 & 0 & 0 & 0 & 1 & 1 \\ 1 & 0 & 0 & 0 & 1 & 1 & 0 \\ 0 & 0 & 0 & 1 & 1 & 0 & 1 \\ 0 & 0 & 1 & 1 & 0 & 1 & 0 \end{bmatrix}.$$

Thus if a symmetric BIB design with parameters $v = b, r = k$, λ is generated by an initial block $(a_0, a_1, ..., a_{k-1})$ mod v, then by adding $+(v-1), +(v-2), ..., +2, +1$ in turn to the initial block, one can form a symmetric incidence matrix of the symmetric BIB design (see also Raghavarao, 1971, Theorem 5.7.3). This corresponds to the discussion following Definition 6.0.1.

Theorem 8.2.10 (Sprott, 1954, Theorem 4.1; Dey, 1986, Theorem 4.18). *If $2m(2\lambda-1)+1$ is a prime or a prime power, then a BIB design with parameters $v = 2m(2\lambda-1)+1, b = mv$, $r = 2m\lambda$, $k = 2\lambda$, λ can be constructed from the m initial blocks*

$$(0, x^i, x^{i+2m}, ..., x^{i+4(\lambda-1)m}), \quad i = 0, 1, ..., m-1,$$

where x is a primitive element of $GF(2m(2\lambda-1)+1)$, i.e., $(0; 2m(2\lambda-1); 0)$-EB with $\varepsilon_1 = \lambda[2m(2\lambda-1)+1]/(4m\lambda^2)$, $\boldsymbol{L}_1 = \boldsymbol{I}_{2m(2\lambda-1)+1} - [2m(2\lambda-1)+1]^{-1}\boldsymbol{1}_{2m(2\lambda-1)+1}\boldsymbol{1}'_{2m(2\lambda-1)+1}$.

Theorem 8.2.11 (Raghavarao, 1971, Theorem 5.7.5). *If $4t+1$ is a prime or a prime power, then there exists a BIB design with parameters $v = 4t+1, b = 8t+2$, $r = 4t, k = 2t$, $\lambda = 2t-1$, having the two initial blocks*

$$(x^0, x^2, x^4, ..., x^{4t-2}), \quad (x, x^3, x^5, ..., x^{4t-1}),$$

where x is a primitive element of $GF(4t+1)$, i.e., $(0; 4t; 0)$-EB with $\varepsilon_1 = (2t-1)(4t+1)/(8t^2)$, $\boldsymbol{L}_1 = \boldsymbol{I}_{4t+1} - (4t+1)^{-1}\boldsymbol{1}_{4t+1}\boldsymbol{1}'_{4t+1}$.

See Example 6.7.2 as an illustration of Theorem 8.2.11.

A BIB design with $\lambda = 1$ is termed a Steiner system (see Doyen and Rosa, 1980; Lindner and Rosa, 1980). In particular, a BIB design with parameters $v, b, r, k = 3$, $\lambda = 1$, called a simple triple system, can be characterized, by using the relations $vr = bk$ and $\lambda(v-1) = r(k-1)$, as

$v = 6t+3, b = (2t+1)(3t+1), r = 3t+1, k = 3, \lambda = 1,$

$v = 6t+1, b = t(6t+1), r = 3t, k = 3, \lambda = 1,$

for a positive integer t. A general solution of the former is given in Theorem 8.2.7, whereas a solution of the latter is presented in the following theorem.

Theorem 8.2.12. (Raghavarao, 1971, Theorem 5.8.2). *If $6t+1$ is a prime or a prime power, then a BIB design with parameters $v = 6t+1, b = t(6t+1)$, $r = 3t, k = 3, \lambda = 1$ can be constructed from t initial blocks $(x^i, x^{i+2t}, x^{i+4t})$, $i = 0, 1, ..., t-1$, where x is a primitive element of GF$(6t+1)$. The resulting design is $(0; 6t; 0)$-EB with $\varepsilon_1 = (6t+1)/(9t)$, $\boldsymbol{L}_1 = \boldsymbol{I}_{6t+1} - (6t+1)^{-1}\mathbf{1}_{6t+1}\mathbf{1}'_{6t+1}$.*

As discussed before, there are several series of symmetric BIB designs, for example, see Corollary 8.2.1. A series has been presented by Takeuchi (1963) through a module (Appendix B). Consider a module $M = \{xy : x, y \in \text{GF}(p)\}$, where p is a prime or a prime power, i.e., the module of residue classes [mod (p,p)] (for illustration of this notation see Example 6.7.3). Here two operations called addition and multiplication are usually defined corresponding to components in xy in M. [For the present convenience, xy is denoted by (x, y).] Further define $p+1$ sets $S_0, S_1, ..., S_p$ of the elements of M as follows:

$$S_0 = \{(0, x_j) : j = 1, 2, ..., p\}, S_i = \{(x_j, x_j x_i) : j = 1, 2, ..., p\}, i = 1, 2, ..., p,$$

where $\{x_1, x_2, ..., x_p\} = \text{GF}(p)$. Now, let the v treatments be denoted by (x, y, z), where $x, y \in \text{GF}(p)$ and z is an element of the residue class mod $(p+2)$. Thus $v = p^2(p+2)$. Then it follows that an initial block

$$\{(S_0, 1), (S_1, 2), ..., (S_p, p+1)\},$$

where $(S_i, j) = \{(x, y, j) : (x, y) \in S_i\}$, produces a symmetric BIB design with parameters $v = b = p^2(p+2)$, $r = k = p(p+1)$, $\lambda = p$, i.e., a $(0; p^2(p+2)-1; 0)$-EB design with $\varepsilon_1 = p(p+2)/(p+1)^2$, $\boldsymbol{L}_1 = \boldsymbol{I}_{p^2(p+2)} - [p^2(p+2)]^{-1}\mathbf{1}_{p^2(p+2)}\mathbf{1}'_{p^2(p+2)}$. For example, when $p = 2$, as GF$(2) = \{x_1, x_2\} = \{0, 1\}$, one gets $S_0 = \{(0,0), (0,1)\}$, $S_1 = \{(0,0), (1,0)\}$, $S_2 = \{(0,0), (1,1)\}$, which give an initial block

$$\{(001), (011), (002), (102), (003), (113)\} \text{ mod } (2, 2, 4),$$

which, by adding to this initial block (000), (001), (002), (003), (010), (011), (012), (013), (100), (101), (102), (103) [mod (2,2,4)], yields a symmetric BIB design with parameters $v = b = 16$, $r = k = 6$, $\lambda = 2$. One can find another solution as No.60 in Table 8.2. When $p = 3$, one can obtain a symmetric BIB design with parameters $v = b = 45$, $r = k = 12$, $\lambda = 3$, by use of an initial block {(001), (011), (021), (002), (102), (202), (003), (113), (223), (004), (124), (214)} mod (3,3, 5). Unfortunately, this design has $r > 10$ which is beyond Table 8.2.

BIB designs generated by an initial block (also called a difference set), i.e., cyclic designs, are listed in Baumert (1971) for $k \leq 100$. Other tables can been found in Jungnickel and Pott (1996; IV 12) and Smith (1996; IV 13). Furthermore, some indications of current research and new developments in the area of difference sets, i.e., initial blocks, can be found in a special issue of the Journal of Statistical Planning and Inference edited by Arasu and Singhi (1997).

Also by some patterned combination of incidence matrices of block designs one can produce other BIB designs. This corresponds to an idea of constructing larger BIB designs from smaller ones. These methods may have close connections with the previous methods.

Theorem 6.6.7 with $c = 1$ can yield the following theorem.

Theorem 8.2.13. *Let N be the $v \times b$ incidence matrix of a BIB design with parameters $v = 2k+1$, b, r, k, λ. Then the incidence matrix*

$$N^* = \begin{bmatrix} N & \mathbf{1}_v\mathbf{1}_b' - N \\ \mathbf{1}_b' & \mathbf{0}' \end{bmatrix}$$

yields a BIB design with parameters $v^* = v+1$, $b^* = 2b$, $r^* = b$, $k^* = k+1$, $\lambda^* = r$, i.e., $(0; 2k+1; 0)$-EB with $\varepsilon_1 = 2k/(2k+1)$, $L_1 = I_{2(k+1)} - [2(k+1)]^{-1}\mathbf{1}_{2(k+1)}\mathbf{1}_{2(k+1)}'$.

Proof. When $v = 2k+1$ in the BIB design, it holds that $b = 3r - 2\lambda$ (see Kageyama and Kuwada, 1985). Hence by Theorem 6.6.7 the proof is accomplished. □

Similarly, one can also take the pattern

$$\begin{bmatrix} \mathbf{1}_v\mathbf{1}_b' - N & N \\ \mathbf{1}_b' & \mathbf{0}' \end{bmatrix}.$$

In this case, if as the basic design a BIB design with parameters $v = 2k-1$, $b, r(= 2\lambda), k, \lambda$ is considered, then another BIB design can be obtained.

Theorem 8.2.14 (Raghavarao, 1971, Theorem 5.2.2). *Let N be the $v \times v$ incidence matrix of a symmetric BIB design with parameters $v = b = 4t^2, r = k = 2t^2 \pm t$, $\lambda = t^2 \pm t$, and let the incidence matrix be written without loss of generality as*

$$N = \begin{bmatrix} 1 & \mathbf{1}_{k-1}' & \mathbf{0}' \\ \mathbf{1}_{k-1} & N_1 & N_2 \\ \mathbf{0} & N_3 & N_4 \end{bmatrix}.$$

Then the following incidence matrix gives a symmetric BIB design with parameters $v^* = b^* = 4t^2 - 1, r^* = k^* = 2t^2 - 1$, $\lambda^* = t^2 - 1$:

$$N^* = \begin{bmatrix} N_1 & \mathbf{1}_{k-1}\mathbf{1}_{v-k}' - N_2 \\ \mathbf{1}_{v-k}\mathbf{1}_{k-1}' - N_3 & N_4 \end{bmatrix},$$

i.e., $(0; 4t^2-2; 0)$-EB with $\varepsilon_1 = (t^2-1)(4t^2-1)/(2t^2-1)^2$, $L_1 = I_{4t^2-1} - (4t^2-1)^{-1}\mathbf{1}_{4t^2-1}\mathbf{1}_{4t^2-1}'$.

Similarly, if N is the $v \times v$ incidence matrix of a symmetric BIB design with parameters $v = b = 4\lambda + 3$, $r = k = 2\lambda + 1$, λ, then

$$N^* = \begin{bmatrix} N & \mathbf{1}_v\mathbf{1}_v' - N & \mathbf{0} \\ \mathbf{1}_v' & \mathbf{0}' & 0 \\ N & N & \mathbf{1}_v \end{bmatrix}$$

is the incidence matrix of a symmetric BIB design with parameters $v^* = b^* = 8\lambda + 7$, $r^* = k^* = 4\lambda + 3$, $\lambda^* = 2\lambda + 1$ (Raghavarao, 1971, Theorem 5.2.4), i.e., $(0; 8\lambda + 6; 0)$-EB with $\varepsilon_1 = (2\lambda + 1)(8\lambda + 7)/(4\lambda + 3)^2$, $L_1 = I_{8\lambda+7} - (8\lambda + 7)^{-1}\mathbf{1}_{8\lambda+7}\mathbf{1}_{8\lambda+7}'$.

Theorem 8.2.15 (Raghavarao, 1971, Theorem 5.3.1). *If N_h, $h = 1, 2$, are the incidence matrices of BIB designs with parameters $v_h, b_h = 4(r_h - \lambda_h), r_h, k_h, \lambda_h$, then*

$$N = N_1 \otimes N_2 + (\mathbf{1}_{v_1}\mathbf{1}_{b_1}' - N_1) \otimes (\mathbf{1}_{v_2}\mathbf{1}_{b_2}' - N_2)$$

is the incidence matrix of a BIB design with parameters $v^* = v_1v_2, b^* = b_1b_2$, $r^* = r_1r_2 + (b_1 - r_1)(b_2 - r_2), k^* = k_1k_2 + (v_1 - k_1)(v_2 - k_2)$, $\lambda^* = r^* - b^*/4$, i.e., $(0; v^* - 1; 0)$-EB with $\varepsilon_1 = \lambda^*v^*/(r^*k^*)$, $L_1 = I_{v^*} - (v^*)^{-1}\mathbf{1}_{v^*}\mathbf{1}_{v^*}'$.

Theorem 8.2.16 (Raghavarao, 1971, Theorem 5.9.2). *If N is the incidence matrix of a BIB design with parameters $v = k(2k-1), b = 4k^2 - 1, r = 2k+1, k$, $\lambda = 1$, then*

$$N'N - kI_b$$

is the incidence matrix of a symmetric BIB design with parameters $v^* = b^* = 4k^2 - 1, r^* = k^* = 2k^2$, $\lambda^* = k^2$, i.e., $(0; 4k^2 - 2; 0)$-EB with $\varepsilon_1 = (4k^2 - 1)/(4k^2)$, $L_1 = I_{4k^2-1} - (4k^2 - 1)^{-1}\mathbf{1}_{4k^2-1}\mathbf{1}_{4k^2-1}'$.

The proofs of these three theorems are straightforward. The reader can also refer to the book by Raghavarao (1971, Sections 5.2, 5.3 and 5.9).

As another recursive construction, by taking a dualization $\mathcal{D}_*^{(2)}$ with respect to pairs (see Remark 6.1.2) for a symmetric BIB design with parameters $v = b$, $r = k$ and λ (≥ 2), one can obtain a BIB design with parameters $v^* = v$, $b^* = \binom{v}{2}$, $r^* = \binom{k}{2}$, $k^* = \lambda$, $\lambda^* = \binom{\lambda}{2}$ (see Vanstone, 1975; Mohan and Kageyama, 1983, Theorem 3.1).

Other combinatorial constructions can be found in some relevant books, e.g., John (1980), Hughes and Piper (1985), Beth, Jungnickel and Lenz (1985, 1999), Rasch and Herrendörfer (1986, Section 4.2), Colbourn and Mathon (1987), and Colbourn and Dinitz (1996).

Taking into account the point of view mentioned at the beginning of this section, a list of BIB designs for $v \leq 100$ and $r \leq 10$, with the reference to their solutions and their common efficiency factor ε_1, is given at the end of this chapter as Table 8.2. Hence, if one wants to use a binary proper equireplicate $(0; v-1; 0)$-EB design for some experiment within the scope of parameters $v \leq 100$ and $r \leq 10$, then a suitable design can be found among the designs given in Table 8.2. BIB designs in this table will also be frequently used in this Volume.

Several tables of BIB designs and their discussions can be found in Hall (1956, 1986), Sprott (1962), Takeuchi (1962), Sillitto (1964), Clatworthy and Lewyckyj (1968), Mullin and Stanton (1968), Raghavarao (1971, Table 5.10.1), Kageyama (1972a), Collens (1973), Di Paola, Wallis and Wallis (1973), Hanani (1975), Mathon and Rosa (1985, 1996), Rasch and Herrendörfer (1986, Tables 4.1 and 8.2), and Colbourn and Dinitz (1996; pp.3-47).

8.2.2 Nonproper designs

BIB designs and RBDs are equireplicate and have equal block sizes. In some practical situations, block designs with block sizes, that are not all equal, may be required, though such designs do not have always clear statistical justification, as indicated in Volume I. The need for blocks of different sizes in biological experiments has been advocated, e.g., by Pearce (1964). Such designs might also be useful in animal experiments for which the available material consists of litters of different sizes.

The statistical implication of nonproper block designs are discussed at several places in Chapter 3. Under the inter-block model (3.2.4) considered in Section 3.2.2, a necessary and sufficient condition for the existence of the BLUE of a contrast $\boldsymbol{c}'\boldsymbol{\tau}$, where $\boldsymbol{c} = \boldsymbol{C}_2\boldsymbol{s}$ and $\boldsymbol{C}_2 = \boldsymbol{N}\boldsymbol{k}^{-\delta}\boldsymbol{N}' - n^{-1}\boldsymbol{r}\boldsymbol{r}'$, is given in Theorem 3.2.2. Furthermore, Corollary 3.2.2 shows that to satisfy the condition of Theorem 3.2.2 by any $\boldsymbol{s}$ it is necessary and sufficient that for any nonzero $b \times 1$ vector $\boldsymbol{t}$,

$$\boldsymbol{N}_0\boldsymbol{t} = \boldsymbol{0} \text{ implies } \boldsymbol{N}_0\boldsymbol{K}_0\boldsymbol{t} = \boldsymbol{0}, \tag{8.2.4}$$

where $\boldsymbol{N}_0 = \boldsymbol{N} - n^{-1}\boldsymbol{r}\boldsymbol{k}'$ and $\boldsymbol{K}_0 = \boldsymbol{k}^{\delta} - n^{-1}\boldsymbol{k}\boldsymbol{k}'$, with $\boldsymbol{N}$ as the incidence matrix of the design. Evidently, this condition automatically holds for a proper design. That is, in case of a proper design there exists the inter-block BLUE for any contrast $\boldsymbol{c}'\boldsymbol{\tau} = \boldsymbol{s}'[k^{-1}\boldsymbol{N}\boldsymbol{N}' - (bk)^{-1}\boldsymbol{r}\boldsymbol{r}']\boldsymbol{\tau}$. In this sense, (8.2.4) is specially important for nonproper designs. Unfortunately, even $(0; v-1; 0)$-EB designs do not always satisfy (8.2.4), as will be seen later (in Example 8.2.13). However, as it follows from Remark 3.2.4(b) and the proof of Theorem 8.2.4, a $(0; v-1; 0)$-EB design with $b = v$ always satisfies (8.2.4). In this section, the condition (8.2.4) may be checked for any of the considered $(0; v-1; 0)$-EB designs.

Thus, among nonproper designs one would recommend designs which satisfy the desirable condition (8.2.4), to have the BLUE of a contrast under both submodels, the intra-block (3.2.2) and the inter-block (3.2.4). In this sense constructions of recommended designs are presented here.

8.2.2.1 Equireplication

Binary equireplicate block designs with block sizes, $k_1, k_2, ..., k_b$ (≥ 2), that are not all equal, will be considered here. As it follows from (8.2.2) and Theorem 8.2.2, such $(0; v-1; 0)$-EB designs have

$$\boldsymbol{C}_1 = \varepsilon_1 r\left(\boldsymbol{I}_v - \frac{1}{v}\boldsymbol{1}_v\boldsymbol{1}_v'\right), \text{ where } \varepsilon_1 = \frac{vr - b}{r(v-1)}.$$

They preserve the estimation properties (1) to (5) given earlier (at the beginning of Section 8.2), with the only change that now $\boldsymbol{r}^{\delta} = r\boldsymbol{I}_v$. In particular, this implies that, according to the property (5), for any set of contrasts $\boldsymbol{U}'\boldsymbol{\tau}$, the covariance matrix of their intra-block BLUEs obtainable in an equireplicate $(0; v-1; 0)$-EB design is of the form $\text{Cov}[(\widehat{\boldsymbol{U}'\boldsymbol{\tau}})_{\text{intra}}] = (\varepsilon_1 r)^{-1}\boldsymbol{U}'\boldsymbol{U}\sigma_1^2$.

Thus, when with regard to a set of contrasts $\boldsymbol{c}_i'\boldsymbol{\tau}$, $i = 1, 2, ...,$ normalized so that $\|\boldsymbol{c}_i\|^2 = \boldsymbol{c}_i'\boldsymbol{c}_i$ is constant for all i, there is an equal interest in estimating them in the intra-block analysis, either in the sense of equal precision (i.e., variance balance) or a common efficiency factor (i.e., efficiency balance) for all these contrasts, then a binary equireplicate $(0; v-1; 0)$-EB design can be recommended. (See also the definitions and discussions in Section 4.1.)

On account of Theorem 8.2.5 and Corollary 2.3.1, it is now enough to discuss the construction of $(0; v-1; 0)$-EB designs with unequal block sizes and $b > v$. Such a $(0; v-1; 0)$-EB design can be characterized, on account of (8.2.1) and Theorem 8.2.1, as follows.

Corollary 8.2.2. *An equireplicate block design with the incidence matrix* $\boldsymbol{N}$ *is a* $(0; v-1; 0)$*-EB design if and only if the off-diagonal elements of* $\boldsymbol{N}\boldsymbol{k}^{-\delta}\boldsymbol{N}'$ *are all equal (in this case equal to* $r\varepsilon_1/v$*).*

This corollary will be used here to derive methods of constructing equireplicate $(0; v-1; 0)$-EB designs. With the application of Corollary 8.2.2, most of their proofs are straightforward and hence they can be omitted.

The following theorem gives the most fundamental obvious technique, generalizing that given in Theorem 6.4.1 (juxtaposition).

Theorem 8.2.17. *If* $\boldsymbol{N}_h$, $h = 1, 2, ..., a$, *are the incidence matrices of equireplicate* $(0; v-1; 0)$*-EB designs with a common number of treatments, then their juxtaposition* $[\boldsymbol{N}_1 : \boldsymbol{N}_2 : \cdots : \boldsymbol{N}_a]$ *is the incidence matrix of an equireplicate* $(0;$ $v-1;$ $0)$*-EB design.*

Note that a slightly more general result of Theorem 8.2.17 was given by Baksalary, Dobek and Kala (1980, p.30). More concrete forms of Theorem 8.2.17 are presented in the following four corollaries. They can be obtained directly from Corollary 8.2.2.

Corollary 8.2.3. *If* $\boldsymbol{N}_h, h = 1, 2,$ *are the* $v \times b_h$ *incidence matrices of BIB designs with parameters* $v, b_h, r_h, k_h, \lambda_h$, *then the incidence matrix* $[\boldsymbol{N}_1 : \boldsymbol{N}_2]$ *yields a* $(0; v-1; 0)$*-EB design with parameters* $v^* = v, b^* = b_1 + b_2, r^* = r_1 + r_2,$ $\boldsymbol{k}^* = [k_1\boldsymbol{1}_{b_1}', k_2\boldsymbol{1}_{b_2}']'$, $\varepsilon_1^* = v(\lambda_1/k_1 + \lambda_2/k_2)/(r_1 + r_2)$, $\boldsymbol{L}_1^* = \boldsymbol{I}_v - v^{-1}\boldsymbol{1}_v\boldsymbol{1}_v'$.

Remark 8.2.4. As a special case of Corollary 8.2.3, the following observation can be obtained: If $\boldsymbol{N}$ is the $v \times b$ incidence matrix of a BIB design with parameters $v = 2k+1, b, r, k, \lambda$, then the incidence matrix $[\boldsymbol{N} : \boldsymbol{1}_v\boldsymbol{1}_v' - \boldsymbol{N}]$ yields a $(0; v-1; 0)$-EB design with parameters $v^* = v, b^* = 2b, r^* = b$, $\boldsymbol{k}^* = [k\boldsymbol{1}_b', (k+1)\boldsymbol{1}_b']'$, $\varepsilon_1^* = v[\lambda/k + (b - 2r + \lambda)/(k+1)]/b$, $\boldsymbol{L}_1^* = \boldsymbol{I}_v - v^{-1}\boldsymbol{1}_v\boldsymbol{1}_v'$. This will be used when one constructs some designs given in Table 8.3.

Corollary 8.2.4. *If $\boldsymbol{N}$ is the $v \times b$ incidence matrix of a BIB design with parameters v, b, r, k, λ, then the incidence matrix $[\boldsymbol{N} : \boldsymbol{1}_v\boldsymbol{1}_s']$ yields a $(0; v-1; 0)$-EB design with parameters $v^* = v, b^* = b + s, r^* = r + s$, $\boldsymbol{k}^* = [k\boldsymbol{1}_b', v\boldsymbol{1}_s']'$, $\varepsilon_1^* = (\lambda v/k + s)/(r + s)$, $\boldsymbol{L}_1^* = \boldsymbol{I}_v - v^{-1}\boldsymbol{1}_v\boldsymbol{1}_v'$.*

Note that designs obtainable by Corollaries 8.2.3 and 8.2.4 satisfy (8.2.4) on account of Corollary 3.2.3.

Example 8.2.3. A BIB design with parameters $v = 4, b = 6, r = 3, k = 2$, $\lambda = 1$ produces, through Corollary 8.2.4 with $s = 2$, a $(0; 3; 0)$-EB design with parameters $v^* = 4, b^* = 8, r^* = 5, \boldsymbol{k}^* = [2\boldsymbol{1}_6', 4\boldsymbol{1}_2']', \varepsilon_1^* = 4/5$, $\boldsymbol{L}_1^* = \boldsymbol{I}_4 - 4^{-1}\boldsymbol{1}_4\boldsymbol{1}_4'$, and with the incidence matrix

$$\begin{bmatrix} 1 & 0 & 1 & 0 & 1 & 0 & 1 & 1 \\ 1 & 0 & 0 & 1 & 0 & 1 & 1 & 1 \\ 0 & 1 & 1 & 0 & 0 & 1 & 1 & 1 \\ 0 & 1 & 0 & 1 & 1 & 0 & 1 & 1 \end{bmatrix}.$$

This design satisfies (8.2.4), as can be shown by taking as $\boldsymbol{t}$ the following four linearly independent vectors: $[1, 1, 0, 0, -1, -1, 1, -1]$, $[0, 0, 1, 1, -1, -1, 1, -1]$, $[3, 3, -2, -2, -1, -1, -1, 1]$ and $[1, 1, 1, 1, 1, 1, -2, -1]$.

Partially balanced incomplete block (PBIB) designs, discussed in Section 6.0.2, can also be used to produce various kinds of $(0; v-1; 0)$-EB designs.

Corollary 8.2.5. *If $\boldsymbol{N}_h, h = 1, 2, ..., a$, are the $v \times b^{(h)}$ incidence matrices of s-associate PBIB designs with the same association scheme and with parameters $v, b^{(h)}, r^{(h)}, k^{(h)}, \lambda_i^{(h)}$ for $i = 1, 2, ..., s$, then the incidence matrix $[\boldsymbol{N}_1 : \boldsymbol{N}_2 : \cdots : \boldsymbol{N}_a]$ yields a $(0; v-1; 0)$-EB design with parameters $v^* = v, b^* = \sum_{h=1}^{a} b^{(h)}$, $r^* = \sum_{h=1}^{a} r^{(h)}$, $\boldsymbol{k}^* = [k^{(1)}\boldsymbol{1}_{b^{(1)}}', k^{(2)}\boldsymbol{1}_{b^{(2)}}', ..., k^{(a)}\boldsymbol{1}_{b^{(a)}}']'$, $\varepsilon_1^* = vw/r^*$, $\boldsymbol{L}_1^* = \boldsymbol{I}_v - v^{-1}\boldsymbol{1}_v\boldsymbol{1}_v'$, provided that*

$$\sum_{h=1}^{a} \frac{\lambda_i^{(h)}}{k^{(h)}} \text{ is a constant } (= w, \text{ say}) \text{ for } i = 1, 2, ..., s.$$

Proof. It is sufficient to note that $\sum_{h=1}^{a} \lambda_i^{(h)}/k^{(h)}$ is in the juxtaposed design the weighted concurrence (see Section 2.4.1) for all pairs of treatments that are ith associates in the component PBIB designs. The equality of these weighted concurrences for all associate classes means the equality of all weighted concurrences in the juxtaposed design. On account of Theorem 8.2.1, this is necessary and sufficient for the design to be $(0; v-1; 0)$-EB, as it is an equireplicate design. It may also be noted that here the constant w is to be equal to $(n-b)/[v(v-1)]$, which allows to give another proof, by referring to (8.2.1). □

Note that if the condition in Corollary 8.2.5 is relaxed, designs with p distinct efficiency factors may be obtained, for $2 \leq p \leq s$.

Some applications of Corollary 8.2.5 have been considered by Gupta and Jones (1983) for two group divisible (GD) PBIB designs satisfying $\lambda_1^{(1)}/k^{(1)} + \lambda_1^{(2)}/k^{(2)} = \lambda_2^{(1)}/k^{(1)} + \lambda_2^{(2)}/k^{(2)}$. They have produced 100 $(0; v-1; 0)$-EB designs with $r \leq 25$. However, their Table 1 is not complete. For example, for the incidence matrix $\boldsymbol{N}_1$ of a regular GD design, R92 in Clatworthy (1973), the matrix $[\boldsymbol{N}_1 : \boldsymbol{I}_4 \otimes \mathbf{1}_6 \mathbf{1}_2']$ gives a $(0; v-1; 0)$-EB design with $v = 24, b = 80, r = 11$, $\boldsymbol{k} = [3\mathbf{1}_{72}', 6\mathbf{1}_8']$, $\varepsilon_1 = 24/33$; for the incidence matrix $\boldsymbol{N}_1$ of a semi-regular GD design, SR75 in Clatworthy (1973), the matrix $[\mathbf{1}_3' \otimes \boldsymbol{N}_1 : \boldsymbol{I}_6 \otimes$ (incidence matrix of an unreduced BIB design with $v = 5$ and $k = 2$)] gives a $(0; v-1; 0)$-EB design with $v = 30, b = 135, r = 19$, $\boldsymbol{k} = [6\mathbf{1}_{75}', 2\mathbf{1}_{60}']$, $\varepsilon_1 = 15/19$; for the incidence matrix $\boldsymbol{N}_1$ of a semi-regular GD design, SR77 in Clatworthy (1973), the matrix $[\mathbf{1}_2' \otimes \boldsymbol{N}_1 : \boldsymbol{I}_6 \otimes$ (incidence matrix of a symmetric BIB design with $v = 7$ and $k = 3$)] gives a $(0; v-1; 0)$-EB design with $v = 42, b = 140, r = 17$, $\boldsymbol{k} = [6\mathbf{1}_{98}', 3\mathbf{1}_{42}']$, $\varepsilon_1 = 14/17$. They are not listed in their Table 1. [These new designs will be quoted at the end of this subsection.] On the other hand, using triangular PBIB designs one can produce further 70 $(0; v-1; 0)$-EB designs with $r \leq 40$ (see Jones, Sinha and Kageyama, 1987). Further tables of such designs can be found in Sinha and Jones (1988), and Agarwal and Kumar (1984). On account of Corollary 3.2.3 all designs obtainable by Corollary 8.2.5 satisfy the desirable condition (8.2.4). Other application of Corollary 8.2.5 can be considered. This is left to the reader.

Table 8.3 presents equireplicate $(0; v-1; 0)$-EB designs with $r \leq 10$, two distinct block sizes $k_1, k_2 = k_1 + 1 \leq 10$ and $v, b \leq 40$. This tabulation is not exhaustive, but Table 8.2 for binary equireplicate proper $(0; v-1; 0)$-EB designs, i.e., BIB designs, is exhaustive under the given scope of parameters. In fact, Table 8.3 is prepared by use of Corollaries 8.2.3 – 8.2.5, and Remark 8.2.4.

Corollary 8.2.6. *The existence of a symmetric BIB design with parameters* $v = b, r = k, \lambda$ *implies the existence of a* $(0; v-1; 0)$*-EB design with parameters* $v^* = v, b^* = v(2v-1), r^* = k(2k-1), \boldsymbol{k}^* = [k\mathbf{1}_v', \lambda\mathbf{1}_{2v(v-1)}']'$, $\varepsilon_1^* = v[k(2k-1) - 2v + 1]/[k(2k-1)(v-1)]$, $\boldsymbol{L}_1^* = \boldsymbol{I}_v - v^{-1}\mathbf{1}_v\mathbf{1}_v'$.

Proof. A symmetric BIB design juxtaposed with its copy yields a BIB design with parameters $v' = v, b' = 2v, r' = 2k, k' = k$, $\lambda' = 2\lambda$. Form a $v' \times \binom{b'}{2}$ matrix consisting of all Schur products of b' vectors given by the columns of the incidence matrix of the new BIB design. Then consider the resulting matrix as the incidence matrix $\boldsymbol{N}^*$ of the final design. Note that it is of the form $\boldsymbol{N}^* = [\boldsymbol{N}_1 : \boldsymbol{N}_2]$, where $\boldsymbol{N}_1$ is the $v \times v$ incidence matrix of the original BIB design and $\boldsymbol{N}_2$ is the $v \times 2v(v-1)$ incidence matrix of the BIB design obtained by the remaining $\binom{2v}{2} - v$ Schur products, with parameters $v_2 = v$, $b_2 = 2v(v-1), r_2 = 2k(k-1)$, $k_2 = \lambda$, $\lambda_2 = 2k(k-1)(\lambda-1)/(v-1)$. Hence the parameters v^*, b^*, r^*, k^*, giving the efficiency factor $\varepsilon^* = (v^*r^* - b^*)/[r^*(v^*-1)]$. Thus, on account of Corollary 8.2.3, this is the incidence matrix of the required design. Here a Schur product of vectors means the componentwise product of the vectors.
□

Note again that Corollary 8.2.6 provides designs that satisfy the desirable condition (8.2.4).

Theorem 8.2.18. *The following incidence matrix gives a* $(0; v-1; 0)$*-EB design with parameters* $v = 2t+1, b = (t+1)(2t+1)$, $r = (t+1)^2$, $\boldsymbol{k} = [2t\mathbf{1}'_{t(t+1)}, 2\mathbf{1}'_{t(t+1)}, (t+1)\mathbf{1}'_{t+1}]'$, $\varepsilon_1 = (2t+1)/[2(t+1)]$, $\boldsymbol{L}_1 = \boldsymbol{I}_{2t+1} - (2t+1)^{-1}\mathbf{1}_{2t+1}\mathbf{1}'_{2t+1}$:

$$\begin{bmatrix} (\mathbf{1}_{t+1}\mathbf{1}'_{t+1} - \boldsymbol{I}_{t+1}) \otimes \mathbf{1}'_t & \boldsymbol{I}_{t+1} \otimes \mathbf{1}'_t & \mathbf{1}_{t+1}\mathbf{1}'_{t+1} \\ \mathbf{1}_t\mathbf{1}'_{t(t+1)} & \mathbf{1}'_{t+1} \otimes \boldsymbol{I}_t & \mathbf{O} \end{bmatrix}.$$

Proof. It follows from a direct application of Corollary 8.2.2. □

Example 8.2.4. When $t = 2$, Theorem 8.2.18 yields a $(0;4;0)$-EB design with parameters $v = 5, b = 15, r = 9$, $\boldsymbol{k} = [4\mathbf{1}'_6, 2\mathbf{1}'_6, 3\mathbf{1}'_3]'$, $\varepsilon_1 = 5/6$, $\boldsymbol{L}_1 = \boldsymbol{I}_5 - 5^{-1}\mathbf{1}_5\mathbf{1}'_5$, which has been given by Jones (1959). One can rearrange columns of the incidence matrix into a 3-resolvable solution, as shown below.

$$\left[\begin{array}{ccccc|ccccc|ccccc} 0&0&1&1&1&1&1&1&0&0&1&1&1&0&0 \\ 1&1&1&0&0&0&0&1&1&1&1&1&1&0&0 \\ 1&1&1&0&0&1&1&1&0&0&0&0&1&1&1 \\ 1&1&0&1&0&1&1&0&1&0&1&1&0&1&0 \\ 1&1&0&0&1&1&1&0&0&1&1&1&0&0&1 \end{array}\right].$$

The design resulting from Theorem 8.2.18 does not satisfy (8.2.4) in general.

Finally, one can consider a design constructed by trial and error.

Example 8.2.5. The following is a (4,2,2)-resolvable or 4-resolvable $(0;5;0)$-EB design with parameters $v = 6, b = 18, r = 8$, $\boldsymbol{k} = [4\mathbf{1}'_6, 2\mathbf{1}'_{12}]'$, $\varepsilon_1 = 3/4$, $\boldsymbol{L}_1 = \boldsymbol{I}_6 - 6^{-1}\mathbf{1}_6\mathbf{1}'_6$ (see Example 7.3.14):

$$\left[\begin{array}{cccccc|cccccc|cccccc} 0&0&1&1&1&1&1&1&0&0&0&0&1&1&0&0&0&0 \\ 1&1&0&0&1&1&1&0&1&0&0&0&0&0&1&1&0&0 \\ 1&1&1&1&0&0&0&1&1&0&0&0&0&0&0&0&1&1 \\ 1&1&1&1&0&0&0&0&0&1&1&0&1&0&1&0&0&0 \\ 1&1&0&0&1&1&0&0&0&1&0&1&0&1&0&0&1&0 \\ 0&0&1&1&1&1&0&0&0&0&1&1&0&0&0&1&0&1 \end{array}\right].$$

This design satisfies the condition (8.2.4), as it follows from Corollary 3.2.3.

As to Examples 8.2.4 and 8.2.5, note that there do not exist BIB designs with parameters $v = 5$ and $r = 9$, and with $v = 6$ and $r = 8$, respectively. There are many such cases of nonexistence, because BIB designs have strong restrictions on parameters, as already mentioned in Section 6.0.1. Furthermore, when $v - 1$

is a prime, as in a BIB design $r = \lambda(v-1)/(k-1)$, it follows that the replication number is at least $v-1$, i.e., $r \geq v-1$. In such cases it is often possible to find $(0; v-1; 0)$-EB designs which are smaller (in the sense of having smaller values of treatment replications) than the existing BIB designs. The following table is from Gupta and Jones (1983; Table 3), where r_1 and r_2 denote the number of treatment replications for a BIB design and a relevant $(0; v-1; 0)$-EB design, respectively.

v	8	12	14	18	20	24	30	32	42
$r_1 \geq$	7	11	13	17	19	23	29	31	41
r_2	5,6	7,10	8,9	7,8,11,15,16	16,18	11	19	9,10,14	17

It should be mentioned that in the original table of Gupta and Jones (1983), $r_2 = 14$ for $v = 18$ appears erroneously. Here three new columns, for $v = 24, 30, 42$, are added. Note that there are no BIB designs for $v = 24, 30, 42$ in available books.

In general, for given values of v and r there may exist both the BIB and another of the $(0; v-1; 0)$-EB designs. Compare such designs given in Tables 8.2 and 8.3. Even in these cases one cannot say which is better in terms of the value of the efficiency factor. That depends on block structures related to the number of blocks and block sizes, which in turn may be restricted by the experimental circumstances.

8.2.2.2 Nonequireplication

Consider now some methods of constructing nonequireplicate $(0; v-1; 0)$-EB designs. Not so many methods of construction of such designs are available (see Kageyama, 1981). The problem of construction is related here, on account of Theorem 8.2.1, to finding such incidence matrix $\boldsymbol{N} = [n_{ij}]$ for which the off-diagonal elements (i, i') of $\boldsymbol{N}\boldsymbol{k}^{-\delta}\boldsymbol{N}'$ are proportional to $r_i r_{i'}$, i.e., that the equality

$$\sum_{j=1}^{b} \frac{n_{ij} n_{i'j}}{k_j} = \frac{\varepsilon_1}{n} r_i r_{i'} \quad \text{for all } i, i' \ (i \neq i') = 1, 2, ..., v, \tag{8.2.5}$$

holds.

Note that the intermediate designs used in the constructions that follow are binary.

Example 8.2.6. As a simple example of the saturated case ($b = v$) considered in Theorem 8.2.4, the following $(0; 3; 0)$-EB design can be taken.

$$\begin{bmatrix} 1 & 0 & 0 & 0 \\ 1 & 1 & 1 & 0 \\ 1 & 1 & 0 & 1 \\ 1 & 0 & 1 & 1 \end{bmatrix} \quad \begin{array}{l} v = b = 4, \ \boldsymbol{r} = [1, 3\boldsymbol{1}_3']', \\ \boldsymbol{k} = [4, 2\boldsymbol{1}_3']', \ n = 10, \\ \varepsilon_1 = 5/6, \end{array} \qquad \boldsymbol{L}_1 = \boldsymbol{I}_4 - \frac{1}{10}\begin{bmatrix} 1 & 3 & 3 & 3 \\ 1 & 3 & 3 & 3 \\ 1 & 3 & 3 & 3 \\ 1 & 3 & 3 & 3 \end{bmatrix}.$$

For obtaining a series of nonequireplicate $(0; v-1; 0)$-EB designs with $b = v$, the following theorem can be applied.

Theorem 8.2.19. *For a positive integer s, the incidence matrix*

$$\boldsymbol{N} = \begin{bmatrix} \mathbf{1}'_s & 0 \\ \boldsymbol{I}_s & \mathbf{1}_s \end{bmatrix}$$

gives a $(0; v-1; 0)$*-EB design with parameters* $v = b = s+1, \boldsymbol{r} = [s, 2\mathbf{1}'_s]'$, $\boldsymbol{k} = [2\mathbf{1}'_s, s]', \varepsilon_1 = 3/4$, $\boldsymbol{L}_1 = \boldsymbol{I}_{s+1} - (3s)^{-1}\mathbf{1}_{s+1}\boldsymbol{r}'$.

Proof. With $\boldsymbol{r} = [s, 2\mathbf{1}'_s]'$, $\boldsymbol{r}^\delta = \text{diag}[s : 2\boldsymbol{I}_s]$ and $\boldsymbol{k}^{-\delta} = \text{diag}[2^{-1}\boldsymbol{I}_s : s^{-1}]$, the above $\boldsymbol{N}$ gives $\boldsymbol{C}_1 = (3/4)[\boldsymbol{r}^\delta - (3s)^{-1}\boldsymbol{r}\boldsymbol{r}']$, which, on account of (8.2.2), shows that the resulting design is a $(0; v-1; 0)$-EB design. □

Note that, in the light of Theorem 8.2.6, the dual of the design considered in Theorem 8.2.19 is again a $(0; v-1; 0)$-EB design. In this case, after some permutations within treatments and also within blocks, the two designs coincide. In fact, Theorem 8.2.19 has more useful usage when $s \geq 3$. Also recall that any $(0; v-1; 0)$-EB design with $b = v$ satisfies the desirable condition (8.2.4) [see the discussion below (8.2.4)].

There is an obvious and useful juxtaposition method of constructing a $(0; v-1; 0)$-EB design from designs with one efficiency factor.

Theorem 8.2.20. *If* $\boldsymbol{N}$ *is the* $v \times b$ *incidence matrix of a* $(0; v-1; 0)$*-EB design with parameters* $v, b, \boldsymbol{r}, \boldsymbol{k}, \varepsilon_1, \boldsymbol{L}_1$*, then any juxtaposition of this design,* $\mathbf{1}'_s \otimes \boldsymbol{N}$*, gives the incidence matrix of a* $(0; v-1; 0)$*-EB design with parameters* $v^* = v, b^* = sb, \boldsymbol{r}^* = s\boldsymbol{r}, \boldsymbol{k}^* = \mathbf{1}_s \otimes \boldsymbol{k}, \varepsilon_1^* = \varepsilon_1, \boldsymbol{L}_1^* = \boldsymbol{L}_1$.

Proof. Let $\boldsymbol{C}_1^*$ be the C-matrix of the design $\boldsymbol{N}^* = \mathbf{1}'_s \otimes \boldsymbol{N}$. Then $\boldsymbol{C}_1^* = s\boldsymbol{C}_1$, where $\boldsymbol{C}_1$ is the C-matrix of the design $\boldsymbol{N}$ [see (8.2.2)], and hence $\boldsymbol{C}_1^* = \varepsilon_1(\boldsymbol{r}^*)^\delta \boldsymbol{L}_1^*$, where $\boldsymbol{L}_1^* = \boldsymbol{L}_1$, as for $\boldsymbol{N}$, which completes the proof. □

Thus, once a $(0; v-1; 0)$-EB design exists, one can take its copies to produce more $(0; v-1; 0)$-EB designs. It may also be noted that if $\boldsymbol{N}$ satisfies (8.2.4), then the condition is also satisfied by $\boldsymbol{N}^* = \mathbf{1}'_s \otimes \boldsymbol{N}$.

Another standard technique is to add a new treatment to some available designs, as shown in Theorem 6.6.13. Its another proof can be obtained directly from the condition (8.2.5). To see this, note that the incidence matrix $\boldsymbol{N}^*$ of Theorem 6.6.13 satisfies (8.2.5) if and only if for some α

$$\frac{xr_1}{k_1+1} = \alpha b_1 x(xr_1 + yr_2) \tag{8.2.6}$$

and

$$\frac{x\lambda_1}{k_1+1} + y\xi r_2^2 = \alpha(xr_1 + yr_2)^2 \quad \text{with} \quad \xi = \frac{\varepsilon_1}{vr_2}. \tag{8.2.7}$$

Relations (8.2.6) and (8.2.7) yield, after some algebra, the ratio $x/y = \{r_2[vr_2\xi(k_1 +1) - k_1]\}/(r_1 - \lambda_1)$, which shows the required condition, as $\xi = (vr_2 - b_2)/[vr_2^2(v-$

1)]. The value of ε_1^* can be obtained as $\varepsilon_1^* = \alpha(vr_2y+vr_1x+xb_1) = 1-yr_2/[(k_1+1)(xr_1+yr_2)]$ from (8.2.6).

Example 8.2.7. For an unreduced BIB design with parameters $v = 5, b = 10, r = 4, k = 2$, $\lambda = 1$, and a (0;4;0)-EB design with parameters $v = 5, b = 15, r = 9$, $\boldsymbol{k} = [4\mathbf{1}_6', 2\mathbf{1}_6', 3\mathbf{1}_3']'$, $\varepsilon_1 = 5/6$ (see Example 8.2.4), Theorem 6.6.13 produces a (0;5;0)-EB design with parameters $v^* = 6, b^* = 10x + 15y$, $\boldsymbol{r}^* = [10x, (4x+9y)\mathbf{1}_5']'$, $\boldsymbol{k}^* = [3\mathbf{1}_{10x}', \mathbf{1}_y' \otimes \boldsymbol{k}']'$, $\varepsilon_1^* = 4/5$, $\boldsymbol{L}_1^* = \boldsymbol{I}_6 - (60x)^{-1}\mathbf{1}_6(\boldsymbol{r}^*)'$ for positive integers x and y such that $x/y = 3/2$. (Compare with Example 7.3.13.)

Now, returning to Theorem 6.6.13, and the comment following it, note that a series of its application can yield nonequireplicate $(0; v^*-1; 0)$-EB designs from equireplicate $(0; v-1; 0)$-EB designs. However, the resulting designs in general may not satisfy the desirable condition (8.2.4), unless $\boldsymbol{N}_2$ is the incidence matrix of a BIB design (as will be seen below).

It can be seen that any design obtained from Corollary 6.6.2 satisfies the condition (8.2.4), on account of Corollary 3.2.4.

Example 8.2.8. For a BIB design with parameters $v = 6, b = 10, r = 5, k = 3$, $\lambda = 2$ (No. 18 in Table 8.2), Corollary 6.6.2 produces a (0;6;0)-EB design with parameters $v^* = 7, b^* = 10(x+y)$, $\boldsymbol{r}^* = [10x, 5(x+y)\mathbf{1}_6']'$, $\boldsymbol{k}^* = [4\mathbf{1}_{10x}', 3\mathbf{1}_{10y}']'$, $\varepsilon_1^* = 13/16$, $\boldsymbol{L}_1^* = \boldsymbol{I}_7 - (130x)^{-1}\mathbf{1}_7(\boldsymbol{r}^*)'$ for positive integers x and y such that $x/y = 1/3$.

Remark 8.2.5. The design given in Remark 6.6.5, obtainable from Corollary 6.6.2, satisfies the desirable condition (8.2.4). For example, a symmetric BIB design with parameters $v = b = 4\lambda+3, r = k = 2\lambda+1$, λ, obtainable in accordance with Corollary 8.2.1, satisfies the above statement. The design is equireplicate and proper. Hence, on account of Theorem 8.2.3, the resulting design is a BIB design with $\lambda^* = k$.

Corollary 8.2.7. *If $\boldsymbol{N}$ is the $v \times b$ incidence matrix of a BIB design with parameters v, b, r, k, λ, then for positive integers x and y such that $x/y = 2\lambda v/k - r$ there exists a $(0; v^*-1; 0)$-EB design with parameters $v^* = v+1, b^* = vx+by$, $\boldsymbol{r}^* = [xv, (yr+x)\mathbf{1}_v']'$, $\boldsymbol{k}^* = [2\mathbf{1}_{xv}', k\mathbf{1}_{yb}']'$, $\varepsilon_1^* = 1-rk/(4\lambda v)$, $\boldsymbol{L}_1^* = \boldsymbol{I}_{v+1} - [v(2x+yr)]^{-1}\mathbf{1}_{v+1}(\boldsymbol{r}^*)'$, whose incidence matrix is given by*

$$\begin{bmatrix} \mathbf{1}_{xv}' & \mathbf{0}' \\ \mathbf{1}_x' \otimes \boldsymbol{I}_v & \mathbf{1}_y' \otimes \boldsymbol{N} \end{bmatrix}.$$

This design satisfies the condition (8.2.4), again on account of Corollary 3.2.4.

Example 8.2.9. When $2\lambda+1$ is a prime or a prime power, for a BIB design with parameters $v = 2(\lambda+1), b = 2(2\lambda+1), r = 2\lambda+1, k = \lambda+1, \lambda > 1$ (see Theorem 8.2.8), Corollary 8.2.7 produces a $(0; v^*-1; 0)$-EB design with parameters $v^* = 2\lambda+3, b^* = 2(\lambda+1)x + 2(2\lambda+1)y$, $\boldsymbol{r}^* = [2(\lambda+1)x, \{x+(2\lambda+1)y\}\mathbf{1}_v']'$, $\boldsymbol{k}^* =$

$[2\mathbf{1}'_{xv}, (\lambda+1)\mathbf{1}'_{yb}]'$, $\varepsilon_1^* = (6\lambda-1)/(8\lambda)$, $\boldsymbol{L}_1^* = \boldsymbol{I}_{2\lambda+3} - [2xv+(\lambda+1)yb]^{-1}\mathbf{1}_{2\lambda+3}(\boldsymbol{r}^*)'$ for positive integers x and y such that $x/y = 2\lambda - 1$.

Corollary 8.2.8. *If $\boldsymbol{N}$ is the $v \times b$ incidence matrix of a BIB design with parameters v, b, r, k, λ, then for positive integers x and y such that $x/y = p/(r-\lambda)$ there exists a $(0; v^*-1; 0)$-EB design with parameters $v^* = v+1, b^* = xb+yp$, $\boldsymbol{r}^* = [xb, (xr+yp)\mathbf{1}'_v]'$, $\boldsymbol{k}^* = [(k+1)\mathbf{1}'_{xb}, v\mathbf{1}'_{yp}]'$, $\varepsilon_1^* = 1-(r-\lambda)/[(k+1)(2r-\lambda)]$, $\boldsymbol{L}_1^* = \boldsymbol{I}_{v+1} - [xb+(xr+yp)v]^{-1}\mathbf{1}_{v+1}(\boldsymbol{r}^*)'$, whose incidence matrix is given by*

$$\begin{bmatrix} \mathbf{1}'_{xb} & \mathbf{0}' \\ \mathbf{1}'_x \otimes \boldsymbol{N} & \mathbf{1}'_y \otimes \mathbf{1}_v \mathbf{1}'_p \end{bmatrix}.$$

This design satisfies also the condition (8.2.4).

Example 8.2.10. Note that when $x = y = 1, p = r - \lambda\ (> 0)$. Then the existence of any BIB design with parameters v, b, r, k, λ implies the existence of a $(0; v^*-1; 0)$-EB design, satisfying (8.2.4), with parameters $v^* = v+1, b^* = b+r-\lambda$, $\boldsymbol{r}^* = [b, (2r-\lambda)\mathbf{1}'_v]'$, $\boldsymbol{k}^* = [(k+1)\mathbf{1}'_b, v\mathbf{1}'_{r-\lambda}]'$, $\varepsilon_1^* = 1-(r-\lambda)/[(k+1)(2r-\lambda)]$, $\boldsymbol{L}_1^* = \boldsymbol{I}_{v+1} - [b+(2r-\lambda)v]^{-1}\mathbf{1}_{v+1}(\boldsymbol{r}^*)'$, and with the incidence matrix given in Corollary 8.2.8 with $x = y = 1$ and $p = r - \lambda$.

As a particular example, for a BIB design with parameters $v = b = 7$, $r = k = 3$, $\lambda = 1$ (No. 24 in Table 8.2), Corollary 8.2.8 produces a $(0; 7; 0)$-EB design with parameters $v^* = 8, b^* = 7x + py$, $\boldsymbol{r}^* = [7x, (3x+py)\mathbf{1}'_7]'$, $\boldsymbol{k}^* = [4\mathbf{1}'_{7x}, 7\mathbf{1}'_{yp}]'$, $\varepsilon_1^* = 9/10$, $\boldsymbol{L}_1^* = \boldsymbol{I}_8 - (42x)^{-1}\mathbf{1}_8(\boldsymbol{r}^*)'$ for positive integers x, y and p satisfying $x/y = p/2$. This design satisfies the condition (8.2.4).

Corollary 8.2.9. *If $\boldsymbol{N}$ is the $v \times b$ incidence matrix of a BIB design with parameters v, b, r, k, λ, then for positive integers x and y such that $x/y = \lambda v^2/k - r(v-1)$ there exists a $(0; v^*-1; 0)$-EB design, satisfying (8.2.4), with parameters $v^* = v+1, b^* = xv+yb$, $\boldsymbol{r}^* = [xv, (x(v-1)+yr)\mathbf{1}'_v]'$, $\boldsymbol{k}^* = [v\mathbf{1}'_{xv}, k\mathbf{1}'_{yb}]'$, $\varepsilon_1^* = 1-k/[v^2(2k-v)]$, $\boldsymbol{L}_1^* = \boldsymbol{I}_{v+1} - [v(xv+yr)]^{-1}\mathbf{1}_{v+1}(\boldsymbol{r}^*)'$, whose incidence matrix is given by*

$$\begin{bmatrix} \mathbf{1}'_{xv} & \mathbf{0}' \\ \mathbf{1}'_x \otimes (\mathbf{1}_v\mathbf{1}'_v - \boldsymbol{I}_v) & \mathbf{1}'_y \otimes \boldsymbol{N} \end{bmatrix}.$$

Example 8.2.11. There exists a symmetric BIB design with parameters $v = b = 4\lambda-1, r = k = 2\lambda, \lambda$, if $4\lambda - 1$ is a prime or a prime power (from Theorem 8.2.10 with $m = 1$). Then Corollary 8.2.9 produces a $(0; 4\lambda-1; 0)$-EB design with parameters $v^* = 4\lambda, b^* = (4\lambda-1)(x+y)$, $\boldsymbol{r}^* = [(4\lambda-1)x, \{2(2\lambda-1)x+2\lambda y\}\mathbf{1}'_v]'$, $\boldsymbol{k}^* = [(4\lambda-1)\mathbf{1}'_{xv}, 2\lambda\mathbf{1}'_{yb}]'$, $\varepsilon_1^* = 1 - 2\lambda/(4\lambda-1)^2$, $\boldsymbol{L}_1^* = \boldsymbol{I}_{4\lambda} - \{x[(4\lambda-1)v + 4\lambda b]\}^{-1}\mathbf{1}_{4\lambda}(\boldsymbol{r}^*)'$ for any positive integers x and y such that $x/y = 1/2$.

Corollary 8.2.10. *If $\boldsymbol{N}$ is the $v \times b$ incidence matrix of a BIB design with parameters v, b, r, k, λ, then for positive integers x and y such that $x/y = (v^2 - 2v - k)/[(v-1)(r-\lambda)]$ there exists a $(0; v^*-1; 0)$-EB design, satisfying (8.2.4), with parameters $v^* = v+1, b^* = xb+yv$, $\boldsymbol{r}^* = [xb, (xr+y(v-1))\mathbf{1}'_v]'$, $\boldsymbol{k}^*$*

$= [(k+1)\mathbf{1}'_{xb}, (v-1)\mathbf{1}'_{yv}]'$, $\varepsilon_1^* = 1-(v-1)^2(r-\lambda)/[vr(k+1)(2v-k-3)]$, $\boldsymbol{L}_1^* = \boldsymbol{I}_{v+1} - [(k+1)xb+(v-1)yv]^{-1}\mathbf{1}_{v+1}(\boldsymbol{r}^*)'$, *whose incidence matrix is given by*

$$\begin{bmatrix} \mathbf{1}'_{xb} & \mathbf{0}' \\ \mathbf{1}'_x \otimes \boldsymbol{N} & \mathbf{1}'_y \otimes (\mathbf{1}_v\mathbf{1}'_v - \boldsymbol{I}_v) \end{bmatrix}.$$

Example 8.2.12. For a BIB design with parameters $v=9, b=12, r=4, k=3$, $\lambda=1$ (No. 34 in Table 8.2), Corollary 8.2.10 produces a (0; 9; 0)-EB design with parameters $v^*=10, b^*=3(4x+3y)$, $\boldsymbol{r}^* = [12x, 4(x+2y)\mathbf{1}'_9]'$, $\boldsymbol{k}^* = [4\mathbf{1}'_{12x}, 8\mathbf{1}'_{9y}]'$, $\varepsilon_1^* = 8/9$, $\boldsymbol{L}_1^* = \boldsymbol{I}_{10} - (192y)^{-1}\mathbf{1}_{10}(\boldsymbol{r}^*)'$ for positive integers x and y such that $2x = 5y$. The resulting design satisfies (8.2.4).

Example 8.2.13. A different example is given here as a counter-example with regard to the condition (8.2.4). The following matrix

$$\boldsymbol{N}^* = \begin{bmatrix} \mathbf{1}'_6 & \mathbf{1}'_6 \\ \boldsymbol{I}_6 & \mathbf{1}_6\mathbf{1}'_6 - \boldsymbol{I}_6 \end{bmatrix}$$

is the incidence matrix of a (0; 6; 0)-EB design with parameters $v^*=7$, $b^*=12$, $\boldsymbol{r}^* = [12, 6\mathbf{1}'_6]'$, $\boldsymbol{k}^* = [2\mathbf{1}'_6, 6\mathbf{1}'_6]'$, $\varepsilon_1^* = 8/9$, $\boldsymbol{L}_1^* = \boldsymbol{I}_7 - (48)^{-1}\mathbf{1}_7(\boldsymbol{r}^*)'$. However, this does not satisfy (8.2.4). In fact, if $\boldsymbol{t} = [1, 1, 1, -1, -1, -1, 1, 1, 1, -1, -1, -1]'$, then $\boldsymbol{N}_0^*\boldsymbol{t} = \mathbf{0}$ but $\boldsymbol{N}_0^*\boldsymbol{K}_0^*\boldsymbol{t} \neq \mathbf{0}$.

As a more simple pattern, the following result applies.

Corollary 8.2.11. *For any positive integer s there exists a* $(0; v^*-1; 0)$*-EB design, satisfying* (8.2.4), *with parameters* $v^* = v+1, b^* = s(v+1)$, $\boldsymbol{r}^* = [sv, 2s\mathbf{1}'_v]'$, $\boldsymbol{k}^* = [2\mathbf{1}'_{sv}, v\mathbf{1}'_s]'$, $\varepsilon_1^* = 3/4$, $\boldsymbol{L}_1^* = \boldsymbol{I}_{v+1} - (3sv)^{-1}\mathbf{1}_{v+1}(\boldsymbol{r}^*)'$, *whose incidence matrix is given by*

$$\begin{bmatrix} \mathbf{1}'_{sv} & \mathbf{0}' \\ \mathbf{1}'_s \otimes \boldsymbol{I}_v & \mathbf{1}_v\mathbf{1}'_s \end{bmatrix}.$$

Note that when $s=1$ Corollary 8.2.11 coincides with Theorem 8.2.19.

The designs following from Theorem 6.6.13 and its corollaries look evidently suitable for testing several new scarcely-available treatments and a single wide-available control treatment.

8.2.3 Illustration

In the present Section 8.2, numerous procedures of constructing designs with a single efficiency factor have been presented and discussed. They make the choice of construction of a suitable design quite easy, if one is searching for a design allowing all contrasts of treatment parameters to be estimated in the intra-block analysis with equal efficiency. Among such designs, the BIB designs are most common. Here, a real example of an experiment in a BIB design will be presented, drawing attention to its construction and to the analysis of the experimental data, worked in detail. To save space, a suitable part of the data from Example 7.3.22 will be used for this purpose. In addition, to illustrate an experiment based on a nonproper design with a single efficiency factor, i.e., such which does not belong to the class of BIB designs, a second example will be discussed shortly. It will refer to an experiment already presented in Chapter 3.

Example 8.2.14. It may be interesting to return to the experiment considered in Example 7.3.22 discarding, however, the last two treatments there, i.e., the two standard varieties. Without them, the design is reduced to a BIB design, represented by the remaining 25×30 incidence matrix $\boldsymbol{N}$. Its construction has been described in Example 7.3.22. As noted there, the design is a resolvable block design, but this has been ignored in the analysis in that example. The same will be done here, leaving the exploration of this property to further analysis in Chapter 9. So, here the BIB design described by the incidence matrix $\boldsymbol{N}$ from Example 7.3.22 will be considered to illustrate the analysis of experimental data obtained under such design. As far as possible, the analysis will follow the same pattern as in the previous example, however, with some differences that result from the reduction of the experiment (going from the supplemented to the basic design, in the terminology of Section 6.3).

First note that while the design considered in Example 7.3.22 belongs to the class of $(\rho_0; \rho_1; 0)$-EB designs, with $\rho_0 > 0$, the design considered in the present example belongs to the class of $(0; v-1; 0)$-EB designs, because any BIB design belongs to this class (as already noted in Section 4.4.2; see also Theorems 8.2.3 and 8.2.5). More precisely, the design considered here is $(0; v-1; 0)$-EB with parameters $v = 25, b = 30, r = 6, k = 5, \lambda = 1$, $\varepsilon_1 = \lambda v/(rk) = v(k-1)/[k(v-1)] = 5/6$ $(= 0.8333)$, $\rho_1 = v - 1 = 24$, $\boldsymbol{L}_1 = \boldsymbol{I}_v - v^{-1}\boldsymbol{1}_v\boldsymbol{1}_v'$ (with $v = 25$). The total number of units (plots) of the design is $n = bk = vr = 150$, of 60 less than in the supplemented design considered in Example 7.3.22.

The same trait, the average diameter of capitulum (head) of sunflower plants, has been taken for the analysis. Thus, the data to be analyzed are those presented in Table 7.1, except observations conserning strains 26th and 27th.

The intra-block analysis is now based on the matrix

$$\boldsymbol{C}_1 = r\boldsymbol{I}_v - k^{-1}\boldsymbol{N}\boldsymbol{N}' = \varepsilon_1 r\boldsymbol{L}_1 = 5[\boldsymbol{I}_{25} - (25)^{-1}\boldsymbol{1}_{25}\boldsymbol{1}_{25}']$$

(see Section 6.0.1), a possible g-inverse of wich is the matrix $(1/5)\boldsymbol{I}_{25}$, and on the vector $\boldsymbol{Q}_1 = \boldsymbol{\Delta y} - k^{-1}\boldsymbol{N}\boldsymbol{D}\boldsymbol{y}$, which here results as

$$\begin{aligned}\boldsymbol{Q}_1 = \; &[0.160, \;\; 0.640, \; -2.980, \;\; 0.440, \;\; 4.660,\\ &-0.260, \; -1.280, \;\; 1.880, \;\; 2.740, \; -2.660,\\ &3.060, \; -0.060, \; -0.060, \;\; 0.860, \;\; 1.220,\\ &0.720, \; -2.020, \;\; 4.360, \;\; 0.240, \;\; 8.920,\\ &-4.260, \; -1.160, \; -4.780, \; -4.460, \; -5.920]'.\end{aligned}$$

Applying the formulae given in Section 3.2.1, one obtains the intra-block analysis of variance as

Source	Degrees of freedom	Sum of squares	Mean square	F
Treatments	$v-1=24$	52.632	2.193	2.692
Residuals	$n-b-v+1=96$	78.200	$s_1^2=0.814583$	
Total	$n-b=120$	130.832		

The P value corresponding to $F = 2.692$ is 0.000343, which certainly leads to the rejection of the hypothesis H_{01} : $\mathrm{E}(\boldsymbol{y}_1) = \boldsymbol{0}$ (see Section 3.2.1), equivalent to $\boldsymbol{L}_1\boldsymbol{\tau} = \boldsymbol{0}$, i.e., to $\tau_1 = \tau_2 = \cdots = \tau_{25}$.

The inter-block analysis is based on the matrix

$$\boldsymbol{C}_2 = r(\boldsymbol{I}_v - v^{-1}\boldsymbol{1}_v\boldsymbol{1}_v') - \boldsymbol{C}_1 = (1-\varepsilon_1)r\boldsymbol{L}_1 = \boldsymbol{I}_{25} - (25)^{-1}\boldsymbol{1}_{25}\boldsymbol{1}_{25}',$$

as $(1-\varepsilon_1)r = 1$, a possible g-inverse of wich is simply the matrix $\boldsymbol{I}_{25}$, and on the vector $\boldsymbol{Q}_2 = k^{-1}\boldsymbol{N}\boldsymbol{D}\boldsymbol{y} - v^{-1}\boldsymbol{1}_v\boldsymbol{1}_v'\boldsymbol{y}$, which here gives

$$\begin{aligned}\boldsymbol{Q}_2 = [&-1.400, & 0.720, & \;\; 0.340, & -0.680, & -0.400,\\ & \;\; 2.120, & 0.140, & -0.820, & 0.020, & \;\; 1.720,\\ & \;\; 2.700, & -0.780, & -1.580, & -1.000, & -1.960,\\ & \;\; 0.440, & 0.180, & \;\; 1.400, & -1.680, & \;\; 4.240,\\ & \;\; 1.620, & -1.780, & -1.160, & -0.180, & -2.220]'.\end{aligned}$$

Using, with these results, the formulae given in Section 3.2.2, one obtains the inter-block analysis of variance as

Source	Degrees of freedom	Sum of squares	Mean square	F
Treatments	$v-1=24$	61.683	2.570	0.018
Residuals	$b-v=\ 5$	708.820	$s_2^2=141.764$	
Total	$b-1=29$	770.503		

This result does not lead to the rejection of the hypothesis H_{02} : $\mathrm{E}(\boldsymbol{y}_2) = \boldsymbol{0}$ (see Section 3.2.2).

Clearly, the results of testing the two hypotheses, H_{01} and H_{02} (which in the present design coincide, both being equivalent to $H_0 : \tau_1 = \tau_2 = \cdots = \tau_{25}$), are here similar to those in Example 7.3.22, though the rejection of H_{01} is now based on a smaller value of F. To see which contrasts are responsible for the rejection, it will be now appropriate to show results corresponding to those reported in Table 7.2.

In the present example, of interest are the basic contrasts denoted in Example 7.3.22 as $\{\boldsymbol{c}_{1j}'\boldsymbol{\tau} = \boldsymbol{s}_{1j}'\boldsymbol{r}^{\delta}\boldsymbol{\tau},\ j = 1, 2, ..., 24\}$. Because the design used here has only one distinct efficiency factor, common for all basic contrasts, there is no need to use the subscript β ($\beta = 1$). Therefore, the basic contrasts of interest are written here as $\{\boldsymbol{c}_i'\boldsymbol{\tau} = \boldsymbol{s}_i'\boldsymbol{r}^{\delta}\boldsymbol{\tau},\ i = 1, 2, ..., 24\}$. Their intra-block and inter-block estimates together with corresponding estimated variances and F test results are given in Table 8.1.

Table 8.1
Intra-block and inter-block estimates of the basic contrasts, together with their estimated variances and the relevant F test results, obtained in Example 8.2.14

Basic	Intra-block			Inter-block		
contrast	estimate	F value	P value	estimate	F value	P value
$c_1'\tau$	0.17	0.028	0.8668	3.67	0.016	0.9047
$c_2'\tau$	1.18	1.436	0.2338	−1.77	0.004	0.9540
$c_3'\tau$	−0.82	0.685	0.4100	0.42	0.000	0.9892
$c_4'\tau$	2.23	5.099	0.0262	−0.32	0.000	0.9917
$c_5'\tau$	−0.35	0.128	0.7216	−3.43	0.014	0.9110
$c_6'\tau$	0.30	0.091	0.7638	1.45	0.002	0.9621
$c_7'\tau$	1.51	2.329	0.1303	−3.75	0.017	0.9028
$c_8'\tau$	−1.50	2.311	0.1318	2.97	0.010	0.9229
$c_9'\tau$	−1.08	1.195	0.2771	−6.03	0.043	0.8444
$c_{10}'\tau$	0.32	0.104	0.7479	1.00	0.001	0.9739
$c_{11}'\tau$	−0.54	0.297	0.5870	−5.51	0.036	0.8575
$c_{12}'\tau$	0.12	0.014	0.9050	−3.93	0.018	0.8980
$c_{13}'\tau$	−0.95	0.922	0.3395	−0.45	0.000	0.9883
$c_{14}'\tau$	−1.43	2.084	0.1521	−5.33	0.033	0.8620
$c_{15}'\tau$	1.45	2.137	0.1471	−1.10	0.001	0.9713
$c_{16}'\tau$	3.55	12.871	0.0005	9.10	0.097	0.7675
$c_{17}'\tau$	1.07	1.180	0.2801	−5.89	0.041	0.8479
$c_{18}'\tau$	0.11	0.013	0.9110	1.70	0.003	0.9558
$c_{19}'\tau$	−0.94	0.896	0.3463	−1.45	0.002	0.9624
$c_{20}'\tau$	−0.99	0.999	0.3201	−4.04	0.019	0.8952
$c_{21}'\tau$	−0.39	0.153	0.6961	3.56	0.015	0.9075
$c_{22}'\tau$	1.12	1.273	0.2621	5.58	0.037	0.8559
$c_{23}'\tau$	1.52	2.372	0.1268	0.11	0.000	0.9971
$c_{24}'\tau$	−5.04	25.997	0.0000	−4.56	0.024	0.8820
Estimated variance	0.978			850.584		

Examining it, first note that

$$\varepsilon_1 \sum_{i=1}^{24} (\widehat{c_i'\tau})_1^2 = 52.6318,$$

equal to the intra-block treatment sum of squares, and that

$$(1-\varepsilon_1)\sum_{i=1}^{24}(\widehat{c_i'\tau})_2^2 = 61.6832,$$

equal to the inter-block treatment sum of squares. These equalities, in turn, imply that the average value of the intra-block individual F statistics is equal to the value of F obtained in the intra-block analysis of vatiance, and the average value of the inter-block individual F statistics is equal to that obtained in the inter-block analysis (the reader may easily check these equalities). Next, looking at the details of Table 8.1, it can be concluded that only the contrasts denoted by $i = 4, 16, 24$ can be considered responsible for rejecting the hypothesis H_{01}, and so the overall hypothesis H_0. Referring to the definitions of the analyzed basic contrasts, as given in Example 7.3.22, note that this conclusion means that there is a significant difference between the first four strains (labeled 1, 2, 3 and 4) on one side and the strain 5th on the other, and that, as in Example 7.3.22, the contrasts 16 and 24 differ significantly from zero. Also, it may be noted that the results of the inter-block individual F tests confirm that the hypothesis H_{02} cannot be rejected in the inter-block analysis of variance. Again, one has to note that the inter-block F test of this hypothesis is rather poor, as based on not more than 5 d.f. for residuals.

Nevertheless, it may be interesting to obtain for all of the 24 considered basic contrasts their combined estimates and tests, based on the information provided by both of the strata, intra-block and inter-block. Using the same approach as in Example 7.3.22, first the classic Yates method has been employed, with formulae (3.9.1) and (3.9.2), to obtain the preliminary estimates of the stratum variances, $\hat{\sigma}_{1(Y)}^2 \equiv s_1^2 = 0.814585$ and $\hat{\sigma}_{2(Y)}^2 = 29.9155$. Then, the iterative procedure described in Section 3.8.2 has provided the following results:

Iteration cycle	1	2	3
d_1'	96.12999	96.14761	96.14751
d_2'	28.87001	28.85239	28.85249
$\hat{w}_1$	0.99458	0.99385	0.99385
$\hat{w}_2$	0.00542	0.00615	0.00615
$\hat{\sigma}_1^2 \equiv \hat{\sigma}_{1(N)}^2$	0.81356	0.81343	0.81344
$\hat{\sigma}_2^2 \equiv \hat{\sigma}_{2(N)}^2$	26.29238	26.30586	26.30579

Coinciding results have been obtained from the REML estimation procedure provided by GENSTAT 5 (1996) software, reaching convergence at the 8th iteration cycle (with $\hat{\sigma}_1^2 = 0.81343$ and $\hat{\sigma}_2^2 = 26.30578$).

Now, applying the formula (3.8.79), one obtains

$$\hat{\zeta} = \frac{2\hat{w}_1\hat{w}_2}{b-v+(v-1)\hat{w}_1^2-(n-v)^{-1}[b-v+(v-1)\hat{w}_1]^2} = 0.0005542,$$

with which it can be seen that the condition (3.8.84) is satisfied, as $\hat{w}_2/\hat{w}_1 = 0.0061881$. This implies that the estimated approximate variance (3.8.77) of the empirical combined estimator (3.8.96) will be smaller than the estimated variance of the intra-block BLUE for any of the contrasts considered. In fact, using (3.8.77), one finds that

$$\widetilde{\mathrm{Var}}(\widetilde{c_i'\boldsymbol{\tau}}) \cong \varepsilon_1^{-1}\hat{w}_1(1+\hat{\zeta})\hat{\sigma}_1^2 = 1.19328\hat{\sigma}_1^2 \ (= 0.97066),$$

which is slightly smaller than

$$\widehat{\mathrm{Var}}[(\widehat{c_i'\boldsymbol{\tau}})_1] = \varepsilon_1^{-1}\hat{\sigma}_1^2 = 1.20000\hat{\sigma}_1^2 \ (= 0.97750, \ \text{ if } \hat{\sigma}_1^2 = s_1^2),$$

obtained for the intra-block BLUE (see Table 8.1). Furthermore, proceeding as in Example 7.3.22, the following results have been obtained:

$$\begin{array}{rlrlrl}
\widetilde{c_1'\boldsymbol{\tau}} = & 0.188, & F_1 = & 0.036, & P = & 0.8493,\\
\widetilde{c_2'\boldsymbol{\tau}} = & 1.167, & F_2 = & 1.400, & P = & 0.2396,\\
\widetilde{c_3'\boldsymbol{\tau}} = & -0.811, & F_3 = & 0.676, & P = & 0.4130,\\
\widetilde{c_4'\boldsymbol{\tau}} = & 2.217, & F_4 = & 5.056, & P = & 0.0268,\\
\widetilde{c_5'\boldsymbol{\tau}} = & -0.372, & F_5 = & 0.143, & P = & 0.7066,\\
\widetilde{c_6'\boldsymbol{\tau}} = & 0.305, & F_6 = & 0.096, & P = & 0.7577,\\
\widetilde{c_7'\boldsymbol{\tau}} = & 1.477, & F_7 = & 2.243, & P = & 0.1375,\\
\widetilde{c_8'\boldsymbol{\tau}} = & -1.475, & F_8 = & 2.240, & P = & 0.1378,\\
\widetilde{c_9'\boldsymbol{\tau}} = & -1.111, & F_9 = & 1.270, & P = & 0.2625,\\
\widetilde{c_{10}'\boldsymbol{\tau}} = & 0.323, & F_{10} = & 0.107, & P = & 0.7440,\\
\widetilde{c_{11}'\boldsymbol{\tau}} = & -0.569, & F_{11} = & 0.334, & P = & 0.5649,\\
\widetilde{c_{12}'\boldsymbol{\tau}} = & 0.093, & F_{12} = & 0.009, & P = & 0.9247,\\
\widetilde{c_{13}'\boldsymbol{\tau}} = & -0.946, & F_{13} = & 0.921, & P = & 0.3397,\\
\widetilde{c_{14}'\boldsymbol{\tau}} = & -1.451, & F_{14} = & 2.167, & P = & 0.1443,\\
\widetilde{c_{15}'\boldsymbol{\tau}} = & 1.430, & F_{15} = & 2.102, & P = & 0.1503,\\
\widetilde{c_{16}'\boldsymbol{\tau}} = & 3.581, & F_{16} = & 13.194, & P = & 0.0005,\\
\widetilde{c_{17}'\boldsymbol{\tau}} = & 1.031, & F_{17} = & 1.094, & P = & 0.2983,\\
\widetilde{c_{18}'\boldsymbol{\tau}} = & 0.121, & F_{18} = & 0.015, & P = & 0.9029,\\
\widetilde{c_{19}'\boldsymbol{\tau}} = & -0.939, & F_{19} = & 0.907, & P = & 0.3434,\\
\widetilde{c_{20}'\boldsymbol{\tau}} = & -1.007, & F_{20} = & 1.043, & P = & 0.3097,\\
\widetilde{c_{21}'\boldsymbol{\tau}} = & -0.363, & F_{21} = & 0.136, & P = & 0.7135,\\
\widetilde{c_{22}'\boldsymbol{\tau}} = & 1.143, & F_{22} = & 1.344, & P = & 0.2493,\\
\widetilde{c_{23}'\boldsymbol{\tau}} = & 1.514, & F_{23} = & 2.358, & P = & 0.1279,\\
\widetilde{c_{24}'\boldsymbol{\tau}} = & -5.038, & F_{23} = & 26.112, & P = & 0.0000.
\end{array}$$

As it is known from Section 3.8.5, when the hypothesis $H_{0,i} : \mathbf{c}_i'\boldsymbol{\tau} = 0$ is true, the distribution of the applied statistic F_i can be approximated by the central F distribution with $d_{(i)}$ and d_1 d.f., where $d_{(i)}$ is as given in (3.8.10) and $d_1 = n - b - v + 1 = 96$, the former being estimated by

$$\hat{d}_{(i)} = \frac{2}{2 + \dfrac{3\hat{\zeta}(\hat{w}_2 - 3\hat{w}_1\hat{\zeta})^2}{(1+\hat{\zeta})^2\hat{w}_1\hat{w}_2}} = 0.999997 \quad \text{for } i = 1, 2, ..., 24.$$

The relevant P values, for each of the obtained F value, are as given above. They are below 0.05 for the basic contrasts denoted by $i = 4, 16, 24$, exactly as in the intra-block analysis. Also, note that the statistic

$$F = (24)^{-1}\sum_{i=1}^{24} F_i = 2.7084,$$

can be used to test the intersection hypothesis

$$H_0 = \bigcap_{i=1}^{24} H_{0,i} : \boldsymbol{U}'\boldsymbol{\tau} = \mathbf{0}, \quad \text{where} \quad \boldsymbol{U} = [\mathbf{c}_1 : \mathbf{c}_2 : \cdots : \mathbf{c}_{24}],$$

which is equivalent to the overall null hypothesis $H_0 : \tau_1 = \tau_2 = \cdots = \tau_{25}$. If it is true, the distribution of the above statistic F can be approximated by the central F distribution with d and d_1 d.f., where (using the same formula as in Example 7.3.22, with $\rho_1 = v - 1$)

$$d = \frac{2(v-1)}{2 + \dfrac{\zeta(w_2 - 3w_1\zeta)^2}{(1+\zeta)^2 w_1 w_2}(v+1)}.$$

Replacing in this formula w_1, w_2 and ζ by their estimates, one obtains $\hat{d} =$ 23.9994. With these d.f., $\hat{d}$ and $d_1 = 96$, the P value relevant to $F = 2.7084$ is equal to 0.000316, which leads to the rejection of the hypothesis H_0 at the significance level almost exactly the same as in the intra-block F test of the hypothesis H_{01}, which actually is equivalent to H_0 (as already indicated).

Finally, it may be desirable to complete the analysis by giving the empirical estimator of $\boldsymbol{\tau}$ and its covariance matrix. This can be done similarly as in Example 7.3.22, applying the formula (3.8.52), which here reduces to

$$\tilde{\boldsymbol{\tau}} = r^{-1}\sum_{i=1}^{v-1} \mathbf{c}_i(\widetilde{\mathbf{c}_i'\boldsymbol{\tau}}) + n^{-1}\mathbf{1}_v\mathbf{1}_n'\boldsymbol{y},$$

where $r = 6$, $v = 25$, $n = 150$ and $\widetilde{\mathbf{c}_j'\boldsymbol{\tau}}$, $j = 1, 2, ..., 24$, are the empirical combined estimators of the chosen basic contrasts. The result is

$$\tilde{\tau} = [15.613,\ 15.722,\ 15.000,\ 15.673,\ 16.514,\\ 15.551,\ 15.336,\ 15.959,\ 16.135,\ 15.072,\\ 16.215,\ 15.573,\ 15.568,\ 15.755,\ 15.820,\\ 15.736,\ 15.190,\ 16.465,\ 15.627,\ 17.389,\\ 14.753,\ 15.348,\ 14.633,\ 14.702,\ 14.400]'.$$

The covariance matrix of $\tilde{\tau}$ (considered in Theorem 3.8.2) can be given in its approximate form as

$$\mathrm{Cov}(\tilde{\tau}) \cong \sigma_1^2 \frac{w_1}{\varepsilon_1}(1+\zeta)\boldsymbol{H}_1 + \sigma_3^2 n^{-1}\mathbf{1}_v\mathbf{1}_v',$$

obtainable as in Example 7.3.22, with

$$\boldsymbol{H}_1 = r^{-2}\sum_{i=1}^{v-1}\boldsymbol{c}_i\boldsymbol{c}_i' = r^{-1}(\boldsymbol{I}_v - v^{-1}\mathbf{1}_v\mathbf{1}_v') = 6^{-1}[\boldsymbol{I}_{25} - (25)^{-1}\mathbf{1}_{25}\mathbf{1}_{25}'].$$

Hence, for any contrast $\boldsymbol{c}'\boldsymbol{\tau}$, the variance of its empirical estimator $\widetilde{\boldsymbol{c}'\boldsymbol{\tau}} = \boldsymbol{c}'\tilde{\boldsymbol{\tau}}$ can be approximated as

$$\mathrm{Var}(\widetilde{\boldsymbol{c}'\boldsymbol{\tau}}) = \boldsymbol{c}'\mathrm{Cov}(\tilde{\boldsymbol{\tau}})\boldsymbol{c} \cong (\varepsilon_1 r)^{-1} w_1(1+\zeta)\boldsymbol{c}'\boldsymbol{c}\sigma_1^2.$$

To estimate it, one has to replace the unknown values of w_1 and ζ by their estimates, given above. For example, for any elementary contrast, $\tau_i - \tau_{i'}$ $(i, i' = 1, 2, ..., 25)$, the estimate of this variance will be

$$\widehat{\mathrm{Var}}(\widetilde{\tau_i - \tau_{i'}}) \cong 2(\varepsilon_1 r)^{-1}\hat{w}_1(1+\hat{\zeta})\hat{\sigma}_1^2 = 0.32355.$$

When ignoring the correction factor $1 + \hat{\zeta} = 1.0005542$, the result becomes 0.32337, exactly as obtainable from the GENSTAT 5 (1996) software. It may also be noted that if confining attention to the intra-block analysis, then the estimated variance of the intra-block BLUE of an elementary contrast would be

$$\widehat{\mathrm{Var}}[(\widehat{\tau_i - \tau_{i'}})_1] = 2(\varepsilon_1 r)^{-1}s_1^2 = (2/5)0.814583 = 0.32583.$$

It may also be interesting to note that the relevant results in Example 7.3.22 would be

$$\widehat{\mathrm{Var}}(\widetilde{\tau_i - \tau_{i'}}) \cong 2(\varepsilon_1^* r)^{-1}\hat{w}_1(1+\hat{\zeta})\hat{\sigma}_1^2 = 0.34662$$

(using $\varepsilon_1^* = 37/42$, $\hat{w}_1 = 0.99662$, $\hat{\zeta} = 0.0002776$ and $\hat{\sigma}_1^2 = 0.91893$) and

$$\widehat{\mathrm{Var}}[(\widehat{\tau_i - \tau_{i'}})_1] = 2(\varepsilon_1^* r)^{-1}s_1^2 = (14/37)0.919390 = 0.34788$$

for $i, i' = 1, 2, ..., 24$. Comparing these results, it should be noted that although the design considered in Example 7.3.22 has better efficiency properties than the design considered in the present example ($\varepsilon_1^* = 37/42 > 5/6 = \varepsilon_1$), the estmates of the variance σ_1^2 are smaller in the latter, both from the REML procedure (0.91893 > 0.81344) and from the intra-block analysis of variance (0.91939 > 0.81458). In effect, the estimated variances of the estimators of elementary contrasts among the strains compared in the experiment are in both approaches smaller here than in the previous example. This shows that not always better efficiency of the design leads to better empirical efficiency of the experiment. Nevertheless, note that the original experiment has the advantage of providing estimates of some important contrasts not available in the experiment considered in the present example.

Example 8.2.15. Ceranka (1975) has considered an agricultural field experiment with $v = 4$ treatments (varieties) of sunflower compared in a design with the incidence matrix $\boldsymbol{N} = [\boldsymbol{N}_1 : \boldsymbol{N}_2]$, where

$$\boldsymbol{N}_1 = \begin{bmatrix} 1 & 0 & 1 & 0 & 1 & 0 \\ 1 & 0 & 0 & 1 & 0 & 1 \\ 0 & 1 & 1 & 0 & 0 & 1 \\ 0 & 1 & 0 & 1 & 1 & 0 \end{bmatrix} \quad \text{and} \quad \boldsymbol{N}_2 = \begin{bmatrix} 1 & 1 & 1 & 0 \\ 1 & 1 & 0 & 1 \\ 1 & 0 & 1 & 1 \\ 0 & 1 & 1 & 1 \end{bmatrix}.$$

Note that $\boldsymbol{N}_1$ represents a BIB design with parameters $v = 4$, $b_1 = 6$, $r_1 = 3$, $k_1 = 2$, $\lambda_1 = 1$, and $\boldsymbol{N}_2$ represents another BIB design, with parameters $v = 4$, $b_2 = 4$, $r_2 = 3$, $k_2 = 3$, $\lambda_2 = 2$, both recorded in Table 8.2, the first at No. 7, the second at No. 6. On account of Corollary 8.2.3, the whole design, represented by $\boldsymbol{N}$, is a $(0; v-1; 0)$-EB design with parameters $v^* = v = 4$, $b^* = b_1 + b_2 = 10$, $r^* = r_1 + r_2 = 6$, $\boldsymbol{k}^* = [k_1\boldsymbol{1}'_{b_1}, k_2\boldsymbol{1}'_{b_2}]' = [2\boldsymbol{1}'_6, 3\boldsymbol{1}'_4]'$, $\varepsilon_1^* = v(\lambda_1/k_1 + \lambda_2/k_2)/(r_1 + r_2) = 7/9$ (= 0.7778). It is recorded in Table 8.3 at No. 24.

For the analysis of this experiment see Examples 3.2.1 (the intra-block analysis), 3.2.6 (the inter-block analysis) and 3.7.3 (the combined analysis).

8.3 Designs with two efficiency factors

In Section 7.3, $(\rho_0; \rho_1; 0)$-EB designs have been discussed in which $\rho_0 \geq 1$. Hence in this section constructions of $(0; \rho_1, \rho_2; 0)$-EB designs will be presented, i.e., those in which both of efficiency factors are less than one, though possibly close to 1.

Similarly as in Section 7.3, following particularly the terminology and notation used there, in the present section various cases of $(0; \rho_1, \rho_2; 0)$-EB designs will be discussed.

8.3.1 Proper designs that are equireplicate

First the most typical design, i.e., a PBIB design, is considered. Referring to Sections 6.0.2 and 7.3.1.1, first note that a connected 2-associate PBIB design with parameters v, b, r, k, λ_1, λ_2, ψ_1, ψ_2, ρ_1, ρ_2, $\boldsymbol{A}_i$, $\boldsymbol{A}_i^{\#}$, $i = 0, 1, 2$, is a $(0; \rho_1, \rho_2; 0)$-EB design with parameters v, b, r, k, $\varepsilon_\beta = 1 - \psi_\beta/(rk)$, ρ_β, $\boldsymbol{L}_\beta = \boldsymbol{A}_\beta^{\#}$, $\beta = 1, 2$, so that $1 > \varepsilon_1 > \varepsilon_2 > 0$, provided that ψ_1 and ψ_2 are both positive.

(A) Regular GD designs. To this class belongs any GD design in which both of the eigenvalues $\psi_1 = rk - v\lambda_2$ and $\psi_2 = r - \lambda_1$ are positive [see Section

6.0.2(A)]. Clatworthy (1973) has tabulated 209 individual plans of regular GD designs. A large number of constructions are also available in the literature (see also Freeman, 1957, 1976; Clatworthy, 1973). The expressions of $A_i^{\#}$ for these designs can be seen in Section 6.0.2(A).

Theorem 8.3.1 (Kageyama and Tanaka, 1981, Corollary 4.1.4). *For $n \geq 2$ there exist symmetric regular GD designs, i.e., $(0; \rho_1, \rho_2; 0)$-EB designs for the common $\rho_1 = 6$, $\rho_2 = 7(n-1)$, $\boldsymbol{L}_1 = (\boldsymbol{I}_7 - 7^{-1}\boldsymbol{1}_7\boldsymbol{1}_7') \otimes n^{-1}\boldsymbol{1}_n\boldsymbol{1}_n'$, $\boldsymbol{L}_2 = \boldsymbol{I}_7 \otimes (\boldsymbol{I}_n - n^{-1}\boldsymbol{1}_n\boldsymbol{1}_n')$, with parameters*

(i) $v = b = 7n, r = k = n+2, \lambda_1 = n-2, \lambda_2 = 1, m = 7, n,\ \varepsilon_1 = 7n/(n+2)^2$, $\varepsilon_2 = n(n+4)/(n+2)^2$;

(ii) $v = b = 7n, r = k = 3n-2, \lambda_1 = 3(n-2), \lambda_2 = n-1, m = 7, n,\ \varepsilon_1 = 7n(n-1)/(3n-2)^2$, $\varepsilon_2 = 3n(3n-4)/(3n-2)^2$.

Proof. The incidence matrices of the designs (i) and (ii) can be shown to be $\boldsymbol{I}_n \otimes \boldsymbol{N} + (\boldsymbol{1}_n\boldsymbol{1}_n' - \boldsymbol{I}_n) \otimes \boldsymbol{I}_7$ and $\boldsymbol{I}_n \otimes \boldsymbol{I}_7 + (\boldsymbol{1}_n\boldsymbol{1}_n' - \boldsymbol{I}_n) \otimes \boldsymbol{N}$, respectively, where

$$\boldsymbol{N} = \begin{bmatrix} 0 & 0 & 0 & 1 & 0 & 1 & 1 \\ 1 & 0 & 0 & 0 & 1 & 0 & 1 \\ 1 & 1 & 0 & 0 & 0 & 1 & 0 \\ 0 & 1 & 1 & 0 & 0 & 0 & 1 \\ 1 & 0 & 1 & 1 & 0 & 0 & 0 \\ 0 & 1 & 0 & 1 & 1 & 0 & 0 \\ 0 & 0 & 1 & 0 & 1 & 1 & 0 \end{bmatrix}. \quad \square$$

General cases of Theorem 8.3.1 are given in Kageyama and Tanaka (1981, Corollaries 4.1.2 and 4.1.3).

Theorem 8.3.2 (Raghavarao, 1971, Theorem 8.6.2). *The existence of a BIB design with parameters $v, b, r, k,\ \lambda = 1$ implies the existence of a regular GD design with parameters $v^* = v-1$, $b^* = b-r$, $r^* = r-1$, $k^* = k$, $\lambda_1^* = 0$, $\lambda_2^* = 1$, $m^* = r$, $n^* = k-1$, i.e., a $(0; \rho_1^*, \rho_2^*; 0)$-EB design with $\varepsilon_1^* = r(k-1)/[k(r-1)]$, $\varepsilon_2^* = (k-1)/k$, $\rho_1^* = r-1$, $\rho_2^* = r(k-2)$, $\boldsymbol{L}_1^* = (\boldsymbol{I}_r - r^{-1}\boldsymbol{1}_r\boldsymbol{1}_r') \otimes (k-1)^{-1}\boldsymbol{1}_{k-1}\boldsymbol{1}_{k-1}'$, $\boldsymbol{L}_2^* = \boldsymbol{I}_r \otimes [\boldsymbol{I}_{k-1} - (k-1)^{-1}\boldsymbol{1}_{k-1}\boldsymbol{1}_{k-1}']$.*

Proof. In the starting BIB design, a treatment θ occurs in the r blocks and on omitting θ, these blocks become disjoint, and the remaining $v - 1 = r(k-1)$ treatments form r groups of $k-1$ treatments each. By omitting from the BIB design the r blocks in which the treatment θ occurs, one can get the required design. Here two treatments belong to the same group if they occur together in the same block as θ in the BIB design. $\square$

By applying Theorem 8.3.2 to a BIB design of Corollary 6.9.2 (i), i.e., a BIB design with parameters $v = q^2$, $b = q(q+1)$, $r = q+1$, $k = q$, $\lambda = 1$, when q is a prime or a prime power, the following result can be obtained.

Corollary 8.3.1 (Raghavarao, 1971, Corollary 8.6.2.1). *When q is a prime or a prime power, there exists a regular GD design with parameters $v = b = q^2 - 1$, $r = k = q$, $\lambda_1 = 0$, $\lambda_2 = 1$, $m = q+1$, $n = q-1$, i.e., a $(0;\rho_1,\rho_2;0)$-EB design with $\varepsilon_1 = 1 - 1/q^2$, $\varepsilon_2 = 1 - 1/q$, $\rho_1 = q$, $\rho_2 = (q+1)(q-2)$, $\boldsymbol{L}_1 = [\boldsymbol{I}_{q+1} - (q+1)^{-1}\boldsymbol{1}_{q+1}\boldsymbol{1}'_{q+1}] \otimes (q-1)^{-1}\boldsymbol{1}_{q-1}\boldsymbol{1}'_{q-1}$, $\boldsymbol{L}_2 = \boldsymbol{I}_{q+1} \otimes [\boldsymbol{I}_{q-1} - (q-1)^{-1}\boldsymbol{1}_{q-1}\boldsymbol{1}'_{q-1}]$.*

There are a large number of BIB designs with $\lambda = 1$. For example, see Section 8.2.1. An application of Theorem 8.3.2 to Corollary 6.9.2 (iii) yields the following result.

Corollary 8.3.2. *When q is a prime or a prime power, there exists a regular GD design with parameters $v = q^3 - 1$, $b = (q^2-1)(q^2+q+1)$, $r = q(q+1)$, $k = q$, $\lambda_1 = 0, \lambda_2 = 1$, $m = q^2+q+1, n = q-1$, i.e., a $(0;\rho_1,\rho_2;0)$-EB design with $\varepsilon_1 = 1 - (q^2+1)/[q^2(q+1)]$, $\varepsilon_2 = 1 - 1/q$, $\rho_1 = q(q+1)$, $\rho_2 = (q-2)(q^2+q+1)$, $\boldsymbol{L}_1 = (\boldsymbol{I}_m - m^{-1}\boldsymbol{1}_m\boldsymbol{1}'_m) \otimes n^{-1}\boldsymbol{1}_n\boldsymbol{1}'_n$, $\boldsymbol{L}_2 = \boldsymbol{I}_m \otimes (\boldsymbol{I}_n - n^{-1}\boldsymbol{1}_n\boldsymbol{1}'_n)$.*

Other cases can be seen below.

Theorem 8.3.3. *There are regular GD designs with the parameters as given below, i.e., $(0;\rho_1,\rho_2;0)$-EB designs with the common $\boldsymbol{L}_1 = (\boldsymbol{I}_m - m^{-1}\boldsymbol{1}_m\boldsymbol{1}'_m) \otimes n^{-1}\boldsymbol{1}_n\boldsymbol{1}'_n$ and $\boldsymbol{L}_2 = \boldsymbol{I}_m \otimes (\boldsymbol{I}_n - n^{-1}\boldsymbol{1}_n\boldsymbol{1}'_n)$:*

(1) $v = 2k$, $b = 8k-6$, $r = 4k-3$, k, $\lambda_1 = 2k-3, \lambda_2 = 2(k-1)$, $m = 2, n = k$, $\varepsilon_1 = 1 - 1/(4k-3)$, $\varepsilon_2 = 1 - 2/(4k-3)$, $\rho_1 = 1, \rho_2 = 2(k-1)$, *when $2k-1$ is a prime or a prime power* (see Kageyama, 1985a).

(2) $v = mn$, $b = \binom{mn}{n} - m$, $r = \binom{mn-1}{n-1} - 1$, $k = n$, $\lambda_1 = \binom{mn-2}{n-2} - 1, \lambda_2 = \binom{mn-2}{n-2}$, $\varepsilon_1 = 1 - n[(m-1)\binom{mn-2}{n-2}/(n-1) - 1]/(rk)$, $\varepsilon_2 = 1 - n(m-1)\binom{mn-2}{n-2}/[rk(n-1)]$, $\rho_1 = m-1$, $\rho_2 = m(n-1)$ *for positive integers m and n (both ≥ 2), unless $n = m = 2$* (see Kageyama, 1985a).

(3) $v^* = 7v$, $b^* = 7b$, $r^* = b + 2r$, $k^* = v + 2k$, $\lambda_1^* = b - 2r + 4\lambda$, $\lambda_2^* = r$, $m^* = 7, n^* = v$, $\varepsilon_1^* = 1 - (bv - 3vr + 4rk)/(r^*k^*)$, $\varepsilon_2^* = 1 - 4(r-\lambda)/(r^*k^*)$, $\rho_1^* = 6, \rho_2^* = 7(v-1)$, *if there exists a BIB design with parameters v, b, r, k, λ* (see Kageyama and Tanaka, 1981).

(4) $v^* = 4k$, $b^* = 4r$, $r^* = 3r$, $k^* = 3k$, $\lambda_1^* = 2r + \lambda$, $\lambda_2^* = 2r$, $m^* = 2, n^* = 2k$, $\varepsilon_1^* = 8/9, \varepsilon_2^* = 1 - (r-\lambda)/(9rk)$, $\rho_1^* = 1, \rho_2^* = 2(2k-1)$, *if there exists a BIB design with parameters $v = 2k, b, r, k$, λ* (see Puri, Mehta and Kageyama, 1987).

(5) $v^* = 3v$, $b^* = 3b$, $r^* = b + r$, $k^* = v + k$, $\lambda_1^* = b + \lambda$, $\lambda_2^* = r$, $m^* = 3, n^* = v$, $\varepsilon_1^* = 1 - [v(b-r) + rk]/(r^*k^*)$, $\varepsilon_2^* = 1 - (r-\lambda)/(r^*k^*)$, $\rho_1^* = 2, \rho_2^* = 3(v-1)$, *if there exists a BIB design with parameters v, b, r, k, λ* (see Puri, Mehta and Kageyama, 1987).

(6) $v = b = s(s-1)(s^2+s+1)$, $r = k = s^2$, $\lambda_1 = 0$, $\lambda_2 = 1$, $m = s^2+s+1, n = s(s-1)$, $\varepsilon_1 = 1 - 1/s^3, \varepsilon_2 = 1 - 1/s^2$, $\rho_1 = s(s+1), \rho_2 = (s^2+s+1)(s^2-s-1)$, *if s is a prime or a prime power* (Sprott, 1959).

(7) $v^* = 8k$, $b^* = 8r$, $r^* = 3r$, $k^* = 3k$, $\lambda_1^* = 3\lambda$, $\lambda_2^* = r$, $m^* = 4, n^* = 2k$, $\varepsilon_1^* = 8/9$, $\varepsilon_2^* = 1 - (r - \lambda)/(3rk)$, $\rho_1^* = 3, \rho_2^* = 4(2k - 1)$, *if there exists a BIB design with parameters* $v = 2k, b, r, k, \lambda$ (see Kageyama and Mohan, 1984a).

(8) $v = 2(4u - 1), b = 2u(4u - 1)$, $r = u(2u - 1), k = 2u - 1$, $\lambda_1 = 0$, $\lambda_2 = u(u-1)/2$, $m = 4u - 1, n = 2$, $\varepsilon_1 = 1 - u/(2u - 1)^2$, $\varepsilon_2 = 1 - 1/(2u - 1)$, $\rho_1 = 2(2u-1), \rho_2 = 4u-1$, *if* $4u-1$ *is a prime or a prime power for* $u \geq 2$ (Mohan and Kageyama, 1983).

(9) $v = 2(2^s - 1), b = 2^{s-1}(2^s - 1)$, $r = 2^{s-2}(2^{s-1} - 1), k = 2^{s-1} - 1$, $\lambda_1 = 0$, $\lambda_2 = 2^{s-3}(2^{s-2} - 1)$, $m = 2^s - 1, n = 2$, $\varepsilon_1 = 1 - 2^{s-2}/(2^{s-1} - 1)^2$, $\varepsilon_2 = 1 - 1/(2^{s-1} - 1)$, $\rho_1 = 2(2^{s-1} - 1), \rho_2 = 2^s - 1$ *for* $s \geq 3$ (Mohan and Kageyama, 1983).

(10) $v = p(p+1), b = p^2(p^2+1)/2$, $r = p(s+1)[p(s+1) - 1]/2, k = s+p+1$, $\lambda_1 = p(p-1)/2$, $\lambda_2 = s(s+2)(s+1)^2/2$, $m = p+1, n = p$, $\varepsilon_1 = 1 - ps^3(s+1)/(2rk)$, $\varepsilon_2 = 1 - ps(ps + 2p - 1)/(2rk)$, $\rho_1 = p, \rho_2 = p^2 - 1$, *if* s *and* $p = s^2 + s + 1$ *are both primes or prime powers* (Mohan and Kageyama, 1983).

(11) $v = b = 2(q^2+q+1), r = k = q^2$, $\lambda_1 = 0$, $\lambda_2 = q(q-1)/2$, $m = q^2+q+1, n = 2$, $\varepsilon_1 = 1 - 1/q^3$, $\varepsilon_2 = 1 - 1/q^2$, $\rho_1 = q(q+1), \rho_2 = q^2 + q + 1$, *if* q *is a prime or a prime power* (Seberry, 1978).

Note that the designs of (1) and (2) in Theorem 8.3.3 are resolvable, whereas the designs of (8), (9) and (10) can be constructed by taking dualizations $\mathcal{D}_*^{(2)}$ with respect to pairs (see Remark 6.1.2) for affine α-resolvable BIB designs $\mathcal{D}$ for $\alpha \geq 1$.

Example 8.3.1. Take $k = 3$ in Theorem 8.3.3(1). Then the existence of a resolvable BIB design with parameters $v = 6, b = 20, r = 10, k = 3, \lambda = 4$ (see a design No. 5 of Table 9.1 in Chapter 9) implies the existence of a (0; 1, 4; 0)-EB design with parameters $v^* = 6$, $b^* = 18$, $r^* = 9$, $k^* = 3$, $\varepsilon_1^* = 8/9$, $\varepsilon_2^* = 7/9$, $\rho_1^* = 1$, $\rho_2^* = 4$, $\boldsymbol{L}_1^* = (\boldsymbol{I}_2 - 2^{-1}\boldsymbol{1}_2\boldsymbol{1}_2') \otimes 3^{-1}\boldsymbol{1}_3\boldsymbol{1}_3'$, $\boldsymbol{L}_2^* = \boldsymbol{I}_2 \otimes (\boldsymbol{I}_3 - 3^{-1}\boldsymbol{1}_3\boldsymbol{1}_3')$, and with the incidence matrix

$$\left[\begin{array}{cc|cc|cc|cc|cc|cc|cc|cc|cc}
1 & 0 & 1 & 0 & 1 & 0 & 1 & 0 & 1 & 0 & 1 & 0 & 0 & 1 & 0 & 1 & 0 & 1 \\
0 & 1 & 0 & 1 & 0 & 1 & 1 & 0 & 0 & 1 & 0 & 1 & 1 & 0 & 0 & 1 & 0 & 1 \\
1 & 0 & 1 & 0 & 1 & 0 & 0 & 1 & 0 & 1 & 0 & 1 & 1 & 0 & 1 & 0 & 1 & 0 \\
0 & 1 & 1 & 0 & 0 & 1 & 0 & 1 & 1 & 0 & 0 & 1 & 0 & 1 & 1 & 0 & 0 & 1 \\
1 & 0 & 0 & 1 & 0 & 1 & 1 & 0 & 1 & 0 & 1 & 0 & 1 & 0 & 1 & 0 & 1 & 0 \\
0 & 1 & 0 & 1 & 1 & 0 & 0 & 1 & 0 & 1 & 1 & 0 & 0 & 1 & 0 & 1 & 1 & 0
\end{array}\right].$$

In fact, this is a regular GD design of $v = 2 \times 3$ treatments which are divided into two groups $\{1, 2, 3\}$ and $\{4, 5, 6\}$ of three treatments. Hence, a basic contrast represented by an eigenvector of $\boldsymbol{L}_1^*$, i.e., a contrast involving not individual treatments but groups of them, obtains for its BLUE in the intra-block analysis the variance $9\sigma_1^2/8$, whereas a basic contrast represented by an eigenvector of

L_2^*, i.e., in particular, a contrast involving individual treatments from the same group, obtains for its intra-block BLUE the variance $9\sigma_1^2/7$.

(B) Triangular designs with $\psi_1 = r + (n-4)\lambda_1 - (n-3)\lambda_2 > 0$ and $\psi_2 = r - 2\lambda_1 + \lambda_2 > 0$, i.e., $\varepsilon_1 < 1$ and $\varepsilon_2 < 1$. Clatworthy (1973) has tabulated 100 triangular designs, among which there are 50 individual plans which satisfy $\psi_1 > 0$ and $\psi_2 > 0$. There are also some methods of constructing such designs. See Section 6.0.2(B) for general expressions of ψ_i and $A_i^{\#}$ in these designs.

Theorem 8.3.4. *There are triangular designs, satisfying $\psi_1 > 0$ and $\psi_2 > 0$, with the parameters as given below, i.e., $(0; \rho_1, \rho_2; 0)$-EB designs, for the common $L_1 = [n(n-2)]^{-1}(2nI_v + nA_1 - 4 1_v 1_v')$, $L_2 = [(n-1)(n-2)]^{-1}[(n-1)(n-4)I_v - (n-1)A_1 + 2 1_v 1_v']$, $\rho_1 = n-1$ and $\rho_2 = n(n-3)/2$, where A_1 is an association matrix on the first associates:*

(1) $v = n(n-1)/2$, $b = n(n-1)(n-2)/2$, $r = 2(n-2)$, $k = 2$, $\lambda_1 = 1$, $\lambda_2 = 0$, $\varepsilon_1 = 1 - (3n-8)/[4(n-2)]$, $\varepsilon_2 = 1 - (n-3)/[2(n-2)]$ *for* $n \geq 5$.

(2) $v = n(n-1)/2$, $b = n(n-1)(n-2)(n-3)/8$, $r = (n-2)(n-3)/2$, $k = 2$, $\lambda_1 = 0$, $\lambda_2 = 1$, $\varepsilon_1 = 1-(n-4)/[2(n-2)]$, $\varepsilon_2 = 1-(n^2-5n+8)/[2(n-2)(n-3)]$ *for* $n \geq 5$.

(3) $v = n(n-2)/2$, $b = n(n-1)(n-2)/6$, $r = n-2$, $k = 3$, $\lambda_1 = 1$, $\lambda_2 = 0$, $\varepsilon_1 = n/[3(n-2)]$, $\varepsilon_2 = 2(n-1)/[3(n-2)]$ *for* $n \geq 5$.

(4) $v^* = n(n-1)/2$, $b^* = nb$, $r^* = 2r$, $k^* = k$, $\lambda_1^* = \lambda$, $\lambda_2^* = 0$, $\varepsilon_1^* = 1 - [2r + (n-4)\lambda]/(2rk)$, $\varepsilon_2^* = 1 - (r-\lambda)/(rk)$, *if there exists a BIB design with parameters* $v = n-1, b, r, k, \lambda$ (Chang, Liu and Liu, 1965; Raghavarao, 1971, Theorem 8.8.4).

(5) $v = n(n-1)/2 = b$, $r = 2(n-2) = k$, $\lambda_1 = n-2$, $\lambda_2 = 4$, $\varepsilon_1 = 1 - (n-4)^2/[4(n-2)^2]$, $\varepsilon_2 = 1 - 1/(n-2)^2$, *for* $n \geq 5$.

(6) $v = n(n-1)/2 = b$, $r = (n-2)(n-3)/2 = k$, $\lambda_1 = (n-3)(n-4)/2$, $\lambda_2 = (n-4)(n-5)/2$, $\varepsilon_1 = 1 - 4/(n-2)^2$, $\varepsilon_2 = 1 - 4/[(n-2)^2(n-3)^2]$, *for* $n \geq 5$.

Proof. (1) The design is obtained, its blocks, by pairing a treatment i $(= 1, 2, ..., v)$ with each of its n_1 first associates exactly once, whereas the design (2) is obtained by pairing a treatment i $(= 1, 2, ..., v)$ with each of its n_2 second associates exactly once. (3) The design is obtained by forming all possible triplets of treatments by first writing a treatment pair lying in the same row of the association scheme, and adding to the pair the treatment lying at the intersection of the other row containing the first treatment of the pair and the column containing the second treatment of the pair. (4) The design is obtained by juxtaposing n corresponding BIB designs with $n-1$ treatments in each of the 1st, 2nd, ... , nth rows of the association scheme of n rows and n columns. (See Example 8.3.2 below.) The incidence matrix of the design (5) is given by the association matrix A_1 of the triangular association scheme, whereas (6) is given by A_2. [See Section 6.0.2(B).] □

Because there exists a BIB design with parameters $v = n-1 = b$, $r = n-2 =$

k, $\lambda = n - 3$ for any integer $n \geq 4$, Theorem 8.3.4(4) produces the following result.

Corollary 8.3.3. *There exists a triangular design with parameters* $v^* = n(n-1)/2$, $b^* = n(n-1)$, $r^* = 2(n-2)$, $k^* = n-2$, $\lambda_1^* = n-3$, $\lambda_2^* = 0$, *i.e., a* $(0; \rho_1^*, \rho_2^*; 0)$*-EB design with* $\varepsilon_1^* = 1 - 1/(n-2)^2$, $\varepsilon_2^* = n(n-3)/[2(n-2)^2]$, $\rho_1^* = n(n-3)/2$, $\rho_2^* = n-1$, $\boldsymbol{L}_1^* = [(n-1)(n-2)]^{-1}[(n-1)(n-4)\boldsymbol{I}_{v^*} - (n-1)\boldsymbol{A}_1 + 2\boldsymbol{1}_{v^*}\boldsymbol{1}_{v^*}']$, $\boldsymbol{L}_2^* = [n(n-2)]^{-1}(2n\boldsymbol{I}_{v^*} + n\boldsymbol{A}_1 - 4\boldsymbol{1}_{v^*}\boldsymbol{1}_{v^*}')$, *for all* $n \geq 4$, *where* $\boldsymbol{A}_1$ *is an association matrix on the first associates.*

Remark 8.3.1. In a nontrivial triangular design with parameters $v = n(n-1)/2$, one usually has $n \geq 5$. Because when $n = 4$, $p_{12}^2 = 0$ and it is the same as the GD type (see Clatworthy, 1973).

Example 8.3.2. Take $n = 5$ in Corollary 8.3.3. Then consider an unreduced BIB design with parameters $v = b = 4$, $r = k = 3$, $\lambda = 2$, whose incidence matrix $\boldsymbol{N}$ is of the form

$$\boldsymbol{N} = \begin{bmatrix} 1 & 1 & 1 & 0 \\ 1 & 1 & 0 & 1 \\ 1 & 0 & 1 & 1 \\ 0 & 1 & 1 & 1 \end{bmatrix}; \qquad \begin{array}{ccccc} & 1 & 2 & 3 & 4 \\ 1 & & 5 & 6 & 7 \\ 2 & 5 & & 8 & 9 \\ 3 & 6 & 8 & & 10 \\ 4 & 7 & 9 & 10 & \end{array}$$

and the triangular association scheme of 5 rows and 5 columns given above, used also in Example 6.0.3. The design provided by Corollary 8.3.3 can now be constructed as follows. The first 4 blocks are formed from $\boldsymbol{N}$ for four treatments being 1, 2, 3, 4 (1st row of the association scheme), the next 4 blocks are formed from $\boldsymbol{N}$ for four treatments being 1, 5, 6, 7 (2nd row of the association scheme), ..., the last 4 blocks are formed from $\boldsymbol{N}$ for four treatments being 4, 7, 9, 10 (5th row of the association scheme). Thus the incidence matrix of the resulting design is of the form

$$\begin{bmatrix}
1 & 1 & 1 & 0 & 1 & 1 & 1 & 0 & 0 & 0 & 0 & 0 & 0 & 0 & 0 & 0 & 0 & 0 & 0 & 0 \\
1 & 1 & 0 & 1 & 0 & 0 & 0 & 0 & 1 & 1 & 1 & 0 & 0 & 0 & 0 & 0 & 0 & 0 & 0 & 0 \\
1 & 0 & 1 & 1 & 0 & 0 & 0 & 0 & 0 & 0 & 0 & 0 & 1 & 1 & 1 & 0 & 0 & 0 & 0 & 0 \\
0 & 1 & 1 & 1 & 0 & 0 & 0 & 0 & 0 & 0 & 0 & 0 & 0 & 0 & 0 & 0 & 1 & 1 & 1 & 0 \\
0 & 0 & 0 & 0 & 1 & 1 & 0 & 1 & 1 & 1 & 0 & 1 & 0 & 0 & 0 & 0 & 0 & 0 & 0 & 0 \\
0 & 0 & 0 & 0 & 1 & 0 & 1 & 1 & 0 & 0 & 0 & 0 & 1 & 1 & 0 & 1 & 0 & 0 & 0 & 0 \\
0 & 0 & 0 & 0 & 0 & 1 & 1 & 1 & 0 & 0 & 0 & 0 & 0 & 0 & 0 & 0 & 1 & 1 & 0 & 1 \\
0 & 0 & 0 & 0 & 0 & 0 & 0 & 0 & 1 & 0 & 1 & 1 & 1 & 0 & 1 & 1 & 0 & 0 & 0 & 0 \\
0 & 0 & 0 & 0 & 0 & 0 & 0 & 0 & 0 & 1 & 1 & 1 & 0 & 0 & 0 & 0 & 1 & 0 & 1 & 1 \\
0 & 0 & 0 & 0 & 0 & 0 & 0 & 0 & 0 & 0 & 0 & 0 & 0 & 1 & 1 & 1 & 0 & 1 & 1 & 1
\end{bmatrix},$$

which yields a (0; 5, 4; 0)-EB design with parameters $v^* = 10, b^* = 20, r^* = 6, k^* = 3$, $\varepsilon_1^* = 8/9$, $\varepsilon_2^* = 5/9$, $\rho_1^* = 5$, $\rho_2^* = 4$, $\boldsymbol{L}_1^* = 6^{-1}(2\boldsymbol{I}_{10} - 2\boldsymbol{A}_1 + \boldsymbol{1}_{10}\boldsymbol{1}_{10}')$,

$L_2^* = (15)^{-1}(10\boldsymbol{I}_{10} + 5\boldsymbol{A}_1 - 4\boldsymbol{1}_{10}\boldsymbol{1}_{10}')$, where $\boldsymbol{A}_1$ is shown in Example 6.0.3.

(C) L_2 designs with $\psi_1 = r+(s-2)\lambda_1-(s-1)\lambda_2 > 0$ and $\psi_2 = r-2\lambda_1+\lambda_2 > 0$, i.e., $\varepsilon_1 < 1$ and $\varepsilon_2 < 1$. Clatworthy (1973) has tabulated 146 L_2 designs, among which there are 89 individual plans for which $\psi_1 > 0$ and $\psi_2 > 0$. The expressions of $\boldsymbol{A}_i^{\#}$, $i = 0, 1, 2$, can be seen in Section 6.0.2(C).

Now some methods of constructing such designs will be considered.

Theorem 8.3.5. *There are L_2 designs, satisfying $\psi_1 > 0$ and $\psi_2 > 0$, with the parameters as given below, i.e., $(0; \rho_1, \rho_2; 0)$-EB designs, for the common $\boldsymbol{L}_1 = (\boldsymbol{I}_s - s^{-1}\boldsymbol{1}_s\boldsymbol{1}_s') \otimes (\boldsymbol{I}_s - s^{-1}\boldsymbol{1}_s\boldsymbol{1}_s')$, $\boldsymbol{L}_2 = s^{-1}\boldsymbol{1}_s\boldsymbol{1}_s' \otimes (\boldsymbol{I}_s - s^{-1}\boldsymbol{1}_s\boldsymbol{1}_s') + (\boldsymbol{I}_s - s^{-1}\boldsymbol{1}_s\boldsymbol{1}_s') \otimes s^{-1}\boldsymbol{1}_s\boldsymbol{1}_s'$, $\rho_1 = (s-1)^2$ and $\rho_2 = 2(s-1)$:*

(1) $v^* = s^2$, $b^* = 2sb$, $r^* = 2r$, $k^* = k$, $\lambda_1^* = \lambda, \lambda_2^* = 0$, $\varepsilon_1^* = 1 - (r-\lambda)/(rk)$, $\varepsilon_2^* = 1 - [2r + (s-2)\lambda]/(2rk)$, *if there exists a BIB design with parameters* $v = s$, b, r, k, λ (Raghavarao, 1971, Theorem 8.10.3).

(2) $v = s^2 = b$, $r = 2s - 1 = k$, $\lambda_1 = s, \lambda_2 = 2$, $\varepsilon_1 = 1 - 1/(2s-1)^2$, $\varepsilon_2 = 1 - (s-1)^2/(2s-1)^2$ *for* $s \geq 3$ (Raghavarao, 1971, Theorem 8.10.4).

Proof. The design (1) is obtained by writing the starting BIB designs with treatments in each of the rows and in each of the columns of an L_2 association scheme (see Example 8.3.3 below). (2) In the L_2 association scheme of s^2 treatments, $\boldsymbol{A}_1 + \boldsymbol{I}_{s^2}$ gives the incidence matrix of the required design. Here $\boldsymbol{A}_1$ is an association matrix of the first associates. [See Section 6.0.2(C).] □

Example 8.3.3. Take $s = 3$ in Theorem 8.3.5(1). Then consider an unreduced BIB design with parameters $v = b = 3$, $r = k = 2$, $\lambda = 1$, whose incidence matrix $\boldsymbol{N}$ is of the form

$$\boldsymbol{N} = \begin{bmatrix} 1 & 1 & 0 \\ 1 & 0 & 1 \\ 0 & 1 & 1 \end{bmatrix},$$

and the following L_2 association scheme of nine treatments, used also in Example 6.0.4.

$$\begin{array}{ccc} 1 & 2 & 3 \\ 4 & 5 & 6 \\ 7 & 8 & 9 \end{array}$$

The required design can now be constructed as follows. The first 3 blocks are formed from $\boldsymbol{N}$ for three treatments being 1, 2, 3 (1st row of the association scheme), the next 3 blocks are formed from $\boldsymbol{N}$ for three treatments being 4, 5, 6 (2nd row of the association scheme), 7th, 8th, 9th blocks are formed from $\boldsymbol{N}$ for three treatments being 7, 8, 9 (3rd row of the association scheme), 10th, 11th, 12th blocks are formed from $\boldsymbol{N}$ for three treatments being 1, 4, 7 (1st column of the association scheme), ..., 16th, 17th, 18th blocks are formed from $\boldsymbol{N}$ for three treatments being 3, 6, 9 (3rd column of the association scheme). Thus the resulting design becomes a (0; 4, 4; 0)-EB design with parameters $v^* = 9, b^* = 18, r^* = 4, k^* = 2$, $\varepsilon_1^* = 3/4$, $\varepsilon_2^* = 3/8$, $\rho_1^* = 4$, $\rho_2^* = 4$, $\boldsymbol{L}_1^* = (\boldsymbol{I}_3 - 3^{-1}\boldsymbol{1}_3\boldsymbol{1}_3') \otimes$

$(I_3 - 3^{-1}\mathbf{1}_3\mathbf{1}_3')$, $L_2^* = 3^{-1}\mathbf{1}_3\mathbf{1}_3' \otimes (I_3 - 3^{-1}\mathbf{1}_3\mathbf{1}_3') + (I_3 - 3^{-1}\mathbf{1}_3\mathbf{1}_3') \otimes 3^{-1}\mathbf{1}_3\mathbf{1}_3'$, whose incidence matrix is of the form

$$\begin{bmatrix} 1&1&0&0&0&0&0&0&0&1&1&0&0&0&0&0&0&0\\ 1&0&1&0&0&0&0&0&0&0&0&0&1&1&0&0&0&0\\ 0&1&1&0&0&0&0&0&0&0&0&0&0&0&0&1&1&0\\ 0&0&0&1&1&0&0&0&0&1&0&1&0&0&0&0&0&0\\ 0&0&0&1&0&1&0&0&0&0&0&0&1&0&1&0&0&0\\ 0&0&0&0&1&1&0&0&0&0&0&0&0&0&0&1&0&1\\ 0&0&0&0&0&0&1&1&0&0&1&1&0&0&0&0&0&0\\ 0&0&0&0&0&0&1&0&1&0&0&0&0&1&1&0&0&0\\ 0&0&0&0&0&0&0&1&1&0&0&0&0&0&0&0&1&1 \end{bmatrix}.$$

Theorem 8.3.6. *Let $N_h, h = 1, 2$, be the $v \times b_h$ incidence matrices of two BIB designs with parameters v, b_h, r_h, k, λ_h. Then $N_1 \otimes N_2$ is the incidence matrix of an L_2 PBIB design, i.e., $(0; \rho_1^*, \rho_2^*; 0)$-EB, with parameters $v^* = v^2$, $b^* = b_1 b_2$, $r^* = r_1 r_2$, $k^* = k^2$, $\lambda_1^* = r_1\lambda_2 = r_2\lambda_1$, $\lambda_2^* = \lambda_1\lambda_2$, $\varepsilon_1^* = 1 - r_2 k(r_1 - \lambda_1)/(r^* k^*)$, $\varepsilon_2^* = 1 - (r_1 - \lambda_1)(r_2 - \lambda_2)/(r^* k^*)$, $\rho_1^* = 2(v-1), \rho_2^* = (v-1)^2$, $L_1^* = v^{-1}\mathbf{1}_v\mathbf{1}_v' \otimes (I_v - v^{-1}\mathbf{1}_v\mathbf{1}_v') + (I_v - v^{-1}\mathbf{1}_v\mathbf{1}_v') \otimes v^{-1}\mathbf{1}_v\mathbf{1}_v'$, $L_2^* = (I_v - v^{-1}\mathbf{1}_v\mathbf{1}_v') \otimes (I_v - v^{-1}\mathbf{1}_v\mathbf{1}_v')$.*

Proof. The direct calculation shows the result (see also Kageyama, 1972c). □

(D) Cyclic designs. Note (Ma, 1984) that this design exists only if $4t + 1$ is a prime [see Section 6.0.2(D)]. Furthermore, in this design both ψ_1 and ψ_2 are positive, i.e. $\varepsilon_1 < 1$ and $\varepsilon_2 < 1$. Thus, all the cyclic designs are well fitted for the discussion in this section. Clatworthy (1973) has tabulated 29 individual cyclic designs and, hence, for such designs the efficiency factors ε_1 and ε_2 have been here calculated and presented in Table 8.4. [Note that the cyclic designs described here are all 2-associate PBIB designs based on cyclic association schemes. Cyclic designs are also discussed in a wide class by John (1987) and generalized cyclic designs with application to factorial designs are considered by Bailey (1990).]

Finally, a construction method of PBIB designs from association matrices of association schemes is to be mentioned. A general argument can be found in Kageyama (1974a). Here association matrices A_0, A_1, A_2 of an association scheme with two associate classes and with parameters $v, n_i, p_{ju}^i, i, j, u = 0, 1, 2$, defined in Section 6.0.2, will be utilized. The following three methods can be taken into account.

Theorem 8.3.7. *The existence of an association scheme with two associate classes implies the following.*

(1) *$A_i, i = 1$ or 2 is the incidence matrix of a symmetric PBIB design with parameters $v = b$, $r = k = n_i$, $\lambda_1 = p_{ii}^1$, $\lambda_2 = p_{ii}^2$, having the same association scheme as the original.*

(2.1) *$A_1 + I_v$ is the incidence matrix of a symmetric PBIB design with parameters $v = b$, $r = k = n_1 + 1$, $\lambda_1 = p_{11}^1 + 2$, $\lambda_2 = p_{11}^2$, having the same association scheme as the original.*

(2.2) *$A_2 + I_v$ is the incidence matrix of a symmetric PBIB design with parameters $v = b$, $r = k = n_2 + 1$, $\lambda_1 = p_{22}^1$, $\lambda_2 = p_{22}^2 + 2$, having the same association scheme as the original.*

(3) *$A_1 + A_2$ is the incidence matrix of a symmetric PBIB design with parameters $v = b$, $r = k = n_1 + n_2 = v - 1$, $\lambda_1 = p_{11}^1 + p_{22}^1 + 2p_{12}^1$, $\lambda_2 = p_{11}^2 + p_{22}^2 + 2p_{12}^2$. Unfortunately, in this case $\lambda_1 = \lambda_2 = n_1 + n_2 - 1 = v - 2$. Then the resulting design becomes a trivial BIB design. But, this method is valid for an association scheme with more than two associate classes.*

The other parameters, $\varepsilon_\beta, \rho_\beta, L_\beta$, depend on the original association scheme employed.

Proof. The proofs are straightforward by calculating $A_i A_i'$, $(A_1 + I_v)(A_1 + I_v)'$, $(A_2 + I_v)(A_2 + I_v)'$, $(A_1 + A_2)(A_1 + A_2)'$, respectively. □

As one can easily see from Theorem 8.3.7, association matrices may also be used to construct BIB designs (see also Shrikhande and Singh, 1962; John, 1967).

Considering other constructions of proper and equireplicate designs, one may be interested also in constructing suitable designs that are not PBIB designs in the strict sense of Section 6.0.2, but still are $(0; \rho_1, \rho_2; 0)$-EB. Usually, a merging method is common and powerful (see also the explanation of the method preceding Corollary 7.3.2). For example, as another case of Corollary 7.3.2 based on a merging method, the following result can be observed for a regular GD design. For positive integers n and s (≥ 2) such that n/s is an integer, the existence of a regular GD design with parameters $v = mn, b, r, k, \lambda_1 = 0, \lambda_2$ implies the existence of a $(0; \rho_1^*, \rho_2^*; 0)$-EB design with parameters $v^* = ms$, $b^* = b$, $r^* = nr/s$, $k^* = k$, $\varepsilon_1^* = v\lambda_2/(rk)$, $\varepsilon_2^* = 1 - 1/k$, $\rho_1^* = m - 1$, $\rho_2^* = m(s-1)$, $L_1^* = (I_m - m^{-1}1_m 1_m') \otimes s^{-1}1_s 1_s'$, $L_2^* = I_m \otimes (I_s - s^{-1}1_s 1_s')$. This can be seen as following from Corollary 6.5.4. However, this design is essentially a regular GD design with $v^* = ms$ (m groups of s treatments), $\lambda_1^* = 0$ and $\lambda_2^* = (n/s)^2\lambda_2$. Thus, to authors' knowledge, a method of constructing a series of block designs in this case is not available, contrary to the results given in Section 7.3.1.2. Further development is left to the reader.

8.3.2 Proper designs that are nonequireplicate

As mentioned in Section 6.5, a merging method may be useful for constructing designs in this category. For example, starting with a regular GD design with $\lambda_1 = 0$, Theorem 6.5.3 and Corollary 6.5.4 can be used to produce $(0; \rho_1, \rho_2; 0)$-EB designs which are proper and nonequireplicate. If these results are applied to a regular GD design with $\lambda_1 = 0$ given in Corollary 8.3.1, the following can be obtained. In the merging method, the treatments in each of the m groups are divided into an arbitrary number of disjoint subgroups. The subgroup sizes for the ith group, $i = 1, 2, ..., m$, are denoted by $a_{i1}, a_{i2}, ..., a_{is_i}$ such that $\sum_{j=1}^{s_i} a_{ij} = n$ and $1 \leq a_{ij} \leq n$. Then the treatments in each group are merged to give a single treatment of the resulting design, which now has $\sum_{i=1}^{m} s_i$ treatments with replications $r[\boldsymbol{a}_1', \boldsymbol{a}_2', ..., \boldsymbol{a}_m']'$, where $\boldsymbol{a}_i = [a_{i1}, a_{i2}, ..., a_{is_i}]'$ for $i = 1, 2, ..., m$. Note that $m = q + 1$ in Corollary 8.3.1.

Corollary 8.3.4. *When q is a prime or a prime power, there are $(0; \rho_1, \rho_2; 0)$-EB designs with parameters*

(1) $v = \sum_{i=1}^{q+1} s_i, b = q^2 - 1$, $\boldsymbol{r} = q[\boldsymbol{a}_1', \boldsymbol{a}_2', ..., \boldsymbol{a}_{q+1}']'$, $k = q$, $\varepsilon_1 = 1 - 1/q^2$, $\varepsilon_2 = 1 - 1/q$, $\rho_1 = q$, $\rho_2 = v - q + 1$, $\boldsymbol{L}_1 = (q-1)^{-1}\text{diag}[\boldsymbol{1}_{s_1}\boldsymbol{a}_1' : \boldsymbol{1}_{s_2}\boldsymbol{a}_2' : \cdots : \boldsymbol{1}_{s_{q+1}}\boldsymbol{a}_{q+1}'] - [q(q^2-1)]^{-1}\boldsymbol{1}_v\boldsymbol{r}'$, $\boldsymbol{L}_2 = \boldsymbol{I}_v - (q-1)^{-1}\text{diag}[\boldsymbol{1}_{s_1}\boldsymbol{a}_1' : \boldsymbol{1}_{s_2}\boldsymbol{a}_2' : \cdots : \boldsymbol{1}_{s_{q+1}}\boldsymbol{a}_{q+1}']$;

(2) $v = (q+1)s, b = q^2 - 1$, $\boldsymbol{r} = q\boldsymbol{1}_{q+1} \otimes \boldsymbol{a}$, $k = q$, $\varepsilon_1 = 1 - 1/q^2$, $\varepsilon_2 = 1 - 1/q$, $\rho_1 = q$, $\rho_2 = v - q + 1$, $\boldsymbol{L}_1 = [\boldsymbol{I}_{q+1} - (q+1)^{-1}\boldsymbol{1}_{q+1}\boldsymbol{1}_{q+1}'] \otimes (q-1)^{-1}\boldsymbol{1}_s\boldsymbol{a}'$, $\boldsymbol{L}_2 = \boldsymbol{I}_{q+1} \otimes [\boldsymbol{I}_s - (q-1)^{-1}\boldsymbol{1}_s\boldsymbol{a}']$, *where* $\boldsymbol{a} = [a_1, a_2, ..., a_s]'$ *and* $\sum_{i=1}^{s} a_i = q - 1$.

Proof. Corollary 8.3.1 together with Theorem 6.5.3 yields (1), and with Corollary 6.5.4 gives (2). □

Note that in Corollary 8.3.4 the design (2) can be obtained from (1) when the number of subgroups is the same among all groups, i.e., $s_1 = s_2 = \cdots = s_m = s$, and $a_1, a_2, ..., a_s$ mutually exclusive treatments are merged in each group, i.e., $\boldsymbol{a}_i = [a_1, a_2, ..., a_s]'$ for all i.

Other regular GD designs with $\lambda_1 = 0$ can be found from Corollary 8.3.2 and Theorem 8.3.3 as well as many individual plans in Clatworthy (1973). These regular GD designs can be applied to Corollary 8.3.4 to get more proper and nonequireplicate $(0; \rho_1, \rho_2; 0)$-EB designs.

8.3.3 Nonproper designs that are equireplicate

A juxtaposition method of construction, discussed in Section 6.4, may be useful here. Theorem 6.4.2 is fundamental to produce $(0; \rho_1, \rho_2; 0)$-EB designs in this category, by use of two 2-associate PBIB designs. See also Section 7.3.3.

Here Corollary 6.4.1(i) will be used to produce a series of $(0; \rho_1, \rho_2; 0)$-EB designs by using the two GD designs given in Theorem 8.3.1, as follows.

Corollary 8.3.5. *For a positive integer $n \geq 2$ there exists a $(0; \rho_1, \rho_2; 0)$-EB design with parameters* $v = 7n, b = 14n$, $r = 4n$, $\boldsymbol{k} = [(n+2)\boldsymbol{1}_{7n}', (3n-2)\boldsymbol{1}_{7n}']'$, $\varepsilon_1 = 1 - (5n^2 - 12n + 12)/[4(n+2)(3n-2)]$, $\varepsilon_2 = 1 - 4/[(n+2)(3n-2)]$, $\rho_1 = 6$, $\rho_2 = 7(n-1)$, $\boldsymbol{L}_1 = (\boldsymbol{I}_7 - 7^{-1}\boldsymbol{1}_7\boldsymbol{1}_7') \otimes n^{-1}\boldsymbol{1}_n\boldsymbol{1}_n'$, $\boldsymbol{L}_2 = \boldsymbol{I}_7 \otimes (\boldsymbol{I}_n - n^{-1}\boldsymbol{1}_n\boldsymbol{1}_n')$.

The design given in Corollary 8.3.5 satisfies the desirable condition (8.2.4) on account of Corollary 3.2.3.

A large number of GD designs with a common number of treatments, $v = mn$, and with different block sizes can be found in abundance in Clatworthy (1973).

Corollary 6.4.1(i) for such purpose can be used as follows.

Corollary 8.3.6. *Let $\boldsymbol{N}_h, h = 1,2$, be the $v \times b^{(h)}$ incidence matrices of two GD designs with parameters $v = mn$ (m groups of n treatments each), $b^{(h)}$, $r^{(h)}$, $k^{(h)}$, $\lambda_1^{(h)}$, $\lambda_2^{(h)}$. Then:*

(1) *If $\boldsymbol{N}_1$ is a singular GD design and $\boldsymbol{N}_2$ is a semi-regular GD design, then $[\boldsymbol{N}_1 : \boldsymbol{N}_2]$ is the incidence matrix of a $(0; \rho_1^*, \rho_2^*; 0)$-EB design, satisfying (8.2.4), with parameters $v^* = mn, b^* = b^{(1)} + b^{(2)}$, $r^* = r^{(1)} + r^{(2)}$, $\boldsymbol{k}^* = [k^{(1)}\mathbf{1}'_{b^{(1)}}, k^{(2)}\mathbf{1}'_{b^{(2)}}]'$, $\varepsilon_1^* = 1 - (r^{(1)}k^{(1)} - v\lambda_2^{(1)})/[k^{(1)}(r^{(1)} + r^{(2)})]$, $\varepsilon_2^* = 1 - (r^{(2)} - \lambda_1^{(2)})/[k^{(2)}(r^{(1)} + r^{(2)})]$, $\rho_1^* = m - 1$, $\rho_2^* = m(n-1)$, $\boldsymbol{L}_1^* = (\boldsymbol{I}_m - m^{-1}\mathbf{1}_m\mathbf{1}'_m) \otimes n^{-1}\mathbf{1}_n\mathbf{1}'_n$, $\boldsymbol{L}_2^* = \boldsymbol{I}_m \otimes (\boldsymbol{I}_n - n^{-1}\mathbf{1}_n\mathbf{1}'_n)$.*

(2) *If $\boldsymbol{N}_1$ is a singular GD design and $\boldsymbol{N}_2$ is a regular GD design, then $[\boldsymbol{N}_1 : \boldsymbol{N}_2]$ is the incidence matrix of a $(0; \rho_1^*, \rho_2^*; 0)$-EB design, satisfying (8.2.4), with parameters $v^* = mn, b^* = b^{(1)} + b^{(2)}$, $r^* = r^{(1)} + r^{(2)}$, $\boldsymbol{k}^* = [k^{(1)}\mathbf{1}'_{b^{(1)}}, k^{(2)}\mathbf{1}'_{b^{(2)}}]'$, $\varepsilon_1^* = 1 - [(r^{(1)}k^{(1)} - v\lambda_2^{(1)})/k^{(1)} + (r^{(2)}k^{(2)} - v\lambda_2^{(2)})/k^{(2)}]/(r^{(1)}+r^{(2)})$, $\varepsilon_2^* = 1 - (r^{(2)} - \lambda_1^{(2)})/[k^{(2)}(r^{(1)}+r^{(2)})]$, $\rho_1^* = m-1$, $\rho_2^* = m(n-1)$, $\boldsymbol{L}_1^* = (\boldsymbol{I}_m - m^{-1}\mathbf{1}_m\mathbf{1}'_m) \otimes n^{-1}\mathbf{1}_n\mathbf{1}'_n$, $\boldsymbol{L}_2^* = \boldsymbol{I}_m \otimes (\boldsymbol{I}_n - n^{-1}\mathbf{1}_n\mathbf{1}'_n)$.*

(3) *If $\boldsymbol{N}_1$ is a semi-regular GD design and $\boldsymbol{N}_2$ is a regular GD design, then $[\boldsymbol{N}_1 : \boldsymbol{N}_2]$ is the incidence matrix of a $(0; \rho_1^*, \rho_2^*; 0)$-EB design, satisfying (8.2.4), with parameters $v^* = mn, b^* = b^{(1)} + b^{(2)}$, $r^* = r^{(1)} + r^{(2)}$, $\boldsymbol{k}^* = [k^{(1)}\mathbf{1}'_{b^{(1)}}, k^{(2)}\mathbf{1}'_{b^{(2)}}]'$, $\varepsilon_1^* = 1 - (r^{(2)}k^{(2)} - v\lambda_2^{(2)})/[k^{(2)}(r^{(1)} + r^{(2)})]$, $\varepsilon_2^* = 1 - [(r^{(1)} - \lambda_1^{(1)})/k^{(1)} + (r^{(2)} - \lambda_1^{(2)})/k^{(2)}]/(r^{(1)}+r^{(2)})$, $\rho_1^* = m-1$, $\rho_2^* = m(n-1)$, $\boldsymbol{L}_1^* = (\boldsymbol{I}_m - m^{-1}\mathbf{1}_m\mathbf{1}'_m) \otimes n^{-1}\mathbf{1}_n\mathbf{1}'_n$, $\boldsymbol{L}_2^* = \boldsymbol{I}_m \otimes (\boldsymbol{I}_n - n^{-1}\mathbf{1}_n\mathbf{1}'_n)$.*

(4) *If both $\boldsymbol{N}_1$ and $\boldsymbol{N}_2$ are regular GD designs, then $[\boldsymbol{N}_1 : \boldsymbol{N}_2]$ is the incidence matrix of a $(0; \rho_1^*, \rho_2^*; 0)$-EB design, satisfying (8.2.4), with parameters $v^* = mn, b^* = b^{(1)} + b^{(2)}$, $r^* = r^{(1)} + r^{(2)}$, $\boldsymbol{k}^* = [k^{(1)}\mathbf{1}'_{b^{(1)}}, k^{(2)}\mathbf{1}'_{b^{(2)}}]'$, $\varepsilon_1^* = 1 - [(r^{(1)}k^{(1)} - v\lambda_2^{(1)})/k^{(1)} + (r^{(2)}k^{(2)} - v\lambda_2^{(2)})/k^{(2)}]/(r^{(1)} + r^{(2)})$, $\varepsilon_2^* = 1 - [(r^{(1)} - \lambda_1^{(1)})/k^{(1)} + (r^{(2)} - \lambda_1^{(2)})/k^{(2)}]/(r^{(1)}+r^{(2)})$, $\rho_1^* = m-1$, $\rho_2^* = m(n-1)$, $\boldsymbol{L}_1^* = (\boldsymbol{I}_m - m^{-1}\mathbf{1}_m\mathbf{1}'_m) \otimes n^{-1}\mathbf{1}_n\mathbf{1}'_n$, $\boldsymbol{L}_2^* = \boldsymbol{I}_m \otimes (\boldsymbol{I}_n - n^{-1}\mathbf{1}_n\mathbf{1}'_n)$.*

Note that Corollary 8.3.5 is an application of the result (4) in Corollary 8.3.6. Next a juxtaposition idea of Corollary 7.3.14 can be used to produce various kinds of designs as shown in Corollary 8.3.7 below, where Corollary 6.4.1(i) is applied with a particular singular GD design having the incidence matrix $\boldsymbol{I}_m \otimes \mathbf{1}_n\mathbf{1}'_s$. In fact, the results (1) and (2) in Corollary 8.3.7 correspond to the cases (2) and (1) in Corollary 8.3.6, respectively.

Corollary 8.3.7. *Let $\boldsymbol{N}$ be the $v \times b$ incidence matrix of a GD design with parameters $v = mn$ (m groups of n treatments each), b, r, k, λ_1, λ_2. Then:*

(1) *If $\boldsymbol{N}$ is a regular GD design, then the incidence matrix $[\boldsymbol{N} : \boldsymbol{I}_m \otimes \mathbf{1}_n\mathbf{1}'_s]$ yields a $(0; \rho_1^*, \rho_2^*; 0)$-EB design, satisfying (8.2.4), with parameters $v^* = mn, b^* = b + ms$, $r^* = r + s$, $\boldsymbol{k}^* = [k\mathbf{1}'_b, n\mathbf{1}'_{ms}]'$, $\varepsilon_1^* = 1 - [k(r+s) -$*

$v\lambda_2]/[k(r+s)]$, $\varepsilon_2^* = 1-(r-\lambda_1)/[k(r+s)]$, $\rho_1^* = m-1$, $\rho_2^* = m(n-1)$, $\boldsymbol{L}_1^* = (\boldsymbol{I}_m - m^{-1}\boldsymbol{1}_m\boldsymbol{1}_m')\otimes n^{-1}\boldsymbol{1}_n\boldsymbol{1}_n'$, $\boldsymbol{L}_2^* = \boldsymbol{I}_m\otimes(\boldsymbol{I}_n - n^{-1}\boldsymbol{1}_n\boldsymbol{1}_n')$ *for* $s\geq 1$.

(2) *If* $\boldsymbol{N}$ *is a semi-regular GD design, then the incidence matrix* $[\boldsymbol{N} : \boldsymbol{I}_m\otimes\boldsymbol{1}_n\boldsymbol{1}_s']$ *yields a* $(0;\rho_1^*,\rho_2^*;0)$*-EB design, satisfying* (8.2.4), *with parameters* $v^* = mn, b^* = b+ms$, $r^* = r+s$, $\boldsymbol{k}^* = [k\boldsymbol{1}_b', n\boldsymbol{1}_{ms}']'$, $\varepsilon_1^* = 1-s/(r+s)$, $\varepsilon_2^* = 1-(r-\lambda_1)/[k(r+s)]$, $\rho_1^* = m-1$, $\rho_2^* = m(n-1)$, $\boldsymbol{L}_1^* = (\boldsymbol{I}_m - m^{-1}\boldsymbol{1}_m\boldsymbol{1}_m')\otimes n^{-1}\boldsymbol{1}_n\boldsymbol{1}_n'$, $\boldsymbol{L}_2^* = \boldsymbol{I}_m\otimes(\boldsymbol{I}_n - n^{-1}\boldsymbol{1}_n\boldsymbol{1}_n')$ *for* $s\geq 1$.

Other statements can be given by using the juxtaposition idea of Corollary 7.3.15, in connection with Corollary 6.2.1 and Theorem 6.4.2, but now with the use of 2-associate PBIB designs, in which both of ε_1 and ε_2 are less than one, as follows.

Corollary 8.3.8. *If* $\boldsymbol{N}$ *is the* $v\times b$ *incidence matrix of a 2-associate PBIB design with parameters* v, b, r, k, λ_1, λ_2, $\varepsilon_\beta(<1)$, ρ_β, $\boldsymbol{L}_\beta$, $\beta = 1,2$, *then the incidence matrix* $[\boldsymbol{N} : \boldsymbol{1}_v\boldsymbol{1}_b' - \boldsymbol{N}]$ *yields a* $(0;\rho_1^*,\rho_2^*;0)$*-EB design with parameters* $v^* = v$, $b^* = 2b$, $r^* = b$, $\boldsymbol{k}^* = [k\boldsymbol{1}_b', (v-k)\boldsymbol{1}_b']'$, $\varepsilon_\beta^* = 1-k(1-\varepsilon_\beta)/(v-k)$, $\rho_\beta^* = \rho_\beta$, $\boldsymbol{L}_\beta^* = \boldsymbol{L}_\beta$.

Similarly, modifying accordingly Corollary 7.3.17, one can present the following result.

Corollary 8.3.9. *If* $\boldsymbol{N}$ *is the* $v\times b$ *incidence matrix of a 2-associate PBIB design with parameters* v, b, r, k, λ_1, λ_2, $\varepsilon_\beta(<1)$, ρ_β, $\boldsymbol{L}_\beta$, $\beta = 1,2$, *then for any positive integer* p *the incidence matrix* $[\boldsymbol{N} : \boldsymbol{1}_v\boldsymbol{1}_p']$ *yields a* $(0;\rho_1^*,\rho_2^*;0)$*-EB design, satisfying* (8.2.4), *with parameters* $v^* = v$, $b^* = b+p$, $r^* = r+p$, $\boldsymbol{k}^* = [k\boldsymbol{1}_b', v\boldsymbol{1}_p']'$, $\varepsilon_\beta^* = 1-r(1-\varepsilon_\beta)/(r+p)$, $\rho_\beta^* = \rho_\beta$, $\boldsymbol{L}_\beta^* = \boldsymbol{L}_\beta$.

Theorem 6.6.7 with $c = 1$ yields the following result.

Corollary 8.3.10. *Let* $\boldsymbol{N}$ *be the* $v\times b$ *incidence matrix of a BIB design with parameters* v, b, r, k, λ. *Then the incidence matrix*

$$\boldsymbol{N}^* = \begin{bmatrix} \boldsymbol{N} & \boldsymbol{1}_v\boldsymbol{1}_b' - \boldsymbol{N} \\ \boldsymbol{1}_b' & \boldsymbol{0}' \end{bmatrix}$$

yields a $(0;\rho_1^*,\rho_2^*;0)$*-EB design with parameters* $v^* = v+1$, $b^* = 2b$, $r^* = b$, $\boldsymbol{k}^* = [(k+1)\boldsymbol{1}_b', (v-k)\boldsymbol{1}_b']'$, $\varepsilon_1^* = 1-(v+1)(r-\lambda)/[b(k+1)(v-k)]$, $\varepsilon_2^* = 1-(b-r)/[b(k+1)]$, $\rho_1^* = v-1$, $\rho_2^* = 1$, $\boldsymbol{L}_1^* = \mathrm{diag}[\boldsymbol{I}_v - v^{-1}\boldsymbol{1}_v\boldsymbol{1}_v' : 0]$ *and* $\boldsymbol{L}_2^*$ *as in* (6.6.4) *for* $c = 1$.

8.3.4 Nonproper designs that are nonequireplicate

As mentioned in Section 7.3.4, nonproper and nonequireplicate $(0;\rho_1,\rho_2;0)$-EB designs are less recommendable than those considered in Section 8.3.2 and are not often used in practice. Furthermore, the designs described in Section 7.3.4 have full efficiency for estimation of some contrasts in the intra-block analysis. This property is reasonable from a practical point of view. However, according to the purpose of this chapter, the present section deals with nonproper and nonequireplicate $(0; \rho_1, \rho_2; 0)$-EB designs in which both of efficiency factors are less than one.

Once there exists such a design, by Theorem 6.4.1 (juxtaposition) one can get some series of this type of designs. Unfortunately, the existence of such designs in general is not so rich.

Two methods can be given. The direct calculation of the $\boldsymbol{M}$ matrix can show the following, by evaluating eigenvalues and eigenvectors of $\boldsymbol{N}\boldsymbol{k}^{-\delta}\boldsymbol{N}'$ with respect to $\boldsymbol{r}^{\delta}$.

Theorem 8.3.8. *Let $\boldsymbol{N}$ be the $v \times b$ incidence matrix of a BIB design with parameters v, b, r, k, λ. Then the incidence matrix*

$$\begin{bmatrix} \boldsymbol{N} & \boldsymbol{1}_v \\ \boldsymbol{1}_b' & 1 \end{bmatrix}$$

yields a $(0; \rho_1^*, \rho_2^*; 0)$-EB design with parameters $v^* = v+1$, $b^* = b+1$, $\boldsymbol{r}^* = [(r+1)\boldsymbol{1}_v', b+1]'$, $\boldsymbol{k}^* = [(k+1)\boldsymbol{1}_b', v+1]'$, $\varepsilon_1^* = 1-(v-k)(b-r)/[(v+1)(k+1)(b+1)(r+1)]$, $\varepsilon_2^* = 1-(r-\lambda)/[(r+1)(k+1)]$, $\rho_1^* = 1$, $\rho_2^* = v-1$, $\boldsymbol{L}_1^* = \boldsymbol{I}_{v+1} - [v(r+1)+b+1]^{-1}\boldsymbol{1}_{v+1}(\boldsymbol{r}^*)' - \boldsymbol{L}_2^*$, $\boldsymbol{L}_2^* = \text{diag}[\boldsymbol{I}_v - v^{-1}\boldsymbol{1}_v\boldsymbol{1}_v' : 0]$.

This design may not satisfy the desirable condition (8.2.4) to get the BLUE of a contrast under both, the intra-block and the inter-block submodel. However, a similar pattern can produce another design, given below, that satisfies (8.2.4), because of Corollary 3.2.4.

Theorem 8.3.9. *Let $\boldsymbol{N}$ be the $v \times b$ incidence matrix of a BIB design with parameters v, b, r, k, λ. Then the incidence matrix*

$$\begin{bmatrix} \boldsymbol{1}_b' & 0 \\ \boldsymbol{N} & \boldsymbol{1}_v \end{bmatrix}$$

yields a $(0; \rho_1^*, \rho_2^*; 0)$-EB design, satisfying (8.2.4), with parameters $v^* = v+1$, $b^* = b+1$, $\boldsymbol{r}^* = [b, (r+1)\boldsymbol{1}_v']'$, $\boldsymbol{k}^* = [(k+1)\boldsymbol{1}_b', v]'$, $\varepsilon_1^* = 1 - 1/[(r+1)(k+1)]$, $\varepsilon_2^* = 1-(r-\lambda)/[(r+1)(k+1)]$, $\rho_1^* = 1$, $\rho_2^* = v-1$,

$$\boldsymbol{L}_1^* = \frac{1}{v(r+1)+b}\begin{bmatrix} v(r+1) & -(r+1)\boldsymbol{1}_v' \\ -b\boldsymbol{1}_v & \frac{b}{v}\boldsymbol{1}_v\boldsymbol{1}_v' \end{bmatrix},$$

$\boldsymbol{L}_2^* = \text{diag}[0 : \boldsymbol{I}_v - v^{-1}\boldsymbol{1}_v\boldsymbol{1}_v']$. Here, $\boldsymbol{L}_1^* + \boldsymbol{L}_2^* = \boldsymbol{I}_{v+1} - [v(r+1)+b]^{-1}\boldsymbol{1}_{v+1}(\boldsymbol{r}^*)'$.

Finally note that, as another patterned method, Theorem 6.6.7 with $c \geq 2$ yields a nonbinary, nonequireplicate and nonproper $(0; \rho_1^*, \rho_2^*; 0)$-EB design.

8.4 Designs with three efficiency factors

This section will proceed similarly to the discussions given in Section 7.4 of Chapter 7. That is, $(0; \rho_1, \rho_2, \rho_3; 0)$-EB designs will be treated.

8.4.1 Proper designs that are equireplicate

Two cases are discussed in this section. Notations follow from Section 7.4.1.

8.4.1.1 PBIB designs

As mentioned in Sections 6.0.2 and 7.4.1.1, one has first to note that a connected 3-associate PBIB design is equireplicate and proper, either $(0; \rho_1, \rho_2, \rho_3; 0)$-EB or $(\rho_0; \rho_1, \rho_2; 0)$-EB, with $\rho_0 \geq 1$ and with $\varepsilon_\beta = 1 - \psi_\beta/(rk)$. Section 7.4.1.1 dealt with a design in which at least one of the eigenvalues ψ_β of $\boldsymbol{NN}'$ was zero, i.e., $\varepsilon_\beta = 1$, which is the case when the incidence matrix $\boldsymbol{N}$ of the PBIB design has a rank smaller than v (see Section 6.0.2). In the present section, the case of all the ψ_β being positive (i.e., $\varepsilon_\beta < 1$ for all β) is considered, according to the purpose of this chapter. Then this means that in the present $(0; \rho_1, \rho_2, \rho_3; 0)$-EB designs no basic contrasts can be estimated in the intra-block analysis with full efficiency.

The most fundamental method of the construction of symmetric PBIB designs is to use the construction idea of Theorem 8.3.7. That is, here, the association matrices $\boldsymbol{A}_0 = \boldsymbol{I}_v$, $\boldsymbol{A}_1$, $\boldsymbol{A}_2$, $\boldsymbol{A}_3$ of an association scheme with three associate classes and with parameters $v, n_i, p^i_{ju}, i, j, u = 0, 1, 2, 3$, defined in Section 6.0.2, are to be taken.

Theorem 8.4.1. *The existence of an association scheme with three associate classes implies the following.*

(1) $\boldsymbol{A}_i, i = 1, 2, 3,$ *is the incidence matrix of a symmetric PBIB design with parameters* $v = b$, $r = k = n_i$, $\lambda_1 = p^1_{ii}$, $\lambda_2 = p^2_{ii}$, $\lambda_3 = p^3_{ii}$, *having the same association scheme as the original.*

(2.1) $\boldsymbol{A}_1 + \boldsymbol{I}_v$ *is the incidence matrix of a symmetric PBIB design with parameters* $v = b$, $r = k = n_1 + 1$, $\lambda_1 = p^1_{11} + 2$, $\lambda_2 = p^2_{11}$, $\lambda_3 = p^3_{11}$, *having the same association scheme as the original.*

(2.2) $\boldsymbol{A}_2 + \boldsymbol{I}_v$ *is the incidence matrix of a symmetric PBIB design with parameters* $v = b$, $r = k = n_2 + 1$, $\lambda_1 = p^1_{22}$, $\lambda_2 = p^2_{22} + 2$, $\lambda_3 = p^3_{22}$, *having the same association scheme as the original.*

(2.3) $\boldsymbol{A}_3 + \boldsymbol{I}_v$ *is the incidence matrix of a symmetric PBIB design with parameters* $v = b$, $r = k = n_3 + 1$, $\lambda_1 = p^1_{33}$, $\lambda_2 = p^2_{33}$, $\lambda_3 = p^3_{33} + 2$, *having the same association scheme as the original.*

(3.1) $\boldsymbol{A}_1 + \boldsymbol{A}_2$ *is the incidence matrix of a symmetric PBIB design with parameters* $v = b$, $r = k = n_1 + n_2$, $\lambda_1 = p^1_{11} + p^1_{22} + 2p^1_{12}$, $\lambda_2 = p^2_{11} + p^2_{22} + 2p^2_{12}$, $\lambda_3 = p^3_{11} + p^3_{22} + 2p^3_{12}$, *having the same association scheme as the original.*

(3.2) $\boldsymbol{A}_2 + \boldsymbol{A}_3$ *is the incidence matrix of a symmetric PBIB design with parameters* $v = b$, $r = k = n_2 + n_3$, $\lambda_1 = p^1_{22} + p^1_{33} + 2p^1_{23}$, $\lambda_2 = p^2_{22} + p^2_{33} + 2p^2_{23}$, $\lambda_3 = p^3_{22} + p^3_{33} + 2p^3_{23}$, *having the same association scheme as the original.*

(3.3) $\boldsymbol{A}_1 + \boldsymbol{A}_3$ *is the incidence matrix of a symmetric PBIB design with parameters* $v = b$, $r = k = n_1 + n_3$, $\lambda_1 = p^1_{11} + p^1_{33} + 2p^1_{13}$, $\lambda_2 = p^2_{11} + p^2_{33} + 2p^2_{13}$, $\lambda_3 = p^3_{11} + p^3_{33} + 2p^3_{13}$, *having the same association scheme as the original.*

The other parameters, $\varepsilon_\beta, \rho_\beta, \boldsymbol{L}_\beta$, depend on the original association scheme employed.

Proof. Each case is easily shown by examining $\boldsymbol{N}\boldsymbol{N}'$ for the incidence matrix $\boldsymbol{N}$ of the resulting design (see also Section 6.0.2 for such expressions). □

Note that the values of parameters of symmetric PBIB designs shown in Theorem 8.4.1 can be realized under existing association schemes with three associate classes, and that in some cases disconnected designs appear. Though several association schemes with three associate classes are available, a rectangular association scheme and a group divisible 3-associate association scheme, as in (E) and (F) of Section 6.0.2, are specially used here. The following can be obtained by Theorem 8.4.1, respectively. Some of them to be presented in Corollaries 8.4.1 and 8.4.2 below have $\varepsilon_\beta = 1$. These cases can be used as designs appropriate for Section 7.4.1.

Corollary 8.4.1. *There are symmetric rectangular PBIB designs with the parameters as given below, for the common $\rho_1 = t-1, \rho_2 = s-1$ and $\rho_3 = (s-1)(t-1)$:*

(1) $v = b = st$, $r = k = (s-1)(t-1)$, $\lambda_1 = (s-1)(t-2)$, $\lambda_2 = (s-2)(t-1)$, $\lambda_3 = (s-2)(t-2)$, *i.e.*, $(0;\rho_1,\rho_2,\rho_3;0)$*-EB with* $\varepsilon_1 = 1-1/(t-1)^2$, $\varepsilon_2 = 1-1/(s-1)^2$, $\varepsilon_3 = 1-1/[(s-1)^2(t-1)^2]$;

(2) $v = b = st$, $r = k = (s-1)(t-1)+1$, $\lambda_1 = (s-1)(t-2)$, $\lambda_2 = (s-2)(t-1)$, $\lambda_3 = (s-2)(t-2)+2$, *i.e.*, $(0;\rho_1,\rho_2,\rho_3;0)$*-EB with* $\varepsilon_1 = 1-[(s-1)(s-3)+1]/(rk)$, $\varepsilon_2 = 1-[(t-1)(t-3)+1]/(rk)$, $\varepsilon_3 = 1-4/(rk)$;

(3) $v = b = st$, $r = k = t+s-2$, $\lambda_1 = t-2$, $\lambda_2 = s-2$, $\lambda_3 = 2$, *i.e.*, $(0;\rho_1,\rho_2,\rho_3;0)$*-EB with* $\varepsilon_1 = 1-(s-2)^2/(rk)$, $\varepsilon_2 = 1-(t-2)^2/(rk)$, $\varepsilon_3 = 1-4/(rk)$.

Here $\boldsymbol{L}_\beta = \boldsymbol{A}^{\#}_\beta, \beta = 1,2,3$, can be found in Section 6.0.2(E).

Note that the two cases (3.2) and (3.3) not appearing in Corollary 8.4.1, corresponding to Theorem 8.4.1, are $(\rho_0^*;\rho_1^*;0)$-EB designs with parameters $v = b = st$, $r = k = t(s-1)$, $\lambda_1 = t(s-1)$, $\lambda_2 = t(s-2)$, $\lambda_3 = t(s-2)$, $\varepsilon_0^* = 1$, $\varepsilon_1^* = 1-1/(s-1)^2$ for $\rho_0^* = s(t-1)$ and $\rho_1^* = s-1$, and with parameters $v = b = st$, $r = k = s(t-1)$, $\lambda_1 = s(t-2)$, $\lambda_2 = s(t-1)$, $\lambda_3 = s(t-2)$, $\varepsilon_0^* = 1$, $\varepsilon_1^* = 1-1/(t-1)^2$ for $\rho_0^* = (s-1)t$ and $\rho_1^* = t-1$, respectively.

The Kronecker product method is also useful to construct rectangular PBIB designs.

Theorem 8.4.2. *Let $\boldsymbol{N}_h, h = 1,2$, be the $v_h \times b_h$ incidence matrices of two BIB designs with parameters v_h, b_h, r_h, k_h, λ_h. Then the incidence matrix $\boldsymbol{N}_1 \otimes \boldsymbol{N}_2$ yields a rectangular PBIB design with parameters $v^* = v_1v_2$, $b^* = b_1b_2$, $r^* = r_1r_2$, $k^* = k_1k_2$, $\lambda_1^* = r_1\lambda_2$, $\lambda_2^* = r_2\lambda_1$, $\lambda_3^* = \lambda_1\lambda_2$, $n_1^* = v_2-1$, $n_2^* = v_1-1$, $n_3^* = (v_1-1)(v_1-1)$, i.e., $(0;\rho_1^*;\rho_2^*;\rho_3^*;0)$-EB with $\varepsilon_1^* = 1-(r_2-\lambda_2)/(r_2k_2)$, $\varepsilon_2^* = 1-(r_1-\lambda_1)/(r_1k_1)$, $\varepsilon_3^* = 1-(r_1-\lambda_1)(r_2-\lambda_2)/(r_1r_2k_1k_2)$, $\rho_1^* = v_2-1$,*

$\rho_2^* = v_1 - 1$, $\rho_3^* = (v_1-1)(v_2-1)$, $\boldsymbol{L}_1^* = (v_1)^{-1}\mathbf{1}_{v_1}\mathbf{1}_{v_1}' \otimes [\boldsymbol{I}_{v_2} - (v_2)^{-1}\mathbf{1}_{v_2}\mathbf{1}_{v_2}']$, $\boldsymbol{L}_2^* = [\boldsymbol{I}_{v_1} - (v_1)^{-1}\mathbf{1}_{v_1}\mathbf{1}_{v_1}'] \otimes (v_2)^{-1}\mathbf{1}_{v_2}\mathbf{1}_{v_2}'$, $\boldsymbol{L}_3^* = [\boldsymbol{I}_{v_1} - (v_1)^{-1}\mathbf{1}_{v_1}\mathbf{1}_{v_1}'] \otimes [\boldsymbol{I}_{v_2} - (v_2)^{-1}\mathbf{1}_{v_2}\mathbf{1}_{v_2}']$.

Proof. The direct calculation shows the result by recalling that the rectangular association scheme is defined in a rectangle of v_1 rows and v_2 columes (see also Kageyama, 1972c). □

As a large number of BIB designs are available in the literature (see also Section 8.2.1), Theorem 8.4.2 can produce many rectangular PBIB designs with all the ε_β being less than one. Their statements are omitted here.

Corollary 8.4.2. *There exist symmetric 3-associate GD designs with the parameters as given below, for the common* $\rho_1 = s_1 - 1, \rho_2 = s_1(s_2-1)$ *and* $\rho_3 = s_1s_2(s_3-1)$:

(2.1) $v = b = s_1s_2s_3$, $r = k = s_3(s_2-1)+1$, $\lambda_1 = s_3(s_2-1)$, $\lambda_2 = s_3(s_2-2)+2$, $\lambda_3 = 0$, *i.e.*, $(0;\rho_1,\rho_2,\rho_3;0)$*-EB with* $\varepsilon_1 = 1-[s_3(s_2-1)(s_2s_3-s_3+2)+1]/(rk)$, $\varepsilon_2 = 1-(s_3-1)^2/(rk)$, $\varepsilon_3 = 1-1/(rk)$;

(3.2) $v = b = s_1s_2s_3$, $r = k = s_3s_2(s_1-1)+s_3-1$, $\lambda_1 = s_3s_2(s_1-1)+s_3-2$, $\lambda_2 = s_3s_2(s_1-1)$, $\lambda_3 = s_3s_2(s_1-2)+s_3-1$, *i.e.*, $(0;\rho_1,\rho_2,\rho_3;0)$*-EB with* $\varepsilon_i = 1-[s_3^2(s_2-s_2+1)+s_2s_3-2s_3+1]/(rk)$, $\varepsilon_2 = 1-(s_3-1)^2/(rk)$, $\varepsilon_3 = 1-1/(rk)$.

Here $\boldsymbol{L}_\beta = \boldsymbol{A}_\beta^\#, \beta = 1,2,3,$ *can be found in Section* 6.0.2(F).

Note that the three cases (1), (2.2) and (3.1) not appearing in Corollary 8.4.2 are, respectively, $(\rho_0^*;\rho_1^*;0)$-EB with parameters $v = b = s_1s_2s_3$, $r = k = s_3s_2(s_1-1)$, $\lambda_1 = s_3s_2(s_1-1)$, $\lambda_2 = s_3s_2(s_1-1)$, $\lambda_3 = s_3s_2(s_1-2)$, $\varepsilon_1^* = 1-1/(s_1-1)^2$ for $\rho_0^* = s_1(s_2s_3-1)$ and $\rho_1^* = s_1-1$; $(0;\rho_1^*,\rho_2^*;0)$-EB with parameters $v = b = s_1s_2s_3$, $r = k = s_3s_2(s_1-1)+1$, $\lambda_1 = s_3s_2(s_1-1)$, $\lambda_2 = s_3s_2(s_1-1)$, $\lambda_3 = s_3s_2(s_1-2)+2$, $\varepsilon_1 = 1-(s_2s_3-1)^2/(rk)$, $\varepsilon_2 = 1-1/(rk)$, $\varepsilon_3 = 1-1/(rk)$, for $\rho_1^* = s_1-1$, $\rho_2^* = s_1(s_2s_3-1)$; and $(\rho_0^*;\rho_1^*,\rho_2^*;0)$-EB with parameters $v = b = s_1s_2s_3$, $r = k = s_3(s_1s_2-1)$, $\lambda_1 = s_3(s_1s_2-1)$, $\lambda_2 = s_3(s_1s_2-2)$, $\lambda_3 = s_3(s_1s_2-2)$, $\varepsilon_1^* = 1-1/(s_1s_2-1)^2$ and $\varepsilon_2^* = 1-1/(s_1s_2-1)^2$ for $\rho_0^* = s_1s_2(s_3-1)$, $\rho_1^* = s_1-1$ and $\rho_2^* = s_1(s_2-1)$.

8.4.1.2 Other than PBIB designs

As mentioned in Section 6.0.2, an association scheme in the classic sense (introduced by Bose and Nair, 1939) does not always exist for any parameters. However, there may be a possibility of getting equireplicate and proper $(0;\rho_1,\rho_2,\rho_3;0)$-EB designs, other than block structures like 3-associate PBIB designs, without taking into account the existence of an association scheme.

The construction of such designs is to find out three eigenvalues and their corresponding eigenvectors of $r^{-1}\boldsymbol{C}_1$ for the incidence matrix $\boldsymbol{N}$ of some design. As $r^{-1}\boldsymbol{C}_1 = \boldsymbol{I}_v - (rk)^{-1}\boldsymbol{N}\boldsymbol{N}'$, the present problem essentially results in evaluating a spectral decomposition of the matrix $(rk)^{-1}\boldsymbol{N}\boldsymbol{N}'(=\boldsymbol{M})$.

Theorem 8.4.3. *For the $v \times b$ incidence matrix $\boldsymbol{N}$ of a 2-associate PBIB design with parameters $v = 2k$, b, r, k, λ_1, λ_2, $\varepsilon_1(<1)$, $\varepsilon_2(<1)$, ρ_1, ρ_2, $\boldsymbol{L}_1$, $\boldsymbol{L}_2$, the incidence matrix*

$$\boldsymbol{N}^* = \begin{bmatrix} \boldsymbol{N} & \mathbf{1}_v\mathbf{1}_b' \\ \mathbf{1}_v\mathbf{1}_b' & \mathbf{1}_v\mathbf{1}_b' - \boldsymbol{N} \end{bmatrix}$$

yields a $(0;\rho_1^*,\rho_2^*,\rho_3^*;0)$-EB design with parameters $v^* = 4k$, $b^* = 4r$, $r^* = 3r$, $k^* = 3k$, $\varepsilon_\beta^* = (\varepsilon_\beta + 8)/9$, $\rho_\beta^* = 2\rho_\beta$, $\boldsymbol{L}_\beta^* = \boldsymbol{I}_2 \otimes \boldsymbol{L}_\beta$, $\beta = 1, 2$, $\varepsilon_3^* = 8/9$, $\rho_3^* = 1$, $\boldsymbol{L}_3^* = (\boldsymbol{I}_2 - 2^{-1}\mathbf{1}_2\mathbf{1}_2') \otimes (2k)^{-1}\mathbf{1}_{2k}\mathbf{1}_{2k}'$.

Proof. It follows from a direct evaluation of $\boldsymbol{N}^*(\boldsymbol{N}^*)'$. (See also Theorems 6.6.3 and 7.4.5.) □

Thus, by Theorem 8.4.3, 2-associate PBIB designs with $\varepsilon_1 < 1$ and $\varepsilon_2 < 1$ yield $(0;\rho_1,\rho_2,\rho_3;0)$-EB designs which belong to the present category of designs. Such 2-associate PBIB designs can be found in abundance in Clatworthy (1973), as discussed in Section 8.3.1. For example, by a regular GD design of Theorem 8.3.3(1), one can get the following series.

Corollary 8.4.3. *When $2k-1$ is a prime or a prime power, there exists a $(0;\rho_1^*,\rho_2^*,\rho_3^*;0)$-EB design with parameters $v^* = 4k$, $b^* = 4(4k-3)$, $r^* = 3(4k-3)$, $k^* = 3k$, $\varepsilon_1^* = 1 - 1/[9(4k-3)]$, $\varepsilon_2^* = 1 - 2/[9(4k-3)]$, $\varepsilon_3^* = 8/9$, $\rho_1^* = 2$, $\rho_2^* = 4(k-1)$, $\rho_3^* = 1$, $\boldsymbol{L}_1^* = \boldsymbol{I}_2 \otimes (\boldsymbol{I}_2 - 2^{-1}\mathbf{1}_2\mathbf{1}_2') \otimes k^{-1}\mathbf{1}_k\mathbf{1}_k'$, $\boldsymbol{L}_2^* = \boldsymbol{I}_2 \otimes \boldsymbol{I}_2 \otimes (\boldsymbol{I}_k - k^{-1}\mathbf{1}_k\mathbf{1}_k')$, $\boldsymbol{L}_3^* = (\boldsymbol{I}_2 - 2^{-1}\mathbf{1}_2\mathbf{1}_2') \otimes (2k)^{-1}\mathbf{1}_{2k}\mathbf{1}_{2k}'$.*

When $k = m$ in a regular GD design, Theorem 6.4.3 can yield a $(0;\rho_1,\rho_2,\rho_3;0)$-EB design in the present category, i.e., any contrast is estimated in the intrablock analysis with efficiency less than one.

Corollary 8.4.4. *Let $\boldsymbol{N}$ be the $v \times b$ incidence matrix of a regular GD design with parameters $v = mn$ (m groups of n treatments each), b, r, $k = m$, λ_1, λ_2. Then the incidence matrix $[\boldsymbol{N} : \mathbf{1}_m\mathbf{1}_s' \otimes \boldsymbol{I}_n]$ yields a $(0;\rho_1^*,\rho_2^*,\rho_3^*;0)$-EB design with parameters $v^* = mn$, $b^* = b + ns$, $r^* = r + s$, $k^* = k$, $\varepsilon_1^* = 1 - (r-\lambda_1)/[k(r+s)]$, $\varepsilon_2^* = 1 - (r - \lambda_1 + sk)/[k(r+s)]$, $\varepsilon_3^* = 1 - (rk - v\lambda_2)/[k(r+s)]$, $\rho_1^* = (m-1)(n-1)$, $\rho_2^* = n-1$, $\rho_3^* = m-1$, $\boldsymbol{L}_1^* = (\boldsymbol{I}_m - m^{-1}\mathbf{1}_m\mathbf{1}_m') \otimes (\boldsymbol{I}_n - n^{-1}\mathbf{1}_n\mathbf{1}_n')$, $\boldsymbol{L}_2^* = m^{-1}\mathbf{1}_m\mathbf{1}_m' \otimes (\boldsymbol{I}_n - n^{-1}\mathbf{1}_n\mathbf{1}_n')$, $\boldsymbol{L}_3^* = (\boldsymbol{I}_m - m^{-1}\mathbf{1}_m\mathbf{1}_m') \otimes n^{-1}\mathbf{1}_n\mathbf{1}_n'$ for a positive integer s.*

Regular GD designs with $k = m$ used in Corollary 8.4.4 can be found in abundance in Clatworthy (1973). For example, take a regular GD design, based on five groups $\{1,2\}, \{3,4\}, \{5,6\}, \{7,8\}, \{9,10\}$ of two treatments, R140 in Clatworthy (1973), with parameters $v = 10$, $b = 14$, $r = 7$, $k = 5$, $\lambda_1 = 4$, $\lambda_2 = 3$, $m = 5$, $n = 2$. Then Corollary 8.4.4 produces a $(0;4,1,4;0)$-EB design with parameters $v^* = 10$, $b^* = 2s + 14$, $r^* = s + 7$, $k^* = 5$, $\varepsilon_1^* = 1 - 3/[5(s+7)]$, $\varepsilon_2^* = 1 - 1/(s+7)$, $\varepsilon_3^* = 1 - (5s+3)/[5(s+7)]$, $\rho_1^* = 4$, $\rho_2^* = 1$, $\rho_3^* = 4$, $\boldsymbol{L}_1^* = (\boldsymbol{I}_5 - 5^{-1}\mathbf{1}_5\mathbf{1}_5') \otimes (\boldsymbol{I}_2 - 2^{-1}\mathbf{1}_2\mathbf{1}_2')$, $\boldsymbol{L}_2^* = (\boldsymbol{I}_5 - 5^{-1}\mathbf{1}_5\mathbf{1}_5') \otimes 2^{-1}\mathbf{1}_2\mathbf{1}_2'$,

$L_3^* = 5^{-1}\mathbf{1}_5\mathbf{1}_5' \otimes (I_2 - 2^{-1}\mathbf{1}_2\mathbf{1}_2')$ for a positive integer s.

Another pattern is also considered as a modification of Theorem 7.4.10 based on Theorem 6.6.4 with $a = 1$ and $c = 0$ [see also the pattern (6.6.2) with $a = 1$ and $c = 0$].

Theorem 8.4.4. *Let N be the $v \times b$ incidence matrix of a 2-associate PBIB design with parameters v, b, r, k, λ_1, λ_2, $\varepsilon_1(< 1)$, $\varepsilon_2(< 1)$, ρ_1, ρ_2, L_1, L_2. Then the incidence matrix*

$$N^* = \begin{bmatrix} N & \mathbf{1}_v\mathbf{1}_b' & O \\ O & N & \mathbf{1}_v\mathbf{1}_b' \\ \mathbf{1}_v\mathbf{1}_b' & O & N \end{bmatrix}$$

yields a $(0; \rho_1^*, \rho_2^*, \rho_3^*; 0)$-EB design with parameters $v^* = 3v$, $b^* = 3b$, $r^* = b + r$, $k^* = v + k$, $\varepsilon_\beta^* = 1 - rk(1 - \varepsilon_\beta)/[(b + r)(v + k)]$, $\rho_\beta^* = 3\rho_\beta$, $\varepsilon_3^* = 1 - [vb - r(v - k)]/[(b + r)(v + k)]$, $\rho_3^* = 2$, $L_\beta^* = I_3 \otimes L_\beta$, $L_3^* = (I_3 - 3^{-1}\mathbf{1}_3\mathbf{1}_3') \otimes v^{-1}\mathbf{1}_v\mathbf{1}_v'$, $\beta = 1, 2$.

Proof. Some direct calculation of $N^*(N^*)'$ can yield the required design. □

This theorem implies that a 2-associate PBIB design with $\varepsilon_1 < 1$ and $\varepsilon_2 < 1$ can yield a $(0; \rho_1, \rho_2, \rho_3; 0)$-EB design in which any contrast is estimated in the intra-block analysis with efficiency less than one. For example, by taking a regular GD design of Corollary 8.3.1, the following result can be obtained.

Corollary 8.4.5. *When q is a prime or a prime power, there exists a $(0; \rho_1, \rho_2, \rho_3; 0)$-EB design with parameters $v = 3(q^2 - 1) = b$, $r = q^2 + q - 1 = k$, $\varepsilon_1 = 1 - 1/(q^2 + q - 1)^2$, $\varepsilon_2 = 1 - q/(q^2 + q - 1)^2$, $\varepsilon_3 = 1 - (q^4 - q^3 - q^2 + q + 1)/(q^2 + q - 1)^2$, $\rho_1 = 3q$, $\rho_2 = 3(q + 1)(q - 2)$, $\rho_3 = 2$, $L_1 = I_3 \otimes [I_{q+1} - (q + 1)^{-1}\mathbf{1}_{q+1}\mathbf{1}_{q+1}'] \otimes (q - 1)^{-1}\mathbf{1}_{q-1}\mathbf{1}_{q-1}'$, $L_2 = I_3 \otimes I_{q+1} \otimes [I_{q-1} - (q - 1)^{-1}\mathbf{1}_{q-1}\mathbf{1}_{q-1}']$, $L_3 = (I_3 - 3^{-1}\mathbf{1}_3\mathbf{1}_3') \otimes (q^2 - 1)^{-1}\mathbf{1}_{q^2-1}\mathbf{1}_{q^2-1}'$.*

The pattern (7.3.3) is also useful to the present section. For example, in Theorem 7.3.5, if the condition $v = 2k$ is relaxed, the resulting design may be a $(0; \rho_1^*, \rho_2^*, \rho_3^*; 0)$-EB design. In fact, a BIB design with parameters $v = b = 7, r = k = 3$, $\lambda = 1$ yields through the pattern (7.3.3) a (0;1,6,6;0)-EB design with the efficiencies $\varepsilon_1^* = 811/882 = 0.9195$, $\varepsilon_2^* = (1595 + \sqrt{337})/1764 = 0.9146$, $\varepsilon_3^* = (1595 - \sqrt{337})/1764 = 0.8938$ with $\rho_1^* = 1$ and $\rho_2^* = \rho_3^* = 6$. General consideration is left to the reader.

8.4.2 Proper designs that are nonequireplicate

This section is prepared similarly as Section 7.4.2. Unfortunately, one cannot take a dual method (Section 6.1) of known designs, because any design constructed by this method has some of ε_β being one. A supplementation method (Section 6.3) also has the same status. Once there exists a nonequireplicate $(0; \rho_1, \rho_2, \rho_3; 0)$-EB design, any its juxtaposition (see Theorem 6.4.1) can produce nonequireplicate $(0; \rho_1, \rho_2, \rho_3; 0)$-EB designs.

The Kronecker product of two $(0; v-1; 0)$-EB designs may yield nonequireplicate designs in the present category (see Puri and Nigam, 1977a) that can also belong to the category of designs considered in Section 8.4.4, when starting designs are nonproper and nonequireplicate. This method is explained in Section 8.4.4 as will be shown in Theorem 8.4.9, Corollaries 8.4.8 and 8.4.9.

A merging method (Section 6.5) may work for the construction of designs in the present category. For example, see Puri and Nigam (1983) who constructed such designs by merging some treatments in the extended group divisible partially efficiency-balanced (PEB) designs constructed by Puri and Nigam (1977a). The reader may try this as an exercise.

Next it will be shown (in Theorem 8.4.5) that a balanced block design with two different numbers of replicates, introduced by Corsten (1962), is a $(0; \rho_1, \rho_2, \rho_3; 0)$-EB design. According to his terminology, a balanced block design with two different numbers of replicates is an incomplete binary proper design with the first set of v_1 treatments occurring r_1 times and the second set of v_2 treatments occurring r_2 times ($r_1 < r_2$) arranged into b blocks of constant block size k such that

(i) any two distinct treatments in the ith set occur together in λ_{ii} blocks, $i = 1, 2$;

(ii) any two treatments from different sets occur together in $\lambda_{12} = \lambda_{21}$ (> 0) blocks.

It can be shown that the parameters satisfy

$$v_1 r_1 + v_2 r_2 = bk, \quad r_1(k-1) = \lambda_{11}(v_1 - 1) + \lambda_{12} v_2,$$
$$r_2(k-1) = \lambda_{12} v_1 + \lambda_{22}(v_2 - 1), r_1 \geq \lambda_{12}, \quad r_2 \geq \lambda_{12},$$
$$r_1 \geq \lambda_{11}, \quad r_2 \geq \lambda_{22}, \quad r_1 r_2 \geq \lambda_{12} b.$$

This type of designs for several sets of treatments were first introduced by Nair and Rao (1942) under the name of an inter- and intra-group balanced design.

Theorem 8.4.5. *A balanced block design with parameters $v_1, v_2, b, r_1, r_2, k, \lambda_{11}, \lambda_{12}, \lambda_{22}$ is a $(0; \rho_1^*, \rho_2^*, \rho_3^*; 0)$-EB design with parameters* $v^* = v_1 + v_2, b^* = b, \boldsymbol{r}^* = [r_1 \mathbf{1}'_{v_1}, r_2 \mathbf{1}'_{v_2}]', k^* = k, \varepsilon_1^* = 1 - (r_1 - \lambda_{11})/(kr_1), \varepsilon_2^* = 1 - (r_2 - \lambda_{22})/(kr_2), \varepsilon_3^* = 1 - (r_1 r_2 - b\lambda_{12})/(r_1 r_2), \rho_1^* = v_1 - 1, \rho_2^* = v_2 - 1, \rho_3^* = 1, \boldsymbol{L}_1^* = \mathrm{diag}[\boldsymbol{I}_{v_1} - (v_1)^{-1}\mathbf{1}_{v_1}\mathbf{1}'_{v_1} : \mathbf{O}], \boldsymbol{L}_2^* = \mathrm{diag}[\mathbf{O} : \boldsymbol{I}_{v_2} - (v_2)^{-1}\mathbf{1}_{v_2}\mathbf{1}'_{v_2}],$

$$\boldsymbol{L}_3^* = \frac{1}{bk}\begin{bmatrix} \frac{v_2 r_2}{v_1}\mathbf{1}_{v_1}\mathbf{1}'_{v_1} & -r_2\mathbf{1}_{v_1}\mathbf{1}'_{v_2} \\ -r_1\mathbf{1}_{v_2}\mathbf{1}'_{v_1} & \frac{v_1 r_1}{v_2}\mathbf{1}_{v_2}\mathbf{1}'_{v_2} \end{bmatrix}.$$

Proof. By the definition, the balanced block design with two different numbers of replicates has the parameters $v^* = v_1 + v_2$, $b^* = b$, $\boldsymbol{r}^* = [r_1\mathbf{1}'_{v_1}, r_2\mathbf{1}'_{v_2}]'$, $k^* = k$, and its incidence matrix $\boldsymbol{N}$ obviously satisfies

$$\boldsymbol{NN}' = \begin{bmatrix} (r_1 - \lambda_{11})\boldsymbol{I}_{v_1} + \lambda_{11}\mathbf{1}_{v_1}\mathbf{1}'_{v_1} & \lambda_{12}\mathbf{1}_{v_1}\mathbf{1}'_{v_2} \\ \lambda_{21}\mathbf{1}_{v_2}\mathbf{1}'_{v_1} & (r_2 - \lambda_{22})\boldsymbol{I}_{v_2} + \lambda_{22}\mathbf{1}_{v_2}\mathbf{1}'_{v_2} \end{bmatrix}.$$

Hence, it follows that, relating to Section 4.4, the matrix $\boldsymbol{M}_0$ is calculated as

$$\frac{1}{k^*}(\boldsymbol{r}^*)^{-\delta}\boldsymbol{N}\boldsymbol{N}' - \frac{1}{bk}\mathbf{1}_{v^*}(\boldsymbol{r}^*)'$$

$$= \frac{1}{bk}\begin{bmatrix} \frac{b(r_1-\lambda_{11})}{r_1}\boldsymbol{I}_{v_1} + \frac{b\lambda_{11}-r_1^2}{r_1}\mathbf{1}_{v_1}\mathbf{1}'_{v_1} & \frac{b\lambda_{12}-r_1r_2}{r_1}\mathbf{1}_{v_1}\mathbf{1}'_{v_2} \\ \frac{b\lambda_{12}-r_1r_2}{r_2}\mathbf{1}_{v_2}\mathbf{1}'_{v_1} & \frac{b(r_2-\lambda_{22})}{r_2}\boldsymbol{I}_{v_2} + \frac{b\lambda_{22}-r_2^2}{r_2}\mathbf{1}_{v_2}\mathbf{1}'_{v_2} \end{bmatrix}$$

$$= \mu_1\boldsymbol{L}_1^* + \mu_2\boldsymbol{L}_2^* + \mu_3\boldsymbol{L}_3^* \text{ with } \boldsymbol{L}_\beta^*\boldsymbol{L}_{\beta'}^* = \delta_{\beta\beta'}\boldsymbol{L}_\beta^*,$$

where

$$\mu_1 = \frac{r_1-\lambda_{11}}{kr_1},\ \rho_1^* = v_1 - 1,\ \mu_2 = \frac{r_2-\lambda_{22}}{kr_2},\ \rho_2^* = v_2 - 1,$$

$$\mu_3 = \frac{r_1r_2 - b\lambda_{12}}{r_1r_2},\ \rho_3^* = 1,$$

and $\boldsymbol{L}_\beta^*, \beta = 1,2,3$, are given as in the theorem. □

Remark 8.4.1. Since $r_1 \geq \lambda_{11}, r_2 \geq \lambda_{22}$ and $r_1r_2 \geq b\lambda_{12}$, there is a possibility of getting designs in which some basic contrasts can be estimated in the intrablock analysis with full efficiency. The present designs are also classified mutually exclusively into five types as follows.

(1) $r_1 = \lambda_{12}$; this implies $r_2 = \lambda_{22}$, $r_1r_2 = \lambda_{12}b$, $r_1 > \lambda_{11}$. Then $\varepsilon_2^* = \varepsilon_3^* = 1$, i.e., the design is $(\rho_0^{**}; \rho_1^{**}; 0)$-EB with $\varepsilon_0^{**} = 1$, $\varepsilon_1^{**} = 1 - (r_1 - \lambda_{11})/(kr_1)$, $\rho_0^{**} = v_2$ and $\rho_1^{**} = v_1 - 1$.

(2) $r_2 = \lambda_{22}$ and $r_1 > \lambda_{12}$; this implies $r_1r_2 > \lambda_{12}b$, $r_1 > \lambda_{11}$. Then $\varepsilon_2^* = 1$, i.e., the design is $(\rho_0^{**}; \rho_1^{**}, \rho_2^{**}; 0)$-EB with $\varepsilon_0^{**} = 1$, $\varepsilon_1^{**} = 1 - (r_1 - \lambda_{11})/(kr_1)$, $\varepsilon_2^{**} = 1 - (r_1r_2 - b\lambda_{12})/(r_1r_2)$, $\rho_0^{**} = v_2 - 1$, $\rho_1^{**} = v_1 - 1$ and $\rho_2^{**} = 1$.

(3) $r_1 = \lambda_{11}$; this implies $r_2 > \lambda_{22}$, $r_1r_2 > \lambda_{12}b$, $r_1 > \lambda_{12}$. Then $\varepsilon_1^* = 1$, i.e., the design is $(\rho_0^{**}; \rho_1^{**}, \rho_2^{**}; 0)$-EB with $\varepsilon_0^{**} = 1$, $\varepsilon_1^{**} = 1 - (r_2 - \lambda_{22})/(kr_2)$, $\varepsilon_2^{**} = 1 - (r_1r_2 - b\lambda_{12})/(r_1r_2)$, $\rho_0^{**} = v_1 - 1$, $\rho_1^{**} = v_2 - 1$ and $\rho_2^{**} = 1$.

(4) $r_1r_2 = \lambda_{12}b$ and $r_1 > \lambda_{12}$; this implies $r_2 > \lambda_{22}$, $r_1 > \lambda_{11}$. Then $\varepsilon_3^* = 1$, i.e., the design is $(\rho_0^{**}; \rho_1^{**}, \rho_2^{**}; 0)$-EB with $\varepsilon_0^{**} = 1$, $\varepsilon_1^{**} = 1 - (r_1 - \lambda_{11})/(kr_1)$, $\varepsilon_2^{**} = 1 - (r_2 - \lambda_{22})/(kr_2)$, $\rho_0^{**} = 1$, $\rho_1^{**} = v_1 - 1$ and $\rho_2^{**} = v_2 - 1$.

(5) $r_1 > \lambda_{11}$, $r_2 > \lambda_{22}$ and $r_1r_2 > \lambda_{12}b$; this implies $r_1 > \lambda_{12}$.

For each type above, Corsten (1962) illustrated some construction methods. Some constructions are also given by Adhikary (1965, 1967) and Agrawal (1963). Kageyama and Sinha (1988), and Sinha and Kageyama (1990) further gave several systematic procedures of constructing such designs and tabulated the parameters of 673 plans along with their construction source within the scope of parameters $2 \leq v_1, v_2, r_1, r_2, k \leq 10$ and $r_1 < r_2$. This tabulation can be used to produce a large number of designs with m efficiency factors for $m \leq 3$. The table also includes values of three distinct variances for treatment differences and degrees of freedom to estimate error variance. The A-optimality of some designs (for comparing test treatments with controls) in this class has been investigated by Majumdar (1986).

In general, a design resulting from Theorem 8.4.5 has the following $\boldsymbol{\Omega}$ matrix as

$$\boldsymbol{\Omega} = \begin{bmatrix} \Omega_{11} & \Omega_{12} \\ \Omega_{21} & \Omega_{22} \end{bmatrix},$$

where

$$\Omega_{11} = \frac{k}{r_1(k-1)+\lambda_{11}}\boldsymbol{I}_{v_1} - \frac{1}{r_1}\left\{\frac{r_1-\lambda_{11}}{v_1[r_1(k-1)+\lambda_{11}]} - \frac{(r_1r_2-b\lambda_{12})v_2r_2}{b^2\lambda_{12}v_1k}\right\}\mathbf{1}_{v_1}\mathbf{1}'_{v_1},$$

$$\Omega_{12} = -\frac{r_1r_2-b\lambda_{12}}{b^2\lambda_{12}k}\mathbf{1}_{v_1}\mathbf{1}'_{v_2} = \Omega'_{21},$$

$$\Omega_{22} = \frac{k}{r_2(k-1)+\lambda_{22}}\boldsymbol{I}_{v_2} - \frac{1}{r_2}\left\{\frac{r_2-\lambda_{22}}{v_2[r_2(k-1)+\lambda_{22}]} - \frac{(r_1r_2-b\lambda_{12})v_1r_1}{b^2\lambda_{12}v_2k}\right\}\mathbf{1}_{v_2}\mathbf{1}'_{v_2}.$$

Then one can evaluate the variances of any elementary contrasts in the intra-block analysis as follows:

(i) for any treatments belonging to the first group

$$\frac{2k}{r_1(k-1)+\lambda_{11}}\sigma_1^2;$$

(ii) for any treatments belonging to the second group

$$\frac{2k}{r_2(k-1)+\lambda_{22}}\sigma_1^2;$$

(iii) for a treatment in the first group and a treatment in the second group

$$\left\{\frac{k}{r_1(k-1)+\lambda_{11}} + \frac{k}{r_2(k-1)+\lambda_{22}} + \frac{2(r_1r_2-b\lambda_{12})}{b^2k\lambda_{12}} - \frac{1}{r_1}\left(\frac{r_1-\lambda_{11}}{v_1[r_1(k-1)+\lambda_{11}]} - \frac{(r_1r_2-b\lambda_{12})v_2r_2}{b^2\lambda_{12}v_1k}\right) - \frac{1}{r_2}\left(\frac{r_2-\lambda_{22}}{v_2[r_2(k-1)+\lambda_{22}]} - \frac{(r_1r_2-b\lambda_{12})v_1r_1}{b^2\lambda_{12}v_2k}\right)\right\}\sigma_1^2.$$

Another class of $(0;\rho_1,\rho_2,\rho_3;0)$-EB designs can also be constructed. At first, recall (see Section 8.2.1) that a BIB design with parameters v, b, r, k, λ is a $(0;v-1;0)$-EB design with the efficiency factor $\lambda v/(rk)$. In many experiments in physical and biological sciences the blocks are natural units and it is quite common that the entire block is missing (see also Section 10.2). Suppose that all the k treatments of a block in the BIB design are missing. In this case, it will be seen that $v-k-1$ basic contrasts are still estimated in the intra-block analysis with the same efficiency $\lambda v/(rk)$ in the framework of designs with three efficiency factors, as the following theorem shows.

Theorem 8.4.6. *The existence of a BIB design with parameters* v, b, r, k, λ *implies, through omitting a block, the existence of a* $(0;\rho_1^*,\rho_2^*,\rho_3^*;0)$*-EB design with parameters* $v^* = v$, $b^* = b-1$, $\boldsymbol{r}^* = [(r-1)\mathbf{1}_k', r\mathbf{1}_{v-k}']'$, $k^* = k$, $\varepsilon_1^* = [r^2(k-1)+\lambda(r-k)]/[rk(r-1)]$, $\varepsilon_2^* = v\lambda/(rk)$, $\varepsilon_3^* = (\lambda v - r)/(\lambda v - \lambda)$, $\rho_1^* = 1$, $\rho_2^* = v-k-1$, $\rho_3^* = k-1$, $\boldsymbol{L}_1^* = \boldsymbol{I}_v - [k(b-1)]^{-1}\mathbf{1}_v(\boldsymbol{r}^*)' - \boldsymbol{L}_2^* - \boldsymbol{L}_3^*$, $\boldsymbol{L}_2^* = \operatorname{diag}[\mathbf{O} : \boldsymbol{I}_{v-k} - (v-k)^{-1}\mathbf{1}_{v-k}\mathbf{1}_{v-k}']$, $\boldsymbol{L}_3^* = \operatorname{diag}[\boldsymbol{I}_k - k^{-1}\mathbf{1}_k\mathbf{1}_k' : \mathbf{O}]$.

Proof. Without loss of generality, let the missing block contain the first k treatments in a BIB design with parameters v, b, r, k, λ. That is, the $v \times b$ incidence matrix $\boldsymbol{N}$ of the BIB design is partitioned as follows.

$$\boldsymbol{N} = \left[\begin{array}{c|c} \mathbf{1}_k & \boldsymbol{N}^* \\ \mathbf{0} & \end{array}\right].$$

Then the resulting design has the $v\times(b-1)$ incidence matrix $\boldsymbol{N}^*$ in which there are v treatments, $b-1$ blocks each of size k, the first k treatments have $r-1$ replications and the remaining $v-k$ treatments have r replications. Consider the matrix

$$\boldsymbol{M}^* = \operatorname{diag}[\frac{1}{r-1}\boldsymbol{I}_k : \frac{1}{r}\boldsymbol{I}_{v-k}]\boldsymbol{N}^*\frac{1}{k}\boldsymbol{I}_{b-1}(\boldsymbol{N}^*)' = \frac{1}{k}\begin{bmatrix} \frac{1}{r-1}\boldsymbol{A} & \frac{1}{r-1}\boldsymbol{B} \\ \frac{1}{r}\boldsymbol{B}' & \frac{1}{r}\boldsymbol{D} \end{bmatrix},$$

where

$$\boldsymbol{A} = (r-\lambda)\boldsymbol{I}_k + (\lambda-1)\mathbf{1}_k\mathbf{1}_k', \ \ \boldsymbol{B} = \lambda\mathbf{1}_k\mathbf{1}_{v-k}', \ \ \boldsymbol{D} = (r-\lambda)\boldsymbol{I}_{v-k} + \lambda\mathbf{1}_{v-k}\mathbf{1}_{v-k}'.$$

It is further shown that, for $\boldsymbol{r}^* = [(r-1)\mathbf{1}_k', r\mathbf{1}_{v-k}']'$,

$$\boldsymbol{M}^* = \frac{1}{k(b-1)}\mathbf{1}_v(\boldsymbol{r}^*)' + \frac{(r-\lambda)(r-k)}{rk(r-1)}\boldsymbol{L}_1^* + \frac{r-\lambda}{rk}\boldsymbol{L}_2^* + \frac{r-\lambda}{r(k-1)}\boldsymbol{L}_3^*,$$

where

$$\boldsymbol{L}_1^* = \boldsymbol{I}_v - \frac{1}{k(b-1)}\mathbf{1}_v(\boldsymbol{r}^*)' - \boldsymbol{L}_2^* - \boldsymbol{L}_3^*,$$

$$\boldsymbol{L}_2^* = \operatorname{diag}[\mathbf{O} : \boldsymbol{I}_{v-k} - \frac{1}{v-k}\mathbf{1}_{v-k}\mathbf{1}_{v-k}'], \ \ \boldsymbol{L}_3^* = \operatorname{diag}[\boldsymbol{I}_k - \frac{1}{k}\mathbf{1}_k\mathbf{1}_k' : \mathbf{O}],$$

$$\operatorname{rank}(\boldsymbol{L}_1^*) = 1, \ \operatorname{rank}(\boldsymbol{L}_2^*) = v-k-1, \ \operatorname{rank}(\boldsymbol{L}_3^*) = k-1.$$

This shows that $\boldsymbol{N}^*$ is the incidence matrix of the required $(0;\rho_1,\rho_2,\rho_3;0)$-EB design. □

Remark 8.4.2. In Theorem 8.4.6, $\varepsilon_2^* = \lambda v/(rk)$ which is equal to the efficiency factor of the original BIB design. Similarly, one can consider a GD design having a complete missing block. However, the resulting design becomes a $(\rho_0; \rho_1, \rho_2, \rho_3, \rho_4; 0)$-EB or $(\rho_0; \rho_1, \rho_2, \rho_3; 0)$-EB design. The detail can be found in Puri (1984).

A truncation method of construction can finally be considered.

Let $\mathcal{D}$ be a binary proper $(\rho_0; \rho_1, ..., \rho_{m-1}; 0)$-EB design with parameters v, b, $\boldsymbol{r} = [r_1, r_2, ..., r_v]'$, $k(\geq 3)$, in which

$$\boldsymbol{F} = \boldsymbol{r}^{-\delta/2}\boldsymbol{C}_1\boldsymbol{r}^{-\delta/2} = \sum_{\beta=0}^{m-1} \varepsilon_\beta \boldsymbol{r}^{\delta/2}\boldsymbol{L}_\beta \boldsymbol{r}^{-\delta/2}, \; 0 < \varepsilon_\beta \leq 1$$

[see (4.4.10)]. Now, along a line of Cheng (1981), the following definition of truncation is introduced. For any k' $(2 \leq k' < k)$ the k'th truncation of $\mathcal{D}$ is a block design $\mathcal{D}^*$ obtained from $\mathcal{D}$ by replacing each block in $\mathcal{D}$ by a set of $\binom{k}{k'}$ blocks considering all possible selections of k' treatments from the k treatments in a block. It is clear that $\mathcal{D}^*$ has the parameters

$$v^* = v, \quad b^* = b\binom{k}{k'}, \quad r_i^* = r_i\binom{k-1}{k'-1}, \quad k^* = k'.$$

Let $\boldsymbol{C}_1$ and $\boldsymbol{C}_1^*$ be the C-matrices of designs $\mathcal{D}$ and $\mathcal{D}^*$ with the incidence matrices $\boldsymbol{N}$ and $\boldsymbol{N}^*$, respectively. Then

$$\boldsymbol{C}_1 = \boldsymbol{r}^\delta - \frac{1}{k}\boldsymbol{N}\boldsymbol{N}' \text{ and } \boldsymbol{C}_1^* = (\boldsymbol{r}^*)^\delta - \frac{1}{k^*}\boldsymbol{N}^*(\boldsymbol{N}^*)'.$$

Note that letting the off-diagonal elements of $\boldsymbol{N}\boldsymbol{N}'$ be λ_{ij}, the off-diagonal elements of $\boldsymbol{N}^*(\boldsymbol{N}^*)'$ become $\lambda_{ij}\binom{k-2}{k'-2}$. This means that

$$\boldsymbol{N}^*(\boldsymbol{N}^*)' = \binom{k-2}{k'-2}\left(\boldsymbol{N}\boldsymbol{N}' + \frac{k-k'}{k'-1}\boldsymbol{r}^\delta\right).$$

Hence

$$\begin{aligned}
\boldsymbol{C}_1^* &= \binom{k-1}{k'-1}\boldsymbol{r}^\delta - \frac{1}{k'}\binom{k-2}{k'-2}\left\{k(\boldsymbol{r}^\delta - \boldsymbol{C}_1) + \frac{k-k'}{k'-1}\boldsymbol{r}^\delta\right\} \\
&= \frac{k}{k'}\binom{k-2}{k'-2}\boldsymbol{C}_1 + \boldsymbol{r}^\delta\left\{\binom{k-1}{k'-1} - \frac{k}{k'}\binom{k-2}{k'-2} - \frac{k-k'}{k'(k'-1)}\binom{k-2}{k'-2}\right\} \\
&= \frac{k}{k'}\binom{k-2}{k'-2}\boldsymbol{C}_1.
\end{aligned}$$

Therefore it follows that

$$\begin{aligned}
(\boldsymbol{r}^*)^{-\delta/2}\boldsymbol{C}_1^*(\boldsymbol{r}^*)^{-\delta/2} &= \left\{k\binom{k-2}{k'-2}/[k'\binom{k-1}{k'-1}]\right\}\boldsymbol{r}^{-\delta/2}\boldsymbol{C}_1\boldsymbol{r}^{-\delta/2} \\
&= \sum_{\beta=0}^{m-1} \frac{k(k'-1)\varepsilon_\beta}{k'(k-1)}(\boldsymbol{r}^*)^{\delta/2}\boldsymbol{L}_\beta(\boldsymbol{r}^*)^{-\delta/2}.
\end{aligned}$$

Thus the following theorem has been established.

Theorem 8.4.7. *The existence of a* $(\rho_0; \rho_1, ..., \rho_{m-1}; 0)$*-EB design with parameters* v, b, r, k (≥ 3), ε_β, ρ_β, $\boldsymbol{L}_\beta$, $\beta = 0, 1, ..., m-1$, *implies, through the* k'*th truncation* $(2 \leq k' < k)$, *the existence of a* $(0; \rho_1^*, \rho_2^*, ..., \rho_{m^*-1}^*; 0)$*-EB design with parameters* $v^* = v$, $b^* = b\binom{k}{k'}$, $\boldsymbol{r}^* = \binom{k-1}{k'-1}\boldsymbol{r}$, $\boldsymbol{k}^* = k'$, $\varepsilon_{\beta^*}^* = \{k(k'-1)/[k'(k-1)]\}\varepsilon_{\beta^*-1}$, $\rho_{\beta^*}^* = \rho_{\beta^*-1}$, $\boldsymbol{L}_{\beta^*}^* = \boldsymbol{L}_{\beta^*-1}$, $\beta^* = 1, 2, ..., m^* - 1$, *where* $m^* = m + 1$.

Remark 8.4.3. Unfortunately, $\varepsilon_\beta^* < \varepsilon_\beta$ always hold for all β, because of $k' < k$. This means that the values of the efficiency factors in every truncation become smaller than the original ones. But this is a very convenient way of constructing new designs from a block design which belongs to the same category of designs.

When $m = 2$ and $\rho_0 = 0$, a binary proper $(0; v-1; 0)$-EB design is a BIB design (see Theorem 8.2.3). Hence Theorem 8.4.7 can yield the following result.

Corollary 8.4.6. *The existence of a BIB design with parameters* v, b, r, k (≥ 3), λ, $\varepsilon_1 = \lambda v/(rk)$, $\rho_1 = v - 1$, $\boldsymbol{L}_1 = \boldsymbol{I}_v - v^{-1}\boldsymbol{1}_v\boldsymbol{1}_v'$ *implies, through the* k'*th truncation* $(2 \leq k' < k)$, *the existence of a BIB design with parameters* $v^* = v$, $b^* = b\binom{k}{k'}$, $r^* = \binom{k-1}{k'-1}r$, $k^* = k'$, $\lambda^* = r^*(k'-1)/(v-1)$, $\varepsilon_1^* = \{k(k'-1)/[k'(k-1)]\}\varepsilon_1 = v(k'-1)/[k'(v-1)]$, $\rho_1^* = \rho_1$, $\boldsymbol{L}_1^* = \boldsymbol{L}_1$.

When $m \geq 2$, one may describe statements on construction of designs with more efficiency factors. But, as seen from Theorem 8.4.7, note that the efficiency factors ε_β^* in the truncation never become one, even if in the original design $\varepsilon_0 = 1$ exists.

8.4.3 Nonproper designs that are equireplicate

The most common method of constructing designs in this category is a juxtaposition technique described in Section 6.4. An idea used in Theorem 6.4.2 can also be generalized into Theorem 7.4.16 and then Corollaries 7.4.18 – 7.4.22, which still work here to construct various kinds of designs in which any contrast is estimated in the intra-block analysis with efficiency factor less than one.

One can use rectangular designs and 3-associate GD PBIB designs given in Corollaries 8.4.1 and 8.4.2, respectively, in which there are many such designs with a common number of treatments.

Corollary 8.4.7. *There are* $(0; \rho_1, \rho_2, \rho_3; 0)$*-EB designs, satisfying* (8.2.4), *with the following parameters:*

(1) $v = st$, $b = 2st$, $r = s(t-1)$, $\boldsymbol{k} = [(t-1)\boldsymbol{1}_{st}', (s-1)(t-1)\boldsymbol{1}_{st}']'$, $\varepsilon_1 = 1 - (t-1)^{-2}$, $\varepsilon_2 = 1 - (s-1)^{-1}$, $\varepsilon_3 = 1 - [(s-1)(t-1)^2]^{-1}$, $\rho_1 = t - 1$, $\rho_2 = s - 1$, $\rho_3 = (s-1)(t-1)$, $\boldsymbol{L}_1 = s^{-1}\boldsymbol{1}_s\boldsymbol{1}_s' \otimes (\boldsymbol{I}_t - t^{-1}\boldsymbol{1}_t\boldsymbol{1}_t')$, $\boldsymbol{L}_2 = (\boldsymbol{I}_s - s^{-1}\boldsymbol{1}_s\boldsymbol{1}_s') \otimes t^{-1}\boldsymbol{1}_t\boldsymbol{1}_t'$, $\boldsymbol{L}_3 = (\boldsymbol{I}_s - s^{-1}\boldsymbol{1}_s\boldsymbol{1}_s') \otimes (\boldsymbol{I}_t - t^{-1}\boldsymbol{1}_t\boldsymbol{1}_t')$;

(2) $v = st$, $b = 2st$, $r = t + 2s - 2$, $\boldsymbol{k} = [s\boldsymbol{1}_{st}', (t+s-2)\boldsymbol{1}_{st}']'$, $\varepsilon_1 = 1 - (st + 2s^2 - 6s + 4)/[(t+s-2)(t+2s-2)]$, $\varepsilon_2 = 1 - (t-2)^2/[(t+s-2)(t+2s-2)]$, $\varepsilon_3 = 1 - 4/[(t+s-2)(t+2s-2)]$, ρ_β *and* $\boldsymbol{L}_\beta$ *are the same as* (1);

(3) $v = s_1s_2s_3$, $b = 2s_1s_2s_3$, $r = s_3s_2(s_1-1)+s_3-1$, $\boldsymbol{k} = [(s_3-1)\mathbf{1}'_{s_1s_2s_3}, s_3s_2(s_1-1)\mathbf{1}'_{s_1s_2s_3}]'$, $\varepsilon_1 = 1-\{s_3-1+s_2^2s_3^2/[s_3s_2(s_1-1)]\}/r$, $\varepsilon_2 = 1-(s_3-1)/r$, $\varepsilon_3 = 1-[(s_3-1)r]^{-1}$, $\rho_1 = s_1-1$, $\rho_2 = s_1(s_2-1)$, $\rho_3 = s_1s_2(s_3-1)$, $\boldsymbol{L}_1 = (\boldsymbol{I}_{s_1} - s_1^{-1}\mathbf{1}_{s_1}\mathbf{1}'_{s_1}) \otimes (s_2s_3)^{-1}\mathbf{1}_{s_2s_3}\mathbf{1}'_{s_2s_3}$, $\boldsymbol{L}_2 = \boldsymbol{I}_{s_1} \otimes (\boldsymbol{I}_{s_2} - s_2^{-1}\mathbf{1}_{s_2}\mathbf{1}'_{s_2}) \otimes s_3^{-1}\mathbf{1}_{s_3}\mathbf{1}'_{s_3}$, $\boldsymbol{L}_3 = \boldsymbol{I}_{s_1s_2} \otimes (\boldsymbol{I}_{s_3} - s_3^{-1}\mathbf{1}_{s_3}\mathbf{1}'_{s_3})$;

(4) $v = s_1s_2s_3$, $b = 2s_1s_2s_3$, $r = s_3(2s_2-1)$, $\boldsymbol{k} = [[s_3(s_2-1)+1]\mathbf{1}'_{s_1s_2s_3}, (s_3s_2-1)\mathbf{1}'_{s_1s_2s_3}]'$, $\varepsilon_1 = 1-\{s_2s_3-1+[s_3(s_2-1)(s_2s_3-s_3+2)+1]/[s_3(s_2-1)+1]\}/r$, $\varepsilon_2 = 1-\{(s_3-1)^2/[s_3(s_2-1)+1]+1/(s_2s_3-1)\}/r$, $\varepsilon_3 = 1-\{1/[s_3(s_2-1)+1]+1/(s_2s_3-1)\}/r$, *$\rho_\beta$ and $\boldsymbol{L}_\beta$ are the same as* (3).

Proof. By Theorem 8.4.1, the designs (1) and (2) follow from juxtapositions $[\boldsymbol{A}_1 : \boldsymbol{A}_3]$ and $[\boldsymbol{A}_2+\boldsymbol{I}_{st} : \boldsymbol{A}_1+\boldsymbol{A}_2]$, respectively, in the rectangular association scheme [Section 6.0.2(E)], whereas the designs (3) and (4) follow from juxtapositions $[\boldsymbol{A}_1 : \boldsymbol{A}_3]$ and $[\boldsymbol{A}_2 + \boldsymbol{I}_{s_1s_2s_3} : \boldsymbol{A}_1 + \boldsymbol{A}_2]$, respectively, in the group divisible 3-associate association scheme [Section 6.0.2(F)]. The condition (8.2.4) is satisfied by all the designs here, because of Corollary 3.2.3. □

A Kronecker product method is also useful to get designs in the present category, as the following theorem shows.

Theorem 8.4.8 (Puri and Nigam, 1977a). *Let $\boldsymbol{N}_h, h = 1, 2$, be the $v_h \times b_h$ incidence matrices of two $(0; v_h-1; 0)$-EB designs with parameters v_h, b_h, r_h, $\boldsymbol{k}_h$, $\varepsilon^{(h)}$, $\rho^{(h)} = v_h - 1$, $\boldsymbol{L}^{(h)} = \boldsymbol{I}_{v_h} - (v_h)^{-1}\mathbf{1}_{v_h}\mathbf{1}'_{v_h}$. Then $\boldsymbol{N}_1 \otimes \boldsymbol{N}_2$ is the incidence matrix of a $(0; \rho_1, \rho_2, \rho_3; 0)$-EB design with parameters $v = v_1v_2$, $b = b_1b_2$, $r = r_1r_2$, $\boldsymbol{k} = \boldsymbol{k}_1 \otimes \boldsymbol{k}_2$, $\varepsilon_1 = \varepsilon^{(2)}$, $\varepsilon_2 = \varepsilon^{(1)}$, $\varepsilon_3 = 1-(1-\varepsilon^{(1)})(1-\varepsilon^{(2)})$, $\rho_1 = v_2-1$, $\rho_2 = v_1 - 1$, $\rho_3 = (v_1-1)(v_2-1)$, $\boldsymbol{L}_1 = (v_1)^{-1}\mathbf{1}_{v_1}\mathbf{1}'_{v_1} \otimes [\boldsymbol{I}_{v_2} - (v_2)^{-1}\mathbf{1}_{v_2}\mathbf{1}'_{v_2}]$, $\boldsymbol{L}_2 = [\boldsymbol{I}_{v_1} - (v_1)^{-1}\mathbf{1}_{v_1}\mathbf{1}'_{v_1}] \otimes (v_2)^{-1}\mathbf{1}_{v_2}\mathbf{1}'_{v_2}$, $\boldsymbol{L}_3 = [\boldsymbol{I}_{v_1} - (v_1)^{-1}\mathbf{1}_{v_1}\mathbf{1}'_{v_1}] \otimes [\boldsymbol{I}_{v_2} - (v_2)^{-1}\mathbf{1}_{v_2}\mathbf{1}'_{v_2}]$.*

Proof. This follows from a direct calculation of the $\boldsymbol{M}$ matrix. □

The existence of equireplicate and nonproper $(0; v-1; 0)$-EB designs can be seen in Section 8.2.2.1, as Corollaries 8.2.3, 8.2.4 and 8.2.5, Theorems 8.2.18 and 8.2.19 show. For example, two such designs in Examples 8.2.3 and 8.2.4 yield the following example by Theorem 8.4.8.

Example 8.4.1. There exists a (0; 12, 4, 3; 0)-EB design with parameters $v = 20$, $b = 120$, $r = 45$, $\boldsymbol{k} = [2\mathbf{1}'_6, 4\mathbf{1}'_2]' \otimes [4\mathbf{1}'_2, 3, 2\mathbf{1}'_2, 4\mathbf{1}'_2, 3, 2\mathbf{1}'_2, 4\mathbf{1}'_2, 3, 2\mathbf{1}'_2]'$, $\varepsilon_1 = 29/30$, $\varepsilon_2 = 5/6$, $\varepsilon_3 = 4/5$, $\rho_1 = 12$, $\rho_2 = 4$, $\rho_3 = 3$, $\boldsymbol{L}_1 = (\boldsymbol{I}_4 - 4^{-1}\mathbf{1}_4\mathbf{1}'_4) \otimes (\boldsymbol{I}_5 - 5^{-1}\mathbf{1}_5\mathbf{1}'_5)$, $\boldsymbol{L}_2 = 4^{-1}\mathbf{1}_4\mathbf{1}'_4 \otimes (\boldsymbol{I}_5 - 5^{-1}\mathbf{1}_5\mathbf{1}'_5)$, $\boldsymbol{L}_3 = (\boldsymbol{I}_4 - 4^{-1}\mathbf{1}_4\mathbf{1}'_4) \otimes 5^{-1}\mathbf{1}_5\mathbf{1}'_5$, and with the incidence matrix

$$\begin{bmatrix} 1&0&1&0&1&0&1&1 \\ 1&0&0&1&0&1&1&1 \\ 0&1&1&0&0&1&1&1 \\ 0&1&0&1&1&0&1&1 \end{bmatrix} \otimes \begin{bmatrix} 0&0&1&1&1&1&1&1&0&0&1&1&1&0&0 \\ 1&1&1&0&0&0&0&1&1&1&1&1&1&0&0 \\ 1&1&1&0&0&1&1&1&0&0&0&0&1&1&1 \\ 1&1&0&1&0&1&1&0&1&0&1&1&0&1&0 \\ 1&1&0&0&1&1&1&0&0&1&1&1&0&0&1 \end{bmatrix}.$$

8.4.4 Nonproper designs that are nonequireplicate

A simple technique is stated, i.e., Kronecker product of two incidence matrices of designs. For example, the usual Kronecker product of p $(0; v-1; 0)$-EB designs yields a design with $2^p - 1$ efficiency factors (see Puri and Nigam, 1977a). A case of $p = 2$ is presented here, because the next case, i.e., $p = 3$, produces a design with seven efficiency factors.

Theorem 8.4.9 (Puri and Nigam, 1977a). *Let $\boldsymbol{N}_h, h = 1,2$, be the $v_h \times b_h$ incidence matrices of two $(0; v_h - 1; 0)$-EB designs with parameters v_h, b_h, $\boldsymbol{r}_h$, $\boldsymbol{k}_h$, $\varepsilon^{(h)}$, $\rho^{(h)} = v_h - 1$, $\boldsymbol{L}^{(h)} = \boldsymbol{I}_{v_h} - (n_h)^{-1}\boldsymbol{1}_{v_h}\boldsymbol{r}'_h$, where $n_h = \boldsymbol{1}'_{v_h}\boldsymbol{r}_h$. Then the incidence matrix $\boldsymbol{N}_1 \otimes \boldsymbol{N}_2$ yields a $(0; \rho_1, \rho_2, \rho_3; 0)$-EB design with parameters $v = v_1v_2$, $b = b_1b_2$, $\boldsymbol{r} = \boldsymbol{r}_1 \otimes \boldsymbol{r}_2$, $\boldsymbol{k} = \boldsymbol{k}_1 \otimes \boldsymbol{k}_2$, $\varepsilon_1 = \varepsilon^{(2)}$, $\varepsilon_2 = \varepsilon^{(1)}$, $\varepsilon_3 = 1 - (1-\varepsilon^{(1)})(1-\varepsilon^{(2)})$, $\rho_1 = v_2 - 1$, $\rho_2 = v_1 - 1$, $\rho_3 = (v_1 - 1)(v_2 - 1)$, $\boldsymbol{L}_1 = (n_1)^{-1}\boldsymbol{1}_{v_1}\boldsymbol{r}'_1 \otimes [\boldsymbol{I}_{v_2} - (n_2)^{-1}\boldsymbol{1}_{v_2}\boldsymbol{r}'_2]$, $\boldsymbol{L}_2 = [\boldsymbol{I}_{v_1} - (n_1)^{-1}\boldsymbol{1}_{v_1}\boldsymbol{r}'_1] \otimes (n_2)^{-1}\boldsymbol{1}_{v_2}\boldsymbol{r}'_2$, $\boldsymbol{L}_3 = [\boldsymbol{I}_{v_1} - (n_1)^{-1}\boldsymbol{1}_{v_1}\boldsymbol{r}'_1] \otimes [\boldsymbol{I}_{v_2} - (n_2)^{-1}\boldsymbol{1}_{v_2}\boldsymbol{r}'_2]$.*

Proof. This follows from a direct calculation of the $\boldsymbol{M}$ matrix. □

Once there exists a $(0; v-1; 0)$-EB design, one can easily produce $(0; \rho_1, \rho_2, \rho_3; 0)$-EB designs which belong to all the categories of Sections 8.4.1 (see Theorem 8.4.2) and 8.4.3 (see Theorem 8.4.8). Here nonproper and nonequireplicate $(0; \rho_1, \rho_2, \rho_3; 0)$-EB designs are produced by Theorem 8.4.9.

As starting $(0; v-1; 0)$-EB designs, one can find many such designs in Section 8.2.2.2. For example, two designs given in Example 8.2.6 and Theorem 8.2.19 can yield the following by use of Theorem 8.4.9.

Corollary 8.4.8. *For a positive integer s there exists a $(0; \rho_1, \rho_2, \rho_3; 0)$-EB design with parameters $v = b = 4(s+1)$, $\boldsymbol{r} = [1, 3\boldsymbol{1}'_3]' \otimes [s, 2\boldsymbol{1}'_s]'$, $\boldsymbol{k} = [4, 2\boldsymbol{1}'_3]' \otimes [2\boldsymbol{1}'_s, s]'$, $\varepsilon_1 = 23/24$, $\varepsilon_2 = 5/6$, $\varepsilon_3 = 3/4$, $\rho_1 = 3s$, $\rho_2 = 3$, $\rho_3 = s$, $\boldsymbol{L}_1 = \{\boldsymbol{I}_4 - (10)^{-1}\boldsymbol{1}_4[1, 3\boldsymbol{1}'_3]'\} \otimes \{\boldsymbol{I}_{s+1} - (3s)^{-1}\boldsymbol{1}_{s+1}[s, 2\boldsymbol{1}'_s]'\}$, $\boldsymbol{L}_2 = \{\boldsymbol{I}_4 - (10)^{-1}\boldsymbol{1}_4[1, 3\boldsymbol{1}'_3]'\} \otimes (3s)^{-1}\boldsymbol{1}_{s+1}[s, 2\boldsymbol{1}'_s]'$, $\boldsymbol{L}_3 = (10)^{-1}\boldsymbol{1}_4[1, 3\boldsymbol{1}'_3]' \otimes \{\boldsymbol{I}_{s+1} - (3s)^{-1}\boldsymbol{1}_{s+1}[s, 2\boldsymbol{1}'_s]'\}$, whose incidence matrix is of the form*

$$\begin{bmatrix} 1 & 0 & 0 & 0 \\ 1 & 1 & 1 & 0 \\ 1 & 1 & 0 & 1 \\ 1 & 0 & 1 & 1 \end{bmatrix} \otimes \begin{bmatrix} \boldsymbol{1}'_s & 0 \\ \boldsymbol{I}_s & \boldsymbol{1}_s \end{bmatrix}.$$

Note that the order of ε_β, $\beta = 1, 2, 3$, given in Corollary 8.4.8 may differ from that in Theorem 8.4.9, because here it is given according to the requirement that $\varepsilon_1 > \varepsilon_2 > \varepsilon_3$ (see Section 4.3.2).

As a starting $(0; v-1; 0)$-EB design, one can also use a BIB design which yield the following, by use of Theorem 8.4.9 with a $(0; v-1; 0)$-EB design resulting from Theorem 8.2.19.

Corollary 8.4.9. *The existence of a BIB design with parameters* v, b, r, k, λ *implies the existence of a* $(0;\rho_1^*,\rho_2^*,\rho_3^*;0)$*-EB design with parameters* $v^* = v(s+1)$, $b^* = b(s+1)$, $\boldsymbol{r}^* = r[s, 2\mathbf{1}_s']'$, $\boldsymbol{k}^* = k[2\mathbf{1}_s', s]'$, $\varepsilon_1^* = 3/4$, $\varepsilon_2^* = \lambda v/(rk)$, $\varepsilon_3^* = 1-(r-\lambda)/(4rk)$, $\rho_1^* = s$, $\rho_2^* = v-1$, $\rho_3^* = s(v-1)$, $\boldsymbol{L}_1^* = v^{-1}\mathbf{1}_v\mathbf{1}_v' \otimes \{\boldsymbol{I}_{s+1} - (3s)^{-1}\mathbf{1}_{s+1}[s, 2\mathbf{1}_s']'\}$, $\boldsymbol{L}_2^* = (\boldsymbol{I}_v - v^{-1}\mathbf{1}_v\mathbf{1}_v') \otimes (3s)^{-1}\mathbf{1}_{s+1}[s, 2\mathbf{1}_s']'$, $\boldsymbol{L}_3^* = (\boldsymbol{I}_v - v^{-1}\mathbf{1}_v\mathbf{1}_v') \otimes \{\boldsymbol{I}_{s+1} - (3s)^{-1}\mathbf{1}_{s+1}[s, 2\mathbf{1}_s']'\}$ *for a positive integer* s.

One can get more $(0;\rho_1,\rho_2,\rho_3;0)$-EB designs with all the ε_β being less than one, by use of Corollary 8.4.9, because a large number of series of BIB designs are available in Section 8.2.1. Such exercises are left to the reader.

8.5 Designs with more efficiency factors

In the previous sections the constructions of $(0;\rho_1,\rho_2,\dots,\rho_{m-1};0)$-EB designs have been discussed mainly for $m-1 \le 3$, that is, designs with up to three distinct efficiency factors. The present section is devoted to the existence of designs with more efficiency factors though it is not sure whether an experimenter wishes to use such plans in practice. In fact, such existence can also be found in Chapter 6 and the previous sections of this chapter.

8.5.1 Proper designs that are equireplicate

At first, as described in Section 6.0.2, a PBIB design in which all the eigenvalues ψ_β are positive, i.e., $\varepsilon_\beta < 1$ for all β, is such. That is,

Theorem 8.5.1. *A connected s-associate PBIB design with parameters* v, b, r, k, λ_1, $\lambda_2,\dots,\lambda_s$, $\varepsilon_1(<1)$, $\varepsilon_2(<1),\dots,\varepsilon_s(<1)$, ρ_1, $\rho_2,\dots,\rho_s$, $\boldsymbol{L}_1$, $\boldsymbol{L}_2,\dots,\boldsymbol{L}_s$ *is a* $(0;\rho_1,\rho_2,\dots,\rho_s;0)$*-EB design with parameters* v, b, r, k, ε_β, $\boldsymbol{L}_\beta$, $\beta = 1,2,\dots,s$.

As will be seen from Theorem 8.4.2, the Kronecker product of p BIB designs produces a (2^p-1)-associate PBIB design. A generalization of this idea will be found in Theorem 8.5.2 for $p=3$.

By Theorem 6.1.1 (dualization) with a symmetric PBIB design, the following result can be obtained.

Corollary 8.5.1. *The dual of a connected s-associate PBIB design with parameters* v, $b(=v)$, r, k, λ_1, $\lambda_2,\dots,\lambda_s$, $\varepsilon_\beta(<1)$, ρ_β, $\boldsymbol{L}_\beta$, $\beta = 1,2,\dots,s$, *is a* $(0;\rho_1^*,\rho_2^*,\dots,\rho_s^*;0)$*-EB design with parameters* $v^* = b$, $b^* = v$, $r^* = k$, $k^* = r$, $\varepsilon_\beta^* = \varepsilon_\beta$, $\rho_\beta^* = \rho_\beta$, $\boldsymbol{L}_\beta^* = [rk(1-\varepsilon_\beta)]^{-1}\boldsymbol{N}'\boldsymbol{L}_\beta\boldsymbol{N}$, *where* $\boldsymbol{N}$ *is the incidence matrix of the original PBIB design.*

By complementation (Theorem 6.2.1) with Theorem 8.5.1, more $(0;\rho_1^*,\rho_2^*,\dots,\rho_s^*;0)$-EB designs can be obtained. Corollary 7.5.2 can also apply here. As a patterned method of construction, note that Theorem 6.6.4 with $a=c=1$ and $\rho_0 = 0$ gives equireplicate and proper $(0;\rho_1^*,\rho_2^*,\dots,\rho_{m^*-1}^*;0)$-EB designs.

8.5.2 Proper designs that are nonequireplicate

At first, once there is a proper and nonequireplicate $(\rho_0; \rho_1, ..., \rho_{m-1}; 0)$-EB design, by Theorem 8.4.7 one can obtain other proper and nonequireplicate $(0; \rho_1^*, \rho_2^*, ..., \rho_m^*; 0)$-EB designs. The existence of such designs can be given by a design resulting from Corollary 8.5.2 below, by use of dualization of Theorem 6.1.1.

8.5.3 Nonproper designs that are equireplicate

At first, by Theorem 6.1.1 the dualization of all the symmetric designs (with $v = b$) given in Section 8.5.2 (if they exist) yields nonproper and equireplicate designs with more efficiency factors which belong to the present category.

Next, by juxtaposition (Theorem 6.4.2) the following result can be obtained.

Corollary 8.5.2. *Let $\boldsymbol{N}_h, h = 1, 2, ..., a$, be the incidence matrices of a connected s-associate PBIB designs, based on the same association scheme, with parameters v, $b^{(h)}$, $r^{(h)}$, $k^{(h)}$, $\lambda_1^{(h)}$, $\lambda_2^{(h)}, ..., \lambda_s^{(h)}$, $\varepsilon_\beta^{(h)}$, ρ_β, $\boldsymbol{L}_\beta$, $\beta = 1, 2, ..., s$. Then the incidence matrix $[\boldsymbol{N}_1 : \boldsymbol{N}_2 : \cdots : \boldsymbol{N}_a]$ yields a $(0; \rho_1^*, \rho_2^*, ..., \rho_s^*; 0)$-EB design with parameters $v^* = v$, $b^* = \sum_{h=1}^a b^{(h)}$, $r^* = \sum_{h=1}^a r^{(h)}$, $\boldsymbol{k}^* = [k^{(1)}\boldsymbol{1}'_{b^{(1)}}, k^{(2)}\boldsymbol{1}'_{b^{(2)}}, ..., k^{(a)}\boldsymbol{1}'_{b^{(a)}}]'$, $\varepsilon_\beta^* = 1 - [\sum_{h=1}^a r^{(h)}(1 - \varepsilon_\beta^{(h)})]/r^*$, $\rho_\beta^* = \rho_\beta$, $\boldsymbol{L}_\beta^* = \boldsymbol{L}_\beta$, $\beta = 1, 2, ..., s$, unless $\varepsilon_\beta^{(h)} = 1$ for all $h = 1, 2, ..., a$ with some common fixed β.*

Theorem 7.4.17 can also be generalized as follows.

Corollary 8.5.3. *If $\boldsymbol{N}$ is the incidence matrix of a binary $(0; \rho_1, \rho_2, ..., \rho_{m-1}; 0)$-EB design with parameters v, b, r, k, ε_β, ρ_β, $\boldsymbol{L}_\beta$, $\beta = 1, 2, ..., m-1$, then the incidence matrix $[\boldsymbol{N} : \boldsymbol{1}_v\boldsymbol{1}'_b - \boldsymbol{N}]$ yields a binary $(0; \rho_1, \rho_2, ..., \rho_{m-1}; 0)$-EB design with parameters $v^* = v$, $b^* = 2b$, $r^* = b$, $\boldsymbol{k}^* = [k\boldsymbol{1}'_b, (v-k)\boldsymbol{1}'_b]'$, $\varepsilon_\beta^* = 1 - k(1-\varepsilon_\beta)/(v-k)$, $\rho_\beta^* = \rho_\beta$, $\boldsymbol{L}_\beta^* = \boldsymbol{L}_\beta$.*

8.5.4 Nonproper designs that are nonequireplicate

At first, once there is a nonproper and nonequireplicate $(0; \rho_1, \rho_2, ..., \rho_{m-1}; \rho_m)$-EB design with $v = b$, by dualization (Theorem 6.1.1) one can get another $(0; \rho_1, \rho_2, ..., \rho_{m-1}; \rho_m)$-EB design. Next, by juxtaposition (Theorem 6.4.1) one can get more such designs.

A generalization of Theorem 8.4.9 can easily produce the following theorem.

Theorem 8.5.2 (Puri and Nigam, 1977a). *Let $\boldsymbol{N}_h$, $h = 1, 2, 3$, be the $v_h \times b_h$ incidence matrices of three $(0; v_h - 1; 0)$-EB designs with parameters v_h, b_h, $\boldsymbol{r}_h$, $\boldsymbol{k}_h$, $\varepsilon^{(h)}$, $\rho^{(h)} = v_h - 1$, $\boldsymbol{L}^{(h)} = \boldsymbol{I}_{v_h} - (n_h)^{-1}\boldsymbol{1}_{v_h}\boldsymbol{r}'_h$, where $n_h = \boldsymbol{1}'_{v_h}\boldsymbol{r}_h$. Then the incidence matrix $\boldsymbol{N}_1 \otimes \boldsymbol{N}_2 \otimes \boldsymbol{N}_3$ yields a $(0; \rho_1^*, \rho_2^*, \rho_3^*, \rho_4^*, \rho_5^*, \rho_6^*, \rho_7^*; 0)$-EB design with parameters $v^* = v_1v_2v_3$, $b^* = b_1b_2b_3$, $\boldsymbol{r}^* = \boldsymbol{r}_1 \otimes \boldsymbol{r}_2 \otimes \boldsymbol{r}_3$, $\boldsymbol{k}^* = \boldsymbol{k}_1 \otimes \boldsymbol{k}_2 \otimes \boldsymbol{k}_3$, $\varepsilon_1^* = \varepsilon^{(3)}$, $\varepsilon_2^* = \varepsilon^{(2)}$, $\varepsilon_3^* = \varepsilon^{(1)}$, $\varepsilon_4^* = 1 - (1-\varepsilon^{(2)})(1-\varepsilon^{(3)})$, $\varepsilon_5^* = 1 - (1-\varepsilon^{(1)})(1-\varepsilon^{(3)})$, $\varepsilon_6^* = 1 - (1-\varepsilon^{(1)})(1-\varepsilon^{(2)})$, $\varepsilon_7^* = 1 - (1-\varepsilon^{(1)})(1-\varepsilon^{(2)})(1-\varepsilon^{(3)})$, $\rho_1^* = v_3 - 1$,*

$\rho_2^* = v_2 - 1$, $\rho_3^* = v_1 - 1$, $\rho_4^* = (v_2 - 1)(v_3 - 1)$, $\rho_5^* = (v_1 - 1)(v_3 - 1)$, $\rho_6^* = (v_1 - 1)(v_2 - 1)$, $\rho_7^* = (v_1 - 1)(v_2 - 1)(v_3 - 1)$, $\boldsymbol{L}_1^* = (n_1)^{-1}\mathbf{1}_{v_1}\boldsymbol{r}_1' \otimes (n_2)^{-1}\mathbf{1}_{v_2}\boldsymbol{r}_2' \otimes [\boldsymbol{I}_{v_3} - (n_3)^{-1}\mathbf{1}_{v_3}\boldsymbol{r}_3']$, $\boldsymbol{L}_2^* = (n_1)^{-1}\mathbf{1}_{v_1}\boldsymbol{r}_1' \otimes [\boldsymbol{I}_{v_2} - (n_2)^{-1}\mathbf{1}_{v_2}\boldsymbol{r}_2'] \otimes (n_3)^{-1}\mathbf{1}_{v_3}\boldsymbol{r}_3'$, $\boldsymbol{L}_3^* = [\boldsymbol{I}_{v_1} - (n_1)^{-1}\mathbf{1}_{v_1}\boldsymbol{r}_1'] \otimes (n_2)^{-1}\mathbf{1}_{v_2}\boldsymbol{r}_2' \otimes (n_3)^{-1}\mathbf{1}_{v_3}\boldsymbol{r}_3'$, $\boldsymbol{L}_4^* = (n_1)^{-1}\mathbf{1}_{v_1}\boldsymbol{r}_1' \otimes [\boldsymbol{I}_{v_2} - (n_2)^{-1}\mathbf{1}_{v_2}\boldsymbol{r}_2'] \otimes [\boldsymbol{I}_{v_3} - (n_3)^{-1}\mathbf{1}_{v_3}\boldsymbol{r}_3']$, $\boldsymbol{L}_5^* = [\boldsymbol{I}_{v_1} - (n_1)^{-1}\mathbf{1}_{v_1}\boldsymbol{r}_1'] \otimes (n_2)^{-1}\mathbf{1}_{v_2}\boldsymbol{r}_2' \otimes [\boldsymbol{I}_{v_3} - (n_3)^{-1}\mathbf{1}_{v_3}\boldsymbol{r}_3']$, $\boldsymbol{L}_6^* = [\boldsymbol{I}_{v_1} - (n_1)^{-1}\mathbf{1}_{v_1}\boldsymbol{r}_1'] \otimes [\boldsymbol{I}_{v_2} - (n_2)^{-1}\mathbf{1}_{v_2}\boldsymbol{r}_2'] \otimes (n_3)^{-1}\mathbf{1}_{v_3}\boldsymbol{r}_3'$, $\boldsymbol{L}_7^* = [\boldsymbol{I}_{v_1} - (n_1)^{-1}\mathbf{1}_{v_1}\boldsymbol{r}_1'] \otimes [\boldsymbol{I}_{v_2} - (n_2)^{-1}\mathbf{1}_{v_2}\boldsymbol{r}_2'] \otimes [\boldsymbol{I}_{v_3} - (n_3)^{-1}\mathbf{1}_{v_3}\boldsymbol{r}_3']$.

Note that the similar Kronecker product of p $(0; v-1; 0)$-EB designs produces a $(0; \rho_1, \dots, \rho_{m-1}; 0)$-EB design for $m = 2^p$. For example, a starting $(0; v-1; 0)$-EB design can be found in Section 8.2.2.2.

8.6 Tables

Table 8.2
Parameters of BIB designs with $v \leq 100, r \leq 10$
and their efficiency factors and solutions

No.	v	r	b	k	λ	$100\varepsilon_1$	Solution
1	3	2	3	2	1	75.00	unreduced type
2		4	6	2	2	75.00	unreduced type
3		6	9	2	3	75.00	unreduced type
4		8	12	2	4	75.00	unreduced type
5		10	15	2	5	75.00	unreduced type
6	4	3	4	3	2	88.89	unreduced type
7		3	6	2	1	66.67	unreduced type
8		6	8	3	4	88.89	unreduced type
9		6	12	2	2	66.67	unreduced type
10		9	12	3	6	88.89	unreduced type
11		9	18	2	3	66.67	unreduced type
12	5	4	5	4	3	93.75	unreduced type
13		4	10	2	1	62.50	unreduced type
14		6	10	3	3	83.33	unreduced type
15		8	10	4	6	93.75	unreduced type
16		8	20	2	2	62.50	unreduced type
17	6	5	6	5	4	96.00	unreduced type
18		5	10	3	2	80.00	(0,1,4)(∞,1,4) mod 5
19		5	15	2	1	60.00	unreduced type
20		10	12	5	8	96.00	unreduced type
21		10	15	4	6	90.00	unreduced type
22		10	20	3	4	80.00	unreduced type or (0,1,4) (0,1,4) (∞,1,4) (∞,1,4) mod 5
23		10	30	2	2	60.00	unreduced type
24	7	3	7	3	1	77.78	(0,1,3) mod 7 ··· PG(2,2):1
25		4	7	4	2	87.85	(2,4,5,6) mod 7
26		6	7	6	5	97.22	unreduced type
27		6	14	3	2	77.78	(0,1,3) (0,1,3) mod 7
28		6	21	2	1	58.33	unreduced type
29		8	14	4	4	87.50	(2,4,5,6) (2,4,5,6) mod 7
30		9	21	3	3	77.78	(0,1,3)(0,1,3)(0,1,3) mod 7
31	8	7	8	7	6	97.96	unreduced type
32		7	14	4	3	85.71	(∞,1,2,4)(0,1,2,4) mod 7 ··· AG(3,2):2
33		7	28	2	1	57.14	unreduced type
34	9	4	12	3	1	75.00	(∞,0,4)PC(4)(0,1,3) mod 8 ··· AG(2,3):1

Table 8.2 (continued-1)

No.	v	r	b	k	λ	$100\varepsilon_1$	Solution
35	9	8	9	8	7	98.44	unreduced type
36		8	12	6	5	93.75	(1,2,3,5,6,7)PC(4)(∞,2,4,5,6,7) mod 8
37		8	18	4	3	84.38	(0,1,2,4)(0,3,4,7) mod 9
38		8	24	3	2	75.00	(∞,0,4)(∞,0,4)PC(4) (0,1,3) (0,1,3) mod 8
39		8	36	2	1	56.25	unreduced type
40		10	18	5	5	90.00	(3,5,6,7,8)(1,2,5,6,8) mod 9
41	10	6	15	4	2	83.33	$(0_0,3_0,0_1,4_1)(0_0,0_1,1_1,3_1)$ $(1_0,2_0,3_0,0_1)$ mod 5
42		9	10	9	8	98.77	unreduced type
43		9	15	6	5	92.59	$(1_0,2_0.4_0,1_1,2_1,3_1)$ $(1_0,2_0,3_0,4_0,2_1,4_1)$ $(0_0,4_0,1_1,2_1,3_1,4_1)$ mod 5
44		9	18	5	4	88.89	(∞,0,1,4,6)(0,1,2,4,8) mod 9
45		9	30	3	2	74.07	(0,3,6)PC(3)(∞,0,2)(0,1,3)(0,1,5) mod 9
46		9	45	2	1	55.56	unreduced type
47	11	5	11	5	2	88.00	(0,2,3,4,8) mod 11
48		10	11	10	9	99.00	unreduced type
49		10	22	5	4	88.00	(0,2,3,4,8)(0,2,3,4,8) mod 11
50		10	55	2	1	55.00	unreduced type
51	13	4	13	4	1	81.25	(0,1,3,9) mod 13 $\cdots$ PG(2,3):1
52		6	26	3	1	72.22	(0,1,4)(0,2,7) mod 13
53		8	26	4	2	81.25	(0,1,3,9)(0,1,3,9) mod 13
54		9	13	9	6	96.30	(2,4,5,6,7,8,10,11,12) mod 13
55	15	7	15	7	3	91.84	(0,1,2,4,5,8,10) mod 15 $\cdots$ PG(3,2):2
56		7	21	5	2	-	nonexistence
57		7	35	3	1	71.43	(0,5,10)PC(5)(0,1,4)(0,2,8) mod 15 $\cdots$ PG(3,2):1
58		8	15	8	4	93.75	(3,6,7,9,11,12,13,14) mod 15
59	16	5	20	4	1	80.00	(∞,0,5,10)PC(5)(0,2,3,11) mod 15 $\cdots$ AG(2,4):1
60		6	16	6	2	88.89	(00,01,02,10,23,30) mod (4,4)
61		9	24	6	3	88.89	(∞, ∞, 0∞,∞0,00, 11,21) (0∞,1∞,00,10,21,∞1) mod (3,3) (∞0,00,∞1,01,∞2,02) (00,10,01,11,02,12) mod (3,-)
62	16	10	16	10	6	96.00	(03,11,12,13,20,21,22,31,32,33) mod (4,4)

Table 8.2 (continued-2)

No.	v	r	b	k	λ	$100\varepsilon_1$	Solution
63		10	40	4	2	80.00	$(\infty,0,5,10)(\infty,0,5,10)$PC(5) (0,2,3,11)(0,2,3,11) mod 15
64	19	9	19	9	4	93.83	(0,3,4,5,6,8,10,15,16) mod 19
65		9	57	3	1	70.37	(0,1,5)(0,2,8)(0,3,10) mod 19
66		10	19	10	5	95.00	(1,2,7,9,11,12,13,14,17,18) mod 19
67	21	5	21	5	1	84.00	(0,1,6,8,18) mod 21 $\cdots$ PG(2,4):1
68		8	28	6	2	-	nonexistence
69		10	30	7	3	90.00	$(0_0,1_0,3_0,0_1,1_1,2_2,6_2)$ $(0_0,1_0,3_0,3_1,5_1,0_2,1_2)$ $(0_0,1_0,3_0,2_1,6_1,3_2,5_2)$ $(4_0,0_1,1_1,5_1,0_2,1_2,5_2)$ mod 7 $(0_1,1_1,2_1,3_1,4_1,5_1,6_1)$ $(0_2,1_2,2_2,3_2,4_2,5_2,6_2)$
70		10	42	5	2	84.00	(0,1,6,8,18)(0,1,6,8,18) mod 21
71		10	70	3	1	70.00	(0,7,14)PC(7)(0,1,3)(0,4,12)(0,5,11) mod 21
72	22	7	22	7	2	-	nonexistence
73	25	6	30	5	1	83.33	$(\infty,0,6,12,18)$PC(6)(0,2,3,7,16) mod 24 $\cdots$ AG(2,5):1
74		8	50	4	1	78.13	(00,01,11,43)(00,02,22,31) mod (5,5)
75		9	25	9	3	92.59	Refer to Fisher and Yates table (1963)
76	28	9	36	7	2	88.89	Residual design of No. 83
77		9	63	4	1	77.78	$(01_1,02_1,10_2,20_2)(01_2,02_2,10_3,20_3)$ $(01_3,02_3,10_1,20_1)(21_1,12_1,22_2,11_2)$ $(21_2,12_2,22_3,11_3)(21_3,12_3,22_1,11_1)$ $(\infty,00_1,00_2,00_3)$ mod (3,3)
78	29	8	29	8	2	-	nonexistence
79	31	6	31	6	1	86.11	(0,1,3,10,14,26) mod 31 $\cdots$ PG(2,5):1
80		10	31	10	3	93.00	$(A,0_0,5_0,1_1,4_1,2_2,3_2,2_3,4_3,5_3)$ $(B,1_0,4_0,2_1,3_1,0_2,5_2,2_3,4_3,5_3)$ $(C,2_0,3_0,0_0,5_1,1_2,4_2,2_3,4_3,5_3)$ $(0_0,1_0,3_0,0_1,1_1,3_1,0_2,1_2,3_2,5_3)$ mod 7 $(A,B,C,0_0,1_0,2_0,3_0,4_0,5_0,6_0)$ $(A,B,C,0_1,1_1,2_1,3_1,4_1,5_1,6_1)$ $(A,B,C,0_2,1_2,2_2,3_2,4_2,5_2,6_2)$
81	36	7	42	6	1	-	nonexistence
82		10	45	8	1	-	nonexistence
83	37	9	37	9	2	91.36	(0,1,3,7,17,24,25,29,35) mod 37
84	41	10	82	5	1	82.00	(0,9,15,17,36)(0,18,30,31,34) mod 41

Table 8.2 (continued-3)

No.	v	r	b	k	λ	$100\varepsilon_1$	Solution
85	43	7	43	7	1	-	nonexistence
86	46	9	69	6	1	85.19	(unknown)
87		10	46	10	2	-	nonexistence
88	49	8	56	7	1	87.50	(∞,0,8,16,24,32,40)PC(8) (0,1,4,10,30,35,37) mod 48 $\cdots$AG(2,7):1
89	51	10	85	6	1	85.00	(unknown)
90	57	8	57	8	1	89.06	(0,3,5,13,14,20,32,36) mod 57
91	64	9	72	8	1	88.89	(∞,0,9,18,27,36,45,54)PC(9) (0,5,7,13,37,47,48,51) mod 63 $\cdots$ AG(2,8):1
92	73	9	73	9	1	90.12	(0,1,3,7,15,31,36,54,63) mod 73 $\cdots$ PG(2,8):1
93	81	10	90	9	1	90.00	$(\infty, 0, 10, 20, 30, 40, 50, 60, 70)PC(10)$ (0,12,34,47,48,65,71,73,76) mod 80 $\cdots$ AG(2,9):1
94	91	10	91	10	1	91.00	(0,1,6,10,23,26,34,41,53,55) $\cdots$ PG(2,9):1

Table 8.2 presents initial block(s) only (see Section 6.7). The "unreduced type" means a BIB design with parameters v, $b = \binom{v}{k}$, $r = \binom{v-1}{k-1}$, k, $\lambda = \binom{v-2}{k-2}$, which can be given by the collection of all possible combinations of k treatments from the set of v treatments (see Section 6.0.1). PG(t, q) : d [or AG(t, q) : d] means that the BIB design can also be constructed by taking all d-dimensional subspaces of a finite geometry PG(t, q) [or AG(t, q)] (see Theorems 6.9.1 and 6.9.2). As to the notation PC(n), e.g., in the design of No. 34 with the solution (∞,0,4)PC(4)(0,1,3) mod 8, PC(4) means a short cycle of order 4, i.e., a cyclic development of initial blocks four times and reducing modulo 8 when necessary. In fact, in this case, the solution shows

(∞,0,4), (∞,1,5), (∞,2,6), (∞,3,7),

(0,1,3), (1,2,4), (2,3,5), (3,4,6), (4,5,7), (5,6,0), (6,7,1), (7,0,2).

Note that, as in Theorem 6.7.2, $\infty + \theta = \infty$ for any θ.

For the use of initial blocks expressed in figures with subscripts like a design of No. 41, see Example 6.7.1. For the use of initial block(s) expressed in figures with double modulus, i.e., mod (,), like a design of No. 60, see Example 6.7.3.

Table 8.3
Parameters of $(0; v-1; 0)$-EB designs with $v, b \leq 40, r, k_1, k_2 = k_1 + 1 \leq 10$ and their efficiency factors and solutions

No.	v	r	b	k_1	k_2	$100\varepsilon_1$	Solution
1	3	3	4	2	3	83.33	Cor8.2.4+No.1+$(s=1)$
2		4	5	2	3	87.50	Cor8.2.4+No.1+$(s=2)$
3		5	6	2	3	90.00	Cor8.2.4+No.1+$(s=3)$
4		5	7	2	3	80.00	Cor8.2.4+No.2+$(s=1)$
5		6	7	2	3	91.67	Cor8.2.4+No.1+$(s=4)$
6		6	8	2	3	83.33	Cor8.2.4+No.2+$(s=2)$
7		7	8	2	3	92.29	Cor8.2.4+No.1+$(s=5)$
8		7	9	2	3	85.71	Cor6.2.4+No.2+$(s=3)$
9		7	10	2	3	78.57	Cor8.2.4+No.3+$(s=1)$
10		8	9	2	3	93.75	Cor8.2.4+No.1+$(s=6)$
11		8	10	2	3	87.50	Cor8.2.4+No.2+$(s=4)$
12		8	11	2	3	81.25	Cor8.2.4+No.3+$(s=2)$
13		9	10	2	3	94.44	Cor8.2.4+No.1+$(s=7)$
14		9	11	2	3	88.89	Cor8.2.4+No.2+$(s=5)$
15		9	12	2	3	83.33	Cor8.2.4+No.3+$(s=3)$
16		9	13	2	3	77.78	Cor8.2.4+No.4+$(s=1)$
17		10	11	2	3	95.00	Cor8.2.4+No.1+$(s=8)$
18		10	12	2	3	90.00	Cor8.2.4+No.2+$(s=6)$
19		10	13	2	3	85.00	Cor8.2.4+No.3+$(s=4)$
20		10	14	2	3	80.00	Cor8.2.4+No.4+$(s=2)$
21	4	4	5	3	4	91.67	Cor8.2.4+No.6+$(s=1)$
22		5	6	3	4	93.33	Cor8.2.4+No.6+$(s=2)$
23		6	7	3	4	94.44	Cor8.2.4+No.6+$(s=3)$
24		6	10	2	3	77.78	Cor8.2.3+No.6+No.7
25		7	8	3	4	95.24	Cor8.2.4+No.6+$(s=4)$
26		7	9	3	4	90.48	Cor8.2.4+No.8+$(s=1)$
27		8	9	3	4	95.83	Cor8.2.4+No.6+$(s=5)$
28		8	10	3	4	91.67	Cor8.2.4+No.8+$(s=2)$
29		9	10	3	4	96.30	Cor8.2.4+No.6+$(s=6)$
30		9	11	3	4	92.59	Cor8.2.4+No.8+$(s=3)$
31		9	14	2	3	81.48	Cor8.2.3+No.7+No.8
32		10	11	3	4	96.67	Cor8.2.4+No.6+$(s=7)$
33		10	12	3	4	93.33	Cor8.2.4+No.8+$(s=4)$
34		10	13	3	4	90.00	Cor8.2.4+No.10+$(s=1)$

Here, for example, Cor8.2.4+No.1+$(s=1)$ means that the design is constructed from a BIB design of No.1 in Table 8.2 by Corollary 8.2.4 for $s=1$. On the other hand, Cor8.2.3+No.6+No.7 means that the design is constructed from two BIB designs of Nos.6 and 7 in Table 8.2 by Corollary 8.2.3.

Table 8.3 (continued)

No.	v	r	b	k_1	k_2	$100\varepsilon_1$	Solution
35	5	5	6	4	5	95.00	Cor8.2.4+No.12+($s=1$)
36		6	7	4	5	95.83	Cor8.2.4+No.12+($s=2$)
37		7	8	4	5	96.43	Cor8.2.4+No.12+($s=3$)
38		8	9	4	5	96.88	Cor8.2.4+No.12+($s=4$)
39		9	10	4	5	97.22	Cor8.2.4+No.12+($s=5$)
40		9	11	4	5	94.44	Cor8.2.4+No.15+($s=1$)
41	5	10	11	4	5	97.50	Cor8.2.4+No.12+($s=6$)
42		10	12	4	5	95.00	Cor8.2.4+No.15+($s=2$)
43		10	15	3	4	87.50	Cor8.2.3+No.12+No.14
44		10	20	2	3	75.00	Cor8.2.3+No.13+No.14
45	6	6	7	5	6	96.67	Cor8.2.4+No.17+($s=1$)
46		7	8	5	6	97.14	Cor8.2.4+No.17+($s=2$)
47		8	9	5	6	97.50	Cor8.2.4+No.17+($s=3$)
48		8	18	2	3	75.00	Cor8.2.5+3SR18+2$\boldsymbol{I}_3 \otimes \mathbf{1}_2$
49		9	10	5	6	97.78	Cor6.2.4+No.17+($s=4$)
50		9	25	2	3	66.67	Cor6.2.5+2SR6+3$\boldsymbol{I}_2 \otimes \mathbf{1}_3$
51		10	11	5	6	98.00	Cor8.2.4+No.17+($s=5$)
52		10	25	2	3	70.00	Cor8.2.4+No.18+No.19
53	7	7	8	6	7	97.62	Cor8.2.4+No.26+($s=1$)
54		7	14	3	4	83.33	Cor8.2.3+No.24+No.25
55		8	9	6	7	97.92	Cor8.2.4+No.26+($s=2$)
56		9	10	6	7	98.15	Cor8.2.4+No.26+($s=3$)
57		9	28	2	3	64.81	Cor8.2.3+No.24+No.28
58		10	11	6	7	98.33	Cor8.2.4+No.26+($s=4$)
59		10	21	3	4	81.67	Cor8.2.3+No.25+No.27
60	8	8	9	7	8	98.21	Cor8.2.4+No.31+($s=1$)
61		9	10	7	8	98.41	Cor8.2.4+No.31+($s=2$)
62		10	11	7	8	98.57	Cor8.2.4+No.31+($s=3$)
63	9	9	10	8	9	98.61	Cor8.2.4+No.35+($s=1$)
64		10	11	8	9	98.75	Cor8.2.4+No.35+($s=2$)
65	10	10	11	9	10	98.89	Cor8.2.4+No.42+($s=1$)
66	13	10	39	3	4	75.83	Cor8.2.3+No.51+No.52

Table 8.4
Existing cyclic designs that are 2-associate PBIB designs

No.	v	r	k	b	λ_1	λ_2	n_1	n_2	ε_1	ε_2	Solution
1	5	2	2	5	1	0	2	2	0.90	0.35	(0,2) mod 5
2	5	4	2	10	2	0	2	2	0.90	0.35	(0,2)(0,2) mod 5
3	5	6	2	15	3	0	2	2	0.90	0.35	(0,2)(0,2)(0,2) mod 5
4	5	8	2	20	4	0	2	2	0.90	0.35	(0,2)(0,2)(0,2) (0,2) mod 5
5	5	10	2	25	5	0	2	2	0.90	0.35	(0,2)(0,2)(0,2) (0,2)(0,2) mod 5
6	5	6	2	15	2	1	2	2	0.72	0.53	(0,1)(0,2)(0,2) mod 5
7	5	8	2	20	3	1	2	2	0.76	0.49	(0,1)(0,2)(0,2) (0,2) mod 5
8	5	10	2	25	4	1	2	2	0.79	0.46	(0,1)(0,2)(0,2) (0,2)(0,2) mod 5
9	5	10	2	25	3	2	2	2	0.68	0.57	(0,1)(0,1)(0,2) (0,2)(0,2) mod 5
10	13	6	2	39	1	0	6	6	0.69	0.39	(0,2)(0,5)(0,6) mod 13
11	17	8	2	68	1	0	8	8	0.66	0.40	(0,3)(0,5)(0,6) (0,7) mod 17
12	5	3	3	5	2	1	2	2	0.96	0.71	(0,1,3) mod 5
13	5	6	3	10	4	2	2	2	0.96	0.71	(0,1,3)(0,1,3) mod 5
14	5	9	3	15	6	3	2	2	0.96	0.71	(0,1,3)(0,1,3) (0,1,3) mod 5
15	5	9	3	15	5	4	2	2	0.96	0.71	(0,1,4)(0,2,4) (0,2,4) mod 5
16	13	3	3	13	1	0	6	6	0.92	0.52	(0,2,8) mod 13
17	13	6	3	26	2	0	6	6	0.92	0.52	(0,2,8)(0,2,8) mod 13
18	13	9	3	39	3	0	6	6	0.93	0.52	(0,2,8)(0,2,8) (0,2,8) mod 13
19	13	9	3	39	1	2	6	6	0.79	0.66	(0,1,4)(0,1,10) (0,2,7) mod 13
20	37	9	3	111	1	0	18	18	0.80	0.57	(0,9,25)(0,10,36) (0,30,33) mod 37

Table 8.4 (continued)

No.	v	r	k	b	λ_1	λ_2	n_1	n_2	ε_1	ε_2	Solution
21	13	8	4	26	1	3	6	6	0.93	0.70	(0,3,9,12) (0,3,11,12) mod 13
22	17	8	4	34	1	2	8	8	0.86	0.73	(0,3,4,16) (1,8,10,16) mod 17
23	13	6	6	13	3	2	6	6	0.95	0.85	(0,1,3,6,8,12) mod 13
24	13	7	7	13	4	3	6	6	0.97	0.89	(2,4,5,7,9,10,11) mod 13
25	29	7	7	29	2	1	14	14	0.94	0.83	(0,6,15,19,22,23, 24) mod 29
26	17	8	8	17	4	3	8	8	0.96	0.90	(0,1,3,7,8,12,14, 15) mod 17
27	29	8	8	29	3	1	14	14	0.99	0.82	(0,1,7,16,20,23, 24,25) mod 29
28	17	9	9	17	5	4	8	8	0.97	0.92	(2,4,5,6,9,10,11, 13,16) mod 17
29	13	10	10	13	8	7	6	6	0.99	0.96	(1,3,4,5,6,7,9,10, 11,12) mod 13

The plans above can also be found in Clatworthy (1973; Table XV). The designs of Nos. 2–11 are 2-resolvable (see Definition 6.0.2), the designs of Nos. 13–15 and 17–20 are 3-resolvable, and the designs of Nos. 21 and 22 are 4-resolvable, whereas the designs of Nos. 1, 12, 16 and 23–29 are nonresolvable.

9
Resolvable Designs

Resolvable block designs, introduced in Section 6.0.3, are important in practice because it is often useful to be able to perform an experiment with replicates one or more at a time. The present chapter is devoted only to those among $(\alpha_1, \alpha_2, ..., \alpha_a)$-resolvable block designs which are α-resolvable for $\alpha \geq 1$, according to the concepts discussed in Section 6.0.3. A 1-resolvable block design is simply called resolvable in the usual sense of Bose (1942a). Note that α-resolvable block designs are necessarily equireplicate. Now, recalling the general form of Definition 6.0.2, one can write as follows. A block design with parameters v, b, r, $\boldsymbol{k}$ is said to be α-resolvable if the blocks can be separated into a sets (superblocks) of b_h blocks each ($b = \sum_{h=1}^{a} b_h$) such that the hth superblock contains every treatment exactly α times, for $h = 1, 2, ..., a$. Further, note that if the block design is proper, i.e., $\boldsymbol{k} = k\mathbf{1}_b$, then $b_1 = b_2 = \cdots = b_a (= b_0$, say). Then one gets $b = b_0 a$, $r = \alpha a$, $v\alpha = kb_0$ and $b\alpha = rb_0$. An α-resolvable proper block design with parameters v, b, r, k is said to be affine α-resolvable if every two distinct blocks from the same superblock intersect in the same number, q_1, of treatments, whereas any two blocks from different superblocks intersect in the same number, q_2, of treatments. Here $q_1 = (\alpha - 1)k/(b_0 - 1)$ and $q_2 = k^2/v$ (see Section 6.0.3). A more general case can be seen in Definition 6.0.3.

The chapter begins with a general consideration on such block designs (Section 9.1), by recalling relevant results discussed in Volume I. Next, methods of constructing α-resolvable block designs which are related to $(\rho_0; \rho_1, ..., \rho_{m-1}; \rho_m)$-EB designs for $m - 1 = 1$ (Section 9.2), 2 (Section 9.3), 3 (Section 9.4), or more (Section 9.5). Some construction methods can already be found in Chapters 6, 7 and 8. Square or rectangular lattice designs are mainly discussed in Section 9.6. Finally, as proper equireplicate $(0; v - 1; 0)$-EB designs, four tables are provided for existing (affine) resolvable BIB designs, unknown resolvable BIB designs, existing α-resolvable BIB designs, and possibly existing affine α-resolvable BIB designs with $\alpha \geq 2$.

9.1 General consideration

A randomization model for experiments in block designs with one stratum of blocks of experimental units has been discussed in Chapter 3. Then Chapter 5 has extended the model to the experimental situation in which the blocks are further grouped into superblocks, forming in that way two strata of blocks. Common examples of relevant designs are the lattice designs (introduced by Yates, 1940a) or, more generally, the resolvable incomplete block designs (see John, 1987, Section 3.4). Whereas in an ordinary block design, as considered in Section 3.2, the stratification of the experimental units leads to three strata, of units within blocks, of blocks within the total experimental area, and of the total area, in the extended situation considered here, four strata are distinguished. These strata, as indicated in Section 5.3, are of units within blocks, of blocks within superblocks, of superblocks within the total experimental area, and of the total area. Consequently, the extended randomization model had to take into account three stages of randomization, i.e., of units within blocks, of blocks within superblocks, and of the latter within the total area.

One can speak in this context, as discussed in Chapter 5, of two systems of blocks, one nested in the other, i.e., of blocks (or "subblocks") nested in the superblocks. As to the arrangement of treatments in blocks and superblocks, usually a kind of balance is desirable. For example, in the class of nested balanced incomplete block (NBIB) designs introduced by Preece (1967), the arrangement of treatments is such that ignoring either system of blocking, of the ordinary blocks (subblocks) or superblocks, one obtains a BIB design whose blocks are those of the system which has not been ignored. However, if by ignoring the system of superblocks one obtains a BIB design, but by ignoring the other, of subblocks, one arrives at a connected proper binary orthogonal block design (an RBD, see Section 2.1), then such nested design can be called a resolvable BIB design in the sense of Bose (1942a) (see Section 6.0.3). In general, a resolvable block design belongs to the class of nested block (NB) designs .

The basic notation and terminology will follow here those of Chapter 5. In the present section the main results of Chapter 5 will be recalled, with particular reference to resolvable block designs (including α-resolvable designs). A block design $\mathcal{D}^*$ for v treatments in b blocks is described by its $v \times b$ incidence matrix $\boldsymbol{N} = \boldsymbol{\Delta}\boldsymbol{D}'$, where $\boldsymbol{\Delta}'$ is the $n \times v$ design matrix for treatments and $\boldsymbol{D}'$ is the $n \times b$ design matrix for blocks. Suppose that the blocks of $\mathcal{D}^*$, also called subblocks, are grouped into a superblocks, leading to the partitions $\boldsymbol{\Delta} = [\boldsymbol{\Delta}_1 : \boldsymbol{\Delta}_2 : \cdots : \boldsymbol{\Delta}_a]$, $\boldsymbol{D} = \text{diag}[\boldsymbol{D}_1 : \boldsymbol{D}_2 : \cdots : \boldsymbol{D}_a]$, and, consequently, $\boldsymbol{N} = [\boldsymbol{N}_1 : \boldsymbol{N}_2 : \cdots : \boldsymbol{N}_a]$, where $\boldsymbol{\Delta}_h$, $\boldsymbol{D}_h$ and $\boldsymbol{N}_h = \boldsymbol{\Delta}_h\boldsymbol{D}_h'$ describe a component design $\mathcal{D}_h$, confined to superblock h $(= 1, 2, ..., a)$. The resulting block design $\mathcal{D}$ is then such that the v treatments are arranged into the a superblocks. Its $v \times a$ incidence matrix is written as $\boldsymbol{\Re} = \boldsymbol{\Delta}\boldsymbol{G}'$, where $\boldsymbol{G}'$ is the $n \times a$ design matrix for superblocks of the form $\boldsymbol{G}' = \boldsymbol{D}'\text{diag}[\mathbf{1}_{b_1} : \mathbf{1}_{b_2} : \cdots : \mathbf{1}_{b_a}] = \text{diag}[\mathbf{1}_{n_1} : \mathbf{1}_{n_2} : \cdots : \mathbf{1}_{n_a}]$, with b_h denoting the number of blocks in superblock h in $\mathcal{D}_h$, and n_h the number of units (plots) in that superblock. Note that $n_h = \mathbf{1}_{b_h}'\boldsymbol{k}_h$, where $\boldsymbol{k}_h = [k_{1(h)}, k_{2(h)}, ..., k_{b_h(h)}]' = \boldsymbol{N}_h'\boldsymbol{1}_v$ is the vector of block sizes in $\mathcal{D}_h$. The matrix $\mathfrak{R}$ can also be written as $\mathfrak{R} = [\boldsymbol{r}_1 : \boldsymbol{r}_2 : \cdots : \boldsymbol{r}_a]$, where $\boldsymbol{r}_h = [r_{1(h)}, r_{2(h)}, ..., r_{v(h)}]' = \boldsymbol{N}_h\boldsymbol{1}_{b_h}$ is the vector of treatment replications in $\mathcal{D}_h$ for $h = 1, 2, ..., a$.

Here, $\boldsymbol{N}\boldsymbol{1}_b = \mathfrak{R}\boldsymbol{1}_a = \boldsymbol{r} = [r_1, r_2, ..., r_v]'$ is the vector of treatment replications in the whole nested block design, as well in $\mathcal{D}^*$ as in $\mathcal{D}$, $\boldsymbol{N}'\boldsymbol{1}_v = \boldsymbol{k} = [\boldsymbol{k}_1', \boldsymbol{k}_2', ..., \boldsymbol{k}_a']'$ is the vector of block sizes in $\mathcal{D}^*$, and $\mathfrak{R}'\boldsymbol{1}_v = \boldsymbol{n} = [n_1, n_2, ..., n_a]'$ is the vector of superblock sizes in $\mathcal{D}$. As $\boldsymbol{k}^\delta = \boldsymbol{D}\boldsymbol{D}'$ is the diagonal matrix of block sizes in $\mathcal{D}^*$, $\boldsymbol{n}^\delta = \boldsymbol{G}\boldsymbol{G}'$ is a diagonal matrix with the numbers n_h on the diagonal representing the superblock sizes in $\mathcal{D}$. The total number of units (plots) used in the experiment is $n = \boldsymbol{1}_v'\boldsymbol{r} = \boldsymbol{1}_b'\boldsymbol{k} = \boldsymbol{1}_a'\boldsymbol{n}$.

The best linear unbiased estimation of a treatment parametric function $\boldsymbol{c}'\boldsymbol{\tau}$ under the overall randomization model (5.2.7), i.e., the model $\boldsymbol{y} = \boldsymbol{\Delta}'\boldsymbol{\tau} + \boldsymbol{G}'\boldsymbol{\alpha} + \boldsymbol{D}'\boldsymbol{\beta} + \boldsymbol{\eta} + \boldsymbol{e}$ derived in Section 5.2.1, is restricted to some special cases, those satisfying the conditions of Theorem 5.2.1. For other cases, not complying with these conditions, the estimation can be performed under submodels related to the stratification of the experimental units. In fact, the units of a design can be grouped according to a nested classification with four strata. The strata are defined as follows: 1st stratum—of units within blocks, called "intra-block," 2nd stratum—of blocks within superblocks, called "inter-block-intra-super-block," 3rd stratum—of superblocks within the experimental area, called "inter-superblock" and the 4th stratum—of the total experimental area.

In accordance to this stratification, the observed vector $\boldsymbol{y}$ can be decomposed as $\boldsymbol{y} = \boldsymbol{y}_1 + \boldsymbol{y}_2 + \boldsymbol{y}_3 + \boldsymbol{y}_4$, where each of the four components is related to one of the strata. In the present chapter, devoted to resolvable designs, the first and the second component vectors, $\boldsymbol{y}_1$ and $\boldsymbol{y}_2$, are important, i.e., the intra-block and inter-block-intra-superblock submodels (see Sections 5.3.1 and 5.3.2). The estimation results under the first submodel (Section 5.3.1) are identical with those of the intra-block analysis of a block design with one stratum of blocks (as in Section 3.2.1). This applies to all NB designs. As to the estimation under the second submodel, note that for any resolvable block design, either 1-resolvable or α-resolvable, the design $\mathcal{D}$ is connected and orthogonal, with its incidence matrix of the form $\mathfrak{R} = v^{-1}\boldsymbol{1}_v\boldsymbol{n}'$. Hence, on account of Remark 5.3.9, the inter-block-intra-superblock analysis is similar to the inter-block analysis of a block design with one stratum of blocks (as in Section 3.2.2). In particular, the BLUE under the inter-block-intra-superblock submodel given in Section 5.3.1 is identical with that under the inter-block submodel considered in Section 3.2.2. Also in the analysis of variance, the treatment sum of squares under the inter-block-intra-superblock submodel, considered in Section 5.3.1, and that under the inter-block submodel, considered in Section 3.2.2, are identical. These identities allow to perform the analysis of an α-resolvable block design, with $\alpha = 1$ or $\alpha > 1$, in a way similar to the analysis of an ordinary block design, with one stratum of blocks. It should, however, be noted that for the BLUE of $\boldsymbol{c}'\boldsymbol{\tau}$ under the inter-block-intra-superblock submodel (5.3.3) to exist, certain condition is to be satisfied, as stated in Theorem 5.3.2. Because in the case of resolvability,

$\tilde{\boldsymbol{N}}_0 = \boldsymbol{N}_0$ (as $\boldsymbol{\Delta}\tilde{\boldsymbol{\phi}}_2 = \boldsymbol{\Delta}\boldsymbol{\phi}_2$, see Remark 5.3.9), the condition is similar to that given in Theorem 3.2.2 for estimation under the submodel (3.2.4) in the case of an ordinary block design. The only difference between (5.3.10) and (3.2.16) concerns the matrix $\tilde{\boldsymbol{K}}_0$ defined in Theorem 5.3.2. Thus, confining attention to resolvable designs, under the submodel (5.3.3), i.e., $\boldsymbol{y}_2 = \tilde{\boldsymbol{\phi}}_2\boldsymbol{y}$, where $\tilde{\boldsymbol{\phi}}_2 = \boldsymbol{D}'\boldsymbol{k}^{-\delta}\boldsymbol{D} - \boldsymbol{G}'\boldsymbol{n}^{-\delta}\boldsymbol{G}$, the BLUE of $\boldsymbol{c}'\boldsymbol{\tau}$, with $\boldsymbol{c} = \tilde{\boldsymbol{C}}_2\boldsymbol{s} = \boldsymbol{C}_2\boldsymbol{s}$, exists if and only if $\boldsymbol{s}$ satisfies the condition

$$[\tilde{\boldsymbol{K}}_0 - \boldsymbol{N}_0'(\boldsymbol{N}_0\boldsymbol{k}^{-\delta}\boldsymbol{N}_0')^{-}\boldsymbol{N}_0]\boldsymbol{N}_0'\boldsymbol{s} = \boldsymbol{0}, \tag{9.1.1}$$

or its equivalence

$$[\tilde{\boldsymbol{K}}_0 - \boldsymbol{N}_0'\boldsymbol{N}_0(\boldsymbol{N}_0'\boldsymbol{N}_0)^{-}\tilde{\boldsymbol{K}}_0]\boldsymbol{N}_0'\boldsymbol{s} = \boldsymbol{0}, \tag{9.1.2}$$

where $\tilde{\boldsymbol{K}}_0 = \text{diag}[\boldsymbol{K}_{01} : \boldsymbol{K}_{02} : \cdots : \boldsymbol{K}_{0a}]$, $\boldsymbol{K}_{0h} = \boldsymbol{k}_h^{\delta} - (n_h)^{-1}\boldsymbol{k}_h\boldsymbol{k}_h'$, and $\boldsymbol{N}_0 = \boldsymbol{N} - n^{-1}\boldsymbol{r}\boldsymbol{k}' = \boldsymbol{N} - v^{-1}\boldsymbol{1}_v\boldsymbol{k}'$ (see Theorems 3.2.2 and 5.3.2). Certainly, if the design is proper, i.e., $k_1 = k_2 = \cdots = k_b = k$ (say), this condition is satisfied automatically by any $\boldsymbol{s}$. If the design is not proper, but still resolvable, then the condition (9.1.1) or (9.1.2) is satisfied by any $\boldsymbol{s}$, i.e., the equality (5.3.14) holds, if and only if for any vector $\boldsymbol{t}$ that satisfies the equality $\boldsymbol{N}_0\boldsymbol{t} = \boldsymbol{0}$, the equality $\boldsymbol{N}_0\tilde{\boldsymbol{K}}_0\boldsymbol{t} = \boldsymbol{0}$ is also satisfied (see Corollary 5.3.2). Furthermore, it can be noted (see Remark 5.3.1) that for the above implication to hold, it is sufficient that the block sizes within the superblock h are all equal for $h = 1, 2, ..., a$ (as required, e.g., in Definition 6.0.3).

An example of a resolvable BIB design with parameters $v = 6, b = 15, r = 5\ (= a), k = 2, \lambda = 1$, based on the theory presented in Sections 5.3.1–5.3.5, is presented as Example 5.3.4 in Section 5.3.6 (see also Example 5.4.2). The incidence matrix $\mathfrak{R} = \boldsymbol{1}_6\boldsymbol{1}_5'$ defines a design $\mathcal{D}$ as an equireplicate connected orthogonal and proper block design, i.e., an RBD. One can show that for any contrast there exists the BLUE under the intra-block submodel and under the inter-block-intra-superblock submodel, whereas no contrast can be estimated under the inter-superblock submodel. This is so, because the design $\mathcal{D}$ is orthogonal and connected [see Corollary 5.3.3(a)]. To find the appropriate formulae for the intra-block BLUEs and the inter-block-intra-superblock BLUEs, and their variances, one can proceed as in Example 5.3.3. In connection with Example 5.3.4, recall again the following (from Remark 5.3.9): If the design $\mathcal{D}$ of an NB design is connected and orthogonal, i.e., $\mathfrak{R} = n^{-1}\boldsymbol{r}\boldsymbol{n}'$, then $\boldsymbol{\Delta}\tilde{\boldsymbol{\phi}}_2 = \boldsymbol{\Delta}\boldsymbol{\phi}_2$ and, hence, the inter-block-intra-superblock analysis of variance becomes similar to the inter-block analysis of variance presented in Section 3.2.2, with the treatment sum of squares identical, $\tilde{\boldsymbol{Q}}_2'\tilde{\boldsymbol{C}}_2^{-}\tilde{\boldsymbol{Q}}_2 = \boldsymbol{Q}_2'\boldsymbol{C}_2^{-}\boldsymbol{Q}_2$, as $\tilde{\boldsymbol{C}}_2 = \boldsymbol{C}_2$ and $\tilde{\boldsymbol{Q}}_2 = \boldsymbol{Q}_2$ then, and with the residual sum of squares $\boldsymbol{y}'\tilde{\boldsymbol{\psi}}_2\boldsymbol{y} = \boldsymbol{y}'\tilde{\boldsymbol{\phi}}_2\boldsymbol{y} - \boldsymbol{y}'\boldsymbol{\phi}_2\boldsymbol{\Delta}'\boldsymbol{C}_2^{-}\boldsymbol{\Delta}\boldsymbol{\phi}_2\boldsymbol{y} = \boldsymbol{y}'\boldsymbol{\psi}_2\boldsymbol{y} - \boldsymbol{y}'\tilde{\boldsymbol{\phi}}_3\boldsymbol{y}$, as in general $\tilde{\boldsymbol{\phi}}_2 = \boldsymbol{\phi}_2 - \tilde{\boldsymbol{\phi}}_3$. Furthermore, because in this case $\tilde{\boldsymbol{C}}_3 = \boldsymbol{O}$, the inter-superblock sum of squares $\boldsymbol{y}'\tilde{\boldsymbol{\phi}}_3\boldsymbol{y}$ has no treatment component, and so the inter-superblock analysis of variance is then immaterial (see also the comment preceding Corollary 5.3.3). As already mentioned, any α-resolvable block design ($\alpha \geq 1$) has $\mathfrak{R} = n^{-1}\boldsymbol{r}\boldsymbol{n}' = v^{-1}\boldsymbol{1}_v\boldsymbol{n}'$.

In sum, for any resolvable or α-resolvable block design the $v \times a$ incidence matrix of the design $\mathcal{D}$ for superblocks is $\mathfrak{R} = n^{-1}\boldsymbol{r}\boldsymbol{n}'$, with $\boldsymbol{r} = r\mathbf{1}_v$ and $\boldsymbol{n}' = \alpha v \mathbf{1}_a$, i.e., $\mathcal{D}$ is equireplicate, connected and orthogonal (see Volume I, pages 219 and 234). This simplifies the analysis presented in Chapter 5. Particularly, there is no information on contrasts of treatment parameters under the inter-superblock submodel (Section 5.3.3). Thus, all variation of experimental material in this stratum is eliminated from estimating and testing these contrasts. This corresponds to the advantages of resolvable designs, lattice designs in particular, indicated by Yates (1940b), who wrote: "Allowance for this must be made in the analysis of variance ... by eliminating complete replications (or groups of replications) from the remainder component of blocks" (his Section 5).

Also it is to be remembered that the intra-block analysis (Sections 3.2.1 and 5.3.1) remains unchanged under the resolvability. "Also, it should be noted that the results presented in Sections 3.4 and 4.1 apply here as well" (Volume I, page 229, lines -2 to -1). This, of course, means that the classification of block designs based on efficiency factors for estimation in the intra-block analysis applies also to resolvable and α-resolvable designs.

9.2 Designs with one efficiency factor

Because the resolvable designs are equireplicate, this section will be discussed by considering separately two cases, of proper designs and nonproper designs.

9.2.1 Proper designs

As it follows from Theorem 7.2.1, the unique connected binary block design providing full efficiency in the intra-block analysis for any contrast, is the RBD, a design most widely used among all experimental designs. See also Section 7.2 for such orthogonal designs.

Though an RBD could be considered as a special case of resolvable block designs, here only binary proper resolvable $(0; v-1; 0)$-EB designs will be considered. This, on account of Theorem 8.2.3, means that of interest here are resolvable BIB designs, with $\varepsilon_1 = \lambda v/(rk), \rho_1 = v - 1$, $\boldsymbol{L}_1 = \boldsymbol{I}_v - v^{-1}\mathbf{1}_v\mathbf{1}_v'$ (see Section 6.0.1) and with their b blocks separated into r sets, each of them providing a single replication for every treatment (see Definition 6.0.2).

Though Yates (1939, 1940b) has pointed out some statistical advantages of resolvable designs and their original form had appeared earlier in the mathematical literature, the interest in resolvable BIB designs was greatly enhanced by a combinatorial paper of Bose (1942a). In using a BIB design for an experiment there is a substantial advantage if its blocks can be divided into disjoint sets (superblocks) such that every treatment occurs exactly once in every set (see Chapter 5, Example 5.3.4). In this case in an agricultural experiment the experimental field would be divided into r superblocks and then each superblock divided further into v/k subblocks. Every treatment would then occur exactly once in each superblock. In this layout the component of variance due to the differences between superblocks will not affect the estimation of the treatment differences, because in this case the design $\mathcal{D}$ (for superblocks) becomes then connected and orthogonal (see Sections 5.1 and 5.3.3).

One of the earliest examples of a resolvable BIB design is the Kirkman school girl problem formulated in Kirkman (1850a) and pursued further in Kirkman (1850b). A teacher wants to arrange $6t+3$ girls in $2t+1$ rows of 3 each for $3t+1$ successive days. The problem is to find different row arrangements such that any two girls appear in the same row exactly at one day. This is equivalent to finding a resolvable solution of the BIB design with parameters $v = 6t+3$, $b = (2t+1)(3t+1)$, $r = 3t+1$, $k = 3$, $\lambda = 1$. Kirkman himself gave some solutions and many mathematicians worked on this problem in the late 19th and early 20th centuries. An excellent bibliography can be found in Eckenstein (1912). However, no complete solution was known until when Ray-Chaudhuri and Wilson (1971) completely solved the problem.

From Section 6.0.1 and Definition 6.0.2, it obviously follows that the necessary conditions for the existence of a resolvable BIB design with parameters v, b, r, k, λ are given by $\lambda(v-1) \equiv 0 \pmod{k-1}$ and $v \equiv 0 \pmod{k}$. The necessary conditions are known to be sufficient for any k and λ if v is large enough (Lu, 1984; Lee and Furino, 1995). For $k = 3$ and 4, the existence of a resolvable BIB design is completely solved (see Hanani, Ray-Chaudhuri and Wilson, 1972; Furino, Miao and Yin, 1996). As the next value of block sizes k, some existence results on resolvable BIB designs with $k = 5$ have been discussed by Miao and Zhu (1995), Furino, Miao and Yin (1996), Abel and Greig (1997), and Abel, Ge, Greig and Zhu (2001). They are still far from a complete solution. [Even in a BIB design the existence problem has not been solved completely for general k, as one can see from Sections 6.0.1 and 8.2.1.]

Some nonexistence results are also known. For example, as in Corollary 6.9.3, when q is a prime or a prime power, there exists an affine resolvable BIB design with parameters $v = q^2$, $b = q(q+1)$, $r = q+1$, $k = q$, $\lambda = 1$. However, this design does not exist for $q = 6(= 2 \times 3), 14(= 2 \times 7), 21(= 3 \times 7), 22(= 2 \times 11)$, ... [$\equiv 1$ or $2 \pmod 4$ and the square free part of q contains a prime $\equiv 3 \pmod 4$] (see Raghavarao, 1971, Theorem 12.3.2).

Direct and/or recursive methods of constructing resolvable BIB designs with their combinatorial properties have been investigated in the literature, for example, by Ball (1942), Bose (1942a, 1947, 1959), Rao (1946), Bose, Shrikhande and Bhattacharya (1953), Sprott (1956), Bose and Shrikhande (1960), Kageyama (1971, 1972b, 1983), Kimberley (1971), Hanani, Ray-Chaudhuri and Wilson (1972), Ray-Chaudhuri and Wilson (1973), Hanani (1974), Baker (1983), Kageyama and Miao (1994, 1995b, 1998), Abel and Furino (1996). In particular, a geometrical method has already been described in Section 6.9, its main results being presented in Theorem 6.9.2, Corollary 6.9.2 (resolvable design) and Corollary 6.9.3 (affine resolvable design). Some other constructions are stated below. They are mainly based on results given in Sections 6.7, 6.8 and 6.9.

Theorem 9.2.1 (Bose, 1947; Sprott, 1956). *When* $4t-1$ *is a prime or a prime* *power, an affine resolvable BIB design with parameters* $v = 4t$, $b = 2(4t-1)$, $r = 4t-1$, $k = 2t$, $\lambda = 2t-1$, $q_1 = 0, q_2 = t$, *i.e., a* $(0;\rho_1;0)$*-EB design with* $\varepsilon_1 = 2(2t-1)/(4t-1)$, $\rho_1 = 4t-1$, $\boldsymbol{L}_1 = \boldsymbol{I}_{4t} - (4t)^{-1}\boldsymbol{1}_{4t}\boldsymbol{1}'_{4t}$, *can be constructed from two initial blocks (which compose a superblock)*

$$(0, x^0, x^2, ..., x^{4(t-1)}),\ (\infty, x, x^3, ..., x^{4t-3}),$$

where x *is a primitive element of Galois field* $GF(4t-1)$.

Proof. A direct application of Theorem 6.7.2 provides the proof. □

Note that the same series as in Theorem 9.2.1 can be obtained from Theorem 6.8.4 through a Hadamard matrix of order $4t$. Recall the definition of a Hadamard matrix as given in Section 6.8. In fact, let $\boldsymbol{H}_0$ be a matrix of order $4t-1$ obtained by deleting the first row and the first column of a normalized Hadamard matrix of order $4t$. Then the incidence matrix

$$\begin{bmatrix} \frac{1}{2}(\boldsymbol{1}_{4t-1}\boldsymbol{1}'_{4t-1} + \boldsymbol{H}_0) & \frac{1}{2}(\boldsymbol{1}_{4t-1}\boldsymbol{1}'_{4t-1} - \boldsymbol{H}_0) \\ \boldsymbol{1}'_{4t-1} & \boldsymbol{0}' \end{bmatrix}$$

produces the required design. Here a pair of the ith and $(i+4t-1)$th columns of the above incidence matrix composes a superblock for $i = 1, 2, ..., 4t-1$.

Incidentally, it is conjectured (Shrikhande, 1976) that an affine resolvable BIB design with the parameters given in Theorem 9.2.1 exists for every positive integer t. A complete solution is not yet given.

Example 9.2.1. Since 3 is a primitive element of GF(7), two initial blocks $(0, 3^0, 3^2, 3^4)$, $(\infty, 3, 3^3, 3^5)$, i.e., (0, 1, 2, 4), $(\infty, 3, 6, 5)$ mod 7, yield an affine resolvable BIB design with parameters $v = 8$, $b = 14$, $r = 7$, $k = 4$, $\lambda = 3$, $q_1 = 0, q_2 = 2$ of No. 7* in Table 9.1, i.e., $(0;7;0)$-EB with $\varepsilon_1 = 6/7, \rho_1 = 7$, $\boldsymbol{L}_1 = \boldsymbol{I}_8 - 8^{-1}\boldsymbol{1}_8\boldsymbol{1}'_8$, whose incidence matrix is given by

$$\begin{array}{c} 0 \\ 1 \\ 2 \\ 3 \\ 4 \\ 5 \\ 6 \\ \infty \end{array} \left[\begin{array}{cc|cc|cc|cc|cc|cc|cc} 1 & 0 & 0 & 1 & 0 & 1 & 1 & 0 & 0 & 1 & 1 & 0 & 1 & 0 \\ 1 & 0 & 1 & 0 & 0 & 1 & 0 & 1 & 1 & 0 & 0 & 1 & 1 & 0 \\ 1 & 0 & 1 & 0 & 1 & 0 & 0 & 1 & 0 & 1 & 1 & 0 & 0 & 1 \\ 0 & 1 & 1 & 0 & 1 & 0 & 1 & 0 & 0 & 1 & 0 & 1 & 1 & 0 \\ 1 & 0 & 0 & 1 & 1 & 0 & 1 & 0 & 1 & 0 & 0 & 1 & 0 & 1 \\ 0 & 1 & 1 & 0 & 0 & 1 & 1 & 0 & 1 & 0 & 1 & 0 & 0 & 1 \\ 0 & 1 & 0 & 1 & 1 & 0 & 0 & 1 & 1 & 0 & 1 & 0 & 1 & 0 \\ 0 & 1 & 0 & 1 & 0 & 1 & 0 & 1 & 0 & 1 & 0 & 1 & 0 & 1 \end{array}\right].$$

Note that for these design parameters Theorem 8.2.8 yields a nonresolvable solution as $(0,1,2,4)$, $(\infty,1,2,4)$ mod 7, i.e., a solution No.32 in Table 8.2.

Theorem 9.2.2. *When* $2k-1$ *is a prime or a prime power, a resolvable BIB design with parameters* $v = 2k$, $b = 4(2k-1)$, $r = 2(2k-1)$, k, $\lambda = 2(k-1)$, *i.e., a* $(0;\rho_1;0)$*-EB design with* $\varepsilon_1 = 2(k-1)/(2k-1)$, $\rho_1 = 2k-1$, $\boldsymbol{L}_1 = \boldsymbol{I}_{2k} -$

$(2k)^{-1}\mathbf{1}_{2k}\mathbf{1}'_{2k}$, *can be constructed from initial blocks (a pair of which composes a superblock)*

$$(0, x^i, x^{i+2}, ..., x^{i+2k-4}),\ (\infty, x^{i+1}, x^{i+3}, ..., x^{i+2k-3}),\ i = 0, 1,$$

where x is a primitive element of Galois field $GF(2k-1)$.

Proof. A direct application of Theorem 6.7.2 provides the proof. □

Example 9.2.2. Let $k = 3$. Then Theorem 9.2.2 yields a resolvable BIB design with parameters $v = 6$, $b = 20$, $r = 10$, $k = 3$, $\lambda = 4$ of No. 5 in Table 9.1, i.e., $(0;5;0)$-EB with $\varepsilon_1 = 4/5, \rho_1 = 5$, $\boldsymbol{L}_1 = \boldsymbol{I}_6 - 6^{-1}\mathbf{1}_6\mathbf{1}'_6$, having initial blocks $[(0, 2^0, 2^2), (\infty, 2^1, 2^3)]$, $[(0, 2^1, 2^3), (\infty, 2^2, 2^4)]$, i.e., $[(0, 1, 4), (\infty, 2, 3)]$, $[(0, 2, 3), (\infty, 4, 1)]$ mod 5, $\boldsymbol{L}_1 = \boldsymbol{I}_8 - 8^{-1}\mathbf{1}_8\mathbf{1}'_8$, whose incidence matrix is given by

$$\begin{array}{c} 0 \\ 1 \\ 2 \\ 3 \\ 4 \\ \infty \end{array}
\left[\begin{array}{cc|cc|cc|cc|cc|cc|cc|cc|cc|cc}
1 & 0 & 1 & 0 & 0 & 1 & 0 & 1 & 1 & 0 & 1 & 0 & 0 & 1 & 1 & 0 & 1 & 0 & 0 & 1 \\
1 & 0 & 1 & 0 & 1 & 0 & 0 & 1 & 0 & 1 & 0 & 1 & 1 & 0 & 0 & 1 & 1 & 0 & 1 & 0 \\
0 & 1 & 1 & 0 & 1 & 0 & 1 & 0 & 0 & 1 & 1 & 0 & 0 & 1 & 1 & 0 & 0 & 1 & 1 & 0 \\
0 & 1 & 0 & 1 & 1 & 0 & 1 & 0 & 1 & 0 & 1 & 0 & 1 & 0 & 0 & 1 & 1 & 0 & 0 & 1 \\
1 & 0 & 0 & 1 & 0 & 1 & 1 & 0 & 1 & 0 & 0 & 1 & 1 & 0 & 1 & 0 & 0 & 1 & 1 & 0 \\
0 & 1 & 0 & 1 & 0 & 1 & 0 & 1 & 0 & 1 & 0 & 1 & 0 & 1 & 0 & 1 & 0 & 1 & 0 & 1
\end{array}\right].$$

Theorem 9.2.3. *Let $\boldsymbol{N}$ be the $v \times b$ incidence matrix of a BIB design with parameters $v = 2k+1$, $b = 3r - 2\lambda$, r, k, λ. Then the incidence matrix*

$$\begin{bmatrix} \boldsymbol{N} & \mathbf{1}_v\mathbf{1}'_b - \boldsymbol{N} \\ \mathbf{1}'_b & \mathbf{0}' \end{bmatrix}$$

yields a resolvable BIB design with parameters $v^* = v+1$, $b^* = 2b$, $r^* = b$, $k^* = k+1$, $\lambda^* = r$, i.e., a $(0;\rho_1;0)$-EB design with $\varepsilon_1 = k(v+1)/[v(k+1)]$, $\rho_1 = v$, $\boldsymbol{L}_1 = \boldsymbol{I}_{v+1} - (v+1)^{-1}\mathbf{1}_{v+1}\mathbf{1}'_{v+1}$. Here a pair of the ith and $(i+b)$th columns of the resulting incidence matrix composes a superblock for $i = 1, 2, ..., b$.

Proof. A direct calculation shows the result. □

Note that Theorem 9.2.3 without resolvability coincides with Theorem 8.2.13.

A symmetric BIB design with parameters $v = b = 4\lambda+3$, $r = k = 2\lambda+1$, λ, discussed in Corollary 8.2.1, satisfies the condition of Theorem 9.2.3. Hence the following result can be obtained through Theorem 9.2.3 with the use of Theorem 6.0.5.

Corollary 9.2.1. *When $4\lambda+3$ is a prime or a prime power, there exists an affine resolvable BIB design with parameters $v^* = 4(\lambda+1)$, $b^* = 2(4\lambda+3)$, $r^* = 4\lambda+3$, $k^* = 2(\lambda+1)$, $\lambda^* = 2\lambda+1$, $q_1 = 0, q_2 = \lambda+1$, i.e., $(0;\rho_1;0)$-EB with $\varepsilon_1 = 2(2\lambda+1)/(4\lambda+3)$, $\rho_1 = 4\lambda+3$, $\boldsymbol{L}_1 = \boldsymbol{I}_{4(\lambda+1)} - [4(\lambda+1)]^{-1}\mathbf{1}_{4(\lambda+1)}\mathbf{1}'_{4(\lambda+1)}$.*

One can say that the existence of a symmetric BIB design with $r = 2\lambda+1$ always implies the existence of an affine resolvable BIB design.

To facilitate a desirable choice, 20 (affine) resolvable BIB designs with their plans within the range of parameters $v \le 100$ and $r \le 10$ have been tabulated in Table 9.1. This tabulation would be useful for practitioners. For more such individual plans within a wide range of parameters the reader can be referred to Kageyama (1972b, 1983), Hanani (1974) and Abel (1994). In fact, within the scope of parameters $v \le 100$ and $r \le 20$ there are 48 cases of admissible parameters (i.e., satisfying necessary conditions), among them 40 designs exist (20 of which, with $r \le 10$, are listed in Table 9.1), one design does not exist and 7 designs are still unknown (see Table 9.2). This wide tabulation may also be useful for practitioners (see Kageyama, 1972b). More tables can be found in a book by Furino, Miao and Yin (1996). The smallest unknown case regarding resolvable BIB designs concerns parameters $v = 15$, $b = 42$, $r = 14$, $k = 5$, $\lambda = 4$ (see Table 9.2).

As mentioned in Section 6.0.3, the concept of resolvability, introduced by Bose (1942a), was generalized to α-resolvability by Shrikhande and Raghavarao (1964). Since then, many constructional methods and combinatorial properties of α-resolvable BIB designs have been given, together with relevant tabulations. See, for example, Shrikhande and Raghavarao (1963, 1964), Sprott (1956), Kageyama (1973a, 1973b, 1976a, 1976b), Mohan (1980), Mohan and Kageyama (1982), Kageyama and Mohan (1983), Rajkundlia (1983). Here, a list of 14 parameter combinations of such α-resolvable BIB designs within the practical range of $\alpha \ge 2, r \le 10$ and $v \le 100$ is presented in Table 9.3. [There are simple methods of constructing a large number of α-resolvable BIB designs for $\alpha \ge 2$ from other α-resolvable BIB designs (with lower α), but these designs are not listed there. More on this will be said at the end of this Section 9.2.1.] Furthermore, a list of 12 parameter combinations of affine α-resolvable BIB designs within the range of $\alpha \ge 2$, $v \le 100$ and $\lambda \le 100$ is presented in Table 9.4.

A series of affine α-resolvable BIB designs can be seen in Shrikhande and Raghavarao (1963) who presented two recursive constructions. One of them is stated below without proof.

Theorem 9.2.4 (Shrikhande and Raghavarao, 1963, Theorem 2; Raghavarao, 1971, Theorem 5.9.4). *The existence of a symmetric BIB design with parameters $v_1 = b_1$, $r_1 = k_1$, λ_1, and an affine resolvable BIB design with parameters $v_2 = v_1k_2$, $b_2 = b_1r_2$, r_2, k_2, λ_2, implies the existence of an affine α-resolvable BIB design with parameters $v = v_2$, $b = b_2$, $r = r_1r_2$, $k = k_1k_2$, $\lambda = r_1\lambda_2 + \lambda_1(r_2 - \lambda_2)$, $\alpha = r_1$, $b_0 = b_1$, $a = r_2$, $q_1 = k_1k_2(k_1 - 1)/(v_1 - 1), q_2 = k_1^2k_2/v_1$, i.e., of a $(0;\rho_1;0)$-EB design with $\varepsilon_1 = v_2[r_1\lambda_2 + \lambda_1(r_2 - \lambda_2)]/(r_1r_2k_1k_2)$, $\rho_1 = v_2 - 1$, $\boldsymbol{L}_1 = \boldsymbol{I}_{v_2} - (v_2)^{-1}\mathbf{1}_{v_2}\mathbf{1}'_{v_2}$.*

Theorem 9.2.4 yields an affine α-resolable $(0;\rho_1;0)$-EB design with larger replication while keeping the numbers of treatments and blocks relatively small. The idea of constructing such design by Theorem 9.2.4 will now be illustrated.

Example 9.2.3. Take a symmetric BIB design with parameters $v_1 = b_1 = 3$, $r_1 = k_1 = 2$, $\lambda_1 = 1$, and with the incidence matrix $\boldsymbol{N}^{(1)}$. Also take an affine resolvable BIB design with parameters $v_2 = 9(= v_1 k_2)$, $b_2 = 12(= b_1 r_2)$, $r_2 = 4, k_2 = 3$, $\lambda_2 = 1$ (see Table 9.1), and with the incidence matrix $\boldsymbol{N}^{(2)}$. Here

$$\boldsymbol{N}^{(1)} = \begin{bmatrix} 0 & 1 & 1 \\ 1 & 0 & 1 \\ 1 & 1 & 0 \end{bmatrix},$$

$$\boldsymbol{N}^{(2)} = \left[\begin{array}{ccc|ccc|ccc|ccc} 0 & 0 & 1 & 1 & 0 & 0 & 1 & 0 & 0 & 0 & 1 & 0 \\ 1 & 0 & 0 & 0 & 0 & 1 & 1 & 0 & 0 & 1 & 0 & 0 \\ 0 & 1 & 0 & 1 & 0 & 0 & 0 & 0 & 1 & 1 & 0 & 0 \\ 0 & 1 & 0 & 0 & 1 & 0 & 1 & 0 & 0 & 0 & 0 & 1 \\ 0 & 0 & 1 & 0 & 1 & 0 & 0 & 1 & 0 & 1 & 0 & 0 \\ 0 & 1 & 0 & 0 & 0 & 1 & 0 & 1 & 0 & 0 & 1 & 0 \\ 1 & 0 & 0 & 0 & 1 & 0 & 0 & 0 & 1 & 0 & 1 & 0 \\ 1 & 0 & 0 & 1 & 0 & 0 & 0 & 1 & 0 & 0 & 0 & 1 \\ 0 & 0 & 1 & 0 & 0 & 1 & 0 & 0 & 1 & 0 & 0 & 1 \end{array}\right]$$

$$= [\boldsymbol{N}_1 : \boldsymbol{N}_2 : \boldsymbol{N}_3 : \boldsymbol{N}_4] \text{ (say).}$$

Let $\boldsymbol{N}_j^{(1)}, j = 1, 2, 3, 4(= r_2)$, be the incidence matrix obtained from $\boldsymbol{N}^{(1)}$ by replacing each treatment i of the design corresponding to $\boldsymbol{N}^{(1)}$ by the set of treatments contained in the ith block of $\boldsymbol{N}_j$ for $i = 1, 2, 3(= v_1)$. Then it follows that the incidence matrix $\boldsymbol{N} = [\boldsymbol{N}_1^{(1)} : \boldsymbol{N}_2^{(1)} : \boldsymbol{N}_3^{(1)} : \boldsymbol{N}_4^{(1)}]$ gives an affine 2-resolvable BIB design with parameters $v = 9$, $b = 12$, $r = 8, k = 6$, $\lambda = 5$, $b_0 = 3, a = 4$, $q_1 = 3, q_2 = 4$, i.e., a (0; 8; 0)-EB design with $\varepsilon_1 = 15/16$, $\rho_1 = 8$, $\boldsymbol{L}_1 = \boldsymbol{I}_9 - 9^{-1}\boldsymbol{1}_9\boldsymbol{1}_9'$. In fact, this design with the said incidence matrix

$$\boldsymbol{N} = \left[\begin{array}{ccc|ccc|ccc|ccc} 1 & 1 & 0 & 0 & 1 & 1 & 0 & 1 & 1 & 1 & 0 & 1 \\ 0 & 1 & 1 & 1 & 1 & 0 & 0 & 1 & 1 & 0 & 1 & 1 \\ 1 & 0 & 1 & 0 & 1 & 1 & 1 & 1 & 0 & 0 & 1 & 1 \\ 1 & 0 & 1 & 1 & 0 & 1 & 0 & 1 & 1 & 1 & 1 & 0 \\ 1 & 1 & 0 & 1 & 0 & 1 & 1 & 0 & 1 & 0 & 1 & 1 \\ 1 & 0 & 1 & 1 & 1 & 0 & 1 & 0 & 1 & 1 & 0 & 1 \\ 0 & 1 & 1 & 1 & 0 & 1 & 1 & 1 & 0 & 1 & 0 & 1 \\ 0 & 1 & 1 & 0 & 1 & 1 & 1 & 0 & 1 & 1 & 1 & 0 \\ 1 & 1 & 0 & 1 & 1 & 0 & 1 & 1 & 0 & 1 & 1 & 0 \end{array}\right]$$

is a complement of the design corresponding to $\boldsymbol{N}^{(2)}$.

As in Example 9.2.3, if $\boldsymbol{1}_{v_1}\boldsymbol{1}_{v_1}' - \boldsymbol{I}_{v_1}$ is taken as the incidence matrix of a starting symmetric BIB design in Theorem 9.2.4, then the resulting design is always a complement of another starting affine resolvable BIB design. This fact is also observed from Theorem 9.2.7 later with $\alpha = 1$. In this sense, to get an affine α-resolvable BIB design, one has to take a nontrivial symmetric BIB design. Such smallest case is a symmetric BIB design with parameters

$v_1 = b_1 = 7$, $r_1 = k_1 = 3$, $\lambda_1 = 1$. Then by taking an affine resolvable BIB design with parameters $v_2 = 49$, $b_2 = 56$, $r_2 = 8, k_2 = 7$, $\lambda_2 = 1$ (see Table 9.1), Theorem 9.2.4 yields an affine 3-resolvable BIB design with parameters $v = 49$, $b = 56$, $r = 24, k = 21$, $\lambda = 10$, $b_0 = 7, a = 8$, $q_1 = 7, q_2 = 9$, i.e., a (0;48;0)-EB design with $\varepsilon_1 = 35/36$, $\rho_1 = 48$, $\boldsymbol{L}_1 = \boldsymbol{I}_{49} - (49)^{-1}\mathbf{1}_{49}\mathbf{1}'_{49}$. This coincides with a design given by Theorem 9.2.5 next with $s = 2$.

Other useful observations in this context are given in the following theorems.

Theorem 9.2.5 (Shrikhande and Raghavarao, 1963, Corollary). *If s and $p = s^2+s+1$ are both primes or prime powers, there exists an affine $(s+1)$-resolvable BIB design with parameters $v = p^2$, $b = p(p+1)$, $r = (p+1)(s+1)$, $k = p(s+1)$, $\lambda = s+p+1$, $b_0 = p$, $a = p+1$, $q_1 = p$, $q_2 = (s+1)^2$, i.e., a $(0;\rho_1;0)$-EB design with $\varepsilon_1 = p(s+p+1)/[(p+1)(s+1)^2]$, $\rho_1 = p^2-1$, $\boldsymbol{L}_1 = \boldsymbol{I}_{p^2} - p^{-2}\mathbf{1}_{p^2}\mathbf{1}'_{p^2}$.*

Proof. When s and $p = s^2+s+1$ are both primes or prime powers, one can take PG$(2,s)$: 1, i.e., use Corollary 6.9.1(i), to get the first BIB design, and AG$(2,p)$: 1, i.e., use Corollary 6.9.2(i), to get the second BIB design, and then apply Theorem 9.2.4 accordingly. □

Theorem 9.2.6. *When $4u-1$ is a prime or a prime power, there exists an affine $(2u-1)$-resolvable BIB design with parameters $v = (4u-1)^2$, $b = 4u(4u-1)$, $r = 4u(2u-1)$, $k = (2u-1)(4u-1)$, $\lambda = u(4u-3)$, $b_0 = 4u-1$, $a = 4u$, $q_1 = (u-1)(4u-1)$, $q_2 = (2u-1)^2$, i.e., a $(0;\rho_1;0)$-EB design with $\varepsilon_1 = (4u-3)(4u-1)/[4(2u-1)^2]$, $\rho_1 = 8u(2u-1)$, $\boldsymbol{L}_1 = \boldsymbol{I}_{(4u-1)^2} - (4u-1)^{-2}\mathbf{1}_{(4u-1)^2}\mathbf{1}'_{(4u-1)^2}$.*

Proof. By Corollary 8.2.1, when $4u-1$ is a prime or a prime power, there exists a symmetric BIB design with parameters $v_1 = b_1 = 4u-1$, $r_1 = k_1 = 2u-1$, $\lambda_1 = u-1$. Furthermore, by Corollary 6.9.3, there exists an affine resolvable BIB design with parameters $v_2 = (4u-1)^2$, $b_2 = 4u(4u-1)$, $r_2 = 4u$, $k_2 = 4u-1$, $\lambda_2 = 1$. Now, the application of Theorem 9.2.4 completes the proof. □

Taking the complement of α-resolvable BIB designs yields the following theorem.

Theorem 9.2.7. *The existence of an (affine) α-resolvable BIB design with parameters v, $b = b_0a$, $r = \alpha a$, k, λ, (q_1, q_2), $\varepsilon_1 = \lambda v/(rk)$, $\rho_1 = v-1$, $\boldsymbol{L}_1 = \boldsymbol{I}_v - v^{-1}\mathbf{1}_v\mathbf{1}'_v$ is equivalent to the existence of an (affine) $(b_0-\alpha)$-resolvable BIB design with parameters $v^* = v$, $b^* = b$, $r^* = (b_0-\alpha)a$, $k^* = v-k$, $\lambda^* = (b_0-2\alpha)a+\lambda$, $(q_1^* = v-2k+q_1$, $q_2^* = v-2k+q_2)$, $\varepsilon_1^* = \lambda^*v^*/(r^*k^*)$, $\rho_1^* = v^*-1$, $\boldsymbol{L}_1^* = \boldsymbol{I}_{v^*} - (v^*)^{-1}\mathbf{1}_{v^*}\mathbf{1}'_{v^*}$.*

Proof. As it is known from Section 6.2 (Corollary 6.2.2), the BIB structure is preserved under the complementation. Hence, it is sufficient to show the (affine) resolvability. Since each treatment occurs α times among the b_0 blocks in each of the a superblocks of an α-resolvable BIB design, each treatment obviously occurs exactly $b_0-\alpha$ times in each of the a superblocks of its complementary BIB design. This shows the $(b_0-\alpha)$-resolvability. Furthermore, Theorem 6.0.5 can be applied here and then it follows that $q_1^* = v-2k+q_1$ and $q_2^* = v-2k+q_2$, which imply the affine $(b_0-\alpha)$-resolvability. For the notation see Definition 6.0.3 and the comment following Theorem 6.0.5. □

Thus, many α-resolvable or affine α-resolvable BIB designs can be constructed through Theorem 9.2.7 by using known resolvable or affine resolvable BIB designs.

Example 9.2.4. An affine resolvable BIB design with parameters $v=9$, $b=12$, $r=4$, $k=3$, $\lambda=1$, $q_1=0, q_2=1$, having a solution (as No. 9* in Table 9.1), i.e., PC(4)[(1, 6, 7) (2, 3, 5) (0, 4, ∞)] (forming a superblock) mod 8, gives an affine 2-resolvable BIB design with parameters $v=9$, $b=12$, $r=8$, $k=6$, $\lambda=5$, $b_0=3, a=4, q_1=3, q_2=4$, $\varepsilon_1=15/16$, $\rho_1=8$, $\boldsymbol{L}_1=\boldsymbol{I}_9-9^{-1}\mathbf{1}_9\mathbf{1}_9'$, having a solution, i.e., PC(4)[(0, 2, 3, 4, 5, ∞) (0, 1, 4, 6, 7, ∞) (1, 2, 3, 5, 6, 7)] (forming a superblock) mod 8. Similarly, from a known affine resolvable BIB design (No. 13* in Table 9.1) with parameters $v=16$, $b=20$, $r=5$, $k=4$, $\lambda=1$, $q_1=0, q_2=1$, one can get an affine 3-resolvable BIB design with parameters $v=16$, $b=20$, $r=15$, $k=12$, $\lambda=11$, $b_0=4, a=5, q_1=8, q_2=9$, $\varepsilon_1=44/45$, $\rho_1=15$, $\boldsymbol{L}_1=\boldsymbol{I}_{16}-(16)^{-1}\mathbf{1}_{16}\mathbf{1}_{16}'$. Here PC(4) means a partial (short) cycle of order 4, i.e., a cyclic development of initial blocks four times and reducing modulo 8 when necessary.

Noting (Kageyama, 1973b) that an affine α-resolvable BIB design with $r\le 15$ does not exist for $\alpha\ge 4$, it is obvious that only two affine α-resolvable BIB designs exist for $\alpha\ge 2$ and $r\le 15$, those given in Example 9.2.4.

Furthermore, using the idea of Rao (1966), the following theorem can be obtained.

Theorem 9.2.8. *The existence of a BIB design with parameters v_1, b_1, r_1, k_1, λ_1, and an α_2-resolvable BIB design with parameters v_2, $b_2=b_0^{(2)}a_2$, $r_2=\alpha_2 a_2$, $k_2=v_1$, λ_2, implies the existence of an α-resolvable BIB design with parameters $v=v_2$, $b=b_1b_2$, $r=r_1r_2$, $k=k_1$, $\lambda=\lambda_1\lambda_2$, $b_0=b_0^{(2)}b_1$, $a=a_2$ for $\alpha=r_1\alpha_2$, i.e., of a $(0;\rho_1;0)$-EB design with $\varepsilon_1=\lambda_1\lambda_2v_2/(r_1r_2k_1)$, $\rho_1=v_2-1$, $\boldsymbol{L}_1=\boldsymbol{I}_{v_2}-(v_2)^{-1}\mathbf{1}_{v_2}\mathbf{1}_{v_2}'$.*

Proof. Let $\boldsymbol{N}^{(1)}$ be the $v_1\times b_1$ incidence matrix of the first starting BIB design. Then one can let $\boldsymbol{N}^{(1)}=[\boldsymbol{n}_1,\boldsymbol{n}_2,...,\boldsymbol{n}_{v_1}]'$ where $\boldsymbol{n}_i'\boldsymbol{n}_j=r_1$ $(i=j)$ or λ_1 $(i\ne j)$. Next let $\boldsymbol{N}^{(2)}$ be the $v_2\times b_2$ incidence matrix of the second starting α_2-resolvable BIB design. Now substitute v_1 distinct $1\times b_1$ row vectors $\boldsymbol{n}_i'$ in place of v_1 distinct units and $\mathbf{0}'$ in place of v_2-v_1 distinct 0 in every block of $\boldsymbol{N}^{(2)}$. Then it follows that the resulting matrix is the incidence matrix of the required design. □

Example 9.2.5. A BIB design, having the incidence matrix

$$\boldsymbol{N}^{(1)}=\begin{bmatrix}0&1&1&1\\1&0&1&1\\1&1&0&1\\1&1&1&0\end{bmatrix},$$

with parameters $v_1 = b_1 = 4$, $r_1 = k_1 = 3$, $\lambda_1 = 2$, and a resolvable BIB design, having the incidence matrix $\boldsymbol{N}^{(2)}$, with parameters $v_2 = 8$, $b_2 = 14$, $r_2 = 7$, $k_2 = 4$, $\lambda_2 = 3$ lead to a 3-resolvable BIB design with parameters $v = 8$, $b = 56$, $r = 21$, $k = 3$, $\lambda = 6$, $b_0 = 8$, $a = 7$, $\varepsilon_1 = 16/21$, $\rho_1 = 7$, $\boldsymbol{L}_1 = \boldsymbol{I}_8 - 8^{-1}\boldsymbol{1}_8\boldsymbol{1}_8'$. This is illustrated below. The second incidence matrix is shown by

$$\boldsymbol{N}^{(2)} = \left[\begin{array}{cc|cc|cc|cc|cc|cc|cc}
1 & 0 & 0 & 1 & 0 & 1 & 1 & 0 & 0 & 1 & 1 & 0 & 1 & 0 \\
1 & 0 & 1 & 0 & 0 & 1 & 0 & 1 & 1 & 0 & 0 & 1 & 1 & 0 \\
1 & 0 & 1 & 0 & 1 & 0 & 0 & 1 & 0 & 1 & 1 & 0 & 0 & 1 \\
0 & 1 & 1 & 0 & 1 & 0 & 1 & 0 & 0 & 1 & 0 & 1 & 1 & 0 \\
1 & 0 & 0 & 1 & 1 & 0 & 1 & 0 & 1 & 0 & 0 & 1 & 0 & 1 \\
0 & 1 & 1 & 0 & 0 & 1 & 1 & 0 & 1 & 0 & 1 & 0 & 0 & 1 \\
0 & 1 & 0 & 1 & 1 & 0 & 0 & 1 & 1 & 0 & 1 & 0 & 1 & 0 \\
0 & 1 & 0 & 1 & 0 & 1 & 0 & 1 & 0 & 1 & 0 & 1 & 0 & 1
\end{array}\right].$$

In $\boldsymbol{N}^{(1)}$, let the four column vectors be $\boldsymbol{n}_1 = [0,1,1,1]'$, $\boldsymbol{n}_2 = [1,0,1,1]'$, $\boldsymbol{n}_3 = [1,1,0,1]'$, $\boldsymbol{n}_4 = [1,1,1,0]'$. Then by applying Theorem 9.2.8 for $\boldsymbol{N}^{(1)}$ and $\boldsymbol{N}^{(2)}$ the incidence matrix of the resulting design is simply given by

$$\left[\begin{array}{cc|cc|cc|cc|cc|cc|cc}
\boldsymbol{n}_1' & \boldsymbol{0}' & \boldsymbol{0}' & \boldsymbol{n}_1' & \boldsymbol{0}' & \boldsymbol{n}_1' & \boldsymbol{n}_1' & \boldsymbol{0}' & \boldsymbol{0}' & \boldsymbol{n}_1' & \boldsymbol{n}_1' & \boldsymbol{0}' & \boldsymbol{n}_1' & \boldsymbol{0}' \\
\boldsymbol{n}_2' & \boldsymbol{0}' & \boldsymbol{n}_1' & \boldsymbol{0}' & \boldsymbol{0}' & \boldsymbol{n}_2' & \boldsymbol{0}' & \boldsymbol{n}_1' & \boldsymbol{n}_1' & \boldsymbol{0}' & \boldsymbol{0}' & \boldsymbol{n}_1' & \boldsymbol{n}_2' & \boldsymbol{0}' \\
\boldsymbol{n}_3' & \boldsymbol{0}' & \boldsymbol{n}_2' & \boldsymbol{0}' & \boldsymbol{n}_1' & \boldsymbol{0}' & \boldsymbol{0}' & \boldsymbol{n}_2' & \boldsymbol{0}' & \boldsymbol{n}_2' & \boldsymbol{n}_2' & \boldsymbol{0}' & \boldsymbol{0}' & \boldsymbol{n}_1' \\
\boldsymbol{0}' & \boldsymbol{n}_1' & \boldsymbol{n}_3' & \boldsymbol{0}' & \boldsymbol{n}_2' & \boldsymbol{0}' & \boldsymbol{n}_2' & \boldsymbol{0}' & \boldsymbol{0}' & \boldsymbol{n}_3' & \boldsymbol{0}' & \boldsymbol{n}_2' & \boldsymbol{n}_3' & \boldsymbol{0}' \\
\boldsymbol{n}_4' & \boldsymbol{0}' & \boldsymbol{0}' & \boldsymbol{n}_2' & \boldsymbol{n}_3' & \boldsymbol{0}' & \boldsymbol{n}_3' & \boldsymbol{0}' & \boldsymbol{n}_2' & \boldsymbol{0}' & \boldsymbol{0}' & \boldsymbol{n}_3' & \boldsymbol{0}' & \boldsymbol{n}_2' \\
\boldsymbol{0}' & \boldsymbol{n}_2' & \boldsymbol{n}_4' & \boldsymbol{0}' & \boldsymbol{0}' & \boldsymbol{n}_3' & \boldsymbol{n}_4' & \boldsymbol{0}' & \boldsymbol{n}_3' & \boldsymbol{0}' & \boldsymbol{n}_3' & \boldsymbol{0}' & \boldsymbol{0}' & \boldsymbol{n}_3' \\
\boldsymbol{0}' & \boldsymbol{n}_3' & \boldsymbol{0}' & \boldsymbol{n}_3' & \boldsymbol{n}_4' & \boldsymbol{0}' & \boldsymbol{0}' & \boldsymbol{n}_3' & \boldsymbol{n}_4' & \boldsymbol{0}' & \boldsymbol{n}_4' & \boldsymbol{0}' & \boldsymbol{n}_4' & \boldsymbol{0}' \\
\boldsymbol{0}' & \boldsymbol{n}_4' & \boldsymbol{0}' & \boldsymbol{n}_4' & \boldsymbol{0}' & \boldsymbol{n}_4' & \boldsymbol{0}' & \boldsymbol{n}_4' & \boldsymbol{0}' & \boldsymbol{n}_4' & \boldsymbol{0}' & \boldsymbol{n}_4' & \boldsymbol{0}' & \boldsymbol{n}_4'
\end{array}\right].$$

Theorem 9.2.9. *Let $\boldsymbol{N}^{(1)}$ be the $v_1 \times b_1$ incidence matrix of an α-resolvable BIB design with parameters v_1, $b_1 = b_0 a$, $r_1 = \alpha a$, k_1, λ_1 satisfying $b_1 = 4(r_1 - \lambda_1)$, and $\boldsymbol{N}^{(2)}$ be the $v_2 \times b_2$ incidence matrix of a BIB design with parameters v_2, b_2, r_2, k_2, λ_2 satisfying $b_2 = 4(r_2 - \lambda_2)$. Then $\boldsymbol{N}^{(1)} \otimes \boldsymbol{N}^{(2)} + (\boldsymbol{1}_{v_1}\boldsymbol{1}_{b_1}' - \boldsymbol{N}^{(1)}) \otimes (\boldsymbol{1}_{v_2}\boldsymbol{1}_{b_2}' - \boldsymbol{N}^{(2)})$ yields an α'-resolvable BIB design with parameters $v = v_1 v_2$, $b = b_1 b_2$, $r = r_1 r_2 + (b_1 - r_1)(b_2 - r_2)$, $k = k_1 k_2 + (v_1 - k_1)(v_2 - k_2)$, $\lambda = r - b/4$ for $\alpha' = \alpha r_2 + (b_0 - \alpha)(b_2 - r_2)$, i.e., a $(0; \rho_1; 0)$-EB design with $\varepsilon_1 = \lambda v/(rk)$, $\rho_1 = v - 1$, $\boldsymbol{L}_1 = \boldsymbol{I}_v - v^{-1}\boldsymbol{1}_v\boldsymbol{1}_v'$.*

Proof. Direct calculation can show the result. In fact, it has been proved by Shrikhande (1962) and Sillitto (1957) that the resulting design is a BIB design with the stated parameters. On the other hand, the α'-resolvability can easily be shown. □

For the BIB designs based on finite geometries (Theorems 6.9.1 and 6.9.2), i.e., PG(t, q) and AG(t, q), explained in Appendix C, the following observations can be summarized. Here, recall (see the comments after Theorems 6.9.1 and

6.9.2) that in PG(t,q) [AG(t,q)], by regarding points and d-dimensional subspaces as treatments and blocks in a design, respectively, a BIB design, denoted by PG$(t,q) : d$ [AG$(t,q) : d$], can be constructed.

Theorem 9.2.10. *The following properties hold:*

(1) *When $t+1$ and $d+1$ are relatively prime, a BIB design $PG(t,q) : d$ is k-resolvable, where $k = \phi(d,0,q)$.*

(2) *There does not exist an affine resolvable BIB design $PG(t,q) : d$.*

(3) *A BIB design $AG(t,q) : d$ is resolvable.*

(4) *A BIB design $AG(t,q) : d$ is affine resolvable if and only if $d = t-1$.*

(5) *An affine α-resolvable BIB design $AG(t,q) : d$ does not exist for $\alpha \geq 2$.*

(6) *A complement of $AG(t,q) : d$ is a $(q^{t-d}-1)$-resolvable BIB design.*

(7) *A complement of $AG(t,q) : t-1$ is an affine $(q-1)$-resolvable BIB design.*

Proof. When $t+1$ and $d+1$ are relatively prime, there does not exist an s-resolution set (superblock) of d-flats in PG(t,q) for $1 \leq s < k = \phi(d,0,q)$. (See Section 3 of Kageyama, 1973b.) Then the first result (1) follows from the Corollary of Section 2 in Kageyama (1976b). The results (2), (4) and (5) can be shown by a direct calculation with a reference to Theorem 6.0.5, whereas (3) follows from Theorem 6.9.2. Finally, (6) and (7) are given by (3) and Theorem 9.2.7. □

Remark 9.2.1. A construction of an affine resolvable BIB design AG$(t,q) : t-1$ is given by Rao (1946, 1961). The existence of some series in this category can be seen in Corollaries 6.9.2(i)(ii) and 6.9.3. In particular, the existence of an affine resolvable BIB design, i.e., a $(0; q^2-1; 0)$-EB design with $\varepsilon_1 = q/(q+1)$ and $\boldsymbol{L}_1 = \boldsymbol{I}_{q^2} - q^{-2}\mathbf{1}_{q^2}\mathbf{1}'_{q^2}$, with parameters $v = q^2$, $b = q(q+1)$, $r = q+1$, $k = q$, $\lambda = 1$, $b_0 = q, a = q+1$, $q_1 = 0, q_2 = 1$ for q being a prime or a prime power [see Corollaries 6.9.2(i) and 6.9.3], should be stressed.

Remark 9.2.2. As stated before, an AG$(t,q) : d$ is a resolvable BIB design. Nevertheless, Rao (1961) carelessly listed an AG(3,3):1 as a nonresolvable BIB design. A noncyclical resolvable geometrical solution of this BIB design AG(3,3):1 with parameters $v = 27$, $b = 117$, $r = 13$, $k = 3$, $\lambda = 1$, however, is easily given by a method of taking groups of parallel lines (corresponding to superblocks in a design) in AG(3,3). Note that the parallelism in 1-flats (i.e., lines) constitutes resolution sets for the resolvability, i.e., superblocks, each with one replicate of every treatment.

As a patterned method of construction, one has the following theorem.

Theorem 9.2.11 (Kageyama and Mohan, 1983). *For a prime v the existence of a BIB design with parameters v, b, r, k, λ implies the existence of an α-resolvable BIB design with parameters $v' = v^2$, $b' = b(v+1)$, $r' = r(v+1)$,*

$k' = kv$, $\lambda' = r + \lambda v$, $b_0 = b$, $a = v + 1$ *for* $\alpha = r$, *i.e., of a* $(0; \rho_1; 0)$*-EB design with* $\varepsilon_1 = v(\lambda v + r)/[rk(v+1)]$, $\rho_1 = v^2 - 1$, $\boldsymbol{L}_1 = \boldsymbol{I}_{v^2} - v^{-2}\boldsymbol{1}_{v^2}\boldsymbol{1}'_{v^2}$.

Proof. Let $\boldsymbol{N}$ be the $v \times b$ incidence matrix of the starting design and π be the permutation of the rows of $\boldsymbol{N}$ defined by $\boldsymbol{N}^{(i)} = \pi^{i-1}\boldsymbol{N}$, $i = 1, 2, ..., v$, where

$$\pi = \begin{bmatrix} 0 & \boldsymbol{I}_{v-1} \\ 1 & \boldsymbol{0}' \end{bmatrix}.$$

Let $\boldsymbol{n}_{ij}$ be the v rows of $\boldsymbol{N}^{(i)}$ for $j = 1, 2, ..., v$. Now $\boldsymbol{n}_{1j}^v$ denotes that v copies of the row $\boldsymbol{n}_{1j}$ arranged one upon the other to form a $v \times b$ array. Then the array

$$\begin{matrix} \boldsymbol{n}_{11}^v & \boldsymbol{N}^{(1)} & \boldsymbol{N}^{(1)} & \boldsymbol{N}^{(1)} & \cdots & \boldsymbol{N}^{(1)} \\ \boldsymbol{n}_{12}^v & \boldsymbol{N}^{(1)} & \boldsymbol{N}^{(2)} & \boldsymbol{N}^{(3)} & \cdots & \boldsymbol{N}^{(v)} \\ \boldsymbol{n}_{13}^v & \boldsymbol{N}^{(1)} & \boldsymbol{N}^{(3)} & \boldsymbol{N}^{(5)} & \cdots & \boldsymbol{N}^{(v-1)} \\ \vdots & \vdots & \vdots & \vdots & & \vdots \\ \boldsymbol{n}_{1v}^v & \boldsymbol{N}^{(1)} & \boldsymbol{N}^{(v)} & \boldsymbol{N}^{(v-1)} & \cdots & \boldsymbol{N}^{(2)} \end{matrix} \qquad (9.2.1)$$

gives the required design. It is obvious that each of the $(v+1)$ groups of b columns in (9.2.1) forms a superblock, i.e., r-resolution set. The values of v', b', r', k' are obvious. The rows can be indexed by (t, j), $1 \le t \le v$, $1 \le j \le v$. The (t, k)th row can be written as

$$[\boldsymbol{N}(j), \boldsymbol{N}(t), \boldsymbol{N}(t + j - 1), \boldsymbol{N}(t + 2(j-1)), ..., \boldsymbol{N}(t + (v-1)(j-1))],$$

where $\boldsymbol{N}(s)$ denotes the sth row of $\boldsymbol{N}^{(1)} = \boldsymbol{N}$ and the numbers in parentheses are read modulo v. Consider the intersection of the rows (t, j) and (t', j'). If $j = j'$ and $t \neq t'$, then the intersection is clearly $r + \lambda v$. If $j \neq j'$, then, as v is a prime, $t + x(j-1) = t' + x(j'-1)$ (mod v) has a unique solution, intersection of $\boldsymbol{N}(t + x(j-1))$ and $\boldsymbol{N}(t' + x(j'-1))$ will be r and all other intersections will be λ and the total intersection will be $r + \lambda v = \lambda'$. Therefore the required α-resolvable BIB design is obtained. □

As a special case of Theorem 9.2.11 when the original design is symmetric, the following result can be obtained.

Corollary 9.2.2 (Mohan, 1980). *For a prime* v *the existence of a symmetric BIB design with parameters* v, b, r, k, λ *implies the existence of an affine* α*-resolvable BIB design with parameters* $v' = v^2$, $b' = b(v+1)$, $r' = r(v+1)$, $k' = kv$, $\lambda' = r + \lambda v$, $b_0 = b$, $a = v + 1$, $q_1 = \lambda v$, $q_2 = r^2$ *for* $\alpha = r$, *i.e., of a* $(0; \rho_1; 0)$*-EB design with* $\varepsilon_1 = v(\lambda v + r)/[rk(v+1)]$, $\rho_1 = v^2 - 1$, $\boldsymbol{L}_1 = \boldsymbol{I}_{v^2} - v^{-2}\boldsymbol{1}_{v^2}\boldsymbol{1}'_{v^2}$.

Example 9.2.6. Take a symmetric BIB design with parameters $v = b = 3$ (a prime), $r = k = 2$, $\lambda = 1$, whose incidence matrix is given by

$$\boldsymbol{N} = \begin{bmatrix} 0 & 1 & 1 \\ 1 & 0 & 1 \\ 1 & 1 & 0 \end{bmatrix}.$$

Then Corollary 9.2.2 yields an affine 2-resolvable BIB design with parameters $v' = 9$, $b' = 12$, $r' = 8, k' = 6$, $\lambda' = 5$, $b_0 = 3, a = 4$, $q_1 = 3, q_2 = 4$, $\varepsilon_1 = 15/16$, $\rho_1 = 8$, $\boldsymbol{L}_1 = \boldsymbol{I}_9 - 9^{-1}\mathbf{1}_9\mathbf{1}_9'$, whose incidence matrix is given by

$$\left[\begin{array}{ccc|ccc|ccc|ccc} 0 & 1 & 1 & 0 & 1 & 1 & 0 & 1 & 1 & 0 & 1 & 1 \\ 0 & 1 & 1 & 1 & 0 & 1 & 1 & 0 & 1 & 1 & 0 & 1 \\ 0 & 1 & 1 & 1 & 1 & 0 & 1 & 1 & 0 & 1 & 1 & 0 \\ \hline 1 & 0 & 1 & 0 & 1 & 1 & 1 & 0 & 1 & 1 & 1 & 0 \\ 1 & 0 & 1 & 1 & 0 & 1 & 1 & 1 & 0 & 0 & 1 & 1 \\ 1 & 0 & 1 & 1 & 1 & 0 & 0 & 1 & 1 & 1 & 0 & 1 \\ \hline 1 & 1 & 0 & 0 & 1 & 1 & 1 & 1 & 0 & 1 & 0 & 1 \\ 1 & 1 & 0 & 1 & 0 & 1 & 0 & 1 & 1 & 1 & 1 & 0 \\ 1 & 1 & 0 & 1 & 1 & 0 & 1 & 0 & 1 & 0 & 1 & 1 \end{array}\right].$$

Note that this block structure is nonisomorphic to the design given after Theorem 9.2.4.

Remark 9.2.3. (1) The α-resolvable BIB designs constructed from a BIB design with v being a prime and $k = 2$ form a separate family of the designs. That is, as there exists in general a BIB design with parameters v (being a prime), $b = v(v-1)/2$, $r = v-1$, $k = 2$, $\lambda = 1$, one can get always a family of the required $(v^2 - 1)$-resolvable BIB designs with parameters $v' = v^2$, $b' = v(v^2 - 1)/2$, $r' = v^2 - 1$, $k' = 2v$, $\lambda' = 2v - 1$ for a prime v, i.e., a $(0;\rho_1;0)$-EB design with $\varepsilon_1 = v(2v-1)/[2(v^2-1)]$, $\rho_1 = v^2 - 1$, $\boldsymbol{L}_1 = \boldsymbol{I}_{v^2} - v^{-2}\mathbf{1}_{v^2}\mathbf{1}_{v^2}'$. (2) As special cases, symmetric BIB designs with parameters $v = b$, $r = k = v - 1$, $\lambda = v - 2$, and with $v = b$, $r = k = 1$, $\lambda = 0$, respectively, yield the required α-resolvable BIB designs with parameters $v' = v^2$, $b' = v(v + 1)$, $r' = v^2 - 1$, $k' = v(v - 1)$, $\lambda' = v^2 - v - 1$, $\alpha = v^2 - 1$, $\varepsilon_1 = v(v^2 - v - 1)/[(v - 1)(v^2 - 1)]$, and with parameters $v' = v^2$, $b' = v(v + 1)$, $r' = v + 1$, $k' = v$, $\lambda' = 1$, $\alpha = v + 1$, $\varepsilon_1 = v/(v + 1)$, where v is a prime, for common $\rho_1 = v^2 - 1$ and $\boldsymbol{L}_1 = \boldsymbol{I}_{v^2} - v^{-2}\mathbf{1}_{v^2}\mathbf{1}_{v^2}'$. (3) Using Theorem 1 (i.e., a general recursive construction method of including Theorems 9.2.4 and 9.2.5) of Shrikhande and Raghavarao (1963), it should be noted that the existence of a BIB design and an affine plane of prime order v implies the existence of an α-resolvable BIB design similar to Theorem 9.2.11. However, the present designs may be different from their designs in the sense that block structures are different in some points. (4) The BIB designs constructed by Theorem 9.2.11 have some interesting combinatorial properties. However, their application in practice may seem somewhat limited, because each of them requires large block size, large number of replications and large number of treatments and, therefore, designs of extremely large parameters are being omitted. A large block size makes it difficult to control the variations among experimental units, thus defeating the purpose of blocking. Nevertheless, the method of Theorem 9.2.11 shows a way of constructing α-resolvable and affine α-resolvable BIB designs which is far easier than that of the methods of constructing designs given by Shrikhande and Raghavarao (1963) and of some other existing methods. These comments are also valid for construction methods which will be given in Theorems 9.3.1, 9.3.2 and 9.4.2.

Some construction methods of α-resolvable BIB designs are available in the literature. For the exhaustive list of BIB designs with a prime v, see Takeuchi (1962) who gives difference sets generating BIB designs for general $v \leq 100$ and $r \leq 20$, and symmetric BIB designs for general $v \leq 100$ and $r \leq 30$ (see also Table 8.2). Furthermore, Kageyama and Mohan (1983) systematically tabulated 237 parameter combinations of α-resolvable BIB designs within the scope of the range as in $v \leq 125, b \leq 250, \lambda \leq 100, 3 \leq k \leq v-3$ with $\alpha \geq 2$, along with construction methods which are also available in the literature. The reader who is interested in designs with larger values of parameters can refer to the tabulation. When $k = 2$ or $v - 2$, α-resolvable designs can be constructed by taking all combinations of k treatments out of v treatments, i.e., unreduced BIB designs (see Section 6.0.1). Their table may cover most of α-resolvable BIB designs not having large values of parameters constructed by known methods, though it is not exhaustive (see Table 9.3). However, for $\alpha \geq 2$ an exhaustive search on existence of affine α-resolvable BIB designs has been done within the above range of parameters (see Table 9.4; Kageyama, 1976c).

The spectrum of α-resolvable BIB designs with parameters $v, b, r = \alpha a, k, \lambda$ has been investigated for $k = 3$ and 4 in the literature. Jungnickel, Mullin and Vanstone (1991) showed the existence of α-resolvable BIB designs with $k = 3$ except for $v = 6$ and $\lambda \equiv 2 \pmod 4$, whereas Vasiga, Furino and Ling (2001) showed the existence of α-resolvable BIB designs with $k = 4$ except for $v = 10$, $\lambda = 2$ and $\alpha = 2$, in which there does not exist a 2-resolvable BIB design with such $v = 10$ and $\lambda = 2$.

It is also interesting to note that if r is a multiple of an integer α, then grouping of α superblocks in a resolvable BIB design leads to an α-resolvable BIB design with the same set of parameters. This idea can frequently be used for the preparation of plans of α-resolvable BIB designs. In general, the larger is the value of $\text{rank}(\tilde{\phi}_3) = a - 1$ the better for α-resolvable block designs (see Section 5.3). Hence there may be no necessity from joining already existing superblocks into more extended superblocks, also from the point of view of eliminating the inter-superblock variation. Thus, designs by this idea of construction are not tabulated in Table 9.3. In this sense, Table 9.3 does not include α-resolvable designs for which α is a multiple of a smaller α^* for which relevant resolvability exists. Only such α-resolvable designs are of real interest for which α^*-resolvable designs with $\alpha^* < \alpha$ do not exist for the same parameters. On the other hand, from two α_i-resolvable BIB designs, $i = 1, 2$, with common parameters v, k and $a_1 = a_2$ (i.e., the numbers of superblocks are the same), an $(\alpha_1 + \alpha_2)$-resolvable BIB design can be constructed. Being based on Theorem 6.7.1, it follows that the BIB designs with parameters v, b, r, k, λ in some series of Bose (1939, 1942b) and of Sprott (1954, 1956), i.e., BIB designs generated by initial blocks (without any partial cycle), are k-resolvable (see also Section 8.2.1 for such series). The last observation can be confirmed by seeing that in such series of designs the v

blocks generated from each of initial blocks form a superblock (see also Examples 6.7.1.2 and 6.7.2), and is used for the preparation of Table 9.3.

Finally, note that a resolvable BIB design may be used in two ways from the point of view of the randomization. Either by taking into account the superblocks or by ignoring them. In most cases the former will be appropriate, but sometimes the latter may be more desirable.

9.2.2 Nonproper designs

In this section α-resolvable $(0; v-1; 0)$-EB designs with unequal block sizes are considered by making references to the previous chapters or available literature. The source for construction of such designs is as follows. For example, Theorem 8.2.17 (juxtaposition method) can yield α-resolvable designs for some α, whereas Corollary 8.2.4 can supply α-resolvable designs, depending on some resolvable structure of the original BIB design for $1 \le \alpha \le r$. Furthermore, Examples 8.2.4 and 8.2.5 show a 3-resolvable design and a 4-resolvable design, respectively. Some other methods are stated below. In particular, designs satisfying the desirable condition (5.3.14) of Corollary 5.3.2 will be recommended to provide the inter-block(-intra-superblock) BLUE of a contrast, in addition to the intra-block BLUE. See Section 5.3.2 and Section 9.1 in this chapter. Note that Example 8.2.5, either as a (4,2,2)-resolvable or a 4-resolvable $(0; v-1; 0)$-EB design, satisfies the condition of Corollary 5.3.2, on account of Remark 5.3.1.

By the idea of Theorem 8.2.17 and Corollary 8.2.4, the following result can be obtained.

Corollary 9.2.3. *Let $\boldsymbol{N} = [\boldsymbol{N}_1 : \boldsymbol{N}_2 : \cdots : \boldsymbol{N}_r]$ be the $v \times b$ incidence matrix of an affine resolvable semi-regular GD design with parameters $v = mn$ (m groups of n treatments each), b, r, $k = n(\lambda_2 - \lambda_1)$, λ_1, λ_2. Then the incidence matrix $[\boldsymbol{N} : \boldsymbol{N}_{r+1}]$ with $\boldsymbol{N}_{r+1} = \boldsymbol{I}_m \otimes \boldsymbol{1}_n$ yields an affine resolvable $(0; v-1; 0)$-EB design, satisfying the condition of Corollary 5.3.2, with parameters $v^* = v$, $b^* = b + m$, $r^* = r + 1$, $\boldsymbol{k}^* = [k\boldsymbol{1}_b', n\boldsymbol{1}_m']'$, $\varepsilon_1^* = \lambda_2 v/[k(r+1)]$, $\rho_1^* = v - 1$, $\boldsymbol{L}_1^* = \boldsymbol{I}_v - v^{-1}\boldsymbol{1}_v\boldsymbol{1}_v'$, $a = r+1$, $b_1 = b_2 = \cdots = b_r = v/k$, $b_{r+1} = m$, $q_{hh} = 0, h = 1, 2, ..., r+1$, $q_{hh'} = k^2/v, h, h'(h \ne h') = 1, 2, ..., r$, $q_{hr+1} = k/m, h = 1, 2, ..., r$.*

Proof. As $k = n(\lambda_2 - \lambda_1)$, a direct calculation through (8.2.5) can show that the resulting design is an affine resolvable $(0; v-1; 0)$-EB design. It also follows from Remark 5.3.1 that the design satisfies the condition of Corollary 5.3.2. □

For example, semi-regular GD designs with $k = n(\lambda_2 - \lambda_1)$, i.e., SR36, SR72, SR92, SR95, SR102, in Clatworthy (1973), are affine resolvable, and hence they produce, respectively, affine resolvable $(0; v-1; 0)$-EB designs with the following parameters:

(1) $v = 8$, $b = 12$, $r = 5$, $\boldsymbol{k} = [4\boldsymbol{1}_8', 2\boldsymbol{1}_4']'$, $\rho_1 = 7$, $\varepsilon_1 = 4/5$, $\boldsymbol{L}_1 = \boldsymbol{I}_8 - 8^{-1}\boldsymbol{1}_8\boldsymbol{1}_8'$.

(2) $v = 18$, $b = 24$, $r = 7$, $\boldsymbol{k} = [6\boldsymbol{1}_{18}', 3\boldsymbol{1}_6']'$, $\rho_1 = 17$, $\varepsilon_1 = 6/7$, $\boldsymbol{L}_1 = \boldsymbol{I}_{18} - (18)^{-1}\boldsymbol{1}_{18}\boldsymbol{1}_{18}'$.

(3) $v = 16$, $b = 24$, $r = 9$, $\boldsymbol{k} = [8\mathbf{1}'_{16}, 2\mathbf{1}'_{8}]'$, $\rho_1 = 15$, $\varepsilon_1 = 8/9$, $\boldsymbol{L}_1 = \boldsymbol{I}_{16} - (16)^{-1}\mathbf{1}_{16}\mathbf{1}'_{16}$.

(4) $v = 32$, $b = 40$, $r = 9$, $\boldsymbol{k} = [8\mathbf{1}'_{32}, 4\mathbf{1}'_{8}]'$, $\rho_1 = 31$, $\varepsilon_1 = 8/9$, $\boldsymbol{L}_1 = \boldsymbol{I}_{32} - (32)^{-1}\mathbf{1}_{32}\mathbf{1}'_{32}$.

(5) $v = 27$, $b = 36$, $r = 10$, $\boldsymbol{k} = [9\mathbf{1}'_{27}, 3\mathbf{1}'_{9}]'$, $\rho_1 = 26$, $\varepsilon_1 = 9/10$, $\boldsymbol{L}_1 = \boldsymbol{I}_{27} - (27)^{-1}\mathbf{1}_{27}\mathbf{1}'_{27}$.

Theorem 9.2.12. *Let $\boldsymbol{N} = [\boldsymbol{N}_1 : \boldsymbol{N}_2 : \cdots : \boldsymbol{N}_r]$ be the $v \times b$ incidence matrix of an affine resolvable BIB design with parameters $v = 2k$, $b = 2(2k-1)$, $r = 2k-1$, k, $\lambda = k-1$. Then*

$$\begin{bmatrix} \boldsymbol{N} & \mathbf{1}_v & \mathbf{0} & \boldsymbol{I}_v \\ \mathbf{1}_v\mathbf{1}'_b - \boldsymbol{N} & \mathbf{0} & \mathbf{1}_v & \boldsymbol{I}_v \end{bmatrix}$$

is the incidence matrix of an affine resolvable $(0; v^* - 1; 0)$-EB design, satisfying the condition of Corollary 5.3.2, with parameters $v^* = 4k$, $b^* = 6k$, $r^* = 2k+1$, $\boldsymbol{k}^* = [2k\mathbf{1}'_{4k}, 2\mathbf{1}'_{v}]'$, $\varepsilon_1^* = 2k/(2k+1)$, $\rho_1^* = 4k-1$, $\boldsymbol{L}_1^* = \boldsymbol{I}_{4k} - (4k)^{-1}\mathbf{1}_{4k}\mathbf{1}'_{4k}$, $a = 2k+1, b_1 = b_2 = \cdots = b_{2k} = 2$, $b_{2k+1} = v$, $q_{hh} = 0, h = 1, 2, ..., 2k+1$, $q_{hh'} = k, h, h'(h \neq h') = 1, 2, ..., 2k-1$, $q_{h2k} = k, q_{h2k+1} = 1, h = 1, 2, ..., 2k-1$, $q_{2k2k+1} = 1$.

Proof. As $v = 2k$ and $r = 2k-1$, a straightforward calculation by use of (8.2.5) shows the result on the resulting design. Furthermore, the condition of Corollary 5.3.2 holds by Remark 5.3.1. □

A BIB design resulting from Theorem 9.2.1 satisfies the assumptions of Theorem 9.2.12. Hence one can get the following.

Corollary 9.2.4. *When $4u-1$ is a prime or a prime power, there exists an affine resolvable $(0; v-1; 0)$-EB design, satisfying the condition of Corollary 5.3.2, with parameters $v = 8u$, $b = 12u$, $r = 4u+1$, $\boldsymbol{k} = [4u\mathbf{1}'_{8u}2\mathbf{1}'_{4u}]'$, $\varepsilon_1 = 4u/(4u+1)$, $\rho_1 = 8u-1$, $\boldsymbol{L}_1 = \boldsymbol{I}_{8u} - (8u)^{-1}\mathbf{1}_{8u}\mathbf{1}'_{8u}$, $a = 4u+1, b_1 = b_2 = \cdots = b_{4u} = 2$, $b_{4u+1} = 4u$, $q_{hh} = 0, h = 1, 2, ..., 4u+1$, $q_{hh'} = 2u, h, h'(h \neq h') = 1, 2, ..., 4u-1$, $q_{h4u} = 2u, q_{h4u+1} = 1, h = 1, 2, ..., 4u-1$, $q_{4u4u+1} = 1$.*

9.2.3 Illustration

Most of the present section is devoted to resolvable BIB designs, and several methods of their construction have been given. It may be interesting now to illustrate their application, by considering a real example of an experiment conducted in a resolvable BIB design. To save space, it will be convenient to use the data already analyzed in Examples 7.3.22 and 8.2.14.

Example 9.2.7. Returning to Example 7.3.22, it will be useful now to note that the incidence matrix $\boldsymbol{N}$ appearing in the description of the design used there represents a resolvable BIB design. For the purpose of the present illustration suppose, like in Example 8.2.14, that the experiment has been conducted in the design given by that matrix $\boldsymbol{N}$, but now taking into account its partition into superblocks, i.e., in a resolvable BIB design with parameters $v = 25$, $b = 30$, $r = 6$, $k = 5$, $\lambda = 1$. In fact, the design is affine resolvable, because it satisfies the condition $b = v + r - 1$ (see Theorem 6.0.5). Certainly, here every two distinct blocks from the same superblock have no treatments in common ($q_1 = 0$), whereas any two blocks from different superblocks have one treatment in common ($q_2 = 1$) (see the comment following Theorem 6.0.5). The construction of the considered BIB design, represented by $\boldsymbol{N}$, has been described in Example 7.3.22. The constructional method used for obtaining this design is particularly suitable for constructing resolvable block designs with $v = q^2$. An alternative equivalent design with the same properties and parameters could be obtained following Theorem 9.2.10(4) and Remark 9.2.1, i.e., as an affine resolvable BIB design of the type $\text{AG}(t, q) : t - 1$, with $t = 2$ and $q = 5$ (see also Corollary 6.9.3). This alternative design can be found in Table 9.1 as No. 16*.

As a resolvable design, the present design is a particular case of nested block (NB) designs. Therefore, it will be assumed that the randomization of superblocks, blocks within the superblocks and plots within the blocks have been implemented according to the procedure described in Section 5.2.1. This means that the order in which the submatrices $\boldsymbol{N}_h, h = 1, 2, ..., a$ $(= r = 6)$, of the matrix

$$\boldsymbol{N} = [\boldsymbol{N}_1 : \boldsymbol{N}_2 : \boldsymbol{N}_3 : \boldsymbol{N}_4 : \boldsymbol{N}_5 : \boldsymbol{N}_6],$$

are assigned to the real superblocks, formed of sets of blocks in the experimental field, has been chosen at random, and that for each superblock, $h = 1, 2, ..., 6$, the order in which the columns of $\boldsymbol{N}_h$ are assigned to the real blocks, forming the superblocks in the field, has been chosen at random, and finally that for each block the order in which the treatments indicated by 1's in the assigned column of $\boldsymbol{N}_h$ are then assigned to the plots of the block, has also been chosen at random. Note that this procedure of randomization differs from that assumed in Examples 7.3.22 and 8.2.14, where the resolvability of the BIB design represented by $\boldsymbol{N}$ has been ignored. On the other hand, the part of the experiment represented in Example 7.3.22 by the 2×30 incidence matrix $\mathbf{1}_2\mathbf{1}'_{30}$ will be deleted here, i.e., the last two treatments (standard varieties) will not be taken into account in the present example (exactly as in Example 8.2.14).

Thus, the data to be analyzed are those presented in Table 7.1, but discarding in each block the treatments (strains) labeled 26 and 27. As it is known from Example 8.2.14, the resulting BIB design is a $(0; v - 1; 0)$-EB design with parameters $v = 25$, $b = 30$, $r = 6$, $k = 5$, $\lambda = 1$, and $\varepsilon_1 = \lambda v/(rk) = 5/6$ $(= 0.833333)$, $\rho_1 = v - 1 = 24$, $\boldsymbol{L}_1 = \boldsymbol{I}_v - v^{-1}\mathbf{1}_v\mathbf{1}'_v = \boldsymbol{I}_{25} - (25)^{-1}\mathbf{1}_{25}\mathbf{1}'_{25}$.

As it follows from Section 5.3.1, the intra-block analysis for an NB design, i.e., with two block strata, can be performed exactly as for a block design with one stratum of blocks. This is so because the matrix $\tilde{\boldsymbol{\phi}}_1 = \boldsymbol{I}_n - \boldsymbol{D}'\boldsymbol{k}^{-\delta}\boldsymbol{D}$ used in Section 5.3.1 is exactly the same as $\boldsymbol{\phi}_1$ in Section 3.2.1, and hence $\tilde{\boldsymbol{C}}_1 = \boldsymbol{C}_1$, $\tilde{\boldsymbol{Q}}_1 = \boldsymbol{Q}_1$ and $\tilde{\psi}_1 = \psi_1$. Thus, the results of the intra-block analysis are here exactly the same as those obtained in Example 8.2.14.

For performing the inter-block-intra-superblock analysis, one has to refer to Section 5.3.2. But note that because in a resolvable block design the $v \times r$ incidence matrix of the design $\mathcal{D}$ for superblocks (see Section 5.1) has the incidence matrix $\Re = \mathbf{1}_v\mathbf{1}'_r$, it follows that $\tilde{\boldsymbol{\Delta\phi}}_2 = \boldsymbol{\Delta\phi}_2$ and, hence, $\tilde{\boldsymbol{C}}_2 = \boldsymbol{C}_2$ and $\tilde{\boldsymbol{Q}}_2 = \boldsymbol{Q}_2$. From this, in turn, the inter-block-intra-superblock treatment sum of squares becomes $\tilde{\boldsymbol{Q}}'_2\tilde{\boldsymbol{C}}_2^-\tilde{\boldsymbol{Q}}_2 = \boldsymbol{Q}'_2\boldsymbol{C}_2^-\boldsymbol{Q}_2$, i.e., equal to the inter-block treatment sum of squares in Section 3.2.2, and the residual sum of squares in the inter-block-intra-superblock analysis gets the form

$$\boldsymbol{y}'\tilde{\boldsymbol{\psi}}_2\boldsymbol{y} = \boldsymbol{y}'\tilde{\boldsymbol{\phi}}_2\boldsymbol{y} - \boldsymbol{y}'\boldsymbol{\phi}_2\boldsymbol{\Delta}'\boldsymbol{C}_2^-\boldsymbol{\Delta\phi}_2\boldsymbol{y} = \boldsymbol{y}'\boldsymbol{\psi}_2\boldsymbol{y} - \boldsymbol{y}'\tilde{\boldsymbol{\phi}}_3\boldsymbol{y},$$

where $\tilde{\boldsymbol{\phi}}_3 = v^{-1}\boldsymbol{G}'\boldsymbol{G} - n^{-1}\mathbf{1}_n\mathbf{1}'_n$ (see Remark 5.3.9). Furthermore, it can be shown that, for the present design,

$$\boldsymbol{\psi}_2 = (\boldsymbol{I}_6 - 6^{-1}\mathbf{1}_6\mathbf{1}'_6) \otimes (25)^{-1}\mathbf{1}_{25}\mathbf{1}'_{25} = \tilde{\boldsymbol{\phi}}_3,$$

which implies that in this example $\boldsymbol{y}'\tilde{\boldsymbol{\psi}}_2\boldsymbol{y} = 0$. This is in agreement with the fact that the number of d.f. for residuals in the inter-block-intra-superblock analysis is $b - a - \text{rank}(\tilde{\boldsymbol{C}}_2) = 30 - 6 - 24 = 0$ here. Thus, in this particular case

$$\boldsymbol{y}'\tilde{\boldsymbol{\phi}}_2\boldsymbol{y} = \boldsymbol{y}'\tilde{\boldsymbol{\phi}}_2\boldsymbol{\Delta}'\tilde{\boldsymbol{C}}_2^-\boldsymbol{\Delta}\tilde{\boldsymbol{\phi}}_2\boldsymbol{y} = \boldsymbol{y}'\boldsymbol{\phi}_2\boldsymbol{\Delta}'\boldsymbol{C}_2^-\boldsymbol{\Delta\phi}_2\boldsymbol{y},$$

which means that the inter-block-intra-superblock analysis of variance is reduced to the treatment sum of squares, leaving no residuals. Using the corresponding result from Example 8.2.14, this analysis can be presented as

Source	Degrees of freedom	Sum of squares	Mean square	F
Treatments	$v - 1 = 24$	61.683	2.570	*
Residuals	$b - r - v + 1 = 0$	0	*	
Total	$b - r = 24$	61.683		

* denoting that the quantity is not available.

As to the inter-superblock analysis (in Section 5.3.3), it should be recalled that for any resolvable block design the incidence matrix of the design $\mathcal{D}$ (for superblocks) is of the form $\Re = \mathbf{1}_v\mathbf{1}'_r$, which implies that for no function $\boldsymbol{c}'\boldsymbol{\tau}$ (with $\boldsymbol{c} \neq \mathbf{0}$) the BLUE under the inter-superblock submodel (5.3.4) exists. Thus, the analysis in this stratum supplies no information on contrasts of treatment parameters. This analysis is interesting only for estimating the stratum variance σ_3^2, which for a proper resolvable block design is defined in (5.3.27). In the present example, for which $\tilde{\boldsymbol{\psi}}_3 = \tilde{\boldsymbol{\phi}}_3 = (\boldsymbol{I}_6 - 6^{-1}\mathbf{1}_6\mathbf{1}'_6) \otimes (25)^{-1}\mathbf{1}_{25}\mathbf{1}'_{25}$, the unbiased estimate of σ_3^2 is obtainable as

$$s_3^2 = \boldsymbol{y}'\tilde{\boldsymbol{\psi}}_3\boldsymbol{y}/(r-1) = 708.820/5 = 141.764,$$

using the corresponding result from Example 8.2.14.

So, it can be concluded that the required information on differences between treatment parameters is provided here by the analyses in the first two strata, the intra-block analysis and the inter-block-intra-superblock analysis. However, only in the former the relevant hypothesis can be tested, i.e., the hypothesis H_{01} : $\mathrm{E}(\boldsymbol{y}_1) = \mathbf{0}$, which (as noted in Example 8.2.14) in the present design is equivalent to the overall null hypothesis $H_0 : \tau_1 = \tau_2 = \cdots = \tau_{25}$. This hypothesis has been rejected by the intra-block F test at the significance level lower than 0.001 (see again Example 8.2.14).

Now, after rejecting H_0, the researcher may be interested to ask, which of the basic contrasts defined originally in Example 7.3.22 are responsible for the rejection. In fact, of interest here are the contrasts considered in Example 8.2.14, written there as $\{\boldsymbol{c}_i'\boldsymbol{\tau} = \boldsymbol{s}_i'\boldsymbol{r}^{\delta}\boldsymbol{\tau},\ i = 1, 2, ..., 24\}$. Because all results under the intra-block submodel (Section 5.3.1) are here exactly the same as in Example 8.2.14, the reader is referred to Table 8.1 there. Also, because of the indicated equalities, $\tilde{\boldsymbol{C}}_2 = \boldsymbol{C}_2$ and $\tilde{\boldsymbol{Q}}_2 = \boldsymbol{Q}_2$, the results of the BLUEs of the contrasts under the inter-block-intra-superblock submodel (Section 5.3.2) are here exactly the same as those presented in Table 8.1. However, as shown above, there is no estimator of σ_2^2 under this submodel in the present example. Therefore, the last two columns of Table 8.1 are not applicable here.

At this occasion, it may be interesting to note that while in Example 8.2.14 there have been $b - v = 5$ d.f. for residuals in the inter-block analysis of variance, here this number is reduced to 0 for residuals in the inter-block-intra-superblock analysis of variance, but has been taken over as such, 5 d.f., for residuals in the inter-superblock analysis of variance. This will happen for any resolvable BIB design with $b - v = r - 1$, i.e., for any affine resolvable BIB design.

Proceeding (as in Example 8.2.14) to the combined analysis, it will be interesting first to note that, because here $\boldsymbol{y}'\tilde{\boldsymbol{\psi}}_2\boldsymbol{y} = 0$, the classic Yates method gives $\hat{\sigma}^2_{1(Y)} \equiv s_1^2 = 0.814585$ and, using the formula (3.9.5),

$$\hat{\sigma}^2_{2(Y)} = \frac{1}{v-1}\{(1-\varepsilon_1)\sum_{i=1}^{v-1}[(\widehat{\boldsymbol{c}_i'\boldsymbol{\tau}})_2 - (\widehat{\boldsymbol{c}_i'\boldsymbol{\tau}})_1]^2 - \frac{1-\varepsilon_1}{\varepsilon_1}\frac{v-1}{d_1}\boldsymbol{y}'\boldsymbol{\psi}_1\boldsymbol{y}\} = 1.95334.$$

With them, the employed iterative procedure (of Section 3.8.2) converges in the first cycle (as indicated in Section 3.9.1), giving as the final results

$$d_1' = 97.84761,\quad d_2' = 22.15239,\quad \hat{w}_1 = 0.92302,\quad \hat{w}_2 = 0.07698,$$

$$\hat{\sigma}_1^2 \equiv \hat{\sigma}^2_{1(N)} = 0.81459 \quad \text{and} \quad \hat{\sigma}_2^2 \equiv \hat{\sigma}^2_{2(N)} = 1.95334.$$

The same results have been obtained from the REML estimation procedure provided by GENSTAT 5 (1996) software (converging at the 9th iteration cycle).

Now, applying the formula (5.5.13) with $w_{1\beta}$ and $w_{2\beta}$ replaced by their estimates, $\hat{w}_1$ and $\hat{w}_2$, respectively, and with $w_{3\beta} = 0$ (as here $\varepsilon_{3\beta} = 0$), combined empirical estimates for all basic contrasts can be obtained, as

$$\widetilde{\boldsymbol{c}_i'\boldsymbol{\tau}} = \hat{w}_1(\widehat{\boldsymbol{c}_i'\boldsymbol{\tau}})_1 + \hat{w}_2(\widehat{\boldsymbol{c}_i'\boldsymbol{\tau}})_2,\quad i = 1, 2, ..., 24,$$

together with their approximate variance and tests. For the approximate variance, the formula (5.5.54) can be used, which here reduces (after some calculations) to

$$\mathrm{Var}(\widetilde{c_i'\boldsymbol{\tau}}) \cong \varepsilon_1^{-1} w_1 (1 + \tilde{\zeta}_i) \sigma_1^2,$$

similar to (3.8.77), with

$$\tilde{\zeta} = \frac{2(n - v - r + 1)}{(b - r)(n - v - b + 1)} \frac{1 - w_1}{w_1}$$

($\tilde{\zeta}$ should not be confused with ζ_j used in the original formula in different sense). With w_1 replaced by $\hat{w}_1$ one obtains $\hat{\tilde{\zeta}} = 0.008688$, and with it, the estimated approximate variance is obtained as

$$\widehat{\mathrm{Var}(\widetilde{c_i'\boldsymbol{\tau}})} \cong \varepsilon_1^{-1} \hat{w}_1 (1 + \hat{\tilde{\zeta}}) \hat{\sigma_1^2} = 1.17325 \hat{\sigma_1^2} \; (= 0.91010),$$

which is smaller than

$$\widehat{\mathrm{Var}[(\widetilde{c_i'\boldsymbol{\tau}})_1]} = \varepsilon_1^{-1} \hat{\sigma_1^2} = 1.20000 \hat{\sigma_1^2} \; (= 0.9775, \;\; \text{if } \hat{\sigma_1^2} = s_1^2)$$

obtained in the intra-block analysis (see Table 8.1). Furthermore, using the statistic

$$F_i = [\hat{w}_1 (1 + \hat{\tilde{\zeta}}_i)]^{-1} \varepsilon_{1i} (\widetilde{c_i'\boldsymbol{\tau}})^2 / s_1^2,$$

(similarly as in Examples 7.3.22 and 8.2.14), the following results have been obtained:

$$
\begin{array}{rrrrrr}
\widetilde{c_1'\boldsymbol{\tau}} = & 0.436, & F_1 = & 0.209, & P = & 0.6486, \\
\widetilde{c_2'\boldsymbol{\tau}} = & 0.958, & F_2 = & 1.007, & P = & 0.3180, \\
\widetilde{c_3'\boldsymbol{\tau}} = & -0.728, & F_3 = & 0.575, & P = & 0.4503, \\
\widetilde{c_4'\boldsymbol{\tau}} = & 2.036, & F_4 = & 4.556, & P = & 0.0354, \\
\widetilde{c_5'\boldsymbol{\tau}} = & -0.590, & F_5 = & 0.383, & P = & 0.5376, \\
\widetilde{c_6'\boldsymbol{\tau}} = & 0.387, & F_6 = & 0.165, & P = & 0.6859, \\
\widetilde{c_7'\boldsymbol{\tau}} = & 1.104, & F_7 = & 1.340, & P = & 0.2500, \\
\widetilde{c_8'\boldsymbol{\tau}} = & -1.159, & F_8 = & 1.475, & P = & 0.2275, \\
\widetilde{c_9'\boldsymbol{\tau}} = & -1.462, & F_9 = & 2.347, & P = & 0.1288, \\
\widetilde{c_{10}'\boldsymbol{\tau}} = & 0.371, & F_{10} = & 0.152, & P = & 0.6978, \\
\widetilde{c_{11}'\boldsymbol{\tau}} = & -0.922, & F_{11} = & 0.933, & P = & 0.3364, \\
\widetilde{c_{12}'\boldsymbol{\tau}} = & -0.194, & F_{12} = & 0.041, & P = & 0.8397, \\
\widetilde{c_{13}'\boldsymbol{\tau}} = & -0.911, & F_{13} = & 0.911, & P = & 0.3421, \\
\widetilde{c_{14}'\boldsymbol{\tau}} = & -1.728, & F_{14} = & 3.281, & P = & 0.0732, \\
\widetilde{c_{15}'\boldsymbol{\tau}} = & 1.249, & F_{15} = & 1.714, & P = & 0.1935, \\
\widetilde{c_{16}'\boldsymbol{\tau}} = & 3.975, & F_{16} = & 17.360, & P = & 0.0001, \\
\widetilde{c_{17}'\boldsymbol{\tau}} = & 0.538, & F_{17} = & 0.318, & P = & 0.5742,
\end{array}
$$

$$\begin{aligned}
\widetilde{c'_{18}\tau} &= 0.233, & F_{18} &= 0.060, & P &= 0.8076,\\
\widetilde{c'_{19}\tau} &= -0.975, & F_{19} &= 1.044, & P &= 0.3094,\\
\widetilde{c'_{20}\tau} &= -1.223, & F_{20} &= 1.644, & P &= 0.2029,\\
\widetilde{c'_{21}\tau} &= -0.083, & F_{21} &= 0.008, & P &= 0.9307,\\
\widetilde{c'_{22}\tau} &= 1.459, & F_{22} &= 2.339, & P &= 0.1295,\\
\widetilde{c'_{23}\tau} &= 1.414, & F_{23} &= 2.197, & P &= 0.1416,\\
\widetilde{c'_{24}\tau} &= -5.004, & F_{23} &= 27.511, & P &= 0.0000.
\end{aligned}$$

The relevant P values, for each of the obtained F values, have been obtained similarly as in Example 8.2.14, with $d_{(i)}$ and d_1 d.f., where, in the present example,

$$\hat{d}_{(i)} = \frac{2}{2 + \dfrac{3\hat{\tilde{\zeta}}(\hat{w}_2 - 3\hat{w}_1\hat{\tilde{\zeta}})^2}{(1+\hat{\tilde{\zeta}})^2\hat{w}_1\hat{w}_2}} = 0.999495 \quad \text{for} \quad i = 1, 2, ..., 24,$$

and $d_1 = 96$. The P values are below 0.05 for the basic contrasts denoted by $i = 4, 16, 24$, which confirms the results obtained in Example 8.2.14. As to the test of the overall hypothesis $H_0 = \bigcap_{i=1}^{24} H_{0,i}$, the appropriate statistic is $F = 2.9820$, the average of the above individual F values. Under H_0, the distribution of this F statistic can be approximated by the central F distribution with d and d_1 d.f., where d is obtainable as in Example 8.2.14, with ζ replaced by $\tilde{\zeta}$. The estimate of d is here $\hat{d} = 23.8954$. The corresponding P value is then equal to 0.000083, smaller than in Example 8.2.14.

Finally, to complete the analysis, the empirical estimator of $\boldsymbol{\tau}$ and its approximate covariance matrix can be obtained, using the same formulae as in Example 8.2.14, changing ζ to $\tilde{\zeta}$ and σ_3^2 to σ_4^2, where applicable. The results here are

$$\begin{aligned}
\tilde{\boldsymbol{\tau}} = [&15.512, && 15.764, && 15.066, && 15.619, && 16.419,\\
&15.705, && 15.364, && 15.874, && 16.097, && 15.231,\\
&16.363, && 15.519, && 15.457, && 15.672, && 15.664,\\
&15.757, && 15.231, && 16.503, && 15.505, && 17.563,\\
&14.928, && 15.239, && 14.618, && 14.753, && 14.326]',
\end{aligned}$$

with the estimated variance for any elementary contrast approximated by

$$\text{Var}(\widetilde{\tau_i - \tau_{i'}}) \cong 2(\varepsilon_1 r)^{-1}\hat{w}_1(1+\hat{\tilde{\zeta}})\hat{\sigma}_1^2 = 0.30337.$$

This can be compared with the corresponding estimated variance obtained in Example 8.2.14, equal to 0.32355, and with that obtained in Example 7.3.22, equal to 0.34662. Evidently, the present result, that follows from taking the resolvability of the design into account, is the smallest, i.e., the best among the three. However, to be honest, it should be mentioned that if in Example 7.3.22 the resolvability of the design considered there were taken into account, the relevant estimated variance would be 0.32978, instead of 0.34662. Moreover, such variance for a contrast between any strain and one of the two standard varieties (in the present example not available) would be in this case equal to 0.19507, smaller than the value obtained in Example 7.3.22, where the resolvability has not been taken into account. This result can easily be checked using the intermediate results, obtained under the resolvability property, $\hat{w}_1 = 0.94243$, $\hat{\sigma}_1^2 = 0.91939$ and $\hat{\hat{\zeta}} = 0.005884$. The reader is encouraged to do this as an exercise.

9.3 Designs with two efficiency factors

This section will also be discussed by considering two cases of proper designs and nonproper designs.

9.3.1 Proper designs

First, some 2-associate PBIB designs which are α-resolvable will be considered. In fact, note that a connected 2-associate PBIB design with parameters v, b, r, k, λ_1, λ_2, ψ_1, ψ_2, ρ_1, ρ_2, $\boldsymbol{A}_1^{\#}$, $\boldsymbol{A}_2^{\#}$ is equireplicate and proper, either (i) a $(\rho_0^*; \rho_1^*; 0)$-EB design with $\varepsilon_0^* = 1, \varepsilon_1^* = 1 - \psi_\beta/(rk)$ and $\psi_{\beta'} = 0 (\beta' \neq \beta)$, $\rho_0^* = \rho_{\beta'}, \rho_1^* = \rho_\beta$, $\boldsymbol{L}_0^* = \boldsymbol{A}_{\beta'}^{\#}$, $\boldsymbol{L}_1^* = \boldsymbol{A}_\beta^{\#}$ for $\beta, \beta' \in \{1, 2\}$, or (ii) a $(0; \rho_1^{**}, \rho_2^{**}; 0)$-EB design with $\varepsilon_\beta^{**} = 1 - \psi_\beta/(rk)$, ρ_β^{**}, $\boldsymbol{L}_\beta^{**} = \boldsymbol{A}_\beta^{\#}$ for $\beta = 1, 2$ (see Section 6.0.2). As indicated in Section 7.3.1.1, Clatworthy (1973) gives comprehensive tables of 2-associate PBIB designs along with possible numbers of nonisomorphic solutions for each plan, within the range of parameters $2 \leq r, k \leq 10$. They also include α-resolvable solutions in many possible cases. In this sense, the reader can also refer to the monograph by Clatworthy (1973) to find α-resolvable PBIB designs that are suitable for the planned experimentation. [Note that the grouping of treatments in a GD association scheme there is different from that given in Section 6.0.2(A) or (9.3.3). As always done in the present Volume II, one has to make some trivial regrouping of the treatments before using such GD designs.] Other individual sources are as follows. First recall that an L_2 design is a 2-associate PBIB design based on an L_2 association scheme [see Section 6.0.2(C)].

Example 9.3.1. A set of blocks, [(1,5,9)(2,6,7)(3,4,8)][(1,6,8)(2,4,9)(3,5,7)], is an affine resolvable L_2 design with parameters $v = 9$, $b = 6$, $r = 2$, $k = 3$, $\lambda_1 = 0$, $\lambda_2 = 1$, $q_1 = 0$, $q_2 = 1$, i.e., $(4; 4; 0)$-EB with $\varepsilon_0 = 1$, $\varepsilon_1 = 1/2$, $\boldsymbol{L}_0 = 3^{-1}\mathbf{1}_3\mathbf{1}_3' \otimes (\boldsymbol{I}_3 - 3^{-1}\mathbf{1}_3\mathbf{1}_3') + (\boldsymbol{I}_3 - 3^{-1}\mathbf{1}_3\mathbf{1}_3') \otimes 3^{-1}\mathbf{1}_3\mathbf{1}_3'$, $\boldsymbol{L}_1 = (\boldsymbol{I}_3 - 3^{-1}\mathbf{1}_3\mathbf{1}_3') \otimes (\boldsymbol{I}_3 - 3^{-1}\mathbf{1}_3\mathbf{1}_3')$. Here, $[\cdots]$ shows a superblock.

Example 9.3.2. A set of blocks, [(1, 2, 3, 4, 5, 6) (1, 2, 3, 7, 8, 9) (4, 5, 6, 7, 8, 9)] [(1, 2, 4, 5, 7, 8) (1, 3, 4, 6, 7, 9) (2, 3, 5, 6, 8, 9)], is an affine 2-resolvable L_2 design with parameters $v = 9$, $b = 6$, $r = 4$, $k = 6$, $\lambda_1 = 3$, $\lambda_2 = 2$, $q_1 = 3$, $q_2 = 4$, i.e., $(4; 4; 0)$-EB with $\varepsilon_0 = 1$, $\varepsilon_1 = 7/8$, $\boldsymbol{L}_0 = (\boldsymbol{I}_3 - 3^{-1}\mathbf{1}_3\mathbf{1}_3') \otimes (\boldsymbol{I}_3 - 3^{-1}\mathbf{1}_3\mathbf{1}_3')$, $\boldsymbol{L}_1 = 3^{-1}\mathbf{1}_3\mathbf{1}_3' \otimes (\boldsymbol{I}_3 - 3^{-1}\mathbf{1}_3\mathbf{1}_3') + (\boldsymbol{I}_3 - 3^{-1}\mathbf{1}_3\mathbf{1}_3') \otimes 3^{-1}\mathbf{1}_3\mathbf{1}_3'$. Here, $[\cdots]$ shows a superblock.

Note that the designs given in Examples 9.3.1 and 9.3.2 are not complementary each other under the same association scheme. Other three resolvable 2-associate PBIB designs can be found in Sinha and Dey (1982).

The same notation $\boldsymbol{N}^{(i)} = \pi^{i-1}\boldsymbol{N}$ follows as in the proof of Theorem 9.2.11. Then consider the following pattern as a substructure of (9.2.1).

$$\begin{array}{ccccc} \boldsymbol{N}^{(1)} & \boldsymbol{N}^{(1)} & \boldsymbol{N}^{(1)} & \cdots & \boldsymbol{N}^{(1)} \\ \boldsymbol{N}^{(1)} & \boldsymbol{N}^{(2)} & \boldsymbol{N}^{(3)} & \cdots & \boldsymbol{N}^{(v)} \\ \boldsymbol{N}^{(1)} & \boldsymbol{N}^{(3)} & \boldsymbol{N}^{(5)} & \cdots & \boldsymbol{N}^{(v-1)} \\ \vdots & \vdots & \vdots & & \vdots \\ \boldsymbol{N}^{(1)} & \boldsymbol{N}^{(v)} & \boldsymbol{N}^{(v-1)} & \cdots & \boldsymbol{N}^{(2)} \end{array} \tag{9.3.1}$$

which can easily produce the following by combining the second and third associate classes in the rectangular association scheme [see Section 6.0.2 (E)] defined in a rectangle of v rows and v columns.

Theorem 9.3.1. *The existence of a BIB design with parameters v (being a prime), b, r, k, λ implies the existence of an α-resolvable semi-regular GD design with parameters $v' = v^2$ (in v groups of v treatments each), $b' = vb$, $r' = vr$, $k' = vk$, $\lambda_1' = v\lambda$, $\lambda_2' = rk$, $a = v$, $b_0 = b$ for $\alpha = r$, i.e., of a $(\rho_0;\rho_1;0)$-EB design with $\varepsilon_0 = 1$, $\varepsilon_1 = 1 - (r-\lambda)/(vrk)$, $\rho_0 = v-1$, $\rho_1 = v(v-1)$, $\boldsymbol{L}_0 = (\boldsymbol{I}_v - v^{-1}\boldsymbol{1}_v\boldsymbol{1}_v') \otimes v^{-1}\boldsymbol{1}_v\boldsymbol{1}_v'$, $\boldsymbol{L}_1 = \boldsymbol{I}_v \otimes (\boldsymbol{I}_v - v^{-1}\boldsymbol{1}_v\boldsymbol{1}_v')$.*

Proof. The values of v', b', r', k', a, b_0 are obvious, with the r-resolvability. For the new $v' = v^2$ treatments the usual rectangular association scheme is defined in a rectangle of v rows and v columns. In this case, it follows from the pattern (9.3.1) that $\lambda_1 = v\lambda$, $\lambda_2 = r + \lambda(v-1) = rk$, $\lambda_3 = \lambda + \lambda(v-2) + r = rk$. Hence let $\lambda_1' = \lambda_1$ and $\lambda_2' = \lambda_2 = \lambda_3$ for the GD association scheme. It is sufficient to show that $r' - \lambda_1' = v(r-\lambda) > 0$ and $r'k' - \lambda_2'v' = 0$, which imply that the resulting design is semi-regular. In (9.3.1), each of the v groups of b columns forms a superblock. □

As a special case of Theorem 9.3.1, one can get the following result.

Corollary 9.3.1. *The existence of a symmetric BIB design with parameters $v = b$ (being a prime), $r = k$, λ implies the existence of an affine r-resolvable semi-regular GD symmetric design with parameters $v' = b' = v^2$, $r' = k' = vk$, $\lambda_1' = v\lambda$, $\lambda_2' = k^2$, $a = v, b_0 = v$, $q_1 = \lambda v (= \lambda_1')$, $q_2 = k^2 (= \lambda_2')$, i.e., of a $(\rho_0;\rho_1;0)$-EB design with $\varepsilon_0 = 1$, $\varepsilon_1 = 1-(k-\lambda)/(vk^2)$, $\rho_0 = v-1$, $\rho_1 = v(v-1)$, $\boldsymbol{L}_0 = (\boldsymbol{I}_v - v^{-1}\boldsymbol{1}_v\boldsymbol{1}_v') \otimes v^{-1}\boldsymbol{1}_v\boldsymbol{1}_v'$, $\boldsymbol{L}_1 = \boldsymbol{I}_v \otimes (\boldsymbol{I}_v - v^{-1}\boldsymbol{1}_v\boldsymbol{1}_v')$.*

The pattern (9.3.1) used in Theorem 9.3.1 makes sense for a prime v in the present case. On the other hand, when v is not a prime, another pattern can be considered, as given below under the same notation as in (9.3.1),

$$\begin{array}{ccccc} \boldsymbol{N}^{(1)} & \boldsymbol{N}^{(2)} & \boldsymbol{N}^{(3)} & \cdots & \boldsymbol{N}^{(v)} \\ \boldsymbol{N}^{(2)} & \boldsymbol{N}^{(3)} & \boldsymbol{N}^{(4)} & \cdots & \boldsymbol{N}^{(1)} \\ \boldsymbol{N}^{(3)} & \boldsymbol{N}^{(4)} & \boldsymbol{N}^{(5)} & \cdots & \boldsymbol{N}^{(2)} \\ \vdots & \vdots & \vdots & & \vdots \\ \boldsymbol{N}^{(v)} & \boldsymbol{N}^{(1)} & \boldsymbol{N}^{(2)} & \cdots & \boldsymbol{N}^{(v-1)}, \end{array} \tag{9.3.2}$$

which can provide the following theorem. Here, each of the v groups of b columns forms a superblock.

Theorem 9.3.2 (Mohan and Kageyama, 1989). *The existence of a BIB design with parameters v, b, r, k, λ implies the existence of an α-resolvable singular GD design with parameters $v' = v^2$ (in v groups of v treatments each), $b' = vb$, $r' = vr$, $k' = vk$, $\lambda_1' = vr$, $\lambda_2' = v\lambda$, $a = v$, $b_0 = b$ for $\alpha = r$, i.e., of a $(\rho_0; \rho_1; 0)$-EB design with $\varepsilon_0 = 1$, $\varepsilon_1 = \lambda v/(rk)$, $\rho_0 = v(v-1)$, $\rho_1 = v - 1$, $\boldsymbol{L}_0 = \boldsymbol{I}_v \otimes (\boldsymbol{I}_v - v^{-1}\boldsymbol{1}_v\boldsymbol{1}_v')$, $\boldsymbol{L}_1 = (\boldsymbol{I}_v - v^{-1}\boldsymbol{1}_v\boldsymbol{1}_v') \otimes v^{-1}\boldsymbol{1}_v\boldsymbol{1}_v'$.*

Next α-resolvable designs with two efficiency factors which are proper, but other than PBIB designs, are considered. Some observations can be found in Chapters 6 and 7, as will be recalled now. Theorems 6.6.1, 6.6.2, 7.3.5 and 7.3.13 produce α-resolvable proper $(\rho_0; \rho_1; 0)$-EB designs. Then by complementation (Theorem 6.2.1) other $(b_0 - \alpha)$-resolvable proper $(\rho_0; \rho_1; 0)$-EB designs can be obtained.

The dual of a block design, defined in Section 6.1, can produce some α-resolvable designs if the original design is a singular or semi-regular GD design. That is, certain refinement of Corollaries 7.3.6 and 7.3.7 will be considered. Two preliminary results are needed (see Ceranka, Kageyama and Mejza, 1986).

Lemma 9.3.1. *If a block design with the incidence matrix $\boldsymbol{N}$ is a singular GD design with parameters $v = mn$ (in m groups of n treatments each), b, r, k, then its dual with the incidence matrix $\boldsymbol{N}'$ is (k/n)-resolvable.*

Proof. Recall that a GD association scheme of $v = mn$ treatments is arranged as

$$\begin{array}{cccc} 1 & 2 & \cdots & n \\ n+1 & n+2 & \cdots & 2n \\ \vdots & \vdots & & \vdots \\ (i-1)n+1 & (i-1)n+2 & \cdots & in \\ \vdots & \vdots & & \vdots \\ (m-1)n+1 & (m-1)n+2 & \cdots & mn \end{array}. \tag{9.3.3}$$

Then it follows from Theorem 7.3.2 that the sets of new blocks, $\{B_1, B_{n+1}, ..., B_{(m-1)n+1}\}$, $\{B_2, B_{n+2}, ..., B_{(m-1)n+2}\}$, ..., $\{B_n, B_{2n}, ..., B_{mn}\}$, constitute the required (k/n)-resolvable design, where a new block, B_i, say, means a block, obtained by the dualization, corresponding to the original treatment i $(= 1, 2, ..., v = mn)$. □

Lemma 9.3.2. *If a block design is a semi-regular GD design with parameters $v = mn$, b, r, k, then its dual is (k/m)-resolvable.*

Proof. By Theorem 7.3.3, each block of the original design contains k/m treatments from each group of the association scheme. Hence each set of the new blocks, $\{B_{(i-1)n+1}, B_{(i-1)n+2}, ..., B_{in}\}$, $i = 1, 2, ..., m$, constitutes a complete (k/m)-replicate, i.e., a superblock. This means that the dual of the resulting design is (k/m)-resolvable. □

By Lemma 9.3.1, Corollary 7.3.6 can now be established and restated as follows.

Corollary 9.3.2. *The dual of a singular GD design with parameters $v = mn$, b, r, k, λ_1, λ_2 is a (k/n)-resolvable $(\rho_0^*; \rho_1^*; 0)$-EB design with parameters $v^* = b$, $b^* = mn$, $r^* = k$, $k^* = r$, $a = n, b_0 = m$, $\varepsilon_0^* = 1$, $\varepsilon_1^* = v\lambda_2/(rk)$, $\rho_0^* = b - m$, $\rho_1^* = m - 1$, $\boldsymbol{L}_0^* = \boldsymbol{I}_b - b^{-1}\boldsymbol{1}_b\boldsymbol{1}_b' - \boldsymbol{L}_1^*$, $\boldsymbol{L}_1^* = (rk - v\lambda_2)^{-1}\boldsymbol{N}'[(\boldsymbol{I}_m - m^{-1}\boldsymbol{1}_m\boldsymbol{1}_m') \otimes n^{-1}\boldsymbol{1}_n\boldsymbol{1}_n']\boldsymbol{N}$, where $\boldsymbol{N}$ is the $v \times b$ incidence matrix of the original singular GD design.*

Next by Lemma 9.3.2, Corollary 7.3.7 can be established and restated below.

Corollary 9.3.3. *The dual of a semi-regular GD design with parameters $v = mn$, b, r, k, λ_1, λ_2 is a (k/m)-resolvable $(\rho_0^*; \rho_1^*; 0)$-EB design with parameters $v^* = b$, $b^* = mn$, $r^* = k$, $k^* = r$, $a = m, b_0 = n$, $\varepsilon_0^* = 1$, $\varepsilon_1^* = 1 - (r - \lambda_1)/(rk)$, $\rho_0^* = b - m(n-1) - 1$, $\rho_1^* = m(n-1)$, $\boldsymbol{L}_0^* = \boldsymbol{I}_b - b^{-1}\boldsymbol{1}_b\boldsymbol{1}_b' - \boldsymbol{L}_1^*$, $\boldsymbol{L}_1^* = (r - \lambda_1)^{-1}\boldsymbol{N}'[\boldsymbol{I}_m \otimes (\boldsymbol{I}_n - n^{-1}\boldsymbol{1}_n\boldsymbol{1}_n')]\boldsymbol{N}$, where $\boldsymbol{N}$ is the $v \times b$ incidence matrix of the original semi-regular GD design.*

Note that Corollaries 9.3.2 and 9.3.3 coincide with Corollaries 7.3.6 and 7.3.7, respectively, in which the resolvability has not been proved. A large number of singular or semi-regular GD designs are tabulated in Clatworthy (1973). Hence one can produce many α-resolvable proper $(\rho_0; \rho_1; 0)$-EB designs through Corollaries 9.3.2 and 9.3.3.

Finally, taking the resolution sets (superblocks) of an affine α-resolvable BIB design for $\alpha \geq 1$ (see Section 6.0.3), a recursive method of constructing α-resolvable GD designs can be given.

Theorem 9.3.3 (Mohan and Kageyama, 1997). *The existence of an affine α-resolvable BIB design with parameters $v, b = b_0 a$, $r = \alpha a, k, \lambda$ implies the existence of a k-resolvable semi-regular GD design with parameters*

$$v' = b,\ b' = va,\ r' = ka,\ k' = r,\ \lambda_1 = \frac{(\alpha - 1)ka}{b_0 - 1},\ \lambda_2 = \frac{ak^2}{v},$$

$$m = a,\ n = b_0,$$

i.e., of a k-resolvable $(\rho_0;\rho_1;0)$-EB design with $\varepsilon_0 = 1, \varepsilon_1 = 1-(r-\lambda_1)/(rk)$, $\rho_0 = a-1, \rho_1 = a(b_0-1)$, $\boldsymbol{L}_0 = (\boldsymbol{I}_a - a^{-1}\boldsymbol{1}_a\boldsymbol{1}_a') \otimes (b_0)^{-1}\boldsymbol{1}_{b_0}\boldsymbol{1}_{b_0}'$, $\boldsymbol{L}_1 = \boldsymbol{I}_a \otimes [\boldsymbol{I}_{b_0} - (b_0)^{-1}\boldsymbol{1}_{b_0}\boldsymbol{1}_{b_0}']$.

Proof. The b blocks are divided into a resolution sets (α-replication superblocks) of b_0 blocks each in the affine α-resolvable BIB design. Let the a sets (superblocks) be denoted by the subincidence matrices $\boldsymbol{N}_1, \boldsymbol{N}_2, ..., \boldsymbol{N}_a$ of size $v \times b_0$ each, and let the transpose of these incidence matrices be $\boldsymbol{N}_1', \boldsymbol{N}_2', ..., \boldsymbol{N}_a'$. Now form the following circulant matrix with these matrices $\boldsymbol{N}_i', i = 1, 2, ..., a$.

$$\begin{bmatrix} \boldsymbol{N}_1' & \boldsymbol{N}_2' & \boldsymbol{N}_3' & \boldsymbol{N}_4' & \cdots & \boldsymbol{N}_a' \\ \boldsymbol{N}_a' & \boldsymbol{N}_1' & \boldsymbol{N}_2' & \boldsymbol{N}_3' & \cdots & \boldsymbol{N}_{a-1}' \\ \boldsymbol{N}_{a-1}' & \boldsymbol{N}_a' & \boldsymbol{N}_1' & \boldsymbol{N}_2' & \cdots & \boldsymbol{N}_{a-2}' \\ \cdots & \cdots & \cdots & \cdots & \cdots & \cdots \\ \boldsymbol{N}_2' & \boldsymbol{N}_3' & \boldsymbol{N}_4' & \boldsymbol{N}_5' & \cdots & \boldsymbol{N}_1' \end{bmatrix}$$

which can lead to the required result. Here, the a sets of b_0 columns each (i.e., corresponding to the submatrices as columns above) show superblocks of the resulting design. □

Example 9.3.3. From an affine resolvable BIB design with parameters $v = 9, b = 12, r = 4, k = 3, \lambda = 1, q_1 = 0, q_2 = 1$ (No. 9* of Table 9.1; Section 6.0.3; Kageyama and Mohan, 1983), one can get a 3-resolvable semi-regular GD design with parameters $v = 12, b = 36, r = 12, k = 4, \lambda_1 = 0, \lambda_2 = 4, m = 4, n = 3$, i.e., a 3-resolvable (3; 8; 0)-EB design with $\varepsilon_0 = 1$, $\varepsilon_1 = 3/4, \rho_0 = 3$, $\rho_1 = 8$, $\boldsymbol{L}_0 = (\boldsymbol{I}_4 - 4^{-1}\boldsymbol{1}_4\boldsymbol{1}_4') \otimes 3^{-1}\boldsymbol{1}_3\boldsymbol{1}_3'$, $\boldsymbol{L}_1 = \boldsymbol{I}_4 \otimes (\boldsymbol{I}_3 - 3^{-1}\boldsymbol{1}_3\boldsymbol{1}_3')$.On the other hand, an affine 3-resolvable BIB design with parameters $v = 49, b = 56, r = 24, k = 21, \lambda = 10, q_1 = 7, q_2 = 9, \beta = 7, t = 8$ (Shrikhande and Raghavarao, 1963) produces a 21-resolvable semi-regular GD design with parameters $v = 56, b = 392, r = 168, k = 24, \lambda_1 = 56, \lambda_2 = 72, m = 8, n = 7$, i.e., a 21-resolvable (7; 48; 0)-EB design with $\varepsilon_0 = 1, \varepsilon_1 = 35/36, \rho_0 = 7, \rho_1 = 48$, $\boldsymbol{L}_0 = (\boldsymbol{I}_8 - 8^{-1}\boldsymbol{1}_8\boldsymbol{1}_8') \otimes 7^{-1}\boldsymbol{1}_7\boldsymbol{1}_7'$, $\boldsymbol{L}_1 = \boldsymbol{I}_8 \otimes (\boldsymbol{I}_7 - 7^{-1}\boldsymbol{1}_7\boldsymbol{1}_7')$.However, this design has larger values on parameters.

As an example of the other structure, the following can be given.

Example 9.3.4 (a continuation of Example 6.6.1). There exists a 3-resolvable (4; 3; 0)-EB design with parameters $v^* = 8$, $b^* = 12$, $r^* = 6$, $k^* = 4$, $\varepsilon_0^* = 1$, $\varepsilon_1^* = 2/3$, $\rho_0^* = 4$, $\rho_1^* = 3$, $\boldsymbol{L}_0^* = (\boldsymbol{I}_2 - 2^{-1}\boldsymbol{1}_2\boldsymbol{1}_2') \otimes \boldsymbol{I}_4$, $\boldsymbol{L}_1^* = 2^{-1}\boldsymbol{1}_2\boldsymbol{1}_2' \otimes (\boldsymbol{I}_4 - 4^{-1}\boldsymbol{1}_4\boldsymbol{1}_4')$. Here all the 4 basic contrasts represented by eigenvectors of $\boldsymbol{L}_0^*$ are estimated in the intra-block analysis with full efficiency and the variance σ_1^2 each, whereas all the 3 basic contrasts represented by eigenvectors of $\boldsymbol{L}_1^*$ are estimated in that analysis with the efficiency 2/3 and the variance $3\sigma_1^2/2$ each. The average efficiency factor of the design is $14/17 = 0.8235$. Furthermore, for the evaluation of the variance of the intra-block BLUE of any contrast of treatment parameters, the matrix

$$\boldsymbol{\Omega}^* = \frac{1}{96}\left[\boldsymbol{I}_2 \otimes (20\boldsymbol{I}_4 - \boldsymbol{1}_4\boldsymbol{1}_4') + (\boldsymbol{1}_2\boldsymbol{1}_2' - \boldsymbol{I}_2) \otimes (4\boldsymbol{I}_4 - \boldsymbol{1}_4\boldsymbol{1}_4')\right]$$

$$= \frac{1}{24}\left[4\boldsymbol{I}_2 \otimes \boldsymbol{I}_4 + \boldsymbol{1}_2\boldsymbol{1}_2' \otimes (\boldsymbol{I}_4 - \frac{1}{4}\boldsymbol{1}_4\boldsymbol{1}_4')\right]$$

is given. Hence, for example, the intra-block BLUE of an elementary contrast which concerns any two of the first four treatments or any two of the last four treatments has the variance $5\sigma_1^2/12$, whereas that which concerns, e.g., the first treatments from each of the two subsets has the variance $\sigma_1^2/3$.

The construction and optimality of affine resolvable proper block designs, which are not always PBIB designs, have been discussed by Bailey, Monod and Morgan (1995), who also have shown (in their Theorem 3.1) that such affine resolvable designs with parameters $v = sk, b = sr$, r and k, via orthogonal arrays of strength two (see Hedayat, Sloane and Stufken, 1999), belong to a class of $(\rho_0; \rho_1; 0)$-EB designs with $\varepsilon_0 = 1$, $\varepsilon_1 = 1 - 1/r$, $\rho_0 = v - 1 - r(s-1)$ and $\rho_1 = r(s-1)$. See Example 9.3.1. This idea can be explained in another way as follows. Because $k = m$ and $\lambda_1 = 0$, from Theorem 7.3.3 and Corollary 7.3.7 (or 9.3.3), it follows that the dual of a semi-regular GD design with parameters $v = mn, b, r, k = m, \lambda_1 = 0, \lambda_2$ is an affine resolvable block design with parameters $v^* = b, b^* = kn, r^* = k, k^* = r, q_1^* = 0$ and $q_2^* = \lambda_2$ (see the comment following Theorem 6.0.5), which is a $(\rho_0^*; \rho_1^*; 0)$-EB design with $\varepsilon_0^* = 1$, $\varepsilon_1^* = 1 - 1/r^*$, $\rho_0^* = v^* - 1 - r^*(n-1)$, $\rho_1^* = r^*(n-1)$, $\boldsymbol{L}_0^* = \boldsymbol{I}_{v^*} - (v^*)^{-1}\boldsymbol{1}_{v^*}\boldsymbol{1}_{v^*}' - \boldsymbol{L}_1^*$, $\boldsymbol{L}_1^* = (k^*)^{-1}\boldsymbol{N}^*[\boldsymbol{I}_{r^*} \otimes (\boldsymbol{I}_n - n^{-1}\boldsymbol{1}_n\boldsymbol{1}_n')](\boldsymbol{N}^*)'$, where $\boldsymbol{N}^*$ is the $v^* \times b^*$ incidence matrix of the affine resolvable design. For example, an affine resolvable design with parameters $v^* = 12, b^* = 8, r^* = 4, k^* = 6, q_1^* = 0, q_2^* = 3$, given in Table 2 of Bailey, Monod and Morgan (1995) (in fact, the dual of which is a semi-regular GD design with parameters $v = 8, b = 12, r = 6, k = m = 4, \lambda_1 = 0, \lambda_2 = 3$, $m = 4, n = 2$), is a $(\rho_0^*; \rho_1^*; 0)$-EB design with $\varepsilon_0^* = 1$, $\varepsilon_1^* = 3/4$, $\rho_0^* = 7$, $\rho_1^* = 4$, $\boldsymbol{L}_0^* = \boldsymbol{I}_{12} - (12)^{-1}\boldsymbol{1}_{12}\boldsymbol{1}_{12}' - \boldsymbol{L}_1^*$, $\boldsymbol{L}_1^* = 6^{-1}\boldsymbol{N}^*[\boldsymbol{I}_4 \otimes (\boldsymbol{I}_2 - 2^{-1}\boldsymbol{1}_2\boldsymbol{1}_2')](\boldsymbol{N}^*)' =$

$$\frac{1}{6}\begin{bmatrix}
2 & -1 & 0 & 1 & -1 & 1 & -1 & 0 & -1 & 1 & -1 & 0 \\
-1 & 2 & -1 & 0 & 2 & 0 & 0 & -1 & 0 & 0 & 0 & -1 \\
0 & -1 & 2 & -1 & -1 & 1 & 1 & 0 & -1 & -1 & 1 & 0 \\
1 & 0 & -1 & 2 & 0 & 0 & 0 & 1 & 0 & 0 & -2 & -1 \\
-1 & 2 & -1 & 0 & 2 & 0 & 0 & -1 & 0 & 0 & 0 & -1 \\
1 & 0 & 1 & 0 & 0 & 2 & 0 & -1 & -2 & 0 & 0 & -1 \\
-1 & 0 & 1 & 0 & 0 & 0 & 2 & 1 & 0 & -2 & 0 & -1 \\
0 & -1 & 0 & 1 & -1 & -1 & 1 & 2 & 1 & -1 & -1 & 0 \\
-1 & 0 & -1 & 0 & 0 & -2 & 0 & 1 & 2 & 0 & 0 & 1 \\
1 & 0 & -1 & 0 & 0 & 0 & -2 & -1 & 0 & 2 & 0 & 1 \\
-1 & 0 & 1 & -2 & 0 & 0 & 0 & -1 & 0 & 0 & 2 & 1 \\
0 & -1 & 0 & -1 & -1 & -1 & -1 & 0 & 1 & 1 & 1 & 2
\end{bmatrix}.$$

Here the incidence matrix $\boldsymbol{N}^*$ of the above affine resolvable design with $a = 4, b_1 = b_2 = b_3 = b_4 = 2$, $q_{hh} = 0, h = 1, 2, 3, 4$, and $q_{hh'} = 3, h \neq h' = 1, 2, 3, 4$

(in the notation of Definition 6.0.3) is of the form

$$N^* = \left[\begin{array}{cc|cc|cc|cc}
1 & 0 & 0 & 1 & 1 & 0 & 0 & 1 \\
1 & 0 & 1 & 0 & 0 & 1 & 1 & 0 \\
0 & 1 & 1 & 0 & 1 & 0 & 0 & 1 \\
1 & 0 & 0 & 1 & 1 & 0 & 1 & 0 \\
1 & 0 & 1 & 0 & 0 & 1 & 1 & 0 \\
1 & 0 & 1 & 0 & 1 & 0 & 0 & 1 \\
0 & 1 & 1 & 0 & 1 & 0 & 1 & 0 \\
0 & 1 & 0 & 1 & 1 & 0 & 1 & 0 \\
0 & 1 & 0 & 1 & 0 & 1 & 1 & 0 \\
1 & 0 & 0 & 1 & 0 & 1 & 0 & 1 \\
0 & 1 & 1 & 0 & 0 & 1 & 0 & 1 \\
0 & 1 & 0 & 1 & 0 & 1 & 0 & 1
\end{array}\right].$$

9.3.2 Nonproper designs

In this section α-resolvable nonproper designs with two efficiency factors will be considered. Note (see Remark 5.3.1) that all the designs in this section satisfy the desirable condition (5.3.14), which allows any contrast not estimated in the intra-block analysis with full efficiency to obtain the inter-block(-intra-superblock) BLUE, in addition to the intra-block BLUE.

The idea of juxtaposition, as used in Theorems 6.4.1 and 6.4.2, Corollaries 6.4.1, 7.3.12 and 7.3.13, can produce α-resolvable nonproper designs with two efficiency factors. Incidentally, Example 7.3.9 shows a 3-resolvable nonproper $(\rho_0; \rho_1; 0)$-EB design with $v = 8$.

Corollary 7.3.14 applied to an s-resolvable GD design $\boldsymbol{N}$ with $k \neq n$ can yield an s-resolvable nonproper $(\rho_0; \rho_1; 0)$-EB design, whereas Corollary 7.3.17 applied to a p-resolvable GD design $\boldsymbol{N}$ can yield a p-resolvable nonproper $(\rho_0; \rho_1; 0)$-EB or $(0; \rho_1, \rho_2; 0)$-EB design for $p \geq 1$.

Ceranka, Kageyama and Mejza (1986) have presented several methods of constructing α-resolvable $(\rho_0; \rho_1; 0)$-EB designs with $\rho_0 \geq 1$, so called C-designs (see Section 4.4.2). Two of those methods are based on "dualization" and "merging of treatments and dualization". One can use Corollary 6.1.3 to construct a nonproper $(\rho_0; \rho_1; 0)$-EB design as the dual of a $(\rho_0^*; \rho_1^*; 0)$-EB design. This fact with some merging of treatments is useful to constuct various kinds of resolvable designs in this category. In general, one may use the procedures of merging of treatments and of dualization applying them to a semi-regular GD design with $v = mn$, b, r, k, $\lambda_1 = 0$, $\lambda_2 \neq 0$ to constuct a resolvable nonproper $(\rho_0; \rho_1; 0)$-EB design with $\rho_0 \geq 1$. From Corollary 6.1.3, nonproper $(\rho_0; \rho_1; 0)$-EB designs can be obtained, and hence it is sufficient to establish the resolvability, which depends on the number of merged treatments. Of course, a larger number of merged treatments yields a smaller number of blocks in the resulting design. (See Ceranka, Kageyama and Mejza, 1986, Section 2.2.)

For example, one possibility of merging is as follows. At first, all n treatments belonging to the same group are merged, and then the design after the merging is dualized. Following (9.3.3), the notation and terms used in the proofs of Lemmas 9.3.1 and 9.3.2, merge the n treatments $1, 2, ..., n$, and dualize the design after the merging. In this case, the new blocks $\{B_1+B_2+\cdots+B_n\}$ (i.e., a union of blocks $B_1, B_2, ..., B_n$ as a superblock), $\{B_{(i-1)n+1}, B_{(i-1)n+2}, ..., B_{in}\}, i = 2, 3, ..., m$, constitute a resolvable nonproper $(\rho_0; \rho_1; 0)$-EB design.

Another is as follows. At first, some of the treatments belonging to the same group are merged and then the resulting design is dualized. For example, merge two treatments $1, 2$, and dualize the design after the merging. In this case, the new blocks $\{B_1 + B_2, B_3, ..., B_n\}$, $\{B_{(i-1)n+1}, B_{(i-1)n+2}, ..., B_{in}\}$, $i = 2, 3, ..., m$, constitute a resolvable nonproper $(\rho_0; \rho_1; 0)$-EB design. Note that the latter approach may yield a design with a lesser number of blocks than one obtained by the former approach.

Thus, one can establish the following theorem.

Theorem 9.3.4. *The existence of a semi-regular GD design with parameters $v = mn$, b, r, k, $\lambda_1 = 0$, $\lambda_2 \neq 0$ implies the existence of a resolvable nonproper binary $(\rho_0^*; \rho_1^*; 0)$-EB design with $\varepsilon_0^* = 1$ and $\varepsilon_1^* = 1 - 1/k$.*

Note that in Theorem 9.3.4 other parameters v^*, b^*, r^*, $\boldsymbol{k}^*$, ρ_β^* and $\boldsymbol{L}_\beta^*$ of the resulting design are given by use of Corollaries 6.1.3 and 6.5.2, after treatments to be merged are known or fixed practically. When $\lambda_1 \neq 0$, it holds that $\varepsilon_1^* = 1 - (r - \lambda_1)/(rk)$ in Theorem 9.3.4, which yields a nonbinary design.

According to the knowledge of the authors, there seems to be no other published work on construction of resolvable $(\rho_0; \rho_1; 0)$-EB designs in literature.

9.4 Designs with three efficiency factors

This section will also be discussed by considering two cases of proper designs and nonproper designs, separately.

9.4.1 Proper designs

As mentioned in Section 7.4.1.1, a connected 3-associate PBIB design with parameters $v, b, r, k, \lambda_1, \lambda_2, \lambda_3, \psi_1, \psi_2, \psi_3$ is equireplicate and proper, either $(0; \rho_1, \rho_2, \rho_3; 0)$-EB or $(\rho_0; \rho_1, \rho_2; 0)$-EB with $\rho_0 \geq 1$ and $\varepsilon_\beta = 1 - \psi_\beta/(rk)$ for $\beta = 0, 1, 2, 3$. Here 3-associate PBIB designs which are α-resolvable are of interest. However, there is almost no published work concerning this topic. Of course, some juxtapositions of a 3-associate PBIB design produce α-resolvable proper $(\rho_0; \rho_1, \rho_2; 0)$-EB or $(0; \rho_1, \rho_2, \rho_3; 0)$-EB designs, which are trivial, for example, by use of Theorem 6.4.1. This idea of construction will be used in Section 9.4.2.

In the previous chapters there are examples of such designs. E.g., Example 7.4.1 in Section 7.4.1.1 shows a 2-resolvable proper $(1; 4, 2; 0)$-EB design with $v = 8$.

Using α-resolvable GD designs with $k = m$ for $\alpha \geq 1$, Theorem 6.4.3 (through

Corollaries 7.4.2 and 7.4.3) with $s = \alpha$ produces α-resolvable proper (ρ_0; ρ_1, ρ_2; 0)-EB (when the original design is a singular or semi-regular GD design) or (0; ρ_1, ρ_2, ρ_3; 0)-EB designs (when the original design is a regular GD design). In Clatworthy (1973), a number of such basic α-resolvable GD designs with $k = m$ can be found. By use of such designs, some parameters of designs constructed as mentioned above are listed in Tables 9.5 – 9.7, where "R," "AR" and "α-R" in the parenthesis mean a resolvable, affine resolvable and α-resolvable solution, respectively, in Clatworthy (1973). The matrices $\boldsymbol{L}_\beta$ for such designs constructed here can be formed as in Theorem 6.4.3 for corresponding values of m and n in the GD designs.

Note that even if two designs have the same parameters, as designs of Nos. 2 and 3 in Table 9.7, they have different matrices $\boldsymbol{L}_\beta$. For example, in the design of No. 2, $\boldsymbol{L}_1 = (\boldsymbol{I}_2 - 2^{-1}\mathbf{1}_2\mathbf{1}_2') \otimes (\boldsymbol{I}_2 - 2^{-1}\mathbf{1}_2\mathbf{1}_2')$, $\boldsymbol{L}_2 = 2^{-1}\mathbf{1}_2\mathbf{1}_2' \otimes (\boldsymbol{I}_2 - 2^{-1}\mathbf{1}_2\mathbf{1}_2')$, $\boldsymbol{L}_3 = (\boldsymbol{I}_2 - 2^{-1}\mathbf{1}_2\mathbf{1}_2') \otimes 2^{-1}\mathbf{1}_2\mathbf{1}_2'$, whereas, in the design of No.3, $\boldsymbol{L}_1 = (\boldsymbol{I}_2 - 2^{-1}\mathbf{1}_2\mathbf{1}_2') \otimes 2^{-1}\mathbf{1}_2\mathbf{1}_2'$, $\boldsymbol{L}_2 = (\boldsymbol{I}_2 - 2^{-1}\mathbf{1}_2\mathbf{1}_2') \otimes (\boldsymbol{I}_2 - 2^{-1}\mathbf{1}_2\mathbf{1}_2')$, $\boldsymbol{L}_3 = 2^{-1}\mathbf{1}_2\mathbf{1}_2' \otimes (\boldsymbol{I}_2 - 2^{-1}\mathbf{1}_2\mathbf{1}_2')$. This can be checked by Theorem 6.4.3. One can see the similar phenomena for the designs of Nos. 9 and 10 in Table 9.7. Thus, the structure of any $\boldsymbol{L}_\beta$ should be carefully formed by use of Theorem 6.4.3.

As another illustration, note that SR6(R) in Table 9.6, i.e., a resolvable semi-regular GD design, SR6 in Clatworthy (1973), with parameters $v = 6$, $b = 9, r = 3$, $k = 2$, $\lambda_1 = 0$, $\lambda_2 = 1$, $m = 2(= k)$, $n = 3$, based on two groups $\{1,2,3\}, \{4,5,6\}$ of three treatments, provides, with $s = 1$ in Theorem 6.4.3, a resolvable (1; 2, 2; 0)-EB design, of No. 7 in Table 9.6, with parameters $v^* = 6, b^* = 12, r^* = 4, k^* = 2$, $\varepsilon_0^* = 1$, $\varepsilon_1^* = 5/8$, $\varepsilon_2^* = 3/8$, $\rho_0^* = 1$, $\rho_1^* = 2$, $\rho_2^* = 2$, $\boldsymbol{L}_0^* = (\boldsymbol{I}_2 - 2^{-1}\mathbf{1}_2\mathbf{1}_2') \otimes 3^{-1}\mathbf{1}_3\mathbf{1}_3'$, $\boldsymbol{L}_1^* = (\boldsymbol{I}_2 - 2^{-1}\mathbf{1}_2\mathbf{1}_2') \otimes (\boldsymbol{I}_3 - 3^{-1}\mathbf{1}_3\mathbf{1}_3')$, $\boldsymbol{L}_2^* = 2^{-1}\mathbf{1}_2\mathbf{1}_2' \otimes (\boldsymbol{I}_3 - 3^{-1}\mathbf{1}_3\mathbf{1}_3')$, whose blocks are given by [(1, 4), (2, 5), (3, 6)], [(1, 6), (2, 4), (3, 5)], [(1, 5), (2, 6), (3, 4)], [(1, 4), (2, 5), (3, 6)]. Note that one replicate is taken twice here.

In Tables 9.5 and 9.7, there are some designs with two distinct efficiency factors only. Note that the designs constructed here are different from their starting designs. For example, see a design of No. 1 in Table 9.5. This is a resolvable $(\rho_0^*; \rho_1^*; 0)$-EB design with parameters $v^* = 8$, $b^* = 8$, $r^* = 4, k^* = 4$, $\varepsilon_0^* = 1$, $\varepsilon_1^* = 3/4$, $\rho_0^* = 3$, $\rho_1^* = 4$, $\boldsymbol{L}_0^* = (\boldsymbol{I}_4 - 4^{-1}\mathbf{1}_4\mathbf{1}_4') \otimes (\boldsymbol{I}_2 - 2^{-1}\mathbf{1}_2\mathbf{1}_2')$, $\boldsymbol{L}_1^* = 4^{-1}\mathbf{1}_4\mathbf{1}_4' \otimes (\boldsymbol{I}_2 - 2^{-1}\mathbf{1}_2\mathbf{1}_2') + (\boldsymbol{I}_4 - 4^{-1}\mathbf{1}_4\mathbf{1}_4') \otimes 2^{-1}\mathbf{1}_2\mathbf{1}_2'$, which can be constructed from an affine resolvable singular GD design, S 6, i.e., a resolvable $(\rho_0; \rho_1; 0)$-EB design with parameters $v = 8, b = 6$, $r = 3, k = 4$, $\varepsilon_0 = 1$, $\varepsilon_1 = 2/3$, $\rho_0^* = 4$, $\rho_1^* = 3$, $\boldsymbol{L}_0^* = \boldsymbol{I}_4 \otimes (\boldsymbol{I}_2 - 2^{-1}\mathbf{1}_2\mathbf{1}_2')$, $\boldsymbol{L}_1^* = (\boldsymbol{I}_4 - 4^{-1}\mathbf{1}_4\mathbf{1}_4') \otimes 2^{-1}\mathbf{1}_2\mathbf{1}_2'$.

Tables 9.5 – 9.7 are provided under the exhaustive search of α-resolvable GD designs with $k = m$ given in Clatworthy (1973).

An array obtained by deleting the first horizontal and vertical blocks in the pattern (9.3.1) can establish the following theorem, similarly to Theorem 9.3.1.

Theorem 9.4.1. *The existence of a BIB design with parameters v (being a prime), b, r, k, λ implies the existence of an r-resolvable rectangular PBIB design* *with parameters* $v' = v(v-1)$ *(in* $v-1$ *rows and* v *columns),* $b' = (v-1)b$, $r' = (v-1)r$, $k' = (v-1)k$, $\lambda_1' = \lambda_2' = (v-1)\lambda$, $\lambda_3' = (v-2)\lambda + r$, $a = v-1$, $b_0 = b$, *i.e., of a* $(\rho_0; \rho_1, \rho_2; 0)$*-EB design with* $\varepsilon_0 = 1$, $\varepsilon_1 = 1-(r-\lambda)/[rk(v-1)^2]$, $\varepsilon_2 = 1 - v(r-\lambda)/[rk(v-1)^2]$, $\rho_0 = v-2$, $\rho_1 = v-1$, $\rho_2 = (v-2)(v-1)$, $\boldsymbol{L}_0 = [\boldsymbol{I}_{v-1} - (v-1)^{-1}\mathbf{1}_{v-1}\mathbf{1}_{v-1}'] \otimes v^{-1}\mathbf{1}_v\mathbf{1}_v'$, $\boldsymbol{L}_1 = (v-1)^{-1}\mathbf{1}_{v-1}\mathbf{1}_{v-1}' \otimes (\boldsymbol{I}_v - v^{-1}\mathbf{1}_v\mathbf{1}_v')$, $\boldsymbol{L}_2 = [\boldsymbol{I}_{v-1} - (v-1)^{-1}\mathbf{1}_{v-1}\mathbf{1}_{v-1}'] \otimes (\boldsymbol{I}_v - v^{-1}\mathbf{1}_v\mathbf{1}_v')$.

Note that this rectangular association scheme is not reducible to an L_2 association scheme, by combining first and second associate classes, though $\lambda_1 = \lambda_2$. (See Kageyama, 1974c, Section 2.)

Returning to the pattern (9.3.2), note that one can add some structure to it, to construct other r-resolvable rectangular PBIB designs when v is not a prime. Recalling that $\boldsymbol{N}(s)$, for $s = 1, 2, ..., v$, denotes the sth row of the incidence matrix $\boldsymbol{N}$ (see the proof of Theorem 9.2.11), the extended pattern can be written as

$$\begin{matrix} \mathbf{1}_v \otimes \boldsymbol{N}(1) & \boldsymbol{N}^{(1)} & \boldsymbol{N}^{(2)} & \boldsymbol{N}^{(3)} & \cdots & \boldsymbol{N}^{(v)} \\ \mathbf{1}_v \otimes \boldsymbol{N}(2) & \boldsymbol{N}^{(2)} & \boldsymbol{N}^{(3)} & \boldsymbol{N}^{(4)} & \cdots & \boldsymbol{N}^{(1)} \\ \mathbf{1}_v \otimes \boldsymbol{N}(3) & \boldsymbol{N}^{(3)} & \boldsymbol{N}^{(4)} & \boldsymbol{N}^{(5)} & \cdots & \boldsymbol{N}^{(2)} \\ \vdots & \vdots & \vdots & \vdots & & \vdots \\ \mathbf{1}_v \otimes \boldsymbol{N}(v) & \boldsymbol{N}^{(v)} & \boldsymbol{N}^{(1)} & \boldsymbol{N}^{(2)} & \cdots & \boldsymbol{N}^{(v-1)} \end{matrix} \tag{9.4.1}$$

which provides the following theorem. Here, each of the $(v+1)$ groups of b columns in (9.4.1) forms a superblock.

Theorem 9.4.2. *The existence of a BIB design with parameters* v, b, r, k, λ *implies the existence of an* r*-resolvable rectangular PBIB design with parameters* $v' = v^2$, $b' = (v+1)b$, $r' = (v+1)r$, $k' = vk$, $\lambda_1' = v\lambda + r$, $\lambda_2' = vr + \lambda$, $\lambda_3' = \lambda(v+1)$, $a = v+1$, $b_0 = b$, *i.e., of a* $(\rho_0; \rho_1, \rho_2; 0)$*-EB design with* $\varepsilon_0 = 1$, $\varepsilon_1 = 1-(r-\lambda)/[rk(v+1)]$, $\varepsilon_2 = 1-v(r-\lambda)/[rk(v+1)]$, $\rho_0 = (v-1)^2$, $\rho_1 = v-1$, $\rho_2 = v-1$, $\boldsymbol{L}_0 = (\boldsymbol{I}_v - v^{-1}\mathbf{1}_v\mathbf{1}_v') \otimes (\boldsymbol{I}_v - v^{-1}\mathbf{1}_v\mathbf{1}_v')$, $\boldsymbol{L}_1 = (\boldsymbol{I}_v - v^{-1}\mathbf{1}_v\mathbf{1}_v') \otimes v^{-1}\mathbf{1}_v\mathbf{1}_v'$, $\boldsymbol{L}_2 = v^{-1}\mathbf{1}_v\mathbf{1}_v' \otimes (\boldsymbol{I}_v - v^{-1}\mathbf{1}_v\mathbf{1}_v')$.

Proof. The association scheme is as follows: In the usual $v \times v$ arrangement of v^2 treatments (its first row is $(1, 2, ..., v)$, the second row $(v+1, v+2, ..., 2v)$, and so on), with respect to each treatment θ, first associates are the other $v-1$ $(= n_1)$ treatments in the same row as θ; second associates are the other $v-1$ $(= n_2)$ treatments which have the same position as θ in the cyclic manner, and the remaining $(v-1)^2$ $(= n_3)$ treatments are third associates. That is, following the $v \times v$ arrangement of v^2 treatments, a treatment 1 has the $v-1$ second associate treatments as $v+2, 2v+3, ..., v^2$, and a treatment 2 has the $v-1$ second associate treatments as $v+3, 2v+4, ..., (v-1)v, (v-1)v+1$, and so on. Now the values of parameters v', b', r', k', a, b_0 are obvious, with the r-resolvability. Following the $v \times v$ arrangement of v^2 treatments, it is shown that the pattern (9.4.1) yields other parameters as stated. □

A 3-associate PBIB design resulting from Theorem 9.4.2 is not reducible, because the parameters λ_i are different. Furthermore, BIB designs are available in abundance (see Section 8.2 and Table 8.2) and hence one can produce a large number of α-resolvable PBIB designs by this theorem.

Once there are α-resolvable proper $(\rho_0; \rho_1, \rho_2; 0)$-EB designs, by complementation (Corollary 7.4.1) other $(b_0 - \alpha)$-resolvable proper $(\rho_0; \rho_1, \rho_2; 0)$-EB designs can be obtained.

As far as the authors are aware of, there seems to exist no published work on construction of resolvable 3-associate PBIB designs, other than those mentioned above.

9.4.2 Nonproper designs

The interest here is in α-resolvable $(\rho_0; \rho_1, \rho_2; 0)$-EB or $(0; \rho_1, \rho_2, \rho_3; 0)$-EB designs with unequal block sizes. The following can be observed in Chapters 6, 7 and 8.

The restricted Kronecker product (Theorem 6.4.3 and Corollary 7.4.18) can produce resolvable nonproper $(\rho_0; \rho_1, \rho_2; 0)$-EB or $(0; \rho_1, \rho_2, \rho_3; 0)$-EB designs, when the starting design is resolvable. In fact, Example 7.4.14 shows a resolvable nonproper $(\rho_0; \rho_1, \rho_2; 0)$-EB design with $v = 12$.

Corollary 7.4.20 can produce α-resolvable nonproper $(\rho_0; \rho_1, \rho_2; 0)$-EB or $(0; \rho_1, \rho_2, \rho_3; 0)$-EB designs, when the starting GD design is α-resolvable. In fact, Example 7.4.15 shows a 2-resolvable nonproper $(4; 1, 4; 0)$-EB design with 10 treatments.

Theorem 7.4.16 (Corollaries 7.4.21 and 7.4.22) also produces α-resolvable nonproper $(\rho_0; \rho_1, \rho_2; 0)$-EB or $(0; \rho_1, \rho_2, \rho_3; 0)$-EB designs, when the starting 3-associate PBIB designs satisfy $r^{(1)} = r^{(2)} = \alpha$.

As far as the authors are aware of, there seems to exist no published work on construction of α-resolvable nonproper $(\rho_0; \rho_1, \rho_2; 0)$-EB or $(0; \rho_1, \rho_2, \rho_3; 0)$-EB designs without using an idea of some type of Kronecker products, other than those mentioned above.

9.5 Designs with more efficiency factors

This section will again be discussed by considering two cases of proper designs and nonproper designs, separately. The designs in this category can be constructed by use of some methods described in Sections 7.5 and 8.5. No other results are available.

9.5.1 Proper designs

As mentioned in Theorems 7.5.1 and 8.5.1, a connected s-associate PBIB design is equireplicate and proper, either $(0; \rho_1, \rho_2, ..., \rho_s; 0)$-EB or $(\rho_0; \rho_1, \rho_2, ..., \rho_{s-1}; 0)$-EB with $\rho_0 \geq 1$. Also, as Corollary 8.5.1, the dualization yields another such design. Here s-associate PBIB designs which are α-resolvable are of interest. At first, as discussed in Corollary 7.5.2, some juxtaposition of an s-associate PBIB design with $v = 2k$ produces α-resolvable proper $(\rho_0; \rho_1, \rho_2, ..., \rho_{s-1}; 0)$-

EB or $(0; \rho_1, \rho_2, ..., \rho_s; 0)$-EB designs, which are trivial, for example, by use of an extended version of Theorem 8.4.1 corresponding to existing s-associate PBIB designs. Furthermore, Corollary 8.5.2 yields an α-resolvable nonproper EB design, when the starting designs are α-resolvable PBIB designs with a common block size. This idea of construction will be used for Section 9.5.2.

As far as the authors are aware of, there seems to exist no published work on construction of resolvable s-associate PBIB designs, other than the results indicated above.

9.5.2 Nonproper designs

The interest of this section is in α-resolvable $(\rho_0; \rho_1, \rho_2, ..., \rho_{s-1}; 0)$-EB designs with unequal block sizes.

The juxtaposition technique is powerful. For example, Corollary 7.5.4 produces α-resolvable nonproper designs, when the starting s-associate PBIB designs satisfy $r^{(1)} = r^{(2)} = \cdots = r^{(a)} = \alpha$. Corollary 8.5.2 also yields an α-resolvable nonproper design, when the starting designs are α-resolvable PBIB designs whose incidence matrices have the same rank smaller than v.

To the knowledge of the authors, there seems to exist no published work on construction of α-resolvable nonproper $(\rho_0; \rho_1, \rho_2, ..., \rho_{s-1}; 0)$-EB designs, other than those mentioned above.

9.6 Lattice designs

In an equireplicate design, when the number of treatments is large, it is desirable from an economical point of view to keep the replication number r as low as possible, provided that the desired precision is obtainable. In general, BIB designs with parameters v, b, r, k, λ discussed mainly in Section 8.2.1 are not suitable for this purpose, because fundamental relations $vr = bk$ and $\lambda(v-1) = r(k-1)$ with Fisher's inequality $r \geq k$ (see Corollary 2.3.1 and Section 6.0.1) imply $r^2 - r + 1 \geq v$.

Lattice designs were introduced for experimental purposes by Yates (1936a, 1936b, 1940a). They have the merit that there is no constraint on the replication number r, though they are available only for limited numbers of treatments. For a lattice design with $r = 2$, k^2 treatments are arranged in a $k \times k$ square scheme as

$$\begin{array}{cccc} 1 & 2 & \cdots & k \\ k+1 & k+2 & \cdots & 2k \\ \vdots & \vdots & & \vdots \\ k(k-1)+1 & k(k-1)+2 & \cdots & k^2 \end{array}$$

[as in Section 6.0.2 (A) and (C)] and the blocks are composed of the sets of treatments which appear in the same row or the same column of this square. (In this sense, this type of design is especially called a square lattice design.) Then one obtains the relevant incidence matrix of the form

$$N = [I_k \otimes 1_k : 1_k \otimes I_k],$$

which can be shown to produce a resolvable $(\rho_0; \rho_1; 0)$-EB design with parameters $v = k^2, b = 2k, r = 2, k$, $\varepsilon_0 = 1, \varepsilon_1 = 1/2$, $\rho_0 = (k-1)^2, \rho_1 = 2(k-1)$,

$$L_0 = (I_k - \frac{1}{k}1_k1_k') \otimes (I_k - \frac{1}{k}1_k1_k'),$$

$$L_1 = \frac{1}{k}1_k1_k' \otimes (I_k - \frac{1}{k}1_k1_k') + (I_k - \frac{1}{k}1_k1_k') \otimes \frac{1}{k}1_k1_k'.$$

For example, when $k = 5$, the scheme is

1	2	3	4	5
6	7	8	9	10
11	12	13	14	15
16	17	18	19	20
21	22	23	24	25

and there are 10 blocks given by rows and columns of the scheme, which yields a resolvable (16;8;0)-EB design with parameters $v = 25, b = 10, r = 2, k = 5$, $\varepsilon_0 = 1, \varepsilon_1 = 1/2$, $\rho_0 = 16, \rho_1 = 8$, $L_0 = (I_5 - 5^{-1}1_51_5') \otimes (I_5 - 5^{-1}1_51_5')$, $L_1 = 5^{-1}1_51_5' \otimes (I_5 - 5^{-1}1_51_5') + (I_5 - 5^{-1}1_51_5') \otimes 5^{-1}1_51_5'$, whose incidence matrix is given by $[I_5 \otimes 1_5 : 1_5 \otimes I_5]$. Note that this design can be seen as described by the incidence matrix $N = [N_1 : N_2]$, where the submatrices N_1 and N_2 are exactly the same as those in Examples 7.3.22 and 9.2.6

As seen from this example, a lattice design is always resolvable. When $r > 2$, one can take a set of $r - 2$ mutually orthogonal Latin squares of order k, if they exist (see Appendix D). Corresponding to each Latin square one takes k blocks obtained by superimposing it on the previously described $k \times k$ square scheme and taking as blocks the sets of treatments which correspond to the same symbol of the superimposed Latin square.

For example, when $k = 4$ and $r = 4$, first prepare a 4×4 square and two mutually orthogonal Latin squares as follows.

$$\begin{matrix} 1 & 2 & 3 & 4 \\ 5 & 6 & 7 & 8 \\ 9 & 10 & 11 & 12 \\ 13 & 14 & 15 & 16 \end{matrix}\ ; \quad \begin{bmatrix} 1 & 2 & 3 & 4 \\ 2 & 1 & 4 & 3 \\ 3 & 4 & 1 & 2 \\ 4 & 3 & 2 & 1 \end{bmatrix} \quad \text{and} \quad \begin{bmatrix} 1 & 2 & 3 & 4 \\ 3 & 4 & 1 & 2 \\ 4 & 3 & 2 & 1 \\ 2 & 1 & 4 & 3 \end{bmatrix}.$$

The above construction procedure yields a resolvable square lattice design with the incidence matrix

$$\left[\begin{array}{cccc|cccc|cccc|cccc}
1 & 0 & 0 & 0 & 1 & 0 & 0 & 0 & 1 & 0 & 0 & 0 & 1 & 0 & 0 & 0 \\
1 & 0 & 0 & 0 & 0 & 1 & 0 & 0 & 0 & 1 & 0 & 0 & 0 & 1 & 0 & 0 \\
1 & 0 & 0 & 0 & 0 & 0 & 1 & 0 & 0 & 0 & 1 & 0 & 0 & 0 & 1 & 0 \\
1 & 0 & 0 & 0 & 0 & 0 & 0 & 1 & 0 & 0 & 0 & 1 & 0 & 0 & 0 & 1 \\
0 & 1 & 0 & 0 & 1 & 0 & 0 & 0 & 0 & 1 & 0 & 0 & 0 & 0 & 1 & 0 \\
0 & 1 & 0 & 0 & 0 & 1 & 0 & 0 & 1 & 0 & 0 & 0 & 0 & 0 & 0 & 1 \\
0 & 1 & 0 & 0 & 0 & 0 & 1 & 0 & 0 & 0 & 0 & 1 & 1 & 0 & 0 & 0 \\
0 & 1 & 0 & 0 & 0 & 0 & 0 & 1 & 0 & 0 & 1 & 0 & 0 & 1 & 0 & 0 \\
0 & 0 & 1 & 0 & 1 & 0 & 0 & 0 & 0 & 0 & 1 & 0 & 0 & 0 & 0 & 1 \\
0 & 0 & 1 & 0 & 0 & 1 & 0 & 0 & 0 & 0 & 0 & 1 & 0 & 0 & 1 & 0 \\
0 & 0 & 1 & 0 & 0 & 0 & 1 & 0 & 1 & 0 & 0 & 0 & 0 & 1 & 0 & 0 \\
0 & 0 & 1 & 0 & 0 & 0 & 0 & 1 & 0 & 1 & 0 & 0 & 1 & 0 & 0 & 0 \\
0 & 0 & 0 & 1 & 1 & 0 & 0 & 0 & 0 & 0 & 0 & 1 & 0 & 1 & 0 & 0 \\
0 & 0 & 0 & 1 & 0 & 1 & 0 & 0 & 0 & 0 & 1 & 0 & 1 & 0 & 0 & 0 \\
0 & 0 & 0 & 1 & 0 & 0 & 1 & 0 & 0 & 1 & 0 & 0 & 0 & 0 & 0 & 1 \\
0 & 0 & 0 & 1 & 0 & 0 & 0 & 1 & 1 & 0 & 0 & 0 & 0 & 0 & 1 & 0
\end{array}\right],$$

which is, in fact, a resolvable (3;12;0)-EB design with parameters $v = b = 16, r = k = 4$, $\varepsilon_0 = 1, \varepsilon_1 = 3/4$, $\rho_0 = 3, \rho_1 = 12$, $\boldsymbol{L}_0 = (\boldsymbol{I}_4 - 4^{-1}\mathbf{1}_4\mathbf{1}_4') \otimes 4^{-1}\mathbf{1}_4\mathbf{1}_4', \boldsymbol{L}_1 = \boldsymbol{I}_4 \otimes (\boldsymbol{I}_4 - 4^{-1}\mathbf{1}_4\mathbf{1}_4')$.

As mentioned in Remark 6.0.1, PBIB designs were discovered by Bose and Nair (1939) in an attempt to generalize a class of lattice designs. The concept of association schemes though inherent in the Bose and Nair paper was explicitly given later by Bose and Shimamoto (1952). It can be seen that lattice designs are examples of 2-associate PBIB designs with $\lambda_1 = 1$ and $\lambda_2 = 0$, by introducing the following association scheme, i.e., L_r [see Section 6.0.2(C)]: In the $k \times k$ scheme, two treatments are first associates if and only if they occur together in the same row or the same column of the square scheme, or correspond to the same symbol in one of the $r-2$ mutually orthogonal Latin squares of order k, and otherwise they are second associates. This association scheme, more precisely denoted by $\mathrm{L}_r(k)$ instead of L_r, has the usual parameters described in Section 6.0.2(C) for $2 \leq r \leq k$, in which the construction mentioned above leads to a resolvable $(\rho_0; \rho_1; 0)$-EB design, having the incidence matrix $\boldsymbol{N}$, with parameters $v = k^2, b = rk, r, k, \varepsilon_0 = 1, \varepsilon_1 = 1 - 1/r$, $\rho_0 = (k - r + 1)(k - 1), \rho_1 = r(k - 1)$,

$$\boldsymbol{L}_0 = \boldsymbol{I}_v - \frac{1}{v}\mathbf{1}_v\mathbf{1}_v' - \boldsymbol{L}_1, \quad \boldsymbol{L}_1 = \frac{1}{k}\left(\boldsymbol{N}\boldsymbol{N}' - \frac{r}{k}\mathbf{1}_v\mathbf{1}_v'\right).$$

However, the concrete forms of $\boldsymbol{L}_0$ and $\boldsymbol{L}_1$ can be given only after having practical structures of $r - 2$ mutually orthogonal Latin squares of order k, i.e., the incidence structure of the design in this case (see John, 1987, Section 3.4.2; John and Williams, 1995, Section 4.2). Especially, when $r = k$, it holds that $p_{12}^2 = 0$. This implies (see Raghavarao, 1971, Section 8.5.1) that the $\mathrm{L}_r(k)$ association scheme coincides with a GD association scheme discussed in Section

6.0.2(A), by exchanging the first associates for the second associates. That is, one can get a resolvable $(\rho_0; \rho_1; 0)$-EB design with parameters $v = b = k^2, r = k$, $\varepsilon_0 = 1, \varepsilon_1 = 1 - 1/k$, $\rho_0 = k - 1, \rho_1 = k(k-1)$,

$$\boldsymbol{L}_0 = (\boldsymbol{I}_k - \frac{1}{k}\mathbf{1}_k\mathbf{1}_k') \otimes \frac{1}{k}\mathbf{1}_k\mathbf{1}_k', \quad \boldsymbol{L}_1 = \boldsymbol{I}_k \otimes (\boldsymbol{I}_k - \frac{1}{k}\mathbf{1}_k\mathbf{1}_k').$$

In general, note that when $r = 2$, one can get all the information on the designs as was already shown. This exactly coincides with the L_2 association scheme discussed in Section 6.0.2(C), when $r = 2$ and $k = s$. Hence, lattice designs can be analyzed as designs with two distinct efficiency factors, i.e., one of them is 1. When a lattice design is used, the difference of a pair of treatments occurring together in the same block is estimated with a slightly greater precision than the difference of a pair which do not occur in the same block.

Thus, in a lattice design the number of treatments is requested to be the square of an integer s. The layout consists of r superblocks (Chapter 5) of s^2 plots, one plot for each treatment; the s^2 plots of each superblock are divided into s blocks of s plots each. Simple lattices ($r = 2$) and triple lattices ($r = 3$) are available for any integer s (> 1), because a Latin square of order s always exists. Quadruple lattices ($r = 4$) are available for many but not all values of s: the exact requirement is that s must be such that two orthogonal Latin squares of order s exist. The typical exception is $s = 6$; in fact there do not exist two orthogonal Latin squares of order 6 (see Tarry, 1900; Appendix D) and hence there is no quadruple lattice design for 36 treatments.

The resolvability is an important property from both statistical and experimental point of view (see Chapter 5). Yates (1940a, 1940b) pointed out that, because they are resolvable, lattice designs can never be less efficient than ordinary RBDs. Indeed, if blocking proves ineffective, lattice designs can be analyzed by conventional randomized block analysis, treating the superblocks as ordinary blocks.

As already known, lattice designs are available only for limited numbers of treatments, as $v = s^2$. To relax this serious restriction, rectangular lattice designs were introduced and discussed by Harshbarger (1946, 1947, 1949, 1951). They are resolvable designs with $v = (s-1)s$ treatments in $b = rs$ blocks of size $k = s - 1$ each such that the blocks can be arranged into r complete replications. Several methods of construction of rectangular lattice designs can be seen in Cochran and Cox (1957) for $r = 2$ or 3 and some also for $r = 4$, and also in Williams (1977b), and Bailey and Speed (1986).

A typical method of constructing rectangular lattice designs is given here by taking an idea similar to that of lattice design described above. When $r \geq 2$, one can take a set of $r - 2$ mutually orthogonal Latin squares of order k with some constraint, if they exist (see Appendix D). The blocks of the first two replicates correspond to the rows and columns of the $s \times s$ sqaure, respectively, with the leading diagonal left blank and the $v = s(s-1)$ treatments set out in the square as follows:

$$\begin{array}{ccccc}
 & 1 & \cdots & s-2 & s-1 \\
s & & \cdots & 2(s-1)-1 & 2(s-1) \\
2s-1 & 2s & \cdots & 3(s-1)-1 & 3(s-1) \\
\vdots & \vdots & & \ddots & \vdots \\
(s-1)(s-2)+1 & (s-1)(s-2)+2 & \cdots & & (s-1)^2 \\
(s-1)^2+1 & (s-1)^2+2 & \cdots & s(s-1) &
\end{array} .$$

Corresponding to each Latin square one takes s blocks obtained by superimposing it on the previously described $s \times s$ square scheme and taking as blocks the sets of treatments which correspond to the same symbol of the superimposed Latin square.

For example, when $s = 4$ and $r = 4$, one has $v = 12$ treatments, first prepare a 4×4 square with the diagonal left blank and two mutually orthogonal Latin squares (with the constraint that the leading diagonal of each Latin square contains every symbol in the same order) as follows.

$$\begin{array}{cccc}
 & 1 & 2 & 3 \\
4 & & 5 & 6 \\
7 & 8 & & 9 \\
10 & 11 & 12 &
\end{array} ;$$

$$L_1 = \begin{bmatrix} 1 & 3 & 4 & 2 \\ 4 & 2 & 1 & 3 \\ 2 & 4 & 3 & 1 \\ 3 & 1 & 2 & 4 \end{bmatrix} \text{ and } L_2 = \begin{bmatrix} 1 & 4 & 2 & 3 \\ 3 & 2 & 4 & 1 \\ 4 & 1 & 3 & 2 \\ 2 & 3 & 1 & 4 \end{bmatrix} .$$

The above construction procedure yields a resolvable rectangular lattice design with the incidence matrix

$$N = \left[\begin{array}{cccc|cccc|cccc|cccc}
1 & 0 & 0 & 0 & 0 & 1 & 0 & 0 & 0 & 0 & 1 & 0 & 0 & 0 & 0 & 1 \\
1 & 0 & 0 & 0 & 0 & 0 & 1 & 0 & 0 & 0 & 0 & 1 & 0 & 1 & 0 & 0 \\
1 & 0 & 0 & 0 & 0 & 0 & 0 & 1 & 0 & 1 & 0 & 0 & 0 & 0 & 1 & 0 \\
0 & 1 & 0 & 0 & 1 & 0 & 0 & 0 & 0 & 0 & 0 & 1 & 0 & 0 & 1 & 0 \\
0 & 1 & 0 & 0 & 0 & 0 & 1 & 0 & 1 & 0 & 0 & 0 & 0 & 0 & 0 & 1 \\
0 & 1 & 0 & 0 & 0 & 0 & 0 & 1 & 0 & 0 & 1 & 0 & 1 & 0 & 0 & 0 \\
0 & 0 & 1 & 0 & 1 & 0 & 0 & 0 & 0 & 1 & 0 & 0 & 0 & 0 & 0 & 1 \\
0 & 0 & 1 & 0 & 0 & 1 & 0 & 0 & 0 & 0 & 0 & 1 & 1 & 0 & 0 & 0 \\
0 & 0 & 1 & 0 & 0 & 0 & 0 & 1 & 1 & 0 & 0 & 0 & 0 & 1 & 0 & 0 \\
0 & 0 & 0 & 1 & 1 & 0 & 0 & 0 & 0 & 0 & 1 & 0 & 0 & 1 & 0 & 0 \\
0 & 0 & 0 & 1 & 0 & 1 & 0 & 0 & 1 & 0 & 0 & 0 & 0 & 0 & 1 & 0 \\
0 & 0 & 0 & 1 & 0 & 0 & 1 & 0 & 0 & 1 & 0 & 0 & 1 & 0 & 0 & 0
\end{array}\right] ,$$

which is, in fact, a resolvable $(\rho_0; \rho_1; 0)$-EB design with parameters $v = 12, b = 16, r = 4, k = 3, \varepsilon_0 = 1, \varepsilon_1 = 2/3, \rho_0 = 2, \rho_1 = 9, \boldsymbol{L}_0 = \boldsymbol{I}_{12} - (12)^{-1}\mathbf{1}_{12}\mathbf{1}'_{12} - \boldsymbol{L}_1$, $\boldsymbol{L}_1 = 4^{-1}(\boldsymbol{N}\boldsymbol{N}' - \mathbf{1}_{12}\mathbf{1}'_{12})$.

In general, when $r = s$, the present method produces a resolvable $(\rho_0; \rho_1; 0)$-EB design, having the incidence matrix $\boldsymbol{N}$, with parameters $v = s(s-1), b = s^2, r = s, k = s-1$, $\varepsilon_0 = 1$, $\varepsilon_1 = (s-2)/(s-1)$, $\rho_0 = s-2, \rho_1 = (s-1)^2$,

$$\boldsymbol{L}_0 = \boldsymbol{I}_{s(s-1)} - \frac{1}{s(s-1)}\boldsymbol{1}_{s(s-1)}\boldsymbol{1}'_{s(s-1)} - \boldsymbol{L}_1, \quad \boldsymbol{L}_1 = \frac{1}{s}\left(\boldsymbol{N}\boldsymbol{N}' - \boldsymbol{1}_{s(s-1)}\boldsymbol{1}'_{s(s-1)}\right).$$

On the other hand, when $r < s$, one gets a resolvable $(\rho_0; \rho_1, \rho_2; 0)$-EB design with parameters $v = s(s-1), b = rs, r, k = s-1$, $\varepsilon_0 = 1$, $\varepsilon_1 = s(r-1)/(rk)$, $\varepsilon_2 = (rk-s)/(rk)$, $\rho_0 = v - 1 - rk$, $\rho_1 = k$, $\rho_2 = k(r-1)$.

For example, when $r = 2 < 3 = s$, for the following 3×3 array

$$\begin{array}{ccc} & 1 & 2 \\ 3 & & 4 \\ 5 & 6 & \end{array},$$

one can obtain a resolvable (1; 2, 2; 0)-EB design, whose incidence matrix is given by

$$\left[\begin{array}{ccc|ccc} 1 & 0 & 0 & 0 & 1 & 0 \\ 1 & 0 & 0 & 0 & 0 & 1 \\ 0 & 1 & 0 & 1 & 0 & 0 \\ 0 & 1 & 0 & 0 & 0 & 1 \\ 0 & 0 & 1 & 1 & 0 & 0 \\ 0 & 0 & 1 & 0 & 1 & 0 \end{array}\right],$$

with parameters $v = b = 6, r = k = 2$, $\varepsilon_0 = 1, \varepsilon_1 = 3/4$, $\varepsilon_2 = 1/4$, $\rho_0 = 1, \rho_1 = 2$, $\rho_2 = 2$, $\boldsymbol{L}_0 = \boldsymbol{I}_6 - 6^{-1}\boldsymbol{1}_6\boldsymbol{1}'_6 - \boldsymbol{L}_1 - \boldsymbol{L}_2$,

$$\boldsymbol{L}_1 = \frac{1}{6}\begin{bmatrix} 2 & -1 & 2 & -1 & -1 & -1 \\ -1 & 2 & -1 & -1 & 2 & -1 \\ 2 & -1 & 2 & -1 & -1 & -1 \\ -1 & -1 & -1 & 2 & -1 & 2 \\ -1 & 2 & -1 & -1 & 2 & -1 \\ -1 & -1 & -1 & 2 & -1 & 2 \end{bmatrix},$$

$$\boldsymbol{L}_2 = \frac{1}{8}\begin{bmatrix} 3 & 1 & -3 & \sqrt{2} & -1 & -\sqrt{2} \\ 1 & 3 & -1 & -\sqrt{2} & -3 & \sqrt{2} \\ -3 & -1 & 3 & -\sqrt{2} & 1 & \sqrt{2} \\ \sqrt{2} & -\sqrt{2} & -\sqrt{2} & 2 & \sqrt{2} & -2 \\ -1 & -3 & 1 & \sqrt{2} & 3 & -\sqrt{2} \\ -\sqrt{2} & \sqrt{2} & \sqrt{2} & -2 & -\sqrt{2} & 2 \end{bmatrix}.$$

Rectangular lattice designs are also constructed by deleting treatments in available square lattice designs. Kempthorne (1952, Chapter 25) pointed out that similar designs are available for other v of the form $s(s-u)$ with $k = s - u$ and $s > u$ but gave few details.

These lattice designs are proper and equireplicate. Hence, as noted at the beginning of Section 9.3.1, one can also use resolvable PBIB designs in Clatworthy (1973), depending on the purpose of the designed experiment and the desired efficiency for basic contrasts of interest, if appropriate plans are available. Note (Nair, 1951) that rectangular lattice designs are in general not partially balanced in the sense of Bose and Nair (1939). However, rectangular lattice designs are generally balanced (see Section 5.4), as shown by Bailey and Speed (1986, Section 5). Cochran and Cox (1957), and Williams (1977b) discussed expressions of the intra-block variances of estimated treatment differences depending on whether or not the compared treatments occur together within blocks.

As an extension of these designs, Williams (1977a) defined generalized lattice designs as resolvable designs with $v = ks$ treatments in $b = rs$ blocks of size k each such that the blocks can be arranged into r complete replications. This definition includes the parameters for the square lattice designs ($k = s$) and the rectangular lattice designs ($k = s - 1$). Here one can allow k to take any value. Generalized lattice designs also include as a special case the α-designs defined by Patterson and Williams (1976) to greatly extend the availability of efficient resolvable designs.

The α-designs are simply resolvable incomplete block designs and were developed specifically for a computer-based application with a wide range of parameter values. In fact, the designs can be generated on a computer, as described by Paterson and Patterson (1983) (see John, 1987, Section 4.8; John and Williams, 1995, Sections 4.4 and 4.5). Under the requirements of agricultural plant variety improvement programs, these designs are extensively used in variety trials in many countries, particularly in the United Kingdom (see Patterson and Silvey, 1980). The practical field conditions dictate that all varieties are equally replicated, the number of replication being small, and all the designs used for these trials are resolvable. The α-designs satisfy such request and include as special cases some lattice and resolvable cyclic designs introduced by Yates (1936a), Bose and Nair (1962), and David (1967). For each design the average efficiency factor ε has been calculated. Williams, Patterson and John (1976) discussed efficient α-designs with $r = 2$ and with the average efficiency factor ε as large as possible. Williams and Patterson (1977) derived upper bounds for ε. It appears that α-designs greatly increased the availability of suitable incomplete block designs, because of high efficiency (see John and Williams, 1995, Section 4.10). Furthermore, tables of efficient α-designs were published by Patterson, Williams and Hunter (1978), and Williams (1976). Further practical developments have been made by Patterson and Silvey (1980), and Patterson and Hunter (1983). More discussions on resolvable cyclic designs, generalized cyclic designs and α-designs can be found in two successive monographs by John (1987; Chapters 3 and 4), and John and Williams (1995; Chapters 3 and 4).

9.7 Tables

Tables 9.1 and 9.3 present α-replication set(s) only (see Definition 6.0.2). For a BIB design, also compare two solutions given in Tables 8.2 and 9.1.

Table 9.1
Existing resolvable BIB designs with $v \leq 100, r \leq 10$,
their efficiency factors and solutions

No.	v	r	b	k	λ	$100\varepsilon_1$	Solution
1*	4	3	6	2	1	66.67	(1,2)(0,∞) mod 3
2	4	6	12	2	2	66.67	[(1,2)(0,∞)][(1,2)(0,∞)] mod 3
3	4	9	18	2	3	66.67	[(1,2)(0,∞)][(1,2)(0,∞)] [(1,2)(0,∞)] mod 3
4	6	5	15	2	1	60.00	(1,4)(2,3)(∞,0) mod 5
5	6	10	20	3	4	80.00	[(0,1,4)(∞,2,3)][(0,2,3)(∞,1,4)] mod 5
6	6	10	30	2	2	60.00	[(1,4)(2,3)(0,∞)] [(1,4)(2,3)(0,∞)] mod 5
7*	8	7	14	4	3	85.71	(0,1,2,4)(3,5,6,∞) mod 7
8	8	7	28	2	1	57.14	(1,6)(2,5)(3,4)(0,∞) mod 7
9*	9	4	12	3	1	75.00	PC(4)[(1,6,7)(2,3,5)(∞,0,4)] mod 8
10	9	8	24	3	2	75.00	PC(4)[(1,6,7)(2,3,5)(∞,0,4)] PC(4)[(1,6,7)(2,3,5)(∞,0,4)] mod 8
11	10	9	45	2	1	55.56	(1,8)(2,7)(3,6)(4,5)(0,∞) mod 9
12	15	7	35	3	1	71.43	[(∞,5,10)(1,6,11)(2,7,12) (3,8,13)(4,9,14)] CT(1,2,3,5,6,11,9)(4,8,14,10,13,12,7)
13*	16	5	20	4	1	80.00	$(1_1,4_1,2_2,3_2)(1_2,4_2,2_3,3_3)$ $(1_3,4_3,2_1,3_1)(\infty,0_1,0_2,0_3)$ mod 5
14	16	10	40	4	2	80.00	$[(1_1,4_1,2_2,3_2)(1_2,4_2,2_3,3_3)$ $(1_3,4_3,2_1,3_1)(\infty,0_1,0_2,0_3)]$ $[(1_1,4_1,2_2,3_2)(1_2,4_2,2_3,3_3)$ $(1_3,4_3,2_1,3_1)(\infty,0_1,0_2,0_3)]$ mod 5
15	21	10	70	3	1	70.00	$(0_1,0_2,0_3)(1_1,2_1,4_1)(1_2,2_2,4_2)$ $(1_3,2_3,4_3)(3_1,5_2,6_3)(3_2,5_3,6_1)$ $(3_3,5_1,6_2)$ Reps I-VII, $(1_1,2_3,4_2)$ Rep. VIII, $(1_2,2_1,4_3)$ Rep. IX, $(1_3,2_2,4_1)$ Reps X mod 7
16*	25	6	30	5	1	83.33	PC(6)[(1,3,16,17,20)(2,7,9,22,23) (4,5,8,13,15)(10,11,14,19,21) (0,6,12,18,∞)] mod 24

Table 9.1 (continued)

No.	v	r	b	k	λ	$100\varepsilon_1$	Solution
17	28	9	63	4	1	77.78	$(0 1_1, 0 2_1, 1 0_2, 2 0_2)(0 1_2, 0 2_2, 1 0_3, 2 0_3)$
							$(0 1_3, 0 2_3, 1 0_1, 2 0_1)$ $(2 1_1, 1 2_1, 2 2_2, 1 1_2)$
							$(2 1_2, 1 2_2, 2 2_3, 1 1_3)(2 1_3, 1 2_3, 2 2_1, 1 1_1)$
							$(\infty\infty, 0 0_1, 0 0_2, 0 0_3)$ mod (3,3)
18*	49	8	56	7	1	87.50	PC(8)[(1,2,5,11,31,36,38)
							(9,10,13,19,39,44,46)
							(4,6,17,18,21,27,47)
							(7,12,14,25,26,29,35)
							(15,20,22,33,34,37,43)
							(3,23,28,30,41,42,45)
							(0,8,16,24,32,40,∞)] mod 48
19*	64	9	72	8	1	88.89	PC(9)[(1,6,8,14,38,48,49,52)
							(10,15,17,23,47,57,58,61)
							(3,4,7,19,24,26,32,56)
							(2,12,13,16,28,33,35,41)
							(11,21,22,25,37,42,44,50)
							(20,30,31,34,46,51,53,59)
							(5,29,39,40,43,55,60,62)
							(0,9,18,27,36,45,54,∞)] mod 63
20*	81	10	90	9	1	90.00	PC(10)[(1,13,35,48,49,66,72,74,77)
							(2,4,7,11,23,45,58,59,76)
							(6,12,14,17,21,33,55,68,69)
							(16,22,24,27,31,43,65,78,79)
							(8,9,26,32,34,37,41,53,75)
							(5,18,19,36,42,44,47,51,63)
							(15,28,29,46,52,54,57,61,73)
							(3,25,38,39,56,62,64,67,71)
							(0,10,20,30,40,50,60,70,∞)] mod 80

The reference number with asterisk ($*$) shows an affine resolvable solution. Note that there does not exist a resolvable BIB design with parameters $v = 6, b = 10, r = 5, k = 3, \lambda = 2$, being the half size of a design of No. 5. Because there $b = v + r - 1$ but $k^2/v = 3/2$ being not an integer (see Theorem 6.0.5 with the integrality of q_1).

For the notation PC(n), e.g., in the design of No. 9* with the solution PC(4)[(1, 6, 7)(2, 3, 5)(∞, 0, 4)] mod 8, the symbol PC(4) means a cyclic development of initial blocks four times and reducing modulo 8 when necessary. In fact, this case shows 4 replicates as

[(1, 6, 7)(2, 3, 5)(∞, 0, 4)], [(2, 7, 0)(3, 4, 6)(∞, 1, 5)],
[(3, 0, 1)(4, 5, 7)(∞, 2, 6)], [(4, 1, 2)(5, 6, 0)(∞, 3, 7)]

(see Example 9.2.3).

Table 9.2
Parameters of unknown resolvable BIB designs
with $v \leq 100$ and $r \leq 20$

No.	v	r	b	k	λ
1	15	14	42	5	4
2	28	18	72	7	4
3	35	17	85	7	3
4	45	11	99	5	1
5	66	13	143	6	1
6	91	15	195	7	1
7	96	19	304	6	1

Table 9.3
Existing α-resolvable BIB designs with $\alpha \geq 2$, $v \leq 100, r \leq 10$

No.	v	r	b	k	λ	α	b_0	a	Solution
1	6	10	15	4	6	2	3	5	[(0,2,3,∞)(0,1,4,∞)(1,2,3,4)] mod 5
2	7	6	14	3	2	3	7	2	[(0,1,3)][(0,1,3)] mod 7
3	7	8	14	4	4	4	7	2	[(2,4,5,6)][(2,4,5,6)] mod 7
4	7	9	21	3	3	3	7	3	[(0,1,3)][(0,1,3)][(0,1,3)] mod 7
5*	9	8	12	6	5	2	3	4	PC(4)[(0,2,3,4,5,∞)(0,1,4,6,7,∞) (1,2,3,5,6,7)] mod 8
6	9	8	18	4	3	4	9	2	[(0,1,2,4)][(0,3,4,7)] mod 9
7	9	10	18	5	5	5	9	2	[(3,5,6,7,8)][(1,2,5,6,8)] mod 9
8	11	10	22	5	4	5	11	2	[(0,2,3,4,8)][(0,2,3,4,8)] mod 11
9	13	6	26	3	1	3	13	2	[(0,1,4)][(0,2,7)] mod 13
10	13	8	26	4	2	4	13	2	[(0,1,3,9)][(0,1,3,9)] mod 13
11	19	9	57	3	1	3	19	3	[(0,1,5)][(0,2,8)][(0,3,10)] mod 19
12	21	10	42	5	2	5	21	2	[(0,1,6,8,18)] [(0,1,6,8,18)] mod 21
13	25	8	50	4	1	4	25	2	[(00,01,11,43)][(00,02,22,31)] mod (5,5)
14	41	10	82	5	1	5	41	2	[(0,9,15,17,36)][(0,18,30,31,34)] mod 41

Within the present scope of parameters there is only one affine resolvable solution, i.e., a design of No. 5*. For more designs see the comments at the end of Section 9.2.1.

Table 9.4
Parameters of existent or possibly existent affine α-resolvable BIB designs with $\alpha \geq 2$, $v \leq 100, \lambda \leq 100$

No.	v	r	b	k	λ	q_1	q_2	α	b_0	a	Construction
1	9	8	12	6	5	3	4	2	3	4	(A)
2	16	15	20	12	11	8	9	3	4	5	(A)
3	25	24	30	20	19	15	16	4	5	6	(A)
4	27	26	39	18	17	9	12	2	3	13	(A)
5	49	24	56	21	10	7	9	3	7	8	(B)
6	49	32	56	28	18	14	16	4	7	8	(A)
7	49	48	56	42	41	35	36	6	7	8	(A)
8	64	63	72	56	55	48	49	7	8	9	(A)
9	64	63	84	48	47	32	36	3	4	21	(A)
10	81	80	90	72	71	63	64	8	9	10	(A)
11	81	80	120	54	53	27	36	2	3	40	(A)
12	99	98	147	66	65	33	44	2	3	49	(U)

Symbols (A) and (B) mean that the designs can be constructed by Theorems 9.2.8 and 9.2.6, respectively. Symbol (U) denotes the case in which a solution is unknown. Furthermore, it is not difficult to show that an affine α-resolvable BIB design with $\lambda \leq 100$ does not exist when $\alpha \geq 10$. The last fact can be shown by the approach described in Kageyama (1973b).

The existence of an affine 9-resolvable BIB design with parameters $v = 100$, $b = 110$, $r = 99$, $k = 90$, $\lambda = 89$, $q_1 = 80$, $q_2 = 81$, $b_0 = 10$, $a = 11$ has been unknown for some time. This complement is an affine plane of order 10, which extends to a projective plane of order 10 (see Section 6.9, Appendix C), and this was eliminated by Lam, Thiel and Swiercz (1989). Hence the affine 9-resolvable BIB design does not exist.

Note that affine α-resolvable BIB designs have strong restrictions on parameters and hence a less chance of their existence reveals. Even if $r \leq 20$ is taken as the range (like Table 9.2), then there are only two designs of Nos. 1 and 6. In this sense, the range on parameters is here extended to know the status of existence of such designs in particular.

For the constructions and related explanations on designs in Tables 9.5 – 9.7 below, see Section 9.4.1.

Table 9.5
Existing α-resolvable $(\rho_0; \rho_1, \rho_2; 0)$-EB designs
from singular GD designs with $k = m$

No.	v	b	r	k	α	ε_0	ε_1	ε_2	ρ_0	ρ_1	ρ_2	Source
1	8	8	4	4	1	1	3/4		3	4	0	S6(AR)
2	8	14	7	4	1	1	6/7	5/7	3	1	3	S7(R)
3	8	20	10	4	1	1	9/10	7/10	3	1	3	S8(R)
4	12	30	15	6	5	1	13/15	2/3	5	5	1	S31(5-R)
5	16	16	8	8	1	1	7/8		7	8	0	S63(AR)
6	18	18	6	6	1	1	5/6	2/3	10	2	5	S38(R)
7	18	33	11	6	1	1	10/11	7/11	10	2	5	S40(R)
8	27	15	5	9	1	1	4/5		16	10	0	S91(AR)
9	27	27	9	9	1	1	8/9	7/9	16	2	8	S92(R)
10	32	32	8	8	1	1	7/8	5/8	21	3	7	S75(R)

Table 9.6
Existing α-resolvable $(\rho_0; \rho_1, \rho_2; 0)$-EB designs
from semi-regular GD designs with $k = m$

No.	v	b	r	k	α	ε_0	ε_1	ε_2	ρ_0	ρ_1	ρ_2	Source
1	4	6	3	2	1	1	2/3	1/3	1	1	1	SR1(AR)
2	4	10	5	2	1	1	3/5	2/5	1	1	1	SR2(R)
3	4	14	7	2	1	1	4/7	3/7	1	1	1	SR3(R)
4	4	18	9	2	1	1	5/9	4/9	1	1	1	SR4(R)
5	4	22	11	2	1	1	6/11	5/11	1	1	1	SR5(R)
6	6	10	5	3	1	1	11/15	8/15	2	2	1	SR19(R)
7	6	12	4	2	1	1	5/8	3/8	1	2	2	SR6(R)
8	6	12	6	3	2	1	7/9	4/9	2	2	1	SR19(2-R)
9	6	16	8	3	2	1	3/4	1/2	2	2	1	SR20(2-R)
10	6	20	10	3	2	1	11/15	8/15	2	2	1	SR21(2-R)
11	6	21	7	2	1	1	4/7	3/7	1	2	2	SR7(R)
12	6	24	12	3	2	1	13/18	5/9	2	2	1	SR22(2-R)
13	6	30	10	2	1	1	11/20	9/20	1	2	2	SR8(R)
14	8	10	5	4	1	1	4/5	3/5	3	3	1	SR36(AR)
15	8	12	6	4	2	1	5/6	1/2	3	3	1	SR36(2-R)
16	8	16	8	4	2	1	13/16	9/16	3	3	1	SR37(2-R)
17	8	18	9	4	1	1	7/9	2/3	3	3	1	SR39(R)

Table 9.6 (continued-1)

No.	v	b	r	k	α	ε_0	ε_1	ε_2	ρ_0	ρ_1	ρ_2	Source
18	8	20	5	2	1	1	3/5	2/5	1	3	3	SR9(R)
19	8	20	10	4	2	1	4/5	3/5	3	3	1	SR39(2-R)
20	8	36	9	2	1	1	5/9	4/9	1	3	3	SR10(R)
21	8	30	15	4	5	1	5/6	1/2	3	3	1	SR40(5-R)
22	9	12	4	3	1	1	3/4	1/2	2	4	2	SR23(AR)
23	9	21	7	3	1	1	5/7	4/7	2	4	2	SR24(R)
24	9	30	10	3	1	1	7/10	3/5	2	4	2	SR25(R)
25	10	12	6	5	2	1	13/15	8/15	4	4	1	SR52(2-R)
26	10	20	10	5	2	1	21/25	16/25	4	4	1	SR54(2-R)
27	10	30	6	2	1	1	7/12	5/12	1	4	4	SR11(R)
28	10	55	11	2	1	1	6/11	5/11	1	4	4	SR12(R)
29	12	12	6	6	2	1	8/9	5/9	5	5	1	SR66(2-R)
30	12	18	9	6	1	1	23/27	20/27	5	5	1	SR69(R)
31	12	20	5	3	1	1	11/15	8/15	2	6	3	SR26(R)
32	12	20	10	6	2	1	13/15	2/3	5	5	1	SR69(2-R)
33	12	24	6	3	2	1	7/9	4/9	2	6	3	SR26(2-R)
34	12	27	9	4	3	1	5/6	1/2	3	6	2	SR42(3-R)
35	12	36	9	3	1	1	19/27	16/27	2	6	3	SR27(R)
36	12	36	12	4	3	1	13/16	9/16	3	6	2	SR43(3-R)
37	12	40	10	3	2	1	11/15	8/15	2	6	3	SR27(2-R)
38	12	42	7	2	1	1	4/7	3/7	1	5	5	SR13(R)
39	14	18	9	7	1	1	55/63	16/21	6	6	1	SR82(R)
40	14	24	12	7	4	1	19/21	4/7	6	6	1	SR82(4-R)
41	14	56	8	2	1	1	9/16	7/16	1	6	6	SR14(R)
42	15	21	7	5	1	1	29/35	24/35	4	8	2	SR56(R)
43	15	30	6	3	1	1	13/18	5/9	2	8	4	SR28(R)
44	15	30	10	5	1	1	41/50	18/25	4	8	2	SR57(R)
45	15	55	11	3	1	1	23/33	20/33	2	8	4	SR29(R)
46	16	18	9	8	1	1	8/9	7/9	7	7	1	SR92(AR)
47	16	20	5	4	1	1	4/5	3/5	3	9	3	SR44(AR)
48	16	36	9	4	1	1	7/9	2/3	3	9	3	SR45(R)
49	16	72	9	2	1	1	5/9	4/9	1	7	7	SR15(R)
50	18	20	10	9	2	1	41/45	32/45	8	8	1	SR100(2-R)
51	18	21	7	6	1	1	6/7	5/7	5	10	2	SR72(AR)
52	18	30	10	6	1	1	17/20	3/4	5	10	2	SR73(R)
53	18	90	10	2	1	1	11/20	9/20	1	8	8	SR16(R)
54	20	20	10	10	2	1	23/25	18/25	9	9	1	SR107(2-R)
55	20	30	6	4	1	1	19/24	5/8	3	12	4	SR46(R)

Table 9.6 (continued-2)

No.	v	b	r	k	α	ε_0	ε_1	ε_2	ρ_0	ρ_1	ρ_2	Source
56	20	48	12	5	4	1	13/15	8/15	4	12	3	SR59(4-R)
57	20	55	11	4	1	1	17/22	15/22	3	12	4	SR47(R)
58	20	110	11	2	1	1	6/11	5/11	1	9	9	SR17(R)
59	21	30	10	7	1	1	61/70	27/35	6	12	2	SR85(R)
60	21	56	8	3	1	1	17/24	7/12	2	12	6	SR31(R)
61	24	30	10	8	1	1	71/80	63/80	7	14	2	SR94(R)
62	24	36	9	6	1	1	23/27	20/27	5	15	3	SR74(R)
63	24	72	9	3	1	1	19/27	16/27	2	14	7	SR32(R)
64	25	30	6	5	1	1	5/6	2/3	4	16	4	SR60(AR)
65	25	55	11	5	1	1	9/11	8/11	4	16	4	SR61(R)
66	27	30	10	9	1	1	9/10	4/5	8	16	2	SR102(AR)
67	27	90	10	3	1	1	7/10	3/5	2	16	8	SR33(R)
68	28	36	9	7	1	1	55/63	16/21	6	18	3	SR86(R)
69	28	56	8	4	1	1	25/32	21/32	3	18	6	SR48(R)
70	30	36	12	10	3	1	37/40	83/120	9	18	2	SR109(3-R)
71	30	75	15	6	5	1	8/9	5/9	5	20	4	SR76(5-R)
72	32	36	9	8	1	1	8/9	7/9	7	21	3	SR95(AR)
73	32	72	9	4	1	1	7/9	2/3	3	21	7	SR49(R)
74	35	56	8	5	1	1	33/40	7/10	4	24	6	SR62(R)
75	36	90	10	4	1	1	31/40	27/40	3	24	8	SR50(R)
76	40	72	9	5	1	1	37/45	32/45	4	28	7	SR63(R)
77	42	56	8	6	1	1	41/48	35/48	5	30	6	SR77(R)
78	45	90	10	5	1	1	41/50	18/25	4	32	8	SR64(R)
79	48	72	9	6	1	1	23/27	20/27	5	35	7	SR78(R)
80	49	56	8	7	1	1	7/8	3/4	6	36	6	SR87(AR)
81	54	90	10	6	1	1	17/20	3/4	5	40	8	SR79(R)
82	56	72	9	7	1	1	55/63	16/21	6	42	7	SR88(R)
83	63	90	10	7	1	1	61/70	27/35	6	48	8	SR89(R)
84	64	72	9	8	1	1	8/9	7/9	7	49	7	SR97(AR)
85	72	90	10	8	1	1	71/80	63/80	7	56	8	SR98(R)
86	81	90	10	9	1	1	9/10	4/5	8	64	8	SR105(AR)

Table 9.7
Existing α-resolvable $(0; \rho_1, \rho_2, \rho_3; 0)$-EB designs
from regular GD designs with $k = m$

No.	v	b	r	k	α	ε_1	ε_2	ε_3	ρ_1	ρ_2	ρ_3	Source
1	4	10	5	2	1	4/5	3/5		1	2	0	R1(R)
2	4	12	6	2	1	5/6	2/3	1/2	1	1	1	R2(R)
3	4	12	6	2	1	5/6	2/3	1/2	1	1	1	R3(R)
4	4	14	7	2	1	6/7	5/7	3/7	1	1	1	R4(R)
5	4	16	8	2	1	7/8	3/4	3/8	1	1	1	R5(R)
6	4	16	8	2	1	3/4	5/8		1	2	0	R6(R)
7	4	16	8	2	1	7/8	5/8	1/2	1	1	1	R7(R)
8	4	18	9	2	1	8/9	7/9	1/3	1	1	1	R8(R)
9	4	18	9	2	1	7/9	2/3	5/9	1	1	1	R9(R)
10	4	18	9	2	1	7/9	2/3	5/9	1	1	1	R10(R)
11	4	20	10	2	1	9/10	4/5	3/10	1	1	1	R11(R)
12	4	20	10	2	1	4/5	7/10	1/2	1	1	1	R12(R)
13	4	20	10	2	1	9/10	3/5	1/2	1	1	1	R13(R)
14	4	22	11	2	1	10/11	9/11	3/11	1	1	1	R14(R)
15	4	22	11	2	1	9/11	8/11	5/11	1	1	1	R15(R)
16	4	22	11	2	1	8/11	7/11		1	2	0	R16(R)
17	4	22	11	2	1	9/11	7/11	6/11	1	1	1	R17(R)
18	6	14	7	3	1	19/21	16/21	5/7	2	1	2	R44(R)
19	6	18	9	3	3	25/27	7/9	16/27	2	2	1	R44(3-R)
20	6	20	10	3	2	13/15	4/5	2/3	2	2	1	R48(2-R)
21	6	24	8	2	1	11/16	9/16	1/2	2	2	1	R20(R)
22	6	24	12	3	3	11/12	3/4	2/3	2	2	1	R50(3-R)
23	6	27	9	2	1	7/9	11/18	1/2	1	2	2	R24(R)
24	6	30	10	2	1	7/10	3/5	2/5	2	2	1	R25(R)
25	8	20	10	4	2	19/20	4/5	3/4	3	3	1	R99(2-R)
26	8	22	11	4	1	10/11	9/11		3	4	0	R103(R)
27	8	24	12	4	2	11/12	5/6	3/4	3	3	1	R103(2-R)
28	8	24	12	4	3	11/12	7/8		3	4	0	R101(3-R)
29	8	30	15	4	5	14/15	13/15	3/5	3	3	1	R103(5-R)
30	8	44	11	2	1	7/11	6/11	5/11	3	3	1	R32(R)
31	9	18	6	3	1	5/6	2/3		4	4	0	R59(R)
32	9	21	7	3	1	6/7	5/7	4/7	4	2	2	R60(R)
33	9	24	8	3	1	7/8	3/4	1/2	4	2	2	R61(R)
34	9	24	8	3	1	7/8	3/4	5/8	2	4	2	R62(R)
35	9	27	9	3	1	8/9	7/9	4/9	4	2	2	R63(R)
36	9	36	12	3	4	11/12	7/12		4	4	0	R63(4-R)
37	9	30	10	3	1	9/10	4/5	2/5	4	2	2	R64(R)
38	9	30	10	3	1	4/5	7/10		4	4	0	R65(R)

Table 9.7 (continued)

No.	v	b	r	k	α	ε_1	ε_2	ε_3	ρ_1	ρ_2	ρ_3	Source
39	9	33	11	3	1	10/11	9/11	4/11	4	2	2	R66(R)
40	9	33	11	3	1	9/11	8/11	7/11	4	2	2	R67(R)
41	9	33	11	3	1	10/11	8/11	7/11	2	4	2	R68(R)
42	9	36	12	3	3	5/6	3/4	7/12	4	2	2	R65(3-R)
43	9	45	15	3	5	14/15	3/5	8/15	4	2	2	R66(5-R)
44	10	30	15	5	5	73/75	13/15	48/75	4	4	1	R142(5-R)
45	12	32	8	3	1	19/24	2/3	5/8	6	3	2	R72(R)
46	15	45	15	5	5	73/75	11/15	16/25	8	4	2	R149(5-R)
47	15	45	15	5	5	14/15	67/75	14/25	4	8	2	R150(5-R)
48	15	60	12	3	3	29/36	2/3	5/9	8	2	4	R83(3-R)
49	16	28	7	4	1	6/7	5/7		9	6	0	R118(R)
50	16	32	8	4	1	7/8	3/4	5/8	9	3	3	R119(R)
51	16	36	9	4	1	8/9	7/9	5/9	9	3	3	R120(R)
52	16	40	10	4	1	9/10	4/5	1/2	9	3	3	R121(R)
53	16	40	10	4	1	9/10	4/5	7/10	3	9	3	R122(R)
54	16	44	11	4	1	10/11	9/11	5/11	9	3	3	R123(R)
55	20	60	12	4	3	7/8	2/3	5/8	12	3	4	R124(3-R)
56	20	60	15	5	5	73/75	16/25	3/5	12	3	4	R152(5-R)
57	25	40	8	5	1	7/8	3/4		16	8	0	R155(R)
58	25	45	9	5	1	8/9	7/9	2/3	16	4	4	R156(R)
59	25	50	10	5	1	9/10	4/5	3/5	16	4	4	R157(R)
60	25	55	11	5	1	10/11	9/11	6/11	16	4	4	R158(R)
61	35	105	15	5	5	67/75	4/5	14/25	24	4	6	R159(5-R)
62	49	70	10	7	1	9/10	4/5		36	12	0	R184(R)
63	49	77	11	7	1	10/11	9/11	8/11	36	6	6	R185(R)
64	64	88	11	8	1	10/11	9/11		49	14	0	R192(R)

10
Special Designs

This chapter briefly presents some types of block designs that are different from the designs discussed before, but are useful in practice as well as in the general theory of block designs. The designs discussed here are variance-balanced designs (Section 10.1), resistant/robust designs (Section 10.2), nonbinary designs (Section 10.3) and disconnected designs (Section 10.4). They are not described in a uniform way. Nevertheless, several concepts and methods introduced and discussed earlier, particularly in Volume I, are applied in various sections of this chapter. A different concept of balance is explored in Section 10.1.

10.1 Variance-balanced designs

There is an additional concept of the characterization of block designs, different from that based on efficiency factors and their multiplicities obtainable by considering eigenvalues and corresponding eigenvectors of the matrix $\boldsymbol{C}_1$ with respect to the diagonal matrix $\boldsymbol{r}^{\delta}$, as discussed in the previous chapters (Chapters 2 and 4 in particular).

The concept of variance balance, or total balance of Type T_0 (according to Pearce, 1976), has been considered in Sections 2.4.1 and 4.1 (see Definitions 2.4.3 and 4.1.1). It depends on the ordinary eigenvalues and eigenvectors of $\boldsymbol{C}_1$ (see Theorem 2.4.1). Of course when the design is equireplicate, both concepts coincide. As in the previous chapters, for equireplicate block designs a systematic explanation of variances and efficiencies for the intra-block BLUEs of some contrasts can be given in terms of these eigenvalues and eigenvectors. However, for nonequireplicate block designs such an attempt may fail. Therefore, for nonequireplicate designs a distinction is usually made between variance balance (VB) and efficiency balance (EB), as explained in Section 4.1. Recall again that only in equireplicate block designs the two concepts of balance in terms of efficiencies and variancies coincide (e.g., see Puri and Nigam, 1975a, 1977b). Both of these concepts are useful from different points of view. (See again the discussion in Section 4.1.)

Consider a block design with parameters v, b, $\boldsymbol{r}$, $\boldsymbol{k}$, and with the incidence matrix $\boldsymbol{N}$. Suppose that it is disconnected of degree $g-1$ (see Definition 2.2.6). Then its C-matrix, $\boldsymbol{C}_1 = \boldsymbol{\Delta}\boldsymbol{\phi}_1\boldsymbol{\Delta}' = \boldsymbol{r}^\delta - \boldsymbol{N}\boldsymbol{k}^{-\delta}\boldsymbol{N}'$, of rank $h = v-g$, can be represented as

$$\boldsymbol{C}_1 = \sum_{i=1}^{v-g} \theta_i \boldsymbol{c}_i \boldsymbol{c}_i',$$

where the vectors $\{\boldsymbol{c}_i\}$ are the orthonormal eigenvectors of $\boldsymbol{C}_1$ corresponding to its nonzero eigenvalues $\{\theta_i\}$, respectively. This decomposition also implies that $\boldsymbol{c}_i'\boldsymbol{1}_v = 0$ for all $i \le v-g$, which means that the vectors $\{\boldsymbol{c}_i\}$ represent contrasts, called the "natural contrasts" by Pearce (1983, p.73).

Now, if the design is variance-balanced (VB), i.e., balanced in the sense of Definition 2.4.4 or, equivalently, Definition 4.1.1, then $\theta_1 = \theta_2 = \cdots = \theta_{v-g} = \theta$ (say) and the C-matrix of the design can be written as

$$\boldsymbol{C}_1 = \theta \sum_{i=1}^{v-g} \boldsymbol{c}_i \boldsymbol{c}_i'$$

in general, and as

$$\boldsymbol{C}_1 = \theta\left(\boldsymbol{I}_v - \frac{1}{v}\boldsymbol{1}_v\boldsymbol{1}_v'\right), \tag{10.1.1}$$

if the design is connected, i.e., when $g = 1$ (see Sections 2.4.1 and 4.1). Furthermore, as noted in Section 4.1, for any contrast $\boldsymbol{c}'\boldsymbol{\tau}$ such that $\boldsymbol{c} = \theta\sum_{i=1}^{v-g} \ell_i \boldsymbol{c}_i$, where $\ell_i = \boldsymbol{c}_i'\boldsymbol{s}$ for some $\boldsymbol{s}$, the intra-block BLUE is of the form

$$(\widehat{\boldsymbol{c}'\boldsymbol{\tau}})_{\text{intra}} = \boldsymbol{s}'\boldsymbol{\Delta}\boldsymbol{\phi}_1\boldsymbol{y} = \sum_{i=1}^{v-g} \ell_i \boldsymbol{c}_i'\boldsymbol{\Delta}\boldsymbol{\phi}_1\boldsymbol{y} = \theta^{-1}\boldsymbol{c}'\boldsymbol{\Delta}\boldsymbol{\phi}_1\boldsymbol{y},$$

with

$$\operatorname{Var}[(\widehat{\boldsymbol{c}'\boldsymbol{\tau}})_{\text{intra}}] = \sigma_1^2\theta\sum_{i=1}^{v-g}\ell_i^2 = \sigma_1^2\theta^{-1}\boldsymbol{c}'\boldsymbol{c},$$

the relation between the vectors $\boldsymbol{c}$ and $\boldsymbol{s}$ being given by $\boldsymbol{c} = \boldsymbol{C}_1\boldsymbol{s}$. In particular, if the contrast $\boldsymbol{c}'\boldsymbol{\tau}$ is normalized, i.e., $\boldsymbol{c}'\boldsymbol{c} = 1$, then

$$\operatorname{Var}[(\widehat{\boldsymbol{c}'\boldsymbol{\tau}})_{\text{intra}}] = \sigma_1^2/\theta.$$

Note that in case of a connected VB design, i.e., when $g = 1$, the representation $\boldsymbol{c} = \theta\sum_{i=1}^{v-1}\ell_i\boldsymbol{c}_i$ holds for any contrast $\boldsymbol{c}'\boldsymbol{\tau}$.

When the VB design is equireplicate, its efficiency factor (in the sense of Remark 3.3.3) is given by $\varepsilon = \theta/r$. For nonequireplicate VB designs, there will be several efficiency factors, ε_β, and $\boldsymbol{L}_\beta$ matrices. This is why VB designs are described in this chapter as some of special designs.

The expression (10.1.1) gives a useful characterization of a VB design in terms of its C-matrix. According to some earlier definitions, a connected block design is called a VB design if every elementary contrast is estimated with the same variance (Rao, 1958; Raghavarao, 1971, Definition 4.3.1). Based on (10.1.1), a characterization has been adopted (following Rao, 1958) that a connected block design is a VB design if and only if the nonzero eigenvalues of $\boldsymbol{C}_1$ are all equal (Theorem 2.4.1).

Methods of constructing VB designs will now be described. Only connected designs will be treated here, i.e., those with $h = v - 1$. As mentioned before, a connected VB design with parameters v, b, $\boldsymbol{r} = [r_1, r_2, ..., r_v]'$, $\boldsymbol{k} = [k_1, k_2, ..., k_b]'$ and the constant θ is given by the $v \times b$ incidence matrix $\boldsymbol{N}$ satisfying the condition

$$(\boldsymbol{C}_1 =) \ \mathrm{diag}[r_1, r_2, ..., r_v] - \boldsymbol{N}\mathrm{diag}[k_1^{-1}, k_2^{-1}, ..., k_b^{-1}]\boldsymbol{N}' = \theta\left(\boldsymbol{I}_v - \frac{1}{v}\boldsymbol{1}_v\boldsymbol{1}_v'\right).$$

It may be recalled (from Remark 2.4.1) that for a connected design $\theta = [n - \mathrm{tr}(\boldsymbol{N}\boldsymbol{k}^{-\delta}\boldsymbol{N}')]/(v-1) \leq n/v$, where $n = \sum_{i=1}^{v} r_i$, which reduces to $\theta = (n - b)/(v-1)$ if the design is, in addition, binary. Also it may be recalled that θ plays the role of the so-called "effective replication" (Pearce, 1983, pp.74-75). As $\boldsymbol{C}_1\boldsymbol{1}_v = \boldsymbol{0}$, it is obvious that a block design with the incidence matrix $\boldsymbol{N}$ is a VB design if and only if the off-diagonal elements of $\boldsymbol{N}\boldsymbol{k}^{-\delta}\boldsymbol{N}'$ are all equal (in this case equal to θ/v). This fact coincides with Definition 2.4.3. It is now only to check the constancy of the off-diagonal elements of such matrix, when presenting constructional methods. Furthermore, it follows that a binary proper VB design is a BIB design (see Section 2.4.1; also Theorem 8.2.3). Note that a BIB design with parameters v, b, r, k, λ is VB with $\theta = b(k-1)/(v-1) = \lambda b/r = \lambda v/k$. Thus, it remains to be interested in constructing VB designs which are nonproper and nonequireplicate. The construction of such designs may be necessary when a BIB design with the corresponding v and b is not available, or when the blocks are naturally not of equal size (see Example 3.1.2). There are several methods of constructing such VB designs, for example, see Agrawal (1963), John (1964), Murty and Das (1967), Das and Rao (1968), Dey (1970), Kulshreshtha, Dey and Saha (1972), Hedayat and Federer (1974), Kageyama (1976d, 1988a, 1988b), Khatri (1982), Gupta and Jones (1983), Agarwal and Kumar (1984, 1986a), Mukerjee and Kageyama (1985), Jones, Sinha and Kageyama (1987), Pal and Pal (1988), Sinha (1988, 1989, 1990), Sinha and Jones (1988), Calvin and Sinha (1989), Hedayat and Stufken (1989), Gupta and Kageyama (1992), Ghosh, Joshi and Kageyama (1993), and Sinha, Jones and Kageyama (1996, 1997). Most of these methods utilize BIB designs and GD designs in their constructions. A fundamental method is to juxtapose known designs or to pattern structures as shown below.

Theorem 10.1.1. *For positive integers s and t such that $s = (t+1)/2$, the* *matrix* $\boldsymbol{N}$ *of the form*

$$\begin{bmatrix} \mathbf{0}' & \mathbf{1}_t' & 1 \\ \mathbf{1}_s' & \mathbf{0}' & 1 \\ \mathbf{1}_t\mathbf{1}_s' & \boldsymbol{I}_t & \mathbf{0} \end{bmatrix}$$

is the incidence matrix of a VB design with parameters $v = t+2$, $b = s+t+1$, $\boldsymbol{r} = [t+1, (s+1)\mathbf{1}_{t+1}']'$, $\boldsymbol{k} = [(t+1)\mathbf{1}_s', 2\mathbf{1}_{t+1}']'$, $\theta = (2s+1)/2$. *This design is also* $(0; \rho_1, \rho_2; 0)$*-EB with* $\rho_1 = t$, $\rho_2 = 1$, $\varepsilon_1 = (2s+1)/[2(s+1)]$, $\varepsilon_2 = (s+2)/[2(s+1)]$, $\boldsymbol{L}_1 = \text{diag}[0 : \boldsymbol{I}_{t+1} - (t+1)^{-1}\mathbf{1}_{t+1}\mathbf{1}_{t+1}']$ *and*

$$\boldsymbol{L}_2 = \frac{1}{(s+2)(t+1)} \begin{bmatrix} (s+1)(t+1) & -(s+1)\mathbf{1}_{t+1}' \\ -(t+1)\mathbf{1}_{t+1} & \mathbf{1}_{t+1}\mathbf{1}_{t+1}' \end{bmatrix}.$$

Proof. Under the assumption $s = (t+1)/2$, the direct calculation can easily show that the design is VB with the required parameters. To prove that the design is a $(0; t, 1; 0)$-EB design, refer to Section 4.4.1 and note that the matrix $\boldsymbol{M} = \boldsymbol{r}^{-\delta}\boldsymbol{N}\boldsymbol{k}^{-\delta}\boldsymbol{N}'$ is here of the form

$$\boldsymbol{M} = \frac{1}{2(s+1)(t+1)} \begin{bmatrix} (s+1)(t+1) & (s+1)\mathbf{1}_{t+1}' \\ (t+1)\mathbf{1}_{t+1} & (t+1)\boldsymbol{I}_{t+1} + 2s\mathbf{1}_{t+1}\mathbf{1}_{t+1}' \end{bmatrix}.$$

It can also be shown that $\boldsymbol{M}$ has eigenvaules $1/[2(s+1)]$, $s/[2(s+1)]$ and 1 with respective multiplicities t, 1 and 1. The matrix $\boldsymbol{L}_1$ is easily given as presented above and then $\boldsymbol{L}_2 = \boldsymbol{I}_v - [(t+1)(s+2)]^{-1}\mathbf{1}_v\boldsymbol{r}' - \boldsymbol{L}_1$. □

The design in Theorem 10.1.1 might be interesting for small t, because then small differences among block sizes appear. Note that in Theorem 10.1.1 the condition $s = (t+1)/2$ is used only to show the variance-balance of the design, and then the other properties are not conditioned by it.

Example 10.1.1. When $t = 3$, one gets $s = 2$ in Theorem 10.1.1. Then the incidence matrix

$$\begin{bmatrix} 0 & 0 & 1 & 1 & 1 & 1 \\ 1 & 1 & 0 & 0 & 0 & 1 \\ 1 & 1 & 1 & 0 & 0 & 0 \\ 1 & 1 & 0 & 1 & 0 & 0 \\ 1 & 1 & 0 & 0 & 1 & 0 \end{bmatrix}, \qquad \begin{array}{l} v = 5,\ b = 6, \\ \boldsymbol{r} = [4, 3\mathbf{1}_4']', \\ \boldsymbol{k} = [4\mathbf{1}_2', 2\mathbf{1}_4']', \end{array}$$

produces a VB design with the above parameters. By Theorem 10.1.1, this is also a (0; 3, 1; 0)-EB design with $\varepsilon_1 = 5/6$ and $\varepsilon_2 = 2/3$. Note that a BIB design with $v = 5$ and $b = 6$ does not exist.

Theorem 10.1.2. *Let* $\boldsymbol{N}_h, h = 1, 2, ..., a$, *be the* $v \times b_h$ *incidence matrices of VB designs with parameters* v, b_h, $\boldsymbol{r}_h$, $\boldsymbol{k}_h$, θ_h. *Then the juxtaposition* $\boldsymbol{N}^* =$

$[N_1 : N_2 : \cdots : N_a]$ is the incidence matrix of a VB design with parameters $v^* = v$, $b^* = \sum_{h=1}^{a} b_h$, $r^* = \sum_{h=1}^{a} r_h$, $k^* = [k_1', k_2', ..., k_a']'$, $\theta^* = \sum_{h=1}^{a} \theta_h$.

Proof. The fact that the C-matrix of N^* equals the sum of C-matrices of the component designs with the incidence matrices N_h (see Section 6.4) gives the proof. □

Remark 10.1.1. For the design given in Theorem 10.1.2, it follows that the matrix M^* is given by

$$M^* = (r^*)^{-\delta} N^* (k^*)^{-\delta} (N^*)' = I_v - \theta^* (r^*)^{-\delta} \left(I_v - \frac{1}{v} 1_v 1_v' \right).$$

Once one has any practical information on the component designs, eigenvalues and eigenvectors of M^* can be evaluated.

Example 10.1.2. Consider an unreduced BIB design with parameters $v = 5$, $b = 10$, $r = 6$, $k = 3$, $\lambda = 3$, $\theta = \lambda v/k = 5$, and a VB design with parameters $v = 5$, $b = 6$, $r = [4, 31_4']'$, $k = [21_4', 41_2']'$, $\theta = 5/2$, having blocks (1,2), (1,3), (1,4), (1,5), (2,3,4,5), (2,3,4,5). Then Theorem 10.1.2 produces a VB design with parameters $v^* = 5$, $b^* = 16$, $r^* = [10, 91_4']'$, $k^* = [31_{10}', 21_4', 41_2']'$, $\theta^* = 15/2$, and with the incidence matrix

$$\begin{bmatrix} 1&1&1&1&1&1&0&0&0&0&1&1&1&1&0&0 \\ 1&1&1&0&0&0&1&1&1&0&1&0&0&0&1&1 \\ 1&0&0&1&1&0&1&1&0&1&0&1&0&0&1&1 \\ 0&1&0&1&0&1&1&0&1&1&0&0&1&0&1&1 \\ 0&0&1&0&1&1&0&1&1&1&0&0&0&1&1&1 \end{bmatrix}.$$

As a natural result, it can be observed that the variances $\sigma_1^2/5$ and $2\sigma_1^2/5$ related to the component designs, respectively, are reduced to $2\sigma_1^2/15$ after the juxtaposition. On the other hand, this design has

$$M^* = \frac{1}{60} \begin{bmatrix} 24&9&9&9&9 \\ 10&20&10&10&10 \\ 10&10&20&10&10 \\ 10&10&10&20&10 \\ 10&10&10&10&20 \end{bmatrix}$$

which has eigenvalues 1/6, 7/30 and 1 with respective multiplicities 3, 1 and 1. It can also be shown that $L_1^* = \mathrm{diag}[0 : I_4 - 4^{-1} 1_4 1_4']$ and

$$L_2^* = \frac{1}{92} \begin{bmatrix} 72 & -181_4' \\ -201_4 & 51_4 1_4' \end{bmatrix}.$$

Thus this design is (0; 3, 1; 0)-EB with $\varepsilon_1 = 5/6$ and $\varepsilon_2 = 23/30$ having the matrices L_β^*, $\beta = 1, 2$, as given above.

Theorem 10.1.3. *Let N_1 and N_2 be, respectively, the incidence matrices of a BIB design with parameters v, b, r, k, λ, and a VB design with parameters v,*

b', $\boldsymbol{r}$, $\boldsymbol{k}$, $n = \boldsymbol{r}'\mathbf{1}_v$ for positive integers x and y such that $y/x = v(v-1)(r-\lambda)/[(n-b')(k+1)]$. Then

$$N^* = \begin{bmatrix} \mathbf{1}'_{xb} & \mathbf{0}' \\ \mathbf{1}'_x \otimes N_1 & \mathbf{1}'_y \otimes N_2 \end{bmatrix}$$

is the incidence matrix of a VB design with parameters $v^* = v+1$, $b^* = xb+yb'$, $\boldsymbol{r}^* = [xb, xr\mathbf{1}'_v + y\boldsymbol{r}']'$, $\boldsymbol{k}^* = [(k+1)\mathbf{1}'_{xb}, \mathbf{1}'_y \otimes \boldsymbol{k}']'$, $\theta^* = xr + y(n-b')/v$.

When the above N_2 is the incidence matrix of a BIB design, the condition reduces to $y/x = v(v-1)(r-\lambda)/[b'(k'-1)(k+1)]$. In this case, note that the condition of Corollary 3.2.2 is satisfied automatically, on account of Corollary 3.2.4. Because a large number of BIB designs are presented in Chapter 8, one can construct VB designs for appropriate choices of integers x and y. In particular, when in Theorem 10.1.3 $x = y = 1$, the following results can be obtained.

Corollary 10.1.1. *Let N_1 and N_2 be the incidence matrices of a BIB design with parameters v, b, r, k, λ, and a VB design with parameters v, b', $\boldsymbol{r}$, $\boldsymbol{k}$, $n = \boldsymbol{r}'\mathbf{1}_v$ satisfying the equality $n - b' = v(v-1)(r-\lambda)/(k+1)$, respectively. Then the incidence matrix*

$$N^* = \begin{bmatrix} \mathbf{1}'_b & \mathbf{0}' \\ N_1 & N_2 \end{bmatrix}$$

yields a VB design with parameters $v^* = v+1$, $b^* = b+b'$, $\boldsymbol{r}^* = [b, r\mathbf{1}'_v + \boldsymbol{r}']'$, $\boldsymbol{k}^* = [(k+1)\mathbf{1}'_b, \boldsymbol{k}']'$, $\theta^* = r + (n-b')/v$.

Remark 10.1.2. In Theorem 10.1.3 and Corollary 10.1.1 when the second design with N_2 is a BIB design, one can evaluate the resulting design as some $(0; \rho_1, \rho_2, ..., \rho_{m-1}; 0)$-EB design in terms of the parameters of the component designs, similarly to Theorem 6.6.13 and Corollary 6.6.2. Otherwise, the replication vector $\boldsymbol{r}$ may make some difficulty of evaluating the matrix M in general. This problem also occurs in Corollary 10.1.3 later.

Corollary 10.1.2. *Let N_1 and N_2 be the incidence matrices of two BIB designs, the first with parameters v, b, r, k, λ, and the second with parameters v, b', r', k', λ', satisfying the equality $b'(k'-1) = v(v-1)(r-\lambda)/(k+1)$. Then the incidence matrix*

$$N^* = \begin{bmatrix} \mathbf{1}'_b & \mathbf{0}' \\ N_1 & N_2 \end{bmatrix}$$

yields a VB design with parameters $v^* = v+1$, $b^* = b+b'$, $\boldsymbol{r}^* = [b, (r+r')\mathbf{1}'_v]'$, $\boldsymbol{k}^* = [(k+1)\mathbf{1}'_b, k'\mathbf{1}'_{b'}]'$, $\theta^* = r + b'(k'-1)/v$. This is also a $(0; \rho_1^*, \rho_2^*; 0)$-EB design with $\rho_1^* = v-1$, $\rho_2^* = 1$, $\varepsilon_1^* = 1-(r+r')^{-1}[(r-\lambda)/(k+1)+(r'-\lambda')/k']$, $\varepsilon_2^* = 1 - r'/[(k+1)(r+r')]$, $L_1^* = \mathrm{diag}[0 : I_v - v^{-1}\mathbf{1}_v\mathbf{1}'_v]$ and

$$L_2^* = \frac{1}{v[b+v(r+r')]} \begin{bmatrix} v^2(r+r') & -v(r+r')\mathbf{1}'_v \\ -vb\mathbf{1}_v & b\mathbf{1}_v\mathbf{1}'_v \end{bmatrix}.$$

Proof. To prove that the design is a $(0; v-1, 1; 0)$-EB design, refer to Section 4.4.1 and note that the matrix $\boldsymbol{M}^* = (\boldsymbol{r}^*)^{-\delta}\boldsymbol{N}^*(\boldsymbol{k}^*)^{-\delta}(\boldsymbol{N}^*)'$ is here of the form

$$\boldsymbol{M}^* = \begin{bmatrix} \frac{1}{k+1} & \frac{r}{b(k+1)}\boldsymbol{1}_v' \\ \frac{r}{(r+r')(k+1)}\boldsymbol{1}_v & \frac{1}{r+r'}\left[\left(\frac{r-\lambda}{k+1}+\frac{r'-\lambda'}{k'}\right)\boldsymbol{I}_v + \left(\frac{\lambda}{k+1}+\frac{\lambda'}{k'}\right)\boldsymbol{1}_v\boldsymbol{1}_v'\right] \end{bmatrix}.$$

It can also be shown that $\boldsymbol{M}^*$ has eigenvaules $(r+r')^{-1}[(r-\lambda)/(k+1)+(r'-\lambda')/k']$, $r'/[(k+1)(r+r')]$ and 1 with respective multiplicities $v-1$, 1 and 1. The matrix $\boldsymbol{L}_1^*$ is easily given as presented above and then $\boldsymbol{L}_2^* = \boldsymbol{I}_{v+1} - [b + v(r+r')]^{-1}\boldsymbol{1}_{v+1}(\boldsymbol{r}^*)' - \boldsymbol{L}_1^*$. □

On account of Corollary 3.2.4, for a design resulting from Corollary 10.1.2 the condition of Corollary 3.2.2 is satisfied automatically.

In particular, Corollary 10.1.1 with $\boldsymbol{N}_1 = \boldsymbol{I}_v$ can yield the following result.

Corollary 10.1.3. *Let* $\boldsymbol{N}$ *be the* $v \times b$ *incidence matrix of a VB design with parameters* v, b, $\boldsymbol{r}$, $\boldsymbol{k}$, $n = \boldsymbol{r}'\boldsymbol{1}_v$ *satisfying the equality* $v(v-1) = 2(n-b)$. *Then the incidence matrix*

$$\boldsymbol{N}^* = \begin{bmatrix} \boldsymbol{1}_v' & \boldsymbol{0}' \\ \boldsymbol{I}_v & \boldsymbol{N} \end{bmatrix} \tag{10.1.2}$$

yields a VB design with parameters $v^* = v+1$, $b^* = v+b$, $\boldsymbol{r}^* = [v, \boldsymbol{1}_v' + \boldsymbol{r}']'$, $\boldsymbol{k}^* = [2\boldsymbol{1}_v', \boldsymbol{k}']'$, $\theta^* = (v+1)/2$.

Example 10.1.3. The VB design with parameters $v = 5$, $b = 6$, $\boldsymbol{r} = [4, 3\boldsymbol{1}_4']'$, $\boldsymbol{k} = [2\boldsymbol{1}_4', 4\boldsymbol{1}_2']'$, $\theta = 5/2$, having blocks (1, 2), (1, 3), (1, 4), (1, 5), (2, 3, 4, 5), (2, 3, 4, 5), used in Example 10.1.2, satisfies the condition $v(v-1) = 2(n-b)$. Then Corollary 10.1.3 produces a VB design with parameters $v^* = 6$, $b^* = 11$, $\boldsymbol{r}^* = [5\boldsymbol{1}_2', 4\boldsymbol{1}_4']'$, $\boldsymbol{k}^* = [2\boldsymbol{1}_9', 4\boldsymbol{1}_2']'$, $\theta^* = 3$, and with the incidence matrix

$$\boldsymbol{N}^* = \begin{bmatrix} 1&1&1&1&1&0&0&0&0&0&0 \\ 1&0&0&0&0&1&1&1&1&0&0 \\ 0&1&0&0&0&1&0&0&0&1&1 \\ 0&0&1&0&0&0&1&0&0&1&1 \\ 0&0&0&1&0&0&0&1&0&1&1 \\ 0&0&0&0&1&0&0&0&1&1&1 \end{bmatrix}.$$

Note that among BIB designs, there exists a BIB design with parameters $v = 6$, $b = 15$, $r = 5$, $k = 2$, $\lambda = 1$ in which $\theta = 3$. It has the same values of v and θ as those resulting here, but requires more experimental units. On the other hand, the present example has

$$\boldsymbol{M}^* = \frac{1}{40}\begin{bmatrix} 16\boldsymbol{I}_2 + 4\boldsymbol{1}_2\boldsymbol{1}_2' & 4\boldsymbol{1}_2\boldsymbol{1}_4' \\ 5\boldsymbol{1}_4\boldsymbol{1}_2' & 10\boldsymbol{I}_4 + 5\boldsymbol{1}_4\boldsymbol{1}_4' \end{bmatrix}$$

which has eigenvalues 1/4, 7/20, 2/5 and 1 with respective multiplicities 3, 1, 1 and 1. Thus this design is (0; 3, 1, 1; 0)-EB with $\varepsilon_1 = 3/4$, $\varepsilon_2 = 13/20$ and $\varepsilon_3 = 3/5$.

When for $\boldsymbol{N}$ in Corollary 10.1.3 a BIB design is considered, the following result can be obtained.

Corollary 10.1.4. *Let $\boldsymbol{N}$ be the $v \times b$ incidence matrix of a BIB design with parameters v, b, r, k, λ satisfying $k = 2\lambda$. Then (10.1.2) is the incidence matrix of a VB design with parameters $v^* = v + 1$, $b^* = v + b$, $\boldsymbol{r}^* = [v, (r+1)\mathbf{1}_v']'$, $\boldsymbol{k}^* = [2\mathbf{1}_v', k\mathbf{1}_b']'$, $\theta^* = (v+1)/2$. This design is also $(0; \rho_1^*, \rho_2^*; 0)$-EB with $\rho_1^* = v - 1$, $\rho_2^* = 1$, $\varepsilon_1^* = 1 - (r+1)^{-1}[1/(k+1) + (r-\lambda)/k]$, $\varepsilon_2^* = 1 - r/[2(r+1)]$, $\boldsymbol{L}_1^* = \mathrm{diag}[0 : \boldsymbol{I}_v - v^{-1}\mathbf{1}_v\mathbf{1}_v']$ and*

$$\boldsymbol{L}_2^* = \frac{1}{v^2(r+2)} \begin{bmatrix} v^2(r+1) & -v(r+1)\mathbf{1}_v' \\ -v^2\mathbf{1}_v & v\mathbf{1}_v\mathbf{1}_v' \end{bmatrix}.$$

Proof. This can be shown by, in Corollary 10.1.2, taking $b = v$, $r = k = 1$, $\lambda = 0$, and $b' = b$, $r' = r$, $k' = k$, $\lambda' = \lambda$. In this case note that the assumption $k = 2\lambda$ is not needed. □

Remark 10.1.3. In Corollaries 10.1.3 and 10.1.4, the parameters v^*, b^* and θ^* are the same. Then their application has the possibility of a wide coverage for the same experimental purpose, provided that they both satisfy the condition of Corollary 3.2.2, which secures obtaining the BLUE in the inter-block analysis for any contrast not estimated in the intra-block analysis with full efficiency. Note that any VB design obtained by Corollary 10.1.4 satisfies this condition, on account of Corollary 3.2.4.

Example 10.1.4. Consider a BIB design with parameters $v = b = 7$, $r = k = 4$, $\lambda = 2$, given in Table 8.2. This satisfies $k = 2\lambda$. Then Corollary 10.1.4 yields a VB design with parameters $v^* = 8$, $b^* = 14$, $\boldsymbol{r}^* = [7, 5\mathbf{1}_7']'$, $\boldsymbol{k}^* = [2\mathbf{1}_7', 4\mathbf{1}_7']'$, $\theta^* = 4$, and with the incidence matrix

$$\begin{bmatrix} 1&1&1&1&1&1&1&0&0&0&0&0&0&0 \\ 1&0&0&0&0&0&0&0&1&1&1&0&1&0 \\ 0&1&0&0&0&0&0&0&0&1&1&1&0&1 \\ 0&0&1&0&0&0&0&1&0&0&1&1&1&0 \\ 0&0&0&1&0&0&0&0&1&0&0&1&1&1 \\ 0&0&0&0&1&0&0&1&0&1&0&0&1&1 \\ 0&0&0&0&0&1&0&1&1&0&1&0&0&1 \\ 0&0&0&0&0&0&1&1&1&1&0&1&0&0 \end{bmatrix}.$$

Note that all BIB designs with $v = 8$ require more experimental units than the design considered here. In particular, there exists an unreduced BIB design with parameters $v = 8$, $b = 28$, $r = 7$, $k = 2$, $\lambda = 1$ (see Table 8.2, No. 33), for which $\theta = 4$. However, it has $n = 56$ experimental units. On the other hand, the present example is also a (0; 6, 1; 0)-EB design with $\varepsilon_1^* = 4/5$, $\varepsilon_2^* = 3/5$, $\boldsymbol{L}_1^* = \text{diag}[0 : \boldsymbol{I}_7 - 7^{-1}\mathbf{1}_7\mathbf{1}_7']$ and

$$\boldsymbol{L}_2^* = \frac{1}{42}\begin{bmatrix} 35 & -5\mathbf{1}_7' \\ -7\mathbf{1}_7 & \mathbf{1}_7\mathbf{1}_7' \end{bmatrix}.$$

Other construction methods of VB designs are available in the literature (see Kageyama, 1974c, 1976d, 1977b; Tyagi, 1979). Most of available methods of constructing VB designs are not essentially based on the method of differences, which is frequently used to construct BIB and PBIB designs as shown in Chapters 6, 7 and 8. Of course, in a juxtaposition of BIB or PBIB designs the method of differences can be utilized indirectly. As the resulting designs have unequal block sizes and/or unequal replications, a direct application of the method of differences becomes difficult. However, Tyagi and Rizwi (1986) have utilized the method of differences for the construction of nonproper VB designs. These designs are not so general, but they require lesser number of replications than unreduced BIB designs for the same number of treatments. Furthermore, even in such case after some calculation it may be possible to represent a VB design as a $(\rho_0; \rho_1, ..., \rho_{m-1}; 0)$-EB design.

10.2 Resistant/Robust designs

Block designs are widely used in many fields of agricultural and physical sciences. From physical or meteorological reasons it may happen that some treatments are missing. A BIB design besides being universally optimal (Theorem 8.1.1; Shah and Sinha, 1989, Chapter 2) has an interesting feature which appeals to many practitioners (Theorem 8.1.1; Shah and Sinha, 1989, Chapter 2). All the BIB designs are both of $(0; v-1; 0)$-EB and VB (see Theorems 8.2.1 and 8.2.3, and Section 10.1). Unfortunately, this desirable feature of BIB designs can easily be destroyed if due to some unforeseen circumstances some or all of the data related to experimental units assigned to one or more treatments are lost in actual experimentation. While in some cases nothing can be done to prevent such an undesirable outcome, in many cases there are ways which one can apply to preserve the $(0; v-1; 0)$-EB property of the remaining design, provided that the experimenter is careful in selecting the design to begin with. It is therefore of interest to see whether a particular design is insensitive, or robust, against the unavailability of observations.

Now take a BIB design as one of starting designs. Suppose in a BIB design with parameters v, b, r, k, λ, a treatment, the first treatment, say, is missing. The problem is to investigate properties of the remaining structure considered as a block design with $v-1$ treatments. Without loss of generality, the $v \times b$ incidence matrix $\boldsymbol{N}$ of the original BIB design can be written as

$$\boldsymbol{N} = \begin{bmatrix} \mathbf{1}_r' & \mathbf{0}' \\ \boldsymbol{N}_1 & \boldsymbol{N}_2 \end{bmatrix}.$$

Then, let

$$N^* = [N_1 : N_2] \tag{10.2.1}$$

be the $(v-1) \times b$ incidence matrix of the remaining structure. This leads to the following theorem.

Theorem 10.2.1. *A symmetric BIB design with parameters $v = b$, $r = k$, λ reduces, through omitting a treatment, to a $(0; \rho_1^*, \rho_2^*; 0)$-EB design with parameters $v^* = v - 1$, $b^* = v$, $r^* = r$, $k^* = [(k-1)1_k', k1_{v-k}']'$, $\varepsilon_1^* = \lambda v/k^2$, $\varepsilon_2^* = (\lambda v - k)/[\lambda(v-1)]$, $\rho_1^* = v - k - 1$, $\rho_2^* = k - 1$, $L_1^* = (r-\lambda)^{-1} N_2 N_2' - [(r-\lambda)/(v-k)] 1_{v-k} 1_{v-k}'$, $L_2^* = (r-\lambda)^{-1} N_1 N_1' - \{\lambda^2/[r(r-\lambda)]\} 1_{v-1} 1_{v-1}'$.*

Proof. First consider the dual $(N^*)' = [N_1 : N_2]'$, which, on account of Theorem 8.4.6, is a $(\tilde{\rho}_0; \tilde{\rho}_1, \tilde{\rho}_2; 0)$-EB design with parameters $\tilde{v} = v$, $\tilde{b} = v - 1$, $\tilde{r} = [(k-1)1_k', k1_{v-k}']'$, $\tilde{k} = k$, $\tilde{\varepsilon}_0 = 1$, $\tilde{\varepsilon}_1 = \lambda v/k^2$, $\tilde{\varepsilon}_2 = (\lambda v - k)/[\lambda(v-1)]$, $\tilde{\rho}_0 = 1$, $\tilde{\rho}_1 = v - k - 1$, $\tilde{\rho}_2 = k - 1$, $\tilde{L}_0 = I_v - [k(b-1)]^{-1} 1_v \tilde{r}' - \tilde{L}_1 - \tilde{L}_2$, $\tilde{L}_1 = \mathrm{diag}[O : I_{v-k} - (v-k)^{-1} 1_{v-k} 1_{v-k}']$, $\tilde{L}_2 = \mathrm{diag}[I_k - k^{-1} 1_k 1_k' : O]$. Then it follows from Corollary 7.4.24 (note that here the multiplicity of L_0^* is zero) that N^* is the incidence matrix of a $(0; \rho_1^*, \rho_2^*; 0)$-EB design with parameters $v^* = v - 1$, $b^* = v$, $r^* = r$, $k^* = [(k-1)1_k', k1_{v-k}']'$, $\varepsilon_1^* = \lambda v/k^2$, $\varepsilon_2^* = (\lambda v - k)/[\lambda(v-1)]$, $\rho_1^* = v-k-1$, $\rho_2^* = k-1$, $L_1^* = [(1-\tilde{\varepsilon}_2)k]^{-1} (N^*)' \tilde{L}_1 \mathrm{diag}[(k-1)^{-1} I_k : k^{-1} I_{v-k}] N^* = (r-\lambda)^{-1} N_2 [I_{v-k} - (v-k)^{-1} 1_{v-k} 1_{v-k}'] N_2'$ and $L_2^* = [(1-\tilde{\varepsilon}_1)k]^{-1} (N^*)' \tilde{L}_2 \mathrm{diag}[(k-1)^{-1} I_k : k^{-1} I_{v-k}] N^* = (r-\lambda)^{-1} N_1 (I_k - k^{-1} 1_k 1_k') N_1'$, which completes the proof. □

Remark 10.2.1. Because the common efficiency factor of the original BIB design considered in Theorem 10.2.1 is $\lambda v/k^2$, equal to ε_1^*, in the design that results from omitting a treatment any of the $v-k-1$ basic contrasts represented by eigenvectors of L_1^* corresponding to its nonzero eigenvalue are still estimated in the intra-block analysis with the same efficiency $\lambda v/k^2$.

Coming back to the original motivation of this section, suppose that at first a BIB design [i.e., proper equireplicate $(0; v-1; 0)$-EB design] is chosen to estimate all normalized contrasts with the same variance (precision), but by some reason a treatment is missing. In that situation, if the experimenter still wants to have the same estimation property, it is required that the remaining structure is still a $(0; v^* - 1; 0)$-EB design. Thus, the following situation is to be considered as desirable. For $v \geq 3$,

(i) the original design is a BIB design with parameters v, b, r, k, λ [i.e., proper equireplicate $(0; v-1; 0)$-EB design], in which $\varepsilon_1 = \lambda v/(rk)$;

(ii) the remaining structure, through omitting a treatment in the BIB design, is a $(0; v^* - 1; 0)$-EB design with parameters $v^* = v - 1$, $b^* = b$, $r^* = r$, $k^* = [(k-1)1_r', k1_{b-r}']'$, $\varepsilon_1^* = [(v-1)r - b]/[r(v-2)]$.

In this case, since

$$\varepsilon_1 - \varepsilon_1^* = \frac{b-r}{r(v-1)(v-2)},$$

the common efficiency factor reduces slightly, but one can get the same estimation property, in the sense of $\rho_1 - \rho_1^* = 1$.

From now on, some characterization of the BIB designs satisfying the above conditions (i) and (ii) will be attempted, mainly from a constructional point of view. Most of the results below have been taken from Kageyama (1987b). At first, some general terminology is to be introduced.

Let $\mathcal{D}$ be a BIB design with parameters v, b, r, k, λ on a set T of v treatments. Let L be a subset of T consisting of s ($\leq v-2$) treatments. Denote by $\overline{\mathcal{D}}$ the remaining design upon the loss of all experimental units in $\mathcal{D}$ assigned to the treatments in L.

Definition 10.2.1. A $(0; v-1; 0)$-EB design $\mathcal{D}$ is said to be globally resistant of degree s [GR(s)] if $\overline{\mathcal{D}}$ is $(0; \rho_1; 0)$-EB due to the loss of any s treatments from $\mathcal{D}$.

Definition 10.2.2. A $(0; v-1; 0)$-EB design $\mathcal{D}$ is said to be locally resistant of degree s [LR(s)] if $\overline{\mathcal{D}}$ is $(0; \rho_1; 0)$-EB due to the loss of some (but not all) s treatments from $\mathcal{D}$.

Definition 10.2.3. A $(0; v-1; 0)$-EB design $\mathcal{D}$ is said to be susceptible if any loss of treatments from $\mathcal{D}$ produces a $\overline{\mathcal{D}}$ which is not $(0; \rho_1; 0)$-EB.

Thus, the investigation of (globally or locally) resistant designs may be useful in a robustness problem when some or all of the experimental units assigned to one or more treatments (or blocks) are lost. For other applications see Hedayat and John (1974).

Hedayat and John (1974), Most (1975), Shah and Gujarathi (1977), Chandak (1980), Kageyama and Saha (1987), and Kageyama (1987b) have discussed several problems concerning resistant BIB designs. It is known that the property of being resistant depends not only on the parameters of the design, but also depends on the way the design has been constructed. This can be seen below.

Definition 10.2.4. A block design with parameters v, b, r, k is called a t-(v, k, λ_t) design or simply a t-design if every t-subset of the set of v treatments is contained in exactly λ_t blocks of the design.

It is well known (see, e.g., Hedayat and Kageyama, 1980) that for each $0 \leq s \leq t$ every t-(v, k, λ_t) design is an s-(v, k, λ_s) design with $\lambda_s = \lambda_t \binom{v-s}{t-s} / \binom{k-s}{t-s}$. Hence, a t-design with $t \geq 2$ is a BIB design with parameters v, b ($= \lambda_0$, say), r ($= \lambda_1$, say), k, λ ($= \lambda_2$, say), defined in Section 6.0.1, whereas any BIB design is a 2-design (for more see Hedayat and Kageyama, 1980; for using another notation, Raghavarao, 1971, Chapter 7).

Users of block designs have paid very little attention to t-designs with $t \geq 3$, and their statistical analysis, perhaps because the standard data analysis of such designs totally ignores the structure of these designs beyond the fact that they are BIB designs (see Calvin, 1954). However, Raghavarao, Federer and Schwager (1986) and Bhaumik (1995) have shown some statistical properties of 3-designs for estimating contrasts of brand effects and of competing effects in a marketing setting. Furthermore, in the context of resistant block designs, t-designs play a key role as Hedayat and John (1974) have shown.

Theorem 10.2.2. *A BIB design is GR(1) if and only if it is a 3-design.*

Proof. When a BIB design is a 3-design, $\boldsymbol{N}_1$ and $\boldsymbol{N}_2$ in (10.2.1) become BIB designs, and hence $\boldsymbol{N}^* = [\boldsymbol{N}_1 : \boldsymbol{N}_2]$ is the incidence matrix of a $(0; v^* - 1; 0)$-EB design (see Theorem 8.2.17). This implies that the design is GR(1). On the other hand, the necessity is shown as follows. Let $\lambda_{ii'}^{(j)}$ denote the concurrence number of the ith and the i'th rows of $\boldsymbol{N}_j, j = 1, 2$, in (10.2.1). Then since the original design is a BIB design, $\lambda_{ii'}^{(1)} + \lambda_{ii'}^{(2)}$ is a constant for all i, i'. Furthermore, if the original design is GR(1), then it has the $(0; v^* - 1; 0)$-EB property and, hence, $\lambda_{ii'}^{(1)}/(k-1) + \lambda_{ii'}^{(2)}/k$ is a constant (see Section 2.4 and Theorem 8.2.1). These two equations imply that $\lambda_{ii'}^{(1)}$ is a constant. Repeating this for any treatment considered as missing, it can be concluded that the design is a 3-design. □

Thus, all t-designs with $t \geq 3$ are GR designs of degree one and possibly more. In many situations one can safely secure the $(0; \rho_1; 0)$-EB property of the remaining designs based on available information if one could arrange to have a suitable LR block designs and thus avoiding the use of t-designs, $t \geq 3$, which require in general a large number of experimental units. Therefore, it is useful to identify and catalog LR block designs of various degrees. For example, Hedayat and John (1974) have pointed out that any symmetric BIB design is LR(k), k being the block size.

Now consider LR(1) BIB designs, i.e., proper equireplicate $(0; v-1; 0)$-EB designs that are LR(1). At first, a trivial result worth mentioning is presented (see again Section 6.0.1).

Theorem 10.2.3. *Any BIB design with parameters v, $b = \binom{v}{k}$, $r = \binom{v-1}{k-1}$, k, $\lambda = \binom{v-2}{k-2}$ of unreduced type is GR(s) with $s \leq k$.*

Proof. By taking as blocks all possible combinations of k out of v treatments, the required result follows. [Because this design is also a k-design, when $k \geq 3$, by Theorem 10.2.2 it is GR(1).] □

Note that a GR(s) BIB design is also LR(s). From Theorem 10.2.3 and the present interest in LR(1) BIB designs, only BIB designs satisfying $k \geq 3$ will be considered.

Let $\mathcal{D}$ be a BIB design with parameters v, b, r, k (≥ 3), λ on a set T of v treatments and let $L = \{x\} \subset T$. Divide $\mathcal{D}$ into two parts, $\mathcal{D}_1$ and $\mathcal{D}_2^*$, where $\mathcal{D}_1$ consists of all the blocks of $\mathcal{D}$ which do not contain the treatment x, whereas $\mathcal{D}_2^*$ is the set of the remaining blocks which contain x. Next, let $\mathcal{D}_2$ be the design obtained by deleting x from the blocks of $\mathcal{D}_2^*$. Then the following theorem can be obtained.

Theorem 10.2.4. *A BIB design $\mathcal{D}$ is LR(1) if and only if $\mathcal{D}_1$, obtained for a* *treatment x, is a BIB design (equivalently, if and only if $\mathcal{D}_2$ is a BIB design).*

Proof. If $\mathcal{D}$ is LR(1), it follows that $\mathcal{D}_1$ and $\mathcal{D}_2$ are BIB designs with the parameters as below.

$$\mathcal{D}_1 : v_1 = v-1,\ b_1 = b-r,\ r_1 = r-\lambda,\ k_1 = k,\ \lambda_1 = \lambda - \lambda(k-2)/(v-2);$$
$$\mathcal{D}_2 : v_2 = v-1,\ b_2 = r,\ r_2 = \lambda,\ k_2 = k-1,\ \lambda_2 = \lambda(k-2)/(v-2).$$

The converse is obvious. This completes the proof. □

Theorem 10.2.4 immediately yields the following result.

Corollary 10.2.1. *If a BIB design $\mathcal{D}$ is LR*(1), *then its parameters satisfy the conditions* (i) $r \geq v-1$ *(or equivalently* $\lambda \geq k-1$), (ii) $\lambda(k-2)/(v-2)$ *is a nonnegative integer, and* (iii) $b \geq v+r-1$.

Remark 10.2.2. Hedayat and John (1974) gave another necessary condition, $\lambda > 1$. However, by use of the Fisher inequality (Corollary 2.3.1) for the parameters of $\mathcal{D}_2$, it holds that $\lambda \geq k-1$, which is equivalent to (i). Therefore, their condition $\lambda > 1$ is superfluous. Also (ii) holds for a 3-design, because $\lambda_3 = \lambda(k-2)/(v-1)$ is an integer in a 3-design, by Definition 10.2.4.

It is known (Kageyama, 1971) that each of the following conditions is sufficient for the validity of $b \geq v+r-1$ in a BIB design $\mathcal{D}$; (a) $\mathcal{D}$ is resolvable; (b) v/k is an integer. This can be shown as follows. (a) As shown in Section 6.0.3, for any α-resolvable BIB design with $r = \alpha a$, $b \geq v+a-1$, and, hence, $b \geq v+r-1$ for any 1-resolvable BIB design. (b) Let $v = pk$ for a positive integer p. Then $r(k-1) = \lambda(v-1)$ implies $(r-p\lambda)k = r-\lambda > 0$. Hence $r-p\lambda \geq 1$ and then $r-\lambda \geq k$. Furthermore, it follows that $b-(v+r-1) = (v-1)(r-\lambda-k)/k$, which, with $r-\lambda \geq k$, implies $b \geq v+r-1$. In the light of the sufficient condition (a), consider next a BIB design with the property of affine α-resolvability (see Section 6.0.3) to characterize the family of affine α-resolvable BIB designs with regard to the LR(1) property. Then as an example of satisfying the condition (iii) in Corollary 10.2.1, one can get the following theorem.

Theorem 10.2.5. *No affine α-resolvable BIB design with $\alpha \geq 2$ is LR*(1).

Proof. This follows from the identity $b = v+a-1$ (see Theorem 6.0.5) and (iii) in Corollary 10.2.1 with $r > a$ if $\alpha \geq 2$. □

Affine resolvable BIB designs can be characterized as the special case of $\alpha = 1$, which is not covered by Theorem 10.2.5.

Theorem 10.2.6. *No affine resolvable BIB design is LR*(1) *except for a series of parameters*

$$v = 4t, b = 2(4t-1), r = 4t-1, k = 2t, \lambda = 2t-1 \quad for \quad t \geq 1. \tag{10.2.2}$$

Proof. It is known (see Raghavarao, 1971, p. 72) that the parameters of an affine resolvable BIB design can be expressed in terms of only two integral parameters s and u as $v = s^2[(s-1)u+1]$, $b = s(s^2u+s+1)$, $r = s^2u+s+1$, $k = s[(s-1)u+1]$,

$\lambda = su + 1$, $s \geq 2, u \geq 0$. In this case, the condition (i) of Corollary 10.2.1, i.e., $r \geq v - 1$ is equivalent to $0 \geq (s-2)(s^2u + s + 1)$ which implies that only for $s = 2$ there may exist an affine resolvable LR(1) BIB design. In this case, letting $s = 2$ and $u = t - 1$, one obtains $v = 4t, b = 2(4t-1), r = 4t - 1$, $k = 2t, \lambda = 2t - 1$, which for $t \geq 1$ satisfy the remaining conditions of Corollary 10.2.1. This completes the proof. □

Remark 10.2.3. From Hedayat and John (1974) and Chandak (1980), it follows that if a Hadamard matrix (see Section 6.8) of order $4t$ exists, then an affine resolvable BIB design with (10.2.2) exists (see also Theorem 6.8.4). Furthermore, because the design is a 3-design, as can be seen from the structure given by Theorem 6.8.4, Theorem 10.2.2 yields that this design is also GR(1). As mentioned in Section 9.2.1, it is conjectured (Shrikhande, 1976) that an affine resolvable BIB design with (10.2.2) exists for every positive integer t. Sprott (1956) gave a difference set for an affine resolvable BIB design with (10.2.2) when $4t - 1$ is a prime or a prime power (see also Theorem 9.2.1). A complete solution for the conjecture is not yet given. Finally, it can easily be shown that a BIB design with $v = 2k$ and $k \geq 3$ satisfying $\lambda(k-2)/(v-2)$ being an integer [i.e., (ii) of Corollary 10.2.1] is completely characterized by the following two series, for integers $s, s' \geq 1$:

$$v = 2k,\ b = 2s(2k-1),\ r = s(2k-1),\ k \text{ (being even)},\ \lambda = s(k-1);$$
$$v = 2k,\ b = 4s'(2k-1),\ r = 2s'(2k-1),\ k,\ \lambda = 2s'(k-1),$$

some practical examples of which will be seen later.

There are some construction methods of LR (or GR) BIB designs.

Theorem 10.2.7 (Hedayat and John, 1974, Theorem 5.1). *The existence of a BIB design with parameters v, b, r, k, λ satisfying $b = 3r - 2\lambda$ implies the existence of a GR(1) BIB design with parameters $\tilde{v} = v + 1$, $\tilde{b} = 2b$, $\tilde{r} = b$, $\tilde{k} = (v+1)/2$, $\tilde{\lambda} = r$.*

Theorem 10.2.8 (Shah and Gujarathi, 1977). *The existence of a BIB design with parameters v, b, r, k, λ satisfying $r = 2\lambda$ implies the existence of an LR(1) BIB design with parameters $v' = v + 1$, $b' = 2b$, $r' = b$, $k' = k$, $\lambda' = b - r$ with respect to two given treatments.*

Now, it can be shown (Kageyama, 1987b) that Theorems 10.2.7 and 10.2.8 produce the same family in the sense of constructing LR(1) BIB designs. However, their methods produce designs having remarkably different structures, as the following shows.

When $\boldsymbol{N}$ is the $v \times b$ incidence matrix of a starting BIB design, it can, without loss of generality, be written in the partitioned form as

$$\boldsymbol{N} = \begin{bmatrix} \mathbf{1}_r' & \mathbf{0}' \\ \boldsymbol{N}_1 & \boldsymbol{N}_2 \end{bmatrix}.$$

Then Theorem 10.2.7 utilizes a construction based on the incidence matrix

$$\begin{bmatrix} N & 1_v 1_b' - N \\ 1_b' & 0' \end{bmatrix} \tag{10.2.3}$$

(see Theorem 9.2.3), whereas Theorem 10.2.8 utilizes a construction given by the incidence matrix

$$\begin{bmatrix} 1_r' & 0' & 0' & 1_{b-r}' \\ N_1 & N_2 & N_1 & 1_{v-1}1_{b-r}' - N_2 \\ 0' & 0' & 1_r' & 1_{b-r}' \end{bmatrix}. \tag{10.2.4}$$

The structure (10.2.3) always yields a 3-design, i.e., the resulting design is GR(1), whereas the structure (10.2.4) does not yield a 3-design. It appears that most of available LR(1) BIB designs have the block structure of a 3-design. Thus, Theorem 10.2.8 is very useful and combinatorially interesting in the sense that one can get systematically an LR(1) BIB design which is not a 3-design.

One can also present all the parameter sets of existing LR(1) BIB designs with $r \leq 30$, constructed by the methods given in Theorems 10.2.7 and 10.2.8, as:

1. $v = 6$, $b = 20$, $r = 10$, $k = 3$, $\lambda = 4$, $\varepsilon_1 = 4/5$;
2. $v = 6$, $b = 40$, $r = 20$, $k = 3$, $\lambda = 8$, $\varepsilon_1 = 4/5$;
3. $v = 6$, $b = 60$, $r = 30$, $k = 3$, $\lambda = 12$, $\varepsilon_1 = 4/5$;
4. $v = 8$, $b = 14$, $r = 7$, $k = 4$, $\lambda = 3$, $\varepsilon_1 = 6/7$;
5. $v = 8$, $b = 28$, $r = 14$, $k = 4$, $\lambda = 6$, $\varepsilon_1 = 6/7$;
6. $v = 8$, $b = 42$, $r = 21$, $k = 4$, $\lambda = 9$, $\varepsilon_1 = 6/7$;
7. $v = 8$, $b = 56$, $r = 28$, $k = 4$, $\lambda = 12$, $\varepsilon_1 = 6/7$;
8. $v = 10$, $b = 36$, $r = 18$, $k = 5$, $\lambda = 8$, $\varepsilon_1 = 8/9$;
9. $v = 12$, $b = 22$, $r = 11$, $k = 6$, $\lambda = 5$, $\varepsilon_1 = 10/11$;
10. $v = 12$, $b = 44$, $r = 22$, $k = 6$, $\lambda = 10$, $\varepsilon_1 = 10/11$;
11. $v = 14$, $b = 52$, $r = 26$, $k = 7$, $\lambda = 12$, $\varepsilon_1 = 12/13$;
12. $v = 16$, $b = 30$, $r = 15$, $k = 8$, $\lambda = 7$, $\varepsilon_1 = 14/15$;
13. $v = 16$, $b = 60$, $r = 30$, $k = 8$, $\lambda = 14$, $\varepsilon_1 = 14/15$;
14. $v = 20$, $b = 38$, $r = 19$, $k = 10$, $\lambda = 9$, $\varepsilon_1 = 18/19$;
15. $v = 24$, $b = 46$, $r = 23$, $k = 12$, $\lambda = 11$, $\varepsilon_1 = 22/23$;
16. $v = 28$, $b = 54$, $r = 27$, $k = 14$, $\lambda = 13$, $\varepsilon_1 = 26/27$.

Note that for each parameter set there are two designs with different block structures, i.e., one is a 3-design, but the other is not a 3-design, and that the designs of Nos. 4, 9, 12, 14, 15 and 16 above belong to series (10.2.2) of affine resolvable BIB designs. Furthermore, as examples of parameters of other existing LR(1) BIB designs, one can present more for the range of $r \leq 20$,

17. $v = 10$, $b = 30$, $r = 12$, $k = 4$, $\lambda = 4$, $\varepsilon_1 = 5/6$;
18. $v = 11$, $b = 33$, $r = 15$, $k = 5$, $\lambda = 6$, $\varepsilon_1 = 22/25$;

19. $v = 17$, $b = 68$, $r = 20$, $k = 5$, $\lambda = 5$, $\varepsilon_1 = 17/20$.

In fact, there exist 3-designs with the parameters given under 17 – 19 above (see Hedayat and Kageyama, 1980).

There are not very many sets of parameters in a practical range for BIB designs $\mathcal{D}$. Theorem 10.2.2 says that $\mathcal{D}$ is GR(1) if and only if it is a 3-design, and any t-design, when $t \geq 3$, is at least GR(1), because the t-design is also a 3-design (see a statement after Definition 10.2.4). As a number of families of t-designs are known in the literature (see Colbourn and Dinitz, 1996, pp.47-75; Hedayat and Kageyama, 1980; Kageyama and Hedayat, 1983; Street, 1982; Raghavarao and Zhou, 1997), there exist a number of families of LR(1) BIB designs. However, these designs mostly have large values of the design parameters.

Further results on this topic are given in Hedayat and John (1974), Most (1975), Shah and Gujarathi (1977), Kageyama and Saha (1987), Kageyama (1987b), and Baksalary and Puri (1990).

Other robust designs can be stated here briefly. Kiefer (1959) introduced the concept of optimum designs. Since then a great deal of research has been done in finding designs satisfying one or more optimality criteria, as can be seen, e.g., in Shah and Sinha (1989). Optimum designs will not normally be optimum when some observations are missing, i.e., are unavailable. Moreover, the unavailability of even a single observation may lead to a situation where for some parametric functions the intra-block BLUEs do not exist. A BIB design is always connected. If some observations in a BIB design are unavailable, then the resulting design may not be another BIB design. Furthermore, the resulting block design may or may not even be connected (see Definition 2.2.5). Under this situation, Ghosh (1979, 1982a) introduced a primary robustness property of designs against the unavailability of any s (a positive integer) observations in the sense that, when any s observations are unavailable, there still exist the intra-block BLUE for any parametric function.

Definition 10.2.5. A connected block design is said to be robust against the unavailability of any s observations if the block design obtained by omitting any s observations remains connected in the sense of Definition 2.2.5.

Recall (from Section 6.0.1, noting the condition $\lambda > 0$) that a BIB design is always connected in the sense of Definition 2.2.5. Ghosh (1982b) has proved the following results based on Definition 10.2.5. Connectedness can also be checked by looking at the rank of the usual C-matrix (see Definition 2.2.5 and Lemma 2.3.2).

Theorem 10.2.9 (Ghosh, 1982b, Theorem 1). *A BIB design with parameters v, b, r, k, λ is robust against the unavailability of any $r - 1$ observations.*

Outline of proof. The idea of the proof is to make an argument in terms of the graph theory. Associate with a BIB design a bipartite graph G such that the vertices of G, treatments and blocks, are partitioned into two subsets $T = \{1, 2, ..., v\}$ and $B = \{1, 2, ..., b\}$ consisting of v treatments and b blocks, respectively, and a point in T is joined to a point in B if the corresponding treatment occurs in the corresponding block. Hence a BIB design is connnected (in the sense of Definition 2.2.5) if and only if all points in T of G are connected with each other. Thus the unavailability of any $r-1$ observations means the removal of the corresponding $r-1$ edges from G. Under such removal, the connectedness of the resulting graph can be shown by carefully looking at the properties of r (replication of each treatment) and λ (concurrence of any two distinct treatments) in the BIB design. □

A BIB design is not robust against the unavailability of any r observations, because if all r observations corresponding to a particular treatment are unavailable then the treatment will be disconnected (in the sense of Definition 2.2.5) from the other treatments. It is clear that a BIB design is robust against the unavailability of any t ($\leq r-1$) observations. Thus, Theorem 10.2.9 shows that the maximum value of s for a BIB design in Definition 10.2.5 is $r-1$.

The same idea as the proof of Theorem 10.2.9 can yield the following result.

Theorem 10.2.10 (Ghosh, 1982b, Theorem 2). *A BIB design with parameters v, b, r, k, λ is robust against the unavailability of all observations in any $r-1$ blocks.*

Note that a BIB design is robust against the unavailability of all observations in any b' ($\leq r-1$) blocks.

The above two theorems are still valid for t-designs with $t \geq 3$. On the other hand, the robustness of PBIB designs has been investigated by Ghosh, Rao and Singhi (1983). Further discussions can be seen in Baksalary and Tabis (1987).

One more setup is finally to be stated. When some observations become unavailable in a designed experiment for some reason, it is of interest to examine the loss of information, defined suitably, that is incurred due to missing data. That is, it is useful to know whether the loss of some observations in a design produces the residual design with a serious loss of efficiency as compared with the original design. Designs for which this loss is "small" may be termed robust. The robustness of designs against the unavailability of data has been investigated in abundance, for example, see Ghosh, Kageyama and Mukerjee (1992), Dey and Dhall (1988), Srivastava, Gupta and Dey (1990), Mukerjee and Kageyama (1990), Das and Kageyama (1992), Gupta and Srivastava (1992), Dey (1993), Kageyama and Setoya (1993, 1998), and Duan and Kageyama (1995). They consider various patterns of unavailability of data as: (i) any number of observations in a block are lost; (ii) all the observations in any two blocks are lost; (iii) any one observation in a design is lost; (iv) any two observations in a design are lost, for a BIB design, GD design, extended BIB design, augmented BIB design. They show that the efficiency of residual designs does not reduce mostly. This means that such designs are fairly robust in the sense of efficiency.

10.3 Nonbinary designs

In a block design, suppose that the number of treatments and the number of blocks are fixed as $v = 5$ and $b = 7$, respectively. Then it is obvious that among block designs with the same values of v and b, a BIB design with parameters $v = 5, b = 7, r, k, \lambda$ does not exist. Hence consider the incidence matrix $\boldsymbol{N} = [n_{ij}]$ of the form

$$\begin{bmatrix} 2 & 1 & 1 & 0 & 0 & 0 & 1 \\ 1 & 0 & 0 & 1 & 1 & 1 & 0 \\ 0 & 1 & 1 & 1 & 1 & 0 & 0 \\ 0 & 1 & 0 & 1 & 0 & 1 & 1 \\ 0 & 0 & 1 & 0 & 1 & 1 & 1 \end{bmatrix}, \quad \text{giving} \quad \boldsymbol{C}_1 = \frac{10}{3}\left(\boldsymbol{I}_5 - \frac{1}{5}\mathbf{1}_5\mathbf{1}_5'\right),$$

which yields a proper variance-balanced (VB) design with parameters $v = 5$, $b = 7$, $\boldsymbol{r} = [5, 4\mathbf{1}_4']'$, $k = 3$, $\theta = 10/3$, according to the notation in Section 10.1. Because $n_{11} = 2$, this design is not optimal in the sense of minimizing the maximum variance of the intra-block BLUE of the normalized contrast. In fact (as recalled in Section 10.1), in a connected VB design, the variance of the intra-block BLUE of any normalized contrast $\boldsymbol{c}'\boldsymbol{\tau}$ (i.e., with $\boldsymbol{c}'\boldsymbol{c} = 1$) is equal to σ_1^2/θ. Evidently, with $n = \sum_{i=1}^{v} r_i$, $\theta = [n - \text{tr}(\boldsymbol{N}\boldsymbol{k}^{-\delta}\boldsymbol{N}')]/(v-1)$ $\leq (n-b)/(v-1)$, the equality holding if and only if the design is binary (in the sense of Definition 2.2.1). Thus, when binary VB designs are available for given values of parameters, it is recommended to use such designs (Kageyama, 1987a). [In particular, for a BIB design $\theta = b(k-1)/(v-1) = \lambda b/r = \lambda v/k$ if it exists.] In such situation, if for given parameters there does not exist a binary plan suitable to the purpose, it may be possible to take nonbinary plans for the present experimentation from a viewpoint of the above optimality. On the other hand, the present design above can be shown to be a proper (0; 3, 1; 0)-EB design with parameters $v = 5$, $b = 7$, $\boldsymbol{r} = [5, 4\mathbf{1}_3']'$, $k = 3$, $\varepsilon_1 = 5/6$, $\varepsilon_2 = 7/10$, $\rho_1 = 3$, $\rho_2 = 1$, $\boldsymbol{L}_1 = \text{diag}[0 : \boldsymbol{I}_4 - 4^{-1}\mathbf{1}_4\mathbf{1}_4']$ and

$$\boldsymbol{L}_2 = \frac{1}{21}\begin{bmatrix} 16 & -4\mathbf{1}_4' \\ -5\mathbf{1}_4' & \frac{5}{4}\mathbf{1}_4\mathbf{1}_4' \end{bmatrix},$$

giving the average efficiency factor $\varepsilon = 70/88 = 0.7955$.

Suppose that one wants to compare $v = 5$ treatments when having $b = 10$ blocks of size $k = 3$ for the experiment. If the researcher wants to have all treatments in the same number of replications, a suitable choice is the BIB design with parameters $v = 5, b = 10, r = 6, k = 3, \lambda = 3$ (see Table 8.2, No. 14). However, if he (she) would like to have one of the treatments replicated twice as much as the other, and at the same time retain the property of $(0; v-1; 0)$-EB of the design, then a possible choice is the design $\boldsymbol{N}^*$ considered in Example 6.5.1. Note that it is a nonbinary (0; 4; 0)-EB design with parameters $v^* = 5, b^* = 10$, $\boldsymbol{r}^* = [10, 5\mathbf{1}_4']', k^* = 3$ and $\varepsilon_1^* = 4/5$, which is only slightly smaller than $\varepsilon_1 = 5/6$ of the BIB design.

Remark 10.3.1. It generally seems that binary block designs are the best. However, Bagchi (1988), Bagchi, Mukhopadhyay and Sinha (1990), Shah and Das (1992), and Morgan and Uddin (1993) have discussed some nonbinary block designs which are better than (superior to) binary block designs under certain optimality criterion. These discussions are not related to the present approach and hence they are not considered further here.

There are several methods of constructing nonbinary block designs in the literature, for example, see John (1964), Murty and Das (1967), Das and Rao (1968), Dey (1970), Ceranka and Mejza (1979), Agarwal and Kumar (1986b), Jacroux and Majumdar (1989), Kageyama (1987a, 1991), Sinha and Kageyama (1991), Ghosh, Joshi and Kageyama (1993).

Here constructions of nonbinary block designs with small number of distinct efficiency factors will be discussed. Most of these constructions will be based on the techniques of merging, patterned arrangement, supplementation and juxtaposition of basic designs, as described in Chapter 6.

10.3.1 Merging

The effect of merging treatments to construct designs was shown in Section 6.5. Especially, nonbinary cases are seen in Theorems 6.5.1 – 6.5.3 as well as Corollaries 6.5.1, 6.5.2 and 6.5.4, whereas binary cases are seen in Corollaries 6.5.3 and 7.3.2, and Section 8.3.2. So far the technique of merging treatments has been applied to BIB designs, GD designs and designs with one distinct efficiency factor (see Sections 6.5, 7.3.1.1, 7.3.2, 8.3.2 and 8.4.2).

But it can also be applied to other block designs having some combinatorial property to produce nonbinary designs. For example, a balanced block design with two different numbers of replicates introduced by Corsten (1962) (see Section 8.4.2 for the definition of the balanced block design in the sense of Corsten) can be applied, because this design is proper. In each case, the efficiency factors remain invariant by merging some treatments in the original design (as will be seen below in the special merging procedure). In fact, as an observation similar to the result by Puri and Nigam (1983), Ceranka and Mejza (1987) have derived a nonbinary design with m efficiency factors for $m \leq 3$ from a balanced block design, defined in Section 8.4.2, with two different numbers of replicates, which is actually a binary design with m efficiency factors for $m \leq 3$ given by Theorem 8.4.5. This can be remarked as follows (see Section 8.4.2 for its derivation).

Let $\mathcal{D}$ be a balanced block design with parameters $v^* = v_1 + v_2$, b^*, $\mathbf{r}^* = [r_1\mathbf{1}'_{v_1}, r_2\mathbf{1}'_{v_2}]'$, k^*, given in Theorem 8.4.5. A block design $\mathcal{D}^*$ with parameters $v^{**} = v_1^* + v_2^*$, b^*, $\mathbf{r}^{**}$, k^*, obtained from the design $\mathcal{D}$ by merging its v_1 treatments (in the first set) and/or its v_2 treatments (in the second set) into v_1^* and v_2^* treatments, respectively, is a $(0; \rho_1^{**}, \rho_2^{**}, \rho_3^{**}; 0)$-EB design. The design $\mathcal{D}^*$ has the same efficiency factors ε_1^*, ε_2^*, ε_3^*, as the original design $\mathcal{D}$, however, with multiplicities $v_1^* - 1$, $v_2^* - 1$ and 1, respectively (as compared with $v_1 - 1$, $v_2 - 1$ and 1, respectively, in the original design).

10.3.2 Patterned method

As described in Section 6.6, the patterned method is very simple and useful to produce more efficient designs, unfortunately however, requiring large number of experimental units.

By Theorem 6.6.4, the following corollaries can be obtained.

Corollary 10.3.1. *Let $\boldsymbol{N}$ be the $v \times b$ incidence matrix of a BIB design with parameters $v,\ b, r, k,\ \lambda,\ \varepsilon = v\lambda/(rk)$, $\boldsymbol{L} = \boldsymbol{I}_v - v^{-1}\boldsymbol{1}_v\boldsymbol{1}_v'$. Then the pattern* (6.6.1) *yields a $(0; \rho_1^*, \rho_2^*, \rho_3^*; 0)$-EB design with parameters $v^* = 3v$, $b^* = 3b$, $r^* = (a+c)b+(1-c)r$, $k^* = (a+c)v+(1-c)k$, $\varepsilon_1^* = 1-(1-c)^2(r-\lambda)/(r^*k^*)$, $\varepsilon_2^* = 1-(1+c+c^2)(r-\lambda)/(r^*k^*)$, $\varepsilon_3^* = 1-[(c^2+c+1)rk+(a^2+c^2-ac)vb-(2c^2+a+c-ac)vr]/(r^*k^*)$, $\rho_1^* = v-1$, $\rho_2^* = 2(v-1)$, $\rho_3^* = 2$, $\boldsymbol{L}_1^* = 3^{-1}\boldsymbol{1}_3\boldsymbol{1}_3' \otimes (\boldsymbol{I}_v - v^{-1}\boldsymbol{1}_v\boldsymbol{1}_v')$, $\boldsymbol{L}_2^* = (\boldsymbol{I}_3 - 3^{-1}\boldsymbol{1}_3\boldsymbol{1}_3') \otimes (\boldsymbol{I}_v - v^{-1}\boldsymbol{1}_v\boldsymbol{1}_v')$, $\boldsymbol{L}_3^* = (\boldsymbol{I}_3 - 3^{-1}\boldsymbol{1}_3\boldsymbol{1}_3') \otimes v^{-1}\boldsymbol{1}_v\boldsymbol{1}_v'$. In particular, when $c = 0$, the design is $(0; \rho_1^*, \rho_2^*; 0)$-EB with $\varepsilon_1^* = 1-(r-\lambda)/(r^*k^*)$, $\varepsilon_2^* = 3avr/[(ab+r)(av+k)]$, $\rho_1^* = 3(v-1)$, $\rho_2^* = 2$, $\boldsymbol{L}_1^* = \boldsymbol{I}_3 \otimes (\boldsymbol{I}_v - v^{-1}\boldsymbol{1}_v\boldsymbol{1}_v')$, $\boldsymbol{L}_2^* = (\boldsymbol{I}_3 - 3^{-1}\boldsymbol{1}_3\boldsymbol{1}_3') \otimes v^{-1}\boldsymbol{1}_v\boldsymbol{1}_v'$.*

Corollary 10.3.2. *Let $\boldsymbol{N}$ be the $v \times b$ incidence matrix of a 2-associate PBIB design with parameters $v, b, r, k, \lambda_1, \lambda_2, \varepsilon_1, \varepsilon_2, \rho_1, \rho_2, \boldsymbol{L}_1, \boldsymbol{L}_2$. Then* (6.6.1) *with $c = 0$ yields a $(\rho_0^*; \rho_1^*, \rho_2^*, \rho_3^*; 0)$-EB design with parameters $v^* = 3v$, $b^* = 3b$, $r^* = ab+r$, $k^* = av+k$, $\varepsilon_1^* = 1-rk(1-\varepsilon_1)/(r^*k^*)$, $\varepsilon_2^* = 1-rk(1-\varepsilon_2)/(r^*k^*)$, $\varepsilon_3^* = 3avr/(r^*k^*)$, $\rho_0^* \geq 0$, $\rho_1^* = 3\rho_1$, $\rho_2^* = 3\rho_2$, $\rho_3^* = 2$, $\boldsymbol{L}_1^* = \boldsymbol{I}_3 \otimes \boldsymbol{L}_1$, $\boldsymbol{L}_2^* = \boldsymbol{I}_3 \otimes \boldsymbol{L}_2$, $\boldsymbol{L}_3^* = (\boldsymbol{I}_3 - 3^{-1}\boldsymbol{1}_3\boldsymbol{1}_3') \otimes v^{-1}\boldsymbol{1}_v\boldsymbol{1}_v'$.*

Remark 10.3.2. In Corollaries 10.3.1 and 10.3.2, when $a, c \in \{0,1\}$ such that $(a,c) \neq (0,0)$, the designs become binary. Otherwise, such designs are nonbinary. If one of the ε_1 and ε_2 in Corollary 10.3.2 is 1, then the resulting design is $(\rho_0^{**}; \rho_1^{**}, \rho_2^{**}; 0)$-EB.

Recall that in Theorem 6.6.7, $\varepsilon_1^* = \varepsilon_2^*$ holds if and only if $c = (r-\lambda)/(b-2r+\lambda)$. Then the theorem yields the following result.

Corollary 10.3.3 (Saha, 1976, Theorem 5). *Let $\boldsymbol{N}$ be the $v \times b$ incidence matrix of a BIB design with parameters $v,\ b, r, k,\ \lambda$ such that $(r-\lambda)/(b-2r+\lambda)$ is an integer, c, say. Then the pattern* (6.6.3) *yields a $(0; v^*-1; 0)$-EB design with parameters $v^* = v+1$, $b^* = 2b$, $\boldsymbol{r}^* = b[\boldsymbol{1}_v', c]'$, $\boldsymbol{k}^* = [(k+c)\boldsymbol{1}_b', (v-k)\boldsymbol{1}_b']'$, $\varepsilon_1^* = r(b-r)/[b(r-\lambda)]$, $\boldsymbol{L}_1^* = \boldsymbol{I}_{v+1} - (v+c)^{-1}\boldsymbol{1}_{v+1}[\boldsymbol{1}_v', c]$.*

Note that $c = 1$ (i.e., a binary design) if and only if $b = 3r - 2\lambda$. It is known (see Remark 8.2.6; Kageyama and Kuwada, 1985) that a BIB design with $b = 3r - 2\lambda$ always satisfies the relations $v = 2k+1, b = s(2k+1), r = sk, k, \lambda = s(k-1)/2$ for a positive integer s such that $s(k-1)$ is even, whose practical plans can be seen in Table 8.2 [see also Theorem 6.8.1(i) and Corollary 6.8.1(i)]. Also note that Corollary 10.3.3 with $c = 1$ corresponds to Theorem 10.2.7, i.e., by (10.2.3). Theorem 9.2.3 itself expresses the binary case of the present corollary. Because examples of Corollary 10.3.3 for $c \geq 2$ are not rich, one can produce a more applicable expression of Corollary 10.3.3. It is given by considering a complement of a BIB design.

Corollary 10.3.4. *Let N be the $v \times b$ incidence matrix of a BIB design with parameters v, b, r, k, λ such that $r/\lambda - 1$ is an integer, c^*, say. Then the incidence matrix*

$$N^* = \begin{bmatrix} \mathbf{1}_v\mathbf{1}_b' - N & N \\ c^*\mathbf{1}_b' & \mathbf{0}' \end{bmatrix}$$

yields a $(0; v^* - 1; 0)$-EB design with parameters $v^* = v + 1$, $b^* = 2b$, $r^* = b[\mathbf{1}_v', c^*]'$, $k^* = [(v - k + c^*)\mathbf{1}_b', k\mathbf{1}_b']'$, $\varepsilon_1^* = r(b-r)/[b(r-\lambda)]$, $L_1^* = I_{v+1} - (v + c^*)^{-1}\mathbf{1}_{v+1}[\mathbf{1}_v', c^*]$.

Remark 10.3.3. There are examples in abundance which satisfy $c^* = r/\lambda - 1$ being an integer. For example, BIB designs with $\lambda = 1$ or $r = s\lambda$ for some integer s satisfy the required condition. In this section the case $c^* \geq 2$ are of interest.

Recall (Definition 2.2.1) that a block design with the usual incidence matrix $N = [n_{ij}]$ is said to be *binary* when $n_{ij} = 0$ or 1 for all i, j. It is called *ternary* when every $n_{ij} = 0$, x or y for some positive integers x and y (usually, $x = 1$ and $y = 2$, see Preece, 1982, p. 87).

Note that the existence of a nonbinary $(0; v-1; 0)$-EB design can be checked by simply taking Theorem 8.2.1 or the formula (8.2.5).

Theorem 10.3.1. *Let N be the $v \times b$ incidence matrix of a BIB design with parameters v, b, r, k, λ for r/λ being an integer (≥ 3). Then, for positive integers p and q such that $p/q = [(2\lambda v - rk)(r+1) - 2(r-\lambda)]/(rk)$, the incidence matrix*

$$N^* = \left[\mathbf{1}_q' \otimes \begin{bmatrix} N & \mathbf{1}_v \\ \mathbf{0}' & (r-\lambda)/\lambda \end{bmatrix}, \begin{matrix} \mathbf{1}_p' \otimes I_v \\ \mathbf{1}_{pv}' \end{matrix}\right]$$

yields a ternary $(0; v^*-1; 0)$-EB design with parameters $v^* = v+1$, $b^* = (b+1)q + vp$, $r^* = [((r+1)q+p)\mathbf{1}_v', q(r-\lambda)/\lambda + vp]'$, $k^* = [k\mathbf{1}_{qb}', ((r-\lambda)/\lambda + v)\mathbf{1}_q', 2\mathbf{1}_{pv}']'$, $\varepsilon_1^* = \{v[(r+1)q+p] + vp + (r-\lambda)q/\lambda\}[(r-\lambda)q/(rk) + p/2]/\{[(r+1)q+p][(r-\lambda)q/\lambda + pv]\}$, $L_1^* = I_{v+1} - [(r^*)'\mathbf{1}_{v+1}]^{-1}\mathbf{1}_{v+1}(r^*)'$.

Proof. It follows from (8.2.5) that in the resulting design for some constant α

$$\frac{\lambda q}{k} + \frac{\lambda q}{rk} = \alpha[(r+1)q+p]^2, \quad \frac{(r-\lambda)q}{rk} + \frac{p}{2} = \alpha[(r+1)q+p][\frac{r-\lambda}{\lambda}q + pv]$$

which yield $p/q = [(2\lambda v - rk)(r+1) - 2(r-\lambda)]/(rk)$. □

Note that in Theorem 10.3.1, $p/q < r+1$.

Example 10.3.1. A BIB design with parameters $v = 4, b = 6, r = 3, k = 2, \lambda = 1$, by Theorem 10.3.1 with $p/q = 2/3$, yields a ternary $(0;4;0)$-EB design, for example, for $p = 2$ and $q = 3$, it has the parameters $v^* = 5, b^* = 29, r^* = 14$, $k^* = [\mathbf{1}_3' \otimes [2\mathbf{1}_6', 6], 2\mathbf{1}_8']'$, $\varepsilon_1^* = 5/7$, $L_1^* = I_5 - 5^{-1}\mathbf{1}_5\mathbf{1}_5'$, whose incidence matrix N^* is of the form

$$\begin{bmatrix} 1&1&1&0&0&0&1&1&1&1&0&0&0&1&1&1&1&0&0&0&1&1&0&0&0&1&0&0&0 \\ 1&0&0&1&1&0&1&1&0&0&1&1&0&1&1&0&0&1&1&0&1&0&1&0&0&0&1&0&0 \\ 0&1&0&1&0&1&1&0&1&0&1&0&1&1&0&1&0&1&0&1&1&0&0&1&0&0&0&1&0 \\ 0&0&1&0&1&1&1&0&0&1&0&1&1&1&0&0&1&0&1&1&1&0&0&0&1&0&0&0&1 \\ 0&0&0&0&0&0&2&0&0&0&0&0&0&2&0&0&0&0&0&0&2&1&1&1&1&1&1&1&1 \end{bmatrix}.$$

Note that the value of ε_1^* is independent of p and q.

Remark 10.3.4. Theorem 10.3.1 can easily be generalized by simply taking the pattern

$$\left[\mathbf{1}_q' \otimes \begin{bmatrix} xN & x\mathbf{1}_v \\ \mathbf{0}' & y \end{bmatrix}, \begin{matrix} x\mathbf{1}_p' \otimes I_v \\ x\mathbf{1}_{pv}' \end{matrix}\right]$$

for some integers x and y. For another pattern, one can consider

$$\begin{bmatrix} x\mathbf{1}_q' \otimes N & x\mathbf{1}_v\mathbf{1}_p' \\ x\mathbf{1}_{bq}' & y\mathbf{1}_p' \end{bmatrix}$$

for positive integers x, y, p and q.

Theorem 10.3.2. *Let N be the $v \times b$ incidence matrix of a BIB design with parameters v, b, r, k, λ for r/λ being an integer (≥ 3). Then, for positive integers p and q such that $p/q = \{\lambda v[2(b\lambda + rk) - (r+1)(2r+k+1)]\}/[\lambda v(2r+k+1) - 2(k+1)(b\lambda + rk)]$, the incidence matrix*

$$N^* = \left[\mathbf{1}_q' \otimes \begin{bmatrix} N & \mathbf{0} & I_v \\ \mathbf{1}_b' & (r-\lambda)/\lambda & \mathbf{1}_v' \end{bmatrix}, \begin{matrix} \mathbf{1}_v\mathbf{1}_p' \\ \mathbf{0}' \end{matrix}\right]$$

yields a ternary $(0; v^*-1; 0)$-EB design with parameters $v^* = v+1, b^* = (b+v+1)q+p$, $r^* = [[(r+1)q+p]\mathbf{1}_v', [b+v+(r-\lambda)/\lambda]q]'$, $k^* = [(k+1)\mathbf{1}_{qb}', [(r-\lambda)/\lambda]\mathbf{1}_q', 2\mathbf{1}_{qv}', v\mathbf{1}_p']'$, $\varepsilon_1^* = \{v[(r+1)q+p] + [b+v+(r-\lambda)/\lambda]q\}[\lambda q/(k+1) + p/v]/[(r+1)q+p]^2$, $L_1^* = I_{v+1} - [(r^*)'\mathbf{1}_{v+1}]^{-1}\mathbf{1}_{v+1}(r^*)'$.

Proof. It follows from (8.2.5) that in the resulting design for some α

$$\frac{\lambda q}{k+1} + \frac{p}{v} = \alpha[(r+1)q+p]^2, \quad \frac{rq}{k+1} + \frac{q}{2} = \alpha[(r+1)q+p](b+v+\frac{r-\lambda}{\lambda})q$$

which can yield $p/q = \{\lambda v[2(b\lambda + rk) - (r+1)(2r+k+1)]\}/[\lambda v(2r+k+1) - 2(k+1)(b\lambda + rk)]$. □

Example 10.3.2. Consider a BIB design with parameters $v = 4, b = 6, r = 3, k = 2, \lambda = 1$. Then using Theorem 10.3.2 with $p/q = 4/3$ yields a ternary $(0; 4; 0)$-EB design, for example, for $p = 4$ and $q = 3$, it has the parameters $v^* = 5, b^* = 37, r^* = [16\mathbf{1}_4', 36]', k^* = [\mathbf{1}_3' \otimes [3\mathbf{1}_6', 2\mathbf{1}_5'], 4\mathbf{1}_4']'$, $\varepsilon_1^* = 25/32$, $L_1^* = I_5 - (100)^{-1}\mathbf{1}_5[16\mathbf{1}_4', 36]$. Note that the value of ε_1^* is independent of p and q.

A large number of ternary VB designs or ternary $(0; v-1; 0)$-EB designs can be constructed through the methods presented here.

Remark 10.3.5. In Theorems 10.3.1 and 10.3.2, when $r/\lambda = 2$, i.e., $v = 2k-1$, the resulting designs are binary. Hence these cases belong to Section 8.2.2. However, a binary design in Theorem 10.3.1 includes a complete block.

10.3.3 Supplementation

Some designs can be constructed by supplementing new treatments to the blocks of a given design (see Section 6.3). The incidence matrix of a supplemented block design can be written in general as in (6.3.1), i.e.,

$$\begin{bmatrix} \boldsymbol{N}_1 \\ \boldsymbol{N}_2 \end{bmatrix},$$

where $\boldsymbol{N}_1$ is the incidence matrix of the basic design and $\boldsymbol{N}_2$ is that of the supplementary treatments. As in Theorem 6.3.1, a new design can be constructed from a given design by adjoining to each of its blocks one or more new treatments in numbers proportional to the original block sizes. The proportionality factors need not be equal for all the new treatments, but they have to depend on their replication numbers.

A set of $v-1$ contrasts of treatment parameters

$$\begin{bmatrix} \boldsymbol{s}_{\beta j} \\ \boldsymbol{0} \end{bmatrix}, \ \beta = 0, 1, ..., m-1; \ j = 1, 2, ..., \rho_\beta$$

are eigenvectors of $\boldsymbol{M}_0^*$ in the proof of Theorem 6.3.1 corresponding to eigenvalues $1-\varepsilon_\beta^*$ and these contrasts are estimated in the intra-block analysis with the efficiency ε_β^*. Originally, before supplementation, contrasts represented by the vectors $\{\boldsymbol{s}_{\beta j}\}$ would be estimated in the intra-block analysis with the efficiency factor $\{\varepsilon_\beta\}$, respectively. Note that $\varepsilon_\beta^* > \varepsilon_\beta$ for $\beta = 0, 1, ..., m-1$. Thus, the addition of supplementary treatments in a large number of replications is likely to result in an increase of efficiency factor associated with contrasts of the basic set of treatments, provided the within block variances of the both designs are of the same order. The $t-1$ contrasts among supplementary treatments, and a contrast between supplementary treatments and the basic set of treatments are estimated in the intra-block analysis with full efficiency.

Similarly, as in Theorem 6.3.1, let v_1 original treatments be allocated to the experimental units within each of b blocks of an orthogonal block design with a constant block size k_1, i.e., $\boldsymbol{N}_1 = b^{-1}\boldsymbol{r}_1\boldsymbol{1}_b'$, and let $v_2 = k_2 b$ new (supplementary) treatments be assigned so that in each block of the b blocks k_2 different treatments occur, i.e., $\boldsymbol{N}_2 = \boldsymbol{I}_b \otimes \boldsymbol{1}_{k_2}$. Then the following theorem can be obtained.

Theorem 10.3.3. *A supplemented design with the incidence matrix*

$$\boldsymbol{N}^* = \begin{bmatrix} b^{-1}\boldsymbol{r}_1\boldsymbol{1}_b' \\ \boldsymbol{I}_b \otimes \boldsymbol{1}_{k_2} \end{bmatrix}$$

is a $(\rho_0^*; \rho_1^*; 0)$*-EB design with parameters* $v^* = v_1 + v_2$, $b^* = b$, $\boldsymbol{r}^* = [\boldsymbol{r}_1', \boldsymbol{1}_{v_2}']'$, $\boldsymbol{k}^* = k\boldsymbol{1}_b$, $\varepsilon_0^* = 1$, $\varepsilon_1^* = k_1/k$, $\rho_0^* = v^* - b$, $\rho_1^* = b-1$, $\boldsymbol{L}_0^* = \boldsymbol{I}_{v^*} - (bk)^{-1}\boldsymbol{1}_{v^*}(\boldsymbol{r}^*)' -$

$\boldsymbol{L}_1^*$, $\boldsymbol{L}_1^* = \text{diag}[\mathbf{O} : (\boldsymbol{I}_b - b^{-1}\mathbf{1}_b\mathbf{1}_b') \otimes k_2{}^{-1}\mathbf{1}_{k_2}\mathbf{1}_{k_2}']$, *where* $v_2 = k_2 b$ *and* $k = k_1 + k_2$, *provided that all elements of the vector* $b^{-1}\boldsymbol{r}_1$ *are integers.*

Proof. The parameters $v^* = v_1 + v_2$, $b^* = b$, $\boldsymbol{r}^* = [\boldsymbol{r}_1', \mathbf{1}_{v_2}']'$, $\boldsymbol{k}^* = (k_1 + k_2)\mathbf{1}_b = k\mathbf{1}_b$, $n^* = bk$ are obvious. As it follows that

$$\boldsymbol{M}_0^* = (\boldsymbol{r}^*)^{-\delta}(\boldsymbol{N}^*)(\boldsymbol{k}^*)^{-\delta}(\boldsymbol{N}^*)' - \frac{1}{n^*}\mathbf{1}_{v^*}(\boldsymbol{r}^*)'$$

$$= \frac{k_2}{k}\text{diag}\left[\mathbf{O} : \left(\boldsymbol{I}_b - \frac{1}{b}\mathbf{1}_b\mathbf{1}_b'\right) \otimes \frac{1}{k_2}\mathbf{1}_{k_2}\mathbf{1}_{k_2}'\right],$$

the proof is established. □

Example 10.3.3. In Theorem 10.3.3, let $\boldsymbol{r}_1 = [r_1, r_2, ..., r_{v_1}]'$, where $r_i = c_i b$ for positive integers c_i, $i = 1, 2, ..., v_1$. Here $k_1 = \sum_{i=1}^{v_1} c_i$. Then the supplemented design is a $(v - b; b - 1; 0)$-EB design with the same parameters as those given in Theorem 10.3.3, having the incidence matrix

$$\begin{bmatrix} c_1\mathbf{1}_b' \\ c_2\mathbf{1}_b' \\ \vdots \\ c_{v_1}\mathbf{1}_b' \\ \boldsymbol{I}_b \otimes \mathbf{1}_{k_2} \end{bmatrix},$$

which is nonbinary unless $c_1 = c_2 = \cdots = c_{v_1} = 1$.

When $k_2 = 1$ Theorem 10.3.3 yields the following result.

Corollary 10.3.5. *A supplemented design with the incidence matrix*

$$\begin{bmatrix} b^{-1}\boldsymbol{r}_1\mathbf{1}_b' \\ \boldsymbol{I}_b \end{bmatrix}$$

yields a $(\rho_0^*; \rho_1^*; 0)$*-EB design with parameters* $v^* = v_1 + b$, $b^* = b$, $\boldsymbol{r}^* = [\boldsymbol{r}_1', \mathbf{1}_b']'$, $\boldsymbol{k}^* = (k_1+1)\mathbf{1}_b$, $\varepsilon_0^* = 1$, $\varepsilon_1^* = k_1/(k_1+1)$, $\rho_0^* = v_1$, $\rho_1^* = b - 1$, $\boldsymbol{L}_0^* = \boldsymbol{I}_{v^*} - [b(k_1 + 1)]^{-1}\mathbf{1}_{v^*}(\boldsymbol{r}^*)' - \boldsymbol{L}_1^*$, $\boldsymbol{L}_1^* = \text{diag}[\mathbf{O} : \boldsymbol{I}_b - b^{-1}\mathbf{1}_b\mathbf{1}_b']$, *provided that all elements of the vector* $b^{-1}\boldsymbol{r}_1$ *are integers.*

10.3.4 Supplementation and juxtaposition

Consider two incidence matrices $\boldsymbol{N}_1$ and $\boldsymbol{N}_2$ of BIB designs with parameters v, b, r, k, λ, and with parameters v, b', r', k', λ', respectively. Then take the following pattern

$$\boldsymbol{N}^* = \begin{bmatrix} \mathbf{1}'_m \otimes \boldsymbol{N}_1 & \mathbf{1}'_p \otimes \boldsymbol{N}_2 \\ k_0 \mathbf{1}'_{mb} & \mathbf{0}' \end{bmatrix} \tag{10.3.1}$$

which has parameters $v^* = v+1$, $b^* = mb + pb'$, $\boldsymbol{r}^* = [(mr+pr')\mathbf{1}'_v, mk_0 b]'$, $\boldsymbol{k}^* = [(k+k_0)\mathbf{1}'_{mb}, k'\mathbf{1}'_{pb'}]'$. It also follows that

$$\boldsymbol{N}^* \mathrm{diag}\left[\frac{1}{k+k_0}\boldsymbol{I}_{mb} : \frac{1}{k'}\boldsymbol{I}_{pb'}\right](\boldsymbol{N}^*)' = \begin{bmatrix} \frac{m}{k+k_0}\boldsymbol{N}_1\boldsymbol{N}'_1 + \frac{p}{k'}\boldsymbol{N}_2\boldsymbol{N}'_2 & \frac{k_0 rm}{k+k_0}\mathbf{1}_v \\ \frac{k_0 rm}{k+k_0}\mathbf{1}'_v & \frac{k_0^2 bm}{k+k_0} \end{bmatrix}$$

which, from off-diagonal elements proportional to the products of corresponding replication numbers (i.e., Theorem 8.2.1), can yield the following result.

Theorem 10.3.4. *If $m/p = [vk_0\lambda' - k(r'-\lambda')]/[(r-\lambda)k']$, then (10.3.1) is the incidence matrix of a $(0; v^*-1; 0)$-EB design with parameters $v^* = v+1$, $b^* = mb+pb'$, $\boldsymbol{r}^* = [(mr+pr')\mathbf{1}'_v, mk_0b]'$, $\boldsymbol{k}^* = [(k+k_0)\mathbf{1}'_{mb}, k'\mathbf{1}'_{pb'}]'$, $\varepsilon_1^* = \{v+1 - v[mrk' + pr'(k+k_0)]/[k'(k+k_0)(mr+pr')] - k_0/(k+k_0)\}/v$, $\boldsymbol{L}_1^* = \boldsymbol{I}_{v+1} - [(\boldsymbol{r}^*)'\mathbf{1}_{v+1}]^{-1}\mathbf{1}_{v+1}(\boldsymbol{r}^*)'$. This design is nonbinary for $k_0 \geq 2$.*

Example 10.3.4. Let $\boldsymbol{N}_1$ and $\boldsymbol{N}_2$ be the incidence matrices of two BIB designs with parameters $v = b = 7$, $r = k = 4$, $\lambda = 2$, and $v = b' = 7$, $r' = k' = 3$, $\lambda' = 1$, respectively (see Table 8.2). When $k_0 = 2$, Theorem 10.3.4 yields $m/p = 1$, and hence one gets a $(0;7;0)$-EB design, whose incidence matrix for $m = p = 1$ is

$$\begin{bmatrix} 0&1&1&1&0&1&0&1&0&0&0&1&0&1 \\ 0&0&1&1&1&0&1&1&1&0&0&0&1&0 \\ 1&0&0&1&1&1&0&0&1&1&0&0&0&1 \\ 0&1&0&0&1&1&1&1&0&1&1&0&0&0 \\ 1&0&1&0&0&1&1&0&1&0&1&1&0&0 \\ 1&1&0&1&0&0&1&0&0&1&0&1&1&0 \\ 1&1&1&0&1&0&0&0&0&0&1&0&1&1 \\ 2&2&2&2&2&2&2&0&0&0&0&0&0&0 \end{bmatrix}, \quad \begin{array}{l} v^* = 8,\ b^* = 14, \\ \boldsymbol{r}^* = [7\mathbf{1}'_7, 14]', \\ \boldsymbol{k}^* = [6\mathbf{1}'_7, 3\mathbf{1}'_7]', \\ n^* = 63, \varepsilon_1^* = 6/7, \\ \boldsymbol{L}_1^* = \boldsymbol{I}_8 - \frac{1}{63}\mathbf{1}_8(\boldsymbol{r}^*)'. \end{array}$$

Remark 10.3.6. It appears that the values of m and p above may get large as the number of treatments and/or block sizes increase, and thus these designs might not find much use in agricultural field experiments. However, they may be of use in some laboratory or industrial research (see the discussion in Tocher, 1952; Kulshreshtha, Dey and Saha, 1972; Puri and Nigam, 1977b). If $\boldsymbol{N}_2 = \mathbf{1}_v\mathbf{1}'_b - \boldsymbol{N}$ is taken with $m = p = 1$ for $k_0 = c$, the resulting pattern yields the incidence matrix of designs resulting from Theorems 6.6.7 – 6.6.10, and Corollaries 10.3.3 and 10.3.4. In this sense, Theorem 10.3.4 has further possibility of generalizations in the construction of designs for $m = p$.

10.4 Disconnected designs

In this section it will be convenient to denote the usual C-matrix by $\boldsymbol{C}$ instead of $\boldsymbol{C}_1$ used previously. In a disconnected block design of degree $g-1$ (see Definition 2.2.6a), not all the contrasts of treatment parameters receive the intra-block BLUEs, as some of the contrasts are totally confounded with blocks. In this case, $\text{rank}(\boldsymbol{C}) = v - g$ for some $g \geq 2$ ($g = 1$ meaning that the design is connected). This results from the fact that the v treatments are divided into g groups, each of them being assigned to different groups of blocks. In accordance with this, after an appropriate ordering of its rows and columns, the $v \times b$ incidence matrix $\boldsymbol{N}$ of the design can be expressed as

$$\boldsymbol{N} = \text{diag}[\boldsymbol{N}_1 : \boldsymbol{N}_2 : \cdots : \boldsymbol{N}_g], \tag{10.4.1}$$

where $\boldsymbol{N}_\ell$, $\ell = 1, 2, ..., g$, are the $v_\ell \times b_\ell$ incidence matrices of its subdesigns corresponding to the partitions $v = \sum_{\ell=1}^{g} v_\ell$ and $b = \sum_{\ell=1}^{g} b_\ell$ obtainable with the maximum g. It obviously follows that the C-matrix of a disconnected design is of the form

$$\boldsymbol{C} = \text{diag}[\boldsymbol{C}_1 : \boldsymbol{C}_2 : \cdots : \boldsymbol{C}_g],$$

where $\text{rank}(\boldsymbol{C}_\ell) = v_\ell - 1$ for the C-matrix, $\boldsymbol{C}_\ell$, of $\boldsymbol{N}_\ell$ (see Lemma 2.3.2). Furthermore, as in (4.4.4), $\boldsymbol{C}_\ell = \sum_{\beta=0}^{m_\ell - 1} \varepsilon_{\ell\beta} \boldsymbol{r}_\ell^\delta \boldsymbol{L}_{\ell\beta}$, $\ell = 1, 2, ..., g$, where $\boldsymbol{r}^\delta = \text{diag}[\boldsymbol{r}_1^\delta : \boldsymbol{r}_2^\delta : \cdots : \boldsymbol{r}_g^\delta]$ corresponds to the partition (10.4.1), i.e., $\boldsymbol{N}_\ell \boldsymbol{1}_{b_\ell} = \boldsymbol{r}_\ell$ and $\boldsymbol{r}_\ell^\delta \boldsymbol{1}_{v_\ell} = \boldsymbol{r}_\ell$ (see Section 4.4.1). Thus, a disconnected design can be constructed by some diagonal juxtaposition of connected subdesigns. See Definition 2.2.6, and the relations (2.4.5) and (2.4.6). The C-matrix can now be written in the form

$$\boldsymbol{C} = \sum_{\beta=0}^{m_1-1} \varepsilon_{1\beta} \begin{bmatrix} \boldsymbol{r}_1^\delta \boldsymbol{L}_{1\beta} & \mathbf{O} \\ \mathbf{O} & \mathbf{O} \end{bmatrix} + \cdots + \sum_{\beta=0}^{m_g-1} \varepsilon_{g\beta} \begin{bmatrix} \mathbf{O} & \mathbf{O} \\ \mathbf{O} & \boldsymbol{r}_g^\delta \boldsymbol{L}_{g\beta} \end{bmatrix},$$

which means that the disconnected block design has at most $m_1 + m_2 + \cdots + m_g$ efficiency factors.

Now two observations can be described.

(i) When $\boldsymbol{N}$ is the $v \times b$ incidence matrix of a $(0; v-1; 0)$-EB design with parameters v, b, $\boldsymbol{r}$, $\boldsymbol{k}$, ε, $\boldsymbol{L}$, its diagonal juxtaposition, $\boldsymbol{I}_2 \otimes \boldsymbol{N}$, yields a disconnected $(0; 2v-2; 1)$-EB design with parameters $v^* = 2v$, $b^* = 2b$, $\boldsymbol{r}^* = \boldsymbol{1}_2 \otimes \boldsymbol{r}$, $\boldsymbol{k}^* = \boldsymbol{1}_2 \otimes \boldsymbol{k}$, $\varepsilon_1^* = \varepsilon$, $\varepsilon_2^* = 0$, $\rho_1^* = 2(v-1)$, $\rho_2^* = 1$, $\boldsymbol{L}_1^* = \boldsymbol{I}_2 \otimes \boldsymbol{L}$, $\boldsymbol{L}_2^* = \boldsymbol{I}_{2v} - [(\boldsymbol{r}^*)'\boldsymbol{1}_{2v}]^{-1}\boldsymbol{1}_{2v}(\boldsymbol{r}^*)' - \boldsymbol{L}_1^*$.

(ii) When $\boldsymbol{N}_\ell$, $\ell = 1, 2$, are the $v_\ell \times b_\ell$ incidence matrices of two $(0; v_\ell - 1; 0)$-EB designs with parameters v_ℓ, b_ℓ, $\boldsymbol{r}_\ell$, $\boldsymbol{k}_\ell$, ε, $\boldsymbol{L}_\ell$, its diagonal juxtaposition, $\text{diag}[\boldsymbol{N}_1 : \boldsymbol{N}_2]$, yields a disconnected $(0; v_1 + v_2 - 2; 1)$-EB design with parameters $v^* = v_1 + v_2$, $b^* = b_1 + b_2$, $\boldsymbol{r}^* = [\boldsymbol{r}_1', \boldsymbol{r}_2']'$, $\boldsymbol{k}^* = [\boldsymbol{k}_1', \boldsymbol{k}_2']'$, $\varepsilon_1^* = \varepsilon$, $\varepsilon_2^* = 0$, $\rho_1^* = v_1 + v_2 - 2$, $\rho_2^* = 1$, $\boldsymbol{L}_1^* = \text{diag}[\boldsymbol{L}_1 : \boldsymbol{L}_2]$, $\boldsymbol{L}_2^* = \boldsymbol{I}_{v^*} - [(\boldsymbol{r}^*)'\boldsymbol{1}_{v^*}]^{-1}\boldsymbol{1}_{v^*}(\boldsymbol{r}^*)' - \boldsymbol{L}_1^*$.

Thus, appropriate diagonal juxtapositions of connected designs may yield designs which are disconnected. For example, consider a block design with the following incidence matrix

$$N = \text{diag}\left[\begin{bmatrix} 1 & 1 & 0 \\ 1 & 0 & 1 \\ 0 & 1 & 1 \end{bmatrix} : \begin{bmatrix} 1 & 1 & 0 \\ 1 & 0 & 1 \\ 0 & 1 & 1 \\ 3 & 3 & 3 \end{bmatrix}\right] \quad (= \text{diag}[N_1 : N_2],\ \text{say}),$$

$$v = 7,\ b = 6,\ r = [2\mathbf{1}'_6, 9]',\ k = [2\mathbf{1}'_3, 5\mathbf{1}'_3]'.$$

It can be shown (Ceranka and Mejza, 1979) that the eigenvalues of the C-matrix of the design,

$$C = \text{diag}\left[15I_3 - 5\mathbf{1}_3\mathbf{1}'_3 : \begin{bmatrix} 16 & -2 & -2 & -12 \\ -2 & 16 & -2 & -12 \\ -2 & -2 & 16 & -12 \\ -12 & -12 & -12 & 36 \end{bmatrix}\right],$$

with respect to the matrix r^δ are 1, 9/10, 3/4, 0 with respective multiplicities 1, 2, 2, 2. This implies that the design is a (1; 2, 2; 2)-EB design. This can also be given through the consideration of the portions N_1 and N_2 separately. In fact, (i) N_1 is the incidence matrix of a BIB design with parameters $v = b = 3$, $r = k = 2$, $\lambda = 1$, and hence it is a (0; 2; 0)-EB design with $\varepsilon_1 = 3/4$ and $L_1 = I_3 - 3^{-1}\mathbf{1}_3\mathbf{1}'_3$, and (ii) for N_2, it can be shown that the corresponding M_0 matrix is

$$r_2^{-\delta}N_2k_2^{-\delta}N'_2 - \frac{1}{15}\mathbf{1}_4[2, 2, 2, 9] = \frac{1}{10}\text{diag}[I_3 - \frac{1}{3}\mathbf{1}_3\mathbf{1}'_3 : 0],$$

which shows that N_2 is the incidence matrix of a $(\rho_0; \rho_1; 0)$-EB design with parameters $v = 4$, $b = 3$, $r = [2\mathbf{1}'_3, 9]'$, $k = 5$, $\varepsilon_0 = 1$, $\varepsilon_1 = 9/10$, $\rho_0 = 1$, $\rho_1 = 2$, $L_0 = I_4 - L_1 - (15)^{-1}\mathbf{1}_4[2, 2, 2, 9]$, $L_1 = \text{diag}[I_3 - 3^{-1}\mathbf{1}_3\mathbf{1}'_3 : 0]$. Thus, the resulting disconnected design is a $(\rho_0^*; \rho_1^*, \rho_2^*; 2)$-EB design with parameters

$$\begin{aligned} &v^* = 7,\ b^* = 6,\ r^* = [2\mathbf{1}'_6, 9]',\ k^* = [2\mathbf{1}'_3, 5\mathbf{1}'_3]',\ \varepsilon_0^* = 1,\ \rho_0^* = 1, \\ &L_0^* = \text{diag}[\mathbf{O} : I_4 - (15)^{-1}\mathbf{1}_4[2, 2, 2, 9] - \text{diag}[I_3 - 3^{-1}\mathbf{1}_3\mathbf{1}'_3 : 0]], \\ &\varepsilon_1^* = 9/10,\ \rho_1^* = 2,\ L_1^* = \text{diag}[\mathbf{O} : I_3 - 3^{-1}\mathbf{1}_3\mathbf{1}'_3 : 0], \\ &\varepsilon_2^* = 3/4,\ \rho_2^* = 2,\ L_2^* = \text{diag}[I_3 - 3^{-1}\mathbf{1}_3\mathbf{1}'_3 : \mathbf{O}], \\ &\varepsilon_3^* = 0,\ \rho_3^* = 2,\ L_3^* = I_7 - (21)^{-1}\mathbf{1}_7(r^*)' - L_1^* - L_2^*. \end{aligned}$$

Recall that in this example the matrix C with respect to the matrix r^δ has two zero eigenvalues. In connection with this note that there exists a contrast between the first three treatments and the second four treatments, for which the intra-block BLUE does not exist. But there exists the inter-block BLUE for it.

Routinely the matrix Ω can be calculated, as mentioned in Section 4.4, as follows.

$$\Omega^* = \frac{1}{9}\text{diag}\left[6I_3 - \frac{1}{2}\mathbf{1}_3\mathbf{1}'_3 : \frac{1}{3}\begin{bmatrix} 29 & -1 & -1 & 0 \\ -1 & 29 & -1 & 0 \\ -1 & -1 & 29 & 0 \\ 0 & 0 & 0 & 3 \end{bmatrix}\right].$$

Though not desirable in general, disconnected designs can be useful when certain contrasts, one or more, have deliberately to be totally confounded with blocks in order to allow the remaining contrasts of interest to gain in precision under the available block sizes. In particular, this may be desirable for some factorial schemes of treatments (see, e.g., Finney, 1960, p.67; Pearce, 1983, Sections 3.7 and 8.3).

An illustration of the analysis of experimental data from an experiment conducted in a nonbinary, proper, nonequireplicate and disconnected block design is given in Example 3.8.1.

Appendix
B. Finite Fields

A certain amount of basic knowledge of some algebra is assumed throughout the book, Volume II in particular. However, it will be useful to recall some notions and results concerning finite fields, particularly those related to constructions of block designs. In fact, finite fields have proved to be very useful in the construction of orthogonal Latin squares, block designs, factorial designs, and many other combinatorial designs. No attempt has been made for completeness of the presentation and the interested reader is referred to the indicated references for proofs and details.

Let a, b and n be integers. Then, if there exists an integer ℓ such that $a - b = \ell n$, the relation is denoted by $a \equiv b \pmod{n}$, i.e., it is said that a is "congruent" to b to modulus n. In this case, the following properties hold:

(1) $a \equiv b \pmod{n}$ if and only if $a \pm c \equiv b \pm c \pmod{n}$ for any integer c;

(2) $a \equiv b \pmod{n}$ implies $ac \equiv bc \pmod{n}$ for any integer c;

(3) $a \equiv b \pmod{n}$ and $c \equiv d \pmod{n}$ imply $ac \equiv bd \pmod{n}$ and $a \pm c \equiv b \pm d \pmod{n}$. In particular, if $g = (c, n)$, i.e., g is the greatest common divisor of c and n, then in (2) $ac \equiv bc \pmod{n}$ implies $a \equiv b \pmod{n/g}$. In particular, when $g = 1$, i.e., $(c, n) = 1$, c and n are said to be relatively prime.

Now modules, rings and fields can be introduced following standard literature, such as, for example, Birkhoff and Maclane (1953), Mirsky (1955), and Hoffman and Kunze (1961).

(A) A "module" or an abelian group M is a set of elements, a, b, c,... for which there is a law of composition, i.e., the addition, denoted by $+$, for which the following axioms are satisfied:

I(i) To any two elements a and b of M, there exists a unique element $a + b$ belonging to M.

I(ii) $a + b = b + a$.
I(iii) $a + (b + c) = (a + b) + c$.
I(iv) To any two elements a and b, there exists an element x belonging to M such that $a + x = b$.

The following theorems can be proved.

Theorem B.1. *In M, the element x in* I(iv) *is unique.*

Theorem B.2. *In M, there exists a unique element zero, denoted by* 0, *such that $c + 0 = c$ for any element c of M.*

(B) A "ring" R is a module for which there exists a second law of composition, i.e., the multiplication, denoted by $\times$ or a dot (replaceable by the juxtaposition of the elements to indicate multiplication), satisfying the following axioms:
II(i) To any two elements a and b in R, there exists a unique element ab belonging to R.
II(ii) $(ab)c = a(bc)$.
II(iii) $a(b + c) = ab + ac, \quad (b + c)a = ba + ca$.

The following theorem can be proved.

Theorem B.3. *In R, $0a = a0 = 0$ for any a in R.*

The ring R is said to be "commutative" if
II(iv) $ab = ba$ for any a and b in R.

(C) A "field" F is a commutative ring for which the following axiom is further satisfied:
III. To any two elements a ($\neq 0$) and b in F, there exists an element y in F such that $ya = b$.

In this case, the following theorems can be shown.

Theorem B.4. *In F, the element y in* (III) *is unique.*

Theorem B.5. *In F, there exists a unique element* 1 ($\neq 0$) *such that $c1 = c$ for any element c of F.*

The element 1 is called the "unit element" of F. In particular, a field with a finite number of elements, n, say, is called a "Galois field." It is denoted by GF(n). Galois fields are of special interest in connexion with the present book. For details of finite field theory the reader is referred to Carmichael (1956), Albert (1956) or Hall (1986). Proofs of some results are given to make them more understandable by the interested reader.

Remark B.1. In a finite field, the axiom II(iv), i.e., commutativity on product, is automatically valid.

Theorem B.6. *In a field F, $ab = 0$ implies $a = 0$ or $b = 0$.*

Theorem B.7. *In a field F, if there exists an integer m such that $m1 = 0$, then*

$mx = 0$ for all x in F.

Theorem B.8. *In a field F, if there exists an integer m such that $mx = 0$ for some $x \neq 0$, then $my = 0$ for all y in F.*

Theorem B.9. *In a field, if p is the least positive integer such that $p1 = 0$ holds, then p is a prime.*

Proof. If p is not a prime, one can let $p = mn$, $1 < m, n < p$. Then $p1 = mn1 = (m1)(n1) = 0$, which, from Theorem B.6, implies $m1 = 0$ or $n1 = 0$. This is in contradiction to the assumption of p. □

Remark B.2. In Theorem B.5, $c \in F$, whereas, in Theorem B.9, p is simply a positive integer, not necessarily belonging to the field.

Definition B.1. The number p in Theorem B.9 is called a characteristic of the field.

Theorem B.10. *In a finite field there always exists a characteristic p and the number of elements in the field is a power of the prime p.*

In view of Theorem B.10, the finite field is denoted by GF(p^m) for some positive integer m. The simplest example (when $m = 1$) of a Galois field is a set of so-called residue classes (mod p) when p is a prime. This is expressed as GF$(p) = \{0, 1, ..., p-1\}$ $(= Z_p$, say), with addition and multiplication defined modulus p. For example, when $p = 5$, the addition $a + b$ and multiplication ab can be given as shown below.

$a + b$:

+	0	1	2	3	4
0	0	1	2	3	4
1	1	2	3	4	0
2	2	3	4	0	1
3	3	4	0	1	2
4	4	0	1	2	3

ab:

·	0	1	2	3	4
0	0	0	0	0	0
1	0	1	2	3	4
2	0	2	4	1	3
3	0	3	1	4	2
4	0	4	3	2	1

Note that a table of addition forms a Latin square of order 5 (see Appendix D).

If x is any nonzero element of GF(p), it follows from the Fermat theorem [i.e., $a^{p-1} \equiv 1 \pmod{p}$ for a and p which are relatively prime] that $x^{p-1} = 1$ where 1 is the unit element of GF(p).

Definition B.2. If m is the least positive integer such that $x^m = 1$, then m is said to be the order of the element x $(\neq 0)$.

Theorem B.11. *The order of each element in GF(p) is a divisor of $p - 1$.*

Proof. Let m be the order of x and let $p - 1 = am + b$, $0 \leq b < m$. If $b \neq 0$ then $1 = x^{p-1} = x^{am+b} = (x^m)^a x^b$, which contradicts that m is the order of x. Hence $b = 0$, i.e., $p - 1$ is divisible by m. □

Definition B.3. If m as the order of x takes the maximum value, $p - 1$, in

Theorem B.11, x is called a primitive element of GF(p).

Since GF(p) is a finite field, by Theorem B.11 there always exists a primitive element for every GF(p). Let $\phi(m)$ be the number of integers m' such that $1 \leq m' < m$ and $(m', m) = 1$, i.e., m' and m are relatively prime. Then one has the following result.

Theorem B.12. *In a GF(p), there are $\phi(p-1)$ primitive elements.*

Proof. Since any nonzero element x of GF(p) satisfies $x^{p-1} \equiv 1 \pmod{p}$ for $(x, p) = 1$, i.e., the Fermat theorem, Theorem B.11 shows that the maximum number of the order of the element in GF(p) is $p-1$. Then there is at least one primitive element, w, say, in GF(p). In this case w^i for $(i, p-1) = 1$ is also a primitive element for some i $(1 \leq i \leq p-2)$. Hence, it follows that the number of such i is given by $\phi(p-1)$. Since the nonzero elements in GF(p) are included in the sequence $w^0 = 1$, w^1, w^2, ... , w^{p-2}, there are exactly $\phi(p-1)$ primitive elements. □

For example, in GF(p=5) = Z_5, first consider 2 as a candidate of primitive elements. Then it is calculated that $2^1 = 1, 2^2 = 4$, $2^3 = 8 \equiv 3 \pmod 5$, $2^4 \equiv 6 \equiv 1 \pmod 5$. This means that 2 is a primitive element of GF(5). Next, as in the proof of Theorem B.12, 2^3 must be another primitive element of GF(5), because $(3, p-1) = 1$, as $p = 5$. In fact, as $2^3 \equiv 3$, it holds that $3^1 = 3, 3^2 = 9$ $\equiv 4$, $3^3 \equiv 12 \equiv 2$, $3^4 \equiv 6 \equiv 1$. Thus, one has two primitive elements, 2 and 3, as also $\phi(p-1) = \phi(4) = 2$. In Table B.1, only one primitive element is listed for each of the Galois fields.

Remark B.3. $\phi(m)$ is called the Euler function. Recall that $\phi(m)$ is the number of integers m' such that $1 \leq m' < m$ and $(m', m) = 1$. It holds that in general

$$\phi(p_1^{s_1} \cdots p_\ell^{s_\ell}) = p_1^{s_1} \cdots p_\ell^{s_\ell}(1 - \frac{1}{p_1}) \cdots (1 - \frac{1}{p_\ell}).$$

As described in the proof of Theorem B.12, every nonzero element in GF(p) is included in the sequence (called the "power cycle" of x) $x^0 = 1$, x^1, x^2, ... , x^{p-2} generated by a primitive element x. That is, the multiplication in GF(p) is cyclic. Here the power cycles of a primitive element of GF(p) for the values $p = 3$, 5, 7, 11, 13, 17, 19, 23, 31, are tabulated in Table B.1.

A Galois field GF(p^s) may also be constructed by using a polynomial of degree s with coefficients in GF(p). Let $\mathrm{GF}_p[x]$ be a set of polynomials whose coefficients are from GF(p). Then it follows from a direct checking of II(i) – II(iv) that $\mathrm{GF}_p[x]$ is a commutative ring under the usual addition and multiplication of polynomials.

Definition B.4. A polynomial $f(x)$ of $\mathrm{GF}_p[x]$ is said to be irreducible if it is impossible to express it as the product of two polynomials of $\mathrm{GF}_p[x]$ of positive degrees.

For example, in GF(2^2), x^2+1 is not irreducible, but x^2+x+1 is irreducible.

Because $x^2 + 1 \equiv (x+1)(x+1) \pmod 2$.

Let $f(x)$ be an irreducible polynomial of degree s of $\mathrm{GF}_p[x]$. Then two polynomials $g_1(x)$ and $g_2(x)$ are said to be "congruent" [mod $f(x)$] if $g_1(x)-g_2(x)$ is divisible by $f(x)$. Consider the classes of polynomials of $\mathrm{GF}_p[x]$ congruent [mod $f(x)$]. It can be shown that these classes form a Galois field, under the usual addition and multiplication of polynomials. Since the coefficients of these polynomials are elements of $\mathrm{GF}(p)$, every element of $\mathrm{GF}(p^s)$, with p^s elements, can be expressed in the form $a_0 + a_1x + \cdots + a_{s-1}x^{s-1}$ whose coefficients a_0, $a_1, ..., a_{s-1}$ are from $\mathrm{GF}(p)$.

Similarly to a case of $\mathrm{GF}(p)$, one can show the following. As a generalization of the Fermat theorem, any nonzero element α of $\mathrm{GF}(p^s)$ satisfies $\alpha^{p^s-1} = 1$. If m is the least positive integer such that $\alpha^m = 1$, then m is said to be the order of x. If t is the order of α in $\mathrm{GF}(p^s)$, then $p^s - 1$ is divisible by t. In particular, when $m = p^s - 1$, α is called a "primitive element" of $\mathrm{GF}(p^s)$ (see also Definition B.3). In $\mathrm{GF}(p^s)$, there are $\phi(p^s - 1)$ primitive elements.

Definition B.5. The function $f(x)$, with the help of which $\mathrm{GF}(p^s)$ is constructed, is called the minimum function.

Thus, a minimum function can be obtained among irreducible polynomials of proper degree.

If the minimum function $f(x)$ is suitably chosen, then x will play a role of a primitive element of $\mathrm{GF}(p^s)$, i.e., $x^{p^s-1} \equiv 1 \pmod{f(x)}$. Then all the nonzero elements are expressed by the sequence $x^0 = 1, x, x^2, ..., x^{p^s-2}$, which is also called the power cycle of x (see Table B.3).

As described above, if there exists an irreducible polynomial of degree s with coefficients in $\mathrm{GF}(p)$, one can construct $\mathrm{GF}(p^s)$. In general, the following result is obtained (see Carmichael, 1956; Albert, 1956).

Theorem B.13. *For any prime p and any positive integer s, there exists a Galois field $\mathrm{GF}(p^s)$.*

For example, in $\mathrm{GF}(2^2) = \{0, 1, x, x^2 = x+1\}$, in which $x^2 + x + 1$ is a minimum function (see Table B.3), the results under two operations, addition and multiplication, can be shown below:

$+$	0	1	x	x^2
0	0	1	x	x^2
1	1	0	x^2	x
x	x	x^2	0	1
x^2	x^2	x	1	0

$\cdot$	0	1	x	x^2
0	0	0	0	0
1	0	1	x	x^2
x	0	x	x^2	1
x^2	0	x^2	1	x

It is finally convenient to know the standard representatives of the successive elements of the power cycle. At first some minimum functions for p^s ($p = 2, 3, 5$; $s = 2,3,4,5,6$) are listed in Table B.2. Next the power cycles for cases $p^s = 2^2$, 2^3, 3^2, 2^4, 5^2 are given in Table B.3.

Remark B.4. It is known (see Alanen and Knuth, 1964) that in $\mathrm{GF}(p^s)$ there are exactly $\phi(p^s - 1)/s$ minimum functions, whose ϕ is the Euler function as in Remark B.3. In fact, Alanen and Knuth (1964) tabulated all minimum functions by hand or by digital computer within the range of $p^s \leq 1024$, and a minimum function for each field with $p^s < 10^9$, where $p < 50$. Some of them can be found in Table B.2. Though it is sufficient to know the existence of a minimum function to construct GF(p^s), it may be useful to know several minimum functions for the investigation of nonisomorphic solutions of BIB designs, discussed in Chapter 8, which are constructed by use of Galois fields.

Thus, a Galois field has two representations of elements. For example, by Table B.3, GF$(2^3) = \{0, 1, x, x^2, x^3, x^4, x^5, x^6\}$ (in the light of a primitive element x), and, by Table B.2, GF$(2^3) = \{0, 1, x, x^2, x^2+1, x^2+x+1, x+1, x^2+x\}$ (in the light of a minimum function $x^3 + x^2 + 1$).

With the help of Tables B.2 and B.3, it is easy to perform the four fundamental operations. For addition and subtraction one may use polynomial expressions, while for multiplication and division one may use power expressions, along with $x^{p^s-1} = 1$ if necessary. For example in GF(3^2), if $\alpha = 1 + 2x$ and $\beta = 1 + x$, then it holds that

$$\alpha + \beta = 2 + 3x = 2,\ \alpha - \beta = x,\ \alpha\beta = x^2x^7 = x^9 = x,$$
$$\alpha/\beta = x^2/x^7 = x^{-5} = x^3 = 2 + 2x$$

(see Table B.3 and Example 6.7.3).

Note that in the field of characteristic p =2, the addition is equivalent to subtraction.

Table B.1. Power cycle I

p	A primitive element	Power cycle
3	2	1, 2
5	2	1, 2, 4, 3
7	3	1, 3, 2, 6, 4, 5
11	2	1, 2, 4, 8, 5, 10, 9, 7, 3, 6
13	2	1, 2, 4, 8, 3, 6, 12, 11, 9, 5, 10, 7
17	3	1, 3, 9, 10, 13, 5, 15, 11, 16, 14, 8, 7, 4, 12, 2, 6
19	2	1, 2, 4, 8, 16, 13, 7, 14, 9, 18, 17, 15, 11, 3, 6, 12, 5, 10
23	5	1, 5, 2, 10, 4, 20, 8, 17, 16, 11, 9, 22, 18, 21, 13, 19, 3, 15, 6, 7, 12, 14
29	2	1, 2, 4, 8, 16, 3, 6, 12, 24, 19, 9, 18, 7, 14, 28, 27 25, 21, 13, 26, 23, 17, 5, 10, 20, 11, 22, 15
31	3	1, 3, 9, 27, 19, 26, 16, 17, 20, 29, 25, 13, 8, 24, 10, 30, 28, 22, 4, 12, 5, 15, 14, 11, 2, 6, 18, 23, 7, 21
37	2	1, 2, 4, 8, 16, 32, 27, 17, 34, 31, 25, 13, 26, 15, 30, 23, 9, 18, 36, 35, 33, 29, 21, 5, 10, 20, 3, 6, 12, 24, 11, 22, 7, 14, 28, 19

Table B.2. Minimum functions

$p\backslash s$	2	3	4	5	6
2	x^2+x+1	x^3+x^2+1, x^3+x+1	x^4+x^3+1, x^4+x+1	x^5+x^3+1, x^5+x^2+1, $x^5+x^4+x^3+x^2+1$, $x^5+x^4+x^3+x+1$, $x^5+x^3+x^2+x+1$, $x^5+x^4+x^2+x+1$,	x^6+x^5+1, x^6+x+1, $x^6+x^5+x^3+x^2+1$, $x^6+x^5+x^4+x+1$, $x^6+x^4+x^3+x+1$, $x^6+x^5+x^2+1$
3	x^2+x+2, x^2+2x+2	x^3+2x^2+1, x^3+2x^2+x+1, x^3+2x+1, x^3+x^2+2x+1	x^4+x^3+2, x^4+2x^3+2, x^4+2x+2, x^4+x+2, $x^4+2x^3+x^2+x+2$, $x^4+2x^3+2x^2+x+2$, (more two functions)	x^5+2x^4+1, $x^5+x^4+2x^3+1$, $x^5+x^4+x^2+1$, $x^5+2x^3+x^2+1$, $x^5+x^3+2x^2+1$, x^5+2x^4+x+1, (more 16 functions)	x^6+x^5+2, x^6+2x^5+2, x^6+x+2, $x^6+x^5+x^3+2$, $x^6+2x^5+2x^3+2$, $x^6+x^5+2x^4+x^2+2$, $x^6+x^5+x^4+2x^2+2$, (more 41 functions)
5	x^2+x+2, x^2+4x+2, x^2+2x+3, x^2+3x+3	x^3+x^2+2, x^3+3x^2+2, x^3+4x^2+3, x^3+2x^2+3, (more 16 functions)	$x^4+x^3+2x^2+2$, $x^4+4x^3+2x^2+2$, (more 46 functions)	x^5+x^2+2, (more 279 functions)	x^6+x^5+2, (more 719 functions)
7	x^2+x+3, x^2+2x+3, x^2+5x+3, x^2+6x+3, (more four functions)	x^3+x^2+x+2, x^3+6x^2+x+2, (more 34 functions)	$x^4+x^3+x^2+3$, (more 799 functions)	x^5+x^4+4, (more 1119 functions)	$x^6+x^5+x^4+3$, (more 6047 functions)

Primitive elements and minimum functions play a vital role to generate Galois fields. In Table B.1, the power cycles of a primitive element of GF(p) for some p are given, whereas several minimum functions for generating the elements of GF(p^s) are given in Table B.2 for some p and s. In Table B.3, the power cycles for some GF(p^s) are presented.

Table B.3. Power cycle II

p^s	A minimum function	Power cycle
2^2	x^2+x+1	$x^0=1, x^1=x, x^2=1+x$
2^3	x^3+x^2+1	$1, x, x^2, x^3=1+x^2, x^4=1+x+x^2, x^5=1+x, x^6=x+x^2$
2^4	x^4+x^3+1	$1, x, x^2, x^3, x^4=1+x^3, x^5=1+x+x^3, x^6=1+x+x^2+x^3,$ $x^7=1+x+x^2, x^8=x+x^2+x^3, x^9=1+x^2, x^{10}=x+x^3,$ $x^{11}=1+x^2+x^3, x^{12}=1+x, x^{13}=x+x^2, x^{14}=x^2+x^3$
2^5	x^5+x^3+1	$1, x, x^2, x^3, x^4, x^5=1+x^3, x^6=x+x^4, x^7=1+x^2+x^3,$ $x^8=x+x^3+x^4, x^9=1+x^2+x^3+x^4, x^{10}=1+x+x^4,$ $x^{11}=1+x+x^2+x^3, x^{12}=x+x^2+x^3+x^4,$ $x^{13}=1+x^2+x^4, x^{14}=1+x, x^{15}=x+x^2, x^{16}=x^2+x^3,$ $x^{17}=x^3+x^4, x^{18}=1+x^3+x^4, x^{19}=1+x+x^3+x^4,$ $x^{20}=1+x+x^2+x^3+x^4, x^{21}=1+x+x^2+x^4,$ $x^{22}=1+x+x^2, x^{23}=x+x^2+x^3, x^{24}=x^2+x^3+x^4,$ $x^{25}=1+x^4, x^{26}=1+x+x^3, x^{27}=x+x^2+x^4,$ $x^{28}=1+x^2, x^{29}=x+x^3, x^{30}=x^2+x^4$
3^2	x^2+x+2	$1, x, x^2=1+2x, x^3=2+2x, x^4=2, x^5=2x, x^6=2+x,$ $x^7=1+x$
3^3	x^3+2x^2+1	$1, x, x^2, x^3=2+x^2, x^4=2+2x+x^2, x^5=2+2x,$ $x^6=2x+2x^2, x^7=1+x^2, x^8=2+x+x^2,$ $x^9=2+2x+2x^2, x^{10}=1+2x+x^2, x^{11}=2+x,$ $x^{12}=2x+x^2, x^{13}=2, x^{14}=2x, x^{15}=2x^2, x^{16}=1+2x^2,$ $x^{17}=1+x+2x^2, x^{18}=1+x, x^{19}=x+x^2, x^{20}=2+2x^2,$ $x^{21}=1+2x+2x^2, x^{22}=1+x+x^2, x^{23}=2+x+2x^2,$ $x^{24}=1+2x, x^{25}=x+2x^2$
5^2	x^2+2x+3	$1, x, x^2=2+3x, x^3=1+x, x^4=2+4x, x^5=3+4x, x^6=3,$ $x^7=3x, x^8=1+4x, x^9=3+3x, x^{10}=1+2x, x^{11}=4+2x,$ $x^{12}=4, x^{13}=4x, x^{14}=3+2x, x^{15}=4+4x, x^{16}=3+x,$ $x^{17}=2+x, x^{18}=2, x^{19}=2x, x^{20}=4+x, x^{21}=2+2x,$ $x^{22}=4+3x, x^{23}=1+3x$
7^2	x^2+x+3	$1, x, x^2=4+6x, x^3=3+5x, x^4=6+5x, x^5=6+x,$ $x^6=4+5x, x^7=6+6x, x^8=3, x^9=3x, x^{10}=5+4x,$ $x^{11}=2+x, x^{12}=4+x, x^{13}=4+3x, x^{14}=5+x,$ $x^{15}=4+4x, x^{16}=2, x^{17}=2x, x^{18}=1+5x, x^{19}=6+3x,$ $x^{20}=5+3x, x^{21}=5+2x, x^{22}=1+3x, x^{23}=5+5x,$ $x^{24}=6, x^{25}=6x, x^{26}=3+x, x^{27}=4+2x, x^{28}=1+2x,$ $x^{29}=1+6x, x^{30}=3+2x, x^{31}=1+x, x^{32}=4, x^{33}=4x,$ $x^{34}=2+3x, x^{35}=5+6x, x^{36}=3+6x, x^{37}=3+4x,$ $x^{38}=2+6x, x^{39}=3+3x, x^{40}=5, x^{41}=5x, x^{42}=6+2x,$ $x^{43}=1+4x, x^{44}=2+4x, x^{45}=2+5x, x^{46}=6+4x,$ $x^{47}=2+2x$

Appendix
C. Finite Geometries

Block designs have close connections with finite geometrical configurations, and a large number of methods of constructing block designs are based on finite geometries over a finite field discussed in Appendix B (see Chapters 6 and 8). Details of these connections can be seen in Bose (1939) or Beth, Jungnickel and Lenz (1985). A detailed explanation about finite geometries can be found in Dembowski (1968) and Raghavarao (1971, Appendix A.7).

C.1. Finite projective geometry

The finite projective geometry of t dimensions over the field $\mathrm{GF}(p^s)$ for a prime p, denoted by $\mathrm{PG}(t, p^s)$, is defined as follows. First, let a function be

$$\phi(t,d,q) = \frac{(q^{t+1}-1)(q^t-1)\cdots(q^{t-d+1}-1)}{(q^{d+1}-1)(q^d-1)\cdots(q-1)},$$

which will be shown to be the number of d-dimensional subspaces in $\mathrm{PG}(t, q)$ with $q = p^s$. It should be noted that $\phi(t, d, q) = \phi(t, t-d-1, q)$. It is also convenient to set formally $\phi(t, -1, q) = 1$.

(i) Any vector of $t+1$ elements $[x_0, x_1, ..., x_t]$ where x_i's belong to $\mathrm{GF}(q)$ and are not all simultaneously zero, is called a "point" of $\mathrm{PG}(t, q)$. Two vectors $[x_0, x_1, ..., x_t]$ and $[y_0, y_1, ..., y_t]$ are said to present the same point when and only when there exists a nonzero element c in $\mathrm{GF}(q)$ such that $y_i = cx_i$ for $i = 0, 1, ..., t$. It is easy to show that the number of points in $\mathrm{PG}(t, q)$ is exactly

$$\frac{q^{t+1}-1}{q-1} \quad [= \phi(t, 0, q)]. \tag{C.1}$$

(ii) A "line" joining the point $P_1 = [x_0, x_1, ..., x_t]$ and $P_2 = [x'_0, x'_1, ..., x'_t]$ is given by a set of the following points

$$\lambda_1 P_1 + \lambda_2 P_2 = [\lambda_1 x_0 + \lambda_2 x_0', \lambda_1 x_1 + \lambda_2 x_1', ..., \lambda_1 x_t + \lambda_2 x_t'],$$

$$[\lambda_1, \lambda_2] \neq [0,0], \quad \lambda_1, \lambda_2 \in \mathrm{GF}(q).$$

As there are $q^2 - 1$ choices of $[\lambda_1, \lambda_2] \neq [0,0]$, and among them $q-1$ choices yield the same nonzero points, it follows that there are

$$\frac{q^2-1}{q-1} = 1 + q \; [= \phi(t, 0, q)] \tag{C.2}$$

points in a line joining two distinct points.

In general, let P_1, P_2, ... , P_{d+1} be $d+1$ linearly independent points in $\mathrm{PG}(t,q)$. Then all the points as

$$\lambda_1 P_1 + \lambda_2 P_2 + \cdots + \lambda_{d+1} P_{d+1}, \; \lambda_i \in \mathrm{GF}(q), \; i = 1, 2, ..., d+1,$$

$$[\lambda_1, \lambda_2, ..., \lambda_{d+1}] \neq [0, 0, ..., 0],$$

are said to form a "d-flat" (or d-dimensional subspace) in $\mathrm{PG}(t,q)$. Here the linear independence of P_1, P_2, ... , P_{d+1} means that $\lambda_1 P_1 + \lambda_2 P_2 + \cdots + \lambda_{d+1} P_{d+1} =$ [0,0,...,0] holds only when $\lambda_1 = \lambda_2 = \cdots = \lambda_{d+1} = 0$. Two points $\sum_{i=1}^{d+1} \lambda_i P_i$ and $\sum_{i=1}^{d+1} \mu_i P_i$ present the same point when and only when $[\lambda_1, \lambda_2, ..., \lambda_{d+1}] = \nu[\mu_1, \mu_2, ..., \mu_{d+1}]$, $\nu \neq 0$. In fact, $\sum_i \lambda_i P_i = \nu \sum_i \mu_i P_i$ implies $\sum_i (\lambda_i - \nu\mu_i) P_i =$ [0,0,...,0], which, from linear independence of P_i, yields $\lambda_i = \nu\mu_i$ for all i. From this observation, as the number of points on a d-flat can be given by counting distinct $d+1$ linearly independent nonzero points in $\mathrm{PG}(t,d)$, it follows that the number of points on a d-flat in $\mathrm{PG}(t,q)$ is given by $(q^{d+1}-1)/(q-1)$ $[= \phi(d,0,q)]$. A 0-flat is, of course, identical with a point, and following the usual nomenclature, a 1-flat can be called a line, and a 2-flat be a plane. Now the number of d-flats satisfying some properties can be counted.

Theorem C.1.1. *The number of d-flats in $\mathrm{PG}(t,q)$ is given by $\phi(t,d,q)$.*

Proof. Consider the way of choosing $d+1$ linearly independent points in $\mathrm{PG}(t,q)$ to define a d-flat. The first point P_1 can be chosen in $1+q+\cdots+q^t$ ways. The second point P_2 can be chosen in $q+\cdots+q^t$ ways. When $\ell(< d+1)$ points are chosen, the $(\ell+1)$th point $P_{\ell+1}$ should be taken to be linearly independent of the previously-chosen ℓ points. This is to take the remaining points after excluding all the points in the $(\ell-1)$-flat spanned by the ℓ points P_1,..., P_ℓ. That is, $P_{\ell+1}$ can be chosen in $q^\ell + \cdots + q^t$ ways. Hence, $d+1$ linearly independent points can be chosen in $(1+q+\cdots+q^t)(q+\cdots+q^t)\cdots(q^d+\cdots+q^t)$ ways. On the other hand, the number of sets of $d+1$ linearly independent points in a d-flat is given by $(1+q+\cdots+q^d)(q+\cdots+q^d)\cdots(q^{d-1}+q^d)q^d$. These determine the same d-flat. Therefore, the number of d-flats in $\mathrm{PG}(t,q)$ is given by

$$\frac{(1+q+\cdots+q^t)(q+\cdots+q^t)\cdots(q^d+\cdots+q^t)}{(1+q+\cdots+q^d)(q+\cdots+q^d)\cdots(q^{d-1}+q^d)q^d}$$

$$= \frac{(q^{t+1}-1)(q^t-1)\cdots(q^{t-d+1}-1)}{(q^{d+1}-1)(q^d-1)\cdots(q-1)} \quad [= \phi(t,d,q)]. \quad \square$$

Theorem C.1.2. *The number of d-flats including a u-flat in PG(t, q) is, for $u < d$, given by $\phi(t-u-1, d-u-1, q)$.*

Proof. Consider a point P_{u+2} outside the u-flat spanned by points P_1, P_2, ... , P_{u+1}. Then P_{u+2} can be chosen in $q^{u+1} + \cdots + q^t$ ways. Form a $(u+1)$-flat spanned by points P_1, P_2, ... , P_{u+1}, P_{u+2}. Next, take a point P_{u+3} which is not included in the $(u+1)$-flat. Then this point can be chosen in $q^{u+2} + \cdots + q^t$ ways. By taking the same procedure one can get a d-flat including the started u-flat. Since a d-flat is obtained in $(q^{u+1} + \cdots + q^d) \cdots (q^{d-1} + q^d)q^d$ ways, the required number can be provided. $\square$

Corollary C.1.2.1. *The number of d-flats including a particular point in PG(t, q) is given by $\phi(t-1, d-1, q)$.*

Corollary C.1.2.2. *The number of d-flats including two distinct points in PG(t, q) is given by $\phi(t-2, d-2, q)$.*

Some application of the above theory to block designs can be seen in Section 6.9, particularly in Theorem 6.9.1. There, points and d-flats correspond to treatments and blocks, respectively. Then by (C.1), (C.2), Theorem C.1.1, and Corollaries C.1.2.1 and C.1.2.2, one has the parameters of a BIB design as $v = \phi(t,0,q)$, $b = \phi(t,d,q)$, $r = \phi(t-1,d-1,q)$, $k = \phi(d,0,q)$, $\lambda = \phi(t-2,d-2,q)$. The design constructed as in Theorem 6.9.1 is denoted by PG(t,q) : d. In particular, when $t = 2$ and $d = 1$, PG$(2,q)$: 1 is called a finite projective plane of order q, which is equivalent to a symmetric BIB design with parameters $v = \phi(2,0,q) = q^2+q+1$, $b = \phi(2,1,q) = q^2+q+1$, $r = \phi(1,0,q) = k = q+1$, $\lambda = \phi(0,-1,q) = 1$ [see Corollary 6.9.1(i)]. In this sense, for each prime or prime power q, there exists a finite projective plane of order q.

C.2. Finite affine geometry

The finite affine geometry of t dimensions over the field GF(p^s) denoted by AG(t,p^s) is defined in a way similar to the case of PG(t,p^s). But here AG(t,p^s) is simply defined as a geometry consisting of the remaining points after excluding a $(t-1)$-flat (with its points) from PG(t,p^s). Points forming the $(t-1)$-flat are called points at infinity, and the other points are called finite points. Of course, both definitions provide the same concept of finite affine geometries. That is, a finite affine geometry is always embedded in a finite projective geometry.

Thus, one can derive several observations, useful to the present book, by use of properties on finite projective geometries described in Section C.1.

(1) The number of points in AG$(t,p^s = q)$ is given by [the number of points in PG(t,q)] minus [the number of points in PG$(t-1,q)$], i.e. $\phi(t,0,q) - \phi(t-1,0,q) = q^t$.

(2) The number of AG(d, q) which is included in AG(t, q) is given by $\phi(t, d, q)$ $-\phi(t-1, d, q) = q^{t-d}\phi(t-1, d-1, q)$.

(3) The number of AG(d, q) including a particular point in AG(t, q) is given by $\phi(t-1, d-1, q)$.

(4) The number of points of AG(d, q) in AG(t, q) is given by $\phi(d, 0, q) - \phi(d-1, 0, q) = q^d$.

(5) The number of AG(d, q) including two distinct points in AG(t, q) is given by $\phi(t-2, d-2, q)$.

Remark C.2.1. Raghavarao (1971, Appendix A.7.2) and most available literature use a term "Finite Euclidean Geometry" EG(t, q) instead of AG(t, q) here. However, usually, a metric axiom is not assumed in the geometry and, hence, the present term "affine geometry" is more recommendable for the use in the experimental design.

Some application of this theory to block designs can be seen in Section 6.9, particularly in Theorem 6.9.2. There, points and d-flats correspond to treatments and blocks, respectively. Then by (1) to (5) above, one has the parameters of a BIB design as $v = q^t$, $b = q^{t-d}\phi(t-1, d-1, q)$, $r = \phi(t-1, d-1, q)$, $k = q^d$, $\lambda = \phi(t-2, d-2, q)$. The design constructed as in Theorem 6.9.2 is denoted by AG$(t, q) : d$. In particular, when $t = 2$ and $d = 1$, AG$(2, q) : 1$ is called a finite affine plane of order q, which is equivalent to an affine resolvable BIB design with parameters $v = q^2$, $b = q(q+1)$, $r = q+1, k = q$, $\lambda = 1$ [see Corollary 6.9.3]. In this sense, for each prime or prime power q, there exists a finite affine plane of order q.

C.3. Illustration

A projective plane PG(2,2) and an affine plane AG(2,2) over GF(2) $= \{0, 1\}$ are illustrated in connection with BIB designs. Here, points and lines correspond to treatments and blocks, respectively, in a BIB design, as discussed in Sections C.1 and C.2.

In PG(2,2), as explained in Section C.1, there are

$$\phi(2, 0, 2) = 7 \ (= v \text{ in the BIB design})$$

points, i.e., $P_1 = [0,0,1], P_2 = [0,1,0], P_3 = [0,1,1]$, $P_4 = [1,0,0], P_5 = [1,0,1]$, $P_6 = [1,1,0], P_7 = [1,1,1]$, where P_1, P_2, P_4 are linearly independent, and there are

$$\phi(2, 1, 2) = 7 \ (= b \text{ in the BIB design})$$

lines, i.e., $\ell_1 = \{P_1, P_2, P_3\}$, $\ell_2 = \{P_2, P_4, P_6\}$, $\ell_3 = \{P_1, P_4, P_5\}$, $\ell_4 = \{P_1, P_6,$

$P_7\}$, $\ell_5 = \{P_3, P_4, P_7\}$, $\ell_6 = \{P_2, P_5, P_7\}$, $\ell_7 = \{P_3, P_5, P_6\}$, since there are

$\phi(1,\, 0, 2) = 3$ (= k in the BIB design)

points in a line joining two distinct points. Furthermore, there are

$\phi(2-1, 1-1, 2) = \phi(1, 0, 2) = 3$ (= r in the BIB design)

lines passing through any one point, and there are

$\phi(2-2, 1-2, 2) = \phi(0, -1, 2) = 1$ (= λ in the BIB design)

line passing through any two distinct points. Also note that any two distinct lines intersect in exactly one point.

This PG(2,2) is drawn below.

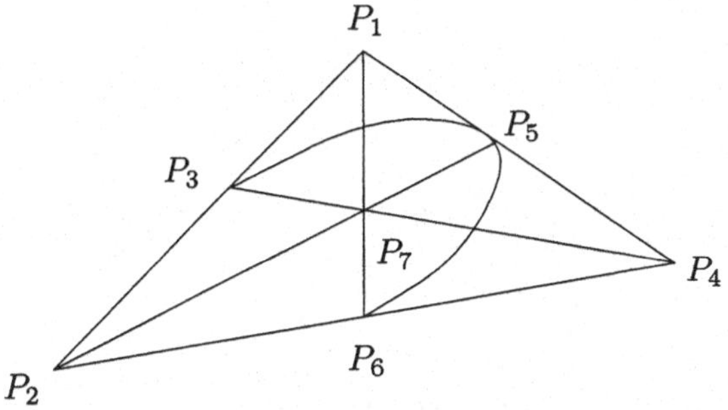

Fig. C.1. PG(2, 2)

This is also expressed in the incidence form as follows.

Matrix C.1

	ℓ_1	ℓ_5	ℓ_3	ℓ_6	ℓ_4	ℓ_2	ℓ_7
P_1	1	0	1	0	1	0	0
P_2	1	0	0	1	0	1	0
P_4	0	1	1	0	0	1	0
P_7	0	1	0	1	1	0	0
P_3	1	1	0	0	0	0	1
P_5	0	0	1	1	0	0	1
P_6	0	0	0	0	1	1	1

This exactly shows the 7×7 incidence matrix of a symmetric BIB design with parameters $v = b = 7$, $r = k = 3$, $\lambda = 1$.

As described in Section C.2, an affine plane AG(2,2) can be obtained by deleting three points at infinity, i.e., P_3, P_5, P_6 (say here), and also deleting a line $\{P_3, P_5, P_6\}$ (called a line at infinity) in the PG(2,2). Thus, in the AG(2,2), there are

$\phi(2, 0, 2) - \phi(1, 0, 2) = 4$ (= v in the BIB design)

points, i.e., P_1, P_2, P_4, P_7, and there are

$\phi(2, 1, 2) - \phi(1, 1, 2) = 6$ (= b in the BIB design)

lines, i.e., $\ell_1^* = \{P_1, P_2\}$, $\ell_2^* = \{P_2, P_4\}$, $\ell_3^* = \{P_1, P_4\}$, $\ell_4^* = \{P_1, P_7\}$, $\ell_5^* =$

$\{P_4, P_7\}$, $\ell_6^* = \{P_2, P_7\}$, and there are

$$\phi(1,0,2) = 3 \ (= r \text{ in the BIB design})$$

lines including a point,

$$\phi(1,0,2) - \phi(0,0,2) = 2 \ (= k \text{ in the BIB design})$$

points in each line,

$$\phi(0,-1,2) = 1 \ (= \lambda \text{ in the BIB design})$$

line passing through any two distinct points.

This AG(2,2) is drawn below.

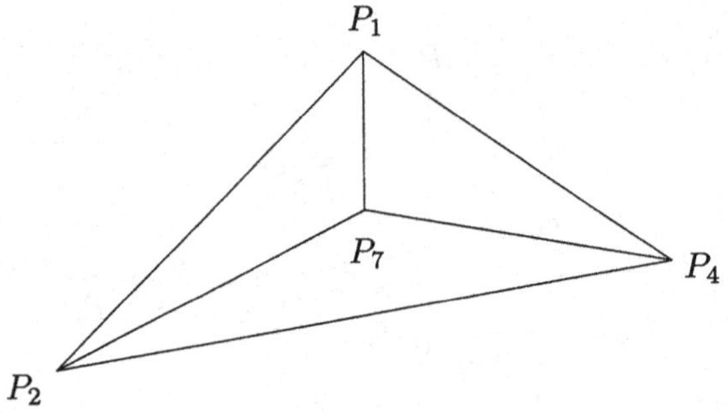

Fig. C.2. AG(2, 2)

This is also expressed in the incidence form as follows.

Matrix C.2

	ℓ_1^*	ℓ_5^*	ℓ_3^*	ℓ_6^*	ℓ_4^*	ℓ_2^*
P_1	1	0	1	0	1	0
P_2	1	0	0	1	0	1
P_4	0	1	1	0	0	1
P_7	0	1	0	1	1	0

This exactly shows the 4×6 incidence matrix of an affine resolvable BIB design with parameters $v = 4, b = 6$, $r = 3, k = 2$, $\lambda = 1$, $q_1 = 0, q_2 = 1$. Also note that pairs of lines, $\{\ell_1^*, \ell_5^*\}, \{\ell_3^*, \ell_6^*\}, \{\ell_2^*, \ell_4^*\}$ are parallel, i.e., no intersection of points occurs.

Finally, note that the matrix in Matrix C.2 is a submatrix of the incidence matrix as in Matrix C.1. This block structure can be generalized as an equivalent relation between a symmetric BIB design with parameters $v = b = q^2 + q + 1$, $r = k = q+1$, $\lambda = 1$ and an affine resolvable BIB design with parameters $v = q^2$, $b = q(q+1)$, $r = q+1, k = q$, $\lambda = 1$ [see Corollaries 6.9.1(i) and 6.9.2(i) with Examples 6.9.1 and 6.9.2]. That is, an affine plane of order q exists if and only if a finite projective plane of order q exists.

Appendix
D. Orthogonal Latin Squares

Latin squares are normally used as statistical designs to remove the heterogeneity of experimental material in two directions. They also find use in constructing some block designs. Detailed explanations about Latin squares and their applications can be found in Raghavarao (1971, Chapter 1), Colbourn and Dinitz (1996, pp. 111-142), Anderson (1997), and Laywine and Mullen (1998).

A Latin square of order h is defined as an arrangement of h symbols in h^2 cells laid out in h rows and h columns, such that every symbol occurs once in each row and in each column.

If in two Latin squares of the same order, when superimposed on one another, every ordered pair of symbols occurs exactly once, the two Latin squares are said to be "orthogonal." If in a set of Latin squares every pair of them is orthogonal, the set is called a set of mutually orthogonal Latin squares (MOLS). For example,

$$L_1 = \begin{bmatrix} 0 & 1 & 2 \\ 1 & 2 & 0 \\ 2 & 0 & 1 \end{bmatrix}, \qquad L_2 = \begin{bmatrix} 0 & 2 & 1 \\ 1 & 0 & 2 \\ 2 & 1 & 0 \end{bmatrix}$$

are MOLS of order three.

A square formed by superimposing two orthogonal Latin squares of order h is called the Euler square (or Graeco-Latin square) of order h. The existence problem of the Euler squares of order $h \equiv 2 \pmod 4$ (i.e., $h = 2, 6, 10, ...$) has been famous since Euler conjectured its existence in 1782. The non-existence of the case $h = 2$ is obvious. The case $h = 6$ is also disproved by Tarry (1900). However, in 1960, Bose, Shrikhande and Parker proved the existence for all $h \equiv 2 \pmod 4 \geq 10$.

When h is odd, the Euler square of order h can easily be constructed as follows: Let the rows and colums are numbered $0, 1, ..., h-1$. Then the (i, j)th position (z, w) of the Euler square is given by

$$z \equiv i + j \pmod{h},\ 0 \le z \le h-1; \quad w \equiv i - j \pmod{h},\ 0 \le w \le h-1,$$

which, for example, when $h = 3$, yields

$$\begin{bmatrix} 00 & 12 & 21 \\ 11 & 20 & 02 \\ 22 & 01 & 10 \end{bmatrix},$$

being a superimposition of the above two orthogonal Latin squares L_1 and L_2 of order 3.

It is known that the cardinality of the set of MOLS of order h is at most $h-1$. A set of $h-1$ MOLS of order h is called a complete set of the MOLS.

A Latin square of order h is easily constructed by writing 1, 2, .., h, cyclically as follows:

$$\begin{bmatrix} 1 & 2 & 3 & \cdots & h \\ h & 1 & 2 & \cdots & h-1 \\ & & \ddots & & \\ & & & \ddots & \\ 2 & 3 & 4 & \cdots & 1 \end{bmatrix}.$$

But, MOLS cannot be constructed in this manner. A standard method of constructing a complete set of $h-1$ MOLS of order h is presented in the following theorem for h being a prime or a prime power.

Theorem D.1. *For* $GF(p^s) = \{g_0 = 0, g_1 = 1, g_2, ..., g_{h-1}\}$ *with* $h = p^s$, *the following arrangements* L_i *form a complete set of* $h-1$ *MOLS of order* h:

$$L_i = \begin{bmatrix} 0 & 1 & \cdots & g_{h-1} \\ g_i & g_i+1 & \cdots & g_i+g_{h-1} \\ g_ig_2 & g_ig_2+1 & \cdots & g_ig_2+g_{h-1} \\ \vdots & \vdots & & \vdots \\ g_ig_{h-1} & g_ig_{h-1}+1 & \cdots & g_ig_{h-1}+g_{h-1} \end{bmatrix}, \quad i = 1, 2, ..., h-1.$$

Proof. (1) First, one has to show that L_i is a Latin square. That is to show that every symbol occurs once in each row and in each column. If not, let in jth row (of kth and rth columns) the same symbol occur. Then

$g_ig_j + g_k = g_ig_j + g_r$ in GF(h),

i.e., $g_k = g_r$ which is a contradiction. Similarly, if the same symbol occurs in the ℓth column, then

$g_ig_j + g_\ell = g_ig_k + g_\ell$ in GF(h),

i.e., $g_j = g_k$ which is again a contradiction. Hence L_i is a Latin square. (2) Next it will be shown that L_i and L_j are mutually orthogonal $(i \ne j)$. Assume that L_i and L_j are not orthogonal. Then to a symbol g_a in L_i there corresponds a symbol g_b in two places in L_j. If these positions are in αth row and βth column, and in γth row and δth column, respectively, it holds that

$g_i g_\alpha + g_\beta = g_i g_\gamma + g_\delta,$
$g_j g_\alpha + g_\beta = g_j g_\gamma + g_\delta,$
which imply $(g_i - g_j)g_\alpha = (g_i - g_j)g_\gamma$, i.e., $g_\alpha = g_\gamma$ and then $g_\beta = g_\delta$. These are in contradiction to the assumption. Thus, the proof is completed. □

When x is a primitive element of GF(h) (see Appendix B), by Theorem D.1 a complete set of MOLS of order h is given by

$$L_i = \begin{bmatrix} 0 & 1 & \cdots & g_{h-1} \\ x^i & x^i + 1 & \cdots & x^i + g_{h-1} \\ x^{i+1} & x^{i+1} + 1 & \cdots & x^{i+1} + g_{h-1} \\ \vdots & \vdots & & \vdots \\ x^{i+h-2} & x^{i+h-2} + 1 & \cdots & x^{i+h-2} + g_{h-1} \end{bmatrix}, \quad i = 0, 1, ..., h-2,$$

with $h = p^s$.

For example, when $h = 2^2$, it holds that GF(h) $= \{0, 1, x, x^2\}$ with $x^2 = x+1$ (and hence $x^3 = 1$) (see Table B.2). In this case, one has

$$L_0 = \begin{bmatrix} 0 & 1 & x & x^2 \\ 1 & 2 & x+1 & x^2+1 \\ x & x+1 & 2x & x^2+x \\ x^2 & x^2+1 & x^2+x & 2x^2 \end{bmatrix} = \begin{bmatrix} 0 & 1 & x & x^2 \\ 1 & 0 & x^2 & x \\ x & x^2 & 0 & 1 \\ x^2 & x & 1 & 0 \end{bmatrix},$$

$$L_1 = \begin{bmatrix} 0 & 1 & x & x^2 \\ x & x^2 & 0 & 1 \\ x^2 & x & 1 & 0 \\ 1 & 0 & x^2 & x \end{bmatrix}, \quad L_2 = \begin{bmatrix} 0 & 1 & x & x^2 \\ x^2 & x & 1 & 0 \\ 1 & 0 & x^2 & x \\ x & x^2 & 0 & 1 \end{bmatrix},$$

which give a set of 3 MOLS of order 4.

Finally, note (Raghavarao, 1971, pp.5-6) that there is a one-to-one correspondence between a complete set of MOLS and PG$(2, p^s)$. That is, a complete set of MOLS of order h exists if and only if a finite projective plane of order h exists (i.e., if and only if an affine plane of order h exists). This means that a complete set of MOLS yields BIB designs (see Appendix C). Though there does not exist a pair of orthogonal Latin squares of order 6, it is known (cf. Dênes and Keedwell, 1974) that there exist at least a pair of orthogonal Latin squares of order h ($\neq$ 2,6). However, it is unknown whether or not there exists a complete set of MOLS of order h when h is neither a prime nor a prime power. The smallest value of h for which the question is open is 12. This is really a challenging problem. It has been proved (see Lam, Thiel and Swiercz, 1989) by a computer search that there does not exist a PG(2,10).

The (orthogonal) Latin squares of orders up to 10000 can be found in Fisher and Yates (1963), Brouwer (1979) or Colbourn and Dinitz (1996, Part II, pp. 97-182).

References

ABEL, R. J. R. (1994). Forty-three balanced incomplete block designs. *J. Combin. Theory Ser. A* **65**, 252-267.

ABEL, R. J. R. and S. C. FURINO (1996). Resolvable and near resolvable designs. In: C. J. Colbourn and J. H. Dinitz (eds.), *C. R. C. Handbook of Combinatorial Designs*. CRC Press, Boca Raton, FL, 87-94.

ABEL, R. J. R. and M. GREIG (1997). Some new RBIBDs with block size 5 and PBDs with block sizes $\equiv 1$ mod 5. *Austral. J. Combin.* **15**, 177-202.

ABEL, R. J. R., G. GE, M. GREIG and L. ZHU (2001). Resolvable balanced incomplete block designs with block size 5. *J. Statist. Plann. Inference* **95**, 49-65.

ADHIKARY, B. (1965). On the properties and construction of balanced block designs with variable replications. *Calcutta Statist. Assoc. Bull.* **14**, 36-64.

ADHIKARY, B. (1967). Group divisible designs with variable replications. *Calcutta Statist. Assoc. Bull.* **16**, 73-92.

AGARWAL, G. G. and S. KUMAR (1984). On a class of variance balanced designs associated with GD designs. *Calcutta Statist. Assoc. Bull.* **33**, 187-190.

AGARWAL, G. G. and S. KUMAR (1986a). On a class of variance balanced incomplete block designs. *Commun. Statist.– Theor. Meth.* **15**, 1529-1533.

AGARWAL, G. G. and S. KUMAR (1986b). Construction of balanced ternary designs. *Austral. J. Statist.* **28**, 251-255.

AGRAWAL, H. L. (1963). On balanced block designs with two different number of replications. *J. Indian Statist. Assoc.* **1**, 145-151.

ALANEN, J. D. and D. E. KNUTH (1964). Tables of finite fields. *Sankhyā Ser. A* **26**, 305-328.

ALBERT, A. A. (1956). *Fundamental Concepts of Higher Algebra.* University of Chicago, Chicago, USA.

ANDERSON, IAN (1997). *Combinatorial Designs and Tournaments.* Clarendon Press, Oxford.

ARASU, K. T. and N. M. SINGHI (editors)(1997). *Special Issue on Difference Sets and Related Topics.* Journal of Statistical Planning and Inference, Volume 62, Number 1, North-Holland, Amsterdam.

BAGCHI, S. (1988). A class of non-binary unequally replicated E-optimal designs. *Metrika* **35**, 1-12.

BAGCHI, S., A. C. MUKHOPADHYAY and B. K. SINHA (1990). A search for optimal nested row-column designs. *Sankhyā Ser. B* **52**, 93-104.

BAILEY, R. A. (1990). Cyclic designs and factorial designs. In: R. R. Bahadur (ed.), *Proceedings of the R. C. Bose Symposium on Probability, Statistics and Design of Experiments*, Delhi, 27-30 December, 1988, Wiley Eastern, New Delhi, 51-74.

BAILEY, R. A., H. MONOD and J. P. MORGAN (1995). Construction and optimality of affine-resolvable designs. *Biometrika* **82**, 187-200.

BAILEY, R. A. and T. P. SPEED (1986). Rectangular lattice designs: Efficiency factors and analysis. *Ann. Statist.* **14**, 874-895.

BAKER, R. D. (1983). Resolvable BIBD and SOLS. *Discrete Math.* **44**, 13-29.

BAKSALARY, J. K., A. DOBEK and R. KALA (1980). Some methods for constructing efficiency-balanced block designs. *J. Statist. Plann. Inference* **4**, 25-32.

BAKSALARY, J. K. and P. D. PURI (1990). Pairwise-balanced, variance-balanced and resistant incomplete block designs revisited. *Ann. Inst. Statist. Math.* **42**, 163-171.

BAKSALARY, J. K. and Z. TABIS (1987). Conditions for the robustness of block designs against the unavailability of data. *J. Statist. Plann. Inference* **16**, 49-54.

BALL, W. W. R. (1942). *Mathematical Recreations and Essays* (revised by H. S. M. COXETER). Macmillan, London.

BAUMERT, L. D. (1971). *Cyclic Difference Sets. Lecture Notes in Mathematics* **182**, Springer, Berlin and New York.

BETH, T., D. JUNGNICKEL and H. LENZ (1985). *Design Theory*, Bibliographisches Institut, Mannheim, Germany. [D. Jungnickel, Design Theory: An update, *Ars Combin.* **28**(1989), 129-199.]

BETH, T., D. JUNGNICKEL and H. LENZ (1999). *Design Theory.* 2nd ed., Volume I, Cambridge Univ. Press, Cambridge.

BHAGWANDAS, S. KAGEYAMA and S. BANERJEE (1985). Patterned constructions of partially balanced incomplete block designs. *Commun. Statist.–Theor. Meth.* **14**, 1259-1267.

BHATTACHARYA, K. N. (1945). On a new symmetrical balanced incomplete block design. *Bull. Calcutta Math. Soc.* **36**, 91-96.

BHAUMIK, D. K. (1995). Optimality in the competing effects models. *Sankhyā Ser. B* **57**, 48-56.

BIRKHOFF, G. and S. MACLANE (1953). *A Survey of Modern Algebra.* Macmillan, New York.

BOGACKA, B. and S. MEJZA (1994). Optimality of generally balanced experimental block designs. In: T. Caliński and R. Kala (eds.), *Proc. Internat. Conf. on Linear Statistical Inference LINSTAT'93*, Kluwer Academic Publishers, Dordrecht, 185-194.

BOSE, R. C. (1939). On the construction of balanced incomplete block designs. *Ann. Eugen.* **9**, 353-399.

BOSE, R. C. (1942a). A note on the resolvability of balanced incomplete block designs. *Sankhyā* **6**, 105-110.

BOSE, R. C. (1942b). On some new series of balanced incomplete block designs. *Bull. Calcutta Math. Soc.* **34**, 17-31.

BOSE, R. C. (1947). On a resolvable series of balanced incomplete block designs. *Sankhyā* **8**, 249-256.

BOSE, R. C. (1959). On the application of finite projective geometry for deriving a certain series of balanced Kirkman arrangements. *Bull. Calcutta Math. Soc.* **51**, 341-354.

BOSE, R. C., W. H. CLATWORTHY and S. S. SHRIKHANDE (1954). Tables of Partially Balanced Designs with Two Associate Classes. *North Carolina Agric. Exp. Station Tech. Bull.* **107**.

BOSE, R. C. and W. S. CONNOR (1952). Combinatorial properties of group divisible incomplete block designs. *Ann. Math. Statist.* **23**, 367-383.

BOSE, R. C. and D. M. MESNER (1959). On linear associative algebras corresponding to association schemes of partially balanced designs. *Ann. Math. Statist.* **30**, 21-38.

BOSE, R. C. and K. R. NAIR (1939). Partially balanced incomplete block designs. *Sankhyā* **4**, 337-372.

BOSE, R. C. and K. R. NAIR (1962). Resolvable incomplete block designs with two replications. *Sankhyā Ser. A* **24**, 9-24.

BOSE, R. C. and T. SHIMAMOTO (1952). Classification and analysis of partially balanced incomplete block designs with two associate classes. *J. Amer. Statist. Assoc.* **47**, 151-184.

BOSE, R. C. and S. S. SHRIKHANDE (1960). On the composition of balanced incomplete block designs. *Canad. J. Math.* **12**, 177-188.

BOSE, R. C., S. S. SHRIKHANDE and K. N. BHATTACHARYA (1953). On the construction of group divisible incomplete block designs. *Ann. Math. Statist.* **24**, 167-195.

BOSE, R. C., S. S. SHRIKHANDE and E. T. PARKER (1960). Further results on the construction of mutually orthogonal Latin squares and the falsity of Euler's conjecture. *Canad. J. Math.* **12**, 189-203.

BROUWER, A. E. (1979). The number of mutually orthogonal Latin squares – a table up to order 10000. *Mathematisch Centrum* ZW 123/79, June, Amsterdam.

BRUCK, R. H. and H. J. RYSER (1949). The non-existence of certain finite projective planes. *Canad. J. Math.* **1**, 88-93.

CALIŃSKI, T. (1971). On some desirable patterns in block designs (with discussion). *Biometrics* **27**, 275-292.

CALIŃSKI, T. (1996). On the existence of BLUEs under a randomization model for the randomized block design (with discussion). *Listy Biometryczne – Biometr. Lett.* **33**, 1-23.

CALIŃSKI, T. and B. CERANKA (1974). Supplemented block designs. *Biom. J.* **16**, 299-305.

CALIŃSKI, T., B. CERANKA and S. MEJZA (1980). On the notion of efficiency of a block design. In: W. Klonecki, A. Kozek and J. Rosiński (eds.), *Mathematical Statistics and Probability Theory. Lecture Notes in Statistics* **2**. Springer, New York, 47-62.

CALVIN, L. D. (1954). Doubly balanced incomplete block designs for expperiments in which the treatment effects are correlated. *Biometrics* **10**, 61-88.

CALVIN, L. D. and K. SINHA (1989). A method of constructing variance balanced designs. *J. Statist. Plann. Inference* **13**, 127-131.

CARMICHAEL, R. D. (1937). *Introduction to the Theory of Groups of Finite Order.* Boston. (Reprint Dover - New York, 1956).

CERANKA, B. (1975). Affine resolvable incomplete block designs. *Zastosowania Matematyki – Applicationes Mathematicae* **14**, 565-572.

CERANKA, B. (1983). Planning of experiments in *C*-designs. *Scientific Dissertations* **136**, *Annals of Poznań Agricultural University*, Poland.

CERANKA, B. and H. CHUDZIK (1984). On construction of some augmented block designs. *Biom. J.* **26**, 849-857.

CERANKA, B., S. KAGEYAMA and S. MEJZA (1986). A new class of *C*-designs. *Sankhyā Ser. B* **48**, 199-206.

CERANKA, B. and M. KOZŁOWSKA (1983). On *C*-property in block designs. *Biom. J.* **25**, 681-687.

CERANKA, B. and M. KOZŁOWSKA (1984). Some methods of constructing *C*-designs. *J. Statist. Plann. Inference* **9**, 253-258.

CERANKA, B. and S. MEJZA (1978). On some properties of the dual of a totally balanced block design. *J. Statist. Plann. Inference* **2**, 93-94.

CERANKA, B. and S. MEJZA (1979). Construction of equi-replicated balanced block designs with block sizes exceeding the number of treatments. *Biom. J.* **21**, 293-296.

CERANKA, B. and S. MEJZA (1987). Merging of treatments in certain partially efficiency balanced block designs with two different numbers of replications. *Calcutta Statist. Assoc. Bull.* **36**, 49-55.

CHANDAK, M. L. (1980). On the theory of resistant block designs. *Calcutta Statist. Assoc. Bull.* **29**, 27-34.

CHANG, C. L., C. W. LIU and W. R. LIU (1965). Incomplete block designs with triangular parameters for which $k \leq 10$ and $r \leq 10$. *Science Sinica* **14**, 329-338.

CHENG, C. S. (1981). Graph and optimum design theories – some connections and examples. *Proc. 43th Session International Statistical Institute*, 580-590.

CHENG, C. S. and R. A. BAILEY (1991). Optimality of some two-associate-class partially balanced incomplete-block designs. *Ann. Statist.* **19**, 1667-1671.

CHOWLA, S. and H. J. RYSER (1950). Combinatorial problems. *Canad. J. Math.* **2**, 93-99.

CLATWORTHY, W. H. (1967). Some new families of partially balanced designs of the Latin square type and related designs. *Technometrics* **9**, 229-243.

CLATWORTHY, W. H. (1973). *Tables of Two-Associate-Class Partially Balanced Designs. NBS Applied Mathematics Series* **63**, U.S. Department of Commerce, National Bureau of Standards, USA.

CLATWORTHY, W. H. and R. J. LEWYCKYJ (1968). Comments on Takeuchi's table of difference sets generating balanced incomplete block designs. *Rev. Internat. Statist. Inst.* **36**, 12-18.

COCHRAN, W. G. and G. M. COX (1957). *Experimental Designs.* 2nd ed., Wiley, New York.

COLBOURN, C. J. and J. H. DINITZ (editors)(1996). *The Handbook of Combinatorial Designs.* CRC Press, New York.

COLBOURN, C. J. and R. A. MATHON (editors)(1987). *Combinatorial Design Theory. Annals of Discrete Mathematics* **34**, North-Holland, Amsterdam.

COLLENS, R. J. (1973). A listing of balanced incomplete block designs. *Proc. Fourth Southeastern Conf. on Combinatorics, Graph Theory and Computing* (Florida Atlantic Univ., Boca Raton, Florida), Utilitas Math., Winnipeg, Manitoba, Canada, 187-231.

CONNOR, W. S. and W. H. CLATWORTHY (1954). Some theorems for partially balanced designs. *Ann. Math. Statist.* **25**, 100-112.

CONSTANTINE, G. M. (1987). *Combinatorial Theory and Statistical Design.* Wiley, New York.

CORSTEN, L. C. A. (1962). Balanced block designs with two different numbers of replicates. *Biometrics* **18**, 499-519.

CORSTEN, L. C. A., B. CERANKA and S. MEJZA (1984). A note on uniform optimality in block designs. *Biom. Praxim.* **24**, 23-26.

DAS, A. and S. KAGEYAMA (1992). Robustness of BIB and extended BIB designs against the unavailability of any number of observations in a block. *Computational Statistics and Data Analysis* **14**, 343-358.

DAS, M. N. (1958). On reinforced incomplete block designs. *J. Ind. Soc. Agric. Statist.* **10**, 73-77.

DAS, M. N. and S. V. S. P. RAO (1968). On balanced n-ary designs. *J. Indian Statist. Assoc.* **6**, 137-146.

DAVID, H. A. (1967). Resolvable cyclic designs. *Sankhyā Ser. A* **29**, 191-198.

DEMBOWSKI, P. (1968). *Finite Geometries.* Springer.

DÊNES, J. and A. D. KEEDWELL (1974). *Latin Squares and Their Applications.* English University Press, London.

DEY, A. (1970). On construction of balanced n-ary block designs. *Ann. Inst. Statist. Math.* **22**, 389-393.

DEY, A. (1986). *Theory of Block Designs.* Wiley Eastern, New Delhi.

DEY, A. (1993). Robustness of block designs against missing data. *Statistica Sinica* **3**, 219-231.

DEY, A. and S. P. DHALL (1988). Robustness of augmented BIB designs. *Sankhyā Ser. B* **50**, 376-381.

DI PAOLA, J. W., J. S. WALLIS and W. D. WALLIS (1973). A list of (v, b, r, k, λ) designs for $r \leq 30$. *Proc. Fourth Southeastern Conf. on Combinatorics, Graph Theory and Computing* (Florida Atlantic Univ., Boca Raton, Florida), Utilitas Math., Manitoba, Canada, 249-258.

DOYEN, J. and A. ROSA (1980). An updated bibliography and survey of Steiner systems. In: C. C. Lindner and A. Rosa (eds), *Topics on Steiner Systems. Ann. Discrete Math.* **7**, 317-349. North-Holland, Amsterdam.

DUAN, X. and S. KAGEYAMA (1995). Robustness of augmented BIB designs against unavailability of some observations. *Sankhyā Ser. B* **57**, 405-419.

ECKENSTEIN, O. (1912). Bibliography of Kirkman's school girl problem. *Messenger of Math.* **41-42**, 33-36.

FEDERER, W. T. (1956). Augmented (or hoonuiaku) designs. *Hawaiian Planters' Record* **55**, 191-208.

FEDERER, W. T. (1961). Augmented designs with one-way elimination of heterogeneity. *Biometrics* **17**, 447-473.

FEDOROV, V. V. (1972). *Theory of Optimal Designs.* Academic Press, New York.

FINNEY, D. J. (1960). *An Introduction to the Theory of Experimental Design.* The University of Chicago Press, Chicago, IL.

FISHER, R. A. (1940). An examination of the different possible solutions of a problem in incomplete blocks. *Ann. Eugen.* **10**, 52-75.

FISHER, R. A. and F. YATES (1963). *Statistical Tables for Biological, Agricultural and Medical Research.* 6th ed. Hafner, New York. (1st ed. 1938).

FREEMAN, G. H. (1957). Some further methods of constructing regular group divisible incomplete block designs. *Ann. Math. Statist.* **28**, 479-487.

FREEMAN, G. H. (1967). The use of cyclic balanced incomplete block designs for directional seed orchards. *Biometrics* **23**, 761-778.

FREEMAN, G. H. (1969). The use of cyclic balanced incomplete block designs for non-directional seed orchards. *Biometrics* **25**, 561-571.

FREEMAN, G. H. (1976). A cyclic method of constructing regular group divisible incomplete block designs. *Biometrika* **63**, 555-558.

FURINO, S., Y. MIAO and J. YIN (1996). *Frames and Resolvable Designs : Uses, Constructions, and Existence.* CRC Press, New York.

GENSTAT 5 COMMITTEE (1996). *Genstat 5 Release 3.2 Command Language Manual.* Numerical Algorithms Group, Oxford.

GHOSH, D. K., G. C. BHIMANI and S. KAGEYAMA (1989). Resolvable semi-regular group divisible designs. *J. Japan Statist. Soc.* **19**, 163-165.

GHOSH, D. K., K. JOSHI and S. KAGEYAMA (1993). Ternary variance-balanced designs through BIB and GD designs. *J. Japan Statist. Soc.* **23**, 75-81.

GHOSH, S. (1979). On robustness of designs against incomplete data. *Sankhyā Ser. B* **40**, 204-208.

GHOSH, S. (1982a). Robustness of designs against the unavailability of data. *Sankhyā Ser. B* **44**, 50-62.

GHOSH, S. (1982b). Robustness of BIBD against the unavailability of data. *J. Statist. Plann. Inference* **6**, 29-32.

GHOSH, S., S. KAGEYAMA and R. MUKERJEE (1992). Efficiency of connected binary block designs when a single observation is unavailable. *Ann. Inst. Statist. Math.* **44**, 593-603.

GHOSH, S., S. B. RAO and N. M. SINGHI (1983). On a robustness property of PBIBD. *J. Statist. Plann. Inference* **8**, 355-364.

GUPTA, S. and B. JONES (1983). Equireplicated balanced block designs with unequal block sizes. *Biometrika* **70**, 433-440.

GUPTA, S. and S. KAGEYAMA (1992). Variance balanced designs with unequal block sizes and unequal replications. *Utilitas Math.* **42**, 15-24.

GUPTA, S. and R. MUKERJEE (1989). *A Calculus for Factorial Arrangements. Lecture Notes in Statistics* **59**. Springer-Verlag, New York.

GUPTA, V. K. and R. SRIVASTAVA (1992). Investigation of robustness of block designs against missing observations. *Sankhyā Ser. B* **54**, 100-105.

HALL, M., JR. (1956). A survey of difference sets. *Proc. Amer. Math. Soc.* **7**, 975-986.

HALL, M., Jr. (1986). *Combinatorial Theory.* 2nd ed. Wiley. (1st ed. 1967, Blaisdell Pub. Co. (Ginn and Co.), Waltham, MA, Toronto, Ontario, London).

HALL, M., JR. and W. S. CONNOR (1953). An embedding theorem for balanced incomplete block designs. *Canad. J. Math.* **6**, 35-41.

HANANI, H. (1974). On resolvable balanced incomplete block designs. *J. Combin. Theory Ser. A* **17**, 275-289.

HANANI, H. (1975). Balanced incomplete block designs and related designs. *Discrete Math.* **11**, 255-369.

HANANI, H., D. K. RAY-CHAUDHURI and R. M. WILSON (1972). On resolvable designs. *Discrete Math.* **3**, 343-357.

HARSHBARGER, B. (1946). Preliminary report on the rectangular lattices. *Biometrics* **2**, 115-119.

HARSHBARGER, B. (1947). Rectangular lattices. *Virginia Agricultural Experiment Station, Memoir 1.*

HARSHBARGER, B. (1949). Rectangular lattices. *Biometrics* **5**, 1-13.

HARSHBARGER, B. (1951). Near balance rectangular lattices. *Virginia J. Sci.* **2**, 13-27.

HEDAYAT, A. and W. T. FEDERER (1974). Pairwise and variance balanced incomplete block designs. *Ann. Inst. Statist. Math.* **26**, 331-338.

HEDAYAT, A. S. and P. W. M. JOHN (1974). Resistant and susceptible BIB designs. *Ann. Statist.* **2**, 148-158.

HEDAYAT, A. S. and S. KAGEYAMA (1980). The family of t-designs – Part I. *J. Statist. Plann. Inference* **4**, 173-212.

HEDAYAT, A. S., N. J. A. SLOANE and J. STUFKEN (1999). *Orthogonal Arrays: Theory and Applications.* Springer, New York.

HEDAYAT, A. S. and J. STUFKEN (1989). A relation between pairwise-balanced and variance-balanced block designs. *J. Amer. Statist. Assoc.* **84**, 753-756.

HEDAYAT, A. S. and W. D. WALLIS (1978). Hadamard matrices and their applications. *Ann. Statist.* **6**, 1184-1238.

HOFFMAN, K. and R. KUNZE (1961). *Linear Algebra.* Prentice-Hall, Englewood Cliffs.

HUGHES, D. R. and F. C. PIPER (1976). On resolutions and Bose's theorem. *Geom. Dedicata* **5**, 129-133.

HUGHES, D. R. and F. C. PIPER (1985). *Design Theory.* Cambridge Univ. Press, Cambridge.

JACROUX, M. and D. MAJUMDAR (1989). Optimal block designs for comparing test treatments with a control when $k > v$. *J. Statist. Plann. Inference* **23**, 381-396.

JOHN, J. A. (1987). *Cyclic Designs.* Monographs on Statistics and Applied Probability, Chapman and Hall, London.

JOHN, J. A. and E. R. WILLIAMS (1995). *Cyclic and Computer Generated Designs.* 2nd ed. Monographs on Statistics and Applied Probability 38, Chapman and Hall, London.

JOHN, P. W. M. (1961). An application of a balanced incomplete block design. *Technometrics* **3**, 51-54.

JOHN, P. W. M. (1964). Balanced designs with unequal numbers of replicates. *Ann. Math. Statist.* **35**, 897-899.

JOHN, P. W. M. (1966). An extension of the triangular association scheme to three associate classes. *J. Roy. Statist. Soc. Ser. B* **28**, 361-365.

JOHN, P. W. M. (1967). On obtaining balanced incomplete block designs from partially balanced association schemes. *Ann. Math. Statist.* **38**, 618-619.

JOHN, P. W. M. (1980). *Incomplete Block Designs.* Marcel Dekker, New York.

JONES B., K. SINHA and S. KAGEYAMA (1987). Further equireplicate variance-balanced designs with unequal block sizes. *Utilitas Math.* **32**, 5-10.

JONES, R. M. (1959). On a property of incomplete blocks. *J. Roy. Statist. Soc. Ser. B* **21**, 172-179.

JUNGNICKEL, D., R. C. MULLIN and S. A. VANSTONE (1991). The spectrum of α-resolvable designs with block size 3. *Discrete Math.* **97**, 269-271.

JUNGNICKEL, D. and A. POTT (1996). Difference sets: Abelian. In: C. J. Colbourn and J. H. Dinitz (eds.), *The CRC Handbook of Combinatorial Designs.* CRC Press, New York, 297-307.

KAGEYAMA, S. (1971). An improved inequality for balanced incomplete block designs. *Ann. Math. Statist.* **42**, 1448-1449.

KAGEYAMA, S. (1972a). Note on Takeuchi's table of difference sets generating balanced incomplete block designs. *Internat. Statist. Rev.* **40**, 275-276.

KAGEYAMA, S. (1972b). A survey of resolvable solutions of balanced incomplete block designs. *Internat. Statist. Rev.* **40**, 269-273.

KAGEYAMA, S. (1972c). On the reduction of associate classes for certain PBIB designs. *Ann. Math. Statist.* **43**, 1528-1540.

KAGEYAMA, S. (1973a). A series of 3-designs. *J. Japan Statist. Soc.* **3**, 67-68.

KAGEYAMA, S. (1973b). On μ-resolvable and affine μ-resolvable balanced incomplete block designs. *Ann. Statist.* **1**, 195-203.

KAGEYAMA, S. (1973c). On the inequality for BIBDs with special parameters. *Ann. Statist.* **1**, 204-207.

KAGEYAMA, S. (1974a). On the incomplete block designs derivable from the association schemes. *J. Combinatorial Theory Ser. A* **17**, 269-272.

KAGEYAMA, S. (1974b). Note on the reduction of associate classes for PBIB designs. *Ann. Inst. Statist. Math.* **26**, 163-170.

KAGEYAMA, S. (1974c). Reduction of associate classes for block designs and related combinatorial arrangements. *Hiroshima Math. J.* **4**, 527-618.

KAGEYAMA, S. (1976a). On μ-resolvable and affine μ-resolvable t-designs. In: S. Ikeda, et al. (eds.), *Essays in Probability and Statistics.* Shinko Tsusho Co. Ltd., Tokyo, 17-31.

KAGEYAMA, S. (1976b). Resolvability of block designs. *Ann. Statist.* **4**, 655-661. Addendum: *Bull. Inst. Math. Statist.* **7**(5)(1978), 312.

KAGEYAMA, S. (1976c). Some observations on affine μ-resolvable incomplete block designs. *Bull. Fac. Educ. Hiroshima Univ.* **25**, 9-14.

KAGEYAMA, S. (1976d). Constructions of balanced block designs. *Utilitas Math.* **9**, 209-229.

KAGEYAMA, S. (1977a). Conditions for α-resolvability and affine α-resolvability of incomplete block designs. *J. Japan Statist. Soc.* **7**, 19-25.

KAGEYAMA, S. (1977b). Note on combinatorial arrangements. *Hiroshima Math. J.* **7**, 449-458.

KAGEYAMA, S. (1980). On properties of efficiency-balanced designs. *Commun. Statist. – Theor. Meth. Ser. A* **9**, 597-616.

KAGEYAMA, S. (1981). Constructions of efficiency-balanced designs. *Commun. Statist. – Theor. Meth. Ser. A* **10**, 559-580.

KAGEYAMA, S. (1983). A resolvable solution of BIBD(18, 51, 17, 6, 5). *Ars Combin.* **15**, 315-316.

KAGEYAMA, S. (1984). Some properties on resolvability of variance-balanced designs. *Geom. Dedicata* **15**, 289-292.

KAGEYAMA, S. (1985a). A construction of group divisible designs. *J. Statist. Plann. Inference* **12**, 123-125.

KAGEYAMA, S. (1985b). A structural classification of SR group divisible designs. *Statistics & Probability Letters* **3**, 25-27.

KAGEYAMA, S. (1987a). Recommendation of binary designs. *Austral. J. Statist.* **29**, 365-366.

KAGEYAMA, S. (1987b). Some characterization of locally resistant BIB designs of degree one. *Ann. Inst. Statist. Math. Ser. A* **39**, 661-669.

KAGEYAMA, S. (1988a). Existence of variance-balanced binary designs with fewer experimental units. *Statistics & Probability Letters* **7**, 27-28.

KAGEYAMA, S. (1988b). Two methods of construction of affine resolvable balanced designs with unequal block sizes. *Sankhyā Ser. B* **50**, 195-199.

KAGEYAMA, S. (1991). An improved upper bound in a proper non-binary variance-balanced design. *Statistics & Probability Letters* **11**, 319-320.

KAGEYAMA, S. and A. S. HEDAYAT (1983). The family of t-designs – Part II. *J. Statist. Plann. Inference* **7**, 257-287.

KAGEYAMA, S. and M. KUWADA (1985). Parametric characterization of BIB designs. *Bull. Fac. Sch. Educ. Hiroshima Univ.* **8**, 137-149.

KAGEYAMA, S. and Y. MIAO (1994). A construction of resolvable designs. *Utilitas Math.* **46**, 49-54.

KAGEYAMA, S. and Y. MIAO (1995a). Some methods of construction of rectangular PBIB designs. *Combinatorics and Graph Theory*, Vol. 1, 186-198.

KAGEYAMA, S. and Y. MIAO (1995b). Difference families with applications to resolvable designs. *Hiroshima Math. J.* **25**, 475-485.

KAGEYAMA, S. and Y. MIAO (1998). A construction for resolvable designs and its generalizations. *Graph and Combin.* **14**, 11-24.

KAGEYAMA, S. and R. N. MOHAN (1983). On μ-resolvable BIB designs. *Discrete Math.* **45**, 113-122.

KAGEYAMA, S. and R. N. MOHAN (1984a). Three methods of constructing PBIB designs. *Commun. Statist.–Theor. Meth.* **13**, 3185-3189.

KAGEYAMA, S. and R. N. MOHAN (1984b). Dualizing with respect to s-tuples. *Proc. Japan Acad., Ser. A* **60**, 266-268.

KAGEYAMA, S. and R. N. MOHAN (1985). Construction of α-resolvable PBIB designs. *Calcutta Statist. Assoc. Bull.* **34**, 221-224.

KAGEYAMA, S. and G. M. SAHA (1987). On resistant t-designs. *Ars Combin.* **23**, 81-92.

KAGEYAMA, S. and H. SETOYA (1993). Bounds on the efficiency of the residual design of extended BIB designs. *Metrika* **40**, 191-201.

KAGEYAMA, S. and H. SETOYA (1998). Robustness of BIB designs against the unavailability of any two observations. *Bull. Fac. Sch. Educ. Hiroshima Univ. Part II* **20**, 23-28.

KAGEYAMA, S. and K. SINHA (1988). Some constructions of balanced bipartite block designs. *Utilitas Math.* **33**, 137-162.

KAGEYAMA, S. and T. TANAKA (1981). Some families of group divisible designs. *J. Statist. Plann. Inference* **5**, 231-141.

KAGEYAMA, S. and T. TSUJI (1977). Characterization of certain incomplete block designs. *J. Statist. Plann. Inference* **1**, 151-161.

KAGEYAMA, S. and T. TSUJI (1979). Inequality for equireplicated n-ary block designs with unequal block sizes. *J. Statist. Plann. Inference* **3**, 101-107.

KEMPTHORNE, O. (1952). *The Design and Analysis of Experiments.* Wiley, New York.

KHATRI, C. G. (1982). A note on variance balanced designs. *J. Statist. Plann. Inference* **6**, 173-177.

KIEFER, J. (1959). Optimum experimental designs (with discussion). *J. Roy. Statist. Soc. Ser. B* **21**, 272-319.

KIEFER, J. (1975). Construction and optimality of generalized Youden designs. In: J. N. Srivastava (ed.), *A Survey of Statistical Design and Linear Models.* North-Holland, Amsterdam, 333-353.

KIMBERLEY, M. E. (1971). On the construction of certain Hadamard designs. *Math. Z.* **119**, 41-59.

KIRKMAN, T. P. (1847). On a problem in combinatorics. *Cambridge Dublin Math. J.* **2**, 191-204.

KIRKMAN, T. P. (1850a). Query. *Ladies and Gentleman's Diary*, 48.

KIRKMAN, T. P. (1850b). Note on an unanswered prize question. *Cambridge Dublin Math. J.* **5**, 191-204.

KULSHRESHTHA, A. C., A. DEY and G. M. SAHA (1972). Balanced designs with unequal replications and unequal block sizes. *Ann. Math. Statist.* **43**, 1342-1345.

LAM, C. W. H., L. THIEL and S. SWIERCZ (1989). The non-existence of finite projective planes of order 10. *Can. J. Math.* **XLI**, 1117-1123.

LAYWINE, C. F. and G. L. MULLEN (1998). *Discrete Mathematics Using Latin Squares.* Wiley, New York.

LEE, T. C. Y. and S. FURINO (1995). A translation of J. X. Lu's existence theory for resolvable block designs. *J. Combin. Designs* **3**, 321-340.

LINDNER, C. C. and C. A. RODGER (1997). *Design Theory.* Boca Raton, New York.

LINDNER, C. C. and A. ROSA (editors) (1980). *Topics on Steiner Systems. Annals of Discrete Mathematics* **7**, North-Holland, Amsterdam.

LISKI, E. P., N. K. MANDAL, K. R. SHAH and B. K. SINHA (2002). *Topics in Optimal Designs. Lecture Notes in Statistics* **163**, Springer, New York.

LU, J. (1984). An existence theory for resolvable block designs. *Acta Math. Sinica* **27**, 458-468 (in Chinese).

MA, S. L. (1984). Partial difference sets. *Discrete Math.* **52**, 75-89.

MAJUMDAR, D. (1986). Optimal designs for comparisons between two sets of treatments. *J. Statist. Plann. Inference* **14**, 359-372.

MATHON, R. A. and A. ROSA (1985). Tables of parameters of BIBDs with $r \leq 41$ including existence, enumeration and resolvability results. In: C. J. Colbourn and M. J. Colbourn (eds.), *Algorithm in Combinatorial Design Theory.* North-Holland, Amsterdam. *Ann. Discrete Math.* **26**, 275-308.

MATHON, R. A. and A. ROSA (1996). 2-(v, k, λ) Designs of small order. In: C. J. Colbourn and J. H. Dinitz (eds.), *The CRC Handbook of Combinatorial Designs.* CRC Press, New York, 3-41.

MIAO, Y. and L. ZHU (1995). On resolvable BIBDs with block size five. *Ars Combin.* **39**, 261-275.

MIRSKY, L. (1955). *An Introduction to Linear Algebra.* Clarendon Press, Oxford.

MOHAN, R. N. (1980). A note on the construction of certain BIB designs. *Discrete Math.* **29**, 209-211.

MOHAN, R. N. and S. KAGEYAMA (1982). On a characterization of affine μ-resolvable BIB designs. *Utilitas Math.* **22**, 17-23.

MOHAN, R. N. and S. KAGEYAMA (1983). A method of construction of group divisible designs. *Utilitas Math.* **24**, 311-318.

MOHAN, R. N. and S. KAGEYAMA (1989). Two constructions of α-resolvable PBIB designs. *Utilitas Math.* **36**, 115-119.

MOHAN, R. N. and S. KAGEYAMA (1997). New series of α-resolvable and almost α-resolvable PBIB designs. *Utilitas Math.* **51**, 21-26.

MORGAN, J. P. and N. UDDIN (1993). Optimality and construction of nested row and column designs. *J. Statist. Plann. Inference* **37**, 81-93.

MOST, B. M. (1975). Resistance of balanced incomplete block designs. *Ann. Statist.* **3**, 1149-1162.

MUKERJEE, R. and S. KAGEYAMA (1985). On resolvable and affine resolvable variance-balanced designs. *Biometrika* **72**, 165-172.

MUKERJEE, R. and S. KAGEYAMA (1990). Robustness of group divisible designs. *Commun. Statist.– Theory Meth.* **19**, 3189-3203.

MUKERJEE, R. and G. M. SAHA (1990). Some optimality results on efficiency-balanced designs. *Sankhyā Ser. B* **52**, 324-331.

MULLIN, R. C. and R. G. STANTON (1968). Classification and embedding of BIBD's. *Sankhyā Ser. A* **30**, 91-100.

MURTY, J. S. and M. N. DAS (1967). Balanced n-ary block designs and their uses. *J. Indian Statist. Assoc.* **5**, 73-82.

NAIR, K. R. (1951). Rectangular lattices and partially balanced incomplete block designs. *Biometrics* **7**, 145-154.

NAIR, K. R. and C. R. RAO (1942). Incomplete block designs for experiments involving several groups of varieties. *Sci. and Culture* **7**, 615-616.

NELDER, J. A. (1968). The combination of information in generally balanced designs. *J. Roy. Statist. Soc. Ser. B* **30**, 303-311.

NIGAM, A. K. (1976). Nearly balanced incomplete block designs. *Sankhyā Ser. B* **38**, 195-198.

PAL, S. and S. N. PAL (1988). Nonproper variance balanced designs and optimality. *Commun. Statist.–Theor. Meth.* **17**, 1685-1695.

PATERSON, L. J. and H. D. PATTERSON (1983). An algorithm for constructing α-lattice designs. *Ars Combin.* **16A**, 87-98.

PATTERSON, H. D. and E. A. HUNTER (1983). The efficiency of incomplete block designs in national list and recommended list cereal variety trials. *J. Agric. Sci.* **101**, 427-433.

PATTERSON, H. D. and V. SILVEY (1980). Statutory and recommended list trials of crop varieties in the United Kingdom. *J. Roy. Statist. Soc. Ser. A* **143**, 219-252.

PATTERSON, H. D. and E. R. WILLIAMS (1976). A new class of resolvable incomplete block designs. *Biometrika* **63**, 83-92.

PATTERSON, H. D., E. R. WILLIAMS and E. A. HUNTER (1978). Block designs for variety trials. *J. Agric. Sci., Camb.* **90**, 395-400.

PEARCE, S. C. (1960). Supplemented balance. *Biometrika* **47**, 263-271.

PEARCE, S. C. (1964). Experimenting with blocks of natural sizes. *Biometrika* **20**, 699-706.

PEARCE, S. C. (1971). Precision in block experiments. *Biometrika* **58**, 161-167.

PEARCE, S. C. (1976). Concurrences and quasi-replication: An alternative approach to precision in designed experiments. *Biom. J.* **18**, 105-116.

PEARCE, S. C. (1983). *The Agricultural Field Experiment: A Statistical Examination of Theory and Practice.* Wiley, Chichester.

PREECE, D. A. (1967). Nested balanced incomplete block designs. *Biometrika* **54**, 479-486.

PREECE, D. A. (1982). Balance and designs: Another terminological tangle. *Utilitas Math.* **21C**, 85-186.

PUKELSHEIM, F. (1993). *Optimal Design of Experiments.* Wiley, New York.

PURI, P. D. (1984). Patterns and analysis of BIB and GD designs having a missing blocks. *Sankhyā Ser. B* **46**, 44-53.

PURI, P. D. and S. KAGEYAMA (1985). Constructions of partially efficiency-balanced designs and their analysis. *Commun. Statist. –Theor. Meth.* **14**, 1315-1342.

PURI, P. D. and A. K. NIGAM (1975a). On patterns of efficiency balanced designs. *J. Roy. Statist. Soc. Ser. B* **37**, 457-458.

PURI, P. D. and A. K. NIGAM (1975b). A note on efficiency balanced designs. *Sankhyā Ser. B* **37**, 457-460.

PURI, P. D. and A. K. NIGAM (1977a). Partially efficiency balanced designs. *Commun. Statist.–Theor. Meth. Ser. A* **6**, 753-771.

PURI, P. D. and A. K. NIGAM (1977b). Balanced block designs. *Commun. Statist. –Theor. Meth. Ser. A* **6**, 1171-1179.

PURI, P. D. and A. K. NIGAM (1983). Merging of treatments in block designs. *Sankhyā Ser. B* **45**, 50-59.

PURI, P. D., B. D. MEHTA and S. KAGEYAMA (1987). Patterned constructions of partially efficiency-balanced designs. *J. Statist. Plann. Inference* **15**, 365-378.

PURI, P. D., A. K. NIGAM and S. KAGEYAMA (1987). Dual designs and their application in genetical experiments. *Biom. J.* **29**, 555-569.

PURI, P. D., A. K. NIGAM and P. NARAIN (1977). Supplemented designs. *Sankhyā Ser. B* **39**, 189-195.

RAGHAVARAO, D. (1960a). A generalization of group divisible designs. *Ann. Math. Statist.* **31**, 756-771.

RAGHAVARAO, D. (1960b). On the block structure of certain PBIB designs with triangular and L_2 association schemes. *Ann. Math. Statist.* **31**, 787-791.

RAGHAVARAO, D. (1962). On balanced unequal block designs. *Biometrika* **49**, 561-562.

RAGHAVARAO, D. (1971). *Constructions and Combinatorial Problems in Design of Experiments.* Wiley, New York. Reprinted (1988) by Dover with some addendum.

RAGHAVARAO, D. and K. CHANDRASEKHARARAO (1964). Cubic designs. *Ann. Math. Statist.* **35**, 389-397.

RAGHAVARAO, D. and B. ZHOU (1997). A method of constructing 3-designs. *Utilitas Math.* **52**, 91-96.

RAGHAVARAO, D., W. T. FEDERER and S. J. SCHWAGER (1986). Characteristics for distinguishing balanced incomplete block designs with repeated blocks. *J. Statist. Plann. Inference* **13**, 151-163.

RAJKUNDLIA, D. P. (1983). Some techniques for constructing infinite families of BIBD's. *Discrete Math.* **44**, 61-96.

RAO, C. R. (1946). Difference sets and combinatorial arrangements derivable from finite geometries. *Proc. Nat. Inst. Sci. India* **12**, 123-135.

RAO, C. R. (1961). A study of BIB designs with replications 11 to 15. *Sankhyā Ser. A* **23**, 117-129.

RAO, M. B. (1966). Group divisible family of PBIB designs. *J. Indian Statist. Assoc.* **4**, 14-28.

RAO, V. R. (1958). A note on balanced designs. *Ann. Math. Statist.* **29**, 290-294.

RASCH, D. and G. HERRENDÖRFER (1986). *Experimental Design : Sample Size Determination and Block Designs.* D. Reidel, Dordrecht.

RAY-CHAUDHURI, D. K. and R. M. WILSON (1971). Solution of Kirkman's school girl problem. *Proc. Symp. in Pure Mathematics, Combinatorics. Amer. Math. Soc.* **19**, 187-203.

RAY-CHAUDHURI, D. K. and R. M. WILSON (1973). The existence of resolvable block designs. In: J. N. Srivastava et al. (ed.), *A Survey of Combinatorial Theory*. North-Holland, Amsterdam, 361-375.

ROY, P. M. (1953-1954). Hierarchical group divisible incomplete block designs with m associate classes. *Sci. and Culture* **19**, 210-211.

SAHA, G. M. (1976). On Caliński's patterns in block designs. *Sankhyā Ser. B* **38**, 383-392.

SAWADA, K. (1985). A Hadamard matrix of order 268. *Graphs and Combin.* **1**, 185-187.

SCHUTZENBERGER, M. P. (1949). A non-existence theorem for an infinite family of symmetrical block designs. *Ann. Eugen.* **14**, 286-287.

SEBERRY, J. (1978). A class of group divisible designs. *Ars Combin.* **6**, 151-152.

SEBERRY, J. and M. YAMADA (1992). Hadamard Matrices, Sequences, and Block Designs. In: J. H. Dinitz and D. R. Stinson (eds.), *Contemporary Design Theory: A Collection of Surveys*. Wiley, New York, 431-560.

SHAH, K. R. and A. DAS (1992). Binary designs are not always the best. *Canad. J. Statist.* **20**, 347-351.

SHAH, K. R. and B. K. SINHA (1989). *Theory of Optimal Designs. Lecture Notes in Statistics* **54**, Springer, New York.

SHAH, S. M. and C. C. GUJARATHI (1977). On a locally resistant BIB design of degree one. *Sankhyā Ser. B* **39**, 406-408.

SHRIKHANDE, S. S. (1950). The impossibility of certain symmetrical balanced incomplete block designs. *Ann. Math. Statist.* **21**, 106-111.

SHRIKHANDE, S. S. (1952). On the dual of some balanced incomplete block designs. *Biometrics* **8**, 66-72.

SHRIKHANDE, S. S. (1960). Relations between certain incomplete block designs. *Contributions to Probability and Statistics.* Stanford University Press, 388-395.

SHRIKHANDE, S. S. (1962). On a two-parameter family of balanced incomplete block designs. *Sankhyā* **24**, 33-40.

SHRIKHANDE, S. S. (1976). Affine resolvable balanced incomplete block designs: a survey. *Aequationes Math.* **14**, 251-269.

SHRIKHANDE, S. S. and D. RAGHAVARAO (1963). A method of construction of incomplete block designs. *Sankhyā Ser. A* **25**, 399-402.

SHRIKHANDE, S. S. and D. RAGHAVARAO (1964). Affine α-resolvable incomplete block designs. In: C. R. Rao (ed.), *Contributions to Statistics.* Pergamon Press, Statistical Publishing Society, Calcutta, 471-480.

SHRIKHANDE, S. S. and N. K. SINGH (1962). On a method of constructing symmetrical balanced incomplete block designs. *Sankhyā Ser. A* **24**, 25-32.

SILLITTO, G. P. (1957). An extension property of a class of balanced incomplete block designs. *Biometrika* **44**, 278-279.

SILLITTO, G. P. (1964). Note on Takeuchi's table of difference sets generating balanced incomplete block designs. *Rev. Inter. Statist. Inst.* **32**, 251.

SINGH, M. and A. DEY (1979). On analysis of some augmented block designs. *Biom. J.* **21**, 87-92.

SINHA, B. K., S. KAGEYAMA and M. K. SINGH (1993). Construction of rectangular designs. *Statistics* **25**, 63-70.

SINHA, K. (1988). A note on equireplicate balanced block designs from BIB designs. *J. Indian Soc. Agric. Statist.* **40**, 150-153.

SINHA, K. (1989). Some new equireplicate balanced block designs. *Statistics & Probability Letters* **7**, 89.

SINHA, K. (1990). Construction of variance-balanced designs. *Utilitas Math.* **38**, 93-96.

SINHA, K. (1991a). A list of new group divisible designs. *J. Res. National Inst. Standards & Technology*, USA, **96**, 613-615.

SINHA, K. (1991b). A construction of rectangular designs. *J. Comb. Math. Comb. Computing* **9**, 199-200.

SINHA, K. and A. DEY (1982). On resolvable PBIB designs. *J. Statist. Plann. Inference* **6**, 179-181.

SINHA, K. and B. JONES (1988). Further equireplicate balanced block designs with unequal block sizes. *Statistics & Probability Letters* **6**, 229-230.

SINHA, K. and S. KAGEYAMA (1990). Further constructions of balanced bipartite block designs. *Utilitas Math.* **38**, 150-160.

SINHA, K. and S. KAGEYAMA (1991). Equiblock-sized efficiency-balanced designs. *Commun. Statist.– Theory Meth.* **20**, 2933-2941.

SINHA, K., B. JONES and S. KAGEYAMA (1996). Some new non-proper variance-balanced designs with unequal replications. *Statistics & Probability Letters* **27**, 149-153.

SINHA, K., B. JONES and S. KAGEYAMA (1997). Constructions of pairwise- and variance-balanced designs. *Statistics* **29**, 241-250.

SINHA, K., M. K. SINGH, S. KAGEYAMA and R. S. SINGH (2002). Some series of rectangular designs. *J. Statist. Plann. Inference*, to appear.

SMITH, K. W. (1996). Difference sets: Nonabelian. In: C. J. Colbourn and J. H. Dinitz (eds.), *The CRC Handbook of Combinatorial Designs.* CRC Press, New York, 308-312.

SPROTT, D. A. (1954). A note on balanced incomplete block designs. *Canad. J. Math.* **6**, 341-346.

SPROTT, D. A. (1955). Some series of partially balanced incomplete block designs. *Canad. J. Math.* **7**, 369-380.

SPROTT, D. A. (1956). Some series of balanced incomplete block designs. *Sankhyā* **17**, 185-192.

SPROTT, D. A. (1959). A series of symmetric group divisible designs. *Ann. Math. Statist.* **30**, 247-251.

SPROTT, D. A. (1962). Listing of BIB designs from $r = 16$ to 20. *Sankhyā Ser. A* **24**, 203-204.

SRIVASTAVA, R., V. K. GUPTA and A. DEY (1990). Robusntess of some designs against missing observations. *Commun. Statist.–Theory Meth.* **19**, 121-126.

STREET, A. P. and D. J. STREET (1987). *Combinatorics of Experimental Design.* Clarendon Press, Oxford.

STREET, D. J. (1982). A difference set construction for inversive plans. *Combinatorial Mathematics IX, Lecture Notes in Math.* **952**, Springer-Verlag, Berlin, 419-422.

SUEN, C. (1989). Some rectangular designs constructed by the method of differences. *J. Statist. Plann. Inference* **21**, 273-276.

TAKEUCHI, K. (1962). A table of difference sets generating balanced incomplete block designs. *Rev. Inst. Internat. Statist.* **30**, 361-366.

TAKEUCHI, K. (1963). On the construction of a series of BIB designs. *Rep. Statist. Appl. Res. Un. Japan Sci. Engrs.* **10**, 226.

TARRY, G. (1900). Le problème des 36 officiers. *C. R. Assoc. Fr. Av. Sci.* **1**, 122-123; **2**(1901), 170-203.

TOCHER, K. D. (1952). The design and analysis of block experiments (with discussion). *J. Roy. Statist. Soc. Ser. B* **14**, 45-100.

TYAGI, B. N. (1979). On a class of variance balanced block designs. *J. Statist. Plann. Inference* **3**, 333-336.

TYAGI, B. N. and S. K. H. RIZWI (1986). Variance balanced designs through difference sets. *Calcutta Statist. Assoc. Bull.* **35**, 25-30.

VANSTONE, S. A. (1975). A note on a construction of BIBD's. *Utilitas Math.* **7**, 321-322.

VARTAK, M. N. (1955). On an application of Kronecker product of matrices to statistical designs. *Ann. Math. Statist.* **26**, 420-438.

VASIGA, T. M. J., S. FURINO and A. C. H. LING (2001). The spectrum of α-resolvable designs with block size four. *J. Combin. Designs* **9**, 1-16.

WILLIAMS, E. R. (1975). Efficiency-balanced designs. *Biometrika* **62**, 686-688.

WILLIAMS, E. R. (1976). Resolvable paired-comparison designs. *J. Roy. Statist. Soc. Ser. B* **38**, 171-174.

WILLIAMS, E. R. (1977a). Iterative analysis of generalized lattice designs. *Austral. J. Statist.* **19**, 39-42.

WILLIAMS, E. R. (1977b). A note on rectangular lattice designs. *Biometrics* **33**, 410-414.

WILLIAMS, E. R. and H. D. PATTERSON (1977). Upper bounds for efficiency factors in block designs. *Austral. J. Statist.* **19**, 194-201.

WILLIAMS, E. R., H. D. PATTERSON and J. A. JOHN (1976). Resolvable designs with two replications. *J. Roy. Statist. Soc. Ser. B* **38**, 296-301.

WOOLHOUSE, W. S. B. (1844). Prize question 1733. *Ladies and Gentleman's Diary.*

YAMAMOTO, S. and Y. FUJII (1963). Analysis of partially balanced incomplete block designs. *J. Sci. Hiroshima Univ. Ser. A-I* **27**, 119-135.

YATES, F. (1936a). Incomplete randomized blocks. *Ann. Eugen.* **7**, 121-140.

YATES, F. (1936b). A new method of arranging variety trials involving a large number of varieties. *J. Agric. Sci.* **26**, 424-455.

YATES, F. (1939). The recovery of inter-block information in variety trials arranged in three-dimensional lattices. *Ann. Eugen.* **9**, 136-156.

YATES, F. (1940a). Lattice squares. *J. Agric. Sci.* **30**, 672-687.

YATES, F. (1940b). The recovery of inter-block information in balanced incomplete block designs. *Ann. Eugen.* **10**, 317-325.

Author Index

Subject Index

Lecture Notes in Statistics

For information about Volumes 1 to 117, please contact Springer-Verlag

118: Radford M. Neal, Bayesian Learning for Neural Networks. xv, 183 pp., 1996.

119: Masanao Aoki and Arthur M. Havenner, Applications of Computer Aided Time Series Modeling. ix, 329 pp., 1997.

120: Maia Berkane, Latent Variable Modeling and Applications to Causality. vi, 288 pp., 1997.

121: Constantine Gatsonis, James S. Hodges, Robert E. Kass, Robert McCulloch, Peter Rossi, and Nozer D. Singpurwalla (Editors), Case Studies in Bayesian Statistics, Volume III. xvi, 487 pp., 1997.

122: Timothy G. Gregoire, David R. Brillinger, Peter J. Diggle, Estelle Russek-Cohen, William G. Warren, and Russell D. Wolfinger (Editors), Modeling Longitudinal and Spatially Correlated Data. x, 402 pp., 1997.

123: D. Y. Lin and T. R. Fleming (Editors), Proceedings of the First Seattle Symposium in Biostatistics: Survival Analysis. xiii, 308 pp., 1997.

124: Christine H. Müller, Robust Planning and Analysis of Experiments. x, 234 pp., 1997.

125: Valerii V. Fedorov and Peter Hackl, Model-Oriented Design of Experiments. viii, 117 pp., 1997.

126: Geert Verbeke and Geert Molenberghs, Linear Mixed Models in Practice: A SAS-Oriented Approach. xiii, 306 pp., 1997.

127: Harald Niederreiter, Peter Hellekalek, Gerhard Larcher, and Peter Zinterhof (Editors), Monte Carlo and Quasi-Monte Carlo Methods 1996. xii, 448 pp., 1997.

128: L. Accardi and C.C. Heyde (Editors), Probability Towards 2000. x, 356 pp., 1998.

129: Wolfgang Härdle, Gerard Kerkyacharian, Dominique Picard, and Alexander Tsybakov, Wavelets, Approximation, and Statistical Applications. xvi, 265 pp., 1998.

130: Bo-Cheng Wei, Exponential Family Nonlinear Models. ix, 240 pp., 1998.

131: Joel L. Horowitz, Semiparametric Methods in Econometrics. ix, 204 pp., 1998.

132: Douglas Nychka, Walter W. Piegorsch, and Lawrence H. Cox (Editors), Case Studies in Environmental Statistics. viii, 200 pp., 1998.

133: Dipak Dey, Peter Müller, and Debajyoti Sinha (Editors), Practical Nonparametric and Semiparametric Bayesian Statistics. xv, 408 pp., 1998.

134: Yu. A. Kutoyants, Statistical Inference For Spatial Poisson Processes. vii, 284 pp., 1998.

135: Christian P. Robert, Discretization and MCMC Convergence Assessment. x, 192 pp., 1998.

136: Gregory C. Reinsel, Raja P. Velu, Multivariate Reduced-Rank Regression. xiii, 272 pp., 1998.

137: V. Seshadri, The Inverse Gaussian Distribution: Statistical Theory and Applications. xii, 360 pp., 1998.

138: Peter Hellekalek and Gerhard Larcher (Editors), Random and Quasi-Random Point Sets. xi, 352 pp., 1998.

139: Roger B. Nelsen, An Introduction to Copulas. xi, 232 pp., 1999.

140: Constantine Gatsonis, Robert E. Kass, Bradley Carlin, Alicia Carriquiry, Andrew Gelman, Isabella Verdinelli, and Mike West (Editors), Case Studies in Bayesian Statistics, Volume IV. xvi, 456 pp., 1999.

141: Peter Müller and Brani Vidakovic (Editors), Bayesian Inference in Wavelet Based Models. xiii, 394 pp., 1999.

142: György Terdik, Bilinear Stochastic Models and Related Problems of Nonlinear Time Series Analysis: A Frequency Domain Approach. xi, 258 pp., 1999.

143: Russell Barton, Graphical Methods for the Design of Experiments. x, 208 pp., 1999.

144: L. Mark Berliner, Douglas Nychka, and Timothy Hoar (Editors), Case Studies in Statistics and the Atmospheric Sciences. x, 208 pp., 2000.

145: James H. Matis and Thomas R. Kiffe, Stochastic Population Models. viii, 220 pp., 2000.

146: Wim Schoutens, Stochastic Processes and Orthogonal Polynomials. xiv, 163 pp., 2000.

147: Jürgen Franke, Wolfgang Härdle, and Gerhard Stahl, Measuring Risk in Complex Stochastic Systems. xvi, 272 pp., 2000.

148: S.E. Ahmed and Nancy Reid, Empirical Bayes and Likelihood Inference. x, 200 pp., 2000.

149: D. Bosq, Linear Processes in Function Spaces: Theory and Applications. xv, 296 pp., 2000.

150: Tadeusz Caliński and Sanpei Kageyama, Block Designs: A Randomization Approach, Volume I: Analysis. ix, 313 pp., 2000.

151: Håkan Andersson and Tom Britton, Stochastic Epidemic Models and Their Statistical Analysis. ix, 152 pp., 2000.

152: David Ríos Insua and Fabrizio Ruggeri, Robust Bayesian Analysis. xiii, 435 pp., 2000.

153: Parimal Mukhopadhyay, Topics in Survey Sampling. x, 303 pp., 2000.

154: Regina Kaiser and Agustín Maravall, Measuring Business Cycles in Economic Time Series. vi, 190 pp., 2000.

155: Leon Willenborg and Ton de Waal, Elements of Statistical Disclosure Control. xvii, 289 pp., 2000.

156: Gordon Willmot and X. Sheldon Lin, Lundberg Approximations for Compound Distributions with Insurance Applications. xi, 272 pp., 2000.

157: Anne Boomsma, Marijtje A.J. van Duijn, and Tom A.B. Snijders (Editors), Essays on Item Response Theory. xv, 448 pp., 2000.

158: Dominique Ladiray and Benoît Quenneville, Seasonal Adjustment with the X-11 Method. xxii, 220 pp., 2001.

159: Marc Moore (Editor), Spatial Statistics: Methodological Aspects and Some Applications. xvi, 282 pp., 2001.

160: Tomasz Rychlik, Projecting Statistical Functionals. viii, 184 pp., 2001.

161: Maarten Jansen, Noise Reduction by Wavelet Thresholding. xxii, 224 pp., 2001.

162: Constantine Gatsonis, Bradley Carlin, Alicia Carriquiry, Andrew Gelman, Robert E. Kass Isabella Verdinelli, and Mike West (Editors), Case Studies in Bayesian Statistics, Volume V. xiv, 448 pp., 2001.

163: Erkki P. Liski, Nripes K. Mandal, Kirti R. Shah, and Bikas K. Sinha, Topics in Optimal Design. xii, 164 pp., 2002.

164: Peter Goos, The Optimal Design of Blocked and Split-Plot Experiments. xiv, 244 pp., 2002.

165: Karl Mosler, Multivariate Dispersion, Central Regions and Depth: The Lift Zonoid Approach. xii, 280 pp., 2002.

166: Hira L. Koul, Weighted Empirical Processes in Dynamic Nonlinear Models, Second Edition. xiii, 425 pp., 2002.

167: Constantine Gatsonis, Alicia Carriquiry, Andrew Gelman, David Higdon, Robert E. Kass, Donna Pauler, and Isabella Verdinelli (Editors), Case Studies in Bayesian Statistics, Volume VI. xiv, 376 pp., 2002.

168: Susanne Rässler, Statistical Matching: A Frequentist Theory, Practical Applications and Alternative Bayesian Approaches. xviii, 238 pp., 2002.

169: Yu. I. Ingster and Irina A. Suslina, Nonparametric Goodness-of-Fit Testing Under Gaussian Models. xiv, 453 pp., 2003.

170: Tadeusz Caliński and Sanpei Kageyama, Block Designs: A Randomization Approach, Volume II: Design. xii, 351 pp., 2003.

ALSO AVAILABLE FROM SPRINGER!

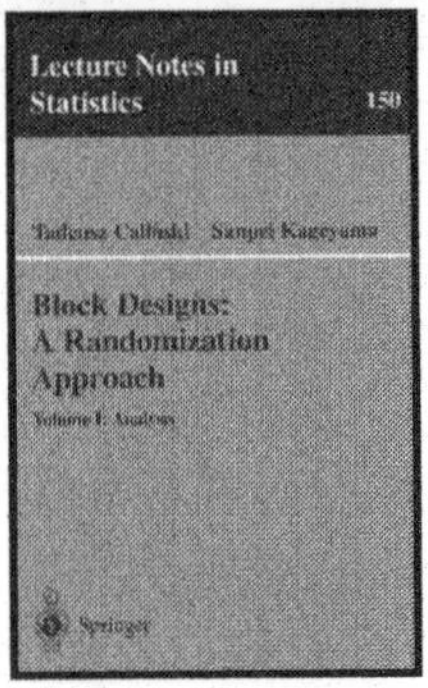

BLOCK DESIGNS: A RANDOMIZATION APPROACH

Volume I: Analysis

TADEUSZ CALIŃSKI and **SANPEI KAGEYAMA**

This volume is devoted to the analysis of experiments in block designs. Chapter 1 introduces the ideas essential to mastering the model building procedures. Chapter 3 is the core of this volume; it presents a randomization model for a general block design and its implications for the analysis. It offers ideas and methods essential for any further study of block designs within the randomization approach. This volume is aimed at an advanced audience, including students at the postgraduate level well acquainted with basic texts on experimental design, and also research workers interested in designing and analyzing experiments with full understanding of the principles.

2000/314 PP./SOFTCOVER/ISBN 0-387-98578-6
LECTURE NOTES IN STATISTICS, VOL. 150

NONPARAMETRIC GOODNESS-OF-FIT TESTING UNDER GAUSSIAN MODELS

YU I. INGSTER and **I.A. SUSLINA**

This book presents the modern theory of nonparametric goodness-of-fit testing. The study is based on an asymptotic version of the minimax approach. The methods for the construction of asymptotically optimal, rate optimal, and optimal adaptive test procedures are developed. The authors present many new results that demonstrate the principal differences between nonparametric goodness-of-fit testing problems with parametric goodness-of-fit testing problems and with non-parametric estimation problems.

2002/448 PP./SOFTCOVER/ISBN 0-387-95531-3
LECTURE NOTES IN STATISTICS, VOL. 169

STATISTICAL MATCHING

A Frequentist Theory, Practical Applications and Alternative Bayesian Approaches

SUSANNE RÄSSLER

This book covers a wide range of different aspects concerning statistical matching that in Europe typically is called data fusion. A theoretical framework is derived to determine the advantages and disadvantages of statistical matching. Its history and practical applications are discussed, and alternative approaches are proposed and evaluated with real world marketing data. Answers to the question of efficiency are provided.

It will be of interest to researchers and practitioners in many data analysis areas, starting with data collection and the production of public use micro files, data banks, and databases.

2002/224 PP./SOFTCOVER/ISBN 0-387-95516-X
LECTURE NOTES IN STATISTICS, VOL. 168

To Order or for Information:

In the Americas: **CALL:** 1-800-SPRINGER or **FAX:** (201) 348-4505 • **WRITE:** Springer-Verlag New York, Inc., Dept. S442, PO Box 2485, Secaucus, NJ 07096-2485 • **VISIT:** Your local technical bookstore • **E-MAIL:** orders@springer-ny.com

Outside the Americas: **CALL:** +49/30/8/27 87-3 73 • +49/30/8 27 87-0 • **FAX:** +49/30 8 27 87 301 • **WRITE:** Springer-Verlag, P.O. Box 140201, D-14302 Berlin, Germany • **E-MAIL:** orders@springer.de

PROMOTION: S442

Springer
www.springer-ny.com

9 780387 954707